Understanding Pure Mathematics

A.J. Sadler, BSc

D.W.S. Thorning, BSc, FIMA

OXFORD
UNIVERSITY PRESS

Great Clarendon Street, Oxford OX2 6DP

Oxford University Press is a department of the University of Oxford. It furthers the University's objective of excellence in research, scholarship, and education by publishing worldwide in

Oxford   New York

Auckland   Cape Town   Dar es Salaam   Hong Kong   Karachi
Kuala Lumpur   Madrid   Melbourne   Mexico City   Nairobi
New Delhi   Shanghai   Taipei   Toronto

With offices in

Argentina   Austria   Brazil   Chile   Czech Republic   France   Greece
Guatemala   Hungary   Italy   Japan   Poland   Portugal   Singapore
South Korea   Switzerland   Thailand   Turkey   Ukraine   Vietnam

Oxford is a registered trade mark of Oxford University Press
in the UK and in certain other countries

First published 1987
Reprinted 1988, 1989, 1990 (twice), 1991, 1992, 1993, 1994, 1995,
      1996 (twice), 1998, 1999 (twice), 2001, 2002 (twice), 2003, 2004, 2005, 2006

British Library Cataloguing in Publication Data

Sadler, A.J.
   Understanding pure mathematics.
   1. Mathematics—Examinations, questions, etc.
   I. Title   II. Thorning, D.W.S.
   510'.76   QA43

ISBN 0-19-914243-2 School Edition
ISBN 0-19-914259-9 Trade Edition
ISBN 9780199142439

30 29 28 27 26 25

Printed by Bell & Bain Ltd., Glasgow

Preface

This book covers the pure mathematics required by candidates taking the single subject GCE advanced level Mathematics course. The syllabuses at this level are varied, even with the introduction of the 'common core', and so teachers and students will find that they can omit certain sections of the book, dependent upon the particular syllabus they are following.

The comprehensive coverage of these syllabuses means that the book also includes a significant part of the work required by those intending to go on to Further Mathematics or advanced level Pure Mathematics. The style and development of the topics should likewise appeal to those preparing for various other intermediate examinations including AS level mathematics.

The authors' aims have been to take each concept and technique at a steady pace. The reader who follows the theory sections and the many worked examples should then be able to make considerable progress through the graded exercises and gain in confidence to attempt the more involved questions that occur later. In particular, an initial chapter on 'Introductory work' covers those basic topics which may not have been studied in pre-advanced level work, while some of the work on vectors in Chapter 2 overlaps material in the authors' companion volume 'Understanding Mechanics'. Likewise, the important work on calculus is introduced in Chapter 10 in order to ensure that the necessary groundwork and algebraic skills have been understood. If this chapter is introduced earlier, the reader is advised to have covered the work on series so that the idea of limits and summations is fully understood.

Each of the twenty-one chapters concludes with an exercise of actual questions from past examination papers. The authors are grateful to the following examination boards for permission to use these questions. The answers provided for these questions are the sole responsibility of the authors.

The Associated Examining Board
University of Cambridge Local Examinations Syndicate
Joint Matriculation Board
University of London School Examinations Council
Oxford Delegacy of Local Examinations
Southern Universities' Joint Board

A.J. Sadler
D.W.S. Thorning

Contents

Introductory work

0.1 The number line

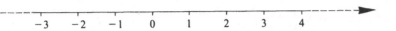

When we first learn to count, we use the numbers, 1, 2, 3, 4, ... These counting numbers form the set of **natural numbers** or **positive integers**, $\mathbb{Z}^+$.

$$\mathbb{Z}^+ = \{1, 2, 3, 4, 5, \ldots\}$$

As our use of number progresses, we need to extend this set backwards to include zero and negative numbers. This gives us the set of all integers $\mathbb{Z}$.

$$\mathbb{Z} = \{\ldots -4, -3, -2, -1, 0, 1, 2, 3, \ldots\}$$

This set of integers is sufficient for many purposes, but we soon need to introduce fractions. Any number that can be expressed in the form a/b, where a and b can take any integer value except $b = 0$, is called a **rational number**. All the integers are rational as they can be expressed in the form $a/1$. If we extend our set of integers $\mathbb{Z}$ to include all rational numbers, we have the set of rational numbers $\mathbb{Q}$.

There is still one group of numbers on our number line that is not included in the set $\mathbb{Q}$. This is the group of numbers that cannot be expressed in the form a/b for integer a and b: these are said to be the **irrational numbers**. One example of an irrational number is $\sqrt{2}$.

To prove that $\sqrt{2}$ is irrational we use the method of **proof by contradiction**, i.e. we assume that what we are trying to prove is not true and then show that this assumption leads to a contradiction.

We initially assume that $\sqrt{2}$ *can* be written as a fraction a/b in its simplest form (i.e. all cancelling carried out), where a and b are integers.

Suppose that $\sqrt{2} = \dfrac{a}{b}$ then $2 = \dfrac{a^2}{b^2}$

$$\text{or}\quad a^2 = 2b^2 \ldots [1]$$

This means that a^2 is even and so a must be even.

Because a is even, we can write $a = 2c$

Substituting for a in [1] gives $(2c)^2 = 2b^2$

$$4c^2 = 2b^2$$

$$\therefore\quad 2c^2 = b^2 \quad \text{and so } b \text{ must be even.}$$

Now we have a contradiction because if both a and b are even, the fraction

a/b was not in its simplest form. Thus $\sqrt{2}$ cannot be written in the form a/b and so $\sqrt{2}$ is irrational.

The set of rational numbers $\mathbb{Q}$ together with the set of irrational numbers form the set of **real numbers** $\mathbb{R}$ and complete the number line.

Note: If we express numbers as decimals, a rational number will either have a finite number of decimal places or will recur. An irrational number will have an infinite number of decimal places without recurring.

0.2 Inequalities

The reader should be familiar with the notation $a > b$ and $a \geqslant b$ standing for 'a is greater than b' and 'a is greater than or equal to b' and with the notation $a < b$ and $a \leqslant b$ standing for 'a is less than b' and 'a is less than or equal to b'.

If we wish to consider a section of the number line, we can define that section by using set notation and the inequality symbols.

The section  can be written

$\{x \in \mathbb{R}: -3 \leqslant x \leqslant 4\}$ and this is read as 'the set of all real x such that x is greater than or equal to -3 and less than or equal to 4.'
[$\{x \in \mathbb{R}: -3 \leqslant x \leqslant 4\}$ can also be written $\{x: x \in \mathbb{R}, -3 \leqslant x \leqslant 4\}$]

If we do not wish to include the integer -3 in our set, we would write $\{x \in \mathbb{R}: -3 < x \leqslant 4\}$.

This would be illustrated as follows:

The open circle indicates that -3 is not included in the section.

If we only want to consider the integer values from -3 to $+4$, we would write this as $\{x \in \mathbb{Z}: -3 \leqslant x \leqslant 4\}$ and the set could be listed: $\{-3, -2, -1, 0, 1, 2, 3, 4\}$.

The notation $\{x: x \text{ obeys some rule}\}$ can also be used for a set with one or more elements missing. For example, the set of all real numbers except the number 1 would be written $\{x \in \mathbb{R}: x \neq 1\}$. In a similar way $\{x \in \mathbb{Z}: x \neq 0, x \neq 1\} = \{\ldots -5, -4, -3, -2, -1, 2, 3, 4, 5, 6 \ldots\}$

Note: The set $\{x \in \mathbb{R}: x < a\}$ is sometimes written $\{x: x < a\}$ or simply as $x < a$. In such cases it should be assumed that the inequalities are for real x.

Inequalities can be simplified using a method similar to that used to solve equations. However, simplifying an inequality will give a range of possible values for the unknown rather than a finite number of solutions.

Example 1

Simplify the following inequality $6x - 3 \leqslant 21 - 2x$.

$$6x - 3 \leqslant 21 - 2x$$

Re-arranging gives $\quad 6x + 2x \leqslant 21 + 3$

$$\therefore \qquad 8x \leqslant 24$$

so $\qquad x \leqslant 3$

The answer may be left in this form or it can be expressed as the solution set $\{x \in \mathbb{R}: x \leqslant 3\}$ or diagrammatically:

One important difference between the techniques used to simplify inequalities and those used when solving equations arises when we multiply or divide the inequality by a negative number. In such cases we must reverse the inequality sign.

Consider the true statements $4 > 3$, $\ 3 > -2$, $\ -4 < -2$. If we multiply each statement by -1 and reverse the inequality sign, we obtain the statements $-4 < -3$, $\ -3 < 2$, $\ 4 > 2$ which are also true. (The reader should verify that the statements obtained by multiplying by -1 and not reversing the inequality signs are not true).

Example 2

Simplify the following inequalities and illustrate the solutions with a diagram: (a) $8 - 2x \leqslant 3$, (b) $-5 < 2x + 3 \leqslant 7$.

(a) $8 - 2x \leqslant 3$
$\quad -2x \leqslant 3 - 8$
$\quad -2x \leqslant -5$
$\quad 2x \geqslant 5$
$\quad x \geqslant 2\frac{1}{2}$

(b) $\qquad -5 < 2x + 3 \leqslant 7$
$\qquad -5 < 2x + 3 \quad \text{and} \quad 2x + 3 \leqslant 7$
$\qquad -5 - 3 < 2x \qquad\qquad 2x \leqslant 7 - 3$
$\qquad -8 < 2x \qquad\qquad\qquad 2x \leqslant 4$
$\qquad -4 < x \qquad \text{and} \qquad x \leqslant 2$
$\qquad\quad \text{i.e.} \quad -4 < x \leqslant 2$

The modulus sign

We write $|x|$ to mean the magnitude or modulus of x, i.e. we are only interested in the size of x and can disregard its sign.

So $\ |-7| = 7; \quad |-3| = 3; \quad |2| = 2 \quad$ etc.

This enables us to write inequalities of the type $-1 < x < 1$ as $|x| < 1$
$$\text{or} \ -3 \leqslant x \leqslant 3 \ \text{as} \ |x| \leqslant 3 \ \text{etc.}$$

Notice that $|x| < a$ means that x must lie between $+a$ and $-a$,
$\qquad$ i.e. $\quad x > -a$ and $x < a$
$\qquad\quad$ whereas $|x| > a$ means that x must either be greater than $+a$ or less than $-a$, i.e. $x > a$ or $x < -a$

Example·3

Express the following as inequalities of x and illustrate the solutions with a diagram:
(a) $|2x - 3| \leqslant 7$,   (b) $|3x + 1| > 8$.

(a) $|2x - 3| \leqslant 7$
$2x - 3 \geqslant -7$ and $2x - 3 \leqslant 7$
$2x \geqslant -4$ and $2x \leqslant 10$
$x \geqslant -2$ and $x \leqslant 5$
So $-2 \leqslant x \leqslant 5$

(b) $|3x + 1| > 8$
$3x + 1 < -8$ or $3x + 1 > 8$
$3x < -9$ or $3x > 7$
$x < -3$ or $x > 2\frac{1}{3}$

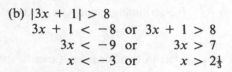

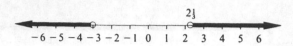

Notice that we do *not* combine the *two* ranges of part (b) in the form $-3 > x > 2\frac{1}{3}$.

Exercise A

1. State whether each of the following statements are true or false.
   (a) $4 > 5$ (b) $5 < 2$ (c) $3 > 1$ (d) $2 < 5$
   (e) $-3 > -2$ (f) $-3 < -2$ (g) $2 > -2$ (h) $-4 < -5$
   (i) $\{x \in \mathbb{Z}^+ : x \neq 2, x \neq 3\} = \{1, 4, 5, 6, 7, 8, 9, \ldots\}$
   (j) $\{x \in \mathbb{Z} : x \neq 2\} = \{\ldots -4, -3, -2, -1, 0, 3, 4, \ldots\}$
   (k) $\{x \in \mathbb{Z} : -1 \leqslant x \leqslant 2\} = \{-1, 0, 1, 2\}$
   (l) $\{x \in \mathbb{Z} : -3 < x \leqslant 4\} = \{-3, -2, -1, 0, 1, 2, 3, 4\}$
   (m) $\{x \in \mathbb{Z} : -2 < x \leqslant 5\} = \{-1, 0, 1, 2, 3, 4, 5\}$
   (n) $\{x \in \mathbb{R} : -2 < x \leqslant 5\} = \{-1, 0, 1, 2, 3, 4, 5\}$
   (o) $\{x \in \mathbb{Z} : x \leqslant 2\} = \{x \in \mathbb{Z} : x < 3\}$
   (p) $\{x \in \mathbb{R} : x \leqslant 2\} = \{x \in \mathbb{R} : x < 3\}$

2. List the elements of the following sets:
   (a) $\{x \in \mathbb{Z} : -2 < x < 4\}$ (b) $\{x \in \mathbb{Z} : -2 \leqslant x \leqslant 4\}$
   (c) $\{x \in \mathbb{Z} : -3 \leqslant x < 2\}$ (d) $\{x \in \mathbb{Z} : -3 < x \leqslant 2\}$
   (e) $\{x \in \mathbb{Z}^+ : x < 5\}$ (f) $\{x \in \mathbb{Z}^+ : x \leqslant 5\}$

3. Give the following sets of numbers in the form
   $\{x \in$ a set of numbers: x satisfies some inequality$\}$.

   (a) $\{-3, -2, -1, 0, 1, 2\}$ (b) $\{-5, -4, -3, -2, -1\}$

   (c) $\{\ldots -4, -3, -2, -1, 0, 1\}$ (d)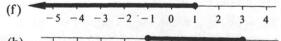

   (e) 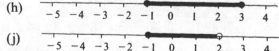   (f)

   (g)   (h)

   (i)   (j)

4. Simplify each of the following inequalities and draw a diagram for each
   solution set for $x \in \mathbb{R}$.
   (a) $2x \leqslant 6$ (b) $2x + 1 < 5$ (c) $3x - 1 \geqslant 11$
   (d) $4 - x < 6$ (e) $4 - 3x > 7$ (f) $2x - 4 \geqslant -1$
   (g) $5x - 3 \leqslant 11 - 2x$ (h) $2x + 1 > 3x - 4$ (i) $-5 \leqslant 2x - 3 \leqslant 7$

(j) $-7 \leqslant 3x + 2 \leqslant 5$ (k) $1 < 2x + 7 < 11$ (l) $1 < 7x + 15 < 36$
(m) $4 \leqslant 2 - x < 7$ (n) $-9 < 3 - 2x \leqslant 1$

5. Find the solution sets of the following inequalities:
(a) $|x - 2| \leqslant 5$ (b) $|2x - 1| \leqslant 5$ (c) $|2x - 5| > 7$
(d) $|4x - 1| < 15$ (e) $|5x - 1| \geqslant 4$ (f) $|3x - 1| > 5$

0.3 Basic algebraic manipulation

The following examples and Exercise B are intended to revise basic algebraic
processes. The ability to manipulate algebraic expressions correctly is an
essential skill for advanced mathematics.

Example 4

Simplify the following:
(a) $12x + 3y - 4x$ (b) $2cd^2 - 5cd^2 + 6d^2c + 4c^2d$

(a) $12x + 3y - 4x$
$= 8x + 3y$

(b) $2cd^2 - 5cd^2 + 6d^2c + 4c^2d$
$= 3cd^2 + 4c^2d$

Example 5

Expand the following and simplify where possible:
(a) $5x(x - 7)$ (b) $14(x - 3) - 5(x - 9)$ (c) $(3x - 2)(5x + 3)$
(d) $(2x + 3)^3$ (e) $3x(x^2 + 4x - 2) - 3(x^2 - 5x)$

(a) $5x(x - 7)$
$= 5x^2 - 35x$

(b) $14(x - 3) - 5(x - 9)$
$= 14x - 42 - 5x + 45$
$= 9x + 3$

(c) $(3x - 2)(5x + 3)$
$= 15x^2 + 9x - 10x - 6$
$= 15x^2 - x - 6$

(d) $(2x + 3)^3$
$= (2x + 3)(2x + 3)^2$
$= (2x + 3)(4x^2 + 12x + 9)$
$= 8x^3 + 24x^2 + 18x + 12x^2 + 36x + 27$
$= 8x^3 + 36x^2 + 54x + 27$

(e) $3x(x^2 + 4x - 2) - 3(x^2 - 5x)$
$= 3x^3 + 12x^2 - 6x - 3x^2 + 15x$
$= 3x^3 + 9x^2 + 9x$

Example 6

Factorise the following:
(a) $12x^2 + 8xy$ (b) $x^2 + 2x - 24$ (c) $2x^2 - x - 6$ (d) $4x^3 - 9x$
(e) $3a + 6 + xa + 2x$

(a) $12x^2 + 8xy$
$= 4x(3x + 2y)$

(b) $x^2 + 2x - 24$
$= (x - 4)(x + 6)$

(c) $2x^2 - x - 6$
$= (2x + 3)(x - 2)$

(d) $4x^3 - 9x$
$= x(4x^2 - 9)$
$= x(2x - 3)(2x + 3)$

(e) $3a + 6 + xa + 2x$
$= 3(a + 2) + x(a + 2)$
$= (a + 2)(3 + x)$

Example 7

Divide $2x^3 - 11x + 6$ by $(x - 2)$

METHOD 1

$$
\begin{array}{r}
2x^2 + 4x - 3 \\
x - 2\overline{)2x^3 \quad\quad - 11x + 6} \\
\underline{2x^3 - 4x^2} \\
4x^2 - 11x \\
\underline{4x^2 - 8x} \\
-3x + 6 \\
\underline{-3x + 6}
\end{array}
$$

The quotient is $2x^2 + 4x - 3$

METHOD 2

$$\frac{2x^3 - 11x + 6}{x - 2} = \frac{2x^2(x - 2) + 4x^2 - 11x + 6}{x - 2}$$

$$= \frac{2x^2(x - 2)}{x - 2} + \frac{4x(x - 2) - 3x + 6}{x - 2}$$

$$= \frac{2x^2(x - 2)}{x - 2} + \frac{4x(x - 2)}{x - 2} - \frac{3(x - 2)}{x - 2}$$

$$= 2x^2 + 4x - 3$$

The quotient is $2x^2 + 4x - 3$

Example 8

Find the remainder when $x^2 + 3x$ is divided by $x^2 + x - 12$

METHOD 1

$$
\begin{array}{r}
1 \\
x^2 + x - 12\overline{)x^2 + 3x} \\
\underline{x^2 + x - 12} \\
2x + 12
\end{array}
$$

Remainder is $2x + 12$

METHOD 2

$$\frac{x^2 + 3x}{x^2 + x - 12} = \frac{1(x^2 + x - 12) + 2x + 12}{x^2 + x - 12}$$

$$= 1 + \frac{2x + 12}{x^2 + x - 12}$$

Remainder is $2x + 12$

Example 9

Express as single fractions:

(a) $\dfrac{1}{2} + \dfrac{2}{(x + 3)}$ (b) $\dfrac{1}{(x + 3)^2} - \dfrac{2}{(x + 3)}$ (c) $\dfrac{(2x + 6)}{7} \times \dfrac{1}{(x^2 - 9)}$

(d) $\dfrac{5x^2}{(x^2 + 6x - 7)} \div \dfrac{x}{(x^2 - 1)}$

(a) $\dfrac{1}{2} + \dfrac{2}{(x + 3)}$

$= \dfrac{1(x + 3) + 2(2)}{2(x + 3)}$

$= \dfrac{x + 7}{2(x + 3)}$

(b) $\dfrac{1}{(x + 3)^2} - \dfrac{2}{(x + 3)}$

$= \dfrac{1 - 2(x + 3)}{(x + 3)^2}$

$= \dfrac{1 - 2x - 6}{(x + 3)^2}$

$= \dfrac{-2x - 5}{(x + 3)^2}$

(c) $\dfrac{(2x + 6)}{7} \times \dfrac{1}{(x^2 - 9)}$

$= \dfrac{2(x + 3)}{7} \times \dfrac{1}{(x + 3)(x - 3)}$

$= \dfrac{2}{7(x - 3)}$

(d) $\dfrac{5x^2}{(x^2 + 6x - 7)} \div \dfrac{x}{(x^2 - 1)}$

$= \dfrac{5x^2}{(x^2 + 6x - 7)} \times \dfrac{(x^2 - 1)}{x}$

$= \dfrac{5x^2}{(x + 7)(x - 1)} \times \dfrac{(x + 1)(x - 1)}{x}$

$= \dfrac{5x(x + 1)}{(x + 7)}$

Improper algebraic fractions

When the highest power of x in the numerator is equal to or greater than the highest power of x in the denominator, the fraction is said to be **improper**. These improper fractions can be rearranged into expressions that are not improper by long division (method 1 above) or by algebraic 'juggling' (method 2 above). The following example shows these methods.

Example 10

Rearrange the following into expressions that do not involve improper algebraic fractions.

(a) $\dfrac{x^2 + 3}{x^2 - 4x + 5}$ (b) $\dfrac{x^2 + x - 4}{x + 3}$

(a) By long division

$$x^2 - 4x + 5 \overline{)\,x^2 \qquad\;\; + 3}$$
$$\underline{x^2 - 4x + 5}$$
$$4x - 2$$

quotient 1

$$\therefore\quad \frac{x^2 + 3}{x^2 - 4x + 5} = 1 + \frac{4x - 2}{x^2 - 4x + 5}$$

By algebraic 'juggling'

$$\frac{x^2 + 3}{x^2 - 4x + 5} = \frac{1(x^2 - 4x + 5) + 4x - 2}{x^2 - 4x + 5}$$

$$= \frac{1(x^2 - 4x + 5)}{x^2 - 4x + 5} + \frac{4x - 2}{x^2 - 4x + 5}$$

$$= 1 + \frac{4x - 2}{x^2 - 4x + 5}$$

(b) By long division

$$x + 3 \overline{)\,x^2 + \;\;x - 4}$$
$$\underline{x^2 + 3x}$$
$$-2x - 4$$
$$\underline{-2x - 6}$$
$$2$$

quotient $x - 2$

$$\therefore\quad \frac{x^2 + x - 4}{x + 3} = x - 2 + \frac{2}{x + 3}$$

By algebraic 'juggling'

$$\frac{x^2 + x - 4}{x + 3} = \frac{x(x + 3) - 2(x + 3) + 2}{x + 3}$$

$$= \frac{x(x + 3)}{x + 3} - \frac{2(x + 3)}{x + 3} + \frac{2}{x + 3}$$

$$= x - 2 + \frac{2}{x + 3}$$

Exercise B

1. Simplify the following:
  (a) $2x + 6y + 9x$ (b) $7x - 4x - 8$ (c) $5p^2 + 8p - 3p^2 - 3$
  (d) $x^2y + 2xy^2 + 3x^2y$ (e) $15a^2 - 2ab - 7a^2$

2. Expand the following and simplify where possible:
  (a) $3x(x - 4)$ (b) $3(x + 4) - 2(x - 6)$ (c) $5(x + 3) + 4(2x - 3)$
  (d) $2x(y - 3) + y(3x - 4)$ (e) $(x + 3)(x - 5)$ (f) $(2x + 3)(x - 4)$
  (g) $(2x - 3)^2$ (h) $(3x - 1)^3$ (i) $5x(x^2 - 3x - 4) - 2(x + 3)^2$

3. Factorise the following:
  (a) $16x^2 + 24xy$ (b) $15x^2 + 10xy + 20x$ (c) $x^2 + 3x - 10$
  (d) $x^2 + 5x - 14$ (e) $2x^2 - 7x - 15$ (f) $6x^2 - x - 2$
  (g) $4x^2 + 4x - 15$ (h) $x^2 - 9$ (i) $16x^2 - 49y^2$
  (j) $2x^3 - 18x$ (k) $ax + 3x + 2a + 6$ (l) $xa - 2xb + ya - 2yb$

4. Divide:
   (a) $x^3 + 2x^2 - x - 2$ by $(x - 1)$ (b) $2x^3 + 9x^2 - 4x - 21$ by $(2x - 3)$
   (c) $x^4 + x^3 + 7x - 3$ by $(x^2 - x + 3)$ (d) $6x^4 + 14x^3 - 9x^2 - 7x + 3$ by $(2x^2 - 1)$
5. Find the remainder when:
   (a) $3x^2 + 13x - 1$ is divided by $x^2 + 4$
   (b) $x^3 + 2x^2 - 6x - 5$ is divided by $x^2 + 4x + 1$
   (c) $x^4 + 2x^3 + 10x^2 + 13x + 11$ is divided by $x^3 + x^2 + x + 1$
   (d) $2x^5 - x^4 - 11x^3 - 15x + 5$ is divided by $2x + 5$
6. Express as single fractions:
   (a) $\dfrac{x}{4} + \dfrac{2x}{3}$

   (b) $\dfrac{x}{5} + \dfrac{x - 1}{3}$

   (c) $\dfrac{2x - 1}{3} - \dfrac{x - 4}{2}$

   (d) $\dfrac{1}{3} + \dfrac{2}{x + 1}$

   (e) $\dfrac{5}{2x + 3} - \dfrac{1}{4}$

   (f) $\dfrac{1}{(x + 2)^2} - \dfrac{3}{(x + 2)}$

   (g) $\dfrac{3}{x} - \dfrac{2}{x + 4}$

   (h) $\dfrac{1}{2(x + 4)} + \dfrac{3}{(x + 4)^2} + \dfrac{1}{2}$

   (i) $\dfrac{1}{2(x + 4)} - \dfrac{3x}{(x + 4)^2} + \dfrac{1}{2}$

   (j) $\dfrac{4x}{3} \div \dfrac{2x}{7}$

   (k) $\dfrac{2(x + 4)}{5} \times \dfrac{1}{(x^2 - 16)}$

   (l) $\dfrac{5x - 20}{3x} \times \dfrac{1}{2x - 8}$

   (m) $\dfrac{3}{2x - 8} + \dfrac{1}{x - 4} + \dfrac{3}{2x}$ (n) $\dfrac{3}{x} - \dfrac{2}{x(x - 1)}$

   (o) $\left(\dfrac{2}{x + 4} - \dfrac{3}{x} \right) \div \dfrac{x}{x^2 - 16}$

7. Using the method of long division (see Example 10), rearrange the
   following into expressions that do not involve improper algebraic fractions:
   (a) $\dfrac{x^2 + 6x - 2}{x^2 + 4x + 1}$ (b) $\dfrac{2x^2 + 5}{x^2 + 1}$ (c) $\dfrac{5x^2 + 2x - 11}{x^2 + x - 2}$ (d) $\dfrac{x^3 - 5x^2 + 9x - 7}{x^2 - 2x + 3}$

8. Using the method of algebraic 'juggling' (see Example 10), rearrange the
   following into expressions that do not involve improper algebraic fractions:
   (a) $\dfrac{x^2 + 3x + 3}{x^2 + 4}$

   (b) $\dfrac{x + 2}{x + 5}$

   (c) $\dfrac{2x^2 - 4x + 11}{x^2 - 2x + 8}$

   (d) $\dfrac{4x + 1}{2x + 3}$

   (e) $\dfrac{x^3 + x^2 + 3x + 5}{x^2 + 3}$

   (f) $\dfrac{x^2 - 2x - 1}{x + 1}$

   (g) $\dfrac{x^3 + 3x^2 - 5x - 4}{x^2 + 2x - 4}$ (h) $\dfrac{x^4 + x^3 + 5x^2 + 10x - 14}{x^2 + 7}$

0.4 Surds

$\sqrt{25} = 5$, $\sqrt[3]{8} = 2$ and $\sqrt{121} = 11$: these are expressions which can be
evaluated exactly.

Surds are such expressions as $\sqrt{7}$, $\sqrt[3]{42}$, $\sqrt{110}$ which cannot be evaluated
exactly. They are **irrational**.

Surds often arise in calculations. For example, using Pythagoras' theorem
to find the hypotenuse of a right-angled triangle with its other sides 5 and
6 cm:

(hypotenuse)^2 = $5^2 + 6^2$
                 = $25 + 36$
hypotenuse   = $\sqrt{61}$ cm

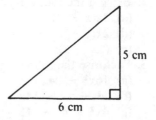

This is an exact answer and it is often preferable to leave an answer in this form, rather than to give a decimal approximation such as 7·8102 ...

Expressions involving surds can often be simplified. The following rules apply:

(i) $\sqrt{a} \times \sqrt{b} = \sqrt{(ab)}$ e.g. $\sqrt{5} \times \sqrt{2} = \sqrt{10}$

(ii) $\sqrt{a} \div \sqrt{b} = \sqrt{\left(\dfrac{a}{b}\right)}$ e.g. $\sqrt{5} \div \sqrt{2} = \sqrt{\left(\dfrac{5}{2}\right)}$

(iii) $a\sqrt{c} \pm b\sqrt{c} = (a \pm b)\sqrt{c}$ e.g. $3\sqrt{2} + 4\sqrt{2} = 7\sqrt{2}$
                           and $3\sqrt{2} - 4\sqrt{2} = -\sqrt{2}$

It must be carefully noted that
$$\sqrt{a} \pm \sqrt{b} \neq \sqrt{(a \pm b)} \text{ for non-zero } a \text{ and } b$$
e.g. $\sqrt{16} + \sqrt{9} \neq \sqrt{25}$, and $\sqrt{16} - \sqrt{9} \neq \sqrt{7}$.

Example 11

Simplify each of the following and hence show that all three expressions are equal:

(a) $\dfrac{\sqrt{18}}{3}$ (b) $3\sqrt{2} - \sqrt{8}$ (c) $\dfrac{\sqrt{6}}{\sqrt{3}}$

(a) $\dfrac{\sqrt{18}}{3}$ (b) $3\sqrt{2} - \sqrt{8}$ (c) $\dfrac{\sqrt{6}}{\sqrt{3}}$

$= \dfrac{\sqrt{9} \times \sqrt{2}}{3}$ $= 3\sqrt{2} - \sqrt{4}\sqrt{2}$ $= \dfrac{\sqrt{3}\sqrt{2}}{\sqrt{3}}$

$= \dfrac{3\sqrt{2}}{3}$ $= 3\sqrt{2} - 2\sqrt{2}$ $= \sqrt{2}$

$= \sqrt{2}$ $= \sqrt{2}$

When fractions are involved with surds, it is normal practice to eliminate the surds from the denominator; this is called rationalising the denominator (i.e. clearing it of irrational numbers.)

For expressions of the type $\dfrac{1}{\sqrt{a}}$,

we multiply the top and bottom by $\sqrt{a}$, giving $\dfrac{1}{\sqrt{a}} - \dfrac{1}{\sqrt{a}} \times \dfrac{\sqrt{a}}{\sqrt{a}} = \dfrac{\sqrt{a}}{a}$

and for expressions of the type $\dfrac{1}{b + \sqrt{a}}$

we multiply the top and bottom by $(b - \sqrt{a})$, giving $\dfrac{1}{b + \sqrt{a}} = \dfrac{1}{b + \sqrt{a}} \times \dfrac{b - \sqrt{a}}{b - \sqrt{a}} = \dfrac{b - \sqrt{a}}{b^2 - a}$

Example 12

Rationalise the denominators of the following fractions (a) $\dfrac{1}{\sqrt{2}}$ (b) $\dfrac{10}{\sqrt{3}-1}$.

(a) $\dfrac{1}{\sqrt{2}} = \dfrac{1}{\sqrt{2}} \times \dfrac{\sqrt{2}}{\sqrt{2}}$

$\qquad = \dfrac{\sqrt{2}}{2}$

(b) $\dfrac{10}{\sqrt{3}-1} = \dfrac{10}{(\sqrt{3}-1)} \times \dfrac{(\sqrt{3}+1)}{(\sqrt{3}+1)}$

$\qquad = \dfrac{10(\sqrt{3}+1)}{3-1}$

$\qquad = 5(\sqrt{3}+1).$

Exercise C

1. Simplify the following:
   (a) $5\sqrt{3} - 3\sqrt{3}$
   (b) $\sqrt{12} + 5\sqrt{3}$
   (c) $\sqrt{200} + 7\sqrt{2} - \sqrt{72}$
   (d) $6\sqrt{3} - \sqrt{12} + \sqrt{48}$
   (e) $\sqrt{45} + \sqrt{20} - \sqrt{80}$
   (f) $(\sqrt{3} + \sqrt{2})(\sqrt{3} - \sqrt{2})$
   (g) $(2\sqrt{5} + \sqrt{7})(2\sqrt{5} - \sqrt{7})$ (h) $(2\sqrt{3} - 3\sqrt{2})(2\sqrt{3} + 3\sqrt{2})$

2. In each of the following state the 'odd one out':
   (a) $3\sqrt{2}, \sqrt{12}, \sqrt{18}$
   (b) $4\sqrt{3}, \sqrt{48}, \sqrt{12}$
   (c) $\sqrt{20}, 2\sqrt{3}, \sqrt{12}$
   (d) $\sqrt{20}, 2\sqrt{5}, \sqrt{18}$
   (e) $5\sqrt{2}, 2\sqrt{5}, \sqrt{50}$
   (f) $\sqrt{6}, 36, \sqrt{2}\sqrt{3}$
   (g) $\sqrt{3}, \dfrac{3}{\sqrt{3}}, \dfrac{1}{\sqrt{3}}$
   (h) $\sqrt{12}, \dfrac{6}{\sqrt{2}}, 3\sqrt{2}$
   (i) $5\sqrt{2}, \dfrac{5}{\sqrt{2}}, \dfrac{10}{\sqrt{2}}$

3. Express each of the following in the form $a + b\sqrt{c}$
   (a) $(\sqrt{3} + \sqrt{2})^2$ (b) $(2\sqrt{5} + \sqrt{3})^2$ (c) $(3\sqrt{3} + \sqrt{2})(\sqrt{3} - \sqrt{2})$

4. Rationalise the denominators of the following:
   (a) $\dfrac{1}{\sqrt{3}}$
   (b) $\dfrac{9}{\sqrt{3}}$
   (c) $\dfrac{16}{3\sqrt{2}}$
   (d) $\dfrac{1}{3-\sqrt{2}}$
   (e) $\dfrac{5}{3-\sqrt{5}}$
   (f) $\dfrac{\sqrt{5}+\sqrt{3}}{\sqrt{5}-\sqrt{3}}$
   (g) $\dfrac{\sqrt{3}+2}{\sqrt{3}-2}$
   (h) $\dfrac{2\sqrt{2}+\sqrt{3}}{2\sqrt{2}-\sqrt{3}}$
   (i) $\dfrac{4-3\sqrt{2}}{5-3\sqrt{2}}$

0.5 Quadratics

Equations of the type $ax^2 + bx + c = 0$, $(a \neq 0)$, are called quadratic equations. There are three ways of solving such equations.

1. By factorisation

This method should only be used if $ax^2 + bx + c$ is readily factorised by inspection.

Example 13

Solve: (a) $x^2 + 2x - 24 = 0$ (b) $2x^2 - x - 6 = 0$

(a) $\qquad x^2 + 2x - 24 = 0$
$\therefore \quad (x + 6)(x - 4) = 0$
So, either $\quad x + 6 = 0$
or $\qquad x - 4 = 0$
giving $\qquad x = -6 \ $ or $\ x = 4$

(b) $\qquad 2x^2 - x - 6 = 0$
$\therefore \quad (2x + 3)(x - 2) = 0$
So, either $\quad 2x + 3 = 0$
or $\qquad x - 2 = 0$
giving $\qquad x = -\tfrac{3}{2} \ $ or $\ x = 2$

2. By the formula $x = \dfrac{-b \pm \sqrt{(b^2 - 4ac)}}{2a}$

Example 14

Solve $x^2 + 4x - 1 = 0$.

Comparing $x^2 - 4x - 1 = 0$ with the general equation
$$ax^2 + bx + c = 0, \text{ gives } a = 1, b = -4 \text{ and } c = -1.$$
Substituting these values in the formula
$$x = \frac{-b \pm \sqrt{(b^2 - 4ac)}}{2a}$$

gives
$$x = \frac{4 \pm \sqrt{(16 + 4)}}{2}$$
$$= 2 + \sqrt{5} \quad \text{or} \quad 2 - \sqrt{5} \quad \text{(these answers are in surds)}$$
$$= 4.24 \quad \text{or} \quad -0.24 \quad \text{(these answers are correct to 2 decimal places)}$$

3. By completing the square

This method uses the expansion $(x + b)^2 = x^2 + 2bx + b^2$.

Notice that the last term, (b^2), is the square of half the coefficient of x, $(2b)$.

Example 15

Solve $\quad x^2 + 3x - 1 = 0$

$$x^2 + 3x - 1 = 0$$
$$\therefore \quad x^2 + 3x \quad\quad = 1$$

Adding the square of half of the coefficient of x to each side of the equation

gives $\quad x^2 + 3x + \left(\dfrac{3}{2}\right)^2 = 1 + \left(\dfrac{3}{2}\right)^2$

So $\quad\quad \left(x + \dfrac{3}{2}\right)^2 = \dfrac{13}{4}$

$\therefore \quad\quad\quad\quad x + \dfrac{3}{2} = \pm\dfrac{\sqrt{13}}{2} \quad \text{giving} \quad x = \dfrac{\sqrt{13} - 3}{2} \quad \text{or} \quad \dfrac{-\sqrt{13} - 3}{2}$

or as decimals $\quad x = 0.30 \quad\quad$ or $\quad -3.30$ (correct to 2 d.p.)

The method of completing the square, used to solve $ax^2 + bx + c = 0$,
can also be used to find the maximum or minimum value of the expression
$ax^2 + bx + c$.
For example, consider the expression $x^2 + 3x + 4$:
$$x^2 + 3x + 4 = x^2 + 3x + (\tfrac{3}{2})^2 - (\tfrac{3}{2})^2 + 4$$
$$= (x + \tfrac{3}{2})^2 + \tfrac{7}{4}$$
Now $(x + \tfrac{3}{2})^2$ cannot be negative for any value of x, i.e. $(x + \tfrac{3}{2})^2 \geqslant 0$.
Thus $x^2 + 3x + 4$ is always positive and will have a minimum value of $\tfrac{7}{4}$
when $x + \tfrac{3}{2} = 0$, i.e. when $x = -\tfrac{3}{2}$.

Example 16

Find the maximum value of $5 - 2x - 4x^2$ for $x \in \mathbb{R}$.

$$5 - 2x - 4x^2 = -4(x^2 + \tfrac{1}{2}x) + 5$$
$$= -4(x^2 + \tfrac{1}{2}x + \tfrac{1}{16}) + \tfrac{4}{16} + 5$$
$$= -4(x + \tfrac{1}{4})^2 + \tfrac{21}{4}$$
$$= \tfrac{21}{4} - 4(x + \tfrac{1}{4})^2$$

Now $(x + \tfrac{1}{4})^2 \geqslant 0$.

Thus $5 - 2x - 4x^2$ has a maximum value of $\tfrac{21}{4}$, or $5\tfrac{1}{4}$, when $x = -\tfrac{1}{4}$.

Exercise D

1. Solve the following equations by factorisation:

(a) $x^2 - x - 12 = 0$ (b) $x^2 + 5x - 24 = 0$ (c) $x^2 + 36 = 13x$

(d) $2x^2 - x - 6 = 0$ (e) $2x^2 - 7x - 4 = 0$ (f) $3x^2 + 5x - 12 = 0$

(g) $x(x + 5) = 6$ (h) $4x^2 - 9x = -3 - x$ (i) $x(x - 1) - 2(x + 5) = 0$

(j) $x - \dfrac{24}{x} + 2 = 0$ (k) $3x - 5 = \dfrac{5x - 3}{x}$ (l) $\dfrac{x(2x - 1)}{(x + 2)} = 12$

2. Solve the following equations by using the formula $x = \dfrac{-b \pm \sqrt{(b^2 - 4ac)}}{2a}$,

giving your answers correct to two decimal places:

(a) $x^2 - 3x + 1 = 0$ (b) $x^2 - x - 4 = 0$ (c) $x^2 + 7x + 5 = 0$

(d) $2x^2 + 3x - 1 = 0$ (e) $3x^2 - 6x + 2 = 0$ (f) $4x^2 - 3x - 2 = 0$

3. Solve the following equations by using the formula $x = \dfrac{-b \pm \sqrt{(b^2 - 4ac)}}{2a}$,

leaving your answers in surd form:

(a) $x^2 - 3x + 1 = 0$ (b) $2x^2 - 6x + 1 = 0$ (c) $3x^2 - 6x + 1 = 0$

4. Solve the following equations by 'completing the square' (leave your answers in surd form):

(a) $x^2 + 6x - 1 = 0$ (b) $x^2 + 4x - 3 = 0$ (c) $x^2 + x - 1 = 0$

(d) $x^2 - x - 3 = 0$ (e) $x^2 - 3x - 5 = 0$ (f) $2x^2 - 6x + 1 = 0$

(g) $3x^2 - 4x - 2 = 0$ (h) $4x^2 - 6x + 1 = 0$ (i) $3x^2 - 6x - 1 = 0$

Each of the expressions given in questions **5–13** has a maximum or minimum value for $x \in \mathbb{R}$.

Find (a) which it is, maximum or minimum, (b) its value, (c) the value of x for which it occurs.

5. $x^2 + 4x - 3$ **6.** $x^2 - 6x + 1$ **7.** $3 - 2x - x^2$

8. $x^2 - 5x + 1$ **9.** $5 + 2x - x^2$ **10.** $6 - x^2 + 8x$

11. $2x^2 + 3x + 1$ **12.** $3x^2 + 2x + 2$ **13.** $1 - 5x - 2x^2$

14. Use the method of 'completing the square' to show that the solutions of

$ax^2 + bx + c = 0$ are given by $x = \dfrac{-b \pm \sqrt{(b^2 - 4ac)}}{2a}$

0.6 Trigonometric ratios of acute and obtuse angles

Acute angles

Trigonometric ratios of acute angles are usually defined by using a right-angled triangle. Denoting the sides of the triangle by a, b and c:

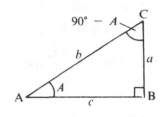

$$\sin A = \frac{a}{b} \qquad \cos A = \frac{c}{b} \qquad \tan A = \frac{a}{c}$$

It follows that $\dfrac{\sin A}{\cos A} = \dfrac{a}{b} \div \dfrac{c}{b} = \dfrac{a}{c} = \tan A$

Also $\sin A = \dfrac{a}{b} = \cos (90° - A)$ and $\cos A = \dfrac{c}{b} = \sin (90° - A)$

These are general results and should be remembered:

$$\frac{\sin x}{\cos x} = \tan x \qquad
\begin{array}{l}
\sin x = \cos (90° - x) \quad \text{(i.e. } \sin x = \text{cosine of complement)} \\
\cos x = \sin (90° - x) \quad (\cos x = \text{sine of complement)}
\end{array}$$

Particular angles

30° and 60°

Suppose $\triangle$PQR is equilateral, with sides 2 units and that PM is the perpendicular bisector of QR.

QM = 1 unit

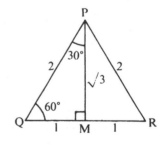

Using Pythagoras' theorem $MP^2 + MQ^2 = PQ^2$

$$\text{or} \qquad MP = \sqrt{(2^2 - 1^2)}$$
$$\text{So} \qquad MP = \sqrt{3}$$

Since $\triangle$PQR is equilateral, $P\hat{Q}M = 60°$ and $Q\hat{P}M = 30°$.

From $\triangle$PQM $\sin 30° = \dfrac{1}{2}$; $\cos 30° = \dfrac{\sqrt{3}}{2}$; $\tan 30° = \dfrac{1}{\sqrt{3}}$ or $\dfrac{\sqrt{3}}{3}$;

and $\sin 60° = \dfrac{\sqrt{3}}{2}$; $\cos 60° = \dfrac{1}{2}$; $\tan 60° = \dfrac{\sqrt{3}}{1} = \sqrt{3}$.

45°

Consider a right-angled triangle which is isosceles and in which the equal sides are 1 unit in length.
The equal angles will each be 45°.
Using Pythagoras' theorem $BC^2 = 1^2 + 1^2$ or $BC = \sqrt{2}$.

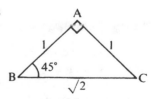

Hence $\sin 45° = \dfrac{1}{\sqrt{2}}$ or $\dfrac{\sqrt{2}}{2}$; $\cos 45° = \dfrac{\sqrt{2}}{2}$; $\tan 45° = \dfrac{1}{1} = 1$.

Although the trigonometric ratios of 30°, 45° and 60° have all been obtained separately, some of these could have been deduced from others:
   to find cos 60°, knowing sin 30°, we could say cos 60° = sin (90° − 60°) = sin 30° = $\frac{1}{2}$.
   to find tan 30°, knowing sin 30° and cos 30° we could say

$$\tan 30° = \frac{\sin 30°}{\cos 30°} = \frac{\frac{1}{2}}{\frac{\sqrt{3}}{2}} = \frac{1}{\sqrt{3}} \quad \text{and so on.}$$

0° and 90°

In the $\triangle ABC$, as the angle x decreases so does the length of the side BC. As x approaches 0°, BC approaches zero and AC approaches AB in length, i.e. as x tends to zero (written $x \to 0$), then BC $\to 0$ and AC $\to$ AB,

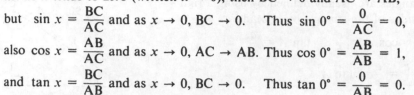

but $\sin x = \dfrac{BC}{AC}$ and as $x \to 0$, BC $\to 0$.   Thus $\sin 0° = \dfrac{0}{AC} = 0$,

also $\cos x = \dfrac{AB}{AC}$ and as $x \to 0$, AC $\to$ AB. Thus $\cos 0° = \dfrac{AB}{AB} = 1$,

and $\tan x = \dfrac{BC}{AB}$ and as $x \to 0$, BC $\to 0$.   Thus $\tan 0° = \dfrac{0}{AB} = 0$.

Similarly, as $x \to 0°$, so $A\hat{C}B \to 90°$. Thus, considering the trigonometric ratios of $A\hat{C}B$ as $x \to 0°$:

$\sin 90° = \dfrac{AB}{AB} = 1$, $\cos 90° = \dfrac{0}{AC} = 0$ and $\tan 90° = \dfrac{\sin 90°}{\cos 90°} = \dfrac{1}{0} = \infty$ (infinity).

These results should be memorised:

Angle	sin	cos	tan
0°	0	1	0
30°	$\dfrac{1}{2}$	$\dfrac{\sqrt{3}}{2}$	$\dfrac{1}{\sqrt{3}}$
45°	$\dfrac{1}{\sqrt{2}}$	$\dfrac{1}{\sqrt{2}}$	1
60°	$\dfrac{\sqrt{3}}{2}$	$\dfrac{1}{2}$	$\sqrt{3}$
90°	1	0	∞

Example 17

Show that $\cos^2 30° + \cos 60° \sin 30° = 1$.

The left-hand side is $\cos^2 30° + \cos 60° \sin 30°$
$$= \cos 30°(\cos 30°) + \cos 60° \sin 30°$$
$$= \frac{\sqrt{3}}{2} \times \frac{\sqrt{3}}{2} + \frac{1}{2} \times \frac{1}{2} = \frac{3}{4} + \frac{1}{4}$$
$$= 1 \text{ as required.}$$

Obtuse angles

Trigonometric ratios of obtuse angles cannot be defined by means of a right-angled triangle.
The sine, cosine or tangent of an obtuse angle is the sine, cosine or tangent of the **supplement** of the angle, with the appropriate sign.
If θ is an obtuse angle:

$\sin \theta = +\sin (180° - \theta)$ i.e. sine of the supplementary angle,
$\cos \theta = -\cos (180° - \theta)$ i.e. $-$cosine of the supplementary angle,
$\tan \theta = -\tan (180° - \theta)$ i.e. $-$tangent of the supplementary angle.

A full definition of the trigonometric ratios of angles of any size is to be found in section 4.1.

Example 18

Write each of the following as a trigonometric ratio of an acute angle
(a) sin 155°, (b) cos 140°, (c) tan 130°.

(a) $\sin 155° = +\sin (180° - 155°)$
$\quad\quad\quad = + \sin 25°$
(b) $\cos 140° = -\cos (180° - 140°)$
$\quad\quad\quad = -\cos 40°$
(c) $\tan 130° = -\tan (180° - 130°)$
$\quad\quad\quad = -\tan 50°$

Example 19

If $\sin 35° = 0.5736$ write down the values of (a) sin 145°, (b) cos 125°

(a) $\sin 145° = +\sin (180° - 145°)$
$\quad\quad\quad = +\sin 35°$
$\quad\quad\quad = +0.5736$

(b) $\cos 125° = -\cos (180° - 125°)$
$\quad\quad\quad = -\cos 55°$
$\quad\quad\quad = -\sin 35°$
$\quad\quad\quad = -0.5736$

Example 20

Given that $\sin \theta = \frac{7}{25}$ and that θ is an acute angle, find: (a) cos θ, (b) tan θ.

First sketch a right-angled triangle containing an angle θ and with two sides of length 7 and 25 units, such that $\sin \theta = \frac{7}{25}$.

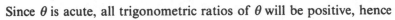

Using Pythagoras' theorem, the third side of the triangle $= \sqrt{(25^2 - 7^2)}$
$\quad\quad\quad\quad\quad\quad\quad\quad\quad\quad\quad\quad\quad = 24$

Since θ is acute, all trigonometric ratios of θ will be positive, hence

(a) $\cos \theta = \frac{24}{25}$ (b) $\tan \theta = \frac{7}{24}$.

Example 21

Given that $\sin \theta = \frac{24}{25}$ and that θ is an obtuse angle, find (a) cos θ, (b) tan θ.

As θ is obtuse, sketch a right-angled triangle containing an angle $(180° - \theta)$ and with two sides of length 24 and 25 units.
As $\sin \theta$ and $\sin (180° - \theta)$ are numerically equal, $\sin (180° - \theta) = \frac{24}{25}$.
Using Pythagoras' theorem, the third side of the triangle $= \sqrt{(25^2 - 24^2)}$
$\quad\quad\quad\quad\quad\quad\quad\quad\quad\quad\quad\quad\quad = 7$

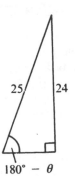

(a) $\cos \theta = -\cos (180° - \theta)$
$\quad\quad\quad = -\frac{7}{25}$

(b) $\tan \theta = -\tan (180° - \theta)$
$\quad\quad\quad = -\frac{24}{7}$

Exercise E

All the questions in this exercise should be answered *without* the use of a calculator or tables

1. Write each of the following as trigonometrical ratios of acute angles:
   (a) sin 130°   (b) cos 130°   (c) tan 130°   (d) sin 140°
   (e) cos 170°   (f) cos 160°   (g) tan 100°   (h) sin 95°
2. Write down the values of the following, leaving surds in your answers:
   (a) sin 30°   (b) cos 30°   (c) tan 45°   (d) sin 45°
   (e) tan 60°   (f) sin 90°   (g) cos 90°   (h) sin 150°
   (i) sin 135°   (j) cos 120°   (k) sin 180°   (l) cos 180°
   (m) tan 135°   (n) cos 150°   (o) tan 120°   (p) sin 120°
3. Show that: (a) $\sin^2 30° + \sin^2 45° + \sin^2 60° = \frac{3}{2}$
             (b) $\sin 60° \cos 30° + \cos 60° \sin 30° = 1$
             (c) $\sin^2 45° + \cos^2 45° = 1$
4. If sin 20° = 0·342 write down the values of:
   (a) sin 160°   (b) cos 70°   (c) cos 110°
5. If sin 40° = 0·643 and cos 40° = 0·766 write down the values of:
   (a) sin 50°   (b) cos 50°   (c) sin 140°
   (d) sin 130°   (e) cos 140°   (f) cos 130°
6. If sin A = 0·98 and cos A = 0·2, find the value of tan A.
7. If sin B = 0·954 and cos B = 0·3, find the value of tan B.
8. If sin $\theta = \frac{2}{3}$ and θ is acute, find the value of (a) cos θ (b) tan θ.
9. If sin $\theta = \frac{2}{3}$ and θ is obtuse, find the value of (a) cos θ (b) tan θ.
10. If sin $\theta = \frac{5}{13}$ and θ is obtuse, find the value of (a) cos θ (b) tan θ.
11. Find the value of x in each of the following, given that x is acute:
    (a) sin 50° = cos x (b) cos 30° = sin x
    (c) cos (40° + x) = sin 30°   (d) sin (20° + x) = cos 50°
    (e) cos (3x − 10°) = sin 10°   (f) cos (2x + 40°) = sin 40°

0.7 Solution of triangles

Cosine rule

Consider $\triangle ABC$ in which CN is the perpendicular from C to BA (produced if necessary—see Case 2). Let the sides of the triangle be a, b and c, the angle CAB be A and AN = x, CN = h.

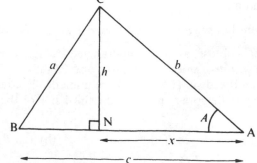

Case 1
Using Pythagoras' theorem
$$\triangle CBN \quad a^2 = h^2 + (c - x)^2$$
$$\triangle CAN \quad b^2 = h^2 + x^2$$
Eliminating h^2
$$a^2 = b^2 - x^2 + (c - x)^2$$
So $\qquad a^2 = b^2 + c^2 - 2cx$
From $\triangle ANC$ $x = b \cos A$
Hence $\quad a^2 = b^2 + c^2 - 2bc \cos A$

Case 2

Using Pythagoras' theorem

$\triangle$CBN $a^2 = h^2 + (c + x)^2$

$\triangle$CAN $b^2 = h^2 + x^2$

Eliminating h^2

$a^2 = b^2 - x^2 + (c + x)^2$

So $a^2 = b^2 + c^2 + 2cx$

From $\triangle$ANC $x = b \cos (180° - A)$

$= -b \cos A$

Hence $a^2 = b^2 + c^2 - 2bc \cos A$

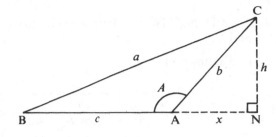

This is the **cosine rule** and it is frequently used to determine a side or angle of a given triangle.

$$a^2 = b^2 + c^2 - 2bc \cos A$$

Example 22

Find the length of the side BC in each of the following triangles:

(a)

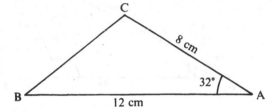

(b)

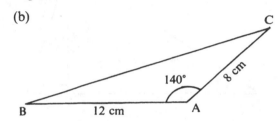

By the cosine rule

$BC^2 = 12^2 + 8^2 - 2(12)(8) \cos 32°$

$= 144 + 64 - 192(0.8480)$

$= 208 - 162.8$

giving BC $= 6.72$ cm

By the cosine rule

$BC^2 = 12^2 + 8^2 - 2(12)(8) \cos 140°$

$= 144 + 64 - 192(-0.7660)$

$= 208 + 147.1$

giving BC $= 18.8$ cm

Example 23

Find the angle θ in the given triangle.

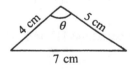

By the cosine rule $7^2 = 5^2 + 4^2 - 2(4)(5) \cos \theta$

$$\cos \theta = \frac{25 + 16 - 49}{40}$$

$$= -0.2$$

We now need to find the angle whose cosine is -0.2. Since the cosine of the angle is negative, the angle must be obtuse.

Using a calculator, press the inverse cosine button (marked $\cos^{-1}$ or arc cos) to give

$\cos^{-1}(-0.2) = 101.54°$ (correct to 2 d.p.)

Alternatively using tables of cosines, search for 0.2 and read off the appropriate angle ($78° 28'$).

As θ must be obtuse $\theta = 180° - 78° 28'$

$= 101.54°$.

Sine rule

Consider the triangle ABC in which the sides are a, b and c and the angles are A, B and C.

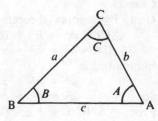

The sine rule states that

$$\frac{a}{\sin A} = \frac{b}{\sin B} = \frac{c}{\sin C}.$$

Proof: Let CD be the perpendicular from C to AB (produced if necessary—see Case 2) and let CD = h.

Case 1

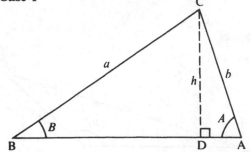

Case 2

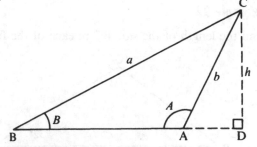

From $\triangle$BCD $\sin B = \dfrac{h}{a}$

or $h = a \sin B$ ... [1]

From $\triangle$ACD $\sin A = \dfrac{h}{b}$

or $h = b \sin A$ ... [2]

From $\triangle$BCD $\sin B = \dfrac{h}{a}$

or $h = a \sin B$ ... [1]

From $\triangle$ACD $\sin (180° - A) = \dfrac{h}{b}$

or $h = b \sin (180° - A)$

$= b \sin A$ ... [2]

Thus, eliminating h from equations [1] and [2] gives $a \sin B = b \sin A$ or $\dfrac{a}{\sin A} = \dfrac{b}{\sin B}$

It can similarly be proved that $\dfrac{a}{\sin A} = \dfrac{c}{\sin C}$ hence $\dfrac{a}{\sin A} = \dfrac{b}{\sin B} = \dfrac{c}{\sin C}.$

Example 24

Find the length of the side BC in the given triangle.

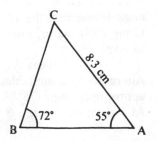

By the sine rule $\dfrac{BC}{\sin 55°} = \dfrac{8\cdot3}{\sin 72°}$

So $BC = \dfrac{8\cdot3 \sin 55°}{\sin 72°}$

$BC = 7\cdot15$ cm.

Example 25

Find the angle x in the given triangle.

By the sine rule $\dfrac{8}{\sin 70°} = \dfrac{6}{\sin x}$

So $\sin x = \dfrac{6 \sin 70°}{8} = 0 \cdot 7048$

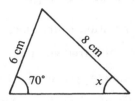

By calculator or tables $\sin^{-1} 0 \cdot 7048 = 44 \cdot 81°$ but we must remember that
$(180° - 44 \cdot 81°)$ would also have a sine of $0 \cdot 7048$, and so

$$x = 44 \cdot 81° \quad \text{or} \quad 180° - 44 \cdot 81°$$
$$= 44 \cdot 81° \quad \text{or} \quad 135 \cdot 19°$$

In this example the obtuse value of the angle can be discarded as it is not
possible when one angle of the triangle is already known to be 70°.

$$\text{Thus} \quad x = 44 \cdot 81°$$

In some cases both answers will be possible.

Example 26

Find the angle x in the given triangle.

By the sine rule $\dfrac{3}{\sin 38°} = \dfrac{4}{\sin x}$

So $\sin x = \dfrac{4 \sin 38°}{3} = 0 \cdot 8209$

hence $x = 55 \cdot 17°$ or $(180° - 55 \cdot 17°)$ i.e. $124 \cdot 83°$

In this case both answers are possible and there are two triangles
that could be drawn from the original information.
If $x_1 = 55 \cdot 17°$ and $x_2 = 124 \cdot 83°$, then these are triangles ABC_1
and ABC_2 as shown on the right.

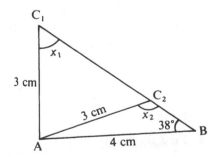

Note on the solution of triangles
Suppose we had to solve the $\triangle ABC$ shown (i.e. find all unknown sides and
angles). We would first find c, using the cosine rule, and then have a choice
as to which angle, A or B, we would find next. If we use the sine rule, we
should then have to decide whether or not the obtuse angle answer is
applicable. It is therefore best to find the angle opposite the smaller side first
(i.e. angle B) as this must be smaller than the angle opposite the larger side
and will therefore be acute.
Alternatively, angle A or B could be found by using the cosine rule.

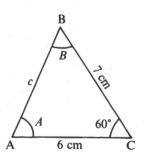

Area of a triangle

In addition to the rule: area $= \dfrac{\text{base} \times \text{perpendicular height}}{2}$ for finding the

area of a triangle, there are two other useful formulae:

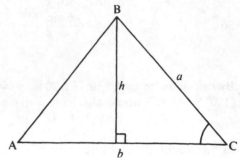

1 Area $= \frac{1}{2}ab \sin C$ (or $\frac{1}{2}bc \sin A$ or $\frac{1}{2}ac \sin B$).

Proof: Consider $\triangle ABC$ in which h is the length
of the perpendicular from B to AC.

Area $= \frac{1}{2}AC \times h$
$ = \frac{1}{2}b \times (BC \sin C)$
$ = \frac{1}{2}ab \sin C$

The other two variations of the formulae can then
be obtained from this by use of the sine rule.

2 Area $= \sqrt{[s(s-a)(s-b)(s-c)]}$

where $s = \dfrac{a+b+c}{2}$, the semi-perimeter.

This second formula was first stated by the Greek Mathematician Hero
and is called Hero's formula. (The proof is not given here).

Example 27

Find the areas of the given triangles:

(a)

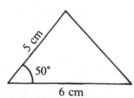

5 cm
50°
6 cm

Area $= \frac{1}{2} \times 5 \times 6 \times \sin 50°$
$ = 11{\cdot}5 \text{ cm}^2$

(b)

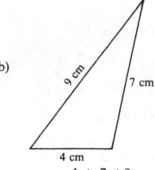

9 cm
7 cm
4 cm

$s = \dfrac{4+7+9}{2} = 10$

Area $= \sqrt{[10(10-4)(10-7)(10-9)]}$
$ = \sqrt{180} = 13{\cdot}4 \text{ cm}^2$

Exercise F

1. Find the length x in each of the following:

(a)

70°
x
50°
8 cm

(b)

5 cm
60°
6 cm
x

(c)

x
75° 55°
4 cm

(d)

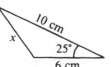

10 cm
x
25°
6 cm

(e)

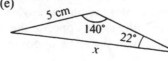

5 cm
140°
22°
x

(f)

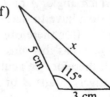

5 cm
x
115°
3 cm

2. Find the angle θ in each of the following:

(a)

(b)

(c)

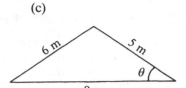

(d)

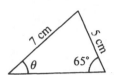

(e)

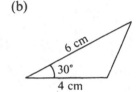

(f)

3. Find the areas of the following triangles:

(a)

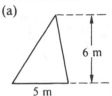

(b)

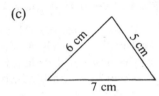

(c)

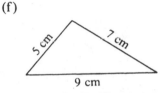

(d)

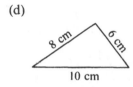

(e)

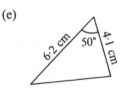

(f)

4. Solve the triangle ABC given that $\hat{A} = 66°$, $\hat{C} = 44°$ and $a = 7$ cm.

5. Solve the triangle ABC given that $\hat{A} = 45°$, $c = 5$ cm and $b = 6$ cm.

6. Solve the triangle ABC given that $\hat{C} = 50°$, $c = 8$ cm and $a = 10$ cm.

7. From a ship A, two other ships B and C, lie on bearings of 320° and 060° respectively. If B is 5 km from A and C is 3 km from A, find the distance from B to C.

8. Three points A, B and C all lie on level ground with B due south of A. The point C lies 250 m from A on a bearing N55°E and C is 400 m from B. Find the bearing of C from B to the nearest degree.

9. A ship A is 7 km away from a lighthouse L on a bearing 080° and a ship B is 5 km away from the lighthouse on a bearing 210°. Find the distance and bearing of A from B.

10. Three points A, B and C lie in a straight line on level ground with B between A and C. A vertical mast stands at A and is supported by wires attached to its top and to points on the ground. One such wire is fastened at B and another at C. The wire to C makes an angle of 40° with CA and the wire to B makes an angle of 60° with BA. If BC = 22 m, find the length of the wires and the height of the mast.

11. A gardener encloses a triangular plot of land by using an existing hedge of length 16 m for one side and fencing, of total length 20 m, for the other two sides. If the area of the plot is $24\sqrt{3}$ m² find the lengths of the sides of the triangle.

12. From a harbour H two ships A and B are situated $2x$ km due north and x km on a bearing N α E respectively (α being an acute angle). Use the cosine rule to show that the distance between the two ships is given by $x\sqrt{(5 - 4\cos\alpha)}$.

0.8 Three-dimensional trigonometry

Angle between a line and a plane

Suppose a line OP meets a plane ABCD at the
point O. The angle between the line OP and
the plane is then defined as the angle between
OP and the line which is the projection of OP
on the plane ABCD. The projection of OP on
the plane ABCD is OQ where PQ is
perpendicular to the plane ABCD. Hence the
required angle is PÔQ.

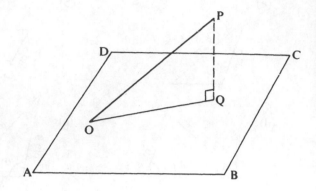

Angle between two planes

Suppose two planes ABCD and ABRS
intersect in the line AB, as shown.
The angle between the planes is the angle
between two lines, both of which are at right-
angles to AB, which intersect on AB, and lie
one in each plane. The required angle is then
MLN.

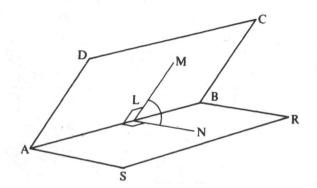

Line of greatest slope

All lines drawn on an inclined plane are not
equally steep. In the diagram ABC is
horizontal and parallel to OX.
The vertical distances of A, B and C above O
are all equal to h.
Hence, since

$$OA < OB < OC$$

then $\dfrac{h}{OA} > \dfrac{h}{OB} > \dfrac{h}{OC}$

thus $\sin\theta_1 > \sin\theta_2 > \sin\theta_3$

i.e. $\theta_1 > \theta_2 > \theta_3$

The steepest line drawn on the inclined plane
will be drawn from O at right-angles to the
line ABC.
Since ABC is parallel to the line OX, this line
of greatest slope will also be at right-angles to
OX.
When we say a plane is inclined at an angle θ
to the horizontal, we mean that the line of
greatest slope on the plane makes an angle θ
with the horizontal plane.

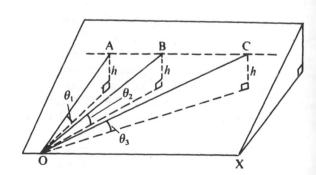

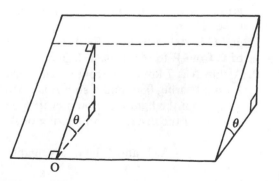

Angle between skew lines

When considering a two-dimensional situation, two lines that are not parallel will intersect. However, in three dimensions two lines which are not parallel to each other will only intersect if the lines themselves are in the same plane, i.e. **coplanar**. Two lines which are not coplanar are said to be **skew lines**.

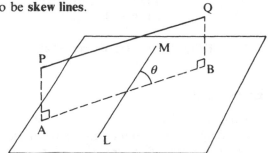

Consider the skew lines LM and PQ:
AB is a line drawn parallel to PQ so that AB and LM intersect (i.e. AB is parallel to PQ and is coplanar with LM).
The angle between the skew lines PQ and LM is then defined as the angle θ between the lines AB and LM.

Example 28

The cuboid shown has PQ = 24 cm, PS = 18 cm and QL = 7 cm. Calculate, to the nearest degree:

(a) the angle between the line SL and the plane PQLM
(b) the angle between the planes LMSR and PQRS
(c) the angle between the skew lines PK and ML.

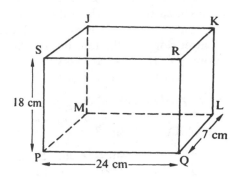

(a) PL is the projection of SL on the plane PQLM. Hence the angle between SL and the plane PQLM is the angle between SL and LP i.e. $S\hat{L}P$

$$\tan S\hat{L}P = \frac{SP}{PL}$$

$$= \frac{18}{\sqrt{(7^2 + 24^2)}}$$

$$\therefore \quad S\hat{L}P = 36° \quad \text{(to nearest degree)}$$

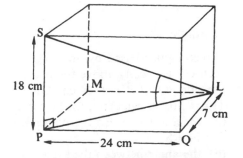

(b) Planes LMSR and PQRS meet in the line RS. RQ is in the plane PQRS and is perpendicular to RS, LR is in the plane LMSR and is perpendicular to RS, so the angle between the planes is $Q\hat{R}L$.

$$\tan Q\hat{R}L = \frac{7}{18}$$

$$\therefore \quad Q\hat{R}L = 21° \quad \text{(to nearest degree)}.$$

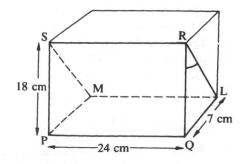

(c) ML is parallel to PQ, so the angle
between the skew lines PK and ML equals
the angle between PK and PQ i.e. KP̂Q.

$$\tan K\hat{P}Q = \frac{KQ}{PQ}$$

$$= \frac{\sqrt{(7^2 + 18^2)}}{24}$$

∴   KP̂Q = 39°   (to nearest degree)

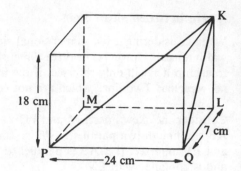

Example 29

A particular hillside may be considered to be a plane inclined at 15° to
the horizontal. A straight path up the hillside makes an angle of
30° with the line of greatest slope. Find, to the nearest degree, the
angle the path makes with the horizontal.

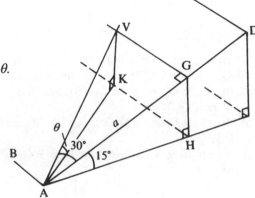

Let ABCD be a portion of the hillside and the path be
represented by AV. AK is the projection of AV in the
horizontal plane, AH is horizontal and AG is a line of
greatest slope on the hillside. Let AG = a and VÂK = θ.

From △AGH   GH = $a \sin 15°$
        hence   VK = $a \sin 15°$

From △AGV $\cos 30° = \dfrac{a}{AV}$ or $AV = \dfrac{a}{\cos 30°}$

In △AKV $\sin \theta = \dfrac{VK}{AV} = a \sin 15° \times \dfrac{\cos 30°}{a}$

∴ $\sin \theta = \sin 15° \cos 30°$

giving $\theta = 13°$ (to nearest degree)

Example 30

A rectangular-based pyramid PQRST has its
vertex T vertically above the mid-point M of
the side PQ. If PQ = 24 cm, QR = 5 cm and
TQ = 20 cm calculate, to the nearest degree:
(a) the angle between the planes RTS and
     PQRS,
(b) the angle between the skew lines RT and QP.

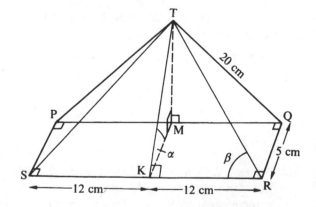

(a) Let K be the mid-point of SR. Planes
     RTS and PQRS meet in RS.
     TK lies in the plane RTS and is
     perpendicular to RS, MK lies in the plane
     PQRS and is perpendicular to RS.

The required angle is TK̂M, which is marked α on the diagram.

$$\tan \alpha = \frac{TM}{MK} = \frac{\sqrt{(20^2 - 12^2)}}{5} = \frac{16}{5}$$

giving $\alpha = 73°$ (to nearest degree)

(b) RS is parallel to QP, so the angle between the skew lines RT and QP equals the angle between RT and RS i.e. the angle TRS, marked β in the diagram.

From $\triangle$TRK $\tan \beta = \dfrac{TK}{12}$

but $TK = \sqrt{(5^2 + TM^2)}$ and $TM^2 = 20^2 - 12^2 = 16^2$

$\therefore$ $\tan \beta = \dfrac{\sqrt{(5^2 + 16^2)}}{12}$

giving $T\hat{R}S = 54°$ (to nearest degree)

Example 31

A particular hillside may be considered to be a plane inclined at 22° to the horizontal. A man walks 60 m due north up a path which follows a line of greatest slope on the hillside. He then walks due west across the hillside to a point P. From P he follows a straight path leading back to his starting point O and the length of this path is 85 m.
Calculate the inclination of the path OP to the horizontal and the total distance the man has walked.

Let OQ be the path due north and let N and M be the feet of the perpendiculars from Q and P to the horizontal plane through the point O. Let $P\hat{O}M = \theta$.

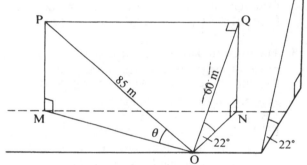

From $\triangle$OQN $QN = 60 \sin 22°$
hence $PM = 60 \sin 22°$

From $\triangle$OPM $\sin \theta = \dfrac{PM}{OP} = \dfrac{60 \sin 22°}{85}$

or $\theta = 15\cdot3°$

Total distance walked $= OQ + QP + PO$
$$= 60 + \sqrt{(85^2 - 60^2)} + 85$$
$$= 205\cdot2 \text{ m}$$

The inclination of the path OP to the horizontal is $15\cdot3°$ and the man walks $205\cdot2$ m.

Exercise G

1. The diagram shows a cuboid ABCDEFGH with CD = 9 cm, BC = 12 cm and FB = 5 cm. Find
   (a) the angle between the line EC and the plane ABCD,
   (b) the angle between the planes EFCD and ABCD,
   (c) the angle between the skew lines EC and AD.

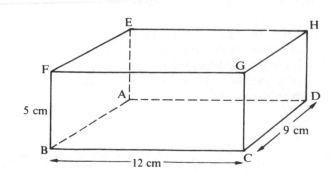

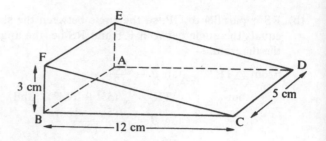

2. The diagram shows a wedge ABCDEF
   with ABCD a horizontal rectangle and
   ABFE a vertical rectangle. BC = 12 cm,
   CD = 5 cm and BF = 3 cm. Find
   (a) the angle between the line FD and the
       base ABCD,
   (b) the angle between the planes EFCD
       and ABCD,
   (c) the angle between the lines BE and
       EC,
   (d) the angle between the skew lines EC
       and FB.

3. ABCDE is a pyramid with ABCD the square horizontal base. AB = 6 cm
   and AE = BE = CE = DE = 8 cm.
   Find (a) the height of the pyramid,
         (b) the angle between the line AE and the base,
         (c) the angle between the plane EAB and the base.

4. ABC is a horizontal triangle, right-angled at B, and BCDE is a vertical
   rectangle. BE = 5 cm, BC = 10 cm and EÂB = 30°.
   Find (a) BÂC, (b) DÂC, (c) DÂE.

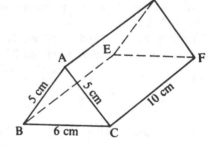

5. The diagram shows a right triangular
   prism of length 10 cm. The rectangular
   base BCFE is horizontal. ABC and DEF
   are vertical isosceles triangles with
   AB = AC = 5 cm and BC = 6 cm.
   Find (a) the height of AD above the base BCFE,
         (b) the angle between the plane DBC and the base,
         (c) the angle between the line DC and the base.

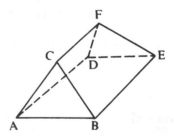

6. The figure ABCDEF shown in the
   diagram has equilateral triangles ABC and
   DEF, of side 8 cm, slanting in towards the
   base such that each triangle makes an
   angle of 60° with the rectangular base ABED.
   Find (a) how much shorter CF is than BE,
         (b) the angle between CB and the base to
             the nearest degree.

7. From the base of a vertical tower of height 30 m point A lies due south
   and point B lies in a direction S60°W with A and B both on the same
   horizontal level as the base of the tower. The top of the tower has angles
   of elevation of 12° from A and 15° from B. Find the distance from A to
   B to the nearest metre.

8. A vertical mast PQ has 4 equal wires attached to its top P. The other ends of the wires are attached to points A, B, C and D on the ground, level with Q. If PA makes an angle of 60° with the ground, find the height of the mast and the angle between the plane PAB and the ground if
(a) ABCD is a square of side 6 m,
(b) ABCD is a rectangle with AB = 6 m and BC = 8 m.

9. A rectangular plot of land ABCD lies in a horizontal plane. A pole of length 3 m is held vertically at C and the top of the pole has an angle of elevation of 16° from B and 25° from D. Find the area of the plot and the elevation of the top of the pole from A (to the nearest degree).

10. VABCD is a pyramid with ABCD as the square base of side 10 m. V is situated vertically above a point P inside ABCD that is 5 m from the line AD and 3 m from the line AB. If VA = VB = 13 m find (a) the angle between the plane VAB and the base, (b) the height of the pyramid, (c) the angle between VA and the base, (d) the angle between the plane VDC and the base.

11. Points A, B and C all lie in the same horizontal plane with B on a bearing N30°W from A and C on a bearing N20°E from A. A vertical tower of height 30 m stands at B and the angle of elevation of its top, from A, is 18°. A vertical tower of height 20 m stands at C and the angle of elevation of the top of the tower from A is 20°. Find the distance from B to C to the nearest metre.

12. From an observation point at sea level, an aircraft is observed on a bearing N30°E, elevation 20° and at height 600 m. The aircraft then flies due east for 1000 m, without altering its height. What will be its bearing and angle of elevation from the observation point then? (Give answers to the nearest degree.)

13. A hillside forms a plane surface inclined at 25° to the horizontal. A straight path up the hillside makes an angle of 60° with the line of greatest slope. Find, to the nearest degree, the inclination of the path to the horizontal.

14. A road is to be constructed up a hillside which may be considered as a plane surface making an angle of 40° with the horizontal. At what angle to the line of greatest slope must the road be constructed if it is to make an angle of 20° with the horizontal? (Give your answer to the nearest degree.)

15. VABCD is a pyramid with the vertex V situated perpendicularly above the centre of the square base ABCD. If θ is the angle between the edge VA and the base, and ϕ is the angle between the plane VAB and the base, show that $\tan \phi = \sqrt{2} \tan \theta$.

16. The diagram below shows a right triangular prism ABCDEF with the rectangular base ABCD horizontal and triangles ABE and DCF vertical. AE = EB, AB = x and BC = $x\sqrt{2}$. The angle between EFCB and the horizontal is θ and the angle between EC and the base is ϕ. Show that $3 \tan \phi = \tan \theta$.

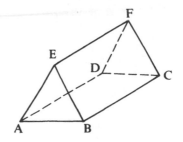

17. The diagram shows a cuboid ABCDEFGH with AB = $4x$, AE = $3x$ and BC = y. If the angle between the skew lines BH and AD is θ show that

$$y \sin \theta - 5x \cos \theta = 0.$$

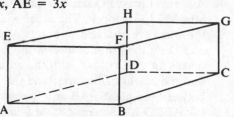

18. Two vertical masts BD and CE, each of height h, have their bases B and C on level ground with C to the east of B.

A point A lies on the same level as B and C and is due south of B. The angle of elevation of D from A is θ and angle DAE = ϕ. If the angle of elevation of E from A is α show that $\sin \alpha = \sin \theta \cos \phi$.

1
Functions

1.1 Basic concepts

Four children, Ann, Bob, Carol and David, are given a spelling test which is marked out of five; their marks for the test are shown in the arrow diagram on the right.

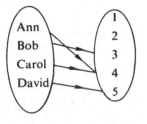

By choosing any one name from the set of names, we can find the mark that relates to it. Any relationship which takes one element of one set and assigns to it one and only one element of a second set is said to be a **function.**

The first set is said to be the **domain** of the function and the second set is the **co-domain.** We say that each element of the first set is **mapped onto** its **image** in the second set. The set of all images will be a subset of the co-domain, and is called the **range.**

Thus in the above example.
the domain is {Ann, Bob, Carol, David}
the co-domain is {1, 2, 3, 4, 5}
the range is {3, 4, 5}

Notice that a function can map more than one element of the domain onto the same element of the range, e.g. Ann → 4 and Carol → 4. Such functions are said to be **many-to-one.** Functions for which each element of the domain is mapped onto a different element of the range are said to be **one-to-one.**

Relationships which are **one-to-many** can occur, but from our definition above, they are *not* functions. The following diagrams illustrate these facts.

one-to-one function   many-to-one function   one-to-many relationship

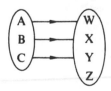   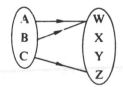   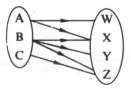

the domain is {A, B, C}   the domain is {A, B, C}   Because this is a
the co-domain is {W, X, Y, Z}   the co-domain is {W, X, Y, Z}   one-to-many relationship,
the range is {W, X, Y}   the range is {W, Z}   it is not a function.

If every element of the co-domain is the image of at least one element of the domain, then the function maps the domain *onto* the co-domain. Otherwise the function maps the domain *into* the co-domain.

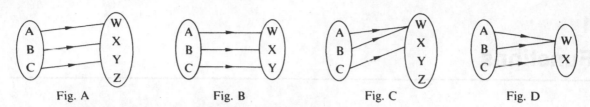

Fig. A   Fig. B   Fig. C   Fig. D

Figure A shows a one-to-one function mapping {A, B, C} into {W, X, Y, Z}
Figure B shows a one-to-one function mapping {A, B, C} onto {W, X, Y}
Figure C shows a many-to-one function mapping {A, B, C} into {W, X, Y, Z}
Figure D shows a many-to-one function mapping {A, B, C} onto {W, X}
For most functions which concern us, the domain will be a set of numbers.

Suppose a function has the set X as the domain and is such that it doubles
any element x of the domain to give the corresponding element of the range.
This function would be written $f: x \rightarrow 2x$ or $f(x) = 2x$, either form being
acceptable. As this function would map the number 2 onto its image, the
number 4, we could write this fact as $f: 2 \rightarrow 4$ or as $f(2) = 4$.

Other letters such as g or h may be used in place of f if we wish to
distinguish between functions.

Example 1

Draw arrow diagrams for the functions
(a) $f: x \rightarrow 2x$ (b) $g: x \rightarrow 3x + 1$ (c) $h: x \rightarrow x^2$
for the domain $\{-1, 0, 1\}$ and state the range of each function.

(a) $f: x \rightarrow 2x$ (b) $g: x \rightarrow 3x + 1$ (c) $h: x \rightarrow x^2$

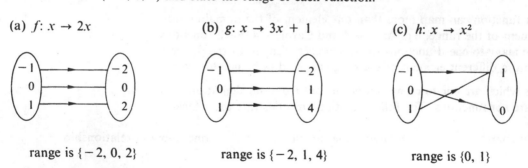

range is $\{-2, 0, 2\}$ range is $\{-2, 1, 4\}$ range is $\{0, 1\}$

It is sometimes helpful to think of functions as 'machines'. A box of
numbers (the domain) is put into the machine. The machine then alters each
number according to some rule and outputs the new numbers into a second
box (the range).

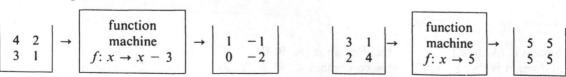

This machine subtracts 3 from   This machine changes each
each number.   number to the number 5.
The domain is {1, 2, 3, 4}   The domain is {1, 2, 3, 4}
The range is $\{-2, -1, 0, 1\}$ The range is {5}

Example 2

State the range of each of the function machines for the domains shown.

(a)

3	2
1	0

$\rightarrow$

function machine
$f: x \rightarrow 2x + 3$

$\rightarrow$ | ? |

(b)

-2		-1
	0	
1		2

$\rightarrow$

function machine
$f: x \rightarrow x^2$

$\rightarrow$ | ? |

(a) $\{0, 1, 2, 3\} \xrightarrow{x \rightarrow 2x + 3} \{3, 5, 7, 9\}$; the range is $\{3, 5, 7, 9\}$

(b) $\{-2, -1, 0, 1, 2\} \xrightarrow{x \rightarrow x^2} \{4, 1, 0\}$; the range is $\{0, 1, 4\}$

Note that it is usual to state the elements of the range in ascending order of magnitude. Therefore the first element of the domain will not necessarily be mapped onto the first element of the range.

Example 3

The functions f and g are given as $f(x) = x + 3$ for $x \geqslant 0$ and $g(x) = x^2$ for $-2 \leqslant x \leqslant 3$. State the range of each of these functions.

If $x \geqslant 0$, then $x + 3 \geqslant 3$. Thus the range of f will be $f(x) \geqslant 3$.
If $-2 \leqslant x \leqslant 3$, then $0 \leqslant x^2 \leqslant 9$. Thus the range of g will be $0 \leqslant g(x) \leqslant 9$.

Example 4

If $f(x) = 4x - 3$ and $g(x) = x^2$ for the domain of all real x, find:
(a) $f(2)$, (b) $f(-2)$, (c) $g(-3)$, (d) the possible values of a if $f(a) = g(a)$.

(a) $f(2) = 4(2) - 3 = 5$

(b) $f(-2) = 4(-2) - 3 = -11$

(c) $g(-3) = (-3)^2 = 9$

(d) $f(a) = 4a - 3$ and $g(a) = a^2$
Thus, if $\quad f(a) = g(a)$
we have $\quad 4a - 3 = a^2$
$0 = a^2 - 4a + 3 \quad$ giving $\quad a = 1$ or 3

If the domain of a function is not stated, it should be assumed to be the set of all real numbers for which the function is defined.

Example 5

The following functions map an element x of the domain onto its image y, i.e. $f: x \rightarrow y$. For each of the three functions below, state (i) the domain for which the function is defined, (ii) the corresponding range of the function, (iii) whether the function is one-to-one or many-to-one.

(a) $f: x \rightarrow x + 3$ (b) $f: x \rightarrow \sqrt{x}$ (c) $f: x \rightarrow \dfrac{1}{x^2}$

(a) $f: x \rightarrow x + 3$
   (i) The function is defined for all real x, so the domain is $\mathbb{R}$.
   (ii) For this domain, the range will contain all elements of $\mathbb{R}$, so the range is $\mathbb{R}$.
   (iii) Each element of the range is obtained from only one element of the domain, so the function is one-to-one.

(b) $f: x \to \sqrt{x}$.

   (i)   The function is not defined for negative x, so the domain is
         $\{x \in \mathbb{R}: x \geqslant 0\}$.

   (ii) For this domain, the range will contain all positive numbers in $\mathbb{R}$.
         Remember that the symbol $\sqrt{}$ is defined to mean the positive square
         root. The range is therefore $\{y \in \mathbb{R}: y \geqslant 0\}$.

   (iii) Each element of the range is obtained from only one element of the
         domain, so the function is one-to-one.

(c) $f: x \to \dfrac{1}{x^2}$

   (i)   The function is defined for all real x except $x = 0$. We write the
         domain as $\{x \in \mathbb{R}: x \neq 0\}$.

   (ii) For this domain, the range will contain neither zero nor any
         negative numbers because x^2 (and hence $1/x^2$) will be positive. The
         range is therefore $\{y \in \mathbb{R}: y > 0\}$.

   (iii) Here the elements of the range can be obtained from more than one
         element of the domain, e.g. $f(3) = \frac{1}{9}$ and $f(-3) = \frac{1}{9}$, so the
         function is many-to-one.

Exercise 1A

1. State which of the following arrow diagrams show functions.

   (a) 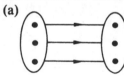   (b) 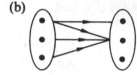   (c) 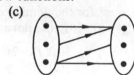   (d)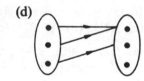

2. State which of the following arrow diagrams show: (i) a one-to-one
   function mapping into the co-domain, (ii) a one-to-one function
   mapping onto the co-domain, (iii) a many-to-one function mapping into
   the co-domain.

   (a) 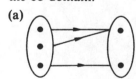   (b) 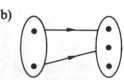   (c) 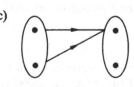   (d)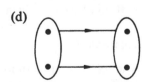

3. State the range of each of the following 'function machines' for the
   domains shown.

   (a)
   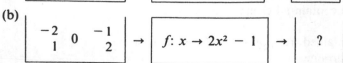

   $$\boxed{\begin{matrix} -2 & & -1 \\ & 0 & \\ 1 & & 2 \end{matrix}} \to \boxed{f: x \to 3x - 2} \to \boxed{?}$$

   (b)

   $$\boxed{\begin{matrix} -2 & & -1 \\ & 0 & \\ 1 & & 2 \end{matrix}} \to \boxed{f: x \to 2x^2 - 1} \to \boxed{?}$$

4. Draw arrow diagrams for the functions $f: x \to x + 2$, $\quad g: x \to x^2 + 1$ and $h: x \to (x + 1)^2$ for the domain $\{-2, -1, 0, 1, 2\}$ and state the range of each function for this domain.

5. Draw arrow diagrams for the functions $f: x \to |x|$, $\quad g: x \to |x| - 1$ and $h: x \to |x - 1|$ for the domain $\{-2, -1, 0, 1, 2\}$ and state the range of each function for this domain.

6. If $f(x) = 2x + 3$ find
(a) $f(2)$, (b) $f(-1)$, (c) $f(6)$, (d) the value of a if $f(a) = a$.

7. If $g(x) = x^2 - 6$ find
(a) $g(4)$, (b) $g(-4)$, (c) $g(2)$, (d) the possible values of a if $g(a) = a$.

8. If $f(x) = 2x^2$ and $g(x) = 3 - x$ find
(a) $f(3)$, (b) $f(-3)$, (c) $g(-3)$, (d) the possible values of a if $f(a) = g(a)$.

9. The function f is given by $f(x) = ax + b$. If $f(3) = 3$ and $f(4) = 5$, find a and b.

10. The function g is given by $g(x) = ax^2 - b$. If $g(2) = 5$ and $g(-1) = 2$, find the values of a and b and hence find $g(-4)$.

11. Each of the following functions maps an element x of the domain onto its image y, i.e. $f(x) = y$. Find the range of each function for the given domains and state whether the function is one-to-one or many-to-one.
(a) $f: x \to x + 3$ with domain $\{x: 0 \leqslant x \leqslant 4\}$,
(b) $f: x \to x - 2$ with domain $\{x: 0 \leqslant x \leqslant 4\}$,
(c) $f: x \to 2x$ with domain $\{x: 0 \leqslant x \leqslant 3\}$,
(d) $f: x \to 2x$ with domain $\{x: -3 \leqslant x \leqslant 3\}$,
(e) $f: x \to x^2$ with domain $\{x: -3 \leqslant x \leqslant 3\}$,
(f) $f: x \to \sqrt{x}$ with domain $\{x: 0 \leqslant x \leqslant 25\}$,
(g) $f: x \to |x|$ with domain $\{x: -3 \leqslant x \leqslant 3\}$,
(h) $f: x \to x^2$ with domain $\mathbb{R}$,
(i) $f: x \to |x|$ with domain $\mathbb{R}$,
(j) $f: x \to \dfrac{1}{x}$ with domain $\{x: x \geqslant 1\}$,
(k) $f: x \to x^2 + 4$ with domain $\mathbb{R}$,
(l) $f: x \to \dfrac{1}{x - 1}$ with domain $\{x \in \mathbb{R}: x \neq 1\}$.

12. The following functions map an element x of the domain onto its image y, i.e. $f: x \to y$. For each function state
(i) the domain for which the function is defined,
(ii) the corresponding range of the function.
(a) $f: x \to 2x$,   (b) $f: x \to |x|$,   (c) $f: x \to \dfrac{1}{x}$,   (d) $f: x \to \dfrac{1}{x - 3}$.

13. f and g are defined as follows: $f: x \to \begin{cases} 2x & \text{for } 0 \leqslant x \leqslant 3 \\ x^3 & \text{for } 3 \leqslant x \leqslant 6 \end{cases}$

$g: x \to \begin{cases} 3x & \text{for } 0 \leqslant x \leqslant 3 \\ x^2 & \text{for } 3 \leqslant x \leqslant 6 \end{cases}$

Explain why g is a function but f is not.

14. f and g are defined as follows: $f: x \to \begin{cases} x^2 + 2 & \text{for } 0 \leqslant x \leqslant 2 \\ 3x & \text{for } 2 \leqslant x \leqslant 4 \end{cases}$

$g: x \to \begin{cases} x^2 & \text{for } 0 \leqslant x \leqslant 2 \\ 3x & \text{for } 2 \leqslant x \leqslant 4 \end{cases}$

Explain why f is a function whereas g is not.

1.2 Composite functions

Consider the functions $f(x) = 2x - 1$ and $g(x) = x^2$. Using the 'machine' analogy, suppose we input the numbers $\{1, 3, 5, 7\}$ into machine f and then take the output from this machine and put it into machine g.

	function machine $f(x) = 2x - 1$			function machine $g(x) = x^2$	
3 7 →		→	5 13 →		→	25 169
1 5			1 9			1 81

This combined or composite function is written $gf(x)$ or simply gf. Notice that the function f is performed first and so is written nearer to the variable x.

The set $\{1, 3, 5, 7\}$ is the domain for the composite function and $\{1, 25, 81, 169\}$ is the range.

Example 6

If $f(x) = 2x$ and $g(x) = x^2 - 1$, find the range of each of the following functions for the domain $\{-2, -1, 0, 1, 2\}$
(a) $f(x)$, (b) $g(x)$, (c) $fg(x)$, (d) $gf(x)$.

(a) $\{-2, -1, 0, 1, 2\} \xrightarrow{f} \{-4, -2, 0, 2, 4\}$; the range is $\{-4, -2, 0, 2, 4\}$

(b) $\{-2, -1, 0, 1, 2\} \xrightarrow{g} \{3, 0, -1\}$; the range is $\{-1, 0, 3\}$

(c) $\{-2, -1, 0, 1, 2\} \xrightarrow{g} \{3, 0, -1\} \xrightarrow{f} \{6, 0, -2\}$; the range is $\{-2, 0, 6\}$

(d) $\{-2, -1, 0, 1, 2\} \xrightarrow{f} \{-4, -2, 0, 2, 4\} \xrightarrow{g} \{15, 3, -1\}$; the range is $\{-1, 3, 15\}$

Example 7

If $f(x) = 2x$ and $g(x) = 3x + 1$, find
(a) $f(2)$, (b) $g(3)$, (c) $fg(2)$, and express $gf(x)$ as a single function $h(x)$.

(a) $f(2) = 2(2)$
$\quad = 4$

(b) $g(3) = 3(3) + 1$
$\quad = 10$

(c) $\quad g(2) = 3(2) + 1$
$\quad\quad = 7;$
so $fg(2) = f(7)$
$\quad\quad = 14$

$f(x) = 2x$, so $gf(x) = g(2x)$
$\quad\quad\quad\quad = 3(2x) + 1$
$\quad\quad\quad\quad = 6x + 1 \quad\quad \therefore \quad h(x) = 6x + 1$

1.3 The inverse of a function

Consider a function f which maps each element x of the domain X onto its image y in the range Y,
  i.e. $f: x \rightarrow y$ where $x \in X$ and $y \in Y$.
Can we find the inverse function, written f^{-1}, that has domain Y and range X, which maps y back to x?
  i.e. $f^{-1}: y \rightarrow x$

First recall the definition of a function. Any one-to-one or many-to-one relationship is a function. However, if we attempted to find the inverse of a many-to-one function, we would obtain a one-to-many relationship which is not a function. Thus, only a one-to-one function can have an inverse function.

The following diagrams illustrate these points.

domain X $\xrightarrow{f}$ range Y

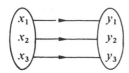

f is a one-to-one function

domain U $\xrightarrow{g}$ range V

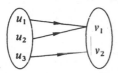

g is a many-to-one function

domain Y $\xrightarrow{f^{-1}}$ range X

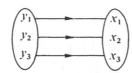

f^{-1} is a one-to-one function

domain V $\xrightarrow{g^{-1}}$ range U

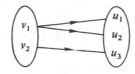

g^{-1} is one-to-many and so it is *not* a function

To find the inverse of one-to-one function, we write the separate operations of the function as a flow chart. We then reverse the flow chart, writing the inverse of each operation.

Example 8

Find the inverses of these functions.

(a) $f(x) = 2x + 3$ (b) $g(x) = 2 - x$ (c) $h(x) = \dfrac{1}{x} - 3$

(a) First write the function as a flow chart with input x and output $2x + 3$

$$x \to \boxed{\times 2} \to \boxed{+3} \to 2x + 3$$

Now reverse the flow chart, and write the inverse of each operation

$$\leftarrow \boxed{\div 2} \leftarrow \boxed{-3} \leftarrow$$

With input x, this will now output the inverse function.

$$\frac{x - 3}{2} \leftarrow \boxed{\div 2} \leftarrow \boxed{-3} \leftarrow x$$

So the inverse function f^{-1} is $x \to \dfrac{x - 3}{2}$

Check: $f(2) = 7$, $f^{-1}(7) = 2$.

(b) First write the function as a flow chart.

$$x \to \boxed{\times (-1)} \to \boxed{+2} \to 2 - x$$

Now reverse the flow chart, writing the inverse of each operation.

$$\frac{x-2}{-1} \leftarrow \boxed{\div (-1)} \leftarrow \boxed{-2} \leftarrow x$$

but $\dfrac{x-2}{-1} = -x + 2 = 2 - x$

So the inverse function g^{-1}, is $x \to 2 - x$, i.e. g is its own inverse.

Check: $g(3) = -1$, $g^{-1}(-1) = 2 - (-1) = 3$.

(c) First write the function as a flow chart

$$x \to \boxed{\text{invert}} \to \boxed{-3} \to \frac{1}{x} - 3$$

Now reverse the flow chart (notice that the operation 'invert' is its own inverse)

$$\frac{1}{x+3} \leftarrow \boxed{\text{invert}} \leftarrow \boxed{+3} \leftarrow x$$

So the inverse function h^{-1} is $x \to \dfrac{1}{x+3}$

Check: $h(2) = -2\frac{1}{2}$, $h^{-1}(-2\frac{1}{2}) = \dfrac{1}{\frac{1}{2}} = 2.$

Example 9

If $f: x \to x + 1$ and $g: x \to \dfrac{3}{x}$ find in similar form:

(a) f^{-1}, (b) g^{-1}, (c) fg, (d) gg, (e) $(fg)^{-1}$.

(a) $f:$ $\quad x \to \boxed{+1} \to x + 1$

$f^{-1}: x - 1 \leftarrow \boxed{-1} \leftarrow x$

So $f^{-1}: x \to x - 1$
Check: $f: 3 \to 4$, $f^{-1}: 4 \to 3$.

(b) $g: x \to \boxed{\text{invert}} \to \boxed{\times 3} \to \dfrac{3}{x}$

$g^{-1}: \dfrac{3}{x} \leftarrow \boxed{\text{invert}} \leftarrow \boxed{\div 3} \leftarrow x$

So $g^{-1}: x \to \dfrac{3}{x}$ (i.e. g is its own inverse)

Check: $g: 6 \to \frac{1}{2}$, $g^{-1}: \frac{1}{2} \to 6$.

(c) $g: x \to \dfrac{3}{x}$

$f: \dfrac{3}{x} \to \dfrac{3}{x} + 1$

So $fg: x \to \dfrac{3}{x} + 1$

(d) $g: x \to \dfrac{3}{x}$

$g: \dfrac{3}{x} \to \dfrac{3}{\frac{3}{x}} = x$

So $gg: x \to x$

(e) fg:

$$x \to \boxed{\text{invert}} \to \boxed{\times 3} \to \boxed{+1} \to \dfrac{3}{x} + 1$$

$(fg)^{-1}: \dfrac{3}{x-1} \leftarrow \boxed{\text{invert}} \leftarrow \boxed{\div 3} \leftarrow \boxed{-1} \leftarrow x$

So $(fg)^{-1}: x \to \dfrac{3}{x-1}$

Check: $fg: 3 \to 2$, $(fg)^{-1}: 2 \to 3$.

The reader may notice similarities between this work on inverse functions and the process of 'changing the subject of a formula' which may have been met in earlier years.

For example, in part (a) of Example 8, we found that for $f(x) = 2x + 3$ then $f^{-1}(x) = \dfrac{x-3}{2}$. This is similar to making x the subject of $y = 2x + 3$ as follows:

$$y = 2x + 3$$

rearranging $y - 3 = 2x$

then $x = \dfrac{y-3}{2}$

Inverse functions can be obtained using this technique.

Example 10

Find the inverses of the functions.

(a) $f(x) = 3x - 1$,   (b) $g(x) = \dfrac{3}{x-1}$.

(a) If $f(x) = y$, we require $f^{-1}(y) = x$

If $y = 3x - 1$,

then $x = \dfrac{y+1}{3}$

So, given y, we can return to x using the expression $\dfrac{y+1}{3}$.

Thus $f^{-1}(x) = \dfrac{x+1}{3}$

(b) If $g(x) = y$, we require $g^{-1}(y) = x$

If $y = \dfrac{3}{x-1}$,

then $\dfrac{1}{y} = \dfrac{x-1}{3}$ or $x = \dfrac{3}{y} + 1$

So, given y, we can return to x using the expression $\dfrac{3}{y} + 1$.

Thus $g^{-1}(x) = \dfrac{3}{x} + 1$

Restricting domains

Although we have said that only one-to-one functions have inverses, we can consider the inverse of a many-to-one function if we restrict the function to a domain for which it is one-to-one.

For example, $f: x \to x^2$ is a many-to-one function for a domain $\mathbb{R}$, but if we restrict the domain to $\{x \in \mathbb{R}: x \geqslant 0\}$, the function is then one-to-one and will have an inverse.

Exercise 1B

1. If $f(x) = 5x$ and $g(x) = x^2 + 3$ find
   (a) $f(3)$,   (b) $g(2)$,   (c) $gf(3)$,   (d) $fg(2)$,   (e) $gg(3)$,   (f) $ff(3)$.
2. If $f(x) = 2x + 1$ and $g(x) = x^2$ find the range of each of the following functions for the domain $\{-3, -2, -1, 0, 1, 2, 3\}$
   (a) $f(x)$,   (b) $g(x)$,   (c) $fg(x)$,   (d) $gf(x)$.
3. If $f(x) = -x$, $g(x) = 1 - x^2$ and $h(x) = |x|$ find the range of each of the following functions for the domain $\{-2, -1, 0, 1, 2\}$:
   (a) fg,   (b) gf,   (c) gh,   (d) fgh,   (e) hgf.
4. If $f: x \to 3x$, $g: x \to 2x - 1$ and $h: x \to x^2$ express the following as single functions:
   (a) fg,   (b) gf,   (c) gh,   (d) hg,   (e) fh,   (f) fgh,   (g) ghf.
5. If $f: x \to 2x$ and $g: x \to x + 4$ state which of the functions ff, fg, gf or gg corresponds to
   (a) $x \to 2x + 4$,   (b) $x \to x + 8$,   (c) $x \to 2x + 8$.
6. If $f: x \to x^2$ and $g: x \to x + 2$ state which of the functions ff, fg, gf or gg corresponds to
   (a) $x \to x + 4$,   (b) $x \to x^2 + 2$,   (c) $x \to x^2 + 4x + 4$.
7. If $f: x \to x + 1$ $g: x \to x^2$ and $h(x) \to \dfrac{2}{x}$ express the following in similar form
   (a) fg,   (b) fh,   (c) gh,   (d) hg,   (e) hfg,   (f) hgf.
8. Find the inverses of the following functions:
   (a) $f: x \to 3x - 2$ (b) $f: x \to \dfrac{x}{2}$ (c) $f: x \to 5 - x$

   (d) $f: x \to (x + 1)^2$ (e) $f: x \to \dfrac{1}{x} + 2$ (f) $f: x \to 2 - \dfrac{1}{x}$

   (g) $f: x \to \dfrac{1}{x + 2}$ (h) $f: x \to \dfrac{1}{2x - 6}$

9. If $f: x \to 2x + 3$, $g: x \to x - 2$ and $h: x \to \dfrac{2}{x}$ find in similar form
   (a) fg,   (b) $(fg)^{-1}$,   (c) hg,   (d) $(hg)^{-1}$,   (e) hgf,   (f) $(hgf)^{-1}$.

1.4 Ordered pairs

Consider the sets A = $\{1, 2\}$ and B = $\{2, 3, 4\}$. The diagram on the right represents function $f: x \to x + 2$ where $x \in A$ and $y \in B$.

Alternatively we could show this relationship by listing the possible values of x and y as **ordered pairs**. For the function above, the ordered pairs would be $(1, 3)$ and $(2, 4)$.

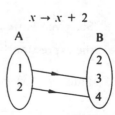

The **cartesian product** of the two sets A and B is written A × B and is the set of all possible ordered pairs (x, y) where $x \in$ A and $y \in$ B. For the sets A = {1, 2} and B = {2, 3, 4}, the cartesian product would be:

$$\{(1, 2), (1, 3), (1, 4), (2, 2), (2, 3), (2, 4)\}$$

Example 11

If A = {2, 3} and B = {4, 5, 6, 7, 8}, write down all the ordered pairs (x, y) such that $x \in$ A, $y \in$ B and x is a factor of y.

The required ordered pairs would be (2, 4), (2, 6), (2, 8), (3, 6).

Example 12

Four members of the cartesian product of two sets A and B are (2, 1), (2, 3), (4, 1) and (4, 5). If there are a total of six such ordered pairs in the cartesian product A × B, find:
(a) the two missing members of A × B,
(b) the elements of A and B,
(c) the ordered pairs (x, y) such that $x \in$ A, $y \in$ B and x is less than y.

(a) From the ordered pairs (2, 1), (2, 3), (4, 1) and (4, 5), set A must contain the elements {2, 4} and set B must contain {1, 3, 5}. Thus the two missing members of A × B are (2, 5) and (4, 3)
(b) A = {2, 4} and B = {1, 3, 5}
(c) The required ordered pairs are (2, 3), (2, 5), (4, 5).

1.5 The graph of a function

Consider the function $f: x \to x + 3$ which maps the domain X onto the range Y. If $x \in$ X and $y \in$ Y such that $f: x \to y$, we can write down the ordered pairs (x, y). If X = ℝ, then Y = ℝ and there would be an infinite number of such ordered pairs (x, y). However, if we restrict the domain X to certain values, say X = {−2, −1, 0, 1, 2}, we can then list the ordered pairs: (−2, 1), (−1, 2), (0, 3), (1, 4), and (2, 5).

This set of ordered pairs can be plotted as points on a graph. If these points are joined, we obtain the graph of the function $f: x \to x + 3$ for the domain $\{x \in ℝ: |x| \leqslant 2\}$.

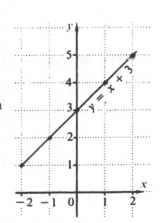

We say that the graph of this function has the **equation** $y = x + 3$; the ordered pair fixing each point gives the x- and y-coordinates of that point on the graph.

As the value of x varies over the whole domain, so each corresponding value of y can be obtained; x and y are called **variables** with x the **independent variable** and y the **dependent variable**.

The equation need not always be given with y as the 'subject'. For example, $y = x + 3$ can be written as $y - x = 3$, $y - x - 3 = 0$, etc.

Clearly, if the domain were the set of all real numbers $\mathbb{R}$, we could not plot the entire graph of the function. Usually we are only interested in certain values of x or in a particular section of $\mathbb{R}$ for which a graph is required.

The ordered pairs are usually written as a table of values. For the ordered pairs obtained above, the table of values would be

x	-2	-1	0	1	2
y	1	2	3	4	5

Example 13

For each of the following functions
(i)   write down the equation of the function,
(ii) construct a table of values for the given domain,
(iii) plot the graph of the function for that domain.
(a) $f: x \rightarrow 2x + 1$ for $|x| \leqslant 2$,   (b) $f: x \rightarrow 3$ for $0 \leqslant x \leqslant 4$,
(c) $f: x \rightarrow x^2 + 2x - 2$ for $-4 \leqslant x \leqslant 2$.

(a) (i)   The equation will be $y = 2x + 1$
    (ii)

x	-2	-1	0	1	2
y	-3	-1	1	3	5

(iii)

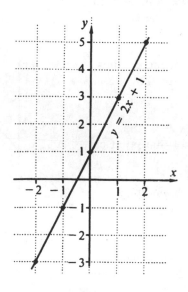

(b) (i)   The equation will be $y = 3$.
    (ii) In this case y will equal 3 for every value of x.

x	0	1	2	3	4
y	3	3	3	3	3

(iii)

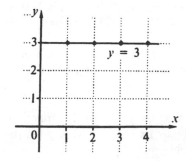

(c) (i) The equation will be
$$y = x^2 + 2x - 2$$

(ii) In this case the equation is more complicated so we will build up each y value gradually.

x	-4	-3	-2	-1	0	1	2
x^2	16	9	4	1	0	1	4
$+2x$	-8	-6	-4	-2	0	2	4
-2	-2	-2	-2	-2	-2	-2	-2
y	6	1	-2	-3	-2	1	6

(iii)

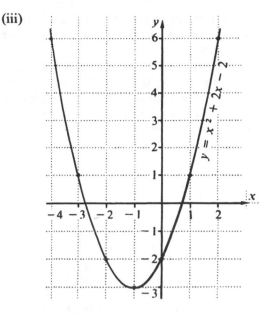

In some cases, the chosen points may be insufficient to draw the graph with certainty, so it will then be necessary to take extra values for x to obtain more points. In Example 13 part (c), we could have included $x = -\frac{1}{2}$ and $x = -1\frac{1}{2}$ to give two more points at the part of the graph which is difficult to draw.

Example 14

State which of the following points lie on the line $y = 2x - 3$:
$(2, 1), (-2, -1), (-1, -5)$.

For a point to lie on a particular line the coordinates of that point must satisfy the equation of that line.

Substituting $x = 2$ into $y = 2x - 3$ gives $y = 1$. $\therefore$ (2, 1) does lie on the line.

Substituting $x = -2$ into $y = 2x - 3$ gives $y = -7$. $\therefore$ $(-2, -1)$ does not lie on the line.

Substituting $x = -1$ into $y = 2x - 3$ gives $y = -5$. $\therefore$ $(-1, -5)$ does lie on the line.

Thus the points $(2, 1)$ and $(-1, -5)$ lie on the line $y = 2x - 3$.

Example 15

Find where the line $y = 2x - 6$ cuts (a) the x-axis (b) the y-axis.

(a) Any point on the x-axis will have a y-coordinate of zero.
Substituting $y = 0$ into $y = 2x - 6$ gives $0 = 2x - 6$
i.e. $x = 3$
$\therefore$ $y = 2x - 6$ cuts the x-axis at (3, 0).

(b) Any point on the y-axis will have an x-coordinate of zero.
Substituting $x = 0$ into $y = 2x - 6$ gives $y = -6$
$\therefore$ $y = 2x - 6$ cuts the y-axis at $(0, -6)$.

It is important to realise that it is possible to write equations
and to draw graphs for relationships between x and y that are
not functions.

The equation $x^2 + y^2 = 25$ is a relationship between x and y
and the graph of all possible ordered pairs (x, y) gives a circle.
However this relationship is not a function because every x
value does not have one and only one associated y value, e.g.
for $x = 0$, y could equal 5 or -5.

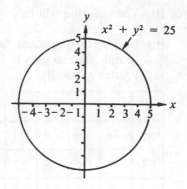

Exercise 1C

1. If $A = \{3, 4, 5, 6\}$ and $B = \{1, 2, 3, 4\}$, write down all the ordered pairs
   (x, y) such that $x \in A$, $y \in B$ and x is twice y.
2. If $A = \{2, 3, 4\}$, write down all the ordered pairs (x, y) such that $x \in A$,
   $y \in A$ and x is greater than y.
3. If $A = \{-2, -1, 0, 1, 2\}$ and $B = \{0, 1, 2, 3, 4\}$, write down all the
   ordered pairs (x, y) such that $x \in A$, $y \in B$ and $y = x^2$.
4. Three elements of the cartesian product $A \times B$ are $(2, 3)$, $(2, 4)$ and
   $(3, 5)$. If there are six such ordered pairs in the cartesian product, find
   (a) the sets A and B,
   (b) the other three elements of $A \times B$,
   (c) set C, a subset of $A \times B$, such that $C = \{(x, y): x \in A, y \in B$ and $x = y\}$.
5. If $A = \{1, 2, 3\}$ and $B = \{1, 2, 3, 4, 5, 6\}$, find the ordered pairs
   of set C given that $C = \{(x, y): x \in A, y \in B$ and $y = 2x\}$.
6. State which of the following points lie on the line $y = 8 - 3x$,
   $(2, 2)$, $(-1, 5)$, $(1, 5)$, $(4, -4)$.
7. If all of the following points lie on the line $y = 2x - 6$, find the values
   of a, b, c, d and e: $(5, a)$, $(2, b)$, $(-2, c)$, $(d, 2)$, $(e, 8)$.
8. If the point $(2, 2)$ lies on the line $y = ax - 4$, find the value of a.
9. If the points $(2, 1)$ and $(-2, -11)$ lie on the line $y = ax + b$, find the
   values of a and b.
10. Find where the following lines cut (i) the y-axis (ii) the x-axis.
    (a) $y = x - 4$ (b) $y = 2x - 4$ (c) $y = 12 - 2x$
    (d) $y = \frac{1}{2}x + 3$ (e) $y + 2x = 8$ (f) $y + 5x = 3$
    (g) $2y - 5x = 12$ (h) $y = x^2 - 3x + 2$ (i) $y = x^2 + x - 6$

For each of the functions in questions **11** to **20**, (a) write down the equation
of the function, (b) construct a table of values for the given domain,
(c) plot the graph of the function for that domain.

11. $f: x \rightarrow x + 1$ for $|x| \leqslant 3$ 12. $f: x \rightarrow x - 2$ for $|x| \leqslant 4$
13. $f: x \rightarrow 2x + 3$ for $|x| \leqslant 3$ 14. $f: x \rightarrow x$ for $-2 \leqslant x \leqslant 4$

15. $f: x \to 5$ for $|x| \leqslant 3$
17. $f: x \to x^2 + 3x - 2$ for $-5 \leqslant x \leqslant 2$
19. $f: x \to 2x^2 - 4x - 5$ for $-2 \leqslant x \leqslant 4$

16. $f: x \to -2$ for $|x| \leqslant 3$
18. $f: x \to x^2 - 2x - 4$ for $-3 \leqslant x \leqslant 5$
20. $f: x \to 10 + x - x^2$ for $-3 \leqslant x \leqslant 4$

1.6 Some further considerations

Odd and even functions

Any function for which $f(-x) = f(x)$ is called an **even** function.
Two examples of even functions are
$$f: x \to x^2 \qquad \text{and} \quad g: x \to |x|$$
$$\text{e.g. } f(-3) \to 9 = f(3) \qquad \text{e.g. } g(-3) \to 3 = g(3)$$

Any function for which $f(-x) = -f(x)$ is called an **odd** function.
Two examples of odd functions are
$$f: x \to x \qquad \text{and} \quad g: x \to x^3$$
$$\text{e.g. } f(-3) \to -3 = -f(3) \qquad \text{e.g. } g(-3) \to -27 = -g(3)$$

(The words even and odd are used because the functions $f: x \to x^n$ have the property that $f(-x) = f(x)$ for even values of n and the property that $f(-x) = -f(x)$ for odd values of n).
   It should be noted that most functions are neither odd nor even.

Example 16

Show that the function $f(x) = 4x^3 - x$ is an odd function.
$$f(-x) = 4(-x)^3 - (-x)$$
$$= -4x^3 + x$$
$$= -(4x^3 - x)$$
$$= -f(x)$$
Thus $f(x) = 4x^3 - x$ is an odd function.

Consider the graph of an even function that maps $x \to y$, i.e. $f(x) = y$. Because the function is even, we know that $f(-x) = y$. Thus for every point (x, y) on the graph of the function, there will also be a point $(-x, y)$. The graph of an even function will therefore be symmetrical about the y-axis.

Consider the graph of an odd function that maps $x \to y$,   i.e. $f(x) = y$. Because the function is odd, we know that $f(-x) = -y$. Thus for every point (x, y) on the graph of the function, there will also be a point $(-x, -y)$. The graph of an odd function will therefore be unchanged under a $180°$ rotation about the origin.

Example 17

For each of the graphs shown below, state whether they are graphs of odd functions, even functions or neither of these.

(a)   (b)   (c)   (d)

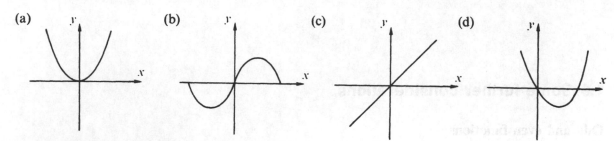

(a) The graph is symmetrical about the y-axis.
   Therefore it is the graph of an even function.
(b) The graph is unchanged under a 180° rotation about the origin.
   Therefore it is the graph of an odd function.
(c) The graph is unchanged under a 180° rotation about the origin.
   Therefore it is the graph of an odd function.
(d) The graph is neither symmetrical about the y-axis nor is it unchanged
   under a 180° rotation about the origin. Therefore the graph is neither an
   odd nor an even function.

Graphs of inverse functions

If the function f maps $x \to y$ then f^{-1} maps $y \to x$. Thus for every
point (x, y) on the graph of the function f, there will exist a point
(y, x) on the graph of the function f^{-1}. For example, a point $(2, 1)$
on the graph of f will mean there is a point $(1, 2)$ on the graph of
f^{-1}. A point $(1, 0)$ on the graph of f will mean there is a point $(0, 1)$
on the graph of f^{-1} and so on.

So we could obtain the graph of f^{-1} by finding the ordered pairs (x, y) for $f: x \to y$ and plotting
them as (y, x). This will give the same line that we would obtain by reflecting the graph of f in the
line $y = x$ as the following graphs illustrate.

(a) $f: x \to x + 3$ for $x \in \mathbb{R}$ (b) $g: x \to 2x + 1$ for $x \in \mathbb{R}$ (c) $h: x \to x^2$ for $\{x \in \mathbb{R}: x \geqslant 0\}$
   $f^{-1}: x \to x - 3$ for $x \in \mathbb{R}$ $g^{-1}: x \to \dfrac{x - 1}{2}$ for $x \in \mathbb{R}$ $h^{-1}: x \to \sqrt{x}$ for $\{x \in \mathbb{R}: x \geqslant 0\}$

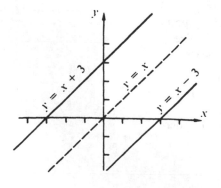
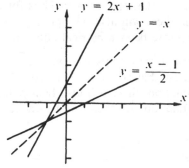
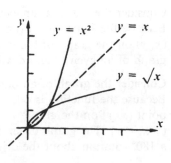
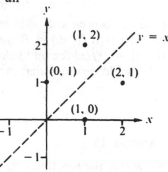

Exercise 1D

1. Show that each of the following functions are odd functions:
   (a) $f(x) = 7x$,   (b) $f(x) = x^3 + x$,   (c) $f(x) = 2x^3 - 3x$.
2. Show that each of the following functions are even functions:
   (a) $f(x) = 4x^2$,   (b) $f(x) = 2 + x^2$,   (c) $f(x) = 3x^2 + 2|x|$.
3. For each of the following functions, state whether they are even, odd or neither of these:
   (a) $f(x) = 4 - 3x^2$,   (b) $f(x) = 3x^2 + x$,   (c) $f(x) = x - \dfrac{1}{x}$,

   (d) $f(x) = x^2 + |x|$,   (e) $f(x) = x^3 + |x|$.
4. For each of the following graphs, state whether they are graphs of odd functions, even functions or neither of these.

(a) 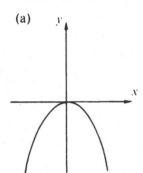   (b)    (c) 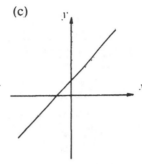   (d)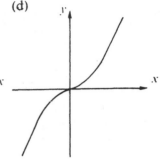

5. Find the equations of the lines obtained if each of the following lines is reflected in the line $y = x$.
   (a) $y = \dfrac{x}{3}$,   (b) $y = 4 - x$,   (c) $y = 2x - 4$,   (d) $y = \dfrac{1}{x + 2}$.

Exercise 1E   Examination questions

1. Find the range of the function $f: x \rightarrow |x - 1|$ corresponding to a domain $-3 \leqslant x \leqslant 3$.
   (Cambridge)

2. The function $f: x \rightarrow \dfrac{a}{x} + b$ is such that $f(-1) = 1\frac{1}{2}$ and $f(2) = 9$.

   (i)   State the value of x for which f is not defined.
   (ii) Find the value of a and of b.
   (iii) Evaluate $f(4)$ and $f^{-1}(4)$.
   (Cambridge)

3. The arrow diagram on the right represents part of the mapping $f: x \rightarrow \dfrac{72}{ax + b}$.

   (i) Find the value of a and of b.
   (ii) Find the element that under this mapping has an image of 4.
   (Cambridge)

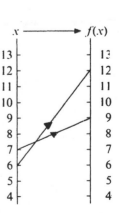

4. Given the functions $f: x \rightarrow 2x - 3$ and $g: x \rightarrow \dfrac{8}{x}$, find in similar form

   (i) f^{-1},   (ii) ff,   (iii) gg,   (iv) fg,   (v) $(fg)^{-1}$.
   (Cambridge)

5. Express in terms of the functions $f: x \to \sqrt{x}$ and $g: x \to x + 5$,

   (i) $x \to \sqrt{(x + 5)}$.   (ii) $x \to x - 5$,   (iii) $x \to x + 10$,

   (iv) $x \to \sqrt{x} + 10$,   (v) $x \to x^2 + 5$.   (Cambridge)

6. Given the functions $f: x \to 2x - 5$ and $g: x \to \dfrac{4}{x}$, express in similar form:

   (i) fg,   (ii) g^2,   (iii) g^{17}, $(g^2 = gg,$ etc.).

   Express in terms of one or both of f and g,

   (iv) $x \to \frac{1}{2}(x + 5)$,   (v) $x \to \dfrac{4}{2x - 5}$,   (vi) $x \to 4x - 15$.   (Cambridge)

2
Vectors

2.1 Basic concepts

Suppose an insect walks directly from a point A to a point B, 30 cm from A, and then to point C, 20 cm from B (see Figure 1). The insect has walked 50 cm altogether but it is clearly not 50 cm from its original position, i.e. AB + BC ≠ AC. However, it is true to say that, in travelling from A to B and then from B to C, the insect arrives at the same point C as it would have done had it travelled directly from A to C. The position of C, relative to A, is not changed by the route taken by the insect. We can write

$$\overrightarrow{AB} + \overrightarrow{BC} = \overrightarrow{AC}$$

where $\overrightarrow{AB}$ indicates that both the length and direction of AB are being considered.

Quantities involving both magnitude and direction are called **vector** quantities. Quantities involving only magnitude are called **scalar** quantities.

Thus $\overrightarrow{AB}$ is the vector whose length is equal to that of the line AB and whose direction is that from A to B. The vector $\overrightarrow{BA}$ would be the same length as $\overrightarrow{AB}$ but will be in the opposite direction, i.e. $\overrightarrow{BA}$ will be in the direction from B to A. We write $|\overrightarrow{AB}|$ or simply AB for the length of the vector $\overrightarrow{AB}$.

Thus, from Figure 1, $|\overrightarrow{AB}| = 30$ cm and $|\overrightarrow{BC}| = 20$ cm or simply AB = 30 cm and BC = 20 cm.

If we wish to find the vector $\overrightarrow{AC}$ we could, given the necessary bearings, find its length and direction by the use of trigonometry. Taking the directions of $\overrightarrow{AB}$ and $\overrightarrow{BC}$ as N60°E and S10°E respectively, we first make a rough sketch:

By the cosine rule
$$AC^2 = 30^2 + 20^2 - 2 \times 30 \times 20 \cos 70°$$
$$\therefore \quad AC = 29 \cdot 83 \text{ cm}$$
By the sine rule
$$\frac{20}{\sin \theta} = \frac{AC}{\sin 70°}$$
$$\therefore \quad \sin \theta = \frac{20 \sin 70°}{29 \cdot 83}$$
$$\therefore \qquad \theta = 39 \cdot 05°$$

Thus $\overrightarrow{AC}$ is of length 29·8 cm and direction S80·95°E

Alternatively, a solution could be obtained by making an accurate scale drawing.

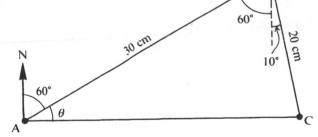
Figure 1

Vectors may also be written using single letters written in heavy type:

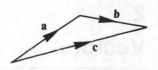

The arrows on the diagram indicate the directions of the vectors. Thus
c = **a** + **b**.
The length of vector **a** is written |**a**| or simply as *a*.

Note: Whilst text books can use heavy type to indicate vectors, this is not
easy when writing by hand. The reader is advised to show these
single letter vectors underlined.
Thus **c** = **a** + **b** is written as c̲ = a̲ + b̲

Equal vectors

For two vectors to be equal, they must have the same magnitude and the
same direction. Thus if we represent vector **a** by a line segment of a certain
magnitude and direction, then any other line segment of the same magnitude
and direction will also equal **a**.

Parallel vectors

Two vectors **a** and **b** are parallel if one is a scalar multiple of the other,
i.e. if **a** = λ**b**.
If λ is positive, the vectors are parallel and in the same direction:
i.e. they are *like* parallel vectors.

If λ is negative, the vectors are parallel and in opposite directions:
i.e. they are *unlike* parallel vectors.

Example 1

In the parallelogram OABC, $\overrightarrow{OA}$ = **a** and $\overrightarrow{OC}$ = **c**. The point D lies on AB
and is such that AD:DB = 1:2. Express the following vectors in terms of **a**
and **c**:
(a) $\overrightarrow{CB}$ (b) $\overrightarrow{BC}$ (c) $\overrightarrow{AB}$ (d) $\overrightarrow{AD}$ (e) $\overrightarrow{OD}$ (f) $\overrightarrow{DC}$.

First draw a diagram:

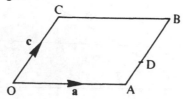

(a) $\overrightarrow{CB}$ is the same length as $\overrightarrow{OA}$ and is in the same direction,
$$\therefore \quad \overrightarrow{CB} = \overrightarrow{OA}$$
$$\text{or} \quad \overrightarrow{CB} = \mathbf{a}$$

(b) $\vec{BC}$ is the same length as $\vec{CB}$ but is in the opposite direction,

$$\therefore \quad \vec{BC} = -\vec{CB}$$

$$\text{or} \quad \vec{BC} = -\mathbf{a}$$

(c) $\vec{AB}$ is the same length as $\vec{OC}$ and is in the same direction,

$$\therefore \quad \vec{AB} = \mathbf{c}$$

(d) D is one third of the way along AB,

$$\therefore \quad \vec{AD} = \tfrac{1}{3}\vec{AB}$$

$$= \tfrac{1}{3}\mathbf{c}$$

(e)
$$\vec{OD} = \vec{OA} + \vec{AD}$$
$$= \mathbf{a} + \tfrac{1}{3}\mathbf{c}$$

(f)
$$\vec{DC} = \vec{DB} + \vec{BC} \qquad \left[\text{or} \quad \vec{DC} = \vec{DA} + \vec{AO} + \vec{OC} \right.$$
$$= \tfrac{2}{3}\mathbf{c} + (-\mathbf{a}) \qquad\qquad = -\tfrac{1}{3}\mathbf{c} - \mathbf{a} + \mathbf{c}$$
$$= \tfrac{2}{3}\mathbf{c} - \mathbf{a} \qquad\qquad\qquad \left. = \tfrac{2}{3}\mathbf{c} - \mathbf{a} \right]$$

Example 2

OABC is a trapezium with $\vec{OA} = \mathbf{a}$, $\vec{OC} = \mathbf{c}$ and $\vec{CB}$ parallel to and twice as long as $\vec{OA}$. The points D and E are the mid-points of AB and CB respectively. Find the following vectors in terms of $\mathbf{a}$ and $\mathbf{c}$:

(a) $\vec{CA}$ (b) $\vec{AB}$ (c) $\vec{ED}$.

Hence show that $\vec{CA}$ is parallel to and twice as long as $\vec{ED}$.

First draw a diagram:

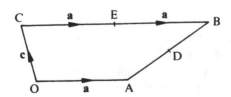

(a) $\quad \vec{CA} = \vec{CO} + \vec{OA}$
$$= -\mathbf{c} + \mathbf{a}$$

(b) $\quad \vec{AB} = \vec{AO} + \vec{OC} + \vec{CB}$
$$= -\mathbf{a} + \mathbf{c} + 2\mathbf{a}$$
$$= \mathbf{a} + \mathbf{c}$$

(c) $\quad \vec{ED} = \vec{EB} + \vec{BD}$
$$= \vec{EB} - \tfrac{1}{2}\vec{AB}$$
$$= \mathbf{a} - \tfrac{1}{2}(\mathbf{a} + \mathbf{c}) \quad \text{from (b)}$$
$$= \tfrac{1}{2}(\mathbf{a} - \mathbf{c})$$

Now $\vec{ED} = \tfrac{1}{2}(\mathbf{a} - \mathbf{c})$
$$= \tfrac{1}{2}(-\mathbf{c} + \mathbf{a}) = \tfrac{1}{2}\vec{CA} \quad \text{or} \quad 2\vec{ED} = \vec{CA}$$

Thus $\vec{CA}$ is parallel to and twice as long as $\vec{ED}$.

Non-parallel vectors

For two non-parallel vectors **a** and **b**, if $\lambda\mathbf{a} + \mu\mathbf{b} = \alpha\mathbf{a} + \beta\mathbf{b}$ then $\lambda = \alpha$
and $\mu = \beta$

i.e. we may equate the coefficients of vector **a** appearing on one side of the equation with those of vector **a** appearing on the other side, and similarly for vector **b**.

Proof: If $\lambda\mathbf{a} + \mu\mathbf{b} = \alpha\mathbf{a} + \beta\mathbf{b}$
then $\lambda\mathbf{a} - \alpha\mathbf{a} = \beta\mathbf{b} - \mu\mathbf{b}$
$\mathbf{a}(\lambda - \alpha) = \mathbf{b}(\beta - \mu)$ ...[1]

But $(\lambda - \alpha)$ and $(\beta - \mu)$ are scalars and so statement [1] means that **a** and **b** are either parallel vectors, which we know they are not,
or $\lambda - \alpha = 0$ and $\beta - \mu = 0$
i.e. $\lambda = \alpha$ and $\beta = \mu$.

Example 3

The diagram shows a parallelogram OABC
with $\overrightarrow{OA} = \mathbf{a}$ and $\overrightarrow{OC} = \mathbf{c}$. D is a point on
AB such that AD:DB = 2:1. OD produced
meets CB produced at E. $\overrightarrow{DE} = h\,\overrightarrow{OD}$ and
$\overrightarrow{BE} = k\overrightarrow{CB}$.

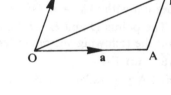

Find   (a) $\overrightarrow{BE}$ in terms of **a** and k,
(b) $\overrightarrow{DE}$ in terms of h, **a** and **c**,
(c) the values of h and k.

(a) $\overrightarrow{BE} = k\overrightarrow{CB}$ (b) $\overrightarrow{DE} = h\overrightarrow{OD}$
$= k\mathbf{a}$ $= h(\mathbf{a} + \tfrac{2}{3}\mathbf{c})$

(c) To determine h and k we need a vector equation containing h and k.
Now $\overrightarrow{BE} = \overrightarrow{BD} + \overrightarrow{DE}$
i.e. $k\mathbf{a} = -\tfrac{1}{3}\mathbf{c} + h(\mathbf{a} + \tfrac{2}{3}\mathbf{c})$
$\therefore\quad k\mathbf{a} = -\tfrac{1}{3}\mathbf{c} + h\mathbf{a} + \tfrac{2}{3}h\mathbf{c}$
equating coefficients of **a** gives $k = h$
equating coefficients of **c** gives $0 = -\tfrac{1}{3} + \tfrac{2}{3}h$
$\therefore\quad h = \tfrac{1}{2}$ and $k = \tfrac{1}{2}$

Exercise 2A

1. A ship is initially at a position A and travels 6 km on a bearing 055° followed by 8 km on a bearing 150° to reach a final position B. Find, by scale drawing, the distance and bearing of B from A.
2. An oil tanker travels 200 km on a bearing 160° followed by 300 km on a bearing 200°. Find, by calculation, the final distance (to the nearest km) and bearing (to the nearest degree) of the tanker from its original position.
3. A hiker leaves a position A and walks 550 m in a direction S60°W followed by 700 m in a direction N10°W. Find, by calculation, the distance that the hiker is then from A and the direction in which she must walk if she is to return directly to A.

4. Four towns A, B, C and D are such that B is 50 km from A on a bearing 080°, C is 70 km from B on a bearing 120° and D is 40 km from C on a bearing 210°. Find, by scale drawing, the distance and bearing of (a) D from A, (b) A from D.

5. If $\overrightarrow{AB}$ = **a** and $\overrightarrow{CD}$ = 3**a** which of the following statements are true?
    (a) $\overrightarrow{AB}$ is parallel to $\overrightarrow{CD}$,
    (b) $\overrightarrow{AB}$ is equal to $\overrightarrow{CD}$,
    (c) $\overrightarrow{AB}$ is three times as long as $\overrightarrow{CD}$,
    (d) $\overrightarrow{CD}$ is three times as long as $\overrightarrow{AB}$.

6. State which of the following vectors are parallel to the vector **a** + 2**b**.
    (a) 3**a** + 6**b**   (b) 3**a**
    (c) ½**a** + **b**   (d) **a** − 2**b**
    (e) $\sqrt{2}$**a** + 2$\sqrt{2}$**b**

7. State which of the following vectors are parallel to the vector 2**a** − 3**b**.
    (a) **a** + 6**b**   (b) **a** − $\frac{3}{2}$**b**
    (c) 4**a** − 6**b**   (d) 2**a** + 3**b**
    (e) 3**a** − 2**b**

8. Find the values of the scalars h and k in the following vector statements given that **a** and **b** are not parallel.
    (a) h**a** + 3**b** = 2**a** + k**b**
    (b) h**a** + k**b** = 3**b** − 2**a**
    (c) h**a** + 2**a** = k**b** − **b**
    (d) **a**(h + 3) = **b**(k − 1)
    (e) h**a** − h**b** = 5**a** − k**a** + **b** − k**b**
    (f) h**a** + k**a** + 3bk = **a** + **b** + 3h**b**
    (g) k**a** + h**a** + k**b** = **a** − 2h**b**
    (h) 2h**a** − **b** + 12 k**b** = 2**a** − 3k**a** + 6h**b**

9. The diagram shows a parallelogram OABC with $\overrightarrow{OA}$ = **a** and $\overrightarrow{OC}$ = **c**. D is a point on CB such that CD:DB = 1:3.

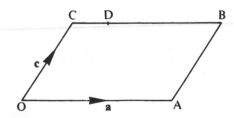

Express the following vectors in terms of **a** and **c**.
    (a) $\overrightarrow{AB}$,   (b) $\overrightarrow{BA}$,   (c) $\overrightarrow{OB}$,   (d) $\overrightarrow{OD}$,
    (e) $\overrightarrow{AD}$.

10. The diagram shows a triangle OAB with $\overrightarrow{OA}$ = **a** and $\overrightarrow{OB}$ = **b**. C is a point on AB such that AC:CB = 3:1.

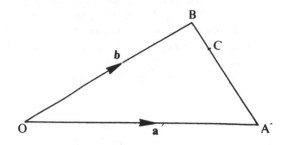

Express the following vectors in terms of **a** and **b**.
    (a) $\overrightarrow{AB}$,   (b) $\overrightarrow{AC}$,   (c) $\overrightarrow{CB}$,   (d) $\overrightarrow{OC}$.

11. OABC is a trapezium with $\overrightarrow{OA}$ = **a**, $\overrightarrow{OC}$ = **c** and $\overrightarrow{CB}$ = 3**a**. D is the mid-point of AB. Express the following vectors in terms of **a** and **c**.
    (a) $\overrightarrow{OB}$,   (b) $\overrightarrow{AB}$,   (c) $\overrightarrow{OD}$,   (d) $\overrightarrow{CD}$.

12. OAB is a triangle with $\overrightarrow{OA}$ = **a** and $\overrightarrow{OB}$ = **b**. C is a point on AB such that AC:CB = 1:2 and D is a point on OB such that OD:DB = 2:1. Express the following vectors in terms of **a** and **b**.
    (a) $\overrightarrow{AB}$,   (b) $\overrightarrow{AC}$,   (c) $\overrightarrow{DC}$.

13. OABC is a parallelogram with $\overrightarrow{OA}$ = **a** and $\overrightarrow{OC}$ = **c**. D is a point on OC such that OD:DC = 1:2 and E is a point on AC such that AE:EC = 2:1. Show that OB is parallel to and three times as long as DE.

14. OABC is a parallelogram with $\overrightarrow{OA}$ = **a** and $\overrightarrow{OC}$ = **c**. D is the mid-point of CB and OD meets AC at E. If $\overrightarrow{OE}$ = $h\overrightarrow{OD}$ and $\overrightarrow{AE}$ = $k\overrightarrow{AC}$ find:
    (a) $\overrightarrow{OE}$ in terms of h, **a** and **c**,
    (b) $\overrightarrow{AE}$ in terms of k, **a** and **c**,
    (c) the values of h and k.

15. OABC is a trapezium with $\overrightarrow{OA}$ = 3**a**, $\overrightarrow{OC}$ = **c** and $\overrightarrow{CB}$ = 2**a**. D is a point on AB such that AD:DB = 3:1 and OD crosses CA at E. If $\overrightarrow{AE}$ = $h\overrightarrow{AC}$ and $\overrightarrow{OE}$ = $k\overrightarrow{OD}$ find the values of h and k.

2.2 Position vectors and unit vectors

As we saw in section 1.5, the position of any
point in a plane can be stated in terms of an
ordered pair (x, y), the coordinates of the
point. This ordered pair gives the
perpendicular distance from the point to each
of the coordinate axes. The point P in the
diagram has coordinates (3, 2) with respect to
the origin O.

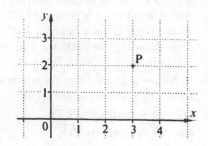

We could also define the point P by giving
the distance and direction of P from the origin
i.e. by stating the vector $\overrightarrow{OP}$. This vector is
called the **position vector** of P.

In addition to writing this position vector
as $\overrightarrow{OP}$ or as a single letter, say **p**, it can also
be expressed in terms of its horizontal and
vertical **components**, in one of the following
ways:

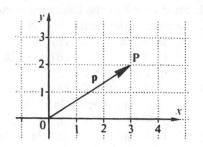

(i) As a column matrix, $\overrightarrow{OP} = \binom{3}{2}$.

(ii) By using **unit vectors**. A unit vector is a vector of length one unit in
a given direction. If we denote a horizontal unit vector by **i** and a
vertical unit vector by **j** then $\overrightarrow{OP} = 3\mathbf{i} + 2\mathbf{j}$.
(If a third dimension is required, we use **k** to represent a unit vector at
right angles to the plane containing **i** and **j**. Chapter 17 covers
three-dimensional vectors.)

If we represent the vector $3\mathbf{i} + 2\mathbf{j}$ by a line
segment of the required length and direction,
then any other line segment of the same length
and direction will also represent the vector
$3\mathbf{i} + 2\mathbf{j}$. Thus in the diagram, **a**, **b**, **c**, **d** and **p**
all equal $3\mathbf{i} + 2\mathbf{j}$. However, with O as the
origin, only **p** is the **position vector** $3\mathbf{i} + 2\mathbf{j}$ as
it is the vector from O to the point P(3, 2).

Note also (from Pythagoras' theorem) that a
vector $a\mathbf{i} + b\mathbf{j}$ has magnitude $\sqrt{(a^2 + b^2)}$.

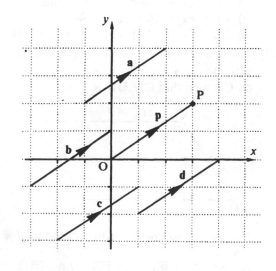

Example 4

For each of the vectors shown below:
- (i) express the vector in terms of **i** and **j**, where **i** is a unit vector in the Ox direction and **j** is a unit vector in the Oy direction,
- (ii) find the length of each vector,
- (iii) find the angle θ.

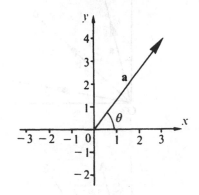 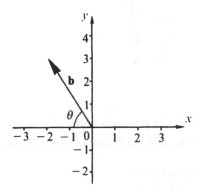 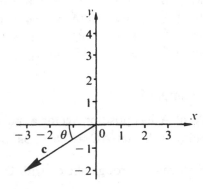

(i) $\mathbf{a} = 3\mathbf{i} + 4\mathbf{j}$

(ii) $|\mathbf{a}| = \sqrt{(3^2 + 4^2)}$
      $= 5$ units

(iii) $\tan \theta = \frac{4}{3}$
    $\therefore$ $\theta = 53 \cdot 1°$

(i) $\mathbf{b} = -2\mathbf{i} + 3\mathbf{j}$

(ii) $|\mathbf{b}| = \sqrt{[(-2)^2 + 3^2]}$
      $= \sqrt{13}$ units

(iii) $\tan \theta = \frac{3}{2}$
    $\therefore$ $\theta = 56 \cdot 3°$

(i) $\mathbf{c} = -3\mathbf{i} - 2\mathbf{j}$

(ii) $|\mathbf{c}| = \sqrt{[(-3)^2 + (-2)^2]}$
      $= \sqrt{13}$ units

(iii) $\tan \theta = \frac{2}{3}$
    $\therefore$ $\theta = 33 \cdot 7°$

Example 5

Find the unit vector in the same direction as the vector $2\mathbf{i} - \mathbf{j}$.

$2\mathbf{i} - \mathbf{j}$ is a vector of length $\sqrt{5}$ units.
Thus a vector parallel to $2\mathbf{i} - \mathbf{j}$ and of length 1 unit will be

$$\frac{1}{\sqrt{5}}(2\mathbf{i} - \mathbf{j}) \text{ or } \frac{\sqrt{5}}{5}(2\mathbf{i} - \mathbf{j})$$

The required unit vector is $\frac{\sqrt{5}}{5}(2\mathbf{i} - \mathbf{j})$.

Note: Whilst we use **i** and **j** for the unit vectors in the direction of the x- and y-axes respectively, a unit vector in the direction of some vector **a** is usually written **â**.

So, $\hat{\mathbf{a}} = \dfrac{\mathbf{a}}{|\mathbf{a}|}$

Example 6

Find a vector that is of magnitude 6 units and is parallel to the vector $\mathbf{i} + \mathbf{j}$.

The vector $\mathbf{i} + \mathbf{j}$ has magnitude $\sqrt{(1^2 + 1^2)} = \sqrt{2}$ units
Thus a vector parallel to $\mathbf{i} + \mathbf{j}$ but of length 6 units will be $\dfrac{6}{\sqrt{2}}(\mathbf{i} + \mathbf{j})$

$$= 3\sqrt{2}\mathbf{i} + 3\sqrt{2}\mathbf{j}$$

Example 7

If A has position vector **a**, B has position vector **b** and C has position vector **c**, find the following vectors in terms of **a**, **b** and **c**:
(a) $\overrightarrow{AB}$ (b) $\overrightarrow{BA}$ (c) $\overrightarrow{BC}$ (d) $\overrightarrow{AC}$.

First draw a sketch showing the position vectors:

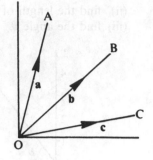

(a) $\overrightarrow{AB} = \overrightarrow{AO} + \overrightarrow{OB} = -\mathbf{a} + \mathbf{b}$ or $\mathbf{b} - \mathbf{a}$
(b) $\overrightarrow{BA} = \overrightarrow{BO} + \overrightarrow{OA} = -\mathbf{b} + \mathbf{a}$ or $\mathbf{a} - \mathbf{b}$
(c) $\overrightarrow{BC} = \overrightarrow{BO} + \overrightarrow{OC} = -\mathbf{b} + \mathbf{c}$ or $\mathbf{c} - \mathbf{b}$
(d) $\overrightarrow{AC} = \overrightarrow{AO} + \overrightarrow{OC} = -\mathbf{a} + \mathbf{c}$ or $\mathbf{c} - \mathbf{a}$

Example 8

Point A has position vector **a**, point B has position vector 2**b** and C is a point such that $\overrightarrow{AC} = 6\mathbf{b} - 3\mathbf{a}$. Show that A, B and C are collinear.

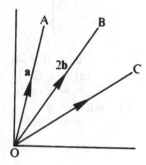

First make a rough sketch with O as this origin:
$$\overrightarrow{AB} = \overrightarrow{AO} + \overrightarrow{OB}$$
$$= -\mathbf{a} + 2\mathbf{b}$$
but $\quad \overrightarrow{AC} = 6\mathbf{b} - 3\mathbf{a}$
$$= 3(-\mathbf{a} + 2\mathbf{b})$$
$$= 3\overrightarrow{AB}$$

Thus $\overrightarrow{AB}$ and $\overrightarrow{AC}$ are parallel and, as the point A is common to both these vectors, the points A, B and C are collinear.

Addition and subtraction of vectors in component vector form

When vectors are expressed in terms of **i** and **j**, it is easy to add two or more vectors to find their combined or **resultant** vector.
From the vector diagram shown, we know that

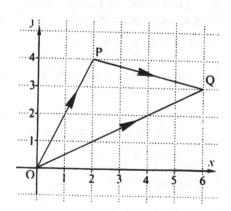

$$\overrightarrow{OP} + \overrightarrow{PQ} = \overrightarrow{OQ} \qquad \ldots[1]$$

Taking **i** and **j** to be the unit vectors in the directions Ox and Oy respectively, we can see from the diagram that $\overrightarrow{OP} = 2\mathbf{i} + 4\mathbf{j}$, $\overrightarrow{PQ} = 4\mathbf{i} - \mathbf{j}$ and $\overrightarrow{OQ} = 6\mathbf{i} + 3\mathbf{j}$.
Substitution into [1] gives
$$(2\mathbf{i} + 4\mathbf{j}) + (4\mathbf{i} - \mathbf{j}) = 6\mathbf{i} + 3\mathbf{j}$$

Thus to add vectors expressed in terms of **i** and **j**, we add the **i** components together, $2 + 4 = 6$, and the **j** components together, $4 + -1 = 3$.

Applying a similar method for subtraction
gives $\quad \overrightarrow{OP} - \overrightarrow{PQ} = (2\mathbf{i} + 4\mathbf{j}) - (4\mathbf{i} - \mathbf{j})$
$$= -2\mathbf{i} + 5\mathbf{j}$$

To interpret $\overrightarrow{OP} - \overrightarrow{PQ}$ we can think of it
as $\overrightarrow{OP} + (-\overrightarrow{PQ})$. The vector $(-\overrightarrow{PQ})$ is equal
in magnitude to $\overrightarrow{PQ}$ but opposite in direction.

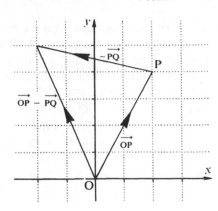

So, $\overrightarrow{OP} - \overrightarrow{PQ}$ means that we must add
to $\overrightarrow{OP}$ a vector equal in length, but opposite
in direction, to that of $\overrightarrow{PQ}$.

Similarly, multiplying a vector expressed in
terms of **i** and **j** by a scalar is straightforward.
If $\mathbf{a} = 3\mathbf{i} + 4\mathbf{j}$ then $2\mathbf{a} = 6\mathbf{i} + 8\mathbf{j}$,
$3\mathbf{a} = 9\mathbf{j} + 12\mathbf{j}$, etc.

Example 9

If $\mathbf{a} = 3\mathbf{i} + 4\mathbf{j}$ and $\mathbf{b} = 2\mathbf{i} + 8\mathbf{j}$ find: (a) $\mathbf{a} + \mathbf{b}$, (b) $|\mathbf{a} + \mathbf{b}|$, (c) $\mathbf{a} - 2\mathbf{b}$, (d) $|\mathbf{a} - 2\mathbf{b}|$.

(a) $\mathbf{a} + \mathbf{b} = (3\mathbf{i} + 4\mathbf{j}) + (2\mathbf{i} + 8\mathbf{j})$
$= 5\mathbf{i} + 12\mathbf{j}$

(b) $|\mathbf{a} + \mathbf{b}| = |5\mathbf{i} + 12\mathbf{j}|$
$= \sqrt{(5^2 + 12^2)} = 13$ units

(c) $\mathbf{a} - 2\mathbf{b} = (3\mathbf{i} + 4\mathbf{j}) - 2(2\mathbf{i} + 8\mathbf{j})$
$= -\mathbf{i} - 12\mathbf{j}$

(d) $|\mathbf{a} - 2\mathbf{b}| = |-\mathbf{i} - 12\mathbf{j}|$
$= \sqrt{[(-1)^2 + (-12^2)]} = \sqrt{145}$ units

Example 10

If a point A has position vector $\mathbf{i} + 2\mathbf{j}$ and B has position vector $5\mathbf{i} + \mathbf{j}$,
find the position vector of the point which divides AB in the ratio $1:-3$.

Note first that the ratio involves two numbers of different sign. This
indicates that the required point divides AB *externally* rather than *internally*.
The difference between these two cases is illustrated below:

Point P dividing AB internally in the ratio $1:3$

Point P dividing AB externally in the ratio
$1:3$. (By writing this as $1:-3$ we need not
state that the division is external).

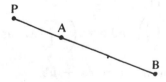

In this case the vectors $\overrightarrow{AP}$ and $\overrightarrow{PB}$ are in the
same direction.

In this case the vectors $\overrightarrow{AP}$ and $\overrightarrow{PB}$ are in
opposite directions. Hence the use of the
minus sign in the ratio.

To solve the given question we first draw a
diagram showing the points A, B and P,
position vectors **a**, **b** and **p** respectively. (Note
that the diagram is an aid to calculation and
need not be drawn accurately).

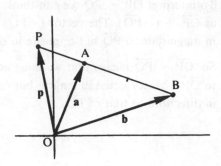

Now $\overrightarrow{OP} = \mathbf{p} = \overrightarrow{OA} + \overrightarrow{AP}$

$\qquad\qquad = \mathbf{a} + \tfrac{1}{2}\overrightarrow{BA}$

$\qquad\qquad = \mathbf{a} + \tfrac{1}{2}(-\mathbf{b} + \mathbf{a})$

$\qquad\qquad = \tfrac{3}{2}\mathbf{a} - \tfrac{1}{2}\mathbf{b}$

But $\mathbf{a} = \mathbf{i} + 2\mathbf{j}$ and $\mathbf{b} = 5\mathbf{i} + \mathbf{j}$

$\therefore \qquad\qquad \mathbf{p} = \tfrac{3}{2}(\mathbf{i} + 2\mathbf{j}) - \tfrac{1}{2}(5\mathbf{i} + \mathbf{j})$

$\qquad\qquad\quad = -\mathbf{i} + \tfrac{5}{2}\mathbf{j}$

The point dividing AB in the ratio $1:-3$ has position vector $-\mathbf{i} + \tfrac{5}{2}\mathbf{j}$.

The general formula for the position vector of a point dividing a line in a
given ratio is given by the section theorem:
The point P which divides the line AB in the ratio $\lambda:\mu$ has position vector
p where $\mathbf{p} = \dfrac{\mu\mathbf{a} + \lambda\mathbf{b}}{\lambda + \mu}$ and **a** and **b** are the position vectors of A and B
respectively.

Applying this formula to example 10 gives $\mathbf{p} = \dfrac{-3(\mathbf{i} + 2\mathbf{j}) + 1(5\mathbf{i} + \mathbf{j})}{1 + (-3)}$

$\qquad\qquad\qquad\qquad\qquad = -\mathbf{i} + \tfrac{5}{2}\mathbf{j} \quad$ as before.

Other base vectors in two dimensions

Any vector lying in the plane of the unit vectors **i** and **j** can be expressed in
the form $x\mathbf{i} + y\mathbf{j}$. The unit vectors **i** and **j** are **base vectors** from which other
coplanar vectors can be built up. Although it is usually convenient to use **i**
and **j** as the base vectors, any pair of non-parallel coplanar vectors **a** and **b**
could be used instead.
Thus for three coplanar vectors **a**, **b** and **c**, it is possible to express any one
of the vectors in terms of the other two.

$\qquad$ i.e.$\quad \mathbf{c} = \lambda_1\mathbf{a} + \mu_1\mathbf{b}$

$\qquad\qquad\quad \mathbf{b} = \lambda_2\mathbf{a} + \mu_2\mathbf{c}$

$\qquad\qquad\quad \mathbf{a} = \lambda_3\mathbf{b} + \mu_3\mathbf{c} \quad$ where λ_1, μ_1, λ_2, μ_2, λ_3 and μ_3 are
$\qquad\qquad\qquad\qquad\qquad\qquad$ suitable scalars.

Example 11

With $\mathbf{a} = \begin{pmatrix} 3 \\ 5 \end{pmatrix}$ and $\mathbf{b} = \begin{pmatrix} 2 \\ -1 \end{pmatrix}$ as base vectors, express $\mathbf{c} = \begin{pmatrix} 5 \\ 17 \end{pmatrix}$ and

$\mathbf{d} = \begin{pmatrix} 3 \\ 2 \end{pmatrix}$ in the form $\lambda\mathbf{a} + \mu\mathbf{b}$.

If $\qquad\quad \mathbf{c} = \lambda\mathbf{a} + \mu\mathbf{b}$ $\qquad\qquad\qquad\qquad$ If $\qquad\quad \mathbf{d} = \lambda\mathbf{a} + \mu\mathbf{b}$

$\quad \begin{pmatrix} 5 \\ 17 \end{pmatrix} = \lambda\begin{pmatrix} 3 \\ 5 \end{pmatrix} + \mu\begin{pmatrix} 2 \\ -1 \end{pmatrix}$ $\qquad\qquad \begin{pmatrix} 3 \\ 2 \end{pmatrix} = \lambda\begin{pmatrix} 3 \\ 5 \end{pmatrix} + \mu\begin{pmatrix} 2 \\ -1 \end{pmatrix}$

Thus $5 = 3\lambda + 2\mu$

and $17 = 5\lambda - \mu$

Solving simultaneously gives $\lambda = 3$ and $\mu = -2$

$\therefore$ $\mathbf{c} = 3\mathbf{a} - 2\mathbf{b}$

Thus $3 = 3\lambda + 2\mu$

and $2 = 5\lambda - \mu$

Solving simultaneously gives $\lambda = \frac{7}{13}$ and $\mu = \frac{9}{13}$

$\therefore$ $\mathbf{d} = \frac{7}{13}\mathbf{a} + \frac{9}{13}\mathbf{b}$

Exercise 2B

1. Taking $\mathbf{i}$ and $\mathbf{j}$ to be unit vectors in the directions Ox and Oy respectively, give the position vectors of the points A to H shown in the diagram. Each square in the grid is a unit square and O is the origin.

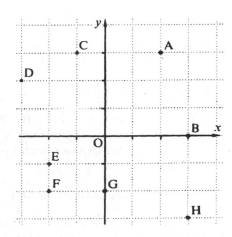

2. For each of the vectors shown below:
   (i) express each vector in terms of $\mathbf{i}$ and $\mathbf{j}$ where $\mathbf{i}$ is a unit vector in the Ox direction and $\mathbf{j}$ is a unit vector in the Oy direction,
   (ii) find the length of each vector, (iii) find the angle θ.

             (a)   (b)   (c)

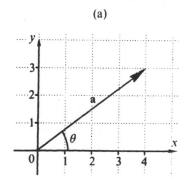 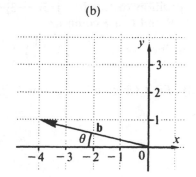 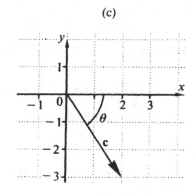

3. If $a\mathbf{i} + 8\mathbf{j}$ is parallel to $2\mathbf{i} + 4\mathbf{j}$, find the value of $\mathbf{a}$.
4. If $\mathbf{a} = \mathbf{i} + 2\mathbf{j}$ find $\hat{\mathbf{a}}$, a unit vector in the direction of $\mathbf{a}$.
5. If $\mathbf{b} = 3\mathbf{i} - \mathbf{j}$ find $\hat{\mathbf{b}}$, a unit vector in the direction of $\mathbf{b}$.
6. Find a vector that is of magnitude 39 units and is parallel to $5\mathbf{i} + 12\mathbf{j}$.
7. Find a vector that is of magnitude $3\sqrt{5}$ units and is parallel to $2\mathbf{i} - \mathbf{j}$.
8. Find a vector that is of magnitude 2 units and is parallel to $4\mathbf{i} - 3\mathbf{j}$.
9. If the point P has position vector $2\mathbf{i} + 3\mathbf{j}$ and point Q has position vector $7\mathbf{i} + 4\mathbf{j}$, find: (a) $\overrightarrow{PQ}$, (b) $\overrightarrow{QP}$.
10. If the point P has position vector $7\mathbf{i} - 3\mathbf{j}$ and point Q has position vector $5\mathbf{i} + 5\mathbf{j}$, find: (a) $\overrightarrow{PQ}$, (b) $\overrightarrow{QP}$.

11. The point P has position vector $-5\mathbf{i} + 3\mathbf{j}$ and Q is a point such that $\overrightarrow{PQ} = 7\mathbf{i} - \mathbf{j}$. Find the position vector of Q.

12. The point P has position vector $3\mathbf{i} - 2\mathbf{j}$ and Q is a point such that $\overrightarrow{QP} = 2\mathbf{i} - 3\mathbf{j}$. Find the position vector of Q.

13. Using $\mathbf{a} = \begin{pmatrix} 3 \\ 4 \end{pmatrix}$ and $\mathbf{b} = \begin{pmatrix} 2 \\ -1 \end{pmatrix}$ as base vectors, express the vectors

$\mathbf{c} = \begin{pmatrix} 7 \\ 2 \end{pmatrix}$ and $\mathbf{d} = \begin{pmatrix} 5 \\ -1 \end{pmatrix}$ in the form $\lambda\mathbf{a} + \mu\mathbf{b}$.

14. Given that $\mathbf{a} = 3\mathbf{i} - \mathbf{j}$ and $\mathbf{b} = 2\mathbf{i} + \mathbf{j}$, find:
(a) $|\mathbf{a}|$,   (b) $|\mathbf{b}|$,   (c) $\mathbf{a} + \mathbf{b}$,   (d) $|\mathbf{a} + \mathbf{b}|$.

15. Given that $\mathbf{a} = \begin{pmatrix} 2 \\ 3 \end{pmatrix}$ and $\mathbf{b} = \begin{pmatrix} 1 \\ -3 \end{pmatrix}$, find:

(a) $\mathbf{a} + 2\mathbf{b}$,   (b) $|\mathbf{a} + 2\mathbf{b}|$,   (c) $2\mathbf{a} + 3\mathbf{b}$,   (d) $|2\mathbf{a} + 3\mathbf{b}|$.

16. The point A has position vector $3\mathbf{i} + \mathbf{j}$ and point B has position vector $10\mathbf{i} + \mathbf{j}$. Find the position vector of the point which divides AB in the ratio $3:4$.

17. The point C has position vector $\begin{pmatrix} 2 \\ 3 \end{pmatrix}$ and point D has position vector $\begin{pmatrix} 1 \\ 2 \end{pmatrix}$.

Find the position vector of the point which divides CD in the ratio $4:-3$.

18. The point K has position vector $3\mathbf{i} + 2\mathbf{j}$ and point L has position vector $\mathbf{i} + 3\mathbf{j}$. Find the position vector of the point which divides KL in the ratio (a) $4:3$, (b) $4:-3$.

19. The three points A, B and C have position vectors $\mathbf{a}$, $\mathbf{b}$ and $\mathbf{c}$ respectively. If $\mathbf{c} = 3\mathbf{b} - 2\mathbf{a}$, show that A, B and C are collinear.

20. The three points A, B and C have position vectors $\mathbf{i} - \mathbf{j}$, $5\mathbf{i} - 3\mathbf{j}$ and $11\mathbf{i} - 6\mathbf{j}$ respectively. Show that A, B and C are collinear.

21. Points A, B, C, D, E and F have position vectors $\begin{pmatrix} 3 \\ -1 \end{pmatrix}, \begin{pmatrix} 1 \\ 3 \end{pmatrix}, \begin{pmatrix} 2 \\ 1 \end{pmatrix},$

$\begin{pmatrix} 5 \\ 3 \end{pmatrix}, \begin{pmatrix} -3 \\ 11 \end{pmatrix}$ and $\begin{pmatrix} 5 \\ -5 \end{pmatrix}$ respectively. Find which of the points C, D, E

and F are collinear with A and B.

2.3 The scalar product

The scalar product of two vectors $\mathbf{a}$ and $\mathbf{b}$ is defined as the product of the magnitudes of the two vectors multiplied by the cosine of the angle between the two vectors.

Writing this scalar product as $\mathbf{a} \cdot \mathbf{b}$ it follows from the definition that

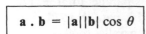

$$\boxed{\mathbf{a} \cdot \mathbf{b} = |\mathbf{a}||\mathbf{b}| \cos \theta}$$

The term **scalar** product is used because $|\mathbf{a}|$, $|\mathbf{b}|$ and $\cos \theta$ are all scalar quantities and so their product will be a scalar quantity.

We read $\mathbf{a} \cdot \mathbf{b}$ as 'a dot b' and for this reason the scalar product may also be referred to as the **dot** product.

Note that the angle between two vectors always refers to the angle between the directions of the vectors when these directions are either both towards their point of intersection or both away from their point of intersection. Thus, in each of the following diagrams, θ is the angle between the two vectors.

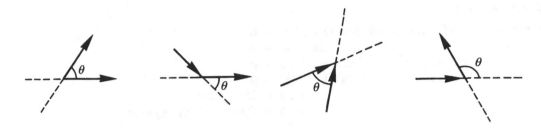

Properties of the scalar product

1. From the definition $\mathbf{a} \cdot \mathbf{b} = |\mathbf{a}||\mathbf{b}| \cos \theta$, it follows that two perpendicular vectors will have a scalar product of zero.

2. $\mathbf{a} \cdot \mathbf{a} = |\mathbf{a}||\mathbf{a}| \cos 0$
$\qquad = |\mathbf{a}||\mathbf{a}|$
This is usually written $\mathbf{a} \cdot \mathbf{a} = |\mathbf{a}|^2$, or just $\mathbf{a} \cdot \mathbf{a} = a^2$

3. The scalar product is commutative i.e. $\mathbf{a} \cdot \mathbf{b} = \mathbf{b} \cdot \mathbf{a}$
Proof: $\mathbf{a} \cdot \mathbf{b} = |\mathbf{a}||\mathbf{b}| \cos \theta$
$\qquad\qquad = |\mathbf{b}||\mathbf{a}| \cos \theta = \mathbf{b} \cdot \mathbf{a}$

4. The scalar product is distributive over addition i.e. $\mathbf{a} \cdot (\mathbf{b} + \mathbf{c}) = \mathbf{a} \cdot \mathbf{b} + \mathbf{a} \cdot \mathbf{c}$

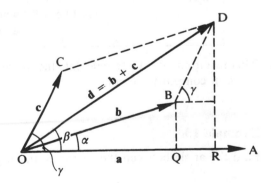

Proof: Consider vectors $\mathbf{a}$, $\mathbf{b}$, $\mathbf{c}$ and $\mathbf{d}$ with
$\mathbf{d} = \mathbf{b} + \mathbf{c}$
It is clear from the diagram that $OR = OQ + QR$
but $OR = |\mathbf{d}| \cos \beta$, $OQ = |\mathbf{b}| \cos \alpha$ and
$QR = |\mathbf{c}| \cos \gamma$
$\therefore \qquad |\mathbf{d}| \cos \beta = |\mathbf{b}| \cos \alpha + |\mathbf{c}| \cos \gamma$
$\therefore \qquad |\mathbf{a}||\mathbf{d}| \cos \beta = |\mathbf{a}||\mathbf{b}| \cos \alpha + |\mathbf{a}||\mathbf{c}| \cos \gamma$
So $\qquad\qquad \mathbf{a} \cdot \mathbf{d} = \mathbf{a} \cdot \mathbf{b} + \mathbf{a} \cdot \mathbf{c}$
giving $\quad \mathbf{a} \cdot (\mathbf{b} + \mathbf{c}) = \mathbf{a} \cdot \mathbf{b} + \mathbf{a} \cdot \mathbf{c}$ as required.

5. The properties for multiplication by a scalar λ are that:
$$\lambda(\mathbf{a} \cdot \mathbf{b}) = \mathbf{a} \cdot (\lambda \mathbf{b}) = (\lambda \mathbf{a}) \cdot \mathbf{b} = \lambda |\mathbf{a}||\mathbf{b}| \cos \theta$$

Example 12

$\overrightarrow{OA}$ and $\overrightarrow{OB}$ are two vectors such that $\overrightarrow{OA} = \mathbf{a} + 2\mathbf{b}$, $\overrightarrow{OB} = 2\mathbf{a} - \mathbf{b}$ and
OA is perpendicular to OB.
(a) Show that $\mathbf{a} \cdot \mathbf{b} = \frac{2}{3}b^2 - a^2$.
(b) If $\mathbf{a}$ has magnitude 4 units and $\mathbf{b}$ has magnitude 5 units, find the angle
between $\mathbf{a}$ and $\mathbf{b}$.

(a) If OA is perpendicular to OB then $\overrightarrow{OA} \cdot \overrightarrow{OB} = 0$

$$\text{i.e.} \qquad (\mathbf{a} + 2\mathbf{b}) \cdot (2\mathbf{a} - \mathbf{b}) = 0$$
$$\therefore \qquad \mathbf{a} \cdot 2\mathbf{a} - \mathbf{a} \cdot \mathbf{b} + 2\mathbf{b} \cdot 2\mathbf{a} - 2\mathbf{b} \cdot \mathbf{b} = 0$$
$$2a^2 - \mathbf{a} \cdot \mathbf{b} + 4\mathbf{a} \cdot \mathbf{b} - 2b^2 = 0$$
$$3\mathbf{a} \cdot \mathbf{b} = 2b^2 - 2a^2$$
$$\mathbf{a} \cdot \mathbf{b} = \tfrac{2}{3}(b^2 - a^2) \text{ as required.}$$

(b) $\qquad \mathbf{a} \cdot \mathbf{b} = \tfrac{2}{3}(b^2 - a^2)$
$$= \tfrac{2}{3}(25 - 16) = 6 \text{ units}$$
but $\qquad \mathbf{a} \cdot \mathbf{b} = |\mathbf{a}||\mathbf{b}| \cos \theta$
$$\therefore \qquad 6 = 4 \times 5 \times \cos \theta$$
giving $\qquad \theta = 72 \cdot 54°$

Thus the angle between $\mathbf{a}$ and $\mathbf{b}$ is $72 \cdot 5°$, correct to 1 d.p.

Consider the vectors $\mathbf{a} = x_1\mathbf{i} + y_1\mathbf{j}$ and $\mathbf{b} = x_2\mathbf{i} + y_2\mathbf{j}$.

Now $\quad \mathbf{a} \cdot \mathbf{b} = (x_1\mathbf{i} + y_1\mathbf{j}) \cdot (x_2\mathbf{i} + y_2\mathbf{j})$
$$= x_1x_2\mathbf{i} \cdot \mathbf{i} + x_1y_2\mathbf{i} \cdot \mathbf{j} + y_1x_2\mathbf{j} \cdot \mathbf{i} + y_1y_2\mathbf{j} \cdot \mathbf{j}$$
But $\quad \mathbf{i} \cdot \mathbf{i} = \mathbf{j} \cdot \mathbf{j} = 1 \quad$ and $\quad \mathbf{i} \cdot \mathbf{j} = \mathbf{j} \cdot \mathbf{i} = 0$
Thus $\quad \mathbf{a} \cdot \mathbf{b} = x_1x_2 + y_1y_2$

Thus, if $\qquad \mathbf{a} = x_1\mathbf{i} + y_1\mathbf{j}$ and $\mathbf{b} = x_2\mathbf{i} + y_2\mathbf{j}$
then $\qquad \mathbf{a} \cdot \mathbf{b} = x_1x_2 + y_1y_2$
$$= |\mathbf{a}||\mathbf{b}| \cos \theta \quad \text{where } \theta \text{ is the angle between}$$
$$\mathbf{a} \text{ and } \mathbf{b}.$$

This result enables us to calculate the angle between two vectors that are
given in component form.

Example 13

Find the angle between the vectors $\mathbf{a}$ and $\mathbf{b}$ given that $\mathbf{a} = 3\mathbf{i} + 4\mathbf{j}$ and $\mathbf{b} = 5\mathbf{i} - 12\mathbf{j}$.

Let θ be the required angle
Then $\quad \mathbf{a} \cdot \mathbf{b} = |\mathbf{a}||\mathbf{b}| \cos \theta$
but $\quad \mathbf{a} \cdot \mathbf{b} = (3 \times 5) + (4 \times (-12)), \quad |\mathbf{a}| = \sqrt{(3^2 + 4^2)} \quad$ and $\quad |\mathbf{b}| = \sqrt{(5^2 + (-12)^2)}$
$$= -33 \qquad\qquad\qquad = 5 \qquad\qquad\qquad\qquad = 13$$
Thus $\quad -33 = 5 \times 13 \cos \theta$
giving $\quad \theta = 120 \cdot 51°$

Thus the angle between $\mathbf{a}$ and $\mathbf{b}$ is $120 \cdot 5°$, correct to 1 d.p.

Example 14

The points A, B, C and D have position vectors $-2i + 3j$, $3i + 8j$, $7i + 6j$ and $7i - 4j$ respectively. Show that AC is perpendicular to BD.

First draw a rough sketch with O as the origin:

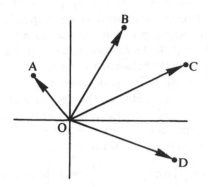

$$\overrightarrow{AC} = \overrightarrow{AO} + \overrightarrow{OC}$$
$$= -(-2i + 3j) + (7i + 6j)$$
$$= 9i + 3j$$
$$\overrightarrow{BD} = \overrightarrow{BO} + \overrightarrow{OD}$$
$$= -(3i + 8j) + (7i - 4j)$$
$$= 4i - 12j$$
$$\therefore \quad \overrightarrow{AC} \cdot \overrightarrow{BD} = (9i + 3j) \cdot (4i - 12j)$$
$$= (9 \times 4) + (3 \times (-12))$$
$$= 0$$

Thus AC is perpendicular to BD.

Projection of one vector onto another

We have already seen that a vector can be expressed in terms of its components in the directions of x- and y-axes. The scalar product can be used to find the component of one vector in the direction of some other vector.

Suppose we wish to find the component of vector **p**, in the direction of vector **q**. The required component will be the projection of vector **p** onto **q** i.e. $|\mathbf{p}| \cos \theta$

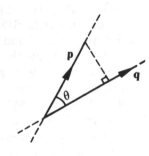

$$\text{but} \quad |\mathbf{p}| \cos \theta = \frac{\mathbf{p} \cdot \mathbf{q}}{|\mathbf{q}|}$$

Thus the component of **p** in the direction of **q** (or the projection of **p** onto **q**) is given by $\frac{\mathbf{p} \cdot \mathbf{q}}{|\mathbf{q}|}$. This is also referred to as the **resolved** part of **p** in the direction of **q**.

Example 15

Find the resolved part of vector $\mathbf{a} = 2i + 5j$ in the direction of the vector $\mathbf{b} = i + j$.

$$\text{Resolved part} = |\mathbf{a}| \cos \theta$$
$$= \frac{\mathbf{a} \cdot \mathbf{b}}{|\mathbf{b}|}$$
$$= \frac{(2 \times 1) + (5 \times 1)}{\sqrt{(1^2 + 1^2)}}$$
$$= \frac{7}{\sqrt{2}} \quad \text{or} \quad \frac{7\sqrt{2}}{2}$$

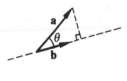

Exercise 2C

1. If $\mathbf{a} = \mathbf{i} + 2\mathbf{j}$, $\mathbf{b} = \mathbf{i} - 2\mathbf{j}$, $\mathbf{c} = 2\mathbf{i} - 3\mathbf{j}$ and $\mathbf{d} = 6\mathbf{i} + 3\mathbf{j}$, find which two of these vectors are perpendicular to each other.

2. If $\mathbf{e} = -\mathbf{i} - 3\mathbf{j}$, $\mathbf{f} = \mathbf{i} + 3\mathbf{j}$, $\mathbf{g} = -3\mathbf{i} - 2\mathbf{j}$ and $\mathbf{h} = 6\mathbf{i} - 9\mathbf{j}$, find which two of these vectors are perpendicular to each other.

3. Find the angle between the vectors $\mathbf{a}$ and $\mathbf{b}$ given that $\mathbf{a} = 3\mathbf{i} + 4\mathbf{j}$ and $\mathbf{b} = 5\mathbf{i} + 12\mathbf{j}$. (Give your answer to the nearest degree.)

4. Find the angle between the vectors $\mathbf{c}$ and $\mathbf{d}$ given that $\mathbf{c} = 5\mathbf{i} - \mathbf{j}$ and $\mathbf{d} = 2\mathbf{i} + 3\mathbf{j}$. (Give your answer to the nearest degree.)

5. Find the angle between the vectors $\mathbf{e}$ and $\mathbf{f}$ given that $\mathbf{e} = -\mathbf{i} - 2\mathbf{j}$ and $\mathbf{f} = 2\mathbf{i} + \mathbf{j}$. (Give your answer to the nearest degree.)

6. If the angle between the vectors $\mathbf{c} = a\mathbf{i} + 2\mathbf{j}$ and $\mathbf{d} = 3\mathbf{i} + \mathbf{j}$ is 45° find the two possible values of a.

7. If $\mathbf{a} = \begin{pmatrix} x \\ y \end{pmatrix}$ and $\mathbf{b} = \begin{pmatrix} 2 \\ 1 \end{pmatrix}$ write down a relationship between x and y for each of the following cases:
   (a) $\mathbf{a}$ is of magnitude 5 units,   (b) $\mathbf{a}$ is perpendicular to $\mathbf{b}$,
   (c) $\mathbf{a}$ is parallel to $\mathbf{b}$,   (d) $(\mathbf{a} - \mathbf{b})$ is perpendicular to the vector $\begin{pmatrix} 4 \\ 1 \end{pmatrix}$.

8. The points A, B, C and D have position vectors $5\mathbf{i} + \mathbf{j}$, $-3\mathbf{i} + 2\mathbf{j}$, $-3\mathbf{i} - 3\mathbf{j}$ and $\mathbf{i} - 6\mathbf{j}$ respectively. Show that AC is perpendicular to BD.

9. The points E, F and G have position vectors $2\mathbf{i} + 2\mathbf{j}$, $\mathbf{i} + 6\mathbf{j}$ and $-7\mathbf{i} + 4\mathbf{j}$. Show that the triangle EFG is right-angled at F.

10. The points A, B, C and D have position vectors $\mathbf{a}$, $\mathbf{b}$, $\mathbf{c}$ and $\mathbf{d}$ respectively where $\mathbf{a} = -2\mathbf{j}$, $\mathbf{b} = -2\mathbf{i} + 4\mathbf{j}$, $\mathbf{c} = 3\mathbf{i} + 4\mathbf{j}$ and $\mathbf{d} = 4\mathbf{i} + y\mathbf{j}$. If AC is perpendicular to BD, find the value of y.

11. The points A, B, C and D have position vectors $-2\mathbf{i} + \mathbf{j}$, $7\mathbf{j}$, $3\mathbf{i} + 6\mathbf{j}$ and $x\mathbf{i} + y\mathbf{j}$ respectively. If $|\overrightarrow{AC}| = |\overrightarrow{BD}|$ and AC is perpendicular to BD, find the two possible values of x and the corresponding values of y.

12. If $\mathbf{a} = \begin{pmatrix} 3 \\ -4 \end{pmatrix}$ find: (a) a unit vector parallel to $\mathbf{a}$,
   (b) a unit vector perpendicular to $\mathbf{a}$.

13. If $\mathbf{b} = \begin{pmatrix} 7 \\ -1 \end{pmatrix}$ find: (a) a vector $\mathbf{c}$, parallel to $\mathbf{b}$ and of magnitude 10 units,
   (b) a vector $\mathbf{d}$, perpendicular to $\mathbf{b}$ and of magnitude $15\sqrt{2}$ units.

14. The points A, B and C have position vectors $4\mathbf{i} - \mathbf{j}$, $\mathbf{i} + 3\mathbf{j}$ and $-5\mathbf{i} + 2\mathbf{j}$ respectively. Find:
   (a) $\overrightarrow{AB}$,   (b) $\overrightarrow{BC}$,   (c) $\overrightarrow{CA}$,   (d) the angles of the triangle ABC giving your answers to the nearest degree.

15. If $\overrightarrow{OA} = 2\mathbf{a} + 3\mathbf{b}$ and $\overrightarrow{OB} = 3\mathbf{a} - 2\mathbf{b}$ show that $\overrightarrow{OA} \cdot \overrightarrow{OB} = 6a^2 + 5\mathbf{a} \cdot \mathbf{b} - 6b^2$.

16. If $\overrightarrow{OC} = 2\mathbf{a} + 3\mathbf{b}$ and $\overrightarrow{OD} = 2\mathbf{a} - 3\mathbf{b}$ show that $\overrightarrow{OC} \cdot \overrightarrow{OD} = 4a^2 - 9b^2$.

17. If $\overrightarrow{OA} = 3\mathbf{a} + 2\mathbf{b}$ and $\overrightarrow{OB} = 2\mathbf{a} - \mathbf{b}$
   (a) show that $\overrightarrow{OA} \cdot \overrightarrow{OB} = 6a^2 + \mathbf{a} \cdot \mathbf{b} - 2b^2$
   (b) find $\overrightarrow{OA} \cdot \overrightarrow{OB}$ given that $\mathbf{a} = 2\mathbf{i} + \mathbf{j}$ and $\mathbf{b} = 2\mathbf{i} - \mathbf{j}$.

18. $\overrightarrow{OC}$ and $\overrightarrow{OD}$ are two vectors such that $\overrightarrow{OC} = \mathbf{a} - \mathbf{b}$ and $\overrightarrow{OD} = 2\mathbf{a} + 3\mathbf{b}$. If OC is perpendicular to $\overrightarrow{OD}$ show that $\mathbf{a} \cdot \mathbf{b} = 3b^2 - 2a^2$.

    If $\mathbf{a}$ has magnitude 7 units and $\mathbf{b}$ has magnitude 5 units, find the angle between $\mathbf{a}$ and $\mathbf{b}$ giving your answer to the nearest degree.

19. Find the resolved part of the vector $\mathbf{a} = 4\mathbf{i} + 2\mathbf{j}$ in the direction of the vector $\mathbf{b} = 3\mathbf{i} + 4\mathbf{j}$.

20. Find the component of the vector $\mathbf{c} = 3\mathbf{i} - 2\mathbf{j}$ in the direction of the vector $\mathbf{d} = 2\mathbf{i} - 3\mathbf{j}$.

21. Find the length of the projection of the vector $\mathbf{e} = 5\mathbf{i} + \mathbf{j}$ onto the vector $\mathbf{f} = 2\mathbf{i} - \mathbf{j}$.

22. The vectors $\mathbf{a}$ and $\mathbf{b}$ are of equal magnitude k $(k \neq 0)$ and the angle between $\mathbf{a}$ and $\mathbf{b}$ is 60°. If $\mathbf{c} = 3\mathbf{a} - \mathbf{b}$ and $\mathbf{d} = 2\mathbf{a} - 10\mathbf{b}$,
    (a) show that $\mathbf{c}$ and $\mathbf{d}$ are perpendicular vectors
    (b) find the magnitudes of $\mathbf{c}$ and $\mathbf{d}$ in terms of k.

2.4 Geometrical proofs by vector methods

The vector properties covered in this chapter and the skills we have now acquired in manipulating and interpreting vector expressions can be used to prove certain geometrical facts as the following example and the questions of Exercise 2D will show.

Example 16

(The mid-points of the sides of any quadrilateral form the vertices of a parallelogram).
O is the origin and points A, B and C have position vectors $\mathbf{a}$, $\mathbf{b}$ and $\mathbf{c}$ respectively. Points P, Q, R and S are the mid-points of OA, AB, BC and CO respectively. Find the following vectors in terms of $\mathbf{a}$, $\mathbf{b}$ and $\mathbf{c}$
(a) $\overrightarrow{AB}$, (b) $\overrightarrow{BC}$, (c) $\overrightarrow{PQ}$, (d) $\overrightarrow{QR}$, (e) $\overrightarrow{SR}$, (f) $\overrightarrow{PS}$.
Hence show that PQRS is parallelogram.

First make a rough sketch:

(a) $\overrightarrow{AB} = \overrightarrow{AO} + \overrightarrow{OB}$
$= -\mathbf{a} + \mathbf{b}$

(b) $\overrightarrow{BC} = \overrightarrow{BO} + \overrightarrow{OC}$
$= -\mathbf{b} + \mathbf{c}$

(c) $\overrightarrow{PQ} = \overrightarrow{PA} + \overrightarrow{AQ}$
$= \frac{1}{2}\mathbf{a} + \frac{1}{2}(-\mathbf{a} + \mathbf{b})$
$= \frac{1}{2}\mathbf{b}$

(d) $\overrightarrow{QR} = \overrightarrow{QB} + \overrightarrow{BR}$
$= \frac{1}{2}(-\mathbf{a} + \mathbf{b}) + \frac{1}{2}(-\mathbf{b} + \mathbf{c})$
$= -\frac{1}{2}\mathbf{a} + \frac{1}{2}\mathbf{c}$

(e) $\overrightarrow{SR} = \overrightarrow{SC} + \overrightarrow{CR}$
$= \frac{1}{2}\mathbf{c} + \frac{1}{2}(-\mathbf{c} + \mathbf{b})$
$= \frac{1}{2}\mathbf{b}$

(f) $\overrightarrow{PS} = \overrightarrow{PO} + \overrightarrow{OS}$
$= -\frac{1}{2}\mathbf{a} + \frac{1}{2}\mathbf{c}$

Parts (c), (d), (e) and (f) above show that $\overrightarrow{PQ} = \overrightarrow{SR}$ and $\overrightarrow{QR} = \overrightarrow{PS}$, i.e. the opposite sides of quadrilateral PQRS are equal and parallel. Thus PQRS is a parallelogram.

Exercise 2D

1. (The diagonals of a parallelogram bisect each other).
   In the diagram, OABC is a parallelogram with $\overrightarrow{OA}$ = **a** and $\overrightarrow{OC}$ = **c**.

   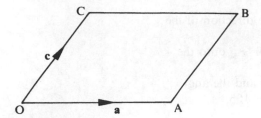

   If G is the mid-point of AC and H is the mid-point of OB, find the following vectors in terms of **a** and **c**
   (a) $\overrightarrow{OB}$, (b) $\overrightarrow{OH}$, (c) $\overrightarrow{AG}$, (d) $\overrightarrow{OG}$.
   Hence show that G and H are concurrent.

2. (The diagonals of a rhombus bisect each other at right angles).
   The diagram shows a parallelogram OABC with $\overrightarrow{OA}$ = **a** and $\overrightarrow{OC}$ = **c**; AC and OB intersect at the point D such that $\overrightarrow{OD}$ = $h\overrightarrow{OB}$ and $\overrightarrow{AD}$ = $k\overrightarrow{AC}$.

   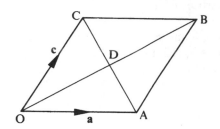

   Show that $h = k = \frac{1}{2}$.
   If the parallelogram is a rhombus, i.e. $|\mathbf{a}| = |\mathbf{c}|$, show that AC is perpendicular to OB. (Hint: find $\overrightarrow{AC} \cdot \overrightarrow{OB}$).

3. (The medians of a triangle intersect at a point that is two-thirds of the way along each median measured from the vertex). (Method one).
   OAB is a triangle with $\overrightarrow{OA}$ = **a** and $\overrightarrow{OB}$ = **b**; C is the mid-point of OB, D is the mid-point of AB and E is the mid-point of OA.
   (a) If F is a point on OD such that $\overrightarrow{OF} = \frac{2}{3}\overrightarrow{OD}$ find $\overrightarrow{OF}$ in terms of **a** and **b**.

   (b) If G is a point on AC such that $\overrightarrow{AG} = \frac{2}{3}\overrightarrow{AC}$ find $\overrightarrow{OG}$ in terms of **a** and **b**.
   (c) If H is a point on EB such that $\overrightarrow{BH} = \frac{2}{3}\overrightarrow{BE}$, find $\overrightarrow{OH}$ in terms of **a** and **b**.
   (d) Hence show that F, G, and H are concurrent.

4. (The medians of a triangle intersect at a point that is two-thirds of the way along each median measured from the vertex). (Method 2).
   OAB is a triangle with $\overrightarrow{OA}$ = **a** and $\overrightarrow{OB}$ = **b**; C is the mid-point of OB, D is the mid-point of AB and E is the mid-point of OA; OD and AC intersect at F.
   If $\overrightarrow{AF} = h\overrightarrow{AC}$ and $\overrightarrow{OF} = k\overrightarrow{OD}$ show that:
   (a) $h = k = \frac{2}{3}$,
   (b) B, F and E are collinear with $\overrightarrow{BF} = \frac{2}{3}\overrightarrow{BE}$.

5. (Cosine rule and Pythagoras' theorem).
   OAB is a triangle and θ is the angle between $\overrightarrow{OA}$ and $\overrightarrow{OB}$.
   (a) By finding $\overrightarrow{AB} \cdot \overrightarrow{AB}$ show that
   $AB^2 = OA^2 + OB^2 - 2(OA)(OB)\cos\theta$
   (b) If triangle OAB is right-angled at O show that $AB^2 = OA^2 + OB^2$

6. (If a straight line is drawn parallel to one side of a triangle, it divides the other two sides in the same ratio).
   In the triangle OAB, a line is drawn parallel to AB, cutting OA at C and OB at D.
   If $\overrightarrow{CD} = \lambda\overrightarrow{AB}$, $\overrightarrow{OC} = h\overrightarrow{OA}$ and $\overrightarrow{OD} = k\overrightarrow{OB}$, show that $h = k = \lambda$.

7. (The angle in a semicircle is a right angle).
   AB is the diameter of a circle centre O; C is a point on the circumference of the circle.
   If $\overrightarrow{OA}$ = **a** and $\overrightarrow{OC}$ = **c** prove that angle A$\hat{C}$B is a right angle.

8. (Apollonius' theorem: the sum of the squares on two sides of a triangle is equal to twice the square on half the third side plus twice the square on the median which bisects the third side).
   OAB is a triangle and point C is the mid-point of AB. Show that:
   (a) $4\overrightarrow{OC} \cdot \overrightarrow{OC} = OA^2 + OB^2 + 2\overrightarrow{OA} \cdot \overrightarrow{OB}$
   (b) $4\overrightarrow{AC} \cdot \overrightarrow{AC} = OA^2 + OB^2 - 2\overrightarrow{OA} \cdot \overrightarrow{OB}$
   (c) $2OC^2 + 2AC^2 = OA^2 + OB^2$

2.5 The vector equation of a straight line

A straight line is uniquely defined if we are given the position of one point that lies on the line and the direction of the line. In terms of vectors this means that a line is defined if we know:

      (i) a vector that is parallel to the line,

and (ii) the position vector of a point on the line.

Consider the line which passes through a point A, position vector **a**, and is parallel to a vector **b**.

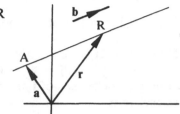

To find the vector equation of this line, we consider some general point R on the line.

Suppose that R has position vector **r**.

$\overrightarrow{AR}$ is parallel to **b**. Thus $\overrightarrow{AR} = \lambda\mathbf{b}$

         but $\overrightarrow{AR} = -\mathbf{a} + \mathbf{r}$

        $\therefore$ $-\mathbf{a} + \mathbf{r} = \lambda\mathbf{b}$

        or $\mathbf{r} = \mathbf{a} + \lambda\mathbf{b}$

The equation $\mathbf{r} = \mathbf{a} + \lambda\mathbf{b}$ is the vector equation for a line that is parallel to a vector **b** and which passes through a point with position vector **a**. The position vector **r** of *any* point lying on the line will satisfy this equation.

Example 17

Find the vector equation of the straight line that is parallel to the vector $2\mathbf{i} - \mathbf{j}$ and which passes through the point with position vector $3\mathbf{i} + 2\mathbf{j}$.

The vector equation is $\mathbf{r} = 3\mathbf{i} + 2\mathbf{j} + \lambda(2\mathbf{i} - \mathbf{j})$ where **r** is the position vector of some general point on the line and λ is a scalar.

Example 18

Find the vector equation of the straight line that passes through the points A, B and C given that the position vectors of A, B and C are $\begin{pmatrix} -1 \\ 1 \end{pmatrix}$, $\begin{pmatrix} 2 \\ 7 \end{pmatrix}$ and $\begin{pmatrix} 3 \\ 9 \end{pmatrix}$ respectively.

To obtain the vector equation, we need a vector that is parallel to the line, for example $\overrightarrow{BC}$, and the position vector of one point on the line, say A. Thus the vector equation of the line is

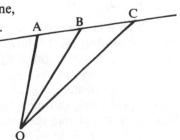

$$\mathbf{r} = \overrightarrow{OA} + \lambda\overrightarrow{BC}$$

but $\overrightarrow{OA} = \begin{pmatrix} -1 \\ 1 \end{pmatrix}$ and $\overrightarrow{BC} = -\begin{pmatrix} 2 \\ 7 \end{pmatrix} + \begin{pmatrix} 3 \\ 9 \end{pmatrix} = \begin{pmatrix} 1 \\ 2 \end{pmatrix}$

Thus the required vector equation is $\mathbf{r} = \begin{pmatrix} -1 \\ 1 \end{pmatrix} + \lambda\begin{pmatrix} 1 \\ 2 \end{pmatrix}$

It should be noticed in Example 18 that $\mathbf{r} = \overrightarrow{OB} + \lambda\overrightarrow{BC}$, $\mathbf{r} = \overrightarrow{OC} + \lambda\overrightarrow{BA}$, $\mathbf{r} = \overrightarrow{OB} + \lambda\overrightarrow{AC}$, etc. would all have given vector equations of the line. Although each equation might appear different, each would give the position vector of any point on the line by the substitution of a suitable value for λ.

Example 19

Show that the point with position vector $\mathbf{i} + 2\mathbf{j}$ lies on the line L, vector equation $\mathbf{r} = 4\mathbf{i} - \mathbf{j} + \lambda(\mathbf{i} - \mathbf{j})$.

The position vector of any point on a line must satisfy the equation of the line.
Thus if $\mathbf{i} + 2\mathbf{j}$ lies on L, there must exist some value of λ such that:
$$\mathbf{i} + 2\mathbf{j} = 4\mathbf{i} - \mathbf{j} + \lambda(\mathbf{i} - \mathbf{j})$$
Equating coefficients of $\mathbf{i}$ and $\mathbf{j}$ gives $1 = 4 + \lambda$ and $2 = -1 - \lambda$ respectively.
The first of these two equations gives $\lambda = -3$ which is consistent with $2 = -1 - \lambda$.
Thus the point with position vector $\mathbf{i} + 2\mathbf{j}$ does lie on the line L.

Example 20

Find the position vector of the point of intersection of the lines L_1 and L_2 given that L_1 and L_2 have vector equations $\mathbf{r} = 2\mathbf{i} + \mathbf{j} + \lambda(\mathbf{i} + 3\mathbf{j})$ and $\mathbf{r} = 6\mathbf{i} - \mathbf{j} + \mu(\mathbf{i} - 4\mathbf{j})$ respectively.

If the required point has position vector $\mathbf{p}$, then this must satisfy the vector equations of both lines
i.e. $\mathbf{p} = 2\mathbf{i} + \mathbf{j} + \lambda(\mathbf{i} + 3\mathbf{j})$
and $\mathbf{p} = 6\mathbf{i} - \mathbf{j} + \mu(\mathbf{i} - 4\mathbf{j})$
$\therefore$ $2\mathbf{i} + \mathbf{j} + \lambda(\mathbf{i} + 3\mathbf{j}) = 6\mathbf{i} - \mathbf{j} + \mu(\mathbf{i} - 4\mathbf{j})$
Thus, equating coefficients: $2 + \lambda = 6 + \mu$ and $1 + 3\lambda = -1 - 4\mu$
Solving these equations simultaneously gives $\lambda = 2$ and $\mu = -2$ and hence $\mathbf{p} = 4\mathbf{i} + 7\mathbf{j}$.
The given lines intersect at the point with position vector $4\mathbf{i} + 7\mathbf{j}$.

Example 21

Find the vector equation of the straight line that passes through the point with position vector $2\mathbf{i} + 3\mathbf{j}$ and which is perpendicular to the line $\mathbf{r} = 3\mathbf{i} + 2\mathbf{j} + \lambda(\mathbf{i} - 2\mathbf{j})$.

The given line is parallel to the vector $\mathbf{i} - 2\mathbf{j}$. If the required line is parallel to a vector $a\mathbf{i} + b\mathbf{j}$, it follows that
$$(a\mathbf{i} + b\mathbf{j}) \cdot (\mathbf{i} - 2\mathbf{j}) = 0 \quad \text{(for the 2 lines to be perpendicular)}$$
so $a - 2b = 0$
hence $a = 2b$
The required line will therefore be parallel to any vector of the form $b(2\mathbf{i} + \mathbf{j})$.
Taking $2\mathbf{i} + \mathbf{j}$ as one such vector, the required vector equation will be:
$$\mathbf{r} = 2\mathbf{i} + 3\mathbf{j} + \lambda(2\mathbf{i} + \mathbf{j}).$$

Example 22

Find the perpendicular distance from the point A, position vector 4i + 5j, to the line L, vector equation **r** = −3i + j + λ(i + 2j).

Let P be the point where the perpendicular from A meets the line L, and let P have position vector **p**. (See diagram on right).
If the value of λ at P is $λ_1$ then

$$\mathbf{p} = -3\mathbf{i} + \mathbf{j} + λ_1(\mathbf{i} + 2\mathbf{j})$$
$$= (-3 + λ_1)\mathbf{i} + (1 + 2λ_1)\mathbf{j}$$

but
$$\overrightarrow{AP} = -\mathbf{a} + \mathbf{p}$$
$$= -(4\mathbf{i} + 5\mathbf{j}) + (-3 + λ_1)\mathbf{i} + (1 + 2λ_1)\mathbf{j}$$
$$= (-7 + λ_1)\mathbf{i} + (-4 + 2λ_1)\mathbf{j}$$

But $\overrightarrow{AP}$ is perpendicular to L and is therefore perpendicular to the vector i + 2j.

So
$$[(-7 + λ_1)\mathbf{i} + (-4 + 2λ_1)\mathbf{j}] \cdot [\mathbf{i} + 2\mathbf{j}] = 0$$
i.e. $-7 + λ_1 - 8 + 4λ_1 = 0$
giving $λ_1 = 3$
∴ $\overrightarrow{AP} = -4\mathbf{i} + 2\mathbf{j}$
and $|\overrightarrow{AP}| = \sqrt{(4^2 + 2^2)} = 2\sqrt{5}$ units

The perpendicular distance from the point A to the line L is $2\sqrt{5}$ units.

Exercise 2E

1. State the vector equation of the straight line which is parallel to 2i + 5j and which passes through the point with position vector 3i − j.
2. State the vector equation of the straight line which passes through the point A, position vector i + 2j, and which is parallel to 4i − j.
3. State the vector equation of the straight line which passes through the point B, position vector $\begin{pmatrix} 1 \\ -2 \end{pmatrix}$, and which is parallel to the vector $\begin{pmatrix} -1 \\ 3 \end{pmatrix}$.
4. State the vector equation of the straight line which is parallel to the vector $\begin{pmatrix} 1 \\ 5 \end{pmatrix}$ and which passes through the point with position vector $\begin{pmatrix} -2 \\ 3 \end{pmatrix}$.
5. Find the vector that is parallel to the line **r** = 2i − j + λ(i + 2j) and is of magnitude $5\sqrt{5}$ units.
6. Find the two vectors that are perpendicular to the line **r** = 2i + 3j + λ(3i − 4j) and are of magnitude 5 units.
7. Show that the point with position vector i + 3j lies on the line with vector equation **r** = 5i + j + λ(2i − j).
8. Points A, B and C have position vectors $\begin{pmatrix} 3 \\ 5 \end{pmatrix}$, $\begin{pmatrix} 1 \\ 9 \end{pmatrix}$ and $\begin{pmatrix} -3 \\ 1 \end{pmatrix}$ respectively.

   Find which of these points lie on the line with vector equation **r** = $\begin{pmatrix} -2 \\ 3 \end{pmatrix} + λ\begin{pmatrix} 1 \\ 2 \end{pmatrix}$.
9. Points D, E and F have position vectors 5i − 5j, i − j and i + 7j respectively. Find which of these points lie on the line with vector equation **r** = 3i + j + λ(−i + 3j).

10. If the point A, position vector $a\mathbf{i} + 3\mathbf{j}$, lies on the line L, vector equation $\mathbf{r} = 2\mathbf{i} + 5\mathbf{j} + \lambda(\mathbf{i} - \mathbf{j})$, find the value of a.

11. Points A and B have position vectors $3\mathbf{i} - 2\mathbf{j}$ and $4\mathbf{i} + \mathbf{j}$ respectively.
    Find: (a) $\overrightarrow{AB}$ in the form $a\mathbf{i} + b\mathbf{j}$
          (b) the vector equation of the straight line that passes through A and B.

12. For each of the following pairs of lines, state whether the lines are parallel and, for those that are not, find the position vector of their point of intersection.
    (a) $\mathbf{r} = -2\mathbf{i} + \mathbf{j} + \lambda(3\mathbf{i} + 2\mathbf{j})$, $\mathbf{r} = \mathbf{i} + 2\mathbf{j} + \mu(\mathbf{i} + \mathbf{j})$
    (b) $\mathbf{r} = 2\mathbf{i} + 3\mathbf{j} + \lambda(\mathbf{i} + 2\mathbf{j})$, $\mathbf{r} = 3\mathbf{i} - \mathbf{j} + \mu(\mathbf{i} + 2\mathbf{j})$
    (c) $\mathbf{r} = 3\mathbf{i} + 5\mathbf{j} + \lambda(-\mathbf{i} + 3\mathbf{j})$, $\mathbf{r} = 2\mathbf{i} + \mathbf{j} + \mu(\mathbf{i} - 3\mathbf{j})$
    (d) $\mathbf{r} = \mathbf{i} + 3\mathbf{j} + \lambda(2\mathbf{i} - \mathbf{j})$, $\mathbf{r} = 5\mathbf{i} - 3\mathbf{j} + \mu(2\mathbf{i} + \mathbf{j})$

13. Show that the lines L_1, vector equation $\mathbf{r} = \begin{pmatrix} 2 \\ 5 \end{pmatrix} + \lambda\begin{pmatrix} 2 \\ -3 \end{pmatrix}$, and L_2,

    vector equation $\mathbf{r} = \begin{pmatrix} 3 \\ -3 \end{pmatrix} + \mu\begin{pmatrix} 3 \\ 2 \end{pmatrix}$, are perpendicular and find the

    position vector of their point of intersection.

14. Find the acute angle between the lines with vector equations
    $\mathbf{r} = 2\mathbf{i} + 3\mathbf{j} + \lambda(\mathbf{i} + 2\mathbf{j})$ and $\mathbf{r} = 3\mathbf{i} - 2\mathbf{j} + \lambda(4\mathbf{i} - \mathbf{j})$, giving your answer to the nearest degree.

15. The lines L_1 and L_2 have vector equations $\mathbf{r} = \mathbf{a} + \lambda(\mathbf{a} - 2\mathbf{b})$ and $\mathbf{r} = 3\mathbf{a} + \mu(-\mathbf{a} + 3\mathbf{b})$ respectively. Given that the lines lie in the same plane and are not parallel, find the position vector of their point of intersection in terms of $\mathbf{a}$ and $\mathbf{b}$.

16. The lines L_1 and L_2 intersect at A and the lines L_1 and L_3 intersect at B.
    Find $\overrightarrow{AB}$ and $|\overrightarrow{AB}|$ given that L_1, L_2 and L_3 have vector equations as
    follows: L_1: $\mathbf{r} = \mathbf{i} + 3\mathbf{j} + \lambda(\mathbf{i} + \mathbf{j})$
               L_2: $\mathbf{r} = \mathbf{i} + \mu(5\mathbf{i} + 2\mathbf{j})$
               L_3: $\mathbf{r} = 4\mathbf{i} + 3\mathbf{j} + \eta(2\mathbf{i} - \mathbf{j})$

17. For each of the following parts, find the perpendicular distance from the given point to the given line:
    (a) the point with position vector $\mathbf{i} + 5\mathbf{j}$ and the line $\mathbf{r} = -2\mathbf{i} + \mathbf{j} + \lambda(2\mathbf{i} + \mathbf{j})$
    (b) the point with position vector $2\mathbf{i} - \mathbf{j}$ and the line $\mathbf{r} = -2\mathbf{j} + \lambda(4\mathbf{i} - 3\mathbf{j})$

    (c) the point with position vector $\begin{pmatrix} 5 \\ 2 \end{pmatrix}$ and the line $\mathbf{r} = \begin{pmatrix} 6 \\ -5 \end{pmatrix} + \lambda\begin{pmatrix} 1 \\ 3 \end{pmatrix}$.

Exercise 2F   Examination questions

1. (a) E is the centre of the rectangle ABCD and $\overrightarrow{AB} = \mathbf{a}$, $\overrightarrow{BC} = \mathbf{b}$.
       Express in terms of $\mathbf{a}$ and $\mathbf{b}$ the vectors:
       (i) $\overrightarrow{AC}$ (ii) $\overrightarrow{CD}$ (iii) $\overrightarrow{BD}$ (iv) $\overrightarrow{EB}$ (v) $\overrightarrow{EA}$.
   (b) The position vectors of P and Q are $5\mathbf{i} + 3\mathbf{j}$ and $\mathbf{i} - 2\mathbf{j}$ respectively referred to an origin $O\mathbf{i} + O\mathbf{j}$.
       Calculate: (i) the position vector of P relative to Q and hence the magnitude of $\overrightarrow{PQ}$:
                  (ii) the magnitude of $2\overrightarrow{OP} - 4\overrightarrow{OQ}$.   (S.U.J.B)

2. The position vectors of three points A, B and C relative to an origin O are **p**, $3\mathbf{q} - \mathbf{p}$, and $9\mathbf{q} - 5\mathbf{p}$ respectively.
Show that the points A, B and C lie on the same straight line, and state the ratio AB:BC.
Given that OBCD is a parallelogram and that E is the point such that $\mathbf{DB} = \frac{1}{3}\mathbf{DE}$, find the position vectors of D and E relative to O.
(Cambridge)

3. Given that $\overrightarrow{OA} = \mathbf{a}$, $\overrightarrow{OB} = \mathbf{b}$, $\overrightarrow{OP} = \frac{4}{5}\overrightarrow{OA}$ and that Q is the midpoint of AB, express $\overrightarrow{AB}$ and $\overrightarrow{PQ}$ in terms of **a** and **b**.
PQ is produced to meet OB produced at R, so that $\overrightarrow{QR} = n\overrightarrow{PQ}$ and $\overrightarrow{BR} = k\mathbf{b}$. Express $\overrightarrow{QR}$ (i) in terms of n, **a** and **b**,
                                          (ii) in terms of k, **a** and **b**.
Hence find the value of n and of k.   (Cambridge)

4. The points A, B and C have position vectors **a**, **b** and **c** respectively referred to an origin O.
   (a) Given that the point X lies on AB produced so that AB:BX = 2:1, find **x**, the position vector of X, in terms of **a** and **b**.
   (b) If Y lies on BC, between B and C so that BY:YC = 1:3, find **y**, the position vector of Y, in terms of **b** and **c**.
   (c) Given that Z is the mid-point of AC, show that X, Y and Z are collinear.
   (d) Calculate XY:YZ.   (London)

5. Given that $\mathbf{a} = \begin{pmatrix} 4 \\ -3 \end{pmatrix}$, $\mathbf{b} = \begin{pmatrix} 2 \\ 4 \end{pmatrix}$ and $\mathbf{c} = \begin{pmatrix} 22 \\ -11 \end{pmatrix}$, find:
   (i) a unit vector perpendicular to **a**,
   (ii) the value of the constants m and n for which $m\mathbf{a} + n\mathbf{b} = \mathbf{c}$.
   Evaluate the scalar product of **a** and **b** and hence find the cosine of the angle between the direction of **a** and the direction of **b**.   (Cambridge)

6. The position vectors of the points A and B with respect to the origin O are $2\mathbf{i} + 3\mathbf{j}$ and $-\mathbf{i} + 5\mathbf{j}$ respectively. Find the position vector of the point C such that $\overrightarrow{AC} = 2\overrightarrow{AB}$.
Calculate the angle between the vectors $\overrightarrow{AB}$ and $\overrightarrow{OB}$.   (A.E.B.)

7. The position vectors of points A, B, C relative to an origin O are $\begin{pmatrix} 1 \\ 3 \end{pmatrix}$, $\begin{pmatrix} 2 \\ 6 \end{pmatrix}$, $\begin{pmatrix} 5 \\ 5 \end{pmatrix}$ respectively.
Write down the vectors representing $\overrightarrow{AB}$, $\overrightarrow{AC}$ and $\overrightarrow{BC}$.
Use vector methods to calculate: (i) $B\hat{A}C$,
                                 (ii) $A\hat{B}C$.
State the special property of $\triangle ABC$ and deduce its area.   (Cambridge)

8. (a) Show that if $(\mathbf{a} + \mathbf{b}) \cdot (\mathbf{a} - \mathbf{b}) = 0$ then $|\mathbf{a}| = |\mathbf{b}|$.
   Give a geometrical example of this fact.
   (b) The diagonals AC, BD of a quadrilateral ABCD bisect each other
   and are represented by the vectors $3\mathbf{i} + 2\mathbf{j}$ and $2\mathbf{i} - 3\mathbf{j}$ respectively.
   Determine, *without scale drawing*, the shape of the quadrilateral
   ABCD.   (A.E.B)

9. The position vectors of A and B relative to an origin O are $\begin{pmatrix} p \\ q \end{pmatrix}$ and $\begin{pmatrix} q \\ r \end{pmatrix}$

   respectively where $q \neq 0$.
   (a) Find an equation connecting p, q and r in each of the cases
       (i) O, A, B are collinear,
       (ii) $|\overrightarrow{OB}| = |\overrightarrow{AB}|$.
   (b) Use the scalar product of two vectors to obtain an expression for
   $\cos A\hat{O}B$ in terms of p, q and r.   (Cambridge)

10. Find the position vector of the point of intersection, P, of the line with

    equation $\mathbf{r} = \begin{pmatrix} 3 \\ 1 \end{pmatrix} + s\begin{pmatrix} 2 \\ 3 \end{pmatrix}$ and the line joining the points with position

    vectors $\begin{pmatrix} 2 \\ 4 \end{pmatrix}$ and $\begin{pmatrix} 1 \\ 4 \end{pmatrix}$. Find the distance between P and the origin.

                                                                 (A.E.B.)

11. The position vectors of the points A and B with respect to the origin O
    are $2\mathbf{i} + 4\mathbf{j}$ and $6\mathbf{i} + 7\mathbf{j}$ respectively.
    (a) Find a vector equation for the line AB, and show that this line
        passes through the point with position vector $14\mathbf{i} + 13\mathbf{j}$.
    (b) The position vector of the point C is $3\mathbf{i} - 4\mathbf{j}$. Find a vector equation
        for the line OC, and show that OC is at right angles to AB.
    (c) Find the position vector of the point of intersection of OC and AB,
        and hence find the perpendicular distance from O to AB.   (A.E.B.)

3

Coordinate geometry I

Coordinate geometry is the study of the geometric properties of points, straight lines and curves using algebraic methods. From earlier work the reader is familiar with the idea of a point P in a plane being defined by stating the perpendicular distances from P to two axes. The point P has coordinates (x, y) or, in vector terms P has position vector $\overrightarrow{OP} = \begin{pmatrix} x \\ y \end{pmatrix}$ or $x\mathbf{i} + y\mathbf{j}$, where O is the origin.

3.1 Loci

Consider the $x–y$ plane, i.e. the plane containing the coordinate axes Ox and Oy. If $P(x, y)$ is a point whose position is fixed by some law, then the set of all such points obeying the law is said to be the **locus** of P.

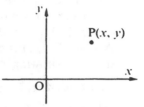

Example 1

The point $P(x, y)$ is 1 cm from the x-axis. Draw a diagram showing the locus of P.

The set of all such points $P(x, y)$ would together form two lines, one on each side of the x-axis and 1 cm from it.

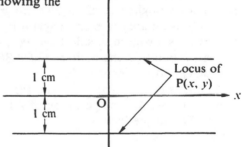

Example 2

The point $P(x, y)$ is equidistant from the origin and the point A(2, 0). Draw a diagram showing the locus of P.

The set of all points $P(x, y)$ that are equidistant from the origin and the point A(2, 0) would together form a line, perpendicular to the x-axis and passing through the point (1, 0). i.e. the locus is the perpendicular bisector of OA.

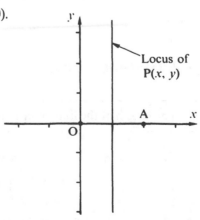

Example 3

ABCD is a square of side 3 cm. The point $P(x, y)$ lies inside the square but is more than 2 cm from D.
Draw a diagram showing ABCD and the locus of P.

The set of all points $P(x, y)$ will form the shaded area shown in the diagram. Notice that broken lines are used for the boundary of the area as the locus does not include these boundary lines.

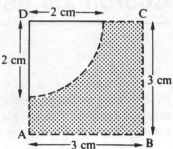

Exercise 3A

1. A point $P(x, y)$ is 2 cm from the origin O. Draw a diagram showing O and the locus of P

2. Point $P(x, y)$ is more than 2 cm but less than 3 cm away from the origin O. Draw a diagram showing O and the locus of P.

3. A point $P(x, y)$ is equidistant from the point A(1, 0) and B(0, 1). Draw a diagram showing A, B and the locus of P.

In questions **4** to **7**, assume that the points A, B, C and D lie in the x–y plane.

4. AB is a straight line of length 3 cm. Draw a diagram showing the line AB and the locus of the point $P(x, y)$ given that P is a distance of 2 cm from the nearest point on AB.

5. AB is a straight line of length 4 cm and point $P(x, y)$ is such that triangle PAB has an area of 6 cm. Explain in words the locus of P.

6. ABCD is a square of side 1 cm. $P(x, y)$ is a point that is 2 cm from the nearest point on the perimeter of ABCD. Draw a diagram showing ABCD and the locus of P.

7. ABCD is a square of side 3 cm. $P(x, y)$ is a point that is 1 cm from the nearest point on the perimeter of ABCD. Draw a diagram showing ABCD and the locus of P.

In the remainder of this chapter we shall obtain certain useful results regarding pairs of points and straight lines lying in the x–y plane.

3.2 Distance between two points

> The length of the line joining the point $A(x_1, y_1)$ to $B(x_2, y_2)$ is given by $\sqrt{[(x_2 - x_1)^2 + (y_2 - y_1)^2]}$

Proof: If A and B have coordinates (x_1, y_1) and (x_2, y_2) respectively, it follows that

$$\mathbf{a} = \begin{pmatrix} x_1 \\ y_1 \end{pmatrix} \text{ and } \mathbf{b} = \begin{pmatrix} x_2 \\ y_2 \end{pmatrix} \text{ where } \mathbf{a} \text{ and } \mathbf{b} \text{ are the position vectors of A and B.}$$

But $\overrightarrow{AB} = -\mathbf{a} + \mathbf{b}$
$$= \mathbf{b} - \mathbf{a}$$

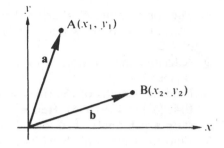

$$= \begin{pmatrix} x_2 \\ y_2 \end{pmatrix} - \begin{pmatrix} x_1 \\ y_1 \end{pmatrix} = \begin{pmatrix} x_2 - x_1 \\ y_2 - y_1 \end{pmatrix}$$

Thus, by Pythagoras' theorem,

$|AB| = \sqrt{[(x_2 - x_1)^2 + (y_2 - y_1)^2]}$ as required.

Example 4

Find the length of the line joining P(5, 1) to Q(9, −3).

Using the above result
$$\begin{aligned} PQ^2 &= (x_2 - x_1)^2 + (y_2 - y_1)^2 \\ &= (9 - 5)^2 + (-3 - 1)^2 \\ &= 16 + 16 \\ &= 32 \\ \therefore \quad PQ &= \sqrt{32} \quad \text{or} \quad 4\sqrt{2} \end{aligned}$$

The length of PQ is $4\sqrt{2}$ units.

Example 5

The points A, B and C have coordinates (2, 1), (7, 3) and (5, k) respectively. If AB and BC are of equal length, find the possible values of k.

$$\begin{aligned} AB^2 &= (7 - 2)^2 + (3 - 1)^2 & \text{and} \quad BC^2 &= (5 - 7)^2 + (k - 3)^2 \\ &= 25 + 4 & &= 4 + k^2 - 6k + 9 \\ &= 29 & &= k^2 - 6k + 13 \end{aligned}$$

Since AB = BC   then $AB^2 = BC^2$

i.e. $29 = k^2 - 6k + 13$

                 $0 = k^2 - 6k - 16$

giving $k = 8$ or -2

The possible values of k are 8 and −2.

Exercise 3B

1. Find the lengths of the straight lines joining each of the following pairs of points:
   (a) (1, 3) and (4, 7)   (b) (1, 8) and (7, 0)
   (c) (4, 0) and (−3, 0)   (d) (1, 5) and (−2, 1)
   (e) (−3, −4) and (2, 8)   (f) (1, 2) and (5, 3)
   (g) (1, 1) and (2, −1)   (h) (3, 3) and (5, 7)
   (i) (3, 3) and (5, −1)   (j) (−1, 0) and (2, 3)
   (k) (7, −3) and (1, −1)   (l) (−7, −1) and (−1, −4)

2. Find the length of the line AB where A is the point with position vector $\begin{pmatrix} -1 \\ 3 \end{pmatrix}$ and B has position vector $\begin{pmatrix} 3 \\ 7 \end{pmatrix}$.

3. Find the length of the line CD where C is the point with position vector $\begin{pmatrix} -3 \\ 2 \end{pmatrix}$ and D has position vector $\begin{pmatrix} 1 \\ -5 \end{pmatrix}$.

4. The three points A, B and C have coordinates (1, 0), (4, 4) and (13, 5) respectively. By how many units does the length of AC exceed the length of AB?

5. A triangle has vertices at (0, 1), (1, 6) and (5, 2). Prove that the triangle is isosceles.

6. Find the lengths of the sides of the triangle with vertices at A(1, 1), B(4, 5) and C(9, −5). Hence prove that the triangle is right-angled, and state which angle, Â, B̂ or Ĉ is the right angle.

7. Find the lengths of the sides of the triangle with vertices at A(2, 7), B(4, 3) and C(10, 6). Hence prove that the triangle is right-angled, and state which angle, Â, B̂ or Ĉ is the right angle.

8. The points A and B have coordinates (1, 1) and (5, −1) respectively. Find, by calculation, which one of the points C(3, 3), D(5, 4) or E(3, −2) lies on the perpendicular bisector of AB.

9. Point A has coordinates (3, 2) and point B has coordinates (9, −4). Find, by calculation, which one of the points C(4, −3), D(10, 3) or E(5, −1) does not lie on the perpendicular bisector of AB.

10. The three points O, A and B have coordinates (0, 0), (a, 7) and (8, 1) respectively. If OA and OB are of equal length, calculate the two possible values of a.

11. The three points D, E and F have coordinates (2, 2) (7, e) and (9, 3) respectively. If DE and DF are of equal length, find the two possible values of e.

12. The three points G(4, 0), H(h, 6) and I(7, 1) are such that GH is twice as long as GI. Calculate the two possible values of h.

3.3 Mid-point of a line joining two points

> The mid-point of the line joining $A(x_1, y_1)$ to $B(x_2, y_2)$ has coordinates $\left(\dfrac{x_1 + x_2}{2}, \dfrac{y_1 + y_2}{2}\right)$

Proof: If A and B have coordinates (x_1, y_1) and (x_2, y_2) respectively, it follows that

$$\mathbf{a} = \begin{pmatrix} x_1 \\ y_1 \end{pmatrix} \text{ and } \mathbf{b} = \begin{pmatrix} x_2 \\ y_2 \end{pmatrix} \text{ where } \mathbf{a} \text{ and } \mathbf{b} \text{ are the position vectors of A and B.}$$

Let M be the mid-point of AB

$$\overrightarrow{OM} = \overrightarrow{OA} + \tfrac{1}{2}\overrightarrow{AB}$$
$$= \mathbf{a} + \tfrac{1}{2}(-\mathbf{a} + \mathbf{b})$$
$$= \tfrac{1}{2}(\mathbf{a} + \mathbf{b})$$
$$= \frac{1}{2}\left[\begin{pmatrix} x_1 \\ y_1 \end{pmatrix} + \begin{pmatrix} x_2 \\ y_2 \end{pmatrix}\right] = \left[\begin{pmatrix} \tfrac{1}{2}(x_1 + x_2) \\ \tfrac{1}{2}(y_1 + y_2) \end{pmatrix}\right]$$

Thus M has coordinates $\left(\dfrac{x_1 + x_2}{2}, \dfrac{y_1 + y_2}{2}\right)$ as required.

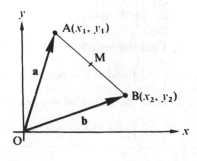

Example 6

Find the coordinates of the mid-point of the straight lines joining the following pairs of points:

(a) (3, 7) and (5, 9)   (b) (−2, 5) and (8, −1)

(a) Mid-point of line joining (3, 7) and (5, 9) has coordinates $\left(\dfrac{3 + 5}{2}, \dfrac{7 + 9}{2}\right)$.

Coordinates of mid-point are (4, 8).

(b) Mid-point of line joining (−2, 5) and (8, −1) has coordinates $\left(\dfrac{-2 + 8}{2}, \dfrac{5 + (-1)}{2}\right)$.

Coordinates of mid-point are (3, 2).

Example 7

The points P(4, −3), Q(−3, 4), R(−2, 7) and S are the vertices of a parallelogram PQRS. Find (a) the coordinates of the mid-point of the diagonal PR, (b) the coordinates of S.

(a) If M is the mid-point of the line joining P(4, −3) to R(−2, 7), M has the coordinates

$$\left(\frac{4 + (-2)}{2}, \frac{(-3) + 7}{2}\right) = (1, 2)$$

Mid-point of PR is the point with coordinates (1, 2)

(b) Since the diagonals of a parallelogram bisect each other, M(1, 2) is also the mid-point of QS.

Suppose the coordinates of S are (x, y),

then $(1, 2) = \left(\dfrac{-3 + x}{2}, \dfrac{4 + y}{2}\right)$

i.e. $1 = \dfrac{-3 + x}{2}$ and $2 = \dfrac{4 + y}{2}$

∴ $x = 5$ ∴ $y = 0$.

The coordinates of S are (5, 0).

In addition to finding the mid-point of a line joining two points it is also possible to determine the coordinates of a point dividing the line in any given ratio. This can either be done using position vectors (see example 10 page 55) or as shown below.

Example 8

Find the coordinates of the point P which divides the line joining A(3, 2) and B(5, 1) in the ratio −2:3.

(Remember that the different signs of the two numbers in the ratio implies that the point P divides AB *externally*, see page 55).

First sketch the situation, with the required
point P having coordinates (x, y).

It then follows that $\dfrac{y - 2}{2 - 1} = \dfrac{2}{1}$

giving $\qquad y = 4$

and $\qquad \dfrac{3 - x}{5 - 3} = \dfrac{2}{1}$

giving $\qquad x = -1$

Thus the point P has coordinates $(-1, 4)$.

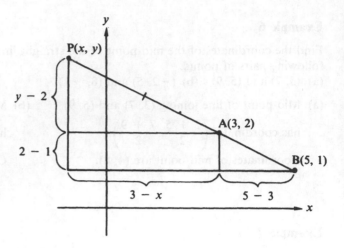

Exercise 3C

1. Find the coordinates of the mid-points of the straight lines joining each
   of the following pairs of points:
   (a) $(3, 4)$ and $(7, 10)$ (b) $(2, 8)$ and $(10, 4)$
   (c) $(0, 5)$ and $(6, 3)$ (d) $(4, 3)$ and $(1, 5)$
   (e) $(-2, 1)$ and $(6, 5)$ (f) $(-3, 5)$ and $(11, -3)$
   (g) $(-3, 8)$ and $(-4, 0)$ (h) $(-2, -1)$ and $(5, 6)$
   (i) $(-3, -9)$ and $(7, -1)$ (j) $(-3, -9)$ and $(-7, -1)$
   (k) $(-2, -6)$ and $(-8, -5)$ (l) $(-3, 4)$ and $(-2, -1)$
2. Find the position vector of the mid-point of the line joining two points
   whose position vectors are as follows:

   (a) $\begin{pmatrix} 3 \\ 5 \end{pmatrix}$ and $\begin{pmatrix} 1 \\ 7 \end{pmatrix}$ (b) $\begin{pmatrix} 2 \\ -1 \end{pmatrix}$ and $\begin{pmatrix} -6 \\ 7 \end{pmatrix}$ (c) $\begin{pmatrix} -2 \\ 3 \end{pmatrix}$ and $\begin{pmatrix} -4 \\ -1 \end{pmatrix}$

   (d) $2\mathbf{i} + 3\mathbf{j}$ and $4\mathbf{i} - \mathbf{j}$ (e) $2\mathbf{i} - \mathbf{j}$ and $10\mathbf{i} + \mathbf{j}$ (f) $\mathbf{i} + \mathbf{j}$ and $\mathbf{i} - 3\mathbf{j}$
3. M is the mid-point of the straight line joining point $A(10, 4)$ to point B.
   If M has coordinates $(7, 3)$, find the coordinates of B.
4. $L(5, 1)$ is the mid-point of the straight line joining point $C(p, -3)$ to
   point $D(7, q)$. Find p and q.
5. Find the co-ordinates of the point E, if the point $F(-4, 3)$ is the mid-
   point of the straight line joining E to $G(-6, 10)$.
6. A triangle has vertices at $A(2, 4)$, $B(4, -2)$ and $C(8, 12)$. L is the mid-
   point of AB and M is the mid-point of BC.
   Find: (a) the coordinates of point L,
             (b) the coordinates of point M,
             (c) the distance LM.
7. A triangle has vertices at $A(9, 9)$, $B(3, 2)$ and $C(9, 4)$.
   Find: (a) the coordinates of M, the mid-point of BC,
             (b) the length of the median from A to M.
8. Find the lengths of the medians of the triangle that has vertices at
   $A(0, 1)$, $B(2, 7)$ and $C(4, -1)$.
9. $A(1, 1)$, $B(6, 9)$, $C(13, 7)$ and D are the vertices of the parallelogram
   ABCD.
   Find: (a) the coordinates of the mid-point of the diagonal AC,
             (b) the coordinates of the mid-point of the diagonal DB,
             (c) the coordinates of D.

10. P(-1, 5), Q(8, 10), R(7, 5) and S are the vertices of the parallelogram PQRS. Calculate the coordinates of S.

11. Points A(2, 5), B(4, -1), C(0, -3) and D(-6, 3) form the quadrilateral ABCD. Points P, Q, R and S are the mid-points of AB, BC, CD and DA respectively.
    (a) Find the coordinates of P, Q, R and S,
    (b) prove that PQ is the same length as SR,
    (c) prove that PS is the same length as QR.

12. Find the coordinates of the point that divides the line joining point K(6, 3) to point L(11, 5) in the ratio (a) $2:3$,　(b) $-2:3$.

13. Find the coordinates of the point that divides the line joining point M(2, 5) to point N(8, 9) in the ratio (a) $3:5$,　(b) $-3:5$.

14. The points A, B, C and D have coordinates (-7, 9), (3, 4), (1, 12) and (-2, -9) respectively. Find the length of the line PQ where P divides AB in the ratio $2:3$ and Q divides CD in the ratio $1:-4$.

3.4 Gradient of a line joining two points

The gradient of the line joining the points A(x_1, y_1) and B(x_2, y_2) is a measure of the steepness of the line AB and it is the ratio of the change in the y-coordinate to the change in the x-coordinate in going from A to B.

From the diagram

$$\frac{\text{change in } y\text{-coordinate from A to B}}{\text{change in } x\text{-coordinate from A to B}} = \frac{BL}{LA} = \frac{y_2 - y_1}{x_2 - x_1}$$

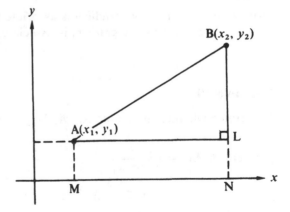

Thus the gradient of the line joining A(x_1, y_1) to B(x_2, y_2) is $\dfrac{y_2 - y_1}{x_2 - x_1}$

Note that $\dfrac{y_2 - y_1}{x_2 - x_1}$ could be written $\dfrac{y_1 - y_2}{x_1 - x_2}$ and the same answer would be obtained.
The point to notice is that the same suffix order is used in both the top and the bottom of the fraction.

Neither $\dfrac{y_2 - y_1}{x_1 - x_2}$ *nor* $\dfrac{y_1 - y_2}{x_2 - x_1}$ would give the correct answer.

Example 9

Find the gradient of the lines joining the following pairs of points:
(a) (4, 1) and (7, 3);　(b) (2, -7) and (4, 3);　(c) (2, 5) and (6, 1).

(a) (4, 1) and (7, 3)

$$\text{gradient} = \frac{y_2 - y_1}{x_2 - x_1}$$
$$= \frac{3 - 1}{7 - 4}$$
$$= \tfrac{2}{3}$$

(b) (2, -7) and (4, 3)

$$\text{gradient} = \frac{y_2 - y_1}{x_2 - x_1}$$
$$= \frac{3 - (-7)}{4 - 2}$$
$$= 5$$

(c) (2, 5) and (6, 1)

$$\text{gradient} = \frac{y_2 - y_1}{x_2 - x_1}$$
$$= \frac{1 - 5}{6 - 2}$$
$$= -1$$

In (a) and (b) the gradient is found to be positive whereas in (c) the gradient is negative.

If the points are plotted, the significance of the negative gradient is seen.

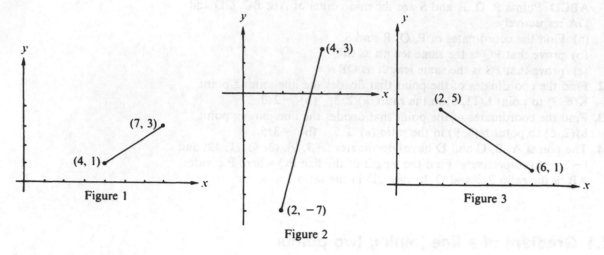

Figure 1

Figure 2

Figure 3

In figures 1 and 2, as we proceed along the line in the direction of increasing x, y also increases.

In figure 3 as we proceed along the line in the direction of increasing x, y decreases.

More simply: a positive gradient is associated with going 'uphill' as x increases,

and   a negative gradient is associated with going 'downhill' as x increases.

Example 10

Determine whether the points A(-4, 3), B(-1, 5) and C(8, 11) are collinear.

gradient of AB $= \dfrac{y_2 - y_1}{x_2 - x_1}$

$= \dfrac{5 - 3}{-1 - (-4)} = \dfrac{2}{3}$

gradient of BC $= \dfrac{y_3 - y_2}{x_3 - x_2}$

$= \dfrac{11 - 5}{8 - (-1)} = \dfrac{2}{3}$

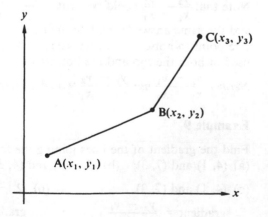

Since the gradient of AB = gradient of BC and B is a point common to the lines AB and BC, the points A, B and C are collinear.

Alternatively, vector methods could be used to show that $\overrightarrow{AB}$ is a scalar multiple of $\overrightarrow{BC}$, as in chapter 2.

Exercise 3D

1. Find the gradients of the straight lines joining each of the following pairs of points:
   (a) (3, 5) and (5, 9)
   (b) (3, 1) and (4, 8)
   (c) (3, 8) and (1, 2)
   (d) (2, 0) and (6, 1)
   (e) (2, 8) and (3, 4)
   (f) (6, 7) and (3, −2)
   (g) (3, 7) and (6, −2)
   (h) (2, 0) and (6, −1)
   (i) (5, −1) and (1, −3)
   (j) (−2, −2) and (0, 2)
   (k) (−1, −4) and (11, 4)
   (l) (−1, −1) and (7, −7)

2. For each of the parts (a) to (e), state whether the three given points are collinear.
   (a) (2, 3), (4, 4), (10, 7).
   (b) (2, 0), (3, 1), (8, 7).
   (c) (1, −3), (−1, 1), (−4, 7).
   (d) (2, −2), (−3, −5), (12, 4).
   (e) (5, 3), (6, 1), (10, −5).

3. If the straight line joining A(3, −5) to B(6, b) has a gradient of 4, find the value of b.

4. If the straight line joining C(c, 5) to D(−3, 2) has a gradient of $\frac{3}{4}$, find the value of c.

5. The three points A(a, −1), B(8, 1) and C(11, 2) are collinear. Find the value of a.

6. The points A(5, 2), B(1, 0), C(c, 5), D(−5, d) and E(e, −2) all lie on the same straight line. Find the values of c, d and e.

3.5 Parallel and perpendicular lines

The gradient of a line is a measure of its 'steepness'. Two parallel lines are equally steep so they will have equal gradients, but what about two perpendicular lines?

Suppose two straight lines (1) and (2) intersect at right angles at the point A and cut the x-axis at the points B and C respectively. If AD is the perpendicular from A to the x-axis and $\hat{ABD} = \theta$, it follows that $\hat{DAC} = \theta$.

Then gradient of line (1) $= \dfrac{AD}{DB} = \tan \theta$

gradient of line (2) $= -\dfrac{AD}{CD}$ (negative as 'downhill' for increasing x)

$$= \dfrac{-1}{CD/AD} = -\dfrac{1}{\tan \theta}$$

∴ gradient of line (1) × gradient of line (2) $= \tan \theta \times \left(\dfrac{-1}{\tan \theta} \right)$

$$= -1$$

> Thus, if two lines are parallel, they have equal gradients
> and if two lines are perpendicular, the product of their gradients is −1.

...his condition for perpendicular lines means that if one line has a gradient m, a perpendicular line will have gradient $-\dfrac{1}{m}$.

Thus, the following pairs of gradients all apply to pairs of perpendicular lines:
$$2 \text{ and } -\tfrac{1}{2}, \quad 3 \text{ and } -\tfrac{1}{3}, \quad \tfrac{2}{3} \text{ and } -\tfrac{3}{2}, \quad \tfrac{3}{5} \text{ and } -\tfrac{5}{3}.$$

Example 11

Given the points A(1, -1), B(5, 2), P(-1, 10), Q(-1, 3) and R(-9, -3), show that AB is parallel to QR and that BP is perpendicular to AB.

gradient AB $= \dfrac{2 - (-1)}{5 - 1}$ gradient QR $= \dfrac{-3 - 3}{-9 - (-1)}$ gradient BP $= \dfrac{10 - 2}{-1 - 5}$

$\qquad\qquad = \dfrac{3}{4}$ $= \dfrac{-6}{-8}$ $= \dfrac{8}{-6}$

$\qquad\qquad\qquad\qquad\qquad\qquad\qquad\qquad = \dfrac{3}{4}$ $= -\dfrac{4}{3}$

Since gradient AB = gradient QR, AB and QR are parallel.
Since (gradient BP) $\times$ (gradient AB) $= -1$, BP is perpendicular to AB.

Example 12

The triangle P(-4, 3), Q(-1, 5), R(0, -3) is right-angled. Find which side is the hypotenuse.

gradient PQ $= \dfrac{5 - 3}{-1 - (-4)}$ gradient QR $= \dfrac{-3 - 5}{0 - (-1)}$ gradient PR $= \dfrac{-3 - 3}{0 - (-4)}$

$\qquad\qquad = \tfrac{2}{3}$ $= -8$ $= -\tfrac{3}{2}$

(gradient PQ) $\times$ (gradient PR) $= \tfrac{2}{3} \times (-\tfrac{3}{2}) = -1$

Hence PQ is perpendicular to PR, so $Q\hat{P}R = 90°$ and QR is the hypotenuse.

Exercise 3E

1. Write down the gradient of a straight line that is perpendicular to a line of gradient:
   (a) 2   (b) 3   (c) -2 (d) $\tfrac{1}{2}$ (e) $-\tfrac{1}{2}$
   (f) $\tfrac{2}{3}$ (g) $\tfrac{3}{2}$ (h) $2\tfrac{1}{2}$ (i) 4   (j) $-\tfrac{2}{3}$

2. Using the points A(2, 4), B(8, 7), C(5, -2) and D(19, 5), prove that
   (a) AB is parallel to CD,   (b) AC is perpendicular to CD.

3. The points A(6, 8), B(11, 9), C(7, -3) and D(4, 2) form the trapezium ABCD. Name the pair of sides that are parallel.

4. A triangle has vertices at A(2, 2), B(5, 7) and C(15, 1). Find the gradients of the lines AB, BC and AC. Hence show that the triangle is right-angled and state which angle, $\hat{A}$, $\hat{B}$ or $\hat{C}$ is the right angle.

5. The line joining point A(a, 3) to point B(2, -3) is perpendicular to the line joining point C(10, 1) to point B. Find the value of a.

6. Points A(0, 3), B(4, 7) and C(10, 5) form the triangle ABC; L is the mid-point of AB and M is the mid-point of BC.
Show that LM is parallel to AC.

7. A triangle has vertices at D(0, 2), E(8, 6) and F(2, f). If the triangle is right-angled at D, use gradients to find the value of f. What would your answer have been had the triangle been right-angled at F?

8. The points A(0, 2), B(8, 0) and C(5, c) form a triangle ABC, right-angled at C. By finding expressions for the gradients of AC and BC, find the two possible values of c.

9. Points A(3, 4), B(5, 0), C(-1, -3) and D(-3, 1) form a quadrilateral ABCD. Find   (a) the gradients of the lines AB, BC, CD and DA,
                 (b) the lengths of the lines AB and AD.
Hence state what type of quadrilateral ABCD is.

3.6 The equation $y = mx + c$

In this chapter we have been discussing points, lines joining points and perpendicular lines, but we saw in chapter 1 that lines on a graph have equations associated with them. We now consider these equations in more detail.

Suppose the point P(x, y) is a point on the straight line which cuts the y-axis at the point A(0, c), and has gradient m.

The gradient of AP is m,

but AP has gradient $\dfrac{y - c}{x - 0}$

$$\therefore \quad \frac{y - c}{x - 0} = m$$

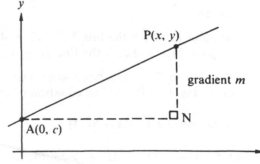

giving $y - c = mx$ and this is usually written $y = mx + c$.

This is the general **cartesian** equation of a straight line. The word cartesian is after René Descartes (1596–1650) the Frenchman who first developed the idea of using an ordered pair of coordinates (x, y) to specify a point in a plane relative to a pair of intersecting straight lines.

> All straight lines have equations which can be written in the form $y = mx + c$ where m is the gradient and (0, c) is the point at which the line cuts the y-axis.

Example 13

Find the gradient of the following straight lines, and in each case find the coordinates of the point where the line cuts the y-axis.

(a) $y = 3x + 4$ (b) $2y = 3x - 4$ (c) $x + 4y = 2$

(a) Comparing $y = mx + c$
and $y = 3x + 4$
gives $m = 3$ and $c = 4$

         Gradient $= 3$ cutting y-axis at $(0, 4)$.

(b) To compare $y = mx + c$
and $2y = 3x - 4$
we write $y = \frac{3}{2}x - 2$
which gives $m = \frac{3}{2}$ and $c = -2$

         Gradient $= \frac{3}{2}$ cutting y-axis at $(0, -2)$.

(c) To compare $y = mx + c$
and $x + 4y = 2$
we write $4y = 2 - x$
or $y = -\frac{1}{4}x + \frac{1}{2}$
which gives $m = -\frac{1}{4}$ and $c = \frac{1}{2}$

         Gradient $= -\frac{1}{4}$ cutting y-axis at $(0, \frac{1}{2})$.

Note: Before comparing the given equation with $y = mx + c$, we must first rearrange the equation so that it is in the form $y = mx + c$. However this initial rearrangement is not necessary if we only require the coordinates of the points of intersection with the axes, as the following example shows.

Example 14

Find the point at which the line $3y - 12 = 4x$ cuts (a) the y-axis, (b) the x-axis. Hence sketch the line $3y - 12 = 4x$.

(a) Any point on the y-axis has x-coordinate equal to zero.
Substituting $x = 0$ into the equation of the line gives $3y - 12 = 0$
                       i.e. $y = 4$

The point on the y-axis is $(0, 4)$.

(b) Any point on the x-axis has y-coordinate equal to zero,
Substituting $y = 0$ into the equation of the line gives $-12 = 4x$
                       i.e. $x = -3$

The point on the x-axis is $(-3, 0)$.

To *sketch* the line, we do not need to make an accurate plot on graph paper. A sketch can be made on plain or lined paper and should show the essential characteristics of the line. For a straight line, a sketch can be made by knowing where the line cuts the axes. The sketch of $3y - 12 = 4x$ is shown on the right.

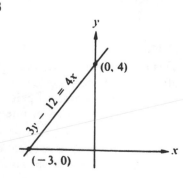

It is also possible to determine the gradient of a line knowing only the angle made with the direction of the positive x-axis.

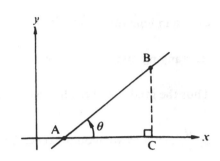

Consider a line which cuts the x-axis at the point A. Suppose B is another point on the line; BC is drawn as shown in the diagram. Assuming the same scale is used on each axis,

$$\text{the gradient of the line} = \frac{\text{BC}}{\text{AC}}$$
$$= \tan \theta$$

Thus the gradient of the line is the tangent of the angle between the direction of the positive x-axis and the line, measured anticlockwise.

Note that by measuring the angle anticlockwise from the direction of the positive x-axis, we still obtain negative gradients for lines that are downhill for increasing x; for such lines, θ is obtuse and $\tan \theta$ is negative for $90° < \theta < 180°$.

It is important to realise that there is no conflict between the cartesian equation of a line and the idea of a vector equation of a line discussed in section 2.5. If a straight line is parallel to some vector $\mathbf{b}$ and passes through a point A having position vector $\mathbf{a}$, the straight line has a vector equation $\mathbf{r} = \mathbf{a} + \lambda\mathbf{b}$.

Consider the straight line $\mathbf{r} = \begin{pmatrix} 0 \\ 1 \end{pmatrix} + \lambda\begin{pmatrix} 2 \\ 1 \end{pmatrix}$.

This line passes through the point having

position vector $\begin{pmatrix} 0 \\ 1 \end{pmatrix}$ and is parallel to the

vector $\begin{pmatrix} 2 \\ 1 \end{pmatrix}$.

To obtain the cartesian equation of this line, consider the general point P on the line having

position vector $\mathbf{r} = \begin{pmatrix} x \\ y \end{pmatrix}$.

Thus $\quad \begin{pmatrix} x \\ y \end{pmatrix} = \begin{pmatrix} 0 \\ 1 \end{pmatrix} + \lambda\begin{pmatrix} 2 \\ 1 \end{pmatrix}$

giving $\qquad x = 0 + 2\lambda \qquad \dots [1]$
and $\qquad y = 1 + \lambda \qquad \dots [2]$

From [1] $\quad \dfrac{x - 0}{2} = \lambda$ and from [2] $\dfrac{y - 1}{1} = \lambda$

Thus $\qquad \dfrac{x - 0}{2} = \dfrac{y - 1}{1} \ (= \lambda)$.

$\therefore \qquad\qquad x = 2(y - 1)$
$\qquad\qquad\qquad = 2y - 2 \quad \text{or} \quad y = \tfrac{1}{2}x + 1$

gradient $= \tfrac{1}{2}$

Thus the line has gradient $\tfrac{1}{2}$ as we would expect for a line parallel to a vector $\begin{pmatrix} 2 \\ 1 \end{pmatrix}$.

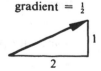

n general terms, the vector equation $\begin{pmatrix} x \\ y \end{pmatrix} = \begin{pmatrix} a \\ b \end{pmatrix} + \lambda \begin{pmatrix} c \\ d \end{pmatrix}$ gives the

cartesian equation $\dfrac{x-a}{c} = \dfrac{y-b}{d}$ (= λ).

rearranging gives $\qquad y = \dfrac{d}{c}x + \dfrac{(bc - ad)}{c}$

Thus the gradient of the line with vector equation $\begin{pmatrix} x \\ y \end{pmatrix} = \begin{pmatrix} a \\ b \end{pmatrix} + \lambda \begin{pmatrix} c \\ d \end{pmatrix}$ is given

by $\dfrac{d}{c}$, again as we would expect for a line parallel to the vector $\begin{pmatrix} c \\ d \end{pmatrix}$:

Thus with the cartesian equation written in the form $\dfrac{x-a}{c} = \dfrac{y-b}{d}$ the

gradient, $\dfrac{d}{c}$, is obtained from the denominators.

Example 15
Find:

(a) the cartesian equation of the line with vector equation $\mathbf{r} = \begin{pmatrix} 2 \\ 3 \end{pmatrix} + \lambda \begin{pmatrix} 1 \\ -2 \end{pmatrix}$

(b) the vector equation of the line with cartesian equation $\dfrac{x-3}{2} = \dfrac{y-1}{4}$

(c) the vector equation of the line with cartesian equation $2y = 3x - 1$.

(a) Considering the general position vector $\mathbf{r} = \begin{pmatrix} x \\ y \end{pmatrix}$ gives $\qquad x = 2 + \lambda$

$$\text{and} \qquad y = 3 - 2\lambda$$
$$\therefore \qquad \frac{x-2}{1} = \frac{y-3}{-2}(= \lambda)$$
$$\text{i.e.} \qquad y = -2x + 7$$

(b) Letting $\dfrac{x-3}{2} = \dfrac{y-1}{4} = \lambda$ gives $\qquad x = 3 + 2\lambda$

$$\text{and} \qquad y = 1 + 4\lambda$$
$$\text{i.e.} \qquad \begin{pmatrix} x \\ y \end{pmatrix} = \begin{pmatrix} 3 \\ 1 \end{pmatrix} + \lambda \begin{pmatrix} 2 \\ 4 \end{pmatrix}$$

The vector equation of the line is $\mathbf{r} = \begin{pmatrix} 3 \\ 1 \end{pmatrix} + \lambda \begin{pmatrix} 2 \\ 4 \end{pmatrix}$.

Alternatively one can simply remember that a cartesian equation of the

form $\dfrac{x-a}{c} = \dfrac{y-b}{d}$ is parallel to the vector $\begin{pmatrix} c \\ d \end{pmatrix}$ and passes through the

point with position vector $\begin{pmatrix} a \\ b \end{pmatrix}$.

(c) We first rearrange $2y = 3x - 1$ into the form $\dfrac{x - a}{c} = \dfrac{y - b}{d}$ $(= \lambda)$.

If

$$2y = 3x - 1$$

then

$$2y + 1 = 3x$$

$$\frac{2y + 1}{3} = x$$

$$\frac{y + \frac{1}{2}}{3} = \frac{x}{2} = \lambda$$

Thus

$$x = 0 + 2\lambda$$

and

$$y = -\tfrac{1}{2} + 3\lambda$$

Thus the vector equation of the line is $\mathbf{r} = \begin{pmatrix} 0 \\ -\frac{1}{2} \end{pmatrix} + \lambda \begin{pmatrix} 2 \\ 3 \end{pmatrix}$ ... [1]

Alternatively, we could say that, as $2y = 3x - 1$ has gradient $\tfrac{3}{2}$, the line is

parallel to the vector $\begin{pmatrix} 2 \\ 3 \end{pmatrix}$ shown on the right. Then, choosing one point on

the line, say $(1, 1)$, we obtain the vector equation

$$\mathbf{r} = \begin{pmatrix} 1 \\ 1 \end{pmatrix} + \lambda \begin{pmatrix} 2 \\ 3 \end{pmatrix}$$ ... [2]

It is not immediately obvious the equations [1] and [2] represent the same
line but each equation does give the position vector of any point on the line
by a suitable choice of λ.
By writing the λ in equation [1] as $(\lambda + \tfrac{1}{2})$ we obtain equation [2].

$$\mathbf{r} = \begin{pmatrix} 0 \\ -\frac{1}{2} \end{pmatrix} + (\lambda + \tfrac{1}{2}) \begin{pmatrix} 2 \\ 3 \end{pmatrix} = \begin{pmatrix} 0 \\ -\frac{1}{2} \end{pmatrix} + \lambda \begin{pmatrix} 2 \\ 3 \end{pmatrix} + \begin{pmatrix} 1 \\ 1\frac{1}{2} \end{pmatrix}$$

$$= \begin{pmatrix} 1 \\ 1 \end{pmatrix} + \lambda \begin{pmatrix} 2 \\ 3 \end{pmatrix}$$

Exercise 3F

1. Find the gradients of the following straight lines:

(a) $y = 3x + 2$ (b) $y = 5x + 4$ (c) $y = -3x + 1$

(d) $y = \tfrac{1}{4}x + 6$ (e) $y = -\tfrac{1}{3}x + 3$ (f) $y = 7 - 3x$

(g) $y = 5 + 2x$ (h) $2y = 4x + 3$ (i) $y - 2x = 4$

(j) $y - 3x + 1 = 0$ (k) $2y + 8x + 5 = 0$ (l) $2y - 3x = 7$

(m) $3x + y = 4$ (n) $\dfrac{y}{3} + \dfrac{x}{2} = 1$ (o) $\dfrac{x}{4} - \dfrac{2y}{3} = 1$

2. Find where the following straight lines cut the y-axis:

(a) $y = 2x + 3$ (b) $y - 3x = 4$

(c) $2y = 3x - 8$ (d) $3x + 4y - 12 = 0$

(e) $\dfrac{x}{3} + \dfrac{y}{2} = 1$ (f) $\dfrac{x}{4} - \dfrac{y}{3} = 1$

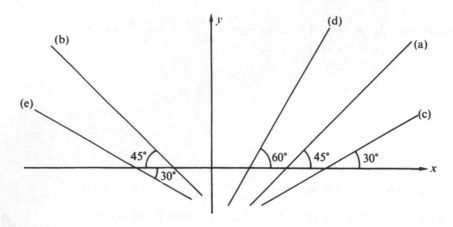

3. Find the gradients of each of the lines (a) to (e) shown on the graph. (Assume the same scale is used on each axis).

4. Find where the line $y = x - 2$ cuts: (a) the x-axis, (b) the y-axis. Hence sketch the line $y = x - 2$.

5. Find where the line $2y = 6 - 3x$ cuts: (a) the x-axis, (b) the y-axis. Hence sketch the line $2y = 6 - 3x$.

6. Which of the following lines are parallel to the line $y = 2x + 4$?
   (a) $y = 2x - 1$, (b) $2y = x + 3$, (c) $3y - 6x = 1$, (d) $2x + 3y = 4$.

7. Which of the following lines are perpendicular to the line $y = 2x + 4$?
   (a) $y = \frac{1}{2}x - 4$, (b) $y = -\frac{1}{2}x + 7$, (c) $2y = 5 - x$, (d) $2y = x + 5$.

8. The line $y = ax + b$ is parallel to the line $y = 2x - 6$ and passes through the point $(-1, 7)$. Find the values of a and b.

9. The line $y = cx + d$ is perpendicular to the line $y + 2x = 4$ and cuts the line $y + x + 3 = 0$ on the y-axis. Find the values of c and d.

10. Find the cartesian equations of the lines whose vector equations are given below. Give your answers in the form $y = mx + c$.

    (a) $\mathbf{r} = \begin{pmatrix} 3 \\ 2 \end{pmatrix} + \lambda \begin{pmatrix} 1 \\ 4 \end{pmatrix}$ (b) $\mathbf{r} = \begin{pmatrix} -1 \\ 2 \end{pmatrix} + \lambda \begin{pmatrix} -2 \\ 3 \end{pmatrix}$

    (c) $\mathbf{r} = 3\mathbf{i} + 2\mathbf{j} + \lambda(3\mathbf{i} - \mathbf{j})$ (d) $\mathbf{r} = 4\mathbf{i} - 5\mathbf{j} + \lambda(2\mathbf{i} + 3\mathbf{j})$

11. Find the vector equations of the lines whose cartesian equations are given below:

    (a) $\dfrac{x - 2}{3} = \dfrac{y - 1}{4}$ (b) $\dfrac{x - 5}{1} = \dfrac{y + 3}{-4}$

    (c) $y = 2x + 3$ (d) $3y = 2x - 4$

12. Two straight lines L_1 and L_2 have equations
    $\dfrac{x - 1}{3} = \dfrac{y + 1}{-2}$ and $\dfrac{x - 4}{4} = \dfrac{y - 1}{6}$ respectively.

    (a) Express each equation in the form $y = mx + c$ and hence show that L_1 and L_2 are perpendicular.
    (b) Express each equation in the form $\mathbf{r} = \mathbf{a} + \lambda\mathbf{b}$ and use the scalar product to show that L_1 and L_2 are perpendicular.

3.7 Finding the cartesian equation of a straight line

In the last section we saw that we can find the gradient of a straight line and the value of its *y*-intercept from its cartesian equation.
This process can be reversed; it is possible to find the equation of a line given sufficient information about the line.

Equation of a line from its gradient *m* and a point (x_1, y_1) on it.

Consider a line having gradient *m* and passing through the point $A(x_1, y_1)$.
Suppose that $P(x, y)$ is any other point on the line.

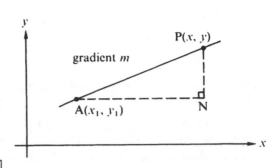

Gradient of line $= m$

$$= \frac{PN}{NA}$$

$$= \frac{y - y_1}{x - x_1}$$

or the equation of the line is $\boxed{y - y_1 = m(x - x_1).}$

Example 16

Find the equation of the straight line that passes through $(-2, 3)$ and has gradient $\frac{4}{5}$.

Let $P(x, y)$ be any point on the line.
Use the equation $\qquad\qquad y - y_1 = m(x - x_1)$
Substituting for x_1, y_1 and *m* gives: $\qquad y - 3 = \frac{4}{5}[x - (-2)]$
$$5y - 15 = 4x + 8$$
So the required equation of the line is $\qquad 5y = 4x + 23$

Equation of a line through $A(x_1, y_1)$ and $B(x_2, y_2)$

Consider a line which passes through the points $A(x_1, y_1)$ and $B(x_2, y_2)$ and suppose $P(x, y)$ is any other point on the line.

The gradient of AP is equal to the gradient of AB as ABP is a straight line.

$$\therefore \quad \frac{PN}{NA} = \frac{BM}{MA}$$

So $\qquad \boxed{\dfrac{y - y_1}{x - x_1} = \dfrac{y_2 - y_1}{x_2 - x_1}}$

This result is easily memorized as the two sides of the equation are both expressions for the gradient of the line.

Example 17

Find the equation of the straight line that passes through $(3, -1)$ and $(7, 2)$.

Let $P(x, y)$ be any point on the line and using

$$\frac{y - y_1}{x - x_1} = \frac{y_2 - y_1}{x_2 - x_1}$$

$$\frac{y - (-1)}{x - 3} = \frac{2 - (-1)}{7 - 3}$$

or $4(y + 1) = 3(x - 3)$

i.e. $4y = 3x - 13$ is the required equation of the line.

Example 18

Find the equation of the perpendicular bisector of the line joining the points $A(2, -3)$ and $B(6, 5)$.

We can find the coordinates of the mid-point of AB. The gradient of AB, and hence that of the perpendicular bisector, can also be found. Thus, knowing the gradient of the perpendicular bisector and one point on it, we can use $y - y_1 = m(x - x_1)$ to obtain the required equation.

Coordinates of M, the mid-point of AB are

$$\left(\frac{2 + 6}{2}, \frac{-3 + 5}{2}\right) = (4, 1)$$

$$\text{gradient of AB} = \frac{5 - (-3)}{6 - 2} = 2$$

$\therefore$ gradient of perpendicular bisector $= -\frac{1}{2}$

The required line has gradient $-\frac{1}{2}$ and passes through M(4, 1)

using $y - y_1 = m(x - x_1)$

$y - 1 = -\frac{1}{2}(x - 4)$

or $2y - 2 = -x + 4$

i.e. $2y = 6 - x$

The equation of the perpendicular bisector is $2y = 6 - x$.

Equation of a line given a law

As we saw in section 3.1, a locus is made up of a set of points which obey some law. We can find the equation of the locus by considering a point $P(x, y)$ on the locus and using the law to derive an equation in x and y. This will be the equation of the locus.

Example 19

Find the equation of the locus of the points which are equidistant from $A(3, -2)$ and $B(-4, 1)$.

Suppose P(x, y) is any point on the required locus.
Since P is equidistant from A and B,

$$PA = PB \text{ and so } PA^2 = PB^2$$

$$\therefore \quad (x - 3)^2 + (y - (-2))^2 = (x - (-4))^2 + (y - 1)^2$$

$$\text{or} \quad (x - 3)^2 + (y + 2)^2 = (x + 4)^2 + (y - 1)^2$$

$$x^2 - 6x + 9 + y^2 + 4y + 4 = x^2 + 8x + 16 + y^2 - 2y + 1$$

which gives

$$3y = 7x + 2$$

The required locus has equation $3y = 7x + 2$.

Note that this question could have been solved by finding the perpendicular bisector of AB, as in Example 18, since that is the required locus.

Exercise 3G

1. Find the equation of the straight line that:
   (a) passes through (1, 7) and has gradient 3,
   (b) passes through $(-1, -6)$ and has gradient 5,
   (c) passes through $(-1, 1)$ and has gradient 2,
   (d) passes through (4, 1) and is parallel to $2y = x - 3$,
   (e) passes through $(-2, 9)$ and is perpendicular to $2y = x - 3$.
2. Find the equation of the straight line that passes through:
   (a) (0, 4) and (3, 10)   (b) (0, 7) and (7, 0)
   (c) (2, 1) and $(-2, -7)$ (d) $(6, -2)$ and (12, 1)
   (e) $(1, -2)$ and $(-1, 4)$.
3. Find the equation of the locus of the points that are equidistant from the points A(1, 1) and B(5, 3).
4. Find the equation of the locus of the points that are equidistant from the points C(2, 7) and D(8, 1).
5. If point E has coordinates (4, 1) and point F has coordinates (2, 7) find:
   (a) the coordinates of the mid-point of EF,
   (b) the gradient of EF,
   (c) the equation of the perpendicular bisector of EF.
6. Find the equation of the straight line that passes through the points L and M where L is the mid-point of the straight line joining A$(-1, 6)$ to B(5, 4) and M is the mid-point of the line joining C$(-5, -1)$ to D$(3, -1)$.
7. A triangle has vertices at A(0, 7), B(9, 4) and C(1, 0). Find:
   (a) the equation of the perpendicular from C to AB,
   (b) the equation of the straight line from A to the mid-point of BC.
8. Find the equations of the medians of the triangle with vertices at A(1, 0), B(5, 2) and C(1, 6).

3.8 Relationships shown as linear graphs

It is frequently desirable to illustrate the results of an experiment graphically. Suppose it is believed that a relationship of the form $P = aQ + b$ exists between the variable quantities P and Q where a and b are constants. By comparing $P = aQ + b$ with $y = mx + c$, we can see that plotting P on the vertical axis against Q on the horizontal axis will lead to a straight line graph if the supposed relationship does exist. If experimental values are used, a close approximation to a straight line may be obtained. Furthermore, the gradient of the straight line will be a and the line will cut the vertical axis at $(0, b)$.

Other suspected relationships can be tested in a similar way. In each case, the relationship is compared with $y = mx + c$ to determine what to plot on the vertical y-axis and the horizontal x-axis. The following table shows some typical cases (a and b are constants in each case).

suspected relationship	plot on y-axis	plot on x-axis	gradient	intercept on vertical axis
$P = aQ^2 + b$	P	Q^2	a	$(0, b)$
$P = a + b\sqrt{Q}$	P	$\sqrt{Q}$	b	$(0, a)$
$P = \dfrac{a}{Q} + b$	P	$\dfrac{1}{Q}$	a	$(0, b)$
$P = aQ + bQ^2$	$\dfrac{P}{Q}$	Q	b	$(0, a)$

Example 20

The cost of producing various quantities of a particular article are shown in the table below.

Cost C	£125	£225	£275	£350	£450	£525	£575
Number of articles N	100	300	400	550	750	900	1000

By plotting a suitable graph, show that a relationship of the form $C = aN + b$ exists and find the constants a and b.

If a relationship of the form $C = aN + b$ exists, plotting C against N should give a straight line graph. The gradient of the line will then be a and the intercept on the C-axis will be $(0, b)$.
The graph is shown on the right.

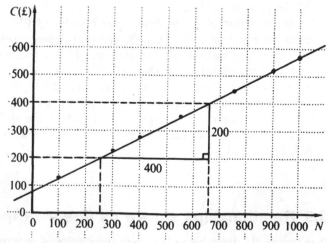

A straight line is obtained so the relationship is of the form $C = aN + b$
The line cuts the C-axis at 75, so $b = 75$.
By drawing a right-angled triangle, the gradient of this line is found to be $\frac{200}{400} = \frac{1}{2}$.
A relationship of the form $C = aN + b$ does exist with $a = \frac{1}{2}$ and $b = 75$.

In some cases the numbers involved may make it impractical to choose scales such that the axes, as drawn on the graph paper, intersect at the point (0, 0). In such cases the gradient can still be found as in example 20 but the intercept made with the axis is determined differently, as the next example shows.

Example 21

In an experiment, the two variables x and y were recorded and the following table obtained.

x	7	8	9	10	11	12
y	40	46	53	60	69	78

By plotting a suitable graph, show that the figures agree approximately with a relationship of the form $y = ax^2 + b$ and find the constants a and b.

Comparing $y = ax^2 + b$ with $y = mx + c$ we see that it is necessary to plot y against x^2. The table below shows values of x, y and x^2.

x	7	8	9	10	11	12
y	40	46	53	60	69	78
x^2	49	64	81	100	121	144

If a scale is chosen so that the axes cross at the origin, the scale will be unnecessarily small giving a less accurate answer.

The straight line graph below shows that the relationship is of the form $y = ax^2 + b$.

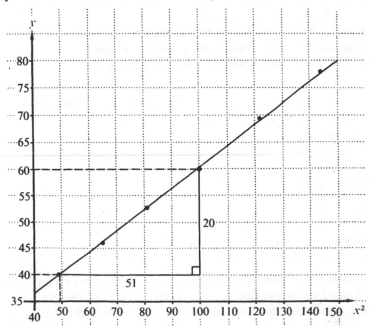

Obtain the gradient from the right-angled triangle.

gradient = $\frac{20}{50} \approx 0.4$

Hence $a = 0.4$

The value of b can be found by substituting any pair of values for y and x^2 together with $a = 0.4$ in the equation, for example $y = 60$ and $x^2 = 100$.

$\therefore$ $60 = 0.4 \times 100 + b$

$b = 60 - 40 = 20$

The relationship $y = ax^2 + b$ does exist; $a = 0.4$, $b = 20$.

Exercise 3H

1. Some Celsius temperatures and the equivalent Fahrenheit temperatures are shown in the table below:

Celsius (C)	50	60	75	95	100	120
Fahrenheit (F)	122	140	167	203	212	248

   By plotting a suitable graph, show that a relationship of the form $F = aC + b$ exists and find the values of the constants a and b.
   Find the Fahrenheit temperature equivalent to
   (a) 80°C,   (b) 40°C,   (c) −40°C.

2. The number of units of gas used in a quarter and the bill received for that quarter were noted for a number of households and the results tabulated:

Bill (B)	£26	£30	£36	£50	£74	£82
Number of units (N)	40	50	65	100	160	180

   By plotting a suitable graph, show that a relationship of the form $B = aN + b$ exists and find the values of the constants a and b.
   Find the quarterly bill for a household that uses 130 units in that quarter.

3. A scientific experiment was carried out to investigate how the electrical resistance of a given wire is dependent on the temperature of the wire. The resistance was measured at various temperatures and the results were tabulated:

Temperature T (°C)	17	28	32	43	47	52	62
Resistance R (ohms)	1·6	1·67	1·69	1·76	1·78	1·81	1·87

   By plotting a suitable graph show that a relationship of the form $R = a + bT$ exists and find the resistance of the wire at 0°C.

4. In an experiment to test the volume-temperature law for gases, a certain mass of gas was heated under constant pressure. The volume of the gas at various temperatures was noted and the results are shown in the following table:

Temperature T (°C)	11	19	41	52	60	86
Volume V (cm³)	104	107	115	119	122	131·5

By plotting a suitable graph, show that a relationship of the form
$V = aT + b$ exists and find approximate values for the constants
a and b. Find the volume of this mass of gas when the
temperature is (a) 0°C, (b) 30°C.

5. The force of wind resistance experienced by a particular car moving with
velocity V was noted for various values of V and the results were
tabulated:

Wind resistance R (in newtons)	26	60	122	135	180	270
Velocity V (in metres/second)	4	10	16	17	20	25

By plotting a suitable graph, show that the figures agree approximately
with a relationship of the form $R = a + bV^2$ and find values for the
constants a and b.
Find the wind resistance experienced by the car when moving with
velocity 15 m/s.

6. In a particular experiment, the two variables x and y were noted and the
results were tabulated.

x	1	1·5	2	3	4	6	8	10
y	6	5·2	4·8	4·4	4·2	4	3·9	3·84

By plotting a suitable graph, show that a relationship of the form
$y = \dfrac{a}{x} + b$ exists and find the values of the constants a and b.

7. An experiment involved noting the time period T of pendulums of
various length l. The results were as follows:

Length l (cm)	20	30	45	50	75	100
Time Period T (secs)	0·90	1·10	1·35	1·42	1·74	2·01

It is suspected that these two quantities T and l are related by a rule of the
form $T = a\sqrt{l}$ where a is a constant. Plot a suitable graph to show this to
be the case.

8. In a particular experiment, the two variables x and y were noted and the
results were tabulated as follows:

x	2	4	7	8	11	12
y	3	3·6	4·4	4·6	5·2	5·4

By plotting a suitable graph show that the figures agree approximately with a relationship of the form $y^2 = ax + b$ and find values for the constants a and b. Find the value of x when $y = 5$.

9. In a particular experiment, the two variables x and y were noted and the results were tabulated as follows.

x	1	1·5	2	4	5	6	10
y	43	23	16	9·3	8·4	8	7·4

By plotting a suitable graph, show that the figures agree approximately with a relationship of the form $y = \dfrac{a}{x^2} + b$ and find values for the constants a and b. Find the value of y when $x = 3$.

10. The sum of the first three even numbers is 12, of the first four even numbers is 20 and of the first seven is 56. These and other similar results are shown in the table below

The first n even numbers where n is:	3	4	7	11	12	15	20
The sum (S) of these first n even numbers	12	20	56	132	156	240	420

It is thought that S and n are related by the formula $S = an + bn^2$. By plotting a suitable graph, show that such a relationship does exist and find the values of the constants a and b. Find the sum of the first fifty even numbers.

3.9 Intersection of lines

Any point on a line has coordinates which will satisfy the equation of that line.
In order to find the point in which two lines intersect we have to find a point with coordinates which satisfy both equations. This is equivalent, from an algebraic point of view, to solving the equations of the lines simultaneously.

Example 22

Find the coordinates of the point in which the lines $x - 3y = 1$ and $2x = 5y + 3$ intersect.

Since the point of intersection satisfies both equations.
$$x - 3y = 1 \qquad \text{i.e.} \qquad 2x = 6y + 2$$
and $\qquad 2x = 5y + 3 \qquad \text{i.e.} \qquad 2x = 5y + 3$
$$\text{subtracting} \quad 0 = y - 1 \quad \text{or} \quad y = 1$$
Substituting in one of the original equations
$$x - 3 = 1 \qquad \text{or} \quad x = 4$$
Point of intersection is (4, 1)

There are various methods which may be used to solve the equations simultaneously. Examples 23 and 24 demonstrate the method of substitution.

Intersection of straight line and curve

A straight line may intersect a curve at more than one point. Thus solving the equations of the line and the curve simultaneously could give more than one answer.

Example 23

The line $y - 2x + 3 = 0$ intersects the curve $y = x^2 - 2x$ at the points A and B. Find the coordinates of A and B.

The coordinates of the points of intersection satisfy both equations.

$$y - 2x + 3 = 0 \qquad \ldots [1]$$
$$y = x^2 - 2x \qquad \ldots [2]$$

Substitute for y from equation [2] into equation [1]:

$$x^2 - 2x - 2x + 3 = 0$$
or $$x^2 - 4x + 3 = 0$$
$\therefore$ $$(x - 3)(x - 1) = 0$$
giving $$x = 3 \quad \text{or} \quad 1$$

Substituting for x in equation [1]:

$$y - 6 + 3 = 0 \quad \text{or} \quad y - 2 + 3 = 0$$
$$\text{i.e.} \quad y = 3 \qquad \qquad \text{i.e.} \quad y = -1$$

The points of intersection are $(3, 3)$ and $(1, -1)$.

Example 24

Find the points at which the curves $y = 3x^2$ and $y = x^2 - 5x - 3$ intersect.

As before, the points of intersection are given by solving simultaneously the equations:

$$\begin{cases} y = 3x^2 \\ y = x^2 - 5x - 3 \end{cases} \qquad \begin{matrix} \ldots [1] \\ \ldots [2] \end{matrix}$$

Substituting for y from equation [1] into equation [2]

$$3x^2 = x^2 - 5x - 3$$
or $$2x^2 + 5x + 3 = 0$$
$\therefore$ $$(2x + 3)(x + 1) = 0$$
giving $$x = -\tfrac{3}{2} \quad \text{or} \quad -1$$

Substituting for x in equation [1]:

$$y = 3(-\tfrac{3}{2})^2 = \tfrac{27}{4} \quad \text{or} \quad y = 3(-1)^2 = 3$$

The points of intersection are $(-\tfrac{3}{2}, \tfrac{27}{4})$ and $(-1, 3)$.

Exercise 3I

1. Find the coordinates of the points where the following pairs of lines intersect:
   (a) $x + 2y = 2$ and $3x - 2y = 14$
   (b) $y + 2x = 5$ and $y = 3x - 5$
   (c) $y = 2x - 1$ and $y = x + 1$
   (d) $y = x + 2$ and $2y = 3x + 1$
   (e) $y = x^2$ and $y + 3x = 4$
   (f) $y = x^2 + 5x$ and $y - x = 12$
   (g) $y = x^2 - 4x + 4$ and $y = 2x - 1$
   (h) $y = x^2 + 1$ and $2y = 5x$
   (i) $y = x^2 + x - 1$ and $2y + 2 = x^2 + 6x$
   (j) $y = 4x^2 + 2x - 1$ and $y = x^2 - 3x + 1$
   (k) $y = x + 1$ and $xy = 6 - 2y$
   (l) $y = x - 3$ and $y + xy = 5$

2. (a) Find the coordinates of the point where the lines $2y - x = 2$ and $y + x = 7$ intersect.
   (b) Prove that the lines $2y - x = 2$, $y + x = 7$ and $y = 2x - 5$ are concurrent.

3. Prove that the three lines $y = 3x$, $y = x + 4$ and $y + 2x = 10$ are concurrent.

4. If the three lines $y = x + 4$, $2y + x = 2$ and $y = ax + 8$ are concurrent, find the value of a.

5. Explain why the lines $y = 2x - 3$ and $2y - 4x = 1$ do not intersect.

6. For each of the following pairs of lines, state whether they intersect or not: (a) $2y + x = 1$ and $4y = 1 - 2x$,
       (b) $x + y = 3$ and $y = x + 7$,
       (c) $y = 2x + 3$ and $2y = 5 + x$

7. Find the coordinates of the mid-point of the line AB where A and B are the points where the line $y = 2x + 4$ and the curve $y = x^2 - 4$ intersect.

8. The line $y = x + 5$ cuts the curve $y = x^2 - 3x$ at two points A and B. Find the coordinates of A and B and the length of the straight line joining these two points.

9. Find:
   (a) the equation of the straight line passing through the point (9, 4) and which is perpendicular to the line $y = 3x - 3$,
   (b) where this line cuts the line $y = 3x - 3$,
   (c) the perpendicular distance from (9, 4) to the line $y = 3x - 3$.

10. Find:
    (a) the equation of the perpendicular from (1, 8) to the line $4y = 3x + 4$,
    (b) the length of the perpendicular from (1, 8) to the line $4y = 3x + 4$.

11. The curves $y = 2x^2 - 3x$ and $y = x^2$ intersect at two points. Find the equation of the straight line joining these points.

12. Find the coordinates of the vertices of the triangle whose sides are given by the equations $y + x = 11$, $y = x - 1$ and $3y = x - 3$.

13. The line $y = ax - 1$ and the curve $y = x^2 + bx - 5$ intersect at the points P(4, -5) and Q. Find the values of a and b and the coordinates of Q.

14. A triangle has vertices at A(0, 8), B(1, 1) and C(5, 3). Show that the triangle is isosceles and find:

(a) the equation of the straight line through A and C,

(b) the coordinates of the foot of the perpendicular from B to AC,

(c) the length of the perpendicular from B to AC.

Exercise 3J   *Examination questions*

1. Three points have coordinates A(1, 7), B(7, 5), and C(0, −2).

Find: (i) the equation of the perpendicular bisector of AB,

(ii) the point of intersection of this perpendicular bisector and BC.

(Cambridge)

2. The perpendicular bisector of the line joining the points (1, 2) and (5, 4) meets the *y*-axis at the point (0, *k*). Calculate *k*.   (S.U.J.B.)

3. Given that the points (4, 2), (1, 6) and (5, *k*) lie on a straight line, calculate the value of *k*.

The line meets the *x*- and *y*-axes at A and B, respectively. If O is the origin, calculate the lengths OA and OB, and the area of the triangle AOB.

(S.U.J.B.)

4. Three points have coordinates A(2, 9), B(4, 3) and C(2, −5). The line through C with gradient $\frac{1}{2}$ meets the line AB produced at D. Calculate

(i) the coordinates of D,

(ii) the equation of the line through D perpendicular to the line $5y - 4x = 17$.

(Cambridge)

5. Find the coordinates of the point C on the line joining the points A(−1, 2) and B(−9, 14) which divides AB internally so that $\frac{AC}{CB} = \frac{1}{3}$.

Find also the equation of the line through A which is perpendicular to AB.   (London)

6. Prove, by calculation, that the triangle formed by the lines $x - 2y + 1 = 0$, $9x + 2y - 11 = 0$ and $7x + 6y - 53 = 0$ is isosceles. Calculate, to the nearest degree, the smallest angle of the triangle.   (A.E.B.)

7. ABCD is a parallelogram in which the coordinates of A, B and C are (1, 2), (7, −1) and (−1, −2) respectively.

(i) Find the equations of AD and CD.

(ii) Find the coordinates of D.

(iii) Prove that BÂC = 90°.

(iv) Calculate the area of the parallelogram.

(v) Find the length of the perpendicular from A to BC, leaving your answer in surd form.   (J.M.B.)

8. P and Q are the points of intersection of the line

$$\frac{x}{a} + \frac{y}{b} = 1, (a > 0, b > 0),$$

with the *x*- and *y*-axes respectively. The distance PQ is 10 and the gradient of PQ is −2. Find the value of *a* and of *b*.   (Cambridge)

9. A point P moves in such a way that its distance from the point (5, 3) is equal to twice its distance from the line $x = 2$. Find the equation of the locus of P. (J.M.B.)

10. Calculate the coordinates of the points of intersection of the straight line $x - y = 3$ and the curve $x^2 - 3xy + y^2 + 19 = 0$. (A.E.B.)

11. Calculate the coordinates of the point of intersection of

$$\frac{x}{y} + \frac{6y}{x} = 5 \text{ and } 2y = x - 2.$$ (Cambridge)

12.

v	5	10	15	20	25
R	149	175	219	280	359

The table shows corresponding values of variables R and v obtained in an experiment. By drawing a suitable linear graph, show that these pairs of values may be regarded as approximations to values satisfying a relation of the form $R = a + bv^2$, where a and b are constants.

Use your graph to estimate the values of a and b, giving your answers to 2 significant figures. (London)

13. The table shows values of the variables x and y which are believed to be related by the equation $ax^2 + y^2 = b$, where a and b are constants.

x	1	2	3	4	4·5
y	2·96	2·83	2·60	2·24	1·98

Show graphically that, for these values, the equation is approximately satisfied. From your graph estimate
(i) values for a and b,   (ii) the value of y when $x = 2·6$. (A.E.B.)

4

Trigonometry 1

4.1 Trigonometric ratios of any angle

In section 0.6 on page 13, we obtained certain facts:

$$\frac{\sin x}{\cos x} = \tan x, \quad \sin x = \cos (90° - x), \text{ etc.}$$

We also saw how to obtain $\sin x$, $\cos x$ and $\tan x$ for $0° \leqslant x \leqslant 180°$. We must now consider the sine, cosine and tangent of angles greater than $180°$.

Consider a rotating arm OP of unit length which traces out any angle θ in an **anticlockwise** direction from the line Ox.

The four positions OP$_1$, OP$_2$, OP$_3$ and OP$_4$ shown below correspond to the four angles θ_1, θ_2, θ_3 and θ_4. The x- and y-axes divide each diagram into four **quadrants**: θ_1 is in the first quadrant, θ_2 in the second, θ_3 in the third and θ_4 in the fourth. Perpendiculars are drawn from each position of P to the x-axis.

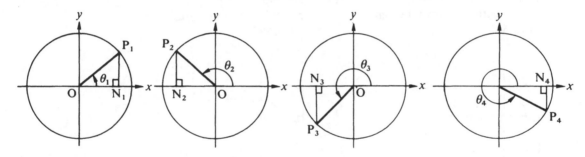

The trigonometric ratio of any angle is defined as the trigonometric ratio of the **acute** angle between the rotating arm OP and the x-axis with the appropriate sign. The appropriate sign is determined by considering the trigonometric ratio of the acute angle.

So $\cos \theta_2 = \cos P_2\hat{O}N_2$ with the appropriate sign

$$= \frac{ON_2}{OP_2} = \frac{ON_2}{1} \text{ as OP}_2 \text{ is of unit length.}$$

But ON$_2$ is negative.

Hence $\cos \theta_2 = - \cos (180° - \theta_2)$.

Similarly, $\tan \theta_3 = \tan P_3\hat{O}N_3$ with the appropriate sign

$$= \frac{P_3N_3}{ON_3}$$

But ON$_3$ and P$_3$N$_3$ are both negative

so $\tan \theta_3 = + \tan (\theta_3 - 180°)$

By considering the trigonometric ratios of angles in each of the four quadrants, we find that all are positive in the first quadrant, only the sine is positive in the second quadrant, only the tangent is positive in the third quadrant and only the cosine is positive in the fourth quadrant. These results are summarised in the diagram on the right.

$\sin > 0$	$\text{all} > 0$
S	A
T	C
$\tan > 0$	$\cos > 0$

Thus $\sin 320° = -\sin 40° = -0.6428$
The result is negative because only the cosine is positive in the fourth quadrant; we use $\sin 40°$ because $40°$ is the **acute** angle that the given angle makes with the x-axis. A calculator gives this answer directly.

Note that if the arm rotates in a clockwise direction, it is said to trace out a negative angle.

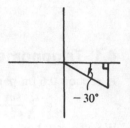

Thus $\tan(-30°) = -\tan 30° = -\dfrac{1}{\sqrt{3}}$
The result is negative because only the cosine is positive in the fourth quadrant.

Also $\cos(-30°) = +\cos 30° = \dfrac{\sqrt{3}}{2}$
The result is positive because the cosine is positive in the fourth quadrant.

Example 1

Write each of the following as trigonometric ratios of acute angles.
(a) $\cos 230°$,   (b) $\sin 140°$.

(a) $\cos 230°$ (3rd quadrant)
     $= \cos 50°$ with appropriate sign
     $= -\cos 50°$

(b) $\sin 140°$
     $= \sin 40°$ with appropriate sign
     $= +\sin 40°$

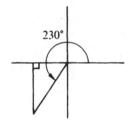

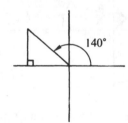

Example 2

Evaluate the following without using tables or a calculator. Leave your answers as surds.
(a) $\cos 135°$,   (b) $\tan 240°$,   (c) $\sin(-45°)$,   (d) $\sin 480°$.

(a) $\cos 135°$ (2nd quadrant) $= -\cos 45° = -\dfrac{1}{\sqrt{2}}$

(b) $\tan 240°$ (3rd quadrant) $= +\tan 60° = +\sqrt{3}$

(c) $\sin(-45°)$ (4th quadrant) $= -\sin 45° = -\dfrac{1}{\sqrt{2}}$

(d) $\sin 480°$ In this case, the rotating arm will be in the same position as
           for $120°$.

     $\sin 480° = \sin 120°$ (2nd quadrant) $= +\sin 60° = +\dfrac{\sqrt{3}}{2}$

Note that the trigonometric ratio of any angle θ greater than $360°$ is the same as the trigonometric ratio of $(\theta - 360°)$.

Maximum and minimum values of sine and cosine

The trigonometric ratios of all angles differ from the trigonometric ratios of acute angles only in sign.
From the definition of the sine and cosine:

the maximum value of $\sin \theta$ is $+1$ (when $\theta = 90°, 450° \ldots$)
and the minimum value of $\sin \theta$ is -1 (when $\theta = 270°, 630° \ldots$)
the maximum value of $\cos \theta$ is $+1$ (when $\theta = 0°, 360°, \ldots$)
and the minimum value of $\cos \theta$ is -1 (when $\theta = 180°, 540°, \ldots$)

Example 3

Write down the maximum and minimum values of each of the following and state the smallest value of θ, from $0°$ to $360°$, for which these values occur.
(a) $1 - 2 \cos \theta$,   (b) $3 \sin \theta - 1$.

(a) $1 - 2 \cos \theta$
Note that $\cos \theta$ varies from $+1$ to -1, hence $2 \cos \theta$ varies from $+2$ to -2.
Thus the maximum value of $1 - 2 \cos \theta$ is $1 - -2 = 3$ and occurs when $\cos \theta = -1$, i.e. $\theta = 180°$.
The minimum value of $1 - 2 \cos \theta$ is $1 - 2 = -1$ and occurs when $\cos \theta = 1$, i.e. $\theta = 0°$.

(b) $3 \sin \theta - 1$
Note that $\sin \theta$ varies from $+1$ to -1, hence $3 \sin \theta$ varies from $+3$ to -3.
Thus the maximum value of $3 \sin \theta - 1$ is $3 - 1 = 2$ and occurs when $\sin \theta = 1$, i.e. $\theta = 90°$.
The minimum value of $3 \sin \theta - 1$ is $-3 - 1 = -4$ and occurs when $\sin \theta = -1$, i.e. $\theta = 270°$.

4.2 Trigonometric graphs

Note that, for any angle θ, a single value of $\sin \theta$ (or $\cos \theta$) can be found. The same applies for $\tan \theta$ with the exception of $\theta = \pm 90°, \pm 270°, \ldots$ for which values of $\tan \theta$ are not defined.
Thus $\sin \theta$ and $\cos \theta$ are functions which are defined for all positive and negative values of θ; $\tan \theta$ is a function which is defined for all positive and negative values of θ except $\pm 90°, \pm 270°, \ldots$
It is therefore possible to construct tables of values giving ordered pairs for these functions and hence to plot their graphs. Such graphs can then be used to find solutions of trigonometric equations.

Example 4

(a) Copy and complete the following table for $y = \sin x$.

x	0°	30°	45°	60°	90°	120°	135°	150°	180°	210°	225°	240°	270°	300°	315°	330°	360°
y	0	0·5	0·71	0·87													

(b) Plot the graph of $y = \sin x$ and use your graph to find approximate solutions to the equation $\sin x = 0.3$, for $0 \leqslant x \leqslant 360°$.

(a) The completed table is as follows:

x	0°	30°	45°	60°	90°	120°	135°	150°	180°	210°	225°	240°	270°	300°	315°	330°	360°
y	0	0.5	0.71	0.87	1.0	0.87	0.71	0.5	0	−0.5	−0.71	−0.87	−1.0	−0.87	−0.71	−0.5	0

(b)

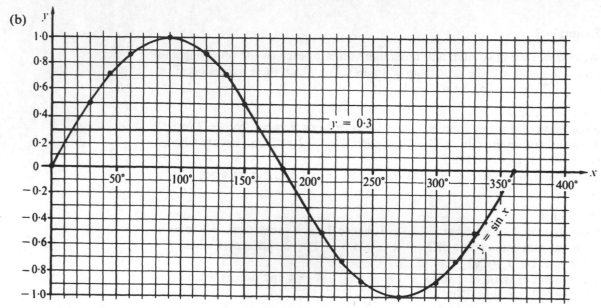

If $y = \sin x$ and $y = 0.3$, solving these simultaneously gives the solution of the equation $\sin x = 0.3$. We can achieve this on the graph by drawing the line $y = 0.3$ as shown above. The points of intersection of the curve with the line give the solutions:

$$x \approx 17° \quad \text{or} \quad x \approx 163°.$$

Example 5

(a) Copy and complete the following table of values for the function $f: x \rightarrow \tan x$.

x	0°	10°	20°	30°	40°	50°	60°	70°
f(x)	0	0.18						

(b) Plot the graph of the function $f: x \rightarrow \tan x$ for the domain $0° \leqslant x \leqslant 70°$ and use your graph to find approximate solutions to the following equations for $0° \leqslant x \leqslant 70°$
(i) $\tan x = 0.3$,   (ii) $\tan x = 2$.

(a) The completed table is as follows:

x	0°	10°	20°	30°	40°	50°	60°	70°
f(x)	0	0.18	0.36	0.58	0.84	1.19	1.73	2.75

(b)

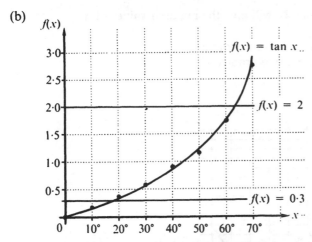

(i) To solve tan x = 0·3 we require the values of x for which $f(x)$ = 0·3.
    From the graph, tan x = 0·3 when $x \approx 17°$.
(ii) To solve tan x = 2 we require the values of x for which $f(x)$ = 2
    From the graph, tan x = 2 when $x \approx 63°$.

Example 6

(a) Plot the graph of $y = 1 + \sin 2x$ for $0° \leqslant x \leqslant 180°$
(b) Obtain approximate solutions from the graph to the equations
    (i) $1 + \sin 2x = 1·2$,   (ii) $\sin 2x = 0·7$,   (iii) $1 + \sin 2x = \sin x$.

(a) The graph of $y = 1 + \sin 2x$ is plotted from a table of values as before.

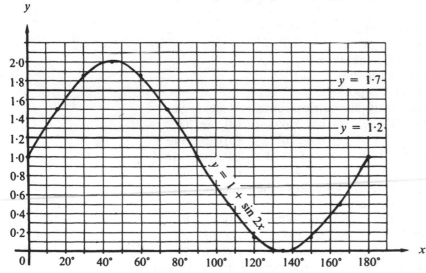

(i) To solve the equation $1 + \sin 2x = 1·2$, we need to draw $y = 1·2$
    on the same graph and find the values of x at the points of
    intersection.
    From the graph $x \approx 6°$ or $x \approx 84°$.
(ii) To solve the equation $\sin 2x = 0·7$ compare it with $y = 1 + \sin 2x$ which we have drawn.
    From $\sin 2x = 0·7$, $1 + \sin 2x = 1 + 0·7 = 1·7$.

So the intersection of $y = 1·7$ with $y = 1 + \sin 2x$ will give the required values of x. From the graph $x \approx 22°$ or $x \approx 68°$.

(iii) To solve $1 + \sin 2x = \sin x$, we need to draw $y = \sin x$ from a table of values. The points of intersection of these two graphs will give the required values of x.

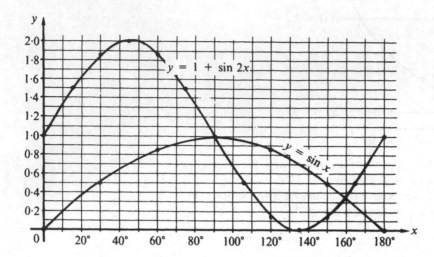

From the graph $x = 90°$ or $x \approx 160°$.

Exercise 4A

1. State in which of the four quadrants each of the angles A to E must lie, given that
   (a) both $\sin A$ and $\cos A$ are positive,
   (b) both $\sin B$ and $\cos B$ are negative,
   (c) $\cos C$ is positive and $\sin C$ is negative,
   (d) both $\cos D$ and $\tan D$ are negative,
   (e) $\sin E$ is negative and $\tan E$ is positive.

2. Write each of the following as trigonometrical ratios of acute angles.
   (a) $\sin 160°$ (b) $\sin 220°$ (c) $\cos 310°$ (d) $\tan 200°$
   (e) $\sin 350°$ (f) $\tan 350°$ (g) $\cos 220°$ (h) $\sin (-40°)$
   (i) $\cos (-50°)$ (j) $\tan (-130°)$ (k) $\sin (-170°)$ (l) $\cos (-160°)$
   (m) $\sin 310°$ (n) $\sin 400°$ (o) $\cos 460°$ (p) $\tan 740°$

3. Without using tables or a calculator, write down the values of the following, leaving surds in your answers.
   (a) $\sin 150°$ (b) $\cos 210°$ (c) $\cos 300°$ (d) $\sin 330°$
   (e) $\tan 300°$ (f) $\tan 225°$ (g) $\sin 270°$ (h) $\cos 270°$
   (i) $\sin (-30°)$ (j) $\tan (-135°)$ (k) $\cos 390°$ (l) $\cos 570°$

4. Without the use of a calculator or tables, write down (i) the maximum value and (ii) the minimum value of each of the following functions and state the smallest value of θ, from $0°$ to $360°$, for which these values occur.
   (a) $\sin \theta$ (b) $2 \sin \theta$ (c) $3 \cos \theta$ (d) $2 + \sin \theta$
   (e) $3 - \cos \theta$ (f) $3 + 2 \sin \theta$ (g) $\sin 2\theta$ (h) $\cos 2\theta - 1$

5. (a) Copy and complete the following table for $y = \cos x$. Tables or a calculator may be used where necessary, giving answers correct to two decimal places.

x	0°	30°	45°	60°	90°	120°	135°	150°	180°	210°	225°	240°	270°	300°	315°	330°	360°
y																	

(b) Plot the graph of $y = \cos x$ for x from 0° to 720° (i.e. twice the interval used in the above table) using a scale of 1 cm to 50° on the x-axis and a scale of 4 cm to 1 unit on the y-axis. Remember that $\cos 390° = \cos 30°$, etc.

6. (a) Copy and complete the following table of values for the function $f: x \to \sin 3x$. Tables or a calculator may be used where necessary, giving answers correct to two decimal places.

x	0°	10°	20°	30°	40°	50°	60°	70°	80°	90°	100°	110°	120°
$f(x)$													

(b) Use a scale of 1 cm for 10° on the horizontal x-axis and a scale of 4 cm to 1 unit on the vertical y-axis, plot the graph of the function for the domain $0° \leqslant x \leqslant 120°$.

(c) From your graph, find approximate solutions to the following equations for $0° \leqslant x \leqslant 120°$,
   (i) $\sin 3x = 0.6$,   (ii) $\sin 3x = -0.6$,   (iii) $\sin 3x = 0.7$.

7. (a) Copy and complete the following table of values for the function $f: x \to 1 + \cos 2x$.

x	0°	15°	30°	45°	60°	75°	90°	105°	120°	135°	150°	165°	180°
$f(x)$													

Tables or a calculator may be used where necessary, giving answers correct to two decimal places.

(b) Plot the graph of the function $f: x \to 1 + \cos 2x$ for the domain $0° \leqslant x \leqslant 180°$ and use your graph to find approximate solutions to the following equations for $0° \leqslant x \leqslant 180°$,
   (i) $1 + \cos 2x = 1.3$,   (ii) $1 + \cos 2x = 0.8$,   (iii) $\cos 2x = -0.7$.

8. (a) Plot the graphs of $y = \cos x$ and $y = 2 \sin x$ on a single pair of axes for $0° \leqslant x \leqslant 360°$.

(b) Use your graph to find approximate solutions of $\cos x = 2 \sin x$ for $0° \leqslant x \leqslant 360°$.

9. (a) Plot the graphs of $y = 2 \cos x - 1$ and $y = \sin x$ on a single pair of axes for $0° \leqslant x \leqslant 360°$.

(b) Use your graph to find approximate solutions to the following equations for $0° \leqslant x \leqslant 360°$,
   (i) $2 \cos x - 1 = -1.5$,   (ii) $2 \cos x = 0.5$,   (iii) $2 \cos x - 1 = \sin x$.

10. (a) Plot the graph of $y = 0.5 + \sin x$ and $y = \cos x$ on a single pair of axes for $0° \leqslant x \leqslant 360°$.

(b) Use your graph to find approximate solutions to the followings equations for $0° \leqslant x \leqslant 360°$.
   (i) $0.5 + \sin x = 1.1$,   (ii) $\sin x = 0.3$,   (iii) $0.5 + \sin x = \cos x$.

4.3 Graph sketching

In the last section some graphs of trigonometric functions were plotted accurately in order to obtain solutions to trigonometric equations. However, the information required by some questions may not require us to make an accurate plot of the graphs. For example we may want to know how many solutions a particular equation has in a given interval and not be required to find the solutions themselves. In such cases a *sketch* of the graph showing its basic characteristics would be sufficient.

Sketch graphs of $y = \sin x$, $y = \cos x$ and $y = \tan x$ are shown below for $0° \leqslant x \leqslant 720°$. They are not drawn on graph paper and could not be used to obtain solutions to equations with any great degree of accuracy but each sketch does show the essential features of each function.

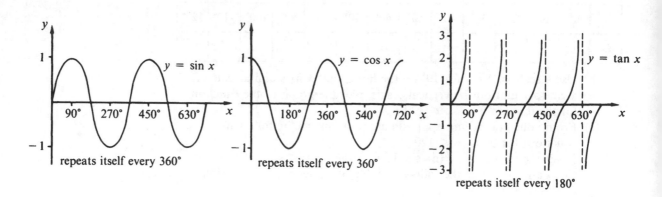

To make sketch graphs of other trigonometric functions we build up information about the graph until sufficient is known to enable a sketch to be made. This will involve far fewer points than are required for an accurate plot of the graph.

We may consider

(a) some easily determined points, for example when sketching $y = \sin x$ we can use the facts that $\sin 0° = 0$, $\sin 90° = 1$ etc.,

(b) the limits (i.e. maximum and minimum values) of the function,

(c) where the graph crosses the axes,

(d) whether the graph repeats itself over a certain interval (i.e. is periodic),

(e) whether the curve has any symmetry.

Example 7

Sketch the graph of $y = \sin 2x$ for $0° \leqslant x \leqslant 360°$.

The values of y will range from $+1$ to -1.
The period of the sine function is $360°$ so our graph should repeat itself after $180°$. The values of y for $x = 0°$, $90°$, $180°$, $270°$, $360°$ are easily marked.

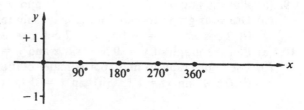

The maximum value of y occurs when
$\sin 2x = 1$
i.e. $2x = 90°, 450° \ldots$ or $x = 45°, 225° \ldots$
The minimum value of y occurs when
$\sin 2x = -1$
i.e. $2x = 270°, 630° \ldots$ or $x = 135°, 315° \ldots$

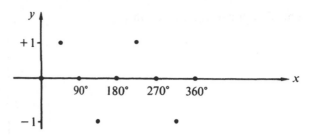

Thus the sketch can be completed:

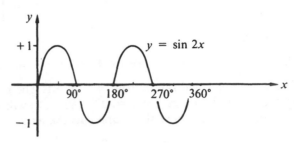

Note that if we continue the sketch backwards
to include negative angles, we would obtain a
sketch for $-180° \leqslant x \leqslant 180°$ as shown:

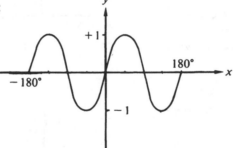

The graph of $y = \sin x$ is symmetrical under a 180° rotation about the
origin as we would expect for an odd function (i.e $\sin(-x) = -\sin x$).
Note that $\tan x$ is also an odd function (i.e. $\tan(-x) = -\tan x$) and so its
graph would also be symmetrical under a 180° rotation about the origin.
However $\cos(-x) = \cos x$ and so the cosine function is even and is
therefore symmetrical for a reflection in the y-axis.

Example 8

Sketch the graph of $y = 2 + \sin x$ for $0° \leqslant x \leqslant 720°$

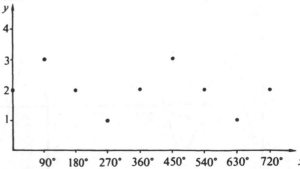

The values of y will range from $+3$ to $+1$,
i.e. the curve does not cut the x-axis.
It crosses the y-axis when $x = 0$, i.e. $y = 2$.
Values of y for $x = 0°, 90°, 180°, \ldots$ can easily
be marked.

The sketch can then be completed:

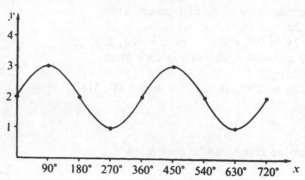

Example 9

Sketch the graph of $y = \cos(x + 30°)$ for $-90° \leqslant x \leqslant 360°$

The values of y will range from $+1$ to -1. The graph will repeat itself every 360°.
In this example the values $x = 90°, 180°, \ldots$ will not necessarily give convenient values for y.
Instead, consider where the graph cuts the axes.

It crosses the x-axis where $y = 0$
i.e. $x + 30° = 90°, 270°, \ldots$
or $x = 60°, 240°, \ldots$
It crosses the y-axis where $x = 0$
i.e. $y = \cos 30° = 0{\cdot}87$
Maximum value of y occurs where
$\cos(x + 30°) = 1$
i.e. $x + 30° = 0°, 360°, \ldots$
or $x = -30°, 330°, \ldots$
Minimum value of y occurs where
$\cos(x + 30°) = -1$
i.e. $x + 30° = 180°, \ldots$
            $x = 150°, \ldots$

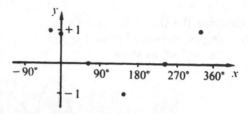

The sketch can then be completed:

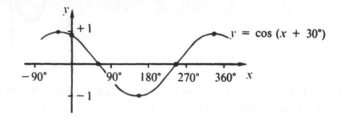

Exercise 4B

1. For each of the equations (a) to (f) listed below, state which one of the sketch graphs (i) to (vi) applies.
   (a) $y = -\tan x$,   (b) $y = -\cos x$,   (c) $y = 1 + \sin x$,   (d) $y = 2 \sin x$,
   (e) $y = 1 - \cos x$,   (f) $y = \sin \frac{1}{2}x$.

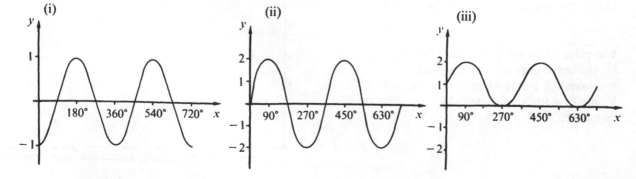

(iv)

(v)

(vi)

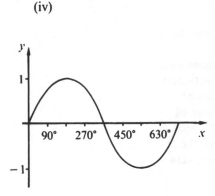
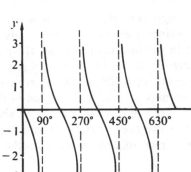
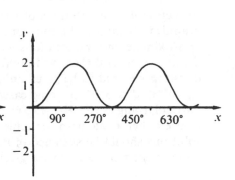

2. For each of the equations (a) to (c) listed below, state which one of the sketch graphs (i) to (iii) applies.
   (a) $y = \tan (x + 30°)$,   (b) $y = \sin (x - 60°)$,   (c) $y = \sin (x + 60°)$.

(i)   (ii)   (iii)

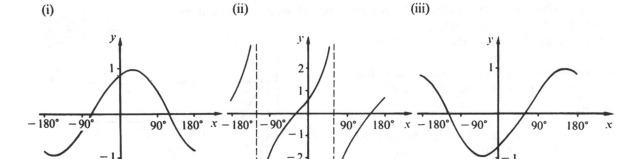

3. Make sketch graphs of the following for $0° \leqslant x \leqslant 360°$
   (a) $y = 2 \cos x$,   (b) $y = 1 + \cos x$,   (c) $y = \cos 2x$.
4. Make sketch graphs of the following for $0° \leqslant x \leqslant 720°$
   (a) $y = -\sin x$,   (b) $y = 1 - \sin x$,   (c) $y = 2 - \cos x$.
5. Make sketch graphs of the following for $-180° \leqslant x \leqslant +180°$
   (a) $y = \sin (x + 30°)$,   (b) $y = \cos (x - 60°)$,   (c) $y = 1 - \sin 2x$.
6. On a single diagram sketch the graphs of $y = \cos x$ and $y = \tan x$ for $0° \leqslant x \leqslant 360°$. Hence state the number of solutions there are to the equation $\cos x = \tan x$ in the interval $0° \leqslant x \leqslant 360°$.
7. On a single diagram sketch the graphs of $y = \cos 2x$ and $y = 2 \cos x$ for $0° \leqslant x \leqslant 360°$. Hence state the number of solutions there are to the equation $\cos 2x = 2 \cos x$ in the interval $0° \leqslant x \leqslant 360°$.
8. On a single diagram sketch the graphs of $y = 1 + \sin x$ and $y = 2 \cos \frac{1}{2}x$ for $0° \leqslant x \leqslant 360°$. Hence state the number of solutions there are to the equation $1 + \sin x = 2 \cos \frac{1}{2}x$ in the interval $0° \leqslant x \leqslant 360°$.

4.4 Solving trigonometric equations

In section 4.2 the solutions of some trigonometric equations were found by graphical methods. The solutions obtained were consequently only approximate and the solutions were in the interval for which the graphs were drawn. In general there will be numerous solutions, so that in any particular question we need to be given an interval for which solutions are required, e.g. $0° \leqslant \theta \leqslant 360°$. In some cases we may recognize the ratios as those of particular angles ($0°$, $30°$, $45°$, $60°$ ... etc.) but in other cases we shall need to use either a calculator or tables to solve the equations. In all cases, the solutions should be sketched in the correct quadrants and then the answers lying in the required interval given.

Example 10

Solve the equation $\cos x = \frac{\sqrt{3}}{2}$ for values of x such that $0° \leqslant x \leqslant 360°$.

The acute angle with a cosine of $\frac{\sqrt{3}}{2}$ is $30°$, so solutions will make an angle

of $30°$ with the x-axis. The fact that the cosine is positive indicates that there are solutions in the 1st and the 4th quadrants. Thus, solutions can be sketched as shown on the right.
For the range $0° \leqslant x \leqslant 360°$, $x = 30°$ or $330°$.

An alternative method is to sketch the graph of $y = \cos x$ for $0° \leqslant x \leqslant 360°$ and of $y = \frac{\sqrt{3}}{2}$.

From the graph we see that the solutions must occur at $30°$ and at $360° - 30° = 330°$.

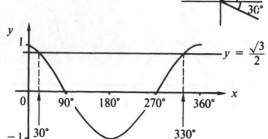

In the following examples this sketch graph method will not be used.

Example 11

Solve the equation $\tan x = -\sqrt{3}$ for values of x such that $-180° \leqslant x \leqslant 180°$.

We first ignore the minus sign and find the acute angle with a tangent of $\sqrt{3}$; this is $60°$ so solutions will make an angle of $60°$ with the x-axis.
Using the fact that the tangent is negative indicates that there are solutions in the 2nd and the 4th quadrants. Thus solutions can be sketched as shown on the right.
For the range $-180° \leqslant x \leqslant 180°$, $x = -60°$ or $120°$.

Example 12

Solve the following equations for $0° \leqslant x \leqslant 360°$.
(a) $\cos x = 0.2$,   (b) $\sin x = -0.2$.

(a) We require the acute angle with a cosine of 0.2. This is found either by using the inverse cosine function (written $\cos^{-1}$ or arccos) on a calculator or by using cosine tables.

$$\cos^{-1} 0.2 = 78.46°$$

i.e. solutions will make an angle of $78.46°$ with the x-axis. Because the cosine is positive, solutions are found in the 1st and the 4th quadrants. Thus solutions can be sketched as shown on the right.
For the range $0° \leqslant x \leqslant 360°$, $x = 78.46°$ or $281.54°$.

(b) There are two possible methods of solution.

METHOD 1 is similar to that used above. We first ignore the minus sign in $\sin x = -0.2$.
From a calculator or tables $\sin^{-1} 0.2 = 11.54°$, i.e. solutions will make an angle of $11.54°$ with the x-axis. Because the sine is negative, solutions are found in the 3rd and the 4th quadrants. Thus, solutions can be sketched as shown on the right.
For the range $0° \leqslant x \leqslant 360°$, $x = 191.54°$ or $348.46°$.

METHOD 2 involves finding $\sin^{-1}(-0.2)$ directly from the calculator:
$\sin^{-1}(-0.2) = -11.54°$ as shown on the sketch. We also know that $\sin x$ is negative in the 3rd quadrant, so giving the second solution.
For the range $0° \leqslant x \leqslant 360°$, $x = 191.54°$ or $348.46°$.

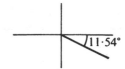

Note Inverse trigonometric functions found on a calculator always give angles that are within a set interval, dependent on the function being used.

For $\sin^{-1}$ and $\tan^{-1}$ the calculator always gives the angle in the range $-90°$ to $+90°$.

For $\cos^{-1}$ the calculator always gives the angle in the range $0°$ to $180°$.

These intervals are used because sine, cosine and tangent are one-to-one functions for these intervals and thus proper inverse functions do exist. Other solutions to an equation can then be found in the usual way using:

S	A
T	C

The inverse functions $\sin^{-1}$, $\cos^{-1}$ and $\tan^{-1}$ are discussed in more detail in chapter 15.

Example 13

Solve $\cos x = -0.3$ for $-180° \leqslant x \leqslant 180°$.

METHOD 1 First ignore the minus sign:
  $\cos^{-1} 0.3 = 72.54°$ from a calculator or tables.
Because the cosine is negative, solutions are in the 2nd and the 3rd quadrants. Thus solutions can be sketched as shown on the right.
For the range $-180° \leqslant x \leqslant 180°$, $x = 107.46°$ or $-107.46°$.

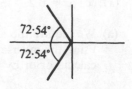

METHOD 2
  $\cos^{-1}(-0.3) = 107.46°$ from a calculator
This is the solution in the 2nd quadrant; because the cosine is negative there is also a solution in the 3rd quadrant of $-107.46°$ as shown on the right.
For the range $-180° \leqslant x \leqslant 180°$, $x = 107.46°$ or $-107.46°$.

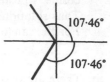

The following examples show the use of method 1 to solve more complicated equations.

Example 14

Solve $\sin(x + 10°) = -0.5$ for $0° \leqslant x \leqslant 360°$.

  $\sin^{-1} 0.5 = 30°$ i.e. solutions will make an angle of $30°$ with the x-axis.
Because the sine is negative, solutions are in the 3rd and the 4th quadrants:
Thus $x + 10° = 210°$ or $x + 10° = 330°$
i.e. $x = 200°$ or $320°$.

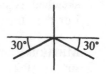

Example 15

Solve $\cos 2x = 0.6$ for $0° \leqslant x \leqslant 360°$.

$\cos^{-1} 0.6 = 53.13°$, i.e. solutions will make an angle of $53.13°$ with the
                        x-axis.
Because the cosine is positive, solutions are in the 1st and the 4th quadrants:
Thus $2x = 53.13°, 306.87°, 413.13°, 666.87°$.
Notice that we must consider solutions for $2x$ from $0°$ to $720°$ as this will give solutions for x in the interval $0°$ to $360°$ as required.
i.e. $x = 26.57°, 153.43°, 206.57°, 333.43°$.

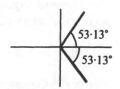

Example 16

Solve $3(\tan x + 1) = 2$ for $-180° \leqslant x \leqslant 180°$.

Expanding $3 \tan x + 3 = 2$
i.e. $\tan x = -\frac{1}{3}$
$\tan^{-1} \frac{1}{3} = 18.43°$, i.e. solutions will make an angle of $18.43°$ with the x-axis.
Because the tangent is negative, solutions are in the 2nd and the 4th quadrants:
For the range $-180° \leqslant x \leqslant 180°$, $x = -18.43°$ or $161.57°$.

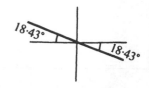

Example 17

Solve $\sin^2 x + \sin x \cos x = 0$ for $0° \leqslant x \leqslant 360°$.

Factorising $\sin x(\sin x + \cos x) = 0$

Thus, either $\sin x = 0$

    Now $\sin^{-1} 0 = 0°$ and so solutions can be sketched:

$x = 0°, 180°, 360°$

or $\sin x + \cos x = 0$

                 $\sin x = -\cos x$

                 $\tan x = -1$ (dividing by $\cos x$)

Now $\tan^{-1} 1 = 45°$ and, because the tangent is negative, solutions will occur in the 2nd and the 4th quadrants:

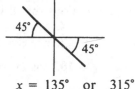

$x = 135°$ or $315°$

Thus for the range $0° \leqslant x \leqslant 360°$, $x = 0°, 135°, 180°, 315°$ or $360°$.

Note: It is important that we factorise $\sin^2 x + \sin x \cos x$ in the last example and do not attempt to cancel by $\sin x$. Cancelling would lead to $\sin x = -\cos x$ and so the solutions arising from $\sin x = 0$ would be lost.

Example 18

Solve $6 \cos^2 x - \cos x - 1 = 0$ for $0° \leqslant x \leqslant 360°$.

Factorising $(3 \cos x + 1)(2 \cos x - 1) = 0$

Thus, either $3 \cos x + 1 = 0$

                 $\cos x = -\tfrac{1}{3}$

Now $\cos^{-1} \tfrac{1}{3} = 70\cdot53°$ and, because the cosine is negative, solutions are in the 2nd and the 3rd quadrants:

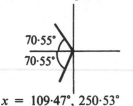

$x = 109\cdot47°, 250\cdot53°$

or $2 \cos x - 1 = 0$

                 $\cos x = \tfrac{1}{2}$

Now $\cos^{-1} \tfrac{1}{2} = 60°$ and, because the cosine is positive, solutions are in the 1st and 4th quadrants:

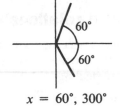

$x = 60°, 300°$

Thus for the range $0° \leqslant x \leqslant 360°$, $x = 60°, 109\cdot47°, 250\cdot53°, 300°$.

Note: In some cases, factorising can give a bracket that does not lead to any solutions. For example, if we had to solve

$$(2 \cos x - 1)(\cos x - 2) = 0 \text{ for } 0° \leqslant x \leqslant 360°$$

either $2 \cos x - 1 = 0$

            $\cos x = \tfrac{1}{2}$

giving $x = 60°, 300°$

or $\cos x - 2 = 0$

        $\cos x = 2$

which has *no* solutions.

Thus $x = 60°$ or $300°$ are the only solutions.

Exercise 4C

1. Solve the following equations for all values of x from $0°$ to $360°$.
   - (a) $\sin x = \frac{1}{2}$
   - (b) $\cos x = -\frac{1}{2}$
   - (c) $\tan x = 1$
   - (d) $\tan x = -1$
   - (e) $\sin x = -\frac{\sqrt{3}}{2}$
   - (f) $\cos x = \frac{1}{\sqrt{2}}$

2. Solve the following equations for all values of x from $-180°$ to $+180°$.
   - (a) $\sin x = -\frac{1}{2}$
   - (b) $\cos x = \frac{1}{2}$
   - (c) $\sin x = \frac{\sqrt{3}}{2}$
   - (d) $\tan x = \sqrt{3}$
   - (e) $\cos x = -\frac{1}{\sqrt{2}}$
   - (f) $\cos x = -\frac{\sqrt{3}}{2}$

Solve the following equations for all values of x from $0°$ to $360°$.

3. $\sin x = 0.3$
4. $\cos x = -0.7$
5. $\tan x = -0.75$
6. $\cos^2 x = \frac{1}{4}$
7. $\sin x = 2 \cos x$
8. $2 \sin x - 3 \cos x = 0$
9. $\sin 2x = -\frac{\sqrt{3}}{2}$
10. $\cos 2x = \frac{1}{2}$
11. $\sin (x + 20°) = -\frac{\sqrt{3}}{2}$
12. $\tan (x - 30°) = 1$
13. $3(\cos x - 1) = -1$
14. $\sin x (1 - 2 \cos x) = 0$
15. $\cos x(2 \sin x + \cos x) = 0$
16. $2 \sin x \cos x + \sin x = 0$
17. $4 \sin x \cos x = 3 \cos x$
18. $4 \cos^2 x + \cos x = 0$
19. $\tan x = 4 \sin x$
20. $(2 \sin x - 1)(\sin x + 1) = 0$
21. $2 \sin^2 x - \sin x - 1 = 0$
22. $2 \tan^2 x - \tan x - 6 = 0$
23. $2 \tan x - \dfrac{1}{\tan x} = 1$

Solve the following equations for all values of x from $-180°$ to $+180°$.

24. $\cos^2 x = \frac{3}{4}$
25. $\sin 2x = 2 \cos 2x$
26. $\cos (x - 20°) = -\dfrac{1}{\sqrt{2}}$
27. $\cos x (\sin x - 1) = 0$
28. $3 \sin^2 x = 2 \sin x \cos x$
29. $2 \cos^2 x - 5 \cos x + 2 = 0$

30. (a) Factorise the expression $6 \sin \theta \cos \theta + 3 \cos \theta + 4 \sin \theta + 2$
    (b) Hence solve the equation $6 \sin \theta \cos \theta + 3 \cos \theta + 4 \sin \theta + 2 = 0$ for $-180° \leqslant \theta \leqslant 180°$.
31. (a) Factorise the expression $3 \sin \theta \cos \theta - 3 \sin \theta + 2 \cos \theta - 2$
    (b) Hence solve the equation $3 \sin \theta \cos \theta - 3 \sin \theta + 2 \cos \theta = 2$ for $0° \leqslant \theta \leqslant 360°$.

4.5 Pythagorean relationships

In section 4.4 we were concerned with the solution of trigonometric equations, i.e. we were finding the particular angles for which the equations were true. In this section we are dealing with trigonometric **identities** which are true for all angles.

In section 0.6 we showed that $\dfrac{\sin x}{\cos x} = \tan x$ and this is an identity. It is true for *all* values of x.

Consider the right-angled triangle ABC.

$$\sin x = \frac{b}{a} \quad \text{and} \quad \cos x = \frac{c}{a}$$

hence $\sin^2 x + \cos^2 x = \dfrac{b^2}{a^2} + \dfrac{c^2}{a^2} = \dfrac{b^2 + c^2}{a^2}$.

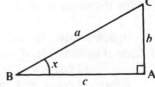

But, by Pythagoras $b^2 + c^2 = a^2$, hence $\sin^2 x + \cos^2 x = \dfrac{a^2}{a^2} = 1$.

This is an *identity* and it is true for *all* values of x.

Example 19

Prove the identity $(\sin \theta + \cos \theta)(1 - \sin \theta \cos \theta) = \sin^3 \theta + \cos^3 \theta$.

Consider the left hand side of the identity

$$
\begin{aligned}
\text{L.H.S.} &= (\sin \theta + \cos \theta)(1 - \sin \theta \cos \theta) \\
&= \sin \theta - \sin^2 \theta \cos \theta + \cos \theta - \sin \theta \cos^2 \theta \\
&= \sin \theta - (1 - \cos^2 \theta)\cos \theta + \cos \theta - \sin \theta(1 - \sin^2 \theta) \quad \text{using} \quad \sin^2 \theta + \cos^2 \theta = 1 \\
&= \sin \theta - \cos \theta + \cos^3 \theta + \cos \theta - \sin \theta + \sin^3 \theta \\
&= \sin^3 \theta + \cos^3 \theta \\
&= \text{R.H.S. of identity}
\end{aligned}
$$

Note: (i) In the proof of an identity we should start from one side of the given identity and try to obtain the other side. It is usually best to start from the more complicated side of the identity and attempt to simplify it.

(ii) The sign '$\equiv$' is sometimes used for identities. Thus Example 19 could have been written:
Prove that $(\sin \theta + \cos \theta)(1 - \sin \theta \cos \theta) \equiv \sin^3 \theta + \cos^3 \theta$.

4.6 Secant, cosecant and cotangent

These three trigonometric ratios can be defined in terms of the sides of a right-angled triangle ABC.

$$\text{secant } x \text{ (written sec } x) = \frac{a}{c} = \frac{1}{\cos x}$$

$$\text{cosecant } x \text{ (written cosec } x) = \frac{a}{b} = \frac{1}{\sin x}$$

$$\text{cotangent } x \text{ (written cot } x) = \frac{c}{b} = \frac{1}{\tan x}$$

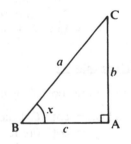

The reciprocal relationships with sine, cosine and tangent should be noted. It follows that the cosecant, secant and cotangent of angles greater than 90° will have the same sign as the sine, cosine and tangent of these angles respectively.

Since $\operatorname{cosec} \theta = \dfrac{1}{\sin \theta}$, the cosec will be positive in the 1st and 2nd quadrants,

$\sec \theta = \dfrac{1}{\cos \theta}$ thus the secant will be positive in the 1st and 4th quadrants,

$\cot \theta = \dfrac{1}{\tan \theta}$ thus the cotangent will be positive in the 1st and 3rd quadrants.

Example 20

Without using tables or a calculator, find the values of
(a) cosec 150°, (b) cot 240°, (c) sec 225°.

$$
\begin{aligned}
\text{(a) } \operatorname{cosec} 150° &= \frac{1}{\sin 150°} \\
&= + \frac{1}{\sin 30°} \\
&= +2
\end{aligned}
\qquad
\begin{aligned}
\text{(b) } \cot 240° &= \frac{1}{\tan 240°} \\
&= + \frac{1}{\tan 60°} \\
&= + \frac{1}{\sqrt{3}}
\end{aligned}
\qquad
\begin{aligned}
\text{(c) } \sec 225° &= \frac{1}{\cos 225°} \\
&= - \frac{1}{\cos 45°} \\
&= - \sqrt{2}
\end{aligned}
$$

In addition to the reciprocal relationships, it also follows that:

$\sec x = \operatorname{cosec}(90° - x)$ and $\cot x = \tan(90° - x)$.

From the identity $\sin^2 x + \cos^2 x = 1$, two further identities can be obtained.

(i) $\qquad\qquad\qquad \sin^2 x + \cos^2 x = 1$ $\qquad$ (ii) $\qquad\qquad\qquad \sin^2 x + \cos^2 x = 1$

divide by $\cos^2 x$ $\qquad \dfrac{\sin^2 x}{\cos^2 x} + 1 = \dfrac{1}{\cos^2 x}$ $\qquad$ divide by $\sin^2 x$ $\qquad 1 + \dfrac{\cos^2 x}{\sin^2 x} = \dfrac{1}{\sin^2 x}$.

$\qquad\qquad\qquad\qquad \Rightarrow \tan^2 x + 1 = \sec^2 x$ $\qquad\qquad\qquad\qquad\qquad \Rightarrow 1 + \cot^2 x = \operatorname{cosec}^2 x$

Thus the three pythagorean identities are

$$\boxed{\begin{aligned} \sin^2 x + \cos^2 x &= 1 \\ 1 + \tan^2 x &= \sec^2 x \\ 1 + \cot^2 x &= \operatorname{cosec}^2 x \end{aligned}}$$

Example 21

Prove the identity $\tan^2 \theta + \sin^2 \theta = (\sec \theta + \cos \theta)(\sec \theta - \cos \theta)$.

$$\begin{aligned} \text{R.H.S.} &= (\sec \theta + \cos \theta)(\sec \theta - \cos \theta) \\ &= \sec^2 \theta - \sec \theta \cos \theta + \cos \theta \sec \theta - \cos^2 \theta \\ &= \sec^2 \theta - \cos^2 \theta \\ &= (1 + \tan^2 \theta) - (1 - \sin^2 \theta) \\ &= 1 + \tan^2 \theta - 1 + \sin^2 \theta \\ &= \tan^2 \theta + \sin^2 \theta = \text{L.H.S.} \end{aligned}$$

Example 22

Prove that $\cot^4 \theta + \cot^2 \theta \equiv \operatorname{cosec}^4 \theta - \operatorname{cosec}^2 \theta$.

$$\begin{aligned} \text{L.H.S.} &= \cot^4 \theta + \cot^2 \theta \\ &= \cot^2 \theta(\cot^2 \theta + 1) \\ &= (\operatorname{cosec}^2 \theta - 1)(\operatorname{cosec}^2 \theta) \quad \text{since} \quad \operatorname{cosec}^2 \theta = 1 + \cot^2 \theta \\ &= \operatorname{cosec}^4 \theta - \operatorname{cosec}^2 \theta = \text{R.H.S.} \end{aligned}$$

Example 23

Prove the identity $\sqrt{\left(\dfrac{1 - \cos \theta}{1 + \cos \theta}\right)} = \operatorname{cosec} \theta - \cot \theta$.

$$\begin{aligned} \text{L.H.S.} &= \sqrt{\left(\frac{1 - \cos \theta}{1 + \cos \theta}\right)} \\ &= \sqrt{\left[\left(\frac{1 - \cos \theta}{1 + \cos \theta}\right)\left(\frac{1 - \cos \theta}{1 - \cos \theta}\right)\right]} \\ &= \sqrt{\left[\frac{(1 - \cos \theta)^2}{(1 - \cos^2 \theta)}\right]} = \sqrt{\left[\frac{(1 - \cos \theta)^2}{\sin^2 \theta}\right]} \\ &= \frac{1 - \cos \theta}{\sin \theta} \\ &= \frac{1}{\sin \theta} - \frac{\cos \theta}{\sin \theta} \\ &= \operatorname{cosec} \theta - \cot \theta = \text{R.H.S.} \end{aligned}$$

Equations involving sec, cosec and cot can be solved in a similar way to those involving sin, cos and tan.

Example 24

Solve the equation $4 \cos \theta - 3 \sec \theta = 2 \tan \theta$ for $-180° \leqslant \theta \leqslant +180°$.

$$4 \cos \theta - 3 \sec \theta = 2 \tan \theta$$

$$4 \cos \theta - \frac{3}{\cos \theta} = \frac{2 \sin \theta}{\cos \theta}$$

Multiplying by $\cos \theta$ gives

$$4 \cos^2 \theta - 3 = 2 \sin \theta$$

i.e. $\quad 4(1 - \sin^2 \theta) - 3 = 2 \sin \theta \qquad$ or $\quad 4 \sin^2 \theta + 2 \sin \theta - 1 = 0$

hence $\qquad\qquad \sin \theta = \dfrac{-2 \pm \sqrt{(4 + 16)}}{8}$

so $\qquad\qquad \sin \theta = 0.3090 \qquad$ or $\quad -0.8090$

Now $\qquad \sin^{-1}(0.3090) = 18° \qquad$ and $\quad \sin^{-1}(0.8090) = 54°$

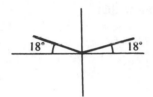   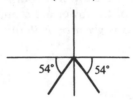

Thus for the range $-180° \leqslant \theta \leqslant +180°$, $\theta = -126°, -54°, 18°, 162°$.

Exercise 4D

1. Without the use of tables or calculator find the values of
   (a) sec 45°   (b) cot 45°   (c) cosec 30°   (d) sec 60°
   (e) cosec 135°   (f) sec 120°   (g) cosec 330°   (h) sec 240°
   (i) cot (−135°)   (j) sec (−60°)   (k) sec (−120°)   (l) cosec 315°
2. Simplify the following expressions
   (a) $\sqrt{[(1 - \sin A)(1 + \sin A)]}$,
   (b) cosec θ tan θ,
   (c) $\dfrac{1}{\sec^2 \theta} + \dfrac{1}{\operatorname{cosec}^2 \theta}$,
   (d) $\cot \theta \sqrt{(1 - \cos^2 \theta)}$.
3. Prove the following identities.
   (a) $\sin \theta \tan \theta + \cos \theta = \sec \theta$, $\qquad\qquad$ (b) $\operatorname{cosec} \theta - \sin \theta = \cot \theta \cos \theta$,
   (c) $(\sin \theta + \cos \theta)^2 + (\sin \theta - \cos \theta)^2 = 2$, $\quad$ (d) $(\sin \theta + \operatorname{cosec} \theta)^2 = \sin^2 \theta + \cot^2 \theta + 3$.
4. If $x = 2 \sin \theta$ and $3y = \cos \theta$ show that $x^2 + 36y^2 = 4$.
5. Eliminate θ from each of the following pairs of relationships.
   (a) $x = \sin \theta$, $y = \cos \theta$, $\qquad\qquad$ (b) $x = 3 \sin \theta$, $y = \operatorname{cosec} \theta$,
   (c) $5x = \sin \theta$, $y = 2 \cos \theta$, $\qquad\qquad$ (d) $x = 3 + \sin \theta$, $y = \cos \theta$,
   (e) $x = 2 + \sin \theta$, $\cos \theta = 1 + y$.
6. Solve the following equations for all values of θ from $-180°$ to $+180°$.
   (a) $4 - \sin \theta = 4 \cos^2 \theta$, $\qquad\qquad$ (b) $\sin^2 \theta + \cos \theta + 1 = 0$,
   (c) $5 - 5 \cos \theta = 3 \sin^2 \theta$, $\qquad\qquad$ (d) $8 \tan \theta = 3 \cos \theta$,
   (e) $\sin^2 \theta + 5 \cos^2 \theta = 3$, $\qquad\qquad$ (f) $1 - \cos^2 \theta = -2 \sin \theta \cos \theta$.

7. Solve the following equations for all values of θ from 0° to 360°.
   (a) $\sec \theta = 2$,
   (b) $\cot 2\theta = -\frac{2}{5}$,
   (c) $3 \cot \theta = \tan \theta$,
   (d) $2 \sin \theta = -3 \cot \theta$,
   (e) $2 \sec^2 \theta - 3 + \tan \theta = 0$,
   (f) $2 \tan \theta = 3 + 5 \cot \theta$,
   (g) $4 \cot \theta + 15 \sec \theta = 0$,
   (h) $\operatorname{cosec}^2 \theta = 3 \cot \theta - 1$,
   (i) $2 \tan \theta = 5 \operatorname{cosec} \theta + \cot \theta$.

8. Prove the following identities.
   (a) $\dfrac{\sin \theta}{1 + \cos \theta} + \dfrac{1 + \cos \theta}{\sin \theta} = 2 \operatorname{cosec} \theta$,

   (b) $\dfrac{\operatorname{cosec} \theta}{\cot \theta + \tan \theta} = \cos \theta$,

   (c) $(1 + \cot \theta - \operatorname{cosec} \theta)(1 + \tan \theta + \sec \theta) = 2$,
   (d) $\cos^4 \theta - \sin^4 \theta + 1 = 2 \cos^2 \theta$,
   (e) $\sec^4 \theta - \sec^2 \theta = \tan^4 \theta + \tan^2 \theta$.

9. A vertical mast CO is of height h and stands with its base, O, on level ground. Points A and B lie on the same level as O and are respectively due south and due east of the tower. From A and B the angles of elevation of the top, C, of the tower are α and β respectively. Show that the distance from A to B is given by $h\sqrt{(\cot^2 \alpha + \cot^2 \beta)}$.

4.7 Compound angles

Sine, cosine, tangent of $(A \pm B)$

In the diagram angles A, and then B, are described in an anticlockwise direction from the line OA.
PN is perpendicular to OB, PM and NR are perpendicular to OA, QN is perpendicular to PM.
Note that $\hat{A} = \hat{RON} = \hat{ONQ}$

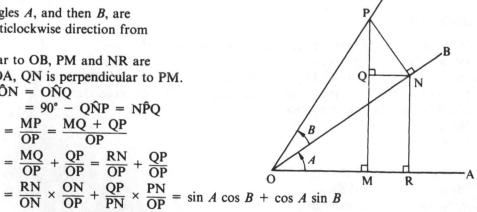

$$= 90° - \hat{QNP} = \hat{NPQ}$$

$$\text{Now } \sin (A + B) = \frac{MP}{OP} = \frac{MQ + QP}{OP}$$

$$= \frac{MQ}{OP} + \frac{QP}{OP} = \frac{RN}{OP} + \frac{QP}{OP}$$

$$= \frac{RN}{ON} \times \frac{ON}{OP} + \frac{QP}{PN} \times \frac{PN}{OP} = \sin A \cos B + \cos A \sin B$$

$$\text{Similarly } \cos (A + B) = \frac{OM}{OP} = \frac{OR - RM}{OP} = \frac{OR - NQ}{OP}$$

$$= \frac{OR}{ON} \times \frac{ON}{OP} - \frac{NQ}{NP} \times \frac{NP}{OP} = \cos A \cos B - \sin A \sin B$$

By replacing B by $-B$ we obtain

$$\sin (A - B) = \sin A \cos (-B) + \cos A \sin (-B) = \sin A \cos B - \cos A \sin B$$

and $$\cos (A - B) = \cos A \cos (-B) - \sin A \sin (-B) = \cos A \cos B + \sin A \sin B$$

Although these results have been obtained for acute angles A and B, they are true for all values of A and B.

Since $\tan (A + B) = \dfrac{\sin (A + B)}{\cos (A + B)}$

$$= \dfrac{\sin A \cos B + \cos A \sin B}{\cos A \cos B - \sin A \sin B}$$

$$= \dfrac{\tan A + \tan B}{1 - \tan A \tan B} \quad \text{(obtained by dividing each term by } \cos A \cos B\text{)}$$

In a similar way

$$\tan (A - B) = \dfrac{\tan A - \tan B}{1 + \tan A \tan B}$$

Thus

$$\sin (A \pm B) = \sin A \cos B \pm \cos A \sin B$$
$$\cos (A \pm B) = \cos A \cos B \mp \sin A \sin B$$
$$\tan (A \pm B) = \dfrac{\tan A \pm \tan B}{1 \mp \tan A \tan B}$$

(Alternatively some of these results can be obtained by matrix methods, see page 183, question 6.)

Example 25

Evaluate the following without the use of tables or a calculator: (a) cos 75°,   (b) tan 15°.

(a) $\cos 75° = \cos (30° + 45°)$

$= \cos 30° \cos 45° - \sin 30° \sin 45°$

$= \dfrac{\sqrt{3}}{2} \times \dfrac{\sqrt{2}}{2} - \dfrac{1}{2} \times \dfrac{\sqrt{2}}{2}$

$\Rightarrow \quad \cos 75° = \dfrac{\sqrt{6} - \sqrt{2}}{4}$

(b) $\tan 15° = \tan (45° - 30°)$

$= \dfrac{\tan 45° - \tan 30°}{1 + \tan 45° \tan 30°}$

$= \dfrac{1 - \dfrac{\sqrt{3}}{3}}{1 + 1 \times \dfrac{\sqrt{3}}{3}}$

$\Rightarrow \quad \tan 15° = \dfrac{3 - \sqrt{3}}{3 + \sqrt{3}}$

Example 26

Given that α and β are acute angles with $\sin \alpha = \frac{7}{25}$ and $\cos \beta = \frac{5}{13}$, find, without using tables or a calculator, $\sin (\alpha + \beta)$ and $\tan (\alpha + \beta)$.

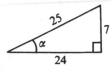

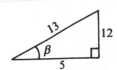

Since $\sin \alpha = \frac{7}{25}$, $\cos \alpha = \frac{24}{25}$ and $\tan \alpha = \frac{7}{24}$

Thus $\sin (\alpha + \beta) = \sin \alpha \cos \beta + \cos \alpha \sin \beta$

$= \frac{7}{25} \times \frac{5}{13} + \frac{24}{25} \times \frac{12}{13}$

$= \frac{323}{325}$

Since $\cos \beta = \frac{5}{13}$, $\sin \beta = \frac{12}{13}$ and $\tan \beta = \frac{12}{5}$

and $\tan (\alpha + \beta) = \dfrac{\tan \alpha + \tan \beta}{1 - \tan \alpha \tan \beta}$

$= \dfrac{\frac{7}{24} + \frac{12}{5}}{1 - (\frac{7}{24})(\frac{12}{5})}$

$= \frac{323}{36}$

Example 27

Solve the equation $\cos \theta \cos 20° + \sin \theta \sin 20° = 0.75$ for values of θ from 0° to 360°.

$$\cos \theta \cos 20° + \sin \theta \sin 20° = 0.75$$
$$\therefore \qquad \cos(\theta + 20°) = 0.75$$

Now $\cos^{-1}(0.75) = 41.41°$ and, because the cosine is positive, solutions are in the 1st and the 4th quadrants.

$\Rightarrow \qquad \theta - 20° = 41.41°$ or $318.59°$
i.e. $\qquad \theta = 61.41°$ or $338.59°$

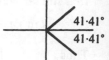

Example 28

Prove that $\cos(A + B)\cos(A - B) \equiv \cos^2 A - \sin^2 B$.

L.H.S. $= \cos(A + B)\cos(A - B)$
$= (\cos A \cos B - \sin A \sin B)(\cos A \cos B + \sin A \sin B)$
$= \cos^2 A \cos^2 B + \cos A \cos B \sin A \sin B - \sin A \sin B \cos A \cos B - \sin^2 A \sin^2 B$
$= \cos^2 A \cos^2 B - \sin^2 A \sin^2 B$
$= \cos^2 A(1 - \sin^2 B) - (1 - \cos^2 A)(\sin^2 B)$
$= \cos^2 A - \cos^2 A \sin^2 B - \sin^2 B + \cos^2 A \sin^2 B$
$= \cos^2 A - \sin^2 B = $ R.H.S.

Exercise 4E

1. Evaluate (a) $\sin 50° \cos 40° + \cos 50° \sin 40°$,
   (b) $\cos 75° \cos 15° + \sin 75° \sin 15°$.
2. Simplify the following expressions
   (a) $\sin A \cos 2A + \cos A \sin 2A$,   (b) $\sin 2\theta \cos \theta - \cos 2\theta \sin \theta$,
   (c) $\cos 4\theta \cos \theta - \sin 4\theta \sin \theta$,   (d) $\sin \theta \sin \theta + \cos \theta \cos \theta$,
   (e) $2 \cos \theta \cos \theta + 2 \sin \theta \sin \theta$.
3. Without using tables or a calculator find the following, leaving surds in your answer.
   (a) $\sin(45° + 30°)$,   (b) $\sin 15°$,   (c) $\tan 105°$.
4. Find (i) the greatest and (ii) the least values that each of the following expressions can take and state the smallest value of θ from 0° to 360° for which these values occur.
   (a) $\sin \theta \cos 20° + \cos \theta \sin 20°$,   (b) $\sin \theta \cos 20° - \cos \theta \sin 20°$,
   (d) $\cos \theta \cos 40° - \sin \theta \sin 40°$.
5. The diagram shows the graph of two lines A and B having gradients of $\frac{1}{2}$ and $\frac{3}{2}$ respectively. With α, β and γ as shown in the diagram, find
   (a) $\tan \alpha$,   (b) $\tan \beta$,   (c) $\tan \gamma$.

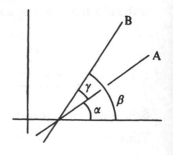

6. A and B are acute angles such that $\sin A = \frac{12}{13}$ and $\cos B = \frac{4}{5}$. Without the use of tables or a calculator find the values of
   (a) $\sin(A - B)$,   (b) $\tan(A - B)$.
7. C and D are both obtuse angles such that $\sin C = \frac{3}{5}$ and $\sin D = \frac{5}{13}$. Without the use of tables or a calculator, find the values of
   (a) $\sin(C + D)$,   (b) $\cos(C - D)$.

8. Solve the following equations for $0° \leqslant \theta \leqslant 360°$.
  (a) $\sin \theta \cos 10° + \cos \theta \sin 10° = -0.5$,
  (b) $\cos 40° \cos \theta - \sin 40° \sin \theta = 0.4$,
  (c) $\sin (\theta + 45°) = \sqrt{2} \cos \theta$.
9. Prove the following identities.
  (a) $\sin (90° - \theta) = \cos \theta$,　　　(b) $\sin (90° + \theta) = \cos \theta$,
  (c) $\cos (90° + \theta) = -\sin \theta$,　　(d) $\cos (180° - \theta) = -\cos \theta$,
  (e) $\sin (A + B) + \sin (A - B) = 2 \sin A \cos B$,
  (f) $\cos (A + B) - \cos (A - B) = -2 \sin A \sin B$,
  (g) $\dfrac{\sin (A + B)}{\sin (A - B)} = \dfrac{\tan A + \tan B}{\tan A - \tan B}$,
  (h) $\dfrac{\cos (A - B)}{\cos (A + B)} = \dfrac{1 + \tan A \tan B}{1 - \tan A \tan B}$,
  (i) $\sin (A + B) \sin (A - B) = \sin^2 A - \sin^2 B$,
  (j) $\dfrac{\sin (A - B)}{\cos A \cos B} + \dfrac{\sin (B - C)}{\cos B \cos C} + \dfrac{\sin (C - A)}{\cos C \cos A} = 0$,
  (k) $\cos (45° - \theta) \cos (45° - \phi) - \sin (45° - \theta) \sin (45° - \phi) = \sin (\theta + \phi)$.

Double-angle formulae

From the identity $\sin (A + B) = \sin A \cos B + \cos A \sin B$
by putting $B = A$ we obtain
$$\sin (A + A) = \sin A \cos A + \cos A \sin A$$

or
$$\boxed{\sin 2A = 2 \sin A \cos A}$$

Similarly by using　$\cos (A + B) = \cos A \cos B - \sin A \sin B$
$$\cos (A + A) = \cos A \cos A - \sin A \sin A$$

$$\boxed{\begin{aligned}\cos 2A &= \cos^2 A - \sin^2 A \\ &= 1 - 2 \sin^2 A \quad (\text{putting } \cos^2 A = 1 - \sin^2 A) \\ &= 2 \cos^2 A - 1 \quad (\text{putting } \sin^2 A = 1 - \cos^2 A)\end{aligned}}$$

Again using　　　$\tan (A + B) = \dfrac{\tan A + \tan B}{1 - \tan A \tan B}$

putting $B = A$　　$\boxed{\tan 2A = \dfrac{2 \tan A}{1 - \tan^2 A}}$

These results are referred to as the **double-angle formulae** because the angle
on the left of the identity is double that on the right.
Thus we also have　$\sin 8\theta = 2 \sin 4\theta \cos 4\theta$
$$\cos 6\theta = \cos^2 3\theta - \sin^2 3\theta \quad \text{etc.}$$

Example 29

Prove that (a) $\dfrac{\sin 2\theta}{1 + \cos 2\theta} \equiv \tan \theta$,   (b) $\operatorname{cosec} 2\theta + \cot 2\theta \equiv \cot \theta$

(a) L.H.S. $= \dfrac{\sin 2\theta}{1 + \cos 2\theta}$

$= \dfrac{2 \sin \theta \cos \theta}{1 + 2 \cos^2 \theta - 1}$

$= \dfrac{2 \sin \theta \cos \theta}{2 \cos^2 \theta}$

$= \dfrac{\sin \theta}{\cos \theta}$

$= \tan \theta =$ R.H.S.

(b) L.H.S $= \operatorname{cosec} 2\theta + \cot 2\theta$

$= \dfrac{1}{\sin 2\theta} + \dfrac{\cos 2\theta}{\sin 2\theta}$

$= \dfrac{1 + \cos 2\theta}{\sin 2\theta}$

$= \dfrac{1 + 2 \cos^2 \theta - 1}{2 \sin \theta \cos \theta}$

$= \dfrac{2 \cos^2 \theta}{2 \sin \theta \cos \theta} = \dfrac{\cos \theta}{\sin \theta}$

$= \cot \theta =$ R.H.S.

Half-angle formulae

From the above double-angle formula for the cosine, it follows that

$$\cos \theta = 2 \cos^2 \frac{\theta}{2} - 1$$

hence $2 \cos^2 \dfrac{\theta}{2} = 1 + \cos \theta$

or $\boxed{\cos^2 \dfrac{\theta}{2} = \tfrac{1}{2}(1 + \cos \theta)}$

and similarly, as

$$\cos \theta = 1 - 2 \sin^2 \frac{\theta}{2}$$

$$2 \sin^2 \frac{\theta}{2} = 1 - \cos \theta$$

or $\boxed{\sin^2 \dfrac{\theta}{2} = \tfrac{1}{2}(1 - \cos \theta)}$

Hence, given the value of $\cos \theta$ we can find $\sin \dfrac{\theta}{2}$ and $\cos \dfrac{\theta}{2}$.

Exercise 4F

In this exercise, a calculator or tables should *not* be used for questions **1** to **9**.

1. Simplify the expressions (a) $\dfrac{\sin 2A}{\cos A}$,   (b) $2 \tan A \cos^2 A$,

   (c) $2 \sin^2 A + \cos 2A$.
2. If $\cos A = -\tfrac{2}{3}$ find $\cos 2A$.
3. Angle B is acute and $\tan B = \tfrac{4}{3}$. Find
   (a) $\sin B$ (b) $\cos B$ (c) $\sin 2B$ (d) $\cos 2B$.
4. If $\sin C = 1/\sqrt{5}$ find $\sin 2C$, $\cos 2C$ and $\tan 2C$ if
   (a) C is acute, (b) C is obtuse.

5. Eliminate θ from each of the following pairs of expressions.
   (a) $x + 1 = \cos 2\theta$, $y = \sin \theta$,   (b) $x = \cos 2\theta$, $y = \cos \theta - 1$,
   (c) $y - 3 = \cos 2\theta$, $x = 2 - \sin \theta$.

6. Angle D is acute and $\tan D = \frac{3}{4}$. Find
   (a) $\sin D$, (b) $\sin 2D$, (c) $\sin (\frac{1}{2}D)$.

7. Angle E lies between $0°$ and $360°$. Find $\sin (\frac{1}{2}E)$ given that $\cos E$ is
   (a) $\frac{3}{8}$, (b) $-\frac{3}{8}$.

8. (a) Obtain an expression for $\sin 3\theta$ involving powers of $\sin \theta$.
   (b) If $\sin \theta = \frac{1}{4}$ find the value of $\sin 3\theta$.

9. (a) Obtain an expression for $\cos 3\theta$ involving power of $\cos \theta$.
   (b) If $\sin \theta = 1/\sqrt{5}$ and θ is obtuse, find the value of $\cos 3\theta$.

10. In triangle ABC, $\hat{BAC} = \theta$ and $\hat{ABC} = 2\theta$. If $BC = x$ show that
   $AC = 2x \cos \theta$.

11. In triangle DEF, $\hat{EDF} = \theta$ and $\hat{DEF} = 4\theta$. If $EF = x$ show that
   $DF = 4x \cos \theta \cos 2\theta$.

12. The three points A, B and C all lie on horizontal ground and form a
triangle right-angled at C with $\hat{BAC} = \alpha$. A point V lies vertically above
B and the angle of elevation of V from A is α. If the length of AV is $2x$,
show that $BC = x \sin 2\alpha$.

13. Solve the following equations for $-180° \leqslant x \leqslant 180°$
   (a) $\sin 2x + \sin x = 0$,   (b) $\sin 2x - 2 \cos^2 x = 0$,
   (c) $3 \cos 2x + 2 + \cos x = 0$,   (d) $\sin 2x = \tan x$.

14. Solve the following equation for $0 \leqslant x \leqslant 360°$
   (a) $2 \sin x \cos x = \cos 2x$,   (b) $3 \sin 2x - \cos x = 0$,
   (c) $3 \sin x + \cos 2x = 2$,   (d) $5 - 13 \sin x = 2 \cos 2x$.

15. Prove the following identities
   (a) $\cos^2 A + \cos 2A = 2 - 3 \sin^2 A$,   (b) $\tan 2A - 2 \tan 2A \sin^2 A = \sin 2A$,

   (c) $\cos^4 A - \sin^4 A = \cos 2A$,   (d) $\dfrac{1 - \cos 2\theta}{1 + \cos 2\theta} = \tan^2 \theta$,

   (e) $\dfrac{\sin \theta + \sin 2\theta}{1 + \cos \theta + \cos 2\theta} = \tan \theta$,   (f) $\tan 3\theta = \dfrac{3 \tan \theta - \tan^3 \theta}{1 - 3 \tan^2 \theta}$

16. Prove the following identities
   (a) $2 \operatorname{cosec} 2\theta = \operatorname{cosec} \theta \sec \theta$,   (b) $\tan A + \cot A = 2 \operatorname{cosec} 2A$,

   (c) $\dfrac{1 + \tan^2 A}{1 - \tan^2 A} = \sec 2A$,   (d) $\cot 2A = \operatorname{cosec} 2A - \tan A$,

   (e) $\dfrac{\sin 2\theta}{1 - \cos 2\theta} = \cot \theta$,   (f) $\tan \theta - \cot \theta = -2 \cot 2\theta$,

   (g) $\operatorname{cosec} 2\theta + \cot 2\theta = \cot \theta$.

17. Each of the following expressions can be written in the form
$a \sin 2\theta + b \cos 2\theta + c$. Find the values of a, b and c in each case.
   (a) $2 \cos^2 \theta + 4 \sin \theta \cos \theta$,   (b) $-2 \sin \theta \cos \theta - 4 \sin^2 \theta$,
   (c) $3 \cos^2 \theta + \sin^2 \theta - 6 \sin \theta \cos \theta$.

18. Prove that
   (a) $\tan \dfrac{\theta}{2} = \dfrac{\sin \theta}{1 + \cos \theta}$,   (b) $\sin \theta = \dfrac{2 \tan (\theta/2)}{1 + \tan^2 (\theta/2)}$.

4.8 Circular measure

Radians

Angles are usually measured in degrees and parts of a degree. A radian is a
larger unit which is often used in trigonometry. It is defined as the angle
subtended at the centre of a circle by an arc equal to the radius of the circle.
In the diagram, suppose that OP is of length 1 unit and that the
arc PQ is also of length 1 unit, then the angle POQ is 1 radian.
Since the circumference of this circle is 2π units, the angle at O
subtended by the whole circumference is 2π radians.

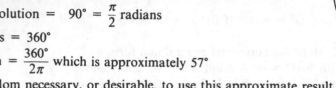

hence   1 revolution $= 360° = 2\pi$ radians

or $\frac{1}{4}$ revolution $= 90° = \dfrac{\pi}{2}$ radians

As 2π radians $= 360°$

   1 radian $= \dfrac{360°}{2\pi}$ which is approximately $57°$

It is very seldom necessary, or desirable, to use this approximate result.
It is better to use one of the exact relationships such as $\pi/2$ radians $= 90°$.

Example 30

Express (a) $\dfrac{5\pi}{6}$ radians in degrees, (b) $210°$ in radians.

(a) Since $\dfrac{\pi}{2}$ radians $= 90°$

   or π radians $= 180°$

   $\dfrac{5\pi}{6}$ radians $= \dfrac{5}{6} \times 180°$

   $= 150°$

(b) Since $90° = \dfrac{\pi}{2}$ radians

   $210° = \dfrac{\pi}{2} \times \dfrac{210}{90}$ radians

   $= \dfrac{7\pi}{6}$ radians

Length of an arc, area of a sector

Suppose the circle centre O has a radius of length r and that the arc PQ
subtends an angle θ rad at O.
The circumference of the circle subtends an angle of 2π rad at O.

Hence $\dfrac{\text{arc PQ}}{\text{circumference of circle}} = \dfrac{\theta}{2\pi}$

$\Rightarrow$ arc PQ $= \dfrac{\theta \times 2\pi r}{2\pi} = r\theta$

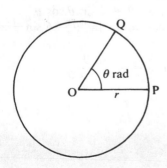

Thus the **length of the arc** PQ is $r\theta$.
The area of a sector may be obtained in a similar way.
The sector POQ subtends an angle θ rad at O,
hence $\dfrac{\text{area of sector POQ}}{\text{area of circle}} = \dfrac{\theta}{2\pi}$

   or   area of sector POQ $= \pi r^2 \times \dfrac{\theta}{2\pi} = \dfrac{1}{2}r^2\theta$

Thus the **area of the sector POQ** is $\frac{1}{2}r^2\theta$.

Example 31

Find (a) the length of the minor arc PQ, (b) the area of the minor sector OPQ of the circle centre O, radius 6 cm, in which angle POQ $= 1\frac{2}{3}$ rad.

(a) arc PQ $= r\theta$
$$= 6 \times \tfrac{5}{3}$$
arc PQ $= 10$ cm

(b) Area $= \tfrac{1}{2}r^2\theta$
$$= \tfrac{1}{2} \times 6^2 \times \tfrac{5}{3}$$
Area $= 30$ cm²

Example 32

Solve the following equations for $0 \leqslant x \leqslant 2\pi$, leaving your answers in terms of π.
(a) $4\cos^2 x = 3$,   (b) $\cos 2x = 3\sin x + 2$.

(a) $4\cos^2 x = 3$
   i.e. $\cos^2 x = \tfrac{3}{4}$
giving $\cos x = \dfrac{\sqrt{3}}{2}$ or $\cos x = -\dfrac{\sqrt{3}}{2}$

giving $x = \dfrac{\pi}{6}, \dfrac{5\pi}{6}, \dfrac{7\pi}{6}$ or $\dfrac{11\pi}{6}$

(b) $\cos 2x = 3\sin x + 2$
$$\therefore \quad 1 - 2\sin^2 x = 3\sin x + 2$$
$$0 = 2\sin^2 x + 3\sin x + 1$$
$$0 = (2\sin x + 1)(\sin x + 1)$$
hence $\sin x = -\tfrac{1}{2}$ or $\sin x = -1$

giving $x = \dfrac{7\pi}{6}, \dfrac{3\pi}{2}$ or $\dfrac{11\pi}{6}$

Example 33

A and B are two points on the circumference of a circle centre O and radius 5 cm. Angle AOB $= \theta$ rad and the length of the chord AB is 6 cm. Find θ and the area of the minor segment bounded by AB and the circle.

$$\sin\frac{\theta}{2} = \frac{3}{5} = 0\cdot6 \quad \text{(see diagram)}$$
$$\theta = 2\sin^{-1}(0\cdot6)$$
$$\therefore \qquad \theta = 1\cdot287 \text{ rad}$$
Area of required minor segment
     $=$ area of sector AOB $-$ area $\triangle$AOB
     $= \tfrac{1}{2}r^2\theta - \tfrac{1}{2}r^2\sin\theta$
     $= \tfrac{1}{2}(25)(1\cdot287) - \tfrac{1}{2}(25)(0\cdot96)$
area of minor segment $= 4\cdot09$ cm²

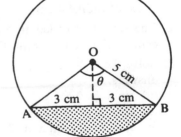

Exercise 4G

1. Express the following angles in degrees.
   (a) π rad,   (b) $\pi/6$ rad,   (c) $3\pi/2$ rad,   (d) $2\pi/3$ rad,
   (e) $11\pi/6$ rad.
2. Express the following angles in radians, leaving π in your answers.
   (a) 360°,   (b) 60°,   (c) 45°,   (d) 150°,   (e) 90°.

3. Without the use of tables or a calculator write down the values of
   (a) $\sin \pi/2$, (b) $\cos 3\pi/2$, (c) $\tan \pi/4$, (d) $\cos \pi$, (e) $\sec \pi$,
   (f) $\sin \pi/6$, (g) $\sin 5\pi/6$, (h) $\cos 5\pi/6$, (i) $\tan 3\pi/4$, (j) $\cos 5\pi/4$.

4. Express (a) 70°, (b) 250° in radians correct to two decimal places.

5. Express (a) 3 rad, (b) 2·5 rad in degrees correct to the nearest degree.

6. Using a calculator or tables find, correct to 2 decimal places, the values
   of (a) $\sin (1 \text{ rad})$, (b) $\tan (3 \text{ rad})$, (c) $\cos (2\cdot4 \text{ rad})$.

7. In each of the following find (i) the length of the minor arc AB,
   (ii) the area of the corresponding sector OAB.
   (a)   (b)   (c)

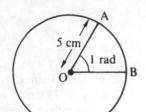
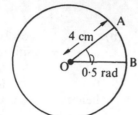
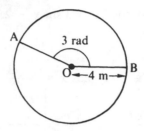

8. Points A and B lie on the circumference of a circle, centre O and radius
   3 cm, and the minor sector OAB has an area equal to one sixth of the
   area of the circle. Find, leaving π in your answers,
   (a) the acute angle AÔB in radians,
   (b) the reflex angle AÔB in radians,
   (c) the length of the minor arc AB,
   (d) the length of the major arc AB.

9. Find the angle subtended at the centre by an arc of length 3 cm on a
   circle of radius 3 cm.

10. Find the area of a sector of a circle of radius 8 cm if the arc of the
    sector subtends an angle of 0·25 radians at the centre.

11. Find the angle subtended at the centre by an arc of length 6 cm on a
    circle of radius 4 cm.

12. The arc AB is of length 6 cm and subtends an angle of 0·5 radians at the
    centre of the circle. Find the radius of the circle.

13. Solve the following equations for $0 \leqslant x \leqslant 2\pi$, leaving π in your
    answers.
    (a) $\sin x = 0\cdot5$ (b) $4 \sin^2 x = 3$
    (c) $\tan 2x = -1$ (d) $2 \sin^2 x + \sin x = 0$
    (e) $\sin^2 x - 3 \sin x + 2 = 0$ (f) $\tan^2 x - 3 \sec x + 3 = 0$

14. A and B are two points on the circumference of a circle, centre O and
    radius 4 cm. If the minor arc AB is of length 3 cm find the area of the
    corresponding sector OAB.

15. A and B are two points on the circumference of a circle centre O, radius
    6 cm. Angle AÔB = $\pi/6$ radians. Find
    (a) the area of the triangle AOB,
    (b) the area of the minor segment bounded by the circle and the chord
        AB (take $\pi = 3\cdot14$).

16. A and B are two points on the circumference of a circle centre O and radius 4 cm. If the chord AB is of length 6 cm, find (a) the angle the minor arc AB subtends at the centre (in radians correct to two decimal places), (b) the area of the corresponding sector OAB.

17. A and B are two points on the circumference of a circle centre O and radius r. The minor arc AB subtends an angle θ radians at O. If the area of the minor segment bounded by the chord AB and the circle is one quarter of the area of the minor sector AOB show that $4 \sin \theta = 3\theta$.

18. In each of the following diagrams find the area of the shaded regions.

(a)

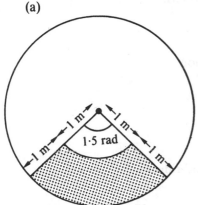

(b)

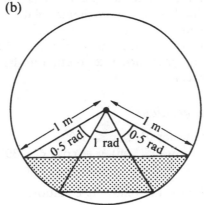

19. The diagram shows a pair of intersecting circles with centres at C and D and of radii 4 cm and 5 cm. AB is the common chord and CD = 7 cm. Find
(a) AĈB in radians,
(b) AD̂B in radians,
(c) the shaded area.

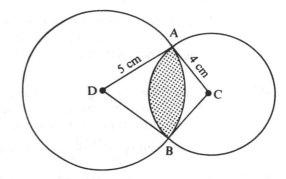

20. (a) Copy and complete the following table for $y = 2 \cos x$ for x from O to π radians.

x in radians	0	$\pi/6$ (0·524)	$\pi/4$ (0·785)	$\pi/3$ (1·047)	$\pi/2$ (1·571)	$2\pi/3$ (2·094)	$3\pi/4$ (2·356)	$5\pi/6$ (2·618)	π (3·142)
y	2	1·73	1·41						

(b) Plot the graph of $y = 2 \cos x$ for $0 \leqslant x \leqslant \pi$ using a scale of 4 cm for 1 radian on the x-axis and 4 cm for 1 unit on the y-axis.

(c) By drawing any additional lines necessary on this graph find approximate solutions to the following equations in the interval $0 \leqslant x \leqslant \pi$.
(i) $2 \cos x = x - 2$ (ii) $4 \cos x = x$

21. (a) Copy and complete the table below for $y = 4 \sin x$ using the values of x shown.

x in radians	0	$\pi/6$ (0·524)	$\pi/4$ (0·785)	$\pi/3$ (1·047)	$\pi/2$ (1·571)	$2\pi/3$ (2·094)	$3\pi/4$ (2·356)	$5\pi/6$ (2·618)	π (3·142)
y	0		2·83	3·46					

(b) Plot the lines $y = 4 \sin x$ and $y = x$ on a single graph using a scale of 4 cm to 1 radian on the x-axis and 4 cm to 1 unit on the y-axis.
(c) From your graph write down an approximate solution to the equation $4 \sin x = x$ in the interval $0 < x \leqslant \pi$.
(d) By plotting a graph of $y = 4 \sin x$ and $y = x$ in the interval $2\cdot4 \leqslant x \leqslant 2\cdot6$, obtain a more accurate solution to the equation $4 \sin x = x$ in this interval.

Exercise 4H   Examination questions

1. Sketch on separate diagrams for $0° \leqslant x \leqslant 360°$
(i) $y = 3 \cos x$,   (ii) $y = \cos 2x$,   (iii) $y = \cos (x - 30°)$.
(Cambridge)

2. Solve, for $-180° \leqslant x \leqslant 180°$, the following equations
(a) $3 \tan x - 2 = 0$,
(b) $3 \tan x - 2 \cos x = 0$.
(A.E.B.)

3. Find all values of θ between $0°$ and $360°$ for which
$$2 \sin \theta + 8 \cos^2 \theta = 5,$$
giving your answers correct to the nearest $0\cdot1°$ where necessary.
(Cambridge)

4. (a) A, B and C are the angles of a triangle such that $\cos A = \frac{2}{3}$ and $\cos B = \frac{5}{13}$. Without using tables or a calculator find the value of
(i) $\tan 2A$, (ii) $\cos (A + B)$, (iii) $\cos C$.
(b) Prove the identity $\dfrac{\sin 2\theta}{1 + \cos 2\theta} = \tan \theta$.
(Cambridge)

5. The angle A lies between $90°$ and $180°$, and $\sin^2 A = 8/9$. Calculate, without using tables or calculator, the value of
(a) $\cos A$,   (b) $\sin 2A$,   (c) $\cos 2A$,   (d) $\cos \frac{1}{2}A$.
(A.E.B.)

6. Find all the angles between $0°$ and $360°$ which satisfy the equations
(i) $(1 + 2 \sin x) \cos 2x = 0$,
(ii) $\sin y = 3 \cos (y - 30°)$.
(Cambridge)

7. Find all the angles between $0°$ and $360°$ which satisfy the equations
(i) $2 \tan x = 1 + 3 \cot x$,
(ii) $5 \sin y \cos y = 2$.
(Cambridge)

8. (a) If $x = \sin^2 2\theta$ and $y = \cos \theta$, eliminate θ and hence express x in terms of y.
(b) Solve the equation
$$2 \sec^2 \theta + \tan \theta - 3 = 0$$
for values of θ in the range $0° \leqslant \theta \leqslant 180°$.
(S.U.J.B.)

9. (a) *A* and *B* are acute angles such that $\sin A = \frac{2}{3}$ and $\tan B = \frac{3}{4}$. Without using tables or calculators, find the value of (i) $\sin (A + B)$, (ii) $\cos (A + B)$, leaving your answers in surd form.
Hence state, with a brief reason, whether angle $A + B$ is acute or obtuse.

   (b) Show that, for all values of θ,
   $$\frac{\cot^2 \theta}{1 + \cot^2 \theta} = \cos^2 \theta.$$
   Hence, or otherwise, find the solutions in the range $-180° \leqslant \theta \leqslant 180°$ of the equation
   $$\frac{\cot^2 \theta}{1 + \cot^2 \theta} = 2 \cos 2\theta. \qquad \text{(J.M.B.)}$$

10. Prove the identities:
    (i) $\tan (45 + A)° \tan (45 - A)° \equiv 1$;
    (ii) $\sin B° + \sin (B + 120)° + \sin (B + 240)° \equiv 0$;
    (iii) $\dfrac{1 - \cos 2C}{1 + \cos 2C} \equiv \tan^2 C.$ \qquad (Oxford)

11. Express $8 \cos^4 \theta$ in the form
    $$a \cos 4\theta + b \cos 2\theta + c,$$
    giving the numerical values of the constants a, b and c. \qquad (Cambridge)

12. (i) Calculate the values of x lying between $0°$ and $360°$ which satisfy the equation,
    $$\tan (x + 35°) = -0.404.$$
    (ii) Show that $\dfrac{\sin \theta}{1 - \cos \theta} + \dfrac{\sin \theta}{1 + \cos \theta} = 2 \csc \theta.$
    Hence, or otherwise, find all the values of θ lying between $0°$ and $360°$ which satisfy the equation,
    $$\frac{\sin \theta}{1 - \cos \theta} + \frac{\sin \theta}{1 + \cos \theta} = 3.76.$$
    (iii) Simplify the expression $f(\theta) = \dfrac{\tan \theta}{1 - \tan \theta} - \dfrac{\cot \theta}{1 - \cot \theta}.$
    Find all the values of α between $0°$ and $360°$ such that $f(\theta) \equiv \tan (\theta + \alpha)$. \qquad (Oxford)

13. (i) Prove the identities
    $$\cos^4 x - \sin^4 x \equiv \cos 2x,$$
    and \qquad $\tan 3x - \tan x \equiv 2 \sin x \sec 3x.$
    (ii) If *A*, *B* and *C* are the angles of a triangle, prove that:
    $$\cos A + \cos (B + C) \equiv 0.$$
    Hence prove that
    $$\cos A + \cos (B - C) \equiv 2 \sin B \sin C. \qquad \text{(Oxford)}$$

14. Sketch on the same diagram the curves $y = |\sin x|$ and $y = \cos 2x$ for the interval $0 \leqslant x \leqslant 2\pi$, labelling each curve clearly.
    State the number of solutions, in this interval, of the equation
    $$|\sin x| = \cos 2x. \qquad \text{(Cambridge)}$$

15. Solve the following equations for values of x between 0 and 2π, giving your answer in radians in terms of π,
(a) $\sec x = -2$,
(b) $\sin x - \sqrt{3} \cos x = 0$,
(c) $2 \sin x \cos x + 1 = 0$.

                                                                      (A.E.B.)

16. A chord PQ of a circle, radius 5 cm, subtends an angle of 2 radians at the centre of the circle. Taking π to be 3·14, calculate, correct to one decimal place,
(i) the length of the **major** arc PQ,
(ii) the area enclosed by the chord PQ and the **minor** arc PQ.

                                                                      (Cambridge)

17. (a) Solve the equation $2 \tan \theta - 4 \cot \theta = \operatorname{cosec} \theta$, giving answers in radians in the range $-\pi < \theta < \pi$.
(b) Prove that the area of the minor segment cut off by a chord subtending an angle of θ radians at the centre of a circle of radius r is $\frac{1}{2}r^2(\theta - \sin \theta)$. A chord which subtends an angle α at the centre of a circle divides the area of the circle into 2 segments in the ratio $1:5$. Prove that $\sin \alpha = \alpha - \pi/3$.
Plot the graphs of $y = \sin \alpha$ and $y = \alpha - \pi/3$ for values of α between 0 and π using the same axes. Hence find the value of α.

                                                                      (S.U.J.B.)

5
Algebra I

5.1 Indices

The reader will know that, using indices, $(a \times a)$ can be written as a^2, $(a \times a \times a)$ can be written as a^3 and so on.
The following particular cases show how expressions involving indices can be simplified.

(i) $a^3 \times a^2 = (a \times a \times a) \times (a \times a) = a^5$

(ii) $a^4 \div a^3 = \dfrac{a \times a \times a \times a}{a \times a \times a} = a^1 = a$

(iii) $(a^3)^2 = (a \times a \times a) \times (a \times a \times a) = a^6$

The three basic rules which include the above particular cases are:
$$a^m \times a^n = a^{m+n}; \quad a^m \div a^n = a^{m-n}; \quad (a^m)^n = a^{mn}.$$
By applying these rules it is possible to interpret other expressions involving indices.

Negative indices

Since $a^5 \div a^2 = \dfrac{a^5}{a^2} = a^{5-2} = a^3$ and $a^5 \times a^{-2} = a^{5-2} = a^3$, it follows

that $a^{-2} = \dfrac{1}{a^2}$

In general, $a^{-m} = \dfrac{1}{a^m}$

Fractional indices

Since $a^{1/2} \times a^{1/2} = a^{1/2+1/2} = a^1$, it follows that $a^{1/2} = \sqrt{a}$.
Similarly from $a^{1/3} \times a^{1/3} \times a^{1/3} = a^{1/3+1/3+1/3} = a^1$, it follows that
$a^{1/3} = \sqrt[3]{a}$, i.e. the cube root of a.
Also $(a^{3/5}) = (a^{1/5})^3 = (\sqrt[5]{a})^3$ or $\sqrt[5]{(a^3)}$.
In general $a^{m/n} = \sqrt[n]{(a^m)} = (\sqrt[n]{a})^m$.
In all cases, $\sqrt{a}$ or $a^{1/2}$ should be taken to mean the *positive* root of a and similarly for $\sqrt[4]{a}$ or $a^{1/4}$, etc.

Zero index

Since $a^m \times a^0 = a^{m+0} = a^m$, it follows that $a^0 = 1$.
Thus we have:

$$\boxed{\begin{array}{l} a^m \times a^n = a^{m+n}, \quad (a^m)^n = a^{mn}, \quad a^{m/n} = \sqrt[n]{a^m} = (\sqrt[n]{a})^m, \\[2mm] a^m \div a^n = a^{m-n}, \quad a^{-m} = \dfrac{1}{a^m}, \quad a^0 = 1. \end{array}}$$

In certain cases it is possible to evaluate numerical expressions involving indices as the following example shows.

Example 1

Simplify each of the following:
(a) 5^3 (b) 13^0 (c) 11^{-2} (d) $8^{2/3}$ (e) $(3\frac{1}{16})^{1/2}$ (f) $16^{-3/4}$.

(a) 5^3
$= 5 \times 5 \times 5$
$= 125$

(b) 13^0
$= 1$

(c) 11^{-2}
$= \dfrac{1}{11^2}$
$= \dfrac{1}{121}$

(d) $8^{2/3}$
$= (2)^2$
$= 4$

(e) $(3\frac{1}{16})^{1/2}$
$= (\frac{49}{16})^{1/2}$
$= \frac{7}{4}$

(f) $16^{-3/4}$
$= \dfrac{1}{16^{3/4}}$
$= \dfrac{1}{2^3}$
$= \dfrac{1}{8}$

Example 2

Simplify the following:
(a) $x^7 \times x^3$ (b) $x^8 \div x^5$ (c) $x^5 \times x^{-3}$ (d) $(x^5)^3$ (e) $8x^4 \div 2x^6$

(a) $x^7 \times x^3$
$= x^{7+3}$
$= x^{10}$

(b) $x^8 \div x^5$
$= x^{8-5}$
$= x^3$

(c) $x^5 \times x^{-3}$
$= x^{5-3}$
$= x^2$

(d) $(x^5)^3$
$= x^{5 \times 3}$
$= x^{15}$

(e) $8x^4 \div 2x^6$
$= \dfrac{8x^4}{2x^6}$
$= \dfrac{4}{x^2}$

Example 3

Find the value of x in each of the following equations:
(a) $5^x = \dfrac{1}{125}$ (b) $\dfrac{3^x \times 3^5}{3^7} = 27$ (c) $(2^x)^3 = 64$

(a) $5^x = \dfrac{1}{125}$
$= \dfrac{1}{5^3}$
$5^x = 5^{-3}$
$\therefore \quad x = -3$

(b) $\dfrac{3^x \times 3^5}{3^7} = 27$
$3^{x+5-7} = 3^3$
$3^{x-2} = 3^3$
$\therefore \quad x = 5$

(c) $(2^x)^3 = 64$
$2^{3x} = 2^6$
$3x = 6$
$\therefore \quad x = 2$

Exercise 5A

Without using a calculator evaluate the following

1. 3^3
2. 2^7
3. $4^2 + 3^2$
4. $(4 + 3)^2$
5. 6^0
6. 8^0
7. $4^{1/2}$
8. $16^{1/2}$
9. $144^{1/2}$
10. $27^{1/3}$
11. $8^{1/3}$
12. $16^{3/2}$
13. 5^{-2}
14. 7^{-1}
15. 10^{-2}
16. $8^{-1/3}$
17. 2^{-3}
18. $16^{-1/4}$
19. $27^{2/3}$
20. $64^{-2/3}$
21. 9^{-2}
22. $16^{-3/2}$
23. $(-8)^{2/3}$
24. $(-8)^{-2/3}$
25. $(\frac{1}{4})^2$
26. $(\frac{2}{3})^2$
27. $(\frac{3}{4})^0$
28. $(1\frac{1}{4})^2$
29. $(2\frac{1}{2})^2$
30. $(\frac{9}{16})^{1/2}$
31. $(1\frac{7}{9})^{1/2}$
32. $(2\frac{1}{4})^{-1/2}$
33. $(6\frac{1}{4})^{3/2}$
34. $(2\frac{1}{4})^{-3/2}$
35. $(3\frac{3}{8})^{-2/3}$

Express each of the following in the form 2^x
36. $2^2 \times 2^4$
37. $2^7 \div 2^3$
38. $2^8 \div 2^2$
39. $2^7 \times 2^4$
40. $\dfrac{2^3 \times 2^4}{2^5}$
41. $\dfrac{2^2 \times 2^4}{2^3}$
42. $4 \times \dfrac{2^9}{2^4}$
43. $\dfrac{32 \times 2^4}{2^2 \times 2}$

Simplify each of the following expressions

44. $x^3 \times x^4$
45. $(x^2)^3$
46. $3x^2 \times 6x^4$
47. $6x^2y \times 4x$
48. $\dfrac{6x^4}{2x^2}$
49. $(15x^5) \div (3x^2)$
50. $x^4 \times x^{-2}$
51. $5x^4y \times 6y^2$
52. $3x^3y \times 4y^2x$

Express each of the following in the form a^x

53. $\dfrac{1}{a}$ **54.** $\dfrac{1}{\sqrt{a}}$ **55.** $\dfrac{a^2}{a^3}$ **56.** $a^2 \times a^3$ **57.** $\dfrac{a}{\sqrt{a}}$

58. $\dfrac{\sqrt{a}}{a}$ **59.** $(a^2)^3$ **60.** $(\sqrt{a})^3$ **61.** $\sqrt{a} \times a^3$ **62.** $\dfrac{a^3 \times \sqrt{a}}{a}$

Find the value of x in each of the following

63. $2^x = 64$ **64.** $2^x = 1$ **65.** $2^x = 0{\cdot}5$

66. $2^x = 0{\cdot}25$ **67.** $3^x = \tfrac{1}{9}$ **68.** $2^4 \times 2^7 = 2^x$

69. $\dfrac{3^5 \times 3^7}{3^9} = 3^x$ **70.** $\dfrac{3^6 \times 3^4}{3^2} = 3^x$ **71.** $\dfrac{2^{15}}{2^7 \times 16} = 2^x$

72. $2^x \times 2^8 = 2^{11}$ **73.** $5^7 \times 5^x = 5^5$ **74.** $7^8 \div 7^x = 49$

75. $\dfrac{6^5 \times 6^x}{36} = 6^9$ **76.** $\dfrac{5^7 \times 5^4}{5^x} = 25$

5.2 Variation

Consider the situation of a person buying a number of copies of a book. As the number of copies bought *increases* then the total cost of buying the books *increases*. We say that the total cost (C) is directly proportional to the number of books (N) bought, or that C varies directly with N. This may be written

$$C \propto N \quad \text{or} \quad C = kN \quad \text{for some constant } k.$$

Now consider the situation of a number of cats eating a given quantity of cat food. In this case, as the number of cats *increases*, so the time for which the cat food lasts *decreases*. We say that the time (T) is inversely proportional to the number of cats (N) or that T varies inversely with N. This may be written

$$T \propto \frac{1}{N} \quad \text{or} \quad T = \frac{k}{N} \text{ for some constant } k.$$

Note that in cases of direct variation we can omit the word 'direct' and simply say C varies with N.

Example 4

Given that y is inversely proportional to the square of x and that when $x = 1{\cdot}5$, $y = 8$, find
(a) a formula giving y in terms of x, (b) the value of y when $x = 2\sqrt{3}$.

(a) y is inversely proportional to the square of x

$$\therefore \quad y \propto \frac{1}{x^2} \quad \text{or} \quad y = \frac{k}{x^2}$$

but $y = 8$ when $x = 1{\cdot}5$,

$$\therefore \quad 8 = \frac{k}{(1{\cdot}5)^2} \text{ which gives } k = 18.$$

$$\therefore \quad y = \frac{18}{x^2} \text{ is the required formula.}$$

(b) Since $y = \dfrac{18}{x^2}$,

when $x = 2\sqrt{3}$, $y = \dfrac{18}{(2\sqrt{3})^2}$

$\therefore$ $\quad y = 1\cdot5$ when $x = 2\sqrt{3}$.

Joint variation

In relationships such as $y = kx$ and $y = \dfrac{k}{x}$, where k is a constant, the variable y is a function of only one variable x. However, in some relationships and formulae a variable may be dependent on two or more other variables. For example, in the formula for the curved surface area of a cone, $A = \pi r l$, and A is dependent *jointly* on r, the base radius, and l the slant height of the cone.

Example 5

Given that y varies jointly with x and with the square root of z, and that $y = 2$ when $x = \frac{1}{8}$ and $z = \frac{1}{4}$, find (a) a formula giving y in terms of x and z, (b) the value of y when $x = \frac{3}{8}$ and $z = \frac{1}{9}$.

(a) y varies jointly with x and with the square root of z

$\quad\therefore \qquad y \propto x\sqrt{z}$

$\quad$or$\qquad y = kx\sqrt{z}$

$\quad$but$\qquad y = 2$ when $x = \frac{1}{8}$ and $z = \frac{1}{4}$,

$\quad\therefore \qquad 2 = k \times \frac{1}{8} \times \sqrt{\frac{1}{4}}$ giving $k = 32$

$\quad\therefore \qquad y = 32x\sqrt{z}$

(b) Using $\quad y = 32x\sqrt{z}$

$\quad$If$\qquad x = \frac{3}{8}$ and $z = \frac{1}{9}$, $y = 32 \times \frac{3}{8} \times \sqrt{\frac{1}{9}}$

$\quad$or$\qquad y = 4$

The required formula is $y = 32x\sqrt{z}$ and $y = 4$ when $x = \frac{3}{8}$ and $z = \frac{1}{9}$.

Note that in this example it was stated that y varies jointly with x and with the square root of z, but usually the word 'jointly' can be omitted without any loss of clarity. Thus we can say 'y varies with x and with the square root of z'.

Partial variation

In relationships of the type $y = k_1 x + k_2 z$, y is said to vary *partly* with x and *partly* with z. The difference between this type of variation and joint variation should be carefully noted.

Example 6

y varies partly with the square root of x and partly with the square of z. Given that $y = 11$ when $x = 9$ and $z = 2$, and that $y = 23$ when $x = 25$ and $z = 4$, find (a) a formula giving y in terms of x and z, (b) the value of x when $y = 30$ and $z = 6$.

(a) $\qquad y = k_1\sqrt{x} + k_2 z^2$

but $\qquad y = 11$ when $x = 9$ and $z = 2$,

$\therefore \qquad 11 = k_1\sqrt{9} + k_2(2)^2$

i.e. $\qquad 11 = 3k_1 + 4k_2 \qquad\qquad\qquad\qquad$ ... [1]

and $\qquad y = 23$ when $x = 25$ and $z = 4$,

$\therefore \qquad 23 = k_1\sqrt{25} + k_2(4)^2$

i.e. $\qquad 23 = 5k_1 + 16k_2 \qquad\qquad\qquad\qquad$ ... [2]

Solving equations [1] and [2] simultaneously gives

$\qquad\qquad k_1 = 3, k_2 = \frac{1}{2}$

thus $\qquad y = 3\sqrt{x} + \frac{1}{2}z^2$

(b) using $\quad y = 3\sqrt{x} + \frac{1}{2}z^2$, with $y = 30$ and $z = 6$,

$\qquad\qquad 30 = 3\sqrt{x} + \frac{1}{2}(6)^2$

$\therefore \qquad \sqrt{x} = 4$

i.e. $\qquad x = 16$

The formula is $y = 3\sqrt{x} + \frac{1}{2}z^2$ and $x = 16$ when $y = 30$ and $z = 6$.

Exercise 5B

1. Using k (or k_1 and k_2) as constants express each of the following statements as equations
   (a) y varies directly with x,
   (b) y is directly proportional to the cube of x,
   (c) y varies directly with the cube root of x,
   (d) y is inversely proportional to x,
   (e) y varies inversely with the square root of x,
   (f) y varies jointly with x and with the square of z,
   (g) y varies with the square of x and inversely with z,
   (h) y varies with w, x and inversely with z,
   (i) y varies partly with x and partly with z,
   (j) y varies partly with the square of x and partly with z,
   (k) The variation of y is in two parts, one part varies directly with the square root of x and the other varies inversely with the square of z.
2. Given that y is inversely proportional to the square of x and that $y = 1\cdot25$ when $x = 2$ find
   (a) a formula giving y in terms of x,   (b) the value of y when $x = \frac{1}{4}$.
3. If y varies partly with x and partly with z^2, and $y = 2$ when $x = 4$ and $z = 2$, and $y = 43$ when $x = 2$ and $z = 4$, find
   (a) a formula giving y in terms of x and z,   (b) the value of x when $z = \frac{1}{2}$ and $y = -\frac{1}{2}$.
4. For a given spring, the extension x varies directly with the magnitude of the applied force F. If $x = 0\cdot3$ metres when $F = 5$ newtons find F as a function of x and hence find x when $F = 8$ newtons.
5. The time period T of a simple pendulum varies directly with the square root of the length (l) of the pendulum. If $T = 1\cdot2$ s when $l = 0\cdot36$ m find T as a function of l and hence find l when $T = 1$ second.
6. For a constant mass moving in a circular path of radius r the force F, acting towards the centre of the circle, varies directly with the square of the speed v and inversely with the radius of the path. If $F = 240$ newtons when $v = 4$ m/s and $r = 0\cdot1$ m express F in terms of v and r and hence find F when $r = 0\cdot5$ m and $v = 2$ m/s.

7. The quarterly telephone charge C is made up of two parts; the first part is a constant and the second part varies directly as the number of units used, u. If the charge is £33 for a quarter in which 400 units are used and £26 for a quarter in which 260 units are used, find C as a function of u and find the charge for a quarter in which 450 units are used.

8. The velocity v of sound waves through metal varies directly with the square root of the modulus of elasticity E of the metal, and inversely with the square root of ρ, the density of the material. For steel $E = 1\cdot95 \times 10^{11}$ N/m², $\rho = 7800$ kg/m³ and the velocity of sound through steel is 5000 m/s. Find v in terms of E and ρ and find the velocity of sound in aluminium for which $E = 7\cdot436 \times 10^{10}$ N/m² and $\rho = 2750$ kg/m³.

5.3 Elementary theory of logarithms

Given any two positive numbers a and b, there exists a third number c such that $a^c = b$.
The number c is said to be the logarithm of b to the base a and we write $\log_a b = c$.
Thus $a^c = b \Leftrightarrow \log_a b = c$.
The logarithm of x to the base a is the power to which a must be raised to give x.
If $p = \log_a x$ and $q = \log_a y$ then $a^p = x$ and $a^q = y$.

$$\text{Thus}\quad xy = a^{(p+q)} \quad \text{i.e.}\quad \log_a(xy) = p + q \Rightarrow \boxed{\log_a(xy) = \log_a x + \log_a y}$$

$$\frac{x}{y} = a^{(p-q)} \quad \text{i.e.}\quad \log_a\left(\frac{x}{y}\right) = p - q \Rightarrow \log_a\left(\frac{x}{y}\right) = \log_a x - \log_a y$$

$$x^n = a^{pn} \quad \text{i.e.}\quad \log_a x^n = pn \quad \Rightarrow \quad \log_a x^n = n \log_a x$$

If $r = \log_a a$ then $a^r = a$, i.e. $r = 1 \quad\Rightarrow\quad \log_a a = 1$

Example 7

(a) Find x if $\log_x 125 = 3$.   (b) Evaluate $\log_3 81$.
(c) Express $7 \log_a 2 - 3 \log_a 12 + 5 \log_a 3$ as a single logarithm.

(a) $\log_x 125 = 3$
$\therefore \quad x^3 = 125$
$\qquad\quad = 5^3$
$\therefore \quad x = 5$

(b) let $\log_3 81 = a$
$\therefore \quad 3^a = 81$
$\qquad\quad = 3^4$
$\therefore \quad a = 4$

(c) $7 \log_a 2 - 3 \log_a 12 + 5 \log_a 3$
$= \log_a 2^7 - \log_a 12^3 + \log_a 3^5$
$= \log_a\left(\dfrac{2^7 \times 3^5}{12^3}\right)$
$= \log_a 18$

Common logarithms

If the base of the logarithm of x is 10, we refer to this as a common logarithm and we write this as $\log_{10} x$. Frequently the base is omitted and we write this as $\log x$. If the base is other than 10 then clearly it is essential that the base is named.

Natural logarithms

We shall see in later chapters that a very important set of logarithms are those to the base e, where e is a constant rather like π. Correct to 5 decimal places, e $= 2{\cdot}718\,28$. These logarithms, to the base e, are referred to as natural logarithms and it is usual to write $\log_e x$ as $\ln x$.

Inverse of logarithmic function

For every positive real number x we can find a unique value y given by $y = \log_a x$.
Thus $f(x) = \log_a x$ is a one-to-one function with domain $\mathbb{R}^+$.
The inverse function of $f : x \to \log_a x$ is the function which maps $\log_a x$ on to x.
This inverse function can be determined by the method shown on page 37, chapter 1.
For the function $f(x) = \log_a x$, let $\quad y = \log_a x$
$$\text{then } x = a^y.$$
Thus, given y, we can return to x using the expression a^y.
The inverse function $f^{-1}(x)$ is $f^{-1}(x) = a^x$.
From section 1.6 on page 44, we would expect the graph of $y = a^x$ to be the reflection of $y = \log_a x$ in the line $y = x$, and question 11 of Exercise 5C verifies this fact.

Example 8

Use your calculator to find, correct to 2 decimal places, the value of x if
(a) $\log 31 = x$,   (b) $\log x = 1{\cdot}6$,   (c) $\ln x = -0{\cdot}25$.

(a) Using the log function   (b) $\log x = 1{\cdot}6$ (c) $\ln x = -0{\cdot}25$
    on the calculator, $\therefore \quad 10^{1{\cdot}6} = x$ $\therefore \quad e^{-0{\cdot}25} = x$
    $x = 1{\cdot}49$ i.e. $x = 39{\cdot}81$ i.e. $x = 0{\cdot}78$

Change of base

It is possible to express the logarithm of a to the base b in terms of logarithms to some other base c.

Suppose $\qquad\qquad\qquad y = \log_b a$
then $\qquad\qquad\qquad\quad a = b^y$
$\therefore \qquad\qquad\qquad \log_c a = \log_c b^y \quad$ by taking $\log_c$ of both sides
$\qquad\qquad\qquad\qquad\quad = y \log_c b$

$\therefore \qquad\qquad\qquad\quad y = \dfrac{\log_c a}{\log_c b} \qquad$ i.e. $\log_b a = \dfrac{\log_c a}{\log_c b}$

putting $\;c = a\;$ gives $\;\log_b a = \dfrac{\log_a a}{\log_a b} \qquad$ i.e. $\log_b a = \dfrac{1}{\log_a b}$

The technique of taking the logarithm of both sides of an equation can be useful when solving equations involving x as a power.

Example 9

Solve, giving your answers correct to 3 decimal places,   (a) $3^{x+1} = 15$,   (b) $(0 \cdot 8)^{2x} = 0 \cdot 45$.

(a)
$$3^{x+1} = 15$$
$$\therefore \quad \log 3^{x+1} = \log 15$$
$$\therefore \quad (x+1) \log 3 = \log 15$$
$$\therefore \quad x+1 = \frac{\log 15}{\log 3}$$
$$x = \frac{\log 15}{\log 3} - 1$$
$$\text{or} \quad x = 1 \cdot 465$$

(b)
$$(0 \cdot 8)^{2x} = 0 \cdot 45$$
$$\therefore \quad \log(0 \cdot 8)^{2x} = \log 0 \cdot 45$$
$$\therefore \quad 2x \log(0 \cdot 8) = \log 0 \cdot 45$$
$$\therefore \quad 2x = \frac{\log 0 \cdot 45}{\log 0 \cdot 8}$$
$$= 3 \cdot 5784$$
$$\text{or} \quad x = 1 \cdot 789$$

Example 10

Solve for x the equations (a) $5^{2x+1} + 4 = 21 \times 5^x$,   (b) $\log_3 x + \log_x 9 = 3$.

(a)
$$5^{2x+1} + 4 = 21 \times 5^x$$
$$\text{or} \quad 5 \times 5^{2x} - 21 \times 5^x + 4 = 0$$
letting $y = 5^x$,
$$5y^2 - 21y + 4 = 0$$
$$(5y - 1)(y - 4) = 0$$
$$y = \tfrac{1}{5} \quad \text{or} \quad y = 4$$
$$\therefore \quad 5^x = \tfrac{1}{5} \quad \text{or} \quad 5^x = 4$$
$$= 5^{-1} \qquad x \log 5 = \log 4$$
$$\text{i.e.} \quad x = -1 \quad \text{or} \quad x = \frac{\log 4}{\log 5} = 0 \cdot 86$$

(b)
$$\log_3 x + \log_x 9 = 3$$
$$\log_3 x + \frac{\log_3 9}{\log_3 x} = 3$$
$$(\log_3 x)^2 + 2 = 3 \log_3 x$$
letting $y = \log_3 x$,
$$y^2 - 3y + 2 = 0$$
$$(y - 1)(y - 2) = 0$$
$$y = 1 \quad \text{or} \quad y = 2$$
$$\therefore \quad \log_3 x = 1 \quad \text{or} \quad \log_3 x = 2$$
$$\text{i.e.} \quad x = 3 \quad \text{or} \quad x = 9$$

Straight line graphs

We know that if y is plotted against x for the equation $y = mx + c$, m and c constant, a straight line is obtained. The gradient of this line is m and it cuts the y-axis at $y = c$.

If we consider the relationship
$$y = ax^n \quad (a, n \text{ constant})$$
then $\log y = \log(ax^n) = \log a + \log x^n$
$$\therefore \quad \log y = n \log x + \log a$$
Since this is of the form $Y = mX + C$, where $\log y = Y$, $\log x = X$, then by plotting values of $\log y$ against $\log x$ we shall obtain a straight line graph. In this way we can verify the existence of such a relationship and, by measuring the gradient and the Y-intercept of the straight line graph, we can obtain approximate values of the constants n and $\log a$.

Similarly, suppose a relationship of the form $y = ab^x$ is believed to exist,
then $\log y = \log(ab^x) = \log a + x \log b$
i.e. $\log y = x \log b + \log a$
which is again of the form $Y = mx + C$. Plotting $\log y$ against x will again produce a straight line graph if the relationship does in fact exist. The

gradient will, in this case, be log b and the line will cut the Y-axis at $Y = \log a$.

This is of course the same technique as was seen in section 3.8 page 89 but now, with our knowledge of logarithms, more involved relationships involving two variables can be tested by seeing whether a straight line results when suitably chosen expressions are plotted against each other.

Exercise 5C

1. Write each of the following statements in the form $a^x = y$.
   (a) $\log_m c = d$ (b) $\log_b p = q$ (c) $\log s = t$ (d) $\ln z = r$

2. Find the value of x in each of the following
   (a) $\log_2 8 = x$ (b) $\log_3 81 = x$ (c) $\log_x 216 = 3$
   (d) $\log_x 16 = 0{\cdot}5$ (e) $\log x = -2$ (f) $\log_4 x = 5\frac{1}{2}$
   (g) $\log_8 16 = x$ (h) $\log_9 27 = x$ (i) $-\log_{16} x = 0{\cdot}75$

3. Evaluate the following
   (a) $\log_2 32$ (b) $\log_6 36$ (c) $\log_5 625$ (d) $\log_2(\frac{1}{2})$
   (e) $\log_2(\frac{1}{4})$ (f) $\log_a 1$ (g) $\log_a a$ (h) $\log_8 2$
   (i) $\log_8(\frac{1}{2})$ (j) $\log_{27} 3$ (k) $\log_8 4$ (l) $\log_{27}(\frac{1}{3})$

4. Express the following in terms of $\log a$, $\log b$ and $\log c$.
   (a) $\log(abc)$ (b) $\log(a^2 bc)$ (c) $\log(a^3 b^2 c)$ (d) $\log(a\sqrt{b})$
   (e) $\log\left(\dfrac{ab}{c}\right)$ (f) $\log\left(\dfrac{a^2}{bc^3}\right)$ (g) $\log\left(\dfrac{1}{ab}\right)$ (h) $\log\left(a\sqrt{\left(\dfrac{b}{c^3}\right)}\right)$

   (i) $\log(10ab^2)$ (j) $\log\left(\dfrac{\sqrt{(10a)}}{b^2}\right)$

5. Express as a single logarithm
   (a) $2\log_a 5 + \log_a 4 - \log_a 10$ (b) $2\log_a 3 - \log_a 15 + \log_a 5$
   (c) $\log_a 12 - (\frac{1}{2}\log_a 9 + \frac{1}{3}\log_a 8)$ (d) $2\log_a(\frac{1}{2}) + 3\log_a 4 - \log_a(\frac{4}{3})$

6. Express each of the following in the form $\log_a[f(x)]$
   (a) $2\log_a x$ (b) $\log_a x + \log_a(x + 3)$
   (c) $\log_a(x + 1) - \log_a 2$ (d) $\log_a(x^2 - 1) - \log_a(x + 1)$
   (e) $2\log_a x - \log_a x(x + 1)$ (f) $3\log_a x + \log_a(x + 1)$
   (g) $4\log_a x - \log_a(x^2 + x^3)$ (h) $\frac{1}{2}\log_a x + \log_a(x + 1) - \log_a \sqrt{x}$

7. Find the value of x in each of the following
   (a) $3\log_2 4 = x$ (b) $2\log_4 8 = x$ (c) $4\log_x 2 = 0{\cdot}5$

8. Use your calculator to find x correct to 2 decimal places
   (a) $\log 23 = x$ (b) $\ln 23 = x$ (c) $\log x = 0{\cdot}32$
   (d) $\log x = 1{\cdot}4$ (e) $\log x = -0{\cdot}65$ (f) $\ln x = 3{\cdot}1$
   (g) $\ln x = 0{\cdot}13$ (h) $\ln x = -0{\cdot}45$

9. Given that $\log 2 = 0{\cdot}3$ show that $\log 5 = 0{\cdot}7$. Using these values find the values of the following:
   (a) $\log(2{\cdot}5)$ (b) $\log 4$ (c) $\log 25$ (d) $\log 32$
   (e) $\log(0{\cdot}5)$ (f) $\log_2 5$ (g) $\log_5 10$

10. Given that $\ln 5 = 1{\cdot}6$ and $\ln 11 = 2{\cdot}4$ find the values of:
    (a) $\ln 625$ (b) $\ln 55$ (c) $\ln(2{\cdot}2)$ (d) $\ln 275$
    (e) $\ln(0{\cdot}2)$ (f) $\ln(0{\cdot}04)$ (g) $\log_5 11$

11. (a) Without the use of a calculator copy and complete the following table of values for $y = 2^x$.

x	-3	-2	-1	0	1	2
y						

(b) Copy and complete the following table of values for $y = \log_2 x$. (Use a calculator and give values correct to 2 decimal places where necessary.)

x	0·2	0·4	0·6	0·8	1	2	3	4
y								

(c) Plot the graphs of $y = 2^x$ and $y = \log_2 x$ on the same set of axes, using a scale of 2 cm to 1 unit on each axis.

12. Show that $\log_a b \times \log_b a = 1$ and hence evaluate
    (a) $\log_3 8 \times \log_2 9$,   (b) $\log_5 64 \times 2 \log_4 25$,   (c) $\log_{10} 8 \times \log_2 1000$.

13. Given that $\log 2 = 0·3$, $\log 3 = 0·48$ and $\log 17 = 1·23$ evaluate
    (a) $\log_2 3$,   (b) $\log_2 17$.

14. Show graphically that the following figures support the belief that a relationship of the form $y = ak^x$ exists and use your graph to find approximate values for the constants a and k.

x	1	2	3	4	6	8	10
y	3·5	4·9	6·9	9·6	18·8	36·9	72·3

15. Show graphically that the following figures support the belief that a relationship of the form $y = ak^{x^2}$ exists and use your graph to find approximate values for the constants a and k.

x	0·5	1	1·5	2	2·5	3
y	3·6	13·8	129·6	2981	$1·679 \times 10^5$	$2·318 \times 10^7$

16. Show graphically that the following figures support the belief that a relationship of the form $y = ax^k$ exists and use your graph to find approximate values for the constants a and k.

x	5	8	12	14	20	30
y	22·4	45·3	83·1	104·8	178·9	328·6

17. Show graphically that the following figures support the belief that a relationship of the form $y = a(2 + x)^k$ exists and use your graph to find approximate values for the constants a and k.

x	1	3	8	10	15	20	24
y	5·2	6·7	9·5	10·4	12·4	14·1	15·3

18. Using the ln or log function on a calculator solve the following equations giving answers correct to 3 significant figures.

(a) $2^x = 5$ (b) $3^x = 17$ (c) $5^x = 10$
(d) $4^x = 12$ (e) $3^x = 365$ (f) $3^{x+1} = 2^x$
(g) $4^{x+1} = 3^{2-x}$ (h) $5^{x-1} = 2^{3x-1}$ (i) $6^{3x+1} = 7^{2-x}$

19. Solve the following equations (for parts (d) and (e) you will need to use a calculator and should give your answers correct to 3 significant figures).

(a) $2^{2x} - 3 \times 2^x - 4 = 0$ (b) $4^{2x} + 4^{x+2} - 80 = 0$
(c) $3^{2x+1} + 3 = 10 \times 3^x$ (d) $4^{2x} - 4^{x+2} = 80$
(e) $2^{2x+1} - 7 \times 2^x + 6 = 0$ (f) $1 + \log_2 x = \dfrac{12}{\log_2 x}$

20. Solve the following logarithmic equations

(a) $\log_a x + \log_a 4 - \log_a 5 = \log_a 12$ (b) $\log_4 x - \log_4 7 = \frac{3}{2}$
(c) $\log_9 x + \log_9 x^2 + \log_9 x^3 + \log_9 x^4 = 5$ (d) $\log_9 4 + \log_3 x = 3$
(e) $\log_2 x - \log_8 x = 4$ (f) $\log_5 x + 2\log_x 5 = 3$

21. Solve the following simultaneous equations

(a) $2\log_x y = 1$ (b) $6\log_3 x + 6\log_{27} y = 7$ (c) $y\log_2 8 = x$
     $xy = 64$ $4\log_9 x + 4\log_3 y = 9$ $2^x + 8^y = 8192$

(d) $\log x - \log 2 = 2\log y$ (e) $5^{x+2} + 7^{y+1} = 3468$ (f) $\log_2 x + 2\log_4 y = 4$
     $x - 5y + 2 = 0$ $7^y = 5^x - 76$ $x + 12y = 52$

(g) $\ln 6 + \ln(x - 3) = 2\ln y$ (h) $2 + \log_2(2x + 1) = 2\log_2 y$
     $2y - x = 3$ $x = 22 - y$

5.4 Quadratics

Sketching

Example 11

For the curve $y = x^2 - 2x - 3$, find (a) where the curve cuts the y-axis, (b) where the curve cuts the x-axis, (c) the coordinates of any maximum or minimum points, and (d) hence sketch the curve.

(a) On the y-axis, $x = 0$,
$\therefore$ $y = 0 - 0 - 3 = -3$
The curve cuts the y-axis at the point $A(0, -3)$.

(b) On the x-axis, $y = 0$,
$\therefore$ $0 = x^2 - 2x - 3$
         $= (x - 3)(x + 1)$
giving $x = +3$ or $x = -1$
The curve cuts the x-axis at the points $B(3, 0)$ and $C(-1, 0)$.

(c) Since $y = x^2 - 2x - 3$
         $= (x - 1)^2 - 4$, (by completing the square, see page 11).
when $x = 1$, the value of y is a minimum of -4.
There is a minimum point at $D(1, -4)$.

(d) Knowing the coordinates of the points A, B, C and D a sketch of the curve can now be made.

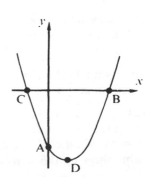

Note   Frequently a satisfactory sketch of a curve can be made without
having the coordinates of any maximum or minimum points. In Example 11
we could have made a reasonable sketch without having the coordinates of
the point D.

Chapter 11 deals more fully with the techniques of sketching curves.

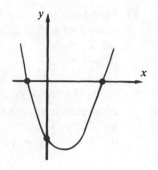

Inequalities

From the introductory chapter, we are familiar with simplifying inequalities
of the form $-9 < 2x - 3 \leqslant 7$, to give a range of possible values for x,
in this case $-3 < x \leqslant 5$.

The following examples show that it is also possible to find a range of values
for x when we are given a quadratic inequality. These ranges can be found
by sketching (Example 12), by examining the signs of the factors (Example 13),
or by completing the square (Example 14).

Example 12

Find, by sketching, the range (or ranges) of values that x can take if
$2x^2 + 5x - 7 \leqslant 0$.

Let $y = 2x^2 + 5x - 7$
        $= (2x + 7)(x - 1)$

Graph of $y = 2x^2 + 5x - 7$ cuts y-axis at $(0, -7)$
and cuts x-axis at $(-3\frac{1}{2}, 0)$, $(1, 0)$.

From this information a sketch can be made.

Thus for $y \leqslant 0$, i.e. for the curve to be 'below' the x-axis, we must
have $-3\frac{1}{2} \leqslant x \leqslant 1$.

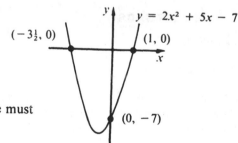

Example 13

By examining the signs of the factors of the expression, find the range of
values x can take for the following inequalities to be true:
(a) $x^2 + 3x - 18 \geqslant 0$, (b) $x^2 - x - 8 \leqslant 4$.

In this method we use the fact that $y = (x - a)(x - b)$ will change sign
when its graph crosses the x-axis, i.e. when $x = a$ and $x = b$.

(a) $x^2 + 3x - 18 \geqslant 0$
     i.e. $(x + 6)(x - 3) \geqslant 0$

$y = (x + 6)(x - 3)$ cuts the x-axis at $x = -6$ and at $x = 3$. Construct a
table showing the signs of $(x + 6)(x - 3)$ on either side of these values.

x	$x < -6$	$-6 < x < 3$	$x > 3$
$x + 6$	$-$ve	$+$ve	$+$ve
$x - 3$	$-$ve	$-$ve	$+$ve
$(x - 3)(x + 6)$	$+$ve	$-$ve	$+$ve

Thus for $(x + 6)(x - 3) \geqslant 0$ (i.e. not $-$ve)
$$x \leqslant -6 \text{ or } x \geqslant 3.$$
(b) $x^2 - x - 8 \leqslant 4$
$$x^2 - x - 12 \leqslant 0$$
   i.e. $(x - 4)(x + 3) \leqslant 0$
$y = (x - 4)(x + 3)$ cuts the x-axis at $x = 4$ and at $x = -3$, and y will change sign at $x = 4$ and at $x = -3$. The table shows the sign of y on either side of these values.

x	$x < -3$	$-3 < x < 4$	$x > 4$
$x - 4$	$-$ve	$-$ve	$+$ve
$x + 3$	$-$ve	$+$ve	$+$ve
$(x - 4)(x + 3)$	$+$ve	$-$ve	$+$ve

Thus for $(x - 4)(x + 3) \leqslant 0$ (i.e. not $+$ve)
$$-3 \leqslant x \leqslant 4.$$
Alternatively, the required ranges could be obtained as follows:-
(a) $(x + 6)(x - 3) \geqslant 0 \Rightarrow$ either $x + 6 \geqslant 0$ and $x - 3 \geqslant 0 \Rightarrow x \geqslant 3$ $\Big\}$ $\Rightarrow x \leqslant -6$ or $x \geqslant 3$.
            or $x + 6 \leqslant 0$ and $x - 3 \leqslant 0 \Rightarrow x \leqslant -6$
(b) $(x - 4)(x + 3) \leqslant 0 \Rightarrow$ either $x - 4 \geqslant 0$ and $x + 3 \leqslant 0 \Rightarrow$ no solution $\Big\}$ $\Rightarrow -3 \leqslant x \leqslant 4$.
            or $x - 4 \leqslant 0$ and $x + 3 \geqslant 0 \Rightarrow -3 \leqslant x \leqslant 4$

Example 14

Find the range of values x can take for the following inequalities to be true:
(a) $x^2 + 2x + 3 \geqslant 0$,   (b) $x^2 - 3x - 5 \geqslant 0$.

(a) $x^2 + 2x + 3 \geqslant 0$
$\therefore$ $x^2 + 2x + 1 - 1 + 3 \geqslant 0$
$\therefore$ $(x + 1)^2 + 2 \geqslant 0$
But this is true for all real x.
$\therefore$ $x^2 + 2x + 3 \geqslant 0$ for all real x.

(b) $x^2 - 3x - 5 \geqslant 0$
$\therefore$ $x^2 - 3x + \dfrac{9}{4} - \dfrac{9}{4} - 5 \geqslant 0$
$\therefore$ $\left(x - \dfrac{3}{2}\right)^2 - \dfrac{29}{4} \geqslant 0$
$\therefore$ $\left(x - \dfrac{3}{2}\right)^2 \geqslant \dfrac{29}{4}$
$\therefore$ $x - \dfrac{3}{2} \leqslant \dfrac{-\sqrt{29}}{2}$ or $x - \dfrac{3}{2} \geqslant \dfrac{\sqrt{29}}{2}$
i.e. $x \leqslant \dfrac{3 - \sqrt{29}}{2}$ or $x \geqslant \dfrac{3 + \sqrt{29}}{2}$

Exercise 5D

Quadratic sketches

1. For the graph of $y = x^2 - 5x + 4$ find the coordinates of
   (a) the point where the line cuts the y-axis,
   (b) the points where the line cuts the x-axis,
   (c) any maximum or minimum points.
   Hence make a sketch of $y = x^2 - 5x + 4$.
2. For the graph of $y = 4 + 3x - x^2$ find the coordinates of
   (a) the point where the line cuts the y-axis,
   (b) the points where the line cuts the x-axis,
   (c) any maximum or minimum points.
   Hence make a sketch of $y = 4 + 3x - x^2$.
3. For the graph of $y = -4x^2 - 12x - 5$ find the coordinates of
   (a) the point where the line cuts the y-axis,
   (b) the points where the line cuts the x-axis,
   (c) any maximum or minimum points.
   Hence make a sketch of $y = -4x^2 - 12x - 5$.

Each of the statements in questions **4, 5** and **6** refer to the graph of
$y = a (x - b)(x - c)$. Find the values of a, b and c in each case ($b \geqslant c$).
4. The curve cuts the x-axis at $(1, 0)$ and at $(2, 0)$ and the y-axis at $(0, 4)$.
5. The curve cuts the x-axis at $(-2, 0)$ and at $(3, 0)$ and the y-axis at $(0, -9)$.
6. The curve *touches* the x-axis at $(2, 0)$ and cuts the y-axis at $(0, 12)$.

Quadratic inequalities

7. Use sketching to determine the range (or ranges) of values x can take for
   each of the following inequalities.
   (a) $(x - 5)(x + 3) \geqslant 0$ (b) $x^2 - 8x + 15 \leqslant 0$
   (c) $x^2 - 4x - 5 \geqslant 0$ (d) $2x^2 - 5x - 3 \geqslant 0$
   (e) $3x^2 - 19x + 6 \leqslant 0$ (f) $5 - 3x - 2x^2 \geqslant 0$
8. By examining the signs of the factors of the expressions, find the range
   (or ranges) of values that x can take in each of the following inequalities.
   (a) $x^2 + 5x - 6 \leqslant 0$ (b) $x^2 + 7x + 12 \geqslant 0$
   (c) $x^2 - 12x + 35 < 0$ (d) $3x^2 - 22x + 35 \geqslant 0$
   (e) $6x^2 - x > 15$ (f) $1 + x - 6x^2 \geqslant 0$
9. For each of the following inequalities use the method of completing the
   square to find the range (or ranges) of values that x can take.
   (a) $x^2 + 4x + 1 \geqslant 0$ (b) $x^2 - 5x + 3 \leqslant 0$ (c) $x^2 - 4x + 5 \geqslant 0$
   (d) $2x^2 - 3x \leqslant 1$ (e) $2x^2 - 3x \geqslant -2$ (f) $3x^2 \geqslant 2 - 2x$
10. Show that $x^2 + x + 1$ is positive for all real x [hint: use 'completing the
    square'].
11. Show that $3x^2 + 4x + 2$ is positive for all real x [hint: use 'completing
    the square'].
12. Use the method of completing the square to find the range of values k
    can take if $x^2 + 4x + k - 1$ is positive for all real x.
13. Use the method of completing the square to find the range of values k
    can take if $x^2 + 2kx + 2k + 3$ is positive for all real x.

The discriminant

On page 11, the formula $x = \dfrac{-b \pm \sqrt{(b^2 - 4ac)}}{2a}$ was used to solve a
quadratic equation. The formula applies to the general quadratic equation
$ax^2 + bx + c = 0$. The value of the expression $(b^2 - 4ac)$ will determine
the nature of the roots of the equation, and it is called the **discriminant** of
the equation.

If the discriminant is zero, i.e. $b^2 - 4ac = 0$, the roots of the equation are $x = \dfrac{-b}{2a}$, $x = \dfrac{-b}{2a}$.

If the discriminant is positive, i.e. $b^2 - 4ac > 0$, the roots are unequal
and $x = \dfrac{-b + \sqrt{(b^2 - 4ac)}}{2a}$ or $x = \dfrac{-b - \sqrt{(b^2 - 4ac)}}{2a}$

If the discriminant is negative, i.e. $b^2 - 4ac < 0$, there are no real roots of the equation.

(i)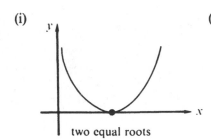

(ii)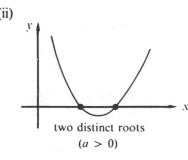

(iii)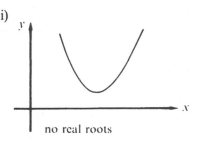

two equal roots   two distinct roots   no real roots
                               $(a > 0)$

The three diagrams of the curve $y = ax^2 + bx + c\ (a > 0)$ show the three possible cases.

In (i) the curve *touches* the x-axis, i.e. for $y = 0$ there are two equal values of x,

In (ii) the curve cuts the x-axis, i.e. for $y = 0$ there are two real distinct values of x,

In (iii) the curve does not cut the x-axis, i.e. for $y = 0$ there is no real value of x.

Similar diagrams exist for the curve $y = ax^2 + bx + c\ (a < 0)$. The value
of the discriminant will again determine the nature of the roots of the
equation $ax^2 + bx + c = 0$.

Example 15

Find the discriminant and state whether the equation will have two real
distinct roots, no real roots or a repeated root in each of the following cases.
(a) $3x^2 - 5x + 2 = 0$,   (b) $4x^2 - 28x + 49 = 0$,   (c) $5x^2 - 7x + 3 = 0$.

(a) Comparing $3x^2 - 5x + 2 = 0$
         with $ax^2 + bx + c = 0$, gives $a = 3, b = -5, c = 2$.

   The discriminant is $b^2 - 4ac$
              $= (-5)^2 - 4(3)2$
              $= +1$
   Since $b^2 - 4ac > 0$ the equation will have 2 real distinct roots.

(b) Comparing $4x^2 - 28x + 49 = 0$
with $ax^2 + bx + c = 0$, gives $a = 4$, $b = -28$, $c = 49$.

The discriminant is $b^2 - 4ac$
$$= (-28)^2 - 4(4)49$$
$$= 784 - 784$$
$$= 0$$

Since $b^2 - 4ac = 0$ the equation will have two equal roots, i.e. a repeated root.

(c) Comparing $5x^2 - 7x + 3 = 0$
with $ax^2 + bx + c = 0$, gives $a = 5$, $b = -7$, $c = 3$.

The discriminant is $b^2 - 4ac$
$$= (-7)^2 - 4(5)3$$
$$= 49 - 60$$
$$= -11$$

Since $b^2 - 4ac < 0$ the equation has no real roots.

Example 16

Find the range of values β can take for the quadratic expression
$x^2 + 4\beta x + \beta$ to be positive for all real values of x.

The discriminant of $x^2 + 4\beta x + \beta$ is $16\beta^2 - 4\beta$.
The graph of $x^2 + 4\beta x + \beta$ will not cut the x-axis if $16\beta^2 - 4\beta < 0$
i.e. $4\beta(4\beta - 1) < 0$

β	$\beta < 0$	$0 < \beta < \frac{1}{4}$	$\beta > \frac{1}{4}$
4β	$-$ ve	$+$ ve	$+$ ve
$4\beta - 1$	$-$ ve	$-$ ve	$+$ ve
$4\beta(4\beta - 1)$	$+$ ve	$-$ ve	$+$ ve

From the table $4\beta(4\beta - 1) < 0$ if $0 < \beta < \frac{1}{4}$.
Thus $x^2 + 4\beta x + \beta$ is always positive or always negative for $0 < \beta < \frac{1}{4}$.
If $x = 0$, $x^2 + 4\beta x + \beta = \beta$ which is positive for $0 < \beta < \frac{1}{4}$.
Thus the required range is $0 < \beta < \frac{1}{4}$.
Alternatively, this example could have been answered by the method of completing the square.

Example 17

Find the range, or ranges, of values K can take for $Kx^2 - 4x + 5 - K = 0$
to have two real distinct roots.

The solution to this question needs care since for $K = 0$, the given equation
is not a quadratic. This special case is considered separately in the working below.

Comparing $Kx^2 - 4x + 5 - K = 0$
with $ax^2 + bx + c = 0$ gives $a = K$, $b = -4$ and $c = (5 - K)$.

If $K \neq 0$, for 2 real distinct roots
$$16 - 4K(5 - K) > 0$$
i.e. $\quad K^2 - 5K + 4 > 0$
or $\quad (K - 1)(K - 4) > 0$

K	$K < 1$	$1 < K < 4$	$K > 4$
$K - 1$	$-ve$	$+ve$	$+ve$
$K - 4$	$-ve$	$-ve$	$+ve$
$(K - 1)(K - 4)$	$+ve$	$-ve$	$+ve$

$\therefore \quad K < 1, K > 4.$

If $\quad K = 0,$
$$-4x + 5 = 0$$
$\therefore \qquad x = 1\tfrac{1}{4}$
i.e.   only one solution.
Thus $K = 0$ does not
give two real distinct roots.

The required ranges are $K < 1$ except $K = 0$ and $K > 4$.

Example 18

Find the range, or ranges, of values K can take for $Kx^2 - 2xK - 3K - 12 = 0$
to have real roots.
(Note: The phrase 'has real roots' should be taken to mean 'has either two real
distinct roots or a repeated root'.)

Comparing $Kx^2 - 2xK - 3K - 12 = 0$
with $\qquad ax^2 + bx + c = 0$ gives $a = K, b = -2K$ and $c = (-3K - 12)$.

If $K \neq 0$, for one or two real roots,
$$4K^2 - 4K(-3K - 12) \geqslant 0$$
i.e. $\qquad\qquad K(K + 3) \geqslant 0$

If $\quad K = 0$
$-12 = 0$ which is impossible.
$\therefore \qquad K \neq 0.$

K	$K < -3$	$-3 < K < 0$	$K > 0$
K	$-ve$	$-ve$	$+ve$
$K + 3$	$-ve$	$+ve$	$+ve$
$K(K + 3)$	$+ve$	$-ve$	$+ve$

$\therefore \quad K \leqslant -3, K \geqslant 0.$

The required ranges are $K \leqslant -3$ and $K > 0$.

The range of a quadratic

The graph of the quadratic function
$f(x) = x^2 + 2x + 3$, is shown in the diagram.
From this we can see that for the domain
$x \in \mathbb{R}$ the function can only take values $\geqslant 2$,
i.e. for real x the range of $f(x)$ is
$\{y \in \mathbb{R}: y \geqslant 2 \}$.
Using the discriminant this range can be
determined without having to make a sketch
of the function.

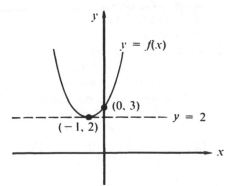

Example 19

Find the range of the function $f(x) = x^2 + 6x - 1$ for the domain $x \in \mathbb{R}$.

Let $\qquad\qquad y = x^2 + 6x - 1$
then $\qquad\qquad 0 = x^2 + 6x - 1 - y$
Comparing with $\quad 0 = ax^2 + bx + c$ gives $a = 1$, $b = 6$, $c = -(1 + y)$.
For real x $\qquad\qquad\quad b^2 - 4ac \geqslant 0$
i.e. $\qquad 36 + 4 \times 1 \times (1 + y) \geqslant 0$
giving $\qquad\qquad\qquad y \geqslant -10$
Thus for real x the function $f(x) = x^2 + 6x - 1$ has range $\{y \in \mathbb{R}: y \geqslant -10\}$.
Alternatively, this range could have been determined by completing the square:

$$\begin{aligned} f(x) &= x^2 + 6x - 1 \\ &= x^2 + 6x + 9 - 9 - 1 \\ &= (x + 3)^2 - 10 \end{aligned}$$

Now $(x + 3)^2 \geqslant 0$,
$\therefore f(x) \geqslant -10$, i.e. for real x, $f(x)$ has range $\{y \in \mathbb{R} : y \geqslant -10\}$.

Exercise 5E

1. The graphs of the quadratic functions $y = f(x)$, $y = g(x)$ and $y = h(x)$ are shown below,

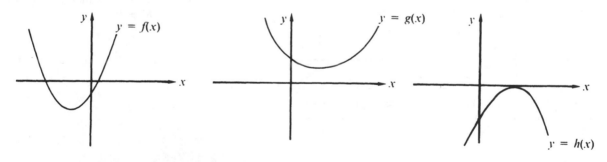

State which of the equations $f(x) = 0$, $g(x) = 0$ and $h(x) = 0$ will have
(a) no real roots, (b) two real distinct roots, (c) a repeated root.
If all of the functions are of the form $ax^2 + bx + c$, find for which of them (d) $a > 0$, (e) $a < 0$.

2. Each of the following quadratic equations are in the form
$ax^2 + bx + c = 0$ (or can be expressed in that form). In each case find the value of $b^2 - 4ac$ and hence state whether the equation will have two real distinct roots, no real roots or a repeated root.
(a) $x^2 + 11x - 2 = 0$ (b) $x^2 - 3x + 3 = 0$ (c) $x^2 + 7x - 2 = 0$
(d) $9x^2 + 6x + 1 = 0$ (e) $3x^2 - 11x + 7 = 0$ (f) $2x^2 + 5x + 4 = 0$
(g) $4x^2 - 12x + 9 = 0$ (h) $7x^2 + 11x + 5 = 0$ (i) $7x^2 - 11x - 5 = 0$

3. Find the range of each of the following functions for the domain $x \in \mathbb{R}$.
   (a) $f(x) = x^2 + 4x - 2$
   (b) $f(x) = x^2 + 3x + 5$
   (c) $f(x) = 5 - 2x - x^2$
   (d) $f(x) = 2x^2 - 8x + 3$
   (e) $f(x) = 5x - 1 - 2x^2$

4. If $ax^2 - 8x + 2 = 0$ has a repeated root find the value of a.

5. If $2x^2 + bx + 50 = 0$ has a repeated root find the possible values of b.

6. Find the range(s) of values b can take for $9x^2 + bx + 4 = 0$ to have two real distinct roots.

7. Find the range(s) of values k can take for $x^2 + (k + 1)x + 1 = 0$ to have real roots.

8. Find the range(s) of values k can take for $2x^2 + (3 - k)x + k + 3 = 0$ to have two real distinct roots.

9. Find the range(s) of values k can take for
   (a) $kx^2 + 4x + 3 + k = 0$
       to have real roots.
   (b) $kx^2 + 3x + k - 4 = 0$
       to have two real distinct roots.
   (c) $kx^2 + 2x + 1 - k = 0$
       to have two real distinct roots.

(d) $kx^2 + 3x(1 - k) + 3 = 0$
    to have real roots.
(e) $kx^2 + k = 8x - 2xk$
    to have two real distinct roots.
(f) $kx^2 + 1 = 4kx - 3k$
    to have real roots.

10. Use the discriminant to find the range of values that β can take if $x^2 + 5\beta x + \beta$ is positive for all real x.

11. Prove that $y = x - 3$ is a tangent to the curve $y = x^2 - 5x + 6$ and find the coordinates of the point of contact.

12. Prove that $y = 4x - 9$ is a tangent to $y = 4x(x - 2)$ and find the coordinates of the point of contact.

13. For each part of this question, state which of the following statements apply:
    (i) the straight line is a tangent to the curve, (ii) the straight line cuts the curve in two distinct points, (iii) the straight line neither cuts nor touches the curve.
    (a) Curve: $y = 3x^2 - 4x - 2$
        Straight line: $y = x - 3$
    (b) Curve: $y = 7x(x - 1)$
        Straight line: $y + 2x + 1 = 0$
    (c) Curve: $y = 9x^2 - 3x + 10$
        Straight line: $y = 3(x + 3)$
    (d) Curve: $y = (3x - 4)(x + 1)$
        Straight line: $y + 10x + 11 = 0$.

Sum and product of the roots of a quadratic equation

If α and β are the roots of the equation $ax^2 + bx + c = 0$ then
$$(x - \alpha)(x - \beta) = 0$$
i.e. $x^2 - x(\alpha + \beta) + \alpha\beta = 0$

Comparing this equation with $ax^2 + bx + c = 0 \left(\text{i.e. } x^2 + \dfrac{bx}{a} + \dfrac{c}{a} = 0 \right)$

we see that $\alpha + \beta = \dfrac{-b}{a}$, $\alpha\beta = \dfrac{c}{a}$.

We can therefore state that if α and β are the roots of the equation $ax^2 + bx + c = 0$,

$$\alpha + \beta \text{(the sum of the roots)} = \frac{-b}{a}$$

$$\alpha\beta \text{(the product of the roots)} = \frac{c}{a}$$

We can use our knowledge of the value of $(\alpha + \beta)$ and $\alpha\beta$ to find the values of other symmetrical expressions in α and β. An expression is symmetrical in α and β if writing α for β and β for α leaves the expression unchanged.

Example 20

If α and β are the roots of the equation $x^2 - x - 3 = 0$, state the values of $(\alpha + \beta)$ and $(\alpha\beta)$ and find the values of
(a) $\alpha^2 + \beta^2$, (b) $(\alpha - \beta)^2$, (c) $\alpha^3 + \beta^3$.

Comparing $x^2 - x - 3 = 0$ with $ax^2 + bx + c = 0$ gives
$a = 1, b = -1, c = -3$.

$$\text{Hence}\quad \alpha + \beta = \frac{-b}{a} = 1 \quad \text{and} \quad \alpha\beta = \frac{c}{a} = -3$$

(a) $\alpha^2 + \beta^2$
$= (\alpha + \beta)^2 - 2\alpha\beta$
$= 1^2 - 2(-3)$
$= 7$

(b) $(\alpha - \beta)^2$
$= \alpha^2 - 2\alpha\beta + \beta^2$
$= (\alpha + \beta)^2 - 2\alpha\beta - 2\alpha\beta$
$= 1^2 - 4(-3)$
$= 13$

(c) $\alpha^3 + \beta^3$
$= (\alpha + \beta)^3 - 3\alpha^2\beta - 3\alpha\beta^2$
$= (\alpha + \beta)^3 - 3\alpha\beta(\alpha + \beta)$
$= 1^3 - 3(-3)(1)$
$= 10$

Example 21

If α and β are the roots of the quadratic equation $x^2 - 3x - 5 = 0$, find the quadratic equation that has roots of

(a) $\alpha - 3, \beta - 3$ (b) $\dfrac{1}{\alpha + 1}, \dfrac{1}{\beta + 1}$ (c) $\dfrac{1}{\alpha^2}, \dfrac{1}{\beta^2}$.

If α and β are the roots of $x^2 - 3x - 5 = 0$ then $\alpha + \beta = 3$ and $\alpha\beta = -5$.
In each case the method will be to find the sum and the product of the required roots and hence to write down the required equation.

(a) sum of new roots $= \alpha - 3 + \beta - 3$
$= \alpha + \beta - 6$
$= 3 - 6$
$= -3$

product of new roots $= (\alpha - 3)(\beta - 3)$
$= \alpha\beta - 3(\alpha + \beta) + 9$
$= -5 - 3(3) + 9$
$= -5$

Thus in the required equation
$\dfrac{-b}{a} = -3$ and $\dfrac{c}{a} = -5$.

With $a = 1, b = 3$ and $c = -5$,
the required equation is
$x^2 + 3x - 5 = 0$.

(b) sum of new roots $= \dfrac{1}{\alpha + 1} + \dfrac{1}{\beta + 1}$
$= \dfrac{\beta + 1 + \alpha + 1}{(\alpha + 1)(\beta + 1)}$
$= \dfrac{(\alpha + \beta) + 2}{\alpha\beta + \beta + \alpha + 1}$
$= -5$

product of new roots $= \left(\dfrac{1}{\alpha + 1}\right)\left(\dfrac{1}{\beta + 1}\right)$
$= \dfrac{1}{\alpha\beta + \alpha + \beta + 1}$
$= \dfrac{1}{-5 + 3 + 1}$
$= -1$

Hence the required equation is
$x^2 + 5x - 1 = 0$.

(c) sum of new roots $= \dfrac{1}{\alpha^2} + \dfrac{1}{\beta^2}$

$$= \frac{\beta^2 + \alpha^2}{\alpha^2 \beta^2}$$

$$= \frac{(\alpha + \beta)^2 - 2\alpha\beta}{\alpha^2 \beta^2}$$

$$= \frac{19}{25}$$

product of new roots $= \dfrac{1}{\alpha^2 \beta^2}$

$$= \frac{1}{25}$$

Hence the required equation is $x^2 - \dfrac{19}{25}x + \dfrac{1}{25} = 0$ or $25x^2 - 19x + 1 = 0$.

Note that, although from any quadratic equation we can write down the sum and the product of the roots of that equation, these two results will not of themselves help us to find the roots of the equation.

If α and β are the roots of the equation $x^2 - 7x + 9 = 0$, then $\alpha + \beta = 7$ and $\alpha\beta = 9$.
Solving these two equations simultaneously:
$\alpha = 7 - \beta$ hence $(7 - \beta)\beta = 9$
$\qquad\qquad$ or $7\beta - \beta^2 = 9$ i.e. $\beta^2 - 7\beta + 9 = 0$ which is the original equation.
However, if some additional information about the roots is available, it is possible to solve the equation by this means as Example 22 shows.

Example 22

Find the roots of the quadratic equation $x^2 - 6x + k = 0$ given that one root is three times the other and that k is a constant. Find also the value of k.

Let the roots of the equation be α and 3α.
Then from the equation $x^2 - 6x + k = 0$,

$$\alpha + 3\alpha = 6 \qquad\qquad \dots [1]$$
and
$$\alpha(3\alpha) = k \qquad\qquad \dots [2]$$

From [1], $\alpha = 1\frac{1}{2}$ and therefore $3\alpha = 4\frac{1}{2}$.
From [2], $k = 3\alpha^2 = 3(\frac{3}{2})^2 = \frac{27}{4}$.
The roots of the equation are $1\cdot5$ and $4\cdot5$ and the constant $k = 6\cdot75$.

Exercise 5F

1. State (i) the sum and (ii) the product of the roots of each of the following quadratics:
   (a) $x^2 + 9x + 4 = 0$
   (b) $x^2 - 2x - 5 = 0$
   (c) $x^2 - 7x + 2 = 0$
   (d) $x^2 - 9x - 3 = 0$
   (e) $2x^2 - 7x + 1 = 0$
   (f) $7x^2 + x - 1 = 0$
   (g) $3x^2 + 10x - 2 = 0$
   (h) $5x^2 + 10x + 1 = 0$
2. In each part of this question, you are given the sum and product of the roots of a quadratic. In each case, find the quadratic equation in the form $ax^2 + bx + c = 0$ with a, b and c taking integer values.

	(a)	(b)	(c)	(d)	(e)	(f)	(g)	(h)
Sum	-3	6	7	$-\frac{2}{3}$	$-\frac{5}{2}$	$-\frac{3}{2}$	$-\frac{1}{4}$	$-1\frac{3}{3}$
Product	-1	-4	-5	$-\frac{7}{3}$	-2	-5	$-\frac{1}{3}$	$\frac{1}{2}$

3. If α and β are the roots of the equation $x^2 + 11x + 2 = 0$ find the values of
   (a) $\alpha\beta$ 　　　　　(b) $\alpha + \beta$
   (c) $(\alpha + \beta)^2$ 　　(d) $\alpha^2 + \beta^2$
   (e) $3 - \alpha - \beta$ 　(f) $\alpha^2\beta + \alpha\beta^2$

4. If α and β are the roots of the equation $2x^2 - 5x + 1 = 0$ find the values of
   (a) $\alpha\beta$ 　　　　　(b) $\alpha + \beta$
   (c) $\alpha^2 + 3\alpha\beta + \beta^2$ (d) $\alpha^2 - 3\alpha\beta + \beta^2$
   (e) $\alpha^3\beta + \alpha\beta^3$ 　(f) $\dfrac{1}{\beta} + \dfrac{1}{\alpha}$.

5. If α and β are the roots of the equation $6x^2 + 2x - 3 = 0$ find the values of
   (a) $\dfrac{\alpha}{\beta} + \dfrac{\beta}{\alpha}$ 　　(b) $\dfrac{1}{\beta^2} + \dfrac{1}{\alpha^2}$
   (c) $\dfrac{2\beta}{1 + \dfrac{\beta}{\alpha}}$ 　　(d) $\dfrac{1}{\alpha\beta} - \dfrac{1}{\beta} - \dfrac{1}{\alpha}$
   (e) $\alpha^3 + \beta^3$ 　　(f) $\dfrac{1}{\alpha^3} + \dfrac{1}{\beta^3}$

6. If the roots of $ax^2 + bx + c = 0$ differ by 3 show that $b^2 = 9a^2 + 4ac$

7. If α and β are roots of the quadratic equation $x^2 - 4x + 2 = 0$ find the quadratic equation that has roots of:
   (a) $\alpha + 2$, $\beta + 2$ 　(b) $\dfrac{1}{\alpha}$, $\dfrac{1}{\beta}$

(c) α^2, β^2 　　(d) $\alpha^2\beta$, $\beta^2\alpha$
(e) α^3, β^3 　　(f) $\dfrac{1}{\alpha^2}$, $\dfrac{1}{\beta^2}$
(g) $(\alpha + 2\beta)$, $(\beta + 2\alpha)$
(h) $\dfrac{\alpha}{\beta}$, $\dfrac{\beta}{\alpha}$

8. Find the roots of the quadratic equation $x^2 - x + k = 0$ and the value of the constant k given that one root of the equation is twice the other.

9. Find the roots of the quadratic equation $x^2 - 3x + k = 0$ and the value of the constant k given that one root of the equation exceeds the other by 2.

10. (a) Find the restrictions on k if the equation $kx^2 + 4x + k - 3 = 0$ has real roots.
    (b) If these roots are of opposite sign find the restrictions on k now.

11. (a) Find the restrictions on k if the equation $kx^2 + (k - 3)x + 1 = 0$ has real roots.
    (b) If these real roots are distinct and are both positive, find the restriction on k now.

5.5 Polynomials in general

If $f(x) = ax^n + bx^{n-1} + cx^{n-2} + \ldots + k$, $a \neq 0$, then $f(x)$ is said to be a polynomial of degree (or order) n in the variable x.

If we consider the equation $f(x) = 0$ this will have up to n real roots.

In particular, if $f(x)$ is a cubic function $ax^3 + bx^2 + cx + d$ then the equation $f(x) = 0$ can have up to three real roots. The number of real roots will depend upon the values of a, b, c and d.

The diagrams show the various possibilities for the number of roots of $f(x) = 0$, for $a > 0$.

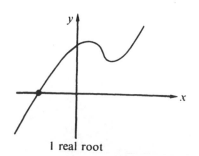

1 real root

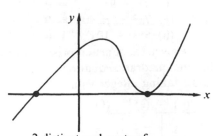
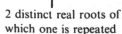

2 distinct real roots of which one is repeated

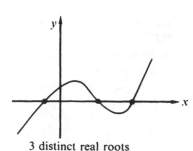

3 distinct real roots

Remainder theorem

If the function $f(x)$, a polynomial in x, is divided by $(x - a)$ until the remainder does not contain x, i.e. it is a constant, then the remainder is equal to $f(a)$.

Proof:

Suppose that when $f(x)$ is divided by $(x - a)$ the quotient is $g(x)$ and the remainder is k, then $f(x) \equiv (x - a) \times g(x) + k$

Substituting $x = a$ in both sides of this identity

$$f(a) = 0 \times g(a) + k$$
$$\text{or} \quad f(a) = k$$

So the remainder when $f(x)$ is divided by $(x - a)$ is equal to $f(a)$.

By a similar argument, when $f(x)$ is divided by $(cx - a)$ the remainder is $f\left(\dfrac{a}{c}\right)$.

Example 23

Find the remainder when $8x^3 - 4x^2 + 6x + 7$ is divided by
(a) $x - 1$, (b) $2x + 1$.

(a) $f(x) = 8x^3 - 4x^2 + 6x + 7$
Using the remainder theorem,
when $f(x)$ is divided by $(x - 1)$,
the remainder is $f(1)$.
$f(1) = 8(1) - 4(1) + 6(1) + 7$
$\quad = +17$
The remainder is $+17$.

(b) $f(x) = 8x^3 - 4x^2 + 6x + 7$
Using the remainder theorem,
when $f(x)$ is divided by $(2x + 1)$
the remainder is $f(-\frac{1}{2})$.
$f(-\frac{1}{2}) = 8(-\frac{1}{2})^3 - 4(-\frac{1}{2})^2 + 6(-\frac{1}{2}) + 7$
$\quad = -1 - 1 - 3 + 7 = +2$
The remainder is $+2$.

Factor theorem

As a direct deduction from the remainder theorem, we can say that for a polynomial $f(x)$, if $f(a) = 0$, then the remainder when $f(x)$ is divided by $(x - a)$ is zero, i.e. $(x - a)$ is a factor of $f(x)$.

This result is known as the factor theorem and can be used to factorise a polynomial.

Example 24

Factorise the expression $x^3 - 5x^2 + 2x + 8$ and hence calculate the ranges of values of x for which $2x + 8 > 5x^2 - x^3$.

Let $\quad f(x) = x^3 - 5x^2 + 2x + 8$
then $\quad f(1) = 1 - 5 + 2 + 8$
$\qquad\qquad = 6 \qquad\qquad\qquad \therefore \quad (x - 1)$ is not a factor
$\quad\quad f(-1) = -1 - 5 - 2 + 8$
$\qquad\qquad = 0 \qquad\qquad\qquad \therefore \quad (x + 1)$ is a factor

Other factors can either be found by further use of the factor theorem or by long division.

factor theorem:

$f(2) = 8 - 20 + 4 + 8$

$\quad = 0 \quad \therefore \quad (x - 2)$ is a factor.

Noticing that if

$f(x) = x^3 - 5x^2 + 2x + 8$

$\quad = (x + 1)(x - 2)(x - p),$

then p must equal 4 so that

$(+1)(-2)(-p)$ gives

the constant term, 8.

long division:

$$
\begin{array}{r}
x^2 - 6x + 8 \\
x + 1{\overline{\smash{\big)}\,x^3 - 5x^2 + 2x + 8}} \\
\underline{x^3 + x^2} \\
-6x^2 + 2x \\
\underline{-6x^2 - 6x} \\
+8x + 8 \\
\underline{+8x + 8}
\end{array}
$$

and $x^2 - 6x + 8 = (x - 2)(x - 4)$

Thus $\quad x^3 - 5x^2 + 2x + 8 = (x + 1)(x - 2)(x - 4)$

For $\qquad\qquad 2x + 8 > 5x^2 - x^3$ we can consider

$\quad x^3 - 5x^2 + 2x + 8 > 0$

or $\quad (x + 1)(x - 2)(x - 4) > 0$

As before, consider the signs of these three factors for ranges of values of x:

x	$x < -1$	$-1 < x < 2$	$2 < x < 4$	$x > 4$
$x + 1$	$-$ve	$+$ve	$+$ve	$+$ve
$x - 2$	$-$ve	$-$ve	$+$ve	$+$ve
$x - 4$	$-$ve	$-$ve	$-$ve	$+$ve
$(x + 1)(x - 2)(x - 4)$	$-$ve	$+$ve	$-$ve	$+$ve

Thus $\quad 2x + 8 > 5x^2 - x^3 \quad$ if $\quad -1 < x < 2 \quad$ or if $\quad x > 4$.

Notes   Consider the cubic $f(x) = ax^3 + bx^2 + cx + d$. For $a = 1$, when searching for values of x for which $f(x) = 0$, we can restrict our attention to those numbers that are factors of d. Thus, in the last example we know immediately that $f(3) \neq 0$ because $(x - p)(x - q)(x - 3)$ will not give 8 as the constant term.

For $a \neq 1$, the process is not as easy and so, having found one factor, it is advisable to use long division to find any others.

Relationships between roots and coefficients

We have already seen on page 149, that if α and β are the roots of the quadratic equation $ax^2 + bx + c = 0$, then $\alpha + \beta = \dfrac{-b}{a}$ and $\alpha\beta = \dfrac{c}{a}$.

Similar relationships exist between the roots and coefficients of polynomials of order higher than two.

Suppose that α, β and γ are the roots of a cubic equation of the type $ax^3 + bx^2 + cx + d = 0$. By the factor theorem, since $x = \alpha$ is a root of the cubic equation $(x - \alpha)$ must be a factor and, in a similar way, so must $(x - \beta)$ and $(x - \gamma)$.

Thus $\quad (x - \alpha)(x - \beta)(x - \gamma) = 0$.

By expanding these brackets and comparing the result with
$ax^3 + bx^2 + cd + d = 0$, the following results are obtained.

$$\alpha + \beta + \gamma = \frac{-b}{a}, \quad \beta\gamma + \alpha\gamma + \alpha\beta = \frac{c}{a} \text{ and } \alpha\beta\gamma = \frac{-d}{a}$$

The details of the proof are left as an exercise for the reader: see question 15 of Exercise 5G.

Exercise 5G

1. Find the remainder when
   (a) $6x^3 + 7x^2 - 15x + 4$ is divided by $x - 1$,
   (b) $2x^3 - 3x^2 + 5x + 3$ is divided by $x + 1$,
   (c) $x^3 - 7x^2 + 6x + 1$ is divided by $x - 3$,
   (d) $5 + 6x + 7x^2 - x^3$ is divided by $x + 2$,
   (e) $x^4 - 3x^3 + 2x^2 + 5$ is divided by $x - 1$,
   (f) $8x^3 - 10x^2 + 7x + 3$ is divided by $2x - 1$,
   (g) $9x^3 + 4$ is divided by $3x + 2$.

2. Factorise the following expressions
   (a) $x^3 - 2x^2 - 5x + 6$ (b) $2x^3 + 7x^2 - 7x - 12$
   (c) $2x^3 + 3x^2 - 17x + 12$ (d) $6x^3 - 5x^2 - 17x + 6$
   (e) $2x^4 + 7x^3 - 17x^2 - 7x + 15$ (f) $6x^4 + 31x^3 + 57x^2 + 44x + 12$

3. Factorise the expression $x^3 - 3x^2 - 4x + 12$. Hence calculate the ranges of values of x for which $x^3 - 3x^2 > 4x - 12$.

4. Factorise the expression $6x^3 - 7x^2 - x + 2$. Hence calculate the ranges of values of x for which $2 - x < 7x^2 - 6x^3$.

5. The remainder obtained when $2x^3 + ax^2 - 6x + 1$ is divided by $(x + 2)$ is twice the remainder obtained when the same expression is divided by $(x - 1)$. Find the value of a.

6. The remainder obtained when $3x^3 - 6x^2 + ax - 1$ is divided by $(x + 1)$ is equal to the remainder obtained when the same expression is divided by $(x - 3)$. Find the value of a.

7. Given that $(x + 2)$ is a factor of $2x^3 + 6x^2 + bx - 5$ find the remainder when the expression is divided by $(2x - 1)$.

8. The expression $3x^3 + 2x^2 - bx + a$ is divisible by $(x - 1)$ but leaves a remainder of 10 when divided by $(x + 1)$. Find the values of a and b.

9. The expression $b + ax - 4x^2 + 8x^3$ gives a remainder of -19 when divided by $(x + 1)$ and a remainder of 2 when divided by $(2x - 1)$. Find the values of a and b.

10. The expression $6x^3 - 23x^2 + ax + b$ gives a remainder of 11 when divided by $(x - 3)$ and a remainder of -21 when divided by $(x + 1)$. Find the values of a and b and hence factorise the expression.

11. If $x^2 + bx + c$ and $x^2 + dx + e$ have a common factor of $(x - p)$ show that $p = \dfrac{e - c}{b - d}$.

12. Find the coordinates of the points where the curve $y = x^3 + 6x^2 + 11x + 6$ cuts
    (a) the y-axis,   (b) the x-axis.
    Hence make a sketch of the curve and state the ranges of values x can take for $x^3 + 6x^2 + 11x + 6 \leqslant 0$.

13. Find the coordinates of the points where the curve $y = x^3 - 5x^2 + 2x + 8$ cuts
    (a) the y-axis,   (b) the x-axis.
    Hence make a sketch of the curve and state the ranges of values x can
    take for $x^3 - 5x^2 + 2x + 8 > 0$.

14. Factorise the cubic expression $2x^3 - 5x^2 - 4x + 12$.
    Give the coordinates of the point where $y = 2x^3 - 5x^2 - 4x + 12$
    (a) cuts the y-axis,   (b) cuts the x-axis,   (c) touches the x-axis.
    Hence sketch the curve $y = 2x^3 - 5x^2 - 4x + 12$ and state the range
    of values x can take for $2x^3 - 5x^2 - 4x + 12 < 0$.

15. If α, β and γ are the three roots of the cubic $ax^3 + bx^2 + cx + d = 0$
    prove that
    $$\alpha + \beta + \gamma = \frac{-b}{a}, \quad \alpha\beta + \beta\gamma + \alpha\gamma = \frac{c}{a}, \quad \alpha\beta\gamma = \frac{-d}{a}.$$

16. If α, β and γ are the three roots of the cubic $x^3 - 2x^2 - 4x + 5 = 0$
    use the results of question 15 to find the values of:
    (a) $\alpha + \beta + \gamma$ (b) $\alpha\beta\gamma$
    (c) $\alpha^2\beta\gamma + \alpha\beta^2\gamma + \alpha\beta\gamma^2$ (d) $\dfrac{1}{\alpha} + \dfrac{1}{\beta} + \dfrac{1}{\gamma}$
    (e) $(\alpha + 1)(\beta + 1)(\gamma + 1)$ (f) $\alpha^2 + \beta^2 + \gamma^2$

17. If α, β and γ are the three roots of the cubic $2x^3 + 5x^2 - 4x - 10 = 0$
    use the results of question 15 to find the values of
    (a) $\dfrac{1}{\alpha\beta\gamma}$ (b) $\dfrac{1}{\alpha} + \dfrac{1}{\beta} + \dfrac{1}{\gamma}$ (c) $\dfrac{1}{\alpha\beta} + \dfrac{1}{\beta\gamma} + \dfrac{1}{\alpha\gamma}$
    Hence find the cubic equation having roots of $\dfrac{1}{\alpha}, \dfrac{1}{\beta}$ and $\dfrac{1}{\gamma}$.

Exercise 5H   *Examination questions*

1. If $5^x \cdot 25^{2y} = 1$ and $3^{5x} \cdot 9^y = \frac{1}{3}$ calculate the value of x and of y.
   (Cambridge)

2. A variable R is directly proportional to the square of P and inversely
   proportional to L, where P and L are two other variables. Given that
   $R = 2$ when $P = 3$ and $L = 18$, determine R when $P = 5$ and
   $L = 50$.
   (S.U.J.B.)

3. A variable, r, is inversely proportional to the square of another
   variable, t. Given that $r = \frac{3}{4}$ when $t = 2$, find the positive value of
   t when $r = \frac{1}{3}$.
   (S.U.J.B.)

4. Find the values of p, q, r if
   (i) $\log_3 81 = p$,   (ii) $\log_q 64 = 2$,   (iii) $\log_2 r = -4$.
   (S.U.J.B.)

5. Use the substitution $y = 2^x$ to solve for x the equation
   $$2^{2x+1} - 2^{x+1} + 1 = 2^x.$$
   (A.E.B.)

6. Find the value of x for which
   $$2^{3x+1} = 3^{x+2},$$
   giving three significant figures in your answer.
   (Cambridge)

7. Solve the equation
   $$e^{\ln x} + \ln e^x = 8.$$
   (London)

8. (i) Simplify:

    (a) $\log x^2 - \log xy$,   (b) $\log z + \log \dfrac{1}{z}$,

    (c) $\dfrac{\log x^3 - \log x}{\log x^2 - \log x}$,   (d) $\log (\log y^2) - \log (\log y)$.

    (ii) Solve the equation $(4 \cdot 381)^{2x-5} = (8 \cdot 032)^x$.   (Oxford)

9. The variables x and y satisfy a law of the form $x^n y = k$, where n and k are constants. Measured values of x and y are given in the table below.

x	1·2	1·6	2·0	2·6	3·5	4·5
y	11·0	6·31	3·98	2·36	1·32	0·79

By plotting a suitable straight-line graph, estimate the values of n and k. Use your graph to estimate the value of y when $x = 3$.   (A.E.B.)

10. The charge q coulombs on a capacitor is given by the formula
$$q = 3 \cdot 16 \times 2 \cdot 72^{-4 \cdot 32t},$$
where t is the time in seconds.
Calculate
(i) q when $t = 0$;   (ii) q when $t = 0 \cdot 2$;   (iii) t when $q = 1 \cdot 58$.
At time $t = T$, q takes the value Q.

Prove that, when q takes the value $\tfrac{1}{2}Q$, $t = T + \dfrac{\log 2}{4 \cdot 32 \times \log 2 \cdot 72}$.

    (Oxford)

11. Given that $\log_2 x + 2 \log_4 y = 4$, show that $xy = 16$. Hence solve for x and y the simultaneous equations
$$\log_{10}(x + y) = 1,$$
$$\log_2 x + 2 \log_4 y = 4.$$
    (A.E.B.)

12. Establish the formula $\log_y x = \dfrac{1}{\log_x y}$.

Solve the simultaneous equations
$\log_x y + 2 \log_y x = 3$,
$\log_9 y + \log_9 x = 3$.
    (J.M.B.)

13. (a) Find the range of values of a for which
$$(2 - 3a)x^2 + (4 - a)x + 2 = 0$$
has no real roots.
(b) Find the values of k for which the line $y = x + k$ is a tangent to the curve $x^2 + xy + 2 = 0$.   (Cambridge)

14. If α^2 and β^2 are the roots of $x^2 - 21x + 4 = 0$ and α and β are both positive, find:

(i) $\alpha\beta$;   (ii) $\alpha + \beta$;   (iii) the equation with roots $\dfrac{1}{\alpha^2}$ and $\dfrac{1}{\beta^2}$.   (S.U.J.B)

15. Given that α and β are the roots of the equation $x^2 - bx + c = 0$,
(a) show that $(\alpha^2 + 1)(\beta^2 + 1) = (c - 1)^2 + b^2$,
(b) find, in terms of b and c, a quadratic equation whose roots are

    $\dfrac{\alpha}{\alpha^2 + 1}$ and $\dfrac{\beta}{\beta^2 + 1}$.   (A.E.B.)

16. Find the real values of k for which the equation $x^2 + (k + 1)x + k^2 = 0$ has (a) real roots, (b) one root double the other.   (A.E.B.)

17. The roots of the quadratic equation
$$(x + 2)^2 - 2kx = 0,$$
where k is a non-zero real constant, are denoted by α and β. Obtain a quadratic equation whose roots are α^2 and β^2. Find a quadratic equation whose roots are $\sqrt{\gamma}$ and $\sqrt{\delta}$, where γ and δ are the roots of
$$x^2 - 28x + 16 = 0.$$ (London)

18. The cubic polynomial $6x^3 + 7x^2 + ax + b$ has a remainder of 72 when divided by $(x - 2)$ and is exactly divisible by $(x + 1)$. Calculate a and b. Show that $(2x - 1)$ is also a factor of the polynomial and obtain the third factor.   (S.U.J.B.)

19. (a) The expression $6x^2 + x + 7$ leaves the same remainder when divided by $x - a$ and by $x + 2a$, where $a \neq 0$.
   Calculate the value of a.
   (b) Given that $x^2 + px + q$ and $3x^2 + q$ have a common factor $x - b$, where p, q, and b are non-zero, show that $3p^2 + 4q = 0$.
                                                                          (Cambridge)

20. Given that $f(x) \equiv 3 - 7x + 5x^2 - x^3$, show that $3 - x$ is a factor of $f(x)$. Factorize $f(x)$ completely and hence state the set of values of x for which $f(x) \leqslant 0$.   (London)

21. Express the polynomial
$$f(x) = 2x^4 + x^3 - x^2 + 8x - 4$$
as the product of two linear factors and a quadratic factor $q(x)$.
Prove that there are no real values of x for which $q(x) = 0$.   (A.E.B.)

6
Matrices

6.1 Basic Concepts

Information can often be conveniently presented as an array of rows and columns.

Bus timetables, football league results and conversion tables often use this form of presentation. Such an arrangement of information is called a matrix. The matrix on the right shows the number of televisions and videos sold in three shops in a particular month. The matrix has 2 rows and 3 columns and is said to be a 2×3 matrix, or a matrix of *order* 2×3. An $n \times m$ matrix would have n rows and m columns.

$$\begin{array}{c} \\ \text{T.V.} \\ \text{video} \end{array} \begin{array}{ccc} \text{shop 1} & \text{shop 2} & \text{shop 3} \\ \begin{pmatrix} 20 & 7 & 13 \\ 31 & 9 & 10 \end{pmatrix} \end{array}$$

It can be useful to combine several matrices. For example, if we had the matrix that showed the numbers of T.V.'s and video's the three shops had sold in the previous month, we could add the matrices to give the numbers sold in each shop over the two-month period. It is assumed that the reader is familiar with the rules for combining matrices but the following statements, examples and exercise should serve to refresh the memory.

If matrices **A**, **B** and **C** are defined as follows:

$$\mathbf{A} = \begin{pmatrix} a & b \\ c & d \end{pmatrix} \quad \mathbf{B} = \begin{pmatrix} e & f \\ g & h \end{pmatrix} \quad \text{and} \quad \mathbf{C} = \begin{pmatrix} i & j & k \\ l & m & n \end{pmatrix}$$

then
$$\left. \begin{array}{l} \mathbf{A} + \mathbf{B} = \begin{pmatrix} a+e & b+f \\ c+g & d+h \end{pmatrix} \\[6pt] \mathbf{A} - \mathbf{B} = \begin{pmatrix} a-e & b-f \\ c-g & d-h \end{pmatrix} \end{array} \right\}$$

Only matrices of the same order can be added or subtracted.
Thus $\mathbf{A} \pm \mathbf{C}$ and $\mathbf{B} \pm \mathbf{C}$ cannot be determined because **A** and **B** are 2×2 matrices and **C** is a 2×3 matrix.

$$s\mathbf{A} = \begin{pmatrix} sa & sb \\ sc & sd \end{pmatrix}$$

$$\left. \begin{array}{l} \mathbf{AB} = \begin{pmatrix} a & b \\ c & d \end{pmatrix}\begin{pmatrix} e & f \\ g & h \end{pmatrix} \\[12pt] \quad = \begin{pmatrix} ae+bg & af+bh \\ ce+dg & cf+dh \end{pmatrix} \\[12pt] \mathbf{AC} = \begin{pmatrix} a & b \\ c & d \end{pmatrix}\begin{pmatrix} i & j & k \\ l & m & n \end{pmatrix} \\[12pt] \quad = \begin{pmatrix} ai+bl & aj+bm & ak+bn \\ ci+dl & cj+dm & ck+dn \end{pmatrix} \end{array} \right\}$$

Matrices can be multiplied together if the number of columns in the first matrix equals the number of rows in the second matrix. Thus **AB**, **BA**, **AC** and **BC** can all be found but **CB** and **CA** cannot.
Matrices that can be multiplied together are said to be *conformable* for multiplication.

The *transpose* of a matrix **A** is written $\mathbf{A}^T$ and is found by interchanging the rows and columns.

Thus $\mathbf{A}^T = \begin{pmatrix} a & c \\ b & d \end{pmatrix}$; $\mathbf{B}^T = \begin{pmatrix} e & g \\ f & h \end{pmatrix}$ and $\mathbf{C}^T = \begin{pmatrix} i & l \\ j & m \\ k & n \end{pmatrix}$

2 × 2 matrices

The *determinant* of a 2 × 2 matrix is obtained as follows:

If $\mathbf{A} = \begin{pmatrix} a & b \\ c & d \end{pmatrix}$ then det $\mathbf{A}\left(\text{or } |\mathbf{A}| \text{ or } \begin{vmatrix} a & b \\ c & d \end{vmatrix}\right) = ad - bc$.

Two particular 2 × 2 matrices are $\mathbf{Z}$, the zero or null matrix $= \begin{pmatrix} 0 & 0 \\ 0 & 0 \end{pmatrix}$

and $\mathbf{I}$, the identity matrix $= \begin{pmatrix} 1 & 0 \\ 0 & 1 \end{pmatrix}$

Note that for any 2 × 2 matrix $\mathbf{A}$, $\mathbf{A} \pm \mathbf{Z} = \mathbf{A}$ and $\mathbf{IA} = \mathbf{AI} = \mathbf{A}$.
The *inverse* of a matrix $\mathbf{A}$ is written $\mathbf{A}^{-1}$ and is such that
$\mathbf{AA}^{-1} = \mathbf{A}^{-1}\mathbf{A} = \mathbf{I}$, the identity matrix.

If $\mathbf{A} = \begin{pmatrix} a & b \\ c & d \end{pmatrix}$ then $\mathbf{A}^{-1} = \dfrac{1}{ad - bc}\begin{pmatrix} d & -b \\ -c & a \end{pmatrix}$ or $\begin{pmatrix} d/\det \mathbf{A} & -b/\det \mathbf{A} \\ -c/\det \mathbf{A} & a/\det \mathbf{A} \end{pmatrix}$

i.e. to determine the inverse of a matrix $\mathbf{A}$, interchange the elements of the leading diagonal of $\mathbf{A}$, change the signs of the elements of the other diagonal and divide by the determinant of the matrix.

Note   If the determinant of a matrix is zero, the inverse cannot be determined because $1/0$ is meaningless. Any 2 × 2 matrix with determinant zero is called a *singular* matrix as it has no inverse.

Example 1

If $\mathbf{A} = \begin{pmatrix} 3 \\ -1 \end{pmatrix}$, $\mathbf{B} = \begin{pmatrix} 1 & -4 \\ 0 & 3 \end{pmatrix}$ and $\mathbf{C} = \begin{pmatrix} 13 & 2 \\ -9 & -3 \end{pmatrix}$ find, where they exist,

(a) $\mathbf{A} + \mathbf{B}$, (b) $\mathbf{BC}$, (c) $\mathbf{BA}$, (d) $\mathbf{AB}$, (e) $\mathbf{CB}$, (f) $\mathbf{B}^{\mathrm{T}}$, (g) det $\mathbf{B}$,
(h) $\mathbf{B}^{-1}$, (i) the 2 × 2 matrix $\mathbf{D}$ such that $\mathbf{BD} = \mathbf{C}$.

(a) $\mathbf{A}$ and $\mathbf{B}$ are not of the same order and so $\mathbf{A} + \mathbf{B}$ cannot be determined.

(b) $\mathbf{BC} = \begin{pmatrix} 1 & -4 \\ 0 & 3 \end{pmatrix}\begin{pmatrix} 13 & 2 \\ -9 & -3 \end{pmatrix} = \begin{pmatrix} 13 + 36 & 2 + 12 \\ 0 - 27 & 0 - 9 \end{pmatrix}$

$$= \begin{pmatrix} 49 & 14 \\ -27 & -9 \end{pmatrix}$$

(c) $\mathbf{BA} = \begin{pmatrix} 1 & -4 \\ 0 & 3 \end{pmatrix}\begin{pmatrix} 3 \\ -1 \end{pmatrix} = \begin{pmatrix} 3 + 4 \\ 0 - 3 \end{pmatrix}$

$$= \begin{pmatrix} 7 \\ -3 \end{pmatrix}$$

(d) $\mathbf{AB} = \begin{pmatrix} 3 \\ -1 \end{pmatrix}\begin{pmatrix} 1 & -4 \\ 0 & 3 \end{pmatrix}$ which cannot be determined.

(e) $\mathbf{CB} = \begin{pmatrix} 13 & 2 \\ -9 & -3 \end{pmatrix}\begin{pmatrix} 1 & -4 \\ 0 & 3 \end{pmatrix} = \begin{pmatrix} 13 + 0 & -52 + 6 \\ -9 + 0 & 36 - 9 \end{pmatrix}$

$$= \begin{pmatrix} 13 & -46 \\ -9 & 27 \end{pmatrix}$$

(f) $\mathbf{B^T} = \begin{pmatrix} 1 & 0 \\ -4 & 3 \end{pmatrix}$

(g) $\det \mathbf{B} = (1 \times 3) - (0 \times -4)$
$= 3$

(h) $\mathbf{B^{-1}} = \frac{1}{3}\begin{pmatrix} 3 & 4 \\ 0 & 1 \end{pmatrix}$ or $\begin{pmatrix} 1 & 1\frac{1}{3} \\ 0 & \frac{1}{3} \end{pmatrix}$

(i) We require $\mathbf{D}$ such that $\begin{pmatrix} 1 & -4 \\ 0 & 3 \end{pmatrix}(\mathbf{D}) = \begin{pmatrix} 13 & 2 \\ -9 & -3 \end{pmatrix}$

Premultiplying both sides by $\mathbf{B^{-1}}$ and remembering that $\mathbf{B^{-1}B = I}$:

$$\begin{pmatrix} 1 & 0 \\ 0 & 1 \end{pmatrix}(\mathbf{D}) = \frac{1}{3}\begin{pmatrix} 3 & 4 \\ 0 & 1 \end{pmatrix}\begin{pmatrix} 13 & 2 \\ -9 & -3 \end{pmatrix}$$

$$\therefore \quad (\mathbf{D}) = \frac{1}{3}\begin{pmatrix} 3 & -6 \\ -9 & -3 \end{pmatrix}$$

Thus the required matrix $\mathbf{D}$ is $\begin{pmatrix} 1 & -2 \\ -3 & -1 \end{pmatrix}$

Example 2

Solve the simultaneous equations $\quad 4x - y = 1$
$-2x + 3y = 12 \quad$ by *matrix* methods.

First we must express the equations in matrix form:
$$\begin{pmatrix} 4 & -1 \\ -2 & 3 \end{pmatrix}\begin{pmatrix} x \\ y \end{pmatrix} = \begin{pmatrix} 1 \\ 12 \end{pmatrix}$$

Premultiplying both sides by the inverse of $\begin{pmatrix} 4 & -1 \\ -2 & 3 \end{pmatrix}$:

$$\begin{pmatrix} 1 & 0 \\ 0 & 1 \end{pmatrix}\begin{pmatrix} x \\ y \end{pmatrix} = \frac{1}{10}\begin{pmatrix} 3 & 1 \\ 2 & 4 \end{pmatrix}\begin{pmatrix} 1 \\ 12 \end{pmatrix}$$

$$\therefore \quad \begin{pmatrix} x \\ y \end{pmatrix} = \frac{1}{10}\begin{pmatrix} 15 \\ 50 \end{pmatrix}$$

Thus $x = 1\cdot5$ and $y = 5$

Exercise 6A

1. If $\mathbf{A} = \begin{pmatrix} -1 & -3 \\ 1 & 1 \end{pmatrix}$, $\mathbf{B} = \begin{pmatrix} 6 & 3 \\ 2 & 1 \end{pmatrix}$, $\mathbf{C} = \begin{pmatrix} 1 & 3 \\ 2 & 0 \\ -3 & 2 \end{pmatrix}$ and

$\mathbf{D} = \begin{pmatrix} 1 & 2 & 1 \\ 0 & 1 & -1 \\ -1 & 0 & 1 \end{pmatrix}$ find, where they exist,

(a) $\mathbf{A + B}$ (b) $\mathbf{A - B}$ (c) $2\mathbf{A}$ (d) $\mathbf{C + D}$
(e) $2\mathbf{B - A}$ (f) $\mathbf{AB}$ (g) $\mathbf{BA}$ (h) $\mathbf{A^2}$
(i) $\mathbf{BC}$ (j) $\mathbf{CB}$ (k) $\mathbf{CD}$ (l) $\mathbf{DC}$
(m) $\mathbf{A^{-1}}$ (n) $\mathbf{B^{-1}}$ (o) $\mathbf{A^T}$ (p) $\mathbf{C^T}$

2. Determine all the possible products that can be formed using any two of the following each time

$$A = \begin{pmatrix} 1 & 2 \\ 3 & -1 \end{pmatrix} \quad B = \begin{pmatrix} 2 \\ 1 \end{pmatrix} \quad C = (-1 \quad 3) \quad D = \begin{pmatrix} 2 & 0 & -1 \\ 1 & 2 & 0 \end{pmatrix}$$

3. Five football teams took part in a league competition in which they each played each other once in the first half of the season and again in the second half of the season. The results matrix for the first half of the season is shown below:

	Won	Drawn	Lost
Aces	2	1	1
Bruisers	2	1	1
City	0	4	0
Defenders	1	1	2
Exiles	1	1	2

The results of the matches played in the second half of the season were as follows:

Bruisers 3 : Aces 4   Exiles 2   : City 2
City 1   : Bruisers 0   Defenders 1 : Aces 0
Exiles 1   : Defenders 0   Exiles 3   : Bruisers 2
Defenders 2 : City 2   City 3   : Aces 0
Exiles 1   : Aces 5   Defenders 2 : Bruisers 3

Construct a results matrix for the whole season.

If points are awarded: 3 for a win, 1 for a draw and 0 for a loss, use the

points matrix $\begin{pmatrix} 3 \\ 1 \\ 0 \end{pmatrix}$ to determine the league positions at the end of the season.

What would the league positions be if points were awarded: 2 for a win, 1 for a draw, 0 for a loss?

4. Find the values of x and y in each of the following matrix equations.

(a) $\begin{pmatrix} 3 & -5 \\ 2 & x \end{pmatrix} + \begin{pmatrix} 1 & y \\ 3 & 2 \end{pmatrix} = \begin{pmatrix} 4 & 6 \\ 5 & -2 \end{pmatrix}$ (b) $\begin{pmatrix} 2 \\ y \end{pmatrix} - \begin{pmatrix} x \\ 3 \end{pmatrix} = \begin{pmatrix} 4 \\ -6 \end{pmatrix}$

(c) $2\begin{pmatrix} 3 \\ y \end{pmatrix} + \begin{pmatrix} x \\ -1 \end{pmatrix} = \begin{pmatrix} 8 \\ 7 \end{pmatrix}$ (d) $\begin{pmatrix} 3 \\ -1 \end{pmatrix} + x\begin{pmatrix} -2 \\ y \end{pmatrix} = \begin{pmatrix} -5 \\ 11 \end{pmatrix}$

(e) $\begin{pmatrix} 3 & x \\ 5 & 4 \end{pmatrix}\begin{pmatrix} 1 & y \\ 2 & 3 \end{pmatrix} = \begin{pmatrix} 7 & 3 \\ 13 & 7 \end{pmatrix}$ (f) $\begin{pmatrix} 2 & 1 \\ x & 3 \end{pmatrix}\begin{pmatrix} 3 & 2 \\ -4 & 3 \end{pmatrix} = \begin{pmatrix} 2 & 7 \\ 3 & y \end{pmatrix}$

(g) $\begin{pmatrix} 2 & x \\ 3 & -1 \end{pmatrix}\begin{pmatrix} 2 & 1 \\ y & -3 \end{pmatrix} = \begin{pmatrix} 0 & 5 \\ 2 & 6 \end{pmatrix}$ (h) $\begin{pmatrix} 3 & 2 \\ 4 & 3 \end{pmatrix}\begin{pmatrix} x \\ y \end{pmatrix} = \begin{pmatrix} 4 \\ 5 \end{pmatrix}$

(i) $\begin{pmatrix} -1 & 3 \\ -2 & 2 \end{pmatrix}\begin{pmatrix} x \\ y \end{pmatrix} = \begin{pmatrix} 9 \\ 10 \end{pmatrix}$ (j) $\begin{pmatrix} 2 & 4 \\ -1 & 2 \end{pmatrix}\begin{pmatrix} x \\ y \end{pmatrix} = \begin{pmatrix} 0 \\ 6 \end{pmatrix}$

5. By letting $A = \begin{pmatrix} a & b \\ c & d \end{pmatrix}$ and $B = \begin{pmatrix} w & x \\ y & z \end{pmatrix}$ prove that $(AB)^T = B^T A^T$ and that $(AB)^{-1} = B^{-1}A^{-1}$.

6. If $A = \begin{pmatrix} 3 & 1 \\ 5 & 2 \end{pmatrix}$, $B = \begin{pmatrix} 6 & 1 \\ 11 & 3 \end{pmatrix}$, $C = \begin{pmatrix} -4 & 3 \\ -5 & 2 \end{pmatrix}$ and $D = \begin{pmatrix} 4 & 7 \\ -2 & 7 \end{pmatrix}$ find the 2×2 matrices X, Y and Z given that $AX = B$, $BY = C$ and $CZ = D$.

7. If $A = \begin{pmatrix} 3 & 2 \\ -4 & 1 \end{pmatrix}$ find the values of m and n given that $A^2 = mA + nI$ where I is the identity matrix $\begin{pmatrix} 1 & 0 \\ 0 & 1 \end{pmatrix}$.

8. Find the possible values x can take given that $A = \begin{pmatrix} x^2 & 3 \\ 1 & 3x \end{pmatrix}$, $B = \begin{pmatrix} 3 & 6 \\ 2 & x \end{pmatrix}$ and $AB = BA$.

9. Solve the following simultaneous equations by matrix methods.
   (a) $x - y = 5$
        $3x + 2y = 5$
   (b) $x - 3y = 3$
        $5x - 9y = 11$
   (c) $x + 3y = 1$
        $2x - 4y = 1$

10. If $A = \begin{pmatrix} -3 & 2 & -1 \\ 2 & -1 & 3 \\ -1 & 1 & 1 \end{pmatrix}$ and $B = \begin{pmatrix} -4 & -3 & 5 \\ -5 & -4 & 7 \\ 1 & 1 & -1 \end{pmatrix}$ find AB.

Hence find the values of x, y and z satisfying the three equations
$$-4x - 3y + 5z = 3$$
$$-5x - 4y + 7z = 4$$
$$x + y - z = 0.$$

6.2 Transformations

A transformation of a plane takes any point A in the plane and maps it onto one and only one image A', also lying in the plane.

We say that the point $A(x, y)$, position vector $\begin{pmatrix} x \\ y \end{pmatrix}$, has an *image* $A'(x', y')$, position vector $\begin{pmatrix} x' \\ y' \end{pmatrix}$, under the transformation.

The nature of certain transformations can best be seen by considering the effect the transformation has on the unit square OABC. Examples of some of the common types of transformation are shown below.

(a) Translations

e.g. OABC → O′A′B′C′ by the translation

vector $\begin{pmatrix} 2 \\ 1 \end{pmatrix}$

e.g. DEF → D′E′F′ by the translation vector

$\begin{pmatrix} 3 \\ -1 \end{pmatrix}$

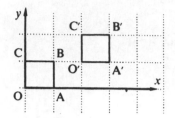

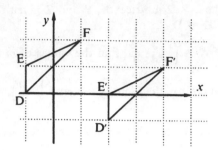

(b) Reflections

e.g. Reflection in the *x*-axis
      OABC → O′A′B′C′

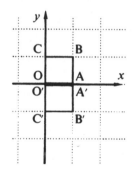

(c) Rotations

e.g. Rotation of 90° anticlockwise about the origin (or 270° clockwise about the origin)
      OABC → O′A′B′C′

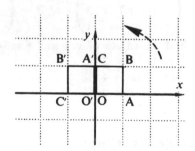

(d) Enlargements (or dilatations)

e.g. Enlargement, scale factor 2, centre (0, 0)
      OABC → O′A′B′C′

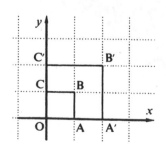

e.g. Enlargement, scale factor 2, centre (0, 0)
      ABC → A′B′C′

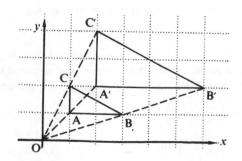

Note   A transformation that reduces the size of a shape, e.g. all sides halved, would be an enlargement, scale factor $\frac{1}{2}$.

(e) Stretch in a particular direction

e.g. Stretch parallel to the *x*-axis, × 3
OABC → O′A′B′C′

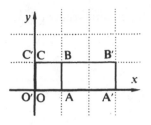

(f) Shear with some line fixed

e.g. Shear with *x*-axis fixed and (0, 1) → (3, 1)
ABCD → A′B′C′D′

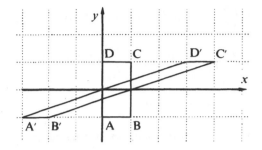

A transformation is said to be linear if any linear relationship between position vectors is conserved under the transformation, i.e. if a linear transformation maps a point A position vector **a** onto its image A′ position vector **a′**, and if **a** = λ**p** + μ**q**, it follows that **a′** = λ**p′** + μ**q′**.

For example, suppose that A and B are two points with position vectors **a** and **b** respectively and Q is a point with position vector **q** where **q** = $\frac{3}{2}$**a** + 2**b**. If under some linear transformation T the images A′, B′ and Q′ have position vectors **a′**, **b′** and **q′**, then **q′** = $\frac{3}{2}$**a′** + 2**b′**.

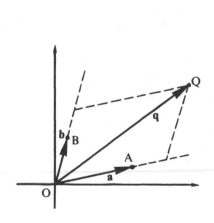

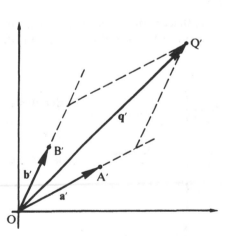

All linear transformations of the plane can be expressed as a pair of equations of the form
$$\begin{cases} x' = ax + by \\ y' = cx + dy \end{cases}$$ or, writing this in matrix form $\begin{pmatrix} x' \\ y' \end{pmatrix} = \begin{pmatrix} a & b \\ c & d \end{pmatrix}\begin{pmatrix} x \\ y \end{pmatrix}$

> Thus, every linear transformation of the plane has an associated 2 × 2 matrix.

Proof: Consider a linear transformation that maps the point with position vector $\begin{pmatrix} 1 \\ 0 \end{pmatrix}$ onto the point with position vector $\begin{pmatrix} a \\ c \end{pmatrix}$, maps the point with position vector $\begin{pmatrix} 0 \\ 1 \end{pmatrix}$ onto the point with position vector $\begin{pmatrix} b \\ d \end{pmatrix}$ and maps the point A with position vector $\begin{pmatrix} x \\ y \end{pmatrix}$ onto A' with position vector $\begin{pmatrix} x' \\ y' \end{pmatrix}$.

Writing $\begin{pmatrix} x \\ y \end{pmatrix}$ as $x\begin{pmatrix} 1 \\ 0 \end{pmatrix} + y\begin{pmatrix} 0 \\ 1 \end{pmatrix}$ it follows that $\begin{pmatrix} x' \\ y' \end{pmatrix} = x\begin{pmatrix} a \\ c \end{pmatrix} + y\begin{pmatrix} b \\ d \end{pmatrix}$

i.e. $\begin{cases} x' = xa + yb \\ y' = xc + yd \end{cases}$ or, in matrix form $\begin{pmatrix} x' \\ y' \end{pmatrix} = \begin{pmatrix} a & b \\ c & d \end{pmatrix}\begin{pmatrix} x \\ y \end{pmatrix}$ as required.

The following points should also be noted:

---

1. The associated 2×2 matrix can be determined by examining the images of $\begin{pmatrix} 1 \\ 0 \end{pmatrix}$ and $\begin{pmatrix} 0 \\ 1 \end{pmatrix}$.

   If $\begin{pmatrix} 1 \\ 0 \end{pmatrix} \to \begin{pmatrix} a \\ c \end{pmatrix}$ and $\begin{pmatrix} 0 \\ 1 \end{pmatrix} \to \begin{pmatrix} b \\ d \end{pmatrix}$, then the transformation matrix is $\begin{pmatrix} a & b \\ c & d \end{pmatrix}$.

---

2. Under any linear transformation, the origin (0, 0) maps onto itself.

3. Under a translation by the vector $\begin{pmatrix} r \\ s \end{pmatrix}$, the transformation equations are $\begin{cases} x' = x + r \\ y' = y + s \end{cases}$.

   Thus a translation is *not* a linear transformation.

Example 3

Find the matrices corresponding to each of the following linear transformations,
(a) a rotation of 90° anticlockwise about the origin,   (b) a reflection in the x-axis.

(a)

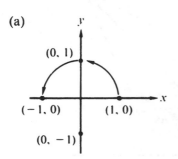

Under this transformation $\begin{pmatrix} 1 \\ 0 \end{pmatrix} \to \begin{pmatrix} 0 \\ 1 \end{pmatrix}$ and $\begin{pmatrix} 0 \\ 1 \end{pmatrix} \to \begin{pmatrix} -1 \\ 0 \end{pmatrix}$

Thus the required matrix is $\begin{pmatrix} 0 & -1 \\ 1 & 0 \end{pmatrix}$

(b)

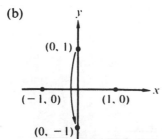

Under this transformation $\begin{pmatrix} 1 \\ 0 \end{pmatrix} \to \begin{pmatrix} 1 \\ 0 \end{pmatrix}$ and $\begin{pmatrix} 0 \\ 1 \end{pmatrix} \to \begin{pmatrix} 0 \\ -1 \end{pmatrix}$

Thus the required matrix is $\begin{pmatrix} 1 & 0 \\ 0 & -1 \end{pmatrix}$

Example 4

A linear transformation **T** has matrix $\begin{pmatrix} 2 & -1 \\ 1 & 1 \end{pmatrix}$. Find (a) the image of the point (2, 3) under **T**, (b) the coordinates of the point having an image of (7, 2) under **T**.

If (x', y') is the image of (x, y) then $\quad \begin{pmatrix} x' \\ y' \end{pmatrix} = \begin{pmatrix} 2 & -1 \\ 1 & 1 \end{pmatrix}\begin{pmatrix} x \\ y \end{pmatrix}$

(a) In this case $(x, y) = (2, 3) \quad \therefore \quad \begin{pmatrix} x' \\ y' \end{pmatrix} = \begin{pmatrix} 2 & -1 \\ 1 & 1 \end{pmatrix}\begin{pmatrix} 2 \\ 3 \end{pmatrix}$

$$= \begin{pmatrix} 1 \\ 5 \end{pmatrix}$$

Thus (1, 5) is the image of the point (2, 3) under **T**.

(b) In this case $(x', y') = (7, 2) \quad \therefore \quad \begin{pmatrix} 7 \\ 2 \end{pmatrix} = \begin{pmatrix} 2 & -1 \\ 1 & 1 \end{pmatrix}\begin{pmatrix} x \\ y \end{pmatrix}$

Premultiplying both sides of this equation by the inverse of $\begin{pmatrix} 2 & -1 \\ 1 & 1 \end{pmatrix}$:

$$\tfrac{1}{3}\begin{pmatrix} 1 & 1 \\ -1 & 2 \end{pmatrix}\begin{pmatrix} 7 \\ 2 \end{pmatrix} = \begin{pmatrix} 1 & 0 \\ 0 & 1 \end{pmatrix}\begin{pmatrix} x \\ y \end{pmatrix}$$

$$\tfrac{1}{3}\begin{pmatrix} 9 \\ -3 \end{pmatrix} = \begin{pmatrix} x \\ y \end{pmatrix}$$

$$\therefore \quad x = 3 \quad \text{and} \quad y = -1$$

Thus (3, −1) is the point having an image of (7, 2) under **T**.

Example 5

Find the 2 × 2 matrix that will transform the point (1, 2) to (3, 3) and the point (−1, 1) to (−3, 3).

Let the matrix be $\begin{pmatrix} a & b \\ c & d \end{pmatrix}$

Thus $\begin{pmatrix} a & b \\ c & d \end{pmatrix}\begin{pmatrix} 1 \\ 2 \end{pmatrix} = \begin{pmatrix} 3 \\ 3 \end{pmatrix}$ and $\begin{pmatrix} a & b \\ c & d \end{pmatrix}\begin{pmatrix} -1 \\ 1 \end{pmatrix} = \begin{pmatrix} -3 \\ 3 \end{pmatrix}$

$\therefore$

$a + 2b = 3$	... [1]	$-a + b = -3$ ... [3]
$c + 2d = 3$	... [2]	$-c + d = 3$ ... [4]

Solving [1] and [3] simultaneously gives $a = 3, b = 0$.
Solving [2] and [4] simultaneously gives $c = -1, d = 2$.

The required matrix is $\begin{pmatrix} 3 & 0 \\ -1 & 2 \end{pmatrix}$

The following facts regarding transformation matrices should also be noted:

1. If a matrix **T** transforms some shape ABC to its image A′B′C′, then the inverse matrix $\mathbf{T}^{-1}$ maps A′B′C′ back onto ABC.

2. If matrix **T** corresponds to a certain linear transformation, then det **T** gives the scale factor for any change of area under the transformation, i.e. if **T** transforms ABC → A'B'C' and det **T** $= t$, then area A'B'C' $= |t| \times$ area ABC.

Thus any matrix with a determinant equal to 1 leaves the area of a shape unchanged. What happens if the determinant is zero, i.e. a singular matrix?

Consider the transformation matrix $\begin{pmatrix} 1 & 2 \\ 2 & 4 \end{pmatrix}$ and the images of various points, say (1, 3) (2, 5) and (−3, 1).

$$\begin{pmatrix} 1 & 2 \\ 2 & 4 \end{pmatrix}\begin{pmatrix} 1 \\ 3 \end{pmatrix} = \begin{pmatrix} 7 \\ 14 \end{pmatrix} \qquad \begin{pmatrix} 1 & 2 \\ 2 & 4 \end{pmatrix}\begin{pmatrix} 2 \\ 5 \end{pmatrix} = \begin{pmatrix} 12 \\ 24 \end{pmatrix} \qquad \begin{pmatrix} 1 & 2 \\ 2 & 4 \end{pmatrix}\begin{pmatrix} -3 \\ 1 \end{pmatrix} = \begin{pmatrix} -1 \\ -2 \end{pmatrix}$$

Notice that all the images lie on the line $y = 2x$.

To show that all points are mapped onto this line, we consider the general point (x, y).

$$\begin{pmatrix} x' \\ y' \end{pmatrix} = \begin{pmatrix} 1 & 2 \\ 2 & 4 \end{pmatrix}\begin{pmatrix} x \\ y \end{pmatrix} = \begin{pmatrix} x + 2y \\ 2x + 4y \end{pmatrix}$$

$$\therefore \begin{pmatrix} x' \\ y' \end{pmatrix} = \begin{pmatrix} x + 2y \\ 2(x + 2y) \end{pmatrix} \quad \text{i.e.} \quad \text{all image points lie on the line } y = 2x.$$

Thus a singular matrix maps all points in the plane onto a line.

3. If matrix **P** transforms (x, y) to (x', y') and matrix **Q** transforms (x', y') to (x'', y''), then the single matrix equivalent to **P** followed by **Q** [i.e. the matrix that will transform $(x, y) \rightarrow (x'', y'')$ direct] is given by **QP**.

Proof: $\begin{pmatrix} x' \\ y' \end{pmatrix} = \mathbf{P}\begin{pmatrix} x \\ y \end{pmatrix}$ and $\begin{pmatrix} x'' \\ y'' \end{pmatrix} = \mathbf{Q}\begin{pmatrix} x' \\ y' \end{pmatrix}$

$$\therefore \begin{pmatrix} x'' \\ y'' \end{pmatrix} = \mathbf{QP}\begin{pmatrix} x \\ y \end{pmatrix} \quad \text{as required.}$$

Exercise 6B

1. The points A(3, 2), B(−1, 4), C(2, 5) and D(1, −1) are transformed to A', B', C', and D' by the transformation matrix $\begin{pmatrix} 3 & 0 \\ 1 & -3 \end{pmatrix}$. Find the coordinates of A', B', C', and D'.

2. Under a certain transformation the image (x', y') of a point (x, y) is obtained by the rule:

$$\begin{pmatrix} x' \\ y' \end{pmatrix} = \begin{pmatrix} 1 & 2 \\ -1 & 3 \end{pmatrix}\begin{pmatrix} x \\ y \end{pmatrix}$$

   Find (a) the image of the point (2, 1),   (b) the image of the point (−1, 3),
   (c) the point with an image of (1, −6),   (d) the point with an image of (−1, 11).

3. Find the 2 × 2 transformation matrix that will map the point (−2, 3) onto (−7, 6) and (1, −1) onto (3, −1).
   Find the image of the point (−1, 3) under this linear transformation and the coordinates of the point that has an image of (6, −2).

4. The square ABCD is transformed to $A_1B_1C_1D_1$ by the transformation

matrix $\begin{pmatrix} 5 & -1 \\ -3 & 1 \end{pmatrix}$. If A, B, C and D have coordinates (1, 0), (3, 0),

(3, 2) and (1, 2) respectively find (a) the coordinates of A_1, B_1, C_1, and D_1,
  (b) the area of $A_1B_1C_1D_1$,
  (c) the matrix that will transform $A_1B_1C_1D_1$ back to ABCD.

5. Find the 2 × 2 transformation matrices **P**, **Q**, **R** and **S** given that
   (a) matrix **P** represents a reflection in the *y*-axis,
   (b) matrix **Q** represents a 90° clockwise rotation about the origin,
   (c) matrix **R** represents a reflection in the line $y = x$,
   (d) matrix **S** transforms the point $(2, -3)$ to $(4, 14)$ and the point $(1, 3)$
       to $(11, -2)$.
   Use your answers to (a), (b) and (c) to show that a reflection in the
   y-axis followed by a 90° clockwise rotation about the origin is equivalent
   to a reflection in the line $y = x$.

6. The linear transformations shown below are shears transforming OABC
   to OA′B′C′. For each transformation, write down
   (i) the associated 2 × 2 matrix,
   (ii) the equations of the transformation in the form $x' = ax + by$
                                                          $y' = cx + dy.$

(a)

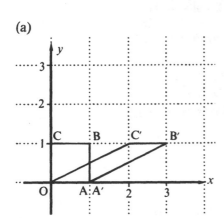

(b)

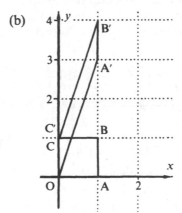

7. A 1st shape is transformed to a 2nd shape by the transformation matrix
   $\begin{pmatrix} 2 & 1 \\ 4 & 3 \end{pmatrix}$ and a 3rd shape is obtained from the 2nd shape using the

   transformation matrix $\begin{pmatrix} -1 & 5 \\ -1 & 2 \end{pmatrix}$. Find the single matrix that would

   transform the 1st shape to the 3rd shape direct. If the 1st shape has an
   area of 5 sq. units, find the areas of the 2nd and 3rd shapes.

8. Find the matrices corresponding to the following linear transformations.
   (a) 180° rotation about the origin,   (b) enlargement scale factor 3, centre (0, 0),
   (c) reflection in the line $y = -x$,   (d) stretch (× 2) parallel to the *y*-axis, *x*-axis fixed,
   (e) shear with *x*-axis fixed and $(0, 1) \rightarrow (1, 1)$,   (f) shear with $y = x$ fixed and $(1, 0) \rightarrow (0, -1)$,
   (g) stretch (× 3) perpendicular to $y = -x$ and with $y = -x$ fixed,
   (h) shear with $y = x$ fixed and $(0, 2) \rightarrow (4, 6)$,
   (i) shear with $y = 2x$ fixed and $(0, 4) \rightarrow (-2, 0)$.

9. Give a geometrical description of the effect of each of the following transformation matrices.

(a) $\begin{pmatrix} 0 & 1 \\ 1 & 0 \end{pmatrix}$ (b) $\begin{pmatrix} 5 & 0 \\ 0 & 5 \end{pmatrix}$ (c) $\begin{pmatrix} 3 & 0 \\ 0 & 1 \end{pmatrix}$ (d) $\begin{pmatrix} 0 & 0 \\ 0 & 0 \end{pmatrix}$ (e) $\begin{pmatrix} 1 & 0 \\ 0 & 0 \end{pmatrix}$

(f) $\begin{pmatrix} 1 & 1 \\ 1 & 1 \end{pmatrix}$ (g) $\begin{pmatrix} 0 & 3 \\ 3 & 0 \end{pmatrix}$ (h) $\begin{pmatrix} 1 & 0 \\ 2 & 1 \end{pmatrix}$ (i) $\begin{pmatrix} 1 & 2 \\ 4 & 8 \end{pmatrix}$ (j) $\begin{pmatrix} -2 & 3 \\ -3 & 4 \end{pmatrix}$

10. Prove that the transformation matrix $\begin{pmatrix} 3 & 6 \\ -2 & -4 \end{pmatrix}$ maps all points of the

    x–y plane onto a straight line and find the equation of that line.

11. The diagram shows the unit square OABC and, under a certain transformation, its image which is the parallelogram OA′B′C′. If A′ and C′ have coordinates (a, c) and (b, d) respectively, prove that the area of OA′B′C′ is $(ad - bc)$ sq. units.

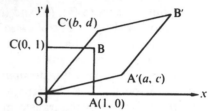

12. A particular transformation takes any point A(x, y) and first reflects the point in the y-axis and then translates this new point by 2 units in the direction of the positive x-axis and 3 units in the direction of the positive y-axis to give the image point A′(x', y'). Represent this transformation in the form

(a) $\begin{pmatrix} x' \\ y' \end{pmatrix} = \begin{pmatrix} * & * \\ * & * \end{pmatrix}\begin{pmatrix} x \\ y \end{pmatrix} + \begin{pmatrix} * \\ * \end{pmatrix}$ (b) $\begin{aligned} x' &= ax + by + c \\ y' &= dx + ey + f \end{aligned}$

13. A certain transformation maps a point A(x, y) onto its image A′(x', y') according to the rule:

$$\begin{pmatrix} x' \\ y' \end{pmatrix} = \begin{pmatrix} 3 & 3 \\ -2 & -1 \end{pmatrix}\begin{pmatrix} x \\ y \end{pmatrix} + \begin{pmatrix} -4 \\ 2 \end{pmatrix}$$

Find (a) the image of the point $(2, -1)$,
     (b) the point with an image of $(-7, 6)$,
     (c) the coordinates of the point that is mapped onto itself by the transformation.

6.3 Further considerations

Invariant points

If a transformation maps some point A(x, y) onto itself, then A is said to be an *invariant point* of the transformation.

Example 6

Find any invariant points of the transformations given by

(a) $\begin{aligned} x' &= 2y - 3 \\ y' &= x + 1 \end{aligned}$

(b) $\begin{pmatrix} x' \\ y' \end{pmatrix} = \begin{pmatrix} 2 & -1 \\ 3 & 4 \end{pmatrix}\begin{pmatrix} x \\ y \end{pmatrix} + \begin{pmatrix} 3 \\ -3 \end{pmatrix}$

(a) If A(x, y) is an invariant point, then $x = 2y - 3$
and $y = x + 1$
Solving simultaneously gives $x = 1$ and $y = 2$

Thus the point (1, 2) is invariant under the transformation $\begin{cases} x' = 2y - 3 \\ y' = x + 1 \end{cases}$

(b) If A(x, y) is an invariant point, then $\begin{pmatrix} x \\ y \end{pmatrix} = \begin{pmatrix} 2 & -1 \\ 3 & 4 \end{pmatrix}\begin{pmatrix} x \\ y \end{pmatrix} + \begin{pmatrix} 3 \\ -3 \end{pmatrix}$

i.e. $x = 2x - y + 3$
and $y = 3x + 4y - 3$
Solving simultaneously gives $x = -1$ and $y = 2$

Thus the point (-1, 2) is invariant under the transformation $\begin{pmatrix} x' \\ y' \end{pmatrix} = \begin{pmatrix} 2 & -1 \\ 3 & 4 \end{pmatrix}\begin{pmatrix} x \\ y \end{pmatrix} + \begin{pmatrix} 3 \\ -3 \end{pmatrix}$

Transformation of a line

A linear transformation will map the straight line with vector equation $\mathbf{r} = \mathbf{a} + \lambda\mathbf{b}$ onto the straight line with vector equation $\mathbf{r} = \mathbf{a'} + \lambda\mathbf{b'}$, i.e. any point lying on $\mathbf{r} = \mathbf{a} + \lambda\mathbf{b}$ will be transformed to a point on the image line $\mathbf{r} = \mathbf{a'} + \lambda\mathbf{b'}$.
All parallel lines will have image lines that are parallel.
Any line that maps onto itself is said to be an **invariant line** of the transformation. It is important to realise that any line which is invariant under a certain transformation need not necessarily be made up of points that are invariant under the transformation. For example, under a stretch parallel to the x-axis, the x-axis itself is an invariant line (as indeed is any line of the form $y = c$). However on the x-axis, only the point (0, 0) is an invariant point under this transformation.

Example 7

All points on the line $y = 2x - 3$ are transformed by the matrix
$\begin{pmatrix} 2 & 1 \\ 3 & -1 \end{pmatrix}$. Find the equation of the image line.

Any point on the line $y = 2x - 3$ will have coordinates of the form
(k, $2k - 3$)

Thus $\begin{pmatrix} x' \\ y' \end{pmatrix} = \begin{pmatrix} 2 & 1 \\ 3 & -1 \end{pmatrix}\begin{pmatrix} k \\ 2k - 3 \end{pmatrix}$

$= \begin{pmatrix} 2k + 2k - 3 \\ 3k - 2k + 3 \end{pmatrix} = \begin{pmatrix} 4k - 3 \\ k + 3 \end{pmatrix}$

i.e. $x' = 4k - 3$ and $y' = k + 3$
Eliminating k from these two equations gives $4y' = x' + 15$. Thus all images (x', y') lie on the line $4y = x + 15$.
The line $4y = x + 15$ is the image of $y = 2x - 3$ under the

transformation $\begin{pmatrix} 2 & 1 \\ 3 & -1 \end{pmatrix}$

Alternatively we could take two points on the line $y = 2x - 3$, find the images of these points under the transformation and then find the equation of the straight line passing through these two image points. The reader should verify that this method does again give the answer $4y = x + 15$.

Example 8

All points on the line $\mathbf{r} = \begin{pmatrix} 1 \\ 3 \end{pmatrix} + \lambda \begin{pmatrix} -1 \\ 4 \end{pmatrix}$ are transformed by the matrix

$\begin{pmatrix} 2 & 1 \\ -1 & 1 \end{pmatrix}$. Find the equation of the image line.

Now $\mathbf{r} = \mathbf{a} + \lambda\mathbf{b}$ will map onto $\mathbf{r} = \mathbf{a}' + \lambda\mathbf{b}'$

In this case $\mathbf{a} = \begin{pmatrix} 1 \\ 3 \end{pmatrix}$ and $\mathbf{b} = \begin{pmatrix} -1 \\ 4 \end{pmatrix}$

$\qquad$ thus $\mathbf{a}' = \begin{pmatrix} 2 & 1 \\ -1 & 1 \end{pmatrix}\begin{pmatrix} 1 \\ 3 \end{pmatrix}$ $\mathbf{b}' = \begin{pmatrix} 2 & 1 \\ -1 & 1 \end{pmatrix}\begin{pmatrix} -1 \\ 4 \end{pmatrix}$

$\qquad\qquad = \begin{pmatrix} 5 \\ 2 \end{pmatrix}$ $\qquad\qquad = \begin{pmatrix} 2 \\ 5 \end{pmatrix}$

Thus the image line will have vector equation $\mathbf{r} = \begin{pmatrix} 5 \\ 2 \end{pmatrix} + \lambda \begin{pmatrix} 2 \\ 5 \end{pmatrix}$

Example 9

Prove that the line $y = 2x - 1$ is mapped onto itself under the transformation $\begin{pmatrix} 3 & -1 \\ 4 & -1 \end{pmatrix}$

Any point on the line $y = 2x - 1$ has coordinates of the form $(k, 2k - 1)$

Thus $\begin{pmatrix} x' \\ y' \end{pmatrix} = \begin{pmatrix} 3 & -1 \\ 4 & -1 \end{pmatrix}\begin{pmatrix} k \\ 2k - 1 \end{pmatrix}$

$\qquad\qquad = \begin{pmatrix} 3k - 2k + 1 \\ 4k - 2k + 1 \end{pmatrix} = \begin{pmatrix} k + 1 \\ 2k + 1 \end{pmatrix}$

i.e. $x' = k + 1$ and $y' = 2k + 1$

Eliminating k from these two equations gives $y' = 2x' - 1$. Thus any image point (x', y') lies on the line $y = 2x - 1$, i.e. $y = 2x - 1$ is mapped onto itself as required.

Example 10

Show that the transformation matrix $\begin{pmatrix} 2 & -1 \\ -4 & 2 \end{pmatrix}$ maps all points on the line

$\mathbf{r} = \begin{pmatrix} 1 \\ 3 \end{pmatrix} + \lambda \begin{pmatrix} 1 \\ 2 \end{pmatrix}$ onto a single point and find the position vector of this point.

$\mathbf{r} = \mathbf{a} + \lambda\mathbf{b}$ will map onto $\mathbf{r} = \mathbf{a}' + \lambda\mathbf{b}'$

In this case $\mathbf{a} = \begin{pmatrix} 1 \\ 3 \end{pmatrix}$ and $\mathbf{b} = \begin{pmatrix} 1 \\ 2 \end{pmatrix}$

thus $\mathbf{a}' = \begin{pmatrix} 2 & -1 \\ -4 & 2 \end{pmatrix}\begin{pmatrix} 1 \\ 3 \end{pmatrix}$ $\mathbf{b}' = \begin{pmatrix} 2 & -1 \\ -4 & 2 \end{pmatrix}\begin{pmatrix} 1 \\ 2 \end{pmatrix}$

$\qquad\qquad = \begin{pmatrix} -1 \\ 2 \end{pmatrix}$ $\qquad = \begin{pmatrix} 0 \\ 0 \end{pmatrix}$

Thus $\mathbf{r} = \begin{pmatrix} 1 \\ 3 \end{pmatrix} + \lambda\begin{pmatrix} 1 \\ 2 \end{pmatrix}$ maps onto $\mathbf{r} = \begin{pmatrix} -1 \\ 2 \end{pmatrix} + \lambda\begin{pmatrix} 0 \\ 0 \end{pmatrix}$

$$= \begin{pmatrix} -1 \\ 2 \end{pmatrix}$$

The matrix $\begin{pmatrix} 2 & -1 \\ -4 & 2 \end{pmatrix}$ maps all points on the line $\mathbf{r} = \begin{pmatrix} 1 \\ 3 \end{pmatrix} + \lambda\begin{pmatrix} 1 \\ 2 \end{pmatrix}$ onto a

single point, position vector $\begin{pmatrix} -1 \\ 2 \end{pmatrix}$.

Example 11

Find the equations of any lines that pass through the origin and map onto

themselves under the transformation whose matrix is $\begin{pmatrix} 3 & 2 \\ 3 & 4 \end{pmatrix}$.

Any straight line passing through the origin has an equation of the form
$y = mx$ and any point on the line will have coordinates of the form (k, mk).

Thus $\begin{pmatrix} x' \\ y' \end{pmatrix} = \begin{pmatrix} 3 & 2 \\ 3 & 4 \end{pmatrix}\begin{pmatrix} k \\ mk \end{pmatrix} = \begin{pmatrix} 3k + 2mk \\ 3k + 4mk \end{pmatrix}$

Now if $y = mx$ maps onto itself then the point $(3k + 2mk, 3k + 4mk)$
must also lie on $y = mx$.

i.e. $3k + 4mk = m(3k + 2mk)$
giving $3 + 4m = 3m + 2m^2$
$\qquad\qquad 0 = 2m^2 - m - 3$
$\qquad\qquad 0 = (2m - 3)(m + 1)$
$\therefore\qquad\qquad m = \tfrac{3}{2}$ or -1.

Thus the required lines are $y = \tfrac{3}{2}x$ and $y = -x$.

Check: $\begin{pmatrix} 3 & 2 \\ 3 & 4 \end{pmatrix}\begin{pmatrix} k \\ \tfrac{3}{2}k \end{pmatrix} = \begin{pmatrix} 6k \\ 9k \end{pmatrix}$ which lies on $y = \tfrac{3}{2}x$

$\qquad\quad \begin{pmatrix} 3 & 2 \\ 3 & 4 \end{pmatrix}\begin{pmatrix} k \\ -k \end{pmatrix} = \begin{pmatrix} k \\ -k \end{pmatrix}$ which lies on $y = -x$

Example 12

Find the equations of any straight lines that are invariant under the
transformation

$$\begin{pmatrix} x' \\ y' \end{pmatrix} = \begin{pmatrix} -6 & 3 \\ 4 & -2 \end{pmatrix}\begin{pmatrix} x \\ y \end{pmatrix} + \begin{pmatrix} 4 \\ -1 \end{pmatrix}$$

Any straight line will be of the form $y = mx + c$ and any point on the line will
have coordinates of the form $(k, mk + c)$.

Thus $\begin{pmatrix} x' \\ y' \end{pmatrix} = \begin{pmatrix} -6 & 3 \\ 4 & -2 \end{pmatrix}\begin{pmatrix} k \\ mk + c \end{pmatrix} + \begin{pmatrix} 4 \\ -1 \end{pmatrix} = \begin{pmatrix} -6k + 3mk + 3c + 4 \\ 4k - 2mk - 2c - 1 \end{pmatrix}$

Now if $y = mx + c$ is invariant, then this point (x', y') must also lie on $y = mx + c$.

i.e. $\quad 4k - 2mk - 2c - 1 = m(-6k + 3mk + 3c + 4) + c$

giving $\quad k(4 + 4m - 3m^2) = 3mc + 4m + 3c + 1$ $\qquad\qquad$ ... [1]

This equation must be true for all values of k (i.e. equation [1] is an identity in k).

Thus $4 + 4m - 3m^2 = 0$ and $3mc + 4m + 3c + 1 = 0$ $\qquad$ ... [2]

i.e. $\quad 3m^2 - 4m - 4 = 0$

giving $\qquad\qquad\qquad m = -\frac{2}{3}$ or 2

From [2], if $m = -\frac{2}{3}$, $c = \frac{5}{3}$ and if $m = 2$, $c = -1$.

This gives the required lines as $y = -\frac{2}{3}x + \frac{5}{3}$ and $y = 2x - 1$.

However we must check this solution:

checking $y = -\frac{2}{3}x + \frac{5}{3}$

$\begin{pmatrix} x' \\ y' \end{pmatrix} = \begin{pmatrix} -6 & 3 \\ 4 & -2 \end{pmatrix}\begin{pmatrix} k \\ -\frac{2}{3}k + \frac{5}{3} \end{pmatrix} + \begin{pmatrix} 4 \\ -1 \end{pmatrix} = \begin{pmatrix} -8k + 9 \\ \frac{16}{3}k - \frac{13}{3} \end{pmatrix}$ which lies on $y = -\frac{2}{3}x + \frac{5}{3}$.

Thus $y = -\frac{2}{3}x + \frac{5}{3}$ is invariant under the transformation

checking $y = 2x - 1$

$\begin{pmatrix} x' \\ y' \end{pmatrix} = \begin{pmatrix} -6 & 3 \\ 4 & -2 \end{pmatrix}\begin{pmatrix} k \\ 2k - 1 \end{pmatrix} + \begin{pmatrix} 4 \\ -1 \end{pmatrix} = \begin{pmatrix} 1 \\ 1 \end{pmatrix}$

Thus all points on the line $y = 2x - 1$ are mapped onto the point $(1, 1)$ and so $y = 2x - 1$ is *not* an invariant line

The line $3y + 2x = 5$ is the only invariant line under the transformation.

Note: The above method will not detect invariant lines of the form $x = c$, a constant (i.e. lines parallel to the y-axis). Such invariant lines may exist if the 2×2 matrix involved has a zero in the 'top right' position. The existence of such invariant lines must then be tested for:

e.g. $\begin{pmatrix} -2 & 0 \\ 4 & 1 \end{pmatrix}\begin{pmatrix} x \\ y \end{pmatrix} + \begin{pmatrix} 6 \\ 2 \end{pmatrix} = \begin{pmatrix} -2x + 6 \\ 4x + y + 2 \end{pmatrix}$

For all values of y the x-coordinate will be unchanged by this transformation if $x = -2x + 6$ i.e. $x = 2$.

Thus $x = 2$ would be an invariant line for the transformation.

Eigenvectors and eigenvalues

Example 11 considered invariant lines that passed through the origin. Now under any linear transformation, the origin itself is an invariant point. If point A lies on an invariant line passing through the origin, then its image A′ must also lie on this line. The position vectors OA and OA′ may be of different magnitudes but *will* be in the same, or exactly opposite directions,

i.e. if $\overrightarrow{OA} = \begin{pmatrix} x \\ y \end{pmatrix}$ and $\overrightarrow{OA'} = \begin{pmatrix} x' \\ y' \end{pmatrix}$

it follows that $\begin{pmatrix} x' \\ y' \end{pmatrix} = \lambda\begin{pmatrix} x \\ y \end{pmatrix}$ for λ a positive or negative constant.

The vector $\begin{pmatrix} x \\ y \end{pmatrix}$ is called an *eigenvector* for the transformation and λ the corresponding *eigenvalue*.

Now for the transformation with matrix $\begin{pmatrix} a & b \\ c & d \end{pmatrix}$, we know that $\begin{pmatrix} x' \\ y' \end{pmatrix} = \begin{pmatrix} a & b \\ c & d \end{pmatrix} \begin{pmatrix} x \\ y \end{pmatrix}$.

Thus if values of x, y and λ can be found such that $\begin{pmatrix} a & b \\ c & d \end{pmatrix} \begin{pmatrix} x \\ y \end{pmatrix} = \lambda \begin{pmatrix} x \\ y \end{pmatrix}$, then $\begin{pmatrix} x \\ y \end{pmatrix}$ is

an eigenvector for the transformation, λ is the associated eigenvalue and the point (x, y) lies on an invariant line passing through the origin.

Furthermore if $\lambda = 1$, then $\begin{pmatrix} a & b \\ c & d \end{pmatrix} \begin{pmatrix} x \\ y \end{pmatrix} = \begin{pmatrix} x \\ y \end{pmatrix}$ and the invariant line consists of invariant points.

Example 13 below is similar to example 11 but the method of solution uses the idea of eigenvectors and eigenvalues.

Example 13

Find the equations of any lines passing through the origin that are mapped

onto themselves by the transformation defined by the matrix $\begin{pmatrix} 4 & 1 \\ 3 & 2 \end{pmatrix}$.

If A(x, y) lies on a line that passes through the origin and that maps onto

itself, then the vector $\begin{pmatrix} x \\ y \end{pmatrix}$ will be an eigenvector. Thus $\begin{pmatrix} 4 & 1 \\ 3 & 2 \end{pmatrix} \begin{pmatrix} x \\ y \end{pmatrix} = \lambda \begin{pmatrix} x \\ y \end{pmatrix}$

i.e. $4x + y = \lambda x$ giving $(4 - \lambda)x = -y$
and $3x + 2y = \lambda y$ giving $3x = (\lambda - 2)y$

dividing: $\dfrac{4 - \lambda}{3} = \dfrac{-1}{\lambda - 2}$

thus $0 = (\lambda - 1)(\lambda - 5)$
and $\lambda = 1$ or 5

With $\lambda = 1$, $4x + y = \lambda x$ becomes $3x + y = 0$
With $\lambda = 5$, $4x + y = \lambda x$ becomes $y = x$

Thus any point with position vector given by the eigenvector $\begin{pmatrix} x \\ y \end{pmatrix}$ must lie

on $3x + y = 0$ or $y = x$, and so these are the required lines.

Notice that the eigenvectors are not unique. In the last example, *any* point lying on the line $3x + y = 0$ will have a position vector that is an eigenvector with eigenvalue 1,

e.g. $\begin{pmatrix} 1 \\ -3 \end{pmatrix}, \begin{pmatrix} 2 \\ -6 \end{pmatrix}, \begin{pmatrix} -1 \\ 3 \end{pmatrix}$ are all eigenvectors with eigenvalue 1:

$\begin{pmatrix} 4 & 1 \\ 3 & 2 \end{pmatrix} \begin{pmatrix} 1 \\ -3 \end{pmatrix} = 1 \begin{pmatrix} 1 \\ -3 \end{pmatrix}$; $\begin{pmatrix} 4 & 1 \\ 3 & 2 \end{pmatrix} \begin{pmatrix} 2 \\ -6 \end{pmatrix} = 1 \begin{pmatrix} 2 \\ -6 \end{pmatrix}$; $\begin{pmatrix} 4 & 1 \\ 3 & 2 \end{pmatrix} \begin{pmatrix} -1 \\ 3 \end{pmatrix} = 1 \begin{pmatrix} -1 \\ 3 \end{pmatrix}$.

Similarly $(1, 1)$, $(3, 3)$ and $(-2, -2)$ all lie on $y = x$ and so the position

vectors $\begin{pmatrix} 1 \\ 1 \end{pmatrix}, \begin{pmatrix} 3 \\ 3 \end{pmatrix}$ and $\begin{pmatrix} -2 \\ -2 \end{pmatrix}$ are all eigenvectors with eigenvalue 5:

$\begin{pmatrix} 4 & 1 \\ 3 & 2 \end{pmatrix} \begin{pmatrix} 1 \\ 1 \end{pmatrix} = 5 \begin{pmatrix} 1 \\ 1 \end{pmatrix}$; $\begin{pmatrix} 4 & 1 \\ 3 & 2 \end{pmatrix} \begin{pmatrix} 3 \\ 3 \end{pmatrix} = 5 \begin{pmatrix} 3 \\ 3 \end{pmatrix}$; $\begin{pmatrix} 4 & 1 \\ 3 & 2 \end{pmatrix} \begin{pmatrix} -2 \\ -2 \end{pmatrix} = 5 \begin{pmatrix} -2 \\ -2 \end{pmatrix}$.

Exercise 6C

1. Find any invariant points of the transformations given by

(a) $x' = 3y + 2$
    $y' = 2x - y + 4$

(b) $\begin{pmatrix} x' \\ y' \end{pmatrix} = \begin{pmatrix} 2 & 3 \\ 1 & 0 \end{pmatrix}\begin{pmatrix} x \\ y \end{pmatrix} + \begin{pmatrix} -3 \\ 5 \end{pmatrix}$

(c) $\begin{pmatrix} x' \\ y' \end{pmatrix} = \begin{pmatrix} -1 & 4 \\ 2 & -5 \end{pmatrix}\begin{pmatrix} x \\ y \end{pmatrix} + \begin{pmatrix} 5 \\ -3 \end{pmatrix}$

2. Two transformations **P** and **T** transform the point (x, y) to its image (x', y') according to the following rules:

$$\mathbf{P}\begin{cases} x' = 3x + y \\ y' = x - 1 \end{cases} \quad \mathbf{T}\begin{cases} x' = 2x - y - 3 \\ y' = 1 - x \end{cases}$$

Express the transformations **P**, **T**, **PT** and **TP** in the form

$$\begin{pmatrix} x' \\ y' \end{pmatrix} = \begin{pmatrix} * & * \\ * & * \end{pmatrix}\begin{pmatrix} x \\ y \end{pmatrix} + \begin{pmatrix} * \\ * \end{pmatrix}$$

Find any invariant points under the transformation
(a) **P**, (b) **T**, (c) **PT**, (d) **TP**.

3. Find the equations of the image lines formed when the lines $y = 2x + 1$ and $3y = 2x - 1$ are transformed using the matrix $\begin{pmatrix} 1 & 3 \\ -1 & 3 \end{pmatrix}$.

4. Find the equation of the image line produced by translating all of the points on the line $y = 3x - 1$ by the vector $\begin{pmatrix} 2 \\ 3 \end{pmatrix}$.

5. Find vector equations of the image lines formed when the lines

$$\mathbf{r} = \begin{pmatrix} 1 \\ -1 \end{pmatrix} + \lambda\begin{pmatrix} 2 \\ 3 \end{pmatrix} \text{ and } \mathbf{r} = \begin{pmatrix} 1 \\ 0 \end{pmatrix} + \lambda\begin{pmatrix} 2 \\ -1 \end{pmatrix}$$ are transformed using the matrix $\begin{pmatrix} 1 & 3 \\ 2 & -1 \end{pmatrix}$.

6. Find the 2×2 transformation matrix **T** that maps $(3, -1)$ onto $(13, -7)$ and $(-1, 3)$ onto $(1, 5)$. Find the equations of the lines obtained when **T** is applied to $y = x$, $y + 2x = 3$ and $y = 2x + 3$.

7. The transformation matrix $\begin{pmatrix} 3 & -2 \\ 3 & -1 \end{pmatrix}$ transforms a line L to the line $y = 2x + 3$. Find the equation of L.

8. Find the vector equation of the line which, when transformed by the matrix $\begin{pmatrix} 2 & 1 \\ 4 & 3 \end{pmatrix}$, has an image line $\mathbf{r} = \begin{pmatrix} 1 \\ 1 \end{pmatrix} + \lambda\begin{pmatrix} -1 \\ 1 \end{pmatrix}$.

9. A transformation **T** assigns to any point (x, y) an image (x', y') according to the rule $\begin{pmatrix} x' \\ y' \end{pmatrix} = \begin{pmatrix} -1 & 2 \\ 0 & 5 \end{pmatrix}\begin{pmatrix} x \\ y \end{pmatrix} + \begin{pmatrix} -2 \\ -4 \end{pmatrix}$.

(a) Find the equation of the image lines obtained when all points on the lines $y = 2x$ and $y = x - 3$ undergo the transformation **T**.
(b) Prove that the line $y = 3x + 1$ maps onto itself under the transformation **T**.

10. Prove that the transformation matrix $\begin{pmatrix} 4 & -1 \\ -3 & 2 \end{pmatrix}$ maps all points on the line $y = 3x$ onto themselves.

11. Prove that all lines of the form $y = x + a$ are mapped onto themselves under the transformation given by the matrix $\begin{pmatrix} 3 & -2 \\ 2 & -1 \end{pmatrix}$.

12. Prove that the lines $y = x + 5$ and $y + 3x = 1$ are mapped onto themselves under the transformation that maps (x, y) onto (x', y') according to the relationship $\begin{pmatrix} x' \\ y' \end{pmatrix} = \begin{pmatrix} 6 & 1 \\ 3 & 4 \end{pmatrix}\begin{pmatrix} x \\ y \end{pmatrix} + \begin{pmatrix} 1 \\ -9 \end{pmatrix}$.

13. Show that the transformation with matrix $\begin{pmatrix} 1 & 3 \\ -1 & -3 \end{pmatrix}$ maps all points on the line $\mathbf{r} = \begin{pmatrix} 4 \\ 1 \end{pmatrix} + \lambda\begin{pmatrix} 3 \\ -1 \end{pmatrix}$ onto a single point and find the position vector of this point.

14. Show that the transformation with matrix $\begin{pmatrix} -4 & 2 \\ 6 & -3 \end{pmatrix}$ maps all points on the line $\mathbf{r} = 2\mathbf{i} + 3\mathbf{j} + \lambda(\mathbf{i} + 2\mathbf{j})$ onto a single point and find the position vector of this point.

15. Prove that all points on the line $y + 3x + 2 = 0$ are mapped onto a single point under the transformation that maps (x, y) onto (x', y') according to the relationship $\begin{pmatrix} x' \\ y' \end{pmatrix} = \begin{pmatrix} 3 & 1 \\ 3 & 1 \end{pmatrix}\begin{pmatrix} x \\ y \end{pmatrix} + \begin{pmatrix} 1 \\ 3 \end{pmatrix}$ and find the coordinates of this single point.

16. Find the equations of any straight lines that pass through the origin and that map onto themselves under the transformation defined by the matrix

    (a) $\begin{pmatrix} -5 & 2 \\ -4 & 1 \end{pmatrix}$ (b) $\begin{pmatrix} -3 & 3 \\ 1 & -5 \end{pmatrix}$ (c) $\begin{pmatrix} a & 1 \\ 8 & a \end{pmatrix}$

17. Find the 2×2 matrix that maps $(1, 2)$ onto $(-3, 0)$ and $(-2, -3)$ onto $(2, -1)$. For the transformation defined by this matrix, find the equations of two invariant straight lines passing through the origin. Which of these two lines is a set of invariant points under the transformation?

18. If $y = 4x$ is mapped onto itself under the transformation $\begin{pmatrix} 3 & -1 \\ a & -3 \end{pmatrix}$, find the value of a and the equation of the other line that passes through the origin and is mapped onto itself by the transformation.

19. Find any straight lines that are invariant under the transformation that maps (x, y) onto (x', y') according to the relationship

    (a) $\begin{pmatrix} x' \\ y' \end{pmatrix} = \begin{pmatrix} 2 & \cdot 3 \\ -2 & 7 \end{pmatrix}\begin{pmatrix} x \\ y \end{pmatrix} + \begin{pmatrix} 6 \\ -12 \end{pmatrix}$,   (b) $\begin{pmatrix} x' \\ y' \end{pmatrix} = \begin{pmatrix} 8 & 1 \\ 4 & 5 \end{pmatrix}\begin{pmatrix} x \\ y \end{pmatrix} + \begin{pmatrix} -1 \\ -4 \end{pmatrix}$,

    (c) $\begin{pmatrix} x' \\ y' \end{pmatrix} = \begin{pmatrix} -1 & 2 \\ 0 & 3 \end{pmatrix}\begin{pmatrix} x \\ y \end{pmatrix} + \begin{pmatrix} 4 \\ -4 \end{pmatrix}$,   (d) $\begin{pmatrix} x' \\ y' \end{pmatrix} = \begin{pmatrix} 2 & -1 \\ -2 & 3 \end{pmatrix}\begin{pmatrix} x \\ y \end{pmatrix} + \begin{pmatrix} -5 \\ -2 \end{pmatrix}$,

    (e) $\begin{pmatrix} x' \\ y' \end{pmatrix} = \begin{pmatrix} 2 & -1 \\ 6 & -3 \end{pmatrix}\begin{pmatrix} x \\ y \end{pmatrix} + \begin{pmatrix} 5 \\ 6 \end{pmatrix}$.

6.4 General reflections and rotations

A general rotation about the origin

The diagram shows the unit square OABC
rotated through an angle θ, anticlockwise
about the origin, to its image OA'B'C'.
As we can see from the diagram

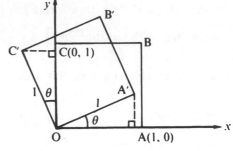

$$\begin{pmatrix} 1 \\ 0 \end{pmatrix} \to \begin{pmatrix} \cos\theta \\ \sin\theta \end{pmatrix} \quad \text{and} \quad \begin{pmatrix} 0 \\ 1 \end{pmatrix} \to \begin{pmatrix} -\sin\theta \\ \cos\theta \end{pmatrix}$$

Thus the matrix representing this

transformation will be $\begin{pmatrix} \cos\theta & -\sin\theta \\ \sin\theta & \cos\theta \end{pmatrix}$.

The determinant of this matrix is $\cos^2\theta + \sin^2\theta$, which equals 1, as we
would expect for a transformation that has no effect on the area of a shape.

> Thus any rotation of the x–y plane about the origin can be represented
>
> by a matrix of the form $\begin{pmatrix} a & -b \\ b & a \end{pmatrix}$ where $a^2 + b^2 = 1$.

A general reflection in a line through the origin

The diagram shows the unit square OABC
reflected in the line $y = mx$, where
$m = \tan\theta$, to its image OA'B'C'.
As can be seen from the diagram

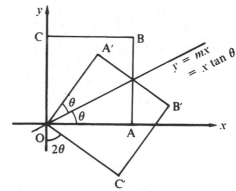

$$\begin{pmatrix} 1 \\ 0 \end{pmatrix} \to \begin{pmatrix} \cos 2\theta \\ \sin 2\theta \end{pmatrix}$$

and $\begin{pmatrix} 0 \\ 1 \end{pmatrix} \to \begin{pmatrix} \sin 2\theta \\ -\cos 2\theta \end{pmatrix}$.

Thus the matrix representing this

transformation will be $\begin{pmatrix} \cos 2\theta & \sin 2\theta \\ \sin 2\theta & -\cos 2\theta \end{pmatrix}$

Again the determinant is as we would expect for such a transformation.

> Thus any reflection of the x–y plane in a straight line passing through the
>
> origin can be represented by a matrix of the form $\begin{pmatrix} a & b \\ b & -a \end{pmatrix}$ where $a^2 + b^2 = 1$.

Suppose now that we wish to find the matrix representation of a rotation
about some point other than the origin.

Example 14

Find the matrix representing a rotation of 90° anticlockwise about the point A (3, 5).

First take some general point with position vector $\begin{pmatrix} x \\ y \end{pmatrix}$.

We now translate the plane so that point A is moved to the origin. Thus our

point $\begin{pmatrix} x \\ y \end{pmatrix}$ becomes $\begin{pmatrix} x - 3 \\ y - 5 \end{pmatrix}$.

We now rotate the plane 90° anticlockwise about the origin: $\begin{pmatrix} 0 & -1 \\ 1 & 0 \end{pmatrix}\begin{pmatrix} x - 3 \\ y - 5 \end{pmatrix}$

and translate the plane so that point A moves back to (3, 5): $\begin{pmatrix} 0 & -1 \\ 1 & 0 \end{pmatrix}\begin{pmatrix} x - 3 \\ y - 5 \end{pmatrix} + \begin{pmatrix} 3 \\ 5 \end{pmatrix}$

Thus if (x, y) has image (x', y') under a 90° anticlockwise rotation about (3, 5).

$$\begin{pmatrix} x' \\ y' \end{pmatrix} = \begin{pmatrix} 0 & -1 \\ 1 & 0 \end{pmatrix}\begin{pmatrix} x - 3 \\ y - 5 \end{pmatrix} + \begin{pmatrix} 3 \\ 5 \end{pmatrix}$$

$$= \begin{pmatrix} -y + 5 + 3 \\ x - 3 + 5 \end{pmatrix}$$

$$= \begin{pmatrix} -y + 8 \\ x + 2 \end{pmatrix}$$

Thus a 90° anticlockwise rotation about the point (3, 5) can be defined by

the equations $\begin{array}{l} x' = -y + 8 \\ y' = x + 2 \end{array}$ or, in matrix form $\begin{pmatrix} x' \\ y' \end{pmatrix} = \begin{pmatrix} 0 & -1 \\ 1 & 0 \end{pmatrix}\begin{pmatrix} x \\ y \end{pmatrix} + \begin{pmatrix} 8 \\ 2 \end{pmatrix}$.

Notice that this is simply a rotation of 90° anticlockwise about the origin
followed by a translation.

If $\begin{pmatrix} a & b \\ c & d \end{pmatrix}$ represents a rotation of angle θ about the origin, then the composite transformation

of this rotation followed by a translation $\begin{pmatrix} e \\ f \end{pmatrix}$, i.e. $\begin{pmatrix} x' \\ y' \end{pmatrix} = \begin{pmatrix} a & b \\ c & d \end{pmatrix}\begin{pmatrix} x \\ y \end{pmatrix} + \begin{pmatrix} e \\ f \end{pmatrix}$, represents

a rotation of angle θ about some other point.

If $\begin{pmatrix} a & b \\ c & d \end{pmatrix}$ represents a reflection in some line $y = mx$, then $\begin{pmatrix} a & b \\ c & d \end{pmatrix}\begin{pmatrix} x \\ y \end{pmatrix} + \begin{pmatrix} e \\ f \end{pmatrix}$ will represent

either a reflection or a glide reflection in some line parallel to $y = mx$
(as shown in the diagrams on the next page).

Diagram showing the effect of

$$\begin{pmatrix} x' \\ y' \end{pmatrix} = \begin{pmatrix} 0 & 1 \\ 1 & 0 \end{pmatrix}\begin{pmatrix} x \\ y \end{pmatrix} + \begin{pmatrix} -2 \\ 2 \end{pmatrix}$$

on the square OABC

$(0, 0) \rightarrow (-2, 2)$ $(1, 0) \rightarrow (-2, 3)$
$(1, 1) \rightarrow (-1, 3)$ $(0, 1) \rightarrow (-1, 2)$

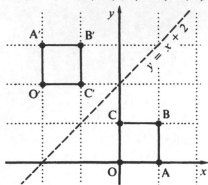

i.e. a reflection in the line $y = x + 2$.

Diagram showing the effect of

$$\begin{pmatrix} x' \\ y' \end{pmatrix} = \begin{pmatrix} 0 & 1 \\ 1 & 0 \end{pmatrix}\begin{pmatrix} x \\ y \end{pmatrix} + \begin{pmatrix} 0 \\ 2 \end{pmatrix}$$

on the square OABC

$(0, 0) \rightarrow (0, 2)$ $(1, 0) \rightarrow (0, 3)$
$(1, 1) \rightarrow (1, 3)$ $(0, 1) \rightarrow (1, 2)$

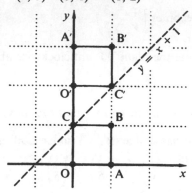

i.e. a glide reflection along the line $y = x + 1$ with $(0, 1) \rightarrow (1, 2)$.

Example 15

Find the matrix equation representing an enlargement, scale factor 3, centre $(-1, 4)$.

First take some general point with position vector $\begin{pmatrix} x \\ y \end{pmatrix}$ and let the point $(-1, 4)$ be A.

Then translate the plane so that A is moved to the origin, i.e. $\begin{pmatrix} x \\ y \end{pmatrix}$ becomes $\begin{pmatrix} x + 1 \\ y - 4 \end{pmatrix}$

Now perform an enlargement, scale factor 3, centre the origin: $\begin{pmatrix} 3 & 0 \\ 0 & 3 \end{pmatrix}\begin{pmatrix} x + 1 \\ y - 4 \end{pmatrix}$

and translate the plane to move A back to $(-1, 4)$: $\begin{pmatrix} 3 & 0 \\ 0 & 3 \end{pmatrix}\begin{pmatrix} x + 1 \\ y - 4 \end{pmatrix} + \begin{pmatrix} -1 \\ 4 \end{pmatrix}$

$$= \begin{pmatrix} 3 & 0 \\ 0 & 3 \end{pmatrix}\begin{pmatrix} x \\ y \end{pmatrix} + \begin{pmatrix} 2 \\ -8 \end{pmatrix}$$

Example 16

Find the matrix equation representing a reflection in the line $y = x + 4$.

Method 1

$y = x + 4$ cuts the y-axis at point B $(0, 4)$.
Translating B to the origin takes (x, y) to $(x, y - 4)$.

Now reflect in the line $y = x$: $\begin{pmatrix} 0 & 1 \\ 1 & 0 \end{pmatrix}\begin{pmatrix} x \\ y - 4 \end{pmatrix}$

and translate B back to $(0, 4)$: $\begin{pmatrix} 0 & 1 \\ 1 & 0 \end{pmatrix}\begin{pmatrix} x \\ y - 4 \end{pmatrix} + \begin{pmatrix} 0 \\ 4 \end{pmatrix}$

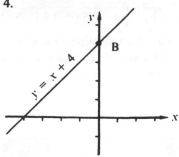

$$\therefore \quad \binom{x'}{y'} = \begin{pmatrix} 0 & 1 \\ 1 & 0 \end{pmatrix}\binom{x}{y-4} + \binom{0}{4}$$

i.e. $\quad \binom{x'}{y'} = \begin{pmatrix} 0 & 1 \\ 1 & 0 \end{pmatrix}\binom{x}{y} + \binom{-4}{4}$

METHOD 2

We know that a reflection in the line $y = x + 4$ will be of the form:

$$\binom{x'}{y'} = \begin{pmatrix} \text{reflection in} \\ y = x \end{pmatrix}\binom{x}{y} + \text{a translation}$$

$$\binom{x'}{y'} = \begin{pmatrix} 0 & 1 \\ 1 & 0 \end{pmatrix}\binom{x}{y} + \binom{a}{b}$$

Now under this reflection, any point on $y = x + 4$ must be invariant.
Considering the point $(0, 4)$

$$\binom{0}{4} = \begin{pmatrix} 0 & 1 \\ 1 & 0 \end{pmatrix}\binom{0}{4} + \binom{a}{b} \quad \therefore \quad a = -4 \quad \text{and} \quad b = 4$$

Thus the required matrix equation is

$$\binom{x'}{y'} = \begin{pmatrix} 0 & 1 \\ 1 & 0 \end{pmatrix}\binom{x}{y} + \binom{-4}{4}$$

Example 17

Give a full geometric description of the transformations defined by the following equations.

(a) $\binom{x'}{y'} = \begin{pmatrix} 0 & 1 \\ -1 & 0 \end{pmatrix}\binom{x}{y} + \binom{7}{-3}$

(b) $x' = y + 3$
$y' = x + 5$

This is a rotation of 90° clockwise about the origin followed by a translation and will be equivalent to a 90° clockwise rotation about some other point (a, b).
The point (a, b) must be invariant under this rotation.

$$\therefore \quad \binom{a}{b} = \begin{pmatrix} 0 & 1 \\ -1 & 0 \end{pmatrix}\binom{a}{b} + \binom{7}{-3}$$

giving (a, b) as the point $(2, -5)$

In matrix form this is $\binom{x'}{y'} = \begin{pmatrix} 0 & 1 \\ 1 & 0 \end{pmatrix}\binom{x}{y} + \binom{3}{5}$

This is a reflection in the line $y = x$ followed by a translation and will be equivalent to a reflection (or glide reflection) in some line $y = x + c$. Thus $y = x + c$ will be an invariant line.

$$\binom{x'}{y'} = \begin{pmatrix} 0 & 1 \\ 1 & 0 \end{pmatrix}\binom{k}{k+c} + \binom{3}{5}$$

$$= \binom{k+c+3}{k+5}$$

Thus $k + 5 = k + c + 3 + c$, i.e. $c = 1$.

Thus $y = x + 1$ is an invariant line with $(k, k + 1) \rightarrow (k + 4, k + 5)$ i.e. a glide reflection in the line $y = x + 1$ with $(0, 1)$ going to $(4, 5)$.

Example 18

Give a geometrical description of the transformation defined by the equations $\begin{aligned} x' &= -2x + 3 \\ y' &= -2y + 6 \end{aligned}$

Writing the equations in matrix form: $\begin{pmatrix} x' \\ y' \end{pmatrix} = \begin{pmatrix} -2 & 0 \\ 0 & -2 \end{pmatrix} \begin{pmatrix} x \\ y \end{pmatrix} + \begin{pmatrix} 3 \\ 6 \end{pmatrix}$

$$= \begin{pmatrix} -1 & 0 \\ 0 & -1 \end{pmatrix} \begin{pmatrix} 2 & 0 \\ 0 & 2 \end{pmatrix} \begin{pmatrix} x \\ y \end{pmatrix} + \begin{pmatrix} 3 \\ 6 \end{pmatrix}$$

$\uparrow$ 180° rotation about origin. $\uparrow$ Enlargement ($\times 2$) centre the origin. $\uparrow$ Translation.

Now a 180° rotation followed by a translation is equivalent to a 180° rotation about some other point. Thus if we can find a point that is invariant under the transformation, this will be the centre of rotation and the centre of the enlargement. If this point is (a, b) then

$$\begin{pmatrix} a \\ b \end{pmatrix} = \begin{pmatrix} -2 & 0 \\ 0 & -2 \end{pmatrix} \begin{pmatrix} a \\ b \end{pmatrix} + \begin{pmatrix} 3 \\ 6 \end{pmatrix} \quad \text{giving } a = 1 \text{ and } b = 2.$$

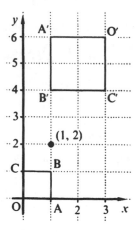

Thus the equations $\begin{aligned} x' &= -2x + 3 \\ y' &= -2y + 6 \end{aligned}$ represent a rotation of 180° about the point $(1, 2)$ and an enlargement, scale factor 2, centre $(1, 2)$.

This result is confirmed by the behaviour of the unit square OABC.

under $\begin{cases} x' = -2x + 3 \\ y' = -2y + 6 \end{cases}$ $\begin{aligned} (0, 0) &\to (3, 6) \\ (1, 0) &\to (1, 6) \\ (1, 1) &\to (1, 4) \\ (0, 1) &\to (3, 4) \end{aligned}$

Exercise 6D

1. Find the matrices which represent the following linear transformations.
   (a) a rotation of 30° anticlockwise about the origin,
   (b) a rotation of 45° anticlockwise about the origin,
   (c) a rotation of 120° anticlockwise about the origin.
2. Find the matrices which represent the following linear transformations,
   (a) a reflection in the line $y = \sqrt{3}x$,
   (b) a reflection in the line $\sqrt{3}y = x$,
   (c) a reflection in the line $y = 2x$.
3. Give a geometrical description of the transformations corresponding to the following matrices
   (a) $\begin{pmatrix} \frac{3}{5} & -\frac{4}{5} \\ \frac{4}{5} & \frac{3}{5} \end{pmatrix}$ (b) $\begin{pmatrix} \frac{3}{5} & \frac{4}{5} \\ \frac{4}{5} & -\frac{3}{5} \end{pmatrix}$ (c) $\begin{pmatrix} 3 & -4 \\ 4 & 3 \end{pmatrix}$
4. Find the matrix equations representing each of the following transformations,
   (a) a rotation of 90° anticlockwise about the point $(-1, 4)$,
   (b) a rotation of 180° about the point $(3, -1)$,

(c) an enlargement, scale factor 3, centre $(2, -1)$,

(d) a reflection in the line $y = x + 5$,

(e) a glide reflection in the line $y = x + 5$ with the point $(0, 5)$ mapped onto $(3, 8)$.

5. Find equations of the form $\begin{array}{l} x' = ax + by + c \\ y' = dx + ey + f \end{array}$ for each of the following transformations.

(a) a rotation of 90° anticlockwise about the point $(1, -1)$,

(b) an enlargement, scale factor 2, centre $(-1, 3)$,

(c) a rotation of 90° clockwise about the point $(4, -1)$,

(d) a reflection in the line $y + x = 3$,

(e) a glide reflection in the line $y + x = 3$ such that $(0, 3)$ maps onto $(3, 0)$.

6. Use the fact that a clockwise rotation of the x–y plane about the origin through an angle θ can be represented by a matrix of the form

$$\begin{pmatrix} \cos\theta & -\sin\theta \\ \sin\theta & \cos\theta \end{pmatrix}$$ to obtain the standard trigonometric identities

$\sin(\theta + \alpha) = \sin\theta\cos\alpha + \cos\theta\sin\alpha$ and
$\cos(\theta + \alpha) = \cos\theta\cos\alpha - \sin\theta\sin\alpha$.

7. Give a full geometric description of the *single* transformation defined by each of the following. For part (g) give your answers as a combination of *two* transformations.

(a) $\begin{pmatrix} x' \\ y' \end{pmatrix} = \begin{pmatrix} 0 & -1 \\ 1 & 0 \end{pmatrix}\begin{pmatrix} x \\ y \end{pmatrix} + \begin{pmatrix} 3 \\ -5 \end{pmatrix}$ (b) $\begin{pmatrix} x' \\ y' \end{pmatrix} = \begin{pmatrix} 0 & 1 \\ 1 & 0 \end{pmatrix}\begin{pmatrix} x \\ y \end{pmatrix} + \begin{pmatrix} 3 \\ -3 \end{pmatrix}$

(c) $\begin{pmatrix} x' \\ y' \end{pmatrix} = \begin{pmatrix} 1 & 0 \\ 0 & -1 \end{pmatrix}\begin{pmatrix} x \\ y \end{pmatrix} + \begin{pmatrix} -6 \\ 2 \end{pmatrix}$ (d) $\begin{array}{l} x' = 5x - 4 \\ y' = 5y - 16 \end{array}$

(e) $\begin{array}{l} x' = 2 - x \\ y' = -y - 3 \end{array}$ (f) $\begin{array}{l} x' = y + 3 \\ y' = -x - 4 \end{array}$ (g) $\begin{array}{l} x' = -3y + 8 \\ y' = 3x + 1 \end{array}$

8. The transformation **T** maps the point (x, y) onto (x', y') according to the equations $x' = -2y - 1$ and $y' = -2x + 4$.

(a) Find the coordinates of the point in the x–y plane that is invariant under **T**.

(b) Find the equations of the two lines that are mapped onto themselves under **T**.

(c) Give a geometric description of the effect of **T**.

Exercise 6E   Examination questions

1. If $\begin{pmatrix} 2 & a & b \\ 1 & 0 & -3 \end{pmatrix}\begin{pmatrix} 1 \\ 3 \\ 2 \end{pmatrix} = \begin{pmatrix} 0 \\ b \end{pmatrix}$, find the values of a and b.   (S.U.J.B.)

2. (a) Arrange the matrices $(2 \quad 0 \quad 3 \quad 4)$, $\begin{pmatrix} 1 \\ 0 \\ 2 \end{pmatrix}$, $\begin{pmatrix} 3 & 2 & -1 \\ 1 & 0 & 3 \end{pmatrix}$ so that they are conformable for multiplication and evaluate their product.

(b) Find the inverse of $\begin{pmatrix} 2 & 1 \\ 4 & 3 \end{pmatrix}$ and **hence** find matrices A and B such that

(i) $\mathbf{A}\begin{pmatrix} 2 & 1 \\ 4 & 3 \end{pmatrix} = \begin{pmatrix} 10 & 6 \\ 10 & 7 \end{pmatrix}$, and (ii) $\begin{pmatrix} 2 & 1 \\ 4 & 3 \end{pmatrix}\mathbf{B} = \begin{pmatrix} 10 & 6 \\ 10 & 7 \end{pmatrix}$.

(Cambridge)

3. (a) Given that $A = \begin{pmatrix} 1 & 2 \\ -1 & 4 \end{pmatrix}$ find (i) $A - 3I$, (ii) $A - 2I$, (iii) $A^2 - 5A + 6I$.

(b) Evaluate all possible products of the matrix A, where $A = \begin{pmatrix} 1 & 0 \\ 2 & -1 \\ 0 & 3 \end{pmatrix}$ with one of the

matrices B, C, D, E where $B = \begin{pmatrix} 1 & 0 & 0 \\ 0 & -1 & 1 \\ 0 & 1 & 1 \end{pmatrix}$, $C = \begin{pmatrix} 2 & 0 \\ 0 & 2 \end{pmatrix}$, $D = (3 \quad 1)$, $E = (1 \quad 2 \quad 1)$.

(Cambridge)

4. (a) Find the value of k for which the simultaneous equations
$$2x - ky = 2,$$
$$3x + (k + 1)y = 4,$$
have no solution.

(b) Given that $A = \begin{pmatrix} 2 & -1 & -2 \\ -1 & 1 & 0 \\ -2 & 1 & 3 \end{pmatrix}$ and $B = \begin{pmatrix} 3 & 1 & 2 \\ 3 & 2 & 2 \\ 1 & 0 & 1 \end{pmatrix}$, evaluate AB.

Hence solve the simultaneous equations
$$3x + y + 2z = 11,$$
$$3x + 2y + 2z = 10,$$
$$x + z = 5.$$
(Cambridge)

5. (a) By means of a matrix method solve the simultaneous equations
$$3x - 2y = -3$$
$$4x - y = 7$$

(b) Given that $A \begin{pmatrix} x \\ y \\ z \end{pmatrix} = 7 \begin{pmatrix} p \\ q \\ r \end{pmatrix}$ and that $\begin{array}{l} x = -2p - 3q + 8r \\ y = p + 5q - 4r, \\ z = 3p + q - 5r \end{array}$ find A^{-1}.

(Cambridge)

6. Write down
(i)   the matrix A which represents reflection in the x-axis,
(ii) the matrix B which represents reflection in the line $y = x$,
(iii) the matrix C which represents rotation through $180°$ about the origin.
Using matrix methods, find
(iv) the image of the point $(5, 2)$ after reflection in the x-axis followed by
      rotation through $180°$ about the origin,
(v)   the point whose image is $(6, -4)$ after reflection in the line $y = x$.

(Cambridge)

7. By using a diagram on graph paper, or otherwise, find the images of the
points $(1, 2)$ and $(2, -1)$ when they are reflected in the line $y = 2x$. Use
your results to find the 2×2 matrix R which carries out this reflection,
and evaluate R^2.
After this reflection a quarter-turn is made clockwise about the origin.
Write down the matrix which represents the quarter-turn, and hence find
the matrix which represents the combined transformation.   (Oxford)

8. Write down the matrices $\mathbf{M}_x$, $\mathbf{M}_y$ and $\mathbf{R}$ which carry out a reflection in the x-axis, a reflection in the y-axis and a half-turn about the origin respectively.

   $\mathbf{T}_x = \begin{pmatrix} 6 \\ 0 \end{pmatrix}$ and $\mathbf{T}_y = \begin{pmatrix} 0 \\ -4 \end{pmatrix}$ are translations parallel to the coordinate axes. P is the point (a, b).

   Find the coordinates of P_1, P_2, P_3, P_4, P_5 which are the images of P under the following five operations:
   (i) $\mathbf{T}_x$ followed by $\mathbf{M}_x$ (ii) $\mathbf{T}_y$ followed by $\mathbf{M}_x$ (iii) $\mathbf{R}$ followed by $\mathbf{T}_x$
   (iv) $\mathbf{R}$ followed by $\mathbf{M}_x$ (v) $\mathbf{M}_x$ followed by $\mathbf{M}_y$ followed by $\mathbf{R}$.
   For each of the five cases find all values of a and b such that P coincides with its image. (Oxford)

9. (i) Calculate the inverse $\mathbf{M}^{-1}$ of the matrix $\mathbf{M} = \begin{pmatrix} x & x-1 \\ y & y \end{pmatrix}$, where $y \neq 0$.

   Find the values of x and y such that (a) $\mathbf{M}^{-1}\begin{pmatrix} 1 \\ 3 \end{pmatrix} = \begin{pmatrix} 0 \\ -2 \end{pmatrix}$;   (b) $\mathbf{M}\begin{pmatrix} 1 \\ 3 \end{pmatrix} = \begin{pmatrix} 0 \\ -2 \end{pmatrix}$.

   (ii) The lines $y = x$ and $x + y = 2$ are transformed by the transformation
   $$\begin{pmatrix} x \\ y \end{pmatrix} \rightarrow \begin{pmatrix} 1 & -1 \\ 2 & 1 \end{pmatrix}\begin{pmatrix} x \\ y \end{pmatrix} + \begin{pmatrix} 1 \\ -1 \end{pmatrix}.$$
   Calculate the equations of their image lines. (Oxford)

10. Find the matrix of the transformation $\mathbf{T}$ which maps $(1, 3)$ and $(2, 1)$ onto $(5, 11)$ and $(5, 7)$ respectively.
    Show that the line $y + x = 0$ is mapped onto itself by $\mathbf{T}$, and find the equation of the other line through the origin that is mapped onto itself. (Cambridge)

11. Find the matrix of
    (i)   the transformation $\mathbf{P}$ that represents enlargement by scale factor 3,
    (ii) the transformation $\mathbf{Q}$ that represents reflection in the line $y = x$,
    (iii) the transformation $\mathbf{R}$ that represents an anticlockwise rotation of θ about the origin where $\tan \theta = \frac{3}{4}$.
    Hence find
    (iv) the matrix of the transformation $\mathbf{T}$ that represents $\mathbf{P}$ followed by $\mathbf{Q}$ followed by $\mathbf{R}$,
    (v)   the image of $(15, 5)$ under $\mathbf{T}$. (Cambridge)

12. The matrices $\mathbf{A}$, $\mathbf{B}$, $\mathbf{C}$ of the single transformations A, B, C are
    $$\mathbf{A} = \begin{pmatrix} \frac{1}{2} & 0 \\ 0 & \frac{1}{2} \end{pmatrix}, \mathbf{B} = \begin{pmatrix} 0 & 1 \\ 1 & 0 \end{pmatrix}, \mathbf{C} = \begin{pmatrix} -\frac{1}{2} & -\frac{\sqrt{3}}{2} \\ \frac{\sqrt{3}}{2} & -\frac{1}{2} \end{pmatrix}.$$
    (i)   Give a geometrical description of each of A, B and C.
    (ii) Find the smallest values of m and n for which $\mathbf{B}^m = \mathbf{C}^n = \mathbf{I}$.
    (iii) Give a geometrical description of the inverse of each transformation. (Cambridge)

13. A reflection **M** in the x-axis is followed by a transformation **R** which is carried out by the matrix $\begin{pmatrix} \frac{1}{2} & -\frac{1}{2}\sqrt{3} \\ \frac{1}{2}\sqrt{3} & \frac{1}{2} \end{pmatrix}$. Write down the matrix which carries out the reflection **M** and find the single matrix which carries out the combined operation **RM**. Show that points of the line $y = x/\sqrt{3}$ are mapped on to themselves by **RM**. Describe the transformation **R**.

Describe also the single transformation which has the same effect as **RM**.

(Oxford)

14. Let $\mathbf{M} = \begin{pmatrix} 7 & -9 \\ 4 & -5 \end{pmatrix}$, let $\mathbf{p} = \begin{pmatrix} x \\ y \end{pmatrix}$ be the position vector of a point P, and let P′ be the point with position vector **Mp**.

Prove that the direction of $\overrightarrow{PP'}$ is constant, and find a fixed vector in this direction.

Let $\mathbf{q} = \begin{pmatrix} 2 \\ -3 \end{pmatrix}$ be the position vector of a point Q and let Q′ be the point with position vector **Mq**. Prove that $\overrightarrow{OQ}$ is perpendicular to $\overrightarrow{PP'}$, where O is the origin.

Calculate the tangent of the angle QOQ′.

(Oxford)

15. Let $\mathbf{P} = \begin{pmatrix} \frac{4}{5} & \frac{2}{5} \\ \frac{2}{5} & \frac{1}{5} \end{pmatrix}$. Verify that $\mathbf{P}^2 = \mathbf{P}$.

Let $\mathbf{r} = \begin{pmatrix} a \\ b \end{pmatrix}$ be the position vector of a point A with respect to the origin O, and let A′ be the point with position vector **Pr**.

Prove that

(i) the vector $\overrightarrow{AA'}$ is a multiple of $\begin{pmatrix} 1 \\ -2 \end{pmatrix}$; (ii) $OA'^2 + A'A^2 = OA^2$.

Prove also that A′ lies on a line of the form $y = cx$, where c is a constant whose value is to be found.

(Oxford)

16. A reflection in the line $y = x - 1$ is followed by an anticlockwise rotation of 90° about the point $(-1, 1)$. Show that the resultant transformation has an invariant line, and give the equation of this line. Describe the resultant transformation in relation to this line.   (Oxford)

17. Find the invariant points of the transformation $\mathbf{T}_1$ of the plane, given by
$$x' = -2y + 1,$$
$$y' = 2x - 3,$$
and give a full geometrical description of $\mathbf{T}_1$.

Express in a similar form the transformation $\mathbf{T}_2$, which consists of a reflection in the line $x + y = 1$. Express also in a similar form the transformation $\mathbf{T}_3$, which consists of $\mathbf{T}_1$ followed by $\mathbf{T}_2$. Show that $\mathbf{T}_3$ may be expressed as a reflection in the line $x = \frac{4}{3}$ followed by an enlargement, and give the centre and scale factor of this enlargement.

(Oxford)

18. The eigenvalues of the matrix
$$\mathbf{T} = \begin{pmatrix} a & b \\ c & d \end{pmatrix} \quad (b > 0, c \geqslant 0)$$
are equal. Prove that $a = d$ and $c = 0$.

If **T** maps the point $(-2, 1)$ into the point $(1, 2)$, prove that
$$\mathbf{T} = \begin{pmatrix} 2 & 5 \\ 0 & 2 \end{pmatrix},$$
and find the image of the line $y + 2x = 0$ under **T**.   (Oxford)

7
Permutations and combinations

7.1 Successive operations

If there are 3 paths joining A to B and 4 paths joining B to C, then there are (3 × 4) or 12 different ways of going from A, through B, to C.

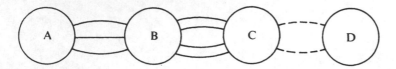

If there are 2 paths joining C to D then there are (3 × 4 × 2) or 24 ways of going from A, through B and C, to D.

This idea can be extended indefinitely so that in general:

If there are r ways of performing one operation, s ways of performing a second operation, t ways of performing a third operation and so on,

then there are ($r × s × t ×$ ...) different ways of performing the operations in succession.

Note that this multiplication rule only applies when the operations are independent, i.e. the choice made for one operation does not affect the choice made for any of the other operations.

Example 1

There are 5 roads joining A to B and 3 roads joining B to C. Find how many different routes there are from A to C via B.

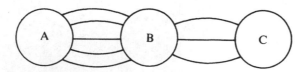

There are two operations to perform in succession:

$$\begin{array}{ll} \text{A to B} & \text{5 ways} \\ \text{B to C} & \text{3 ways} \end{array}$$

Number of routes from A to C = 5 × 3
= 15

Example 2

A man has three choices of the way in which he travels to work; he can walk, go by car or go by train. In how many different ways can he arrange his travel for the 5 working days in the week?

On Monday he has 3 choices: walk, car or train.
On Tuesday he has 3 choices: walk, car or train.
Similarly on each of Wednesday, Thursday and Friday he has 3 choices. Hence there are 5 successive operations, each of which can be performed in 3 ways.

$$\text{Total number of arrangements} = 3 \times 3 \times 3 \times 3 \times 3$$
$$= 3^5 \text{ (or 243) ways.}$$

7.2 Permutations

Consider the three letters A, B and C. If these letters are written in a row, one after another, there are six different possible arrangements:

ABC   ACB   BAC   BCA   CAB   CBA

Each arrangement is a possible *permutation* of the letters A, B and C and so there are six permutations altogether.

The following reasoning shows how to calculate the number of permutations without having to list them all:
The first letter to be written down can be chosen in 3 ways. The second letter can then be chosen in 2 ways and the remaining letter is written down in the third position. Thus the three operations can be performed in $(3 \times 2 \times 1)$ or 6 ways.

This can be stated in general terms as follows:
The number of ways of arranging n different things in a row is $n(n - 1)(n - 2) \ldots \times 2 \times 1$. A useful shorthand way of writing this expression is $n!$ (read as n factorial).

Thus $4! = 4 \times 3 \times 2 \times 1$, $5! = 5 \times 4 \times 3 \times 2 \times 1$ and so on.

Example 3

Evaluate (a) $\dfrac{6!}{2 \times 4!}$, (b) $\dfrac{7!}{4! \times 2!}$.

(a) $\dfrac{6!}{2 \times 4!} = \dfrac{6 \times 5 \times 4 \times 3 \times 2 \times 1}{(2) \times 4 \times 3 \times 2 \times 1}$

$= 15.$

(b) $\dfrac{7!}{4! \times 2!} = \dfrac{7 \times 6 \times 5 \times 4 \times 3 \times 2 \times 1}{4 \times 3 \times 2 \times 1 \times (2 \times 1)}$

$= 105.$

Example 4

Five children are to be seated on a bench. Find (a) how many ways the children can be seated, (b) how many arrangements are possible if the youngest child is to sit at the left-hand end of the bench.

(a) Since there are 5 children, the child to sit at the left-hand end can be chosen in 5 ways.
    The child to be seated next can be chosen in 4 ways, the next child can be chosen in 3 ways, and so on.
    There are 5 operations and the total number of arrangements is

$$5 \times 4 \times 3 \times 2 \times 1 = 5! \text{ or 120 ways.}$$

(b) If the youngest child is to sit at the left-hand end, this place can be filled
in 1 way only.

The next child can then be chosen in 4 ways, the next in 3 ways and so on.
There are 5 operations and the total number of arrangements is

$$1 \times 4 \times 3 \times 2 \times 1 = 1 \times 4! \text{ or } 24 \text{ ways.}$$

.	.	.	.	.
1	4	3	2	1

Example 5

Three different mathematics books and five other different books are to be
arranged on a bookshelf. Find (a) the number of possible arrangements of
the books, (b) the number of possible arrangements if the three mathematics
books must be kept together.

(a) Since there are 8 books altogether, the one to be placed at the left-hand
end can be chosen in 8 ways, the next book can be chosen in 7 ways,
and so on.

There are 8 operations and the total number of arrangements is

$$8 \times 7 \times 6 \times 5 \times 4 \times 3 \times 2 \times 1 = 8! \text{ or } 40\,320 \text{ ways.}$$

(b) Since the 3 mathematics books are to be together, consider these bound
together as one book. There are now 6 books to be arranged and this
can be performed in $6 \times 5 \times 4 \times 3 \times 2 \times 1 = 6! = 720$ ways.

Now in each of these arrangements the 3 mathematics books are bound
together; these mathematics books can be arranged in $3 \times 2 \times 1 = 3!$ ways
$= 6$ ways.

Total number of arrangements $= 720 \times 6$ or 4320 ways.

Circular arrangements

In Example 4(a), we saw that five children can be arranged in a row in
5! ways. Suppose, instead, that we wished to arrange the children around a
circular table. The number of possible arrangements will no longer be 5!
because there is now no distinction between certain arrangements that were
distinct when written in a row.

For example, A B C D E is a different arrangement from E A B C D,

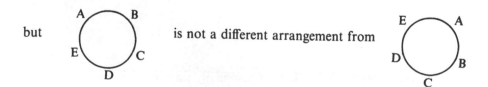

but   is not a different arrangement from

With circular arrangements of this type, it is the relative positions of the
items being arranged which is important. One item can therefore be fixed
and the remaining items arranged around it. The number of arrangements of
n unlike things in a circle will therefore be $(n - 1)!$ In those cases where
clockwise and anticlockwise arrangements are not considered to be different,
this reduces to $\frac{1}{2}(n - 1)!$ (See Example 7).

Example 6

Four men Peters, Rogers, Smith and Thomas are to be seated at a circular table. In how many ways can this be done?

Suppose Peters is seated at some particular place.
The seat on his left can be filled in 3 ways, i.e. by Rogers, by Smith or by Thomas. The next seat on his left can then be filled in 2 ways and the remaining seat in 1 way, i.e. by the remaining man.

Total number of arrangements $= 3 \times 2 \times 1 = 3!$ or 6 ways.

Example 7

Nine beads, all of different colours are to be arranged on a circular wire. Two arrangements are *not* considered to be different if they appear the same when the ring is turned over. How many different arrangements are possible?

When the ring is turned over, the arrangement would appear as

When viewed from one side, these arrangements are only different in that one is a clockwise arrangement and the other is anticlockwise.
If one bead is fixed, there are $(9 - 1)!$ ways of arranging the remaining beads relative to the fixed one, i.e. 8! ways. But half of these arrangements will appear the same as the other half when the ring is turned over, because for every clockwise arrangement there is a similar anticlockwise arrangement. Hence,

number of arrangements is $\frac{1}{2}(8!) = 20\,160$ ways.

Mutually exclusive situations

When two situations A and B are mutually exclusive then, if situation A occurs, situation B cannot occur. Likewise, if situation B occurs, situation A cannot occur.
In such cases, the number of permutations of either situation A or situation B occurring can be obtained by adding the number of permutations of situation A to the number of permutations of situation B.

Example 8

How many different four-digit numbers can be formed from the figures 3, 4, 5, 6 if each figure is used only once in each number?
How many of these numbers (a) end in a 4, (b) end in a 3, (c) end in a 3 or a 4?

The first digit can be chosen in 4 ways, the second in 3 ways, and so on. Thus there are $4 \times 3 \times 2 \times 1 = 24$ different four-digit numbers that can be formed.

(a) The last digit can be chosen in 1 way as it must be a 4; the first digit can be chosen in 3 ways, the second in 2 ways and the third in 1 way. Thus there are $1 \times 3 \times 2 \times 1 = 6$ of the numbers that end in a 4.

(b) By reasoning similar to that used in (a), there will be $1 \times 3 \times 2 \times 1 = 6$ of the numbers that end in a 3.

(c) The numbers that end in a 3 cannot also end in a 4, so these are mutually exclusive situations. Thus $6 + 6 = 12$ of the numbers end either in a 3 or a 4.

Alternatively (c) could be solved as follows.
The last digit can be chosen in 2 ways (a 3 or a 4); the first digit can then be chosen in 3 ways, the second in 2 ways and the third in 1 way, i.e. $2 \times 3 \times 2 \times 1 = 12$ of the numbers end in a 3 or 4.

The permutations in which a certain event A does occur will clearly be mutually exclusive with those permutations in which that event does *not* occur. Thus:

$$\begin{pmatrix} \text{number of permutations in} \\ \text{which event A does occur} \end{pmatrix} + \begin{pmatrix} \text{number of permutations in} \\ \text{which event A does not occur} \end{pmatrix} = \begin{pmatrix} \text{number of permutations in which} \\ \text{event A either does or does not occur} \end{pmatrix}$$

$$= \begin{pmatrix} \text{total number of} \\ \text{permutations} \end{pmatrix}$$

which can be arranged thus

$$\begin{pmatrix} \text{number of permutations in} \\ \text{which event A does not occur} \end{pmatrix} = (\text{total number of permutations}) - \begin{pmatrix} \text{number of permutations in} \\ \text{which event A does occur} \end{pmatrix}$$

Example 9

In how many ways can five people, Smith, Jones, Clark, Brown and White, be arranged around a circular table if
(a) Smith must sit next to Brown,   (b) Smith must *not* sit next to Brown?

(a) Since Smith and Brown must sit next to each other, consider these two bound together as one person. There are now 4 people to seat. Fix one of these, and then the remaining 3 can be seated in $3 \times 2 \times 1 = 6$ ways relative to the one that was fixed.
In each of these arrangements, Brown and Smith are seated together in a particular way. Brown and Smith could now change seats giving another 6 ways of arranging the 5 people.
Total number of arrangements $= 2 \times 6 = 12$ ways.

(b) If Smith is not to sit next to Brown, then this situation is mutually exclusive with the situation in (a).
Hence

$$\begin{pmatrix} \text{number of permutations in which} \\ \text{Smith does not sit next to Brown} \end{pmatrix} = \begin{pmatrix} \text{total number} \\ \text{of permutations} \end{pmatrix} - \begin{pmatrix} \text{number of permutations in which} \\ \text{Smith does sit next to Brown} \end{pmatrix}$$

Total number of arrangements of 5 people at a circular table $= (5 - 1)!$
$$= 4!$$
Required number of arrangements $= 4! - 12$
$$= 24 - 12 = 12$$
Number of arrangements in which Smith does not sit next to Brown is 12.

Exercise 7A

1. Evaluate   (a) $\dfrac{8!}{6!}$ (b) $\dfrac{9!}{3 \times 5!}$ (c) $6! - 5!$ (d) $\dfrac{5! \times 4!}{6!}$.

2. Express the following in terms of $5!$
   (a) $\dfrac{6!}{6}$ (b) $\dfrac{6!}{2!}$ (c) $\dfrac{6!}{3!}$ (d) $4!$ (e) $6! - 5!$
   (f) $2 \times 6! - 3 \times 5!$ (g) $3 \times 4! + 7 \times 4!$ (h) $7! - 6!$ (i) $\dfrac{8! - 6!}{3!}$

3. There are three roads joining town X to town Y; three more roads join Y to Z and two roads join Z to A. How many different routes are there from X to A passing through Y and Z?

4. In how many ways can a group of ten children be arranged in a line?

5. In how many ways can eight different books be arranged on a bookshelf?

6. The letters a, b, c and d are to be arranged in a row with each letter being used once and once only. In how many ways can this be done?

7. The front doors of five houses in a terrace are to be painted blue, brown, black, green and red. In how many different ways can the painting be done if no two doors are to be the same colour?

8. How many different four-digit numbers can be formed from the digits 5, 7, 8 and 9 if each digit is used once only in each number? If repetitions of the digits were allowed, how many arrangements would be possible?

9. With his breakfast a man sometimes has tea, sometimes coffee and sometimes fruit juice, but never more than one of these on any one day. Find the number of possible arrangements he can have in a period of 4 days if (a) he always has a drink of some sort,
   (b) he may choose not to have a drink on certain days.

10. In how many different ways can the letters of the word THURSDAY be arranged?

11. The letters of the word TUESDAY are arranged in a line, each arrangement ending with the letter S. How many different arrangements are possible? How many of these arrangements also start with the letter D?

12. In how many ways can five women be seated at a circular table?

13. How many numbers greater than 50 000 can be formed using the digits 2, 3, 4, 5 and 6 if each digit is used only once in each number?

14. How many numbers greater than 40 000 can be formed using the digits 2, 3, 4, 5 and 6 if each digit is used only once in each number?

15. How many odd numbers greater than 60 000 can be formed using the digits 2, 3, 4, 5 and 6 if each digit is used only once in each number?

16. Three boys and five girls are to be seated on a bench so that the youngest boy and the youngest girl sit next to each other. In how many ways can this be done?

17. There are nine different books on a shelf, one of which is a dictionary and one an atlas. In how many ways can the books be arranged on the shelf if the dictionary and the atlas are to be next to each other?

18. On a bookshelf I have a four-volume encyclopaedia and ten other books that are all different. In how many ways can the fourteen books be arranged on the shelf if the four volumes of the encyclopaedia
    (a) need not be kept together,
    (b) must be kept together and must be in the order 1, 2, 3, 4; (the reverse order is not acceptable),
    (c) must be kept together, but not in any particular order?

19. Seven boys are to be seated on a bench so that the oldest boy and the youngest boy sit next to each other. In how many ways can the boys be seated? If, instead, the boys were seated around a circular table, with the same restriction as on the bench, how many arrangements would be possible?

20. In how many different ways can eight differently coloured beads be arranged on a ring if a particular order arranged clockwise is not considered different from the same order arranged anticlockwise?

21. In how many ways can six boys and two girls be arranged in a line if the two girls must not sit together?

22. How many even numbers greater than 40 000 can be formed using the digits 2, 3, 4, 5 and 6 if each digit is used only once in each number?

23. Mr and Mrs X, Mr A, Mr B and Mr C are to be seated at a circular table. In how many ways can this be done if
    (a) the married couple sit next to each other,
    (b) the married couple do not sit next to each other?

24. Find the number of ways that Mr and Mrs Smith, Mr and Mrs Brown and Mr and Mrs White can be seated around a circular table.
    Find in how many of these arrangements
    (a) Mr and Mrs Smith sit next to each other,
    (b) Mr Smith is not sitting next to Mrs Smith,
    (c) Mrs Smith is sitting between Mr Brown and Mrs Brown.

25. To a meeting involving four companies, each company sends three representatives—the managing director, the chief accountant and the company secretary. In how many ways can the twelve people be arranged around a circular table if the three people from each company sit together, with the managing director between the accountant and the secretary in each case?

26. Sheila, Ann and Harvey are fifth year pupils; Sarah, Jeff and Heather are fourth year pupils; Alan, Rosemary and John are third year pupils. In how many ways can these pupils be arranged in a line if the pupils from each year are kept next to each other?

Permutations of objects selected from a group

In how many ways can we arrange two different letters chosen from the five letters A, B, C, D and E?

The first letter can be chosen in 5 ways and the second letter can be chosen in 4 ways.

Hence there are (5 × 4) or 20 possible arrangements and these are:

    AB AC AD AE BC BD BE CD CE DE
    BA CA DA EA CB DB EB DC EC ED

Thus we say there are 20 possible permutations of two letters taken from five different letters.

Suppose we wish to arrange r objects chosen from n unlike objects:

The first object can be chosen in n ways, the second object can be chosen in $(n - 1)$ ways, the third object in $(n - 2)$ ways and so on until the rth item which can be chosen in $(n - r + 1)$ ways.

Thus the number of permutations of r objects chosen from n unlike objects is

$$n(n - 1)(n - 2)(n - 3) \ldots (n - r + 1)$$

$$= \frac{n(n - 1)(n - 2)(n - 3) \ldots (n - r + 1)(n - r)(n - r - 1) \ldots 2 \times 1}{(n - r)(n - r - 1) \ldots 2 \times 1}$$

$$= \frac{n!}{(n - r)!}$$

We usually say that the number of permutations of r objects selected from n unlike objects is ^nP_r, where $^nP_r = \dfrac{n!}{(n - r)!}$.

Thus in the case of arranging two letters chosen from the five letters A, B, C, D and E mentioned above, this gives $^5P_2 = \dfrac{5!}{3!}$

$$= 20 \text{ as required.}$$

Note that we already know that there are $n!$ arrangements of n objects chosen from n objects. Thus ^nP_n must equal $n!$, i.e. $\dfrac{n!}{(n - n)!} = n!$

and we therefore define $0!$ as 1.

Example 10

Ten athletes are to take part in a race. In how many different ways can the 1st, 2nd and 3rd places be filled?

1st place can be filled in 10 ways	or, using ^nP_r
2nd place can be filled in 9 ways	Number of permutations of three things
3rd place can be filled in 8 ways	from ten is $^{10}P_3 = \dfrac{10!}{7!}$

$\therefore$ number of permutations of 1st, 2nd and $= 720$
3rd places $= 10 \times 9 \times 8 = 720$

There are 720 ways in which the 1st, 2nd and 3rd places can be filled.

Arrangements of like and unlike things

Suppose the n unlike letters $a_1\ a_2\ a_3\ \ldots\ b_1\ b_2\ \ldots\ c_1\ c_2\ \ldots$ are arranged in a row, then the number of possible arrangements is $n!$

If we consider these $n!$ arrangements and remove the suffixes from the letters 'a' we shall have written down the same arrangement several times, for example the arrangements

$$a_1 \ldots a_3 \ldots a_2 \ldots$$
$$a_2 \ldots a_1 \ldots a_3 \ldots$$
$$a_3 \ldots a_1 \ldots a_2 \ldots$$

will all be the same once the suffixes are removed. Indeed, we shall have repeated this particular arrangement $r!$ times where r is the number of a's. In a similar way we can consider the b's and c's.

Hence the number of arrangements of n things, p of one kind, q of another, r of another ... is $$\frac{n!}{p!\,q!\,r!\,\ldots}$$

Example 11

In how many ways can 4 red, 3 yellow and 2 green discs be arranged in a row, if discs of the same colour are indistinguishable?

There are $4 + 3 + 2 = 9$ discs

$\therefore$ the number of arrangements $= \dfrac{9!}{4!\,3!\,2!}$

$\hspace{5.5cm} = 1260$

The discs can be arranged in a row in 1260 ways.

Example 12

Find (a) in how many different ways the letters of the word ALGEBRA can be arranged in a row, (b) in how many of these arrangements the two A's are together, (c) in how many of the arrangements the two A's are not together.

(a) There are 7 letters including two A's

$\therefore$ total number of arrangements $= \dfrac{7!}{2!}$

$\hspace{5.5cm} = 2520$

(b) If the A's are kept together, there are effectively 6 letters to arrange, hence
     number of arrangements $= 6!$
     $\hspace{3.2cm} = 720$

(c) Number of arrangements when A's are not together $= 2520 - 720$
     $\hspace{7.4cm} = 1800$

Example 13

How many different arrangements are there of 3 letters chosen from the word COMBINATION?

There are 11 letters including 2 O's, 2 I's and 2 N's.
To find the total number of different arrangements we consider the possible arrangements as four mutually exclusive situations.

(i)   Arrangements in which all 3 letters are different.
      There are $8 \times 7 \times 6$ (or $^8P_3$) of these, i.e. 336.
(ii) Arrangements containing two O's and one other letter.
      The other letter can be one of seven letters (C, M, B, I, N, A or T) and can appear in any of the three positions (before the two O's, between the two O's, or after the two O's),
      i.e. 3×7 or 21 arrangements that have two O's and one other letter.

(iii) Arrangements containing two I's and one other letter.

By the same reasoning as in (ii) there will be 21 arrangements that have two I's and one other letter.

(iv) Arrangements containing two N's and one other letter.

Similarly there will be 21 arrangements that have two N's and one other letter.

Thus the total number of arrangements of three letters chosen from the word COMBINATION will be $336 + 21 + 21 + 21 = 399$.

Exercise 7B

1. Find the number of permutations of two different letters taken from the letters A, B, C, D, E and F.
2. In how many ways can six books be arranged on a shelf when the books are selected from 10 different books?
3. How many code words, each consisting of five different letters, can be formed from the letters A, B, C, D, E, F, G and H?
4. How many three-digit numbers, with all three digits different, can be formed from the figures 7, 6, 5, 4, 3 and 2?
5. Three-digit numbers are formed from the figures 3, 5, 6 and 7. How many different numbers can be formed if
   (a) no figure is repeated in a number, (b) repetitions are allowed?
6. How many different three-digit even numbers can be formed from the figures 2, 5, 7 and 9 if repetitions (a) are not allowed, (b) are allowed?
7. How many numbers greater than 300 can be formed from the figures 4, 3, 2 and 1 if each figure can be used no more than once in each number and all the figures need not be used each time?
8. In how many different ways can a first, second and third prize be awarded to a group of 15 pupils if no pupil may win more than one prize?
9. A box contains 14 coloured discs which are identical except for their colour. There are 5 red discs, 4 green, 3 blue and 2 yellow. In how many ways can the 14 discs be arranged in a row?
10. In how many ways can the letters of the word PARALLEL be arranged in a row?
11. In how many ways can the letters of the word PHOTOGRAPH be arranged in a row? How many of these arrangements start and finish with an H?
12. (a) In how many ways can the letters of the word GEOMETRY be arranged in a row?
    (b) In how many of these arrangements are the two E's together?
    (c) In how many of these arrangements are the two E's not together?
13. How many even numbers greater than 300 can be formed from the figures 4, 3, 2 and 1 if each figure can be used no more than once in each number and all the figures do not have to be used each time?
14. In how many ways can the letters of the word BANANA be arranged in a row? In how many of these arrangements are the N's separated?
15. Ten coloured beads, 4 red, 3 blue, 1 orange, 1 white and 1 green, are to be threaded onto a circular wire. Find the number of ways in which this can be done if beads of the same colour are indistinguishable and a particular order arranged clockwise is not considered different from the same order arranged anticlockwise.

16. Find the number of different three-letter arrangements that can be made from the letters of the word PYTHAGORAS.

17. Find the number of different three-letter arrangements that can be made from the letters of the word COMMON.

18. Find the number of different three-letter arrangements that can be made from the letters of the word ISOSCELES.
    How many of these arrangements will contain
    (a) no E's at all,   (b) at least one E,
    (c) no vowels at all,   (d) at least one vowel?

19. In how many ways can the letters of the word PERMUTATION be arranged? In how many of these arrangements are
    (a) the T's together,   (b) the vowels together?

20. In how many ways can the letters of the word ARRANGEMENT be arranged? In how many of these arrangements are
    (a) the A's together,   (b) the vowels together?

21. Find the total number of permutations of four letters selected from the word PERMUTATION.

22. Find the total number of permutations of four letters selected from the word ARRANGEMENT. How many of these permutations contain two R's?

7.3 Combinations

A combination is a selection. In making a selection from a number of items, only the contents of the group selected are important, not the order in which the items are selected.

We saw on page 194 that there were 20 possible *arrangements* of two letters chosen from the letters A, B, C, D and E:

$$\text{AB} \quad \text{AC} \quad \text{AD} \quad \text{AE} \quad \text{BC} \quad \text{BD} \quad \text{BE} \quad \text{CD} \quad \text{CE} \quad \text{DE}$$
$$\text{BA} \quad \text{CA} \quad \text{DA} \quad \text{EA} \quad \text{CB} \quad \text{DB} \quad \text{EB} \quad \text{DC} \quad \text{EC} \quad \text{ED}$$

However, the number of *combinations* of two letters chosen from A, B, C, D and E is only 10 because AB and BA are the same combination, as also are AC and CA etc.

Example 15

Find the number of selections of 2 letters which can be made from the three letters a, b and c.

We can in this simple case write out the possible selections:

$$\text{ab; bc; ca}$$

There are 3 possible selections.

If, in the above example, we considered the number of permutations of the 3 letters taken 2 at a time, there would be

$$^3P_2 = \frac{3!}{(3-2)!} = 6 \text{ possible arrangements.}$$

We can see that the number of permutations, or arrangements, is twice the number of combinations, or selections, because there are 2 ways of arranging each pair of letters selected.

For the general case, we know from page 195 that the number of arrangements of n things taken r at a time is

$$^nP_r = \frac{n!}{(n-r)!}.$$

But each selection of r things can be arranged in $r!$ ways, so each selection is repeated $r!$ times in the ^nP_r arrangements, hence

$$\text{number of selections} = \frac{^nP_r}{r!} = \frac{n!}{(n-r)!\,r!}$$

Thus the number of possible combinations of n different objects, taken r at a time, is given by ^nC_r, also written as $\binom{n}{r}$ where $^nC_r = \dfrac{n!}{(n-r)!\,r!}$.

Example 16

How many selections of 4 letters can be made from the 6 letters a, b, c, d, e and f?

The number of selections is ^nC_r where r is the number of things selected from a group of n. Hence for this case $n = 6$ and $r = 4$.

$$^6C_4 = \frac{6!}{2!\,4!} = 15$$

There are 15 selections of 4 letters which can be made from the 6 letters.

Example 17

How many different committees, each consisting of 3 boys and 2 girls, can be chosen from 7 boys and 5 girls?

Number of ways of choosing 3 boys from 7 $= {}^7C_3 = \dfrac{7!}{4!\,3!} = 35$

Number of ways of choosing 2 girls from 5 $= {}^5C_2 = \dfrac{5!}{3!\,2!} = 10$

Number of committees which can be chosen $= 35 \times 10 = 350$

Note that 35 is multiplied by 10 since the choice of the boys and the choice of the girls are independent operations.

Example 18

(a) Find the number of different selections of 4 letters that can be made from the letters of the word SPHERICAL.
(b) How many of these selections do not contain a vowel?

(a) There are 9 letters, all of which are different.

Number of selections of 4 letters $= {}^9C_4 = \dfrac{9!}{5!\,4!} = 126$

(b) There are 3 vowels in the word SPHERICAL.
Each selection either does or does not contain a vowel.
If the selections are not to contain vowels, the selection of 4 letters must be made from the 6 consonants.

Number of selections of 4 letters, not containing a vowel $= {}^6C_4 = \dfrac{6!}{2!\,4!}$

$$= 15$$

The following examples illustrate other techniques which may be used in questions involving selections.

Example 19

(a) Find the number of different selections of 3 letters that can be made from the letters of the word SUCCESSFUL.
(b) How many of these selections contain at least one vowel?

(a) There are 2 C's, 2 U's, 3 S's and 3 other different letters. We must consider all the mutually exclusive selections of 3 letters; these are:

number of selections containing 3 S's	= 1 selection
number of selections containing 2 C's + 1 other letter	$= 1 \times {}^5C_1 = 5$
number of selections containing 2 U's + 1 other letter	$= 1 \times {}^5C_1 = 5$
number of selections containing 2 S's + 1 other (non S)	$= 1 \times {}^5C_1 = 5$
number of selections containing 3 different letters (from 6)	$= {}^6C_3 = 20$
Total number of selections of 3 letters	$= 36$

(b) Without the vowels, there are 3 S's, 2 C's, 1 F and 1 L.

As in (a) 3 S's	= 1 selection	
2 C's + 1 other from 3 (S, F, L)	$= 1 \times {}^3C_1$	= 3
2 S's + 1 other from 3 (C, F, L)	$= 1 \times {}^3C_1$	= 3
3 different letters from 4 (S, C, F, L)	$= {}^4C_3$	= 4
Number of selections without any vowels	= 11	
Hence number of selections with at least 1 vowel $= 36 - 11$	= 25	

Example 20

A team of 7 players is to be chosen from a group of 12 players. One of the 7 is then to be elected as captain and another as vice-captain. In how many ways can this be done?

Number of ways of choosing 7 from 12 is ${}^{12}C_7 = \dfrac{12!}{5!\,7!}$

$$= 792$$

In each chosen team there are 7 ways of choosing a captain, and
there are 6 ways of choosing a vice-captain
(alternatively we say the captain and vice-captain can be elected in ${}^7P_2$ ways).
Total number of selections is $7 \times 6 \times 792 = 33\,264$.

Example 21

A group consists of 4 boys and 7 girls. In how many ways can a team of five be selected if it is to contain (a) no boys, (b) at least one member of each sex, (c) 2 boys and 3 girls, (d) at least 3 boys?

(a) No boys are selected, so the team is chosen from the 7 girls.

Number of ways of choosing 5 girls from 7 is ${}^7C_5 = \dfrac{7!}{2!\,5!}$

$$= 21.$$

(b) Considering mutually exclusive groups:

Number of selections with no boys $= 21$

Number of selections with no girls $= 0$ (as there are only 4 boys)

Total number of possible selections $= {}^{11}C_5$

$= 462$

Thus, number of selections with at least one of each sex $= 462 - 21$

$= 441.$

(c) 2 boys can be chosen from 4, in ${}^4C_2 = 6$ ways

3 girls can be chosen from 7, in ${}^7C_3 = 35$ ways

These are independent events, so

number of teams with 2 boys and 3 girls $= 6 \times 35 = 210.$

(d) If the team is to have at least 3 boys, then there must be either 3 or 4 boys.

Number of teams with 3 boys and 2 girls $= {}^4C_3 \times {}^7C_2 = 84$

Number of teams with 4 boys and 1 girl $= {}^4C_4 \times {}^7C_1 = 7$

These are mutually exclusive events, so

number of teams with at least 3 boys $= 84 + 7 = 91.$

Example 22

How many permutations are there of the letters of the word PARALLELOGRAM? In how many of these are the A's separated?

The word has 3 A's, 3 L's, 2 R's and 5 other letters.

Number of permutations of these 13 letters $= \dfrac{13!}{3!\,3!\,2!} = 86\,486\,400$

Number of permutations of the 10 letters (omitting the A's) $= \dfrac{10!}{3!\,2!}$

The A's have now to be placed in any 3 of the available 11 spaces; the order in which the three spaces are selected is not important, so this can be done in ${}^{11}C_3$ ways.

Thus, number of permutations in which A's are separated $= {}^{11}C_3 \times \dfrac{10!}{3!\,2!}$

$= 49\,896\,000.$

Exercise 7C

1. How many different teams of 7 players can be chosen from 10 girls?

2. How many selections of 4 books can be made from 9 different books?

3. A gardener has space to plant 3 trees. In how many ways can he make his selection from 5 different trees?

4. A boy can choose any 7 stamps from a set of 11 stamps. How many different selections can he make?

5. Three pupils are to be promoted from a particular form. If five pupils are under consideration for promotion, in how many ways can the group to be promoted be selected?

6. A librarian chooses 16 books from a group of 20 books that need to be rebound. In how many ways can the 16 books be selected?

7. How many different hands of three cards can be chosen from a pack of 52 cards?

8. Four boys are to be selected from a group of 10 boys. How many different groups can be selected? If instead 6 boys were chosen, how many groups could be selected?

9. From the numbers 1, 2, 3, 4, ... 15, five odd numbers and three even numbers are to be selected. Find the number of different groups of eight numbers which can be chosen.

10. From a pack of 52 cards, how many different groups of 3 spades can be chosen?

11. A man decides to plant 3 shrubs and 4 trees. In how many ways can he make his selection if he has 5 shrubs and 6 trees from which to choose?

12. How many different groups can be selected from the letters a, b, c, d, e, f, g, h, i if each group is to include 2 vowels and 3 consonants?

13. A librarian has to make a selection of 5 newspapers and 7 magazines from the 8 newspapers and 9 magazines which are available. In how many ways can she make her selection?

14. (a) Find the number of different selections of 5 letters that can be made from the letters of the word CHEMISTRY.
    (b) How many of these selections contain at least one vowel?

15. (a) Find the number of different selections of 2 letters that can be made from the letters of the word METHOD.
    (b) How many of these selections contain no consonants?

16. (a) Find the number of different selections of 3 letters that can be made from the letters of the word METHOD.
    (b) How many of these selections contain no vowels?

17. (a) Find the number of different selections of 2 letters that can be made from the letters of the word STATISTICS.
    (b) How many of these selections contain no vowels?

18. (a) Find the number of different selections of 3 letters that can be made from the letters of the word STATISTICS.
    (b) How many of these selections contain no vowels?
    (c) How many of these selections contain at least one T?

19. How many different selections of 2 beads can be made from a bag containing ten beads, two of which are red and the other eight are of different colours?

20. A team of 6 players is to be selected from a group of ten with one of the six then being nominated as captain, and another as vice-captain. In how many ways can this be done?

21. A group consists of 5 boys and 8 girls. In how many ways can a team of four be chosen, if the team contains
    (a) no girls,   (b) no more than one girl,   (c) at least two boys?

22. A team of 5 managers is to be selected from a group of ten managers— 5 from company A, 3 from company B and 2 from company C. In how many ways can this be done if the team must contain at least one manager from each company?

23. In how many ways may a committee of 5 people be selected from 7 men and 3 women, if it must contain
    (a) 3 men and 2 women,   (b) 3 women and 2 men,   (c) at least 1 woman?

24. In how many ways can a committee of 7 people be selected from 4 men and 6 women if the committee must have at least 4 women on it?

25. A group consists of 5 boys and 8 girls. In how many ways can a team of five be chosen if it is to contain
    (a) no girls,   (b) no boys,
    (c) at least one boy,   (d) at least one member of each sex?
26. A tennis club has to select 2 mixed double pairs from a group of 5 men and 4 women. In how many ways can this be done?
27. How many permutations are there of the letters of the word AARDVARK? In how many of these permutations are the A's separated?
28. How many permutations are there of the letters of the word VILLIFIED? In how many of these permutations are the I's separated?
29. How many permutations are there of the letters of the word NONILLION? In how many of these permutations are the N's separated?

7.4 Selections of any size from a group

The number of possible selections of any size that can be made from a group of unlike things deserves special consideration.

Example 23

How many different selections can be made from the five letters a, b, c, d, e?

First method
Number of selections of 1 letter $= {}^5C_1 = 5$
Number of selections of 2 letters $= {}^5C_2 = 10$
Number of selections of 3 letters $= {}^5C_3 = 10$
Number of selections of 4 letters $= {}^5C_4 = 5$
Number of selections of 5 letters $= {}^5C_5 = 1$
Total number of possible selections $= 5 + 10 + 10 + 5 + 1 = 31.$

With a larger number of objects to select from, the above method is tedious.

Second method
In any given selection, the letter a is either included or not included, i.e. there are 2 ways of dealing with this letter.
Similarly the letter b is either included or not included; again there are 2 ways of dealing with this letter.
Extending this to all five letters, we see there are $2 \times 2 \times 2 \times 2 \times 2$ ways of dealing with the letters, but this includes the case in which none of the letters is included, and this is not a selection.

$$\text{Thus, number of selections} = 2^5 - 1$$
$$= 31 \text{ as obtained in method 1.}$$

In general there are $2^r - 1$ selections which can be made from r unlike items.

Group containing repeated items

It may be that the group from which the selections are to be made includes some repeated items. Suppose, for example that a group of letters includes 3 a's. These can be dealt with in 4 ways: either no a's, 1 a, 2 a's or 3 a's are included in a particular selection. The different letters can then be considered as before.

Example 24

How many different selections can be made from the letters of the word OSMOSIS?

There are 3 S's, 2 O's and 2 other different letters.
The S's can be dealt with in 4 ways.
The O's can be dealt with in 3 ways.
The M can be dealt with in 2 ways and the I can be dealt with in 2 ways.

$$\text{Total number of selections} = 4 \times 3 \times 2^2 - 1$$
$$= 47$$

7.5 Division into groups

We now consider examples in which a number of letters are divided into two or more groups of differing sizes.

Example 25

The letters a, b, c, d, e, f, g, h and i are to be divided into three groups containing 2, 3 and 4 letters respectively. In how many ways can this be done?

The 2 letters for the first group can be chosen in $^9C_2$ ways
The 3 letters for the second group can then be chosen in $^7C_3$ ways and the remaining 4 letters form the third group in 1 way

$$\text{Thus number of ways} = {}^9C_2 \times {}^7C_3 \times 1 = \frac{9!}{7!\,2!} \times \frac{7!}{4!\,3!} \times 1$$

$$= \frac{9!}{2!\,3!\,4!}$$

$$= 1260$$

In general, the number of ways of dividing $(p + q + r)$ unlike things into three groups containing p, q and r things respectively is

$$\frac{(p + q + r)!}{p!\,q!\,r!}.$$

It may be that the groups into which the things are to be divided are not of different sizes.

Example 26

Find the number of ways that 12 people can be arranged into groups if there are to be (a) 2 groups of 6 people, (b) 3 groups of 4 people.

(a) The first group of 6 people can be selected in $^{12}C_6$ ways, but the second group of 6 people is then automatically made up of those not in the first group. So there is only 1 way of selecting the second group.
But these $^{12}C_6$ ways will include the case when $a_1, a_2, a_3, \ldots a_6$ are in the first group and $a_7, a_8, \ldots a_{12}$ are in the second group and *vice versa*. So we must divide by the number of ways in which the 2 groups

can be permutated amongst themselves, i.e. 2!

Number of ways of dividing into two groups of 6 $= \dfrac{^{12}C_6 \times 1}{2!}$

$$= 462$$

(b) The first group of 4 people can be selected in $^{12}C_4$ ways.
The second group of 4 people can then be selected in $^8C_4$ ways.
The third group can then be selected in 1 way.
The 3 groups can be permutated amongst themselves in 3! ways,

thus number of ways of dividing into 3 groups $= \dfrac{^{12}C_4 \times {}^8C_4 \times 1}{3!}$

$$= \dfrac{12!}{(4!)^3 \times 3!} = 5775.$$

Exercise 7D

1. A group of 15 children are to be divided into three groups of 4, 5 and 6 children. In how many ways can this be done?
2. A gardener makes a selection from 7 different shrubs he is shown at a nursery. In how many ways can he make his selection?
3. How many selections can be made from the letters a, b, c, d, e, f?
4. Of 11 girls, 4 are to play tennis and 7 are to play netball. How many ways are there of forming the two groups?
5. How many ways are there of dividing 12 people into three groups containing 2 people, 7 people and 3 people?
6. For a group of 13 people, whilst 3 people are swimming, 4 will be playing tennis and 6 will be spectating. Find in how many ways the 13 people can be organised.
7. A boy has one of each of the following coins: 1p, 2p, 5p, 10p, 20p and 50p. In how many different ways can he make a contribution to a charity?
8. How many different whole numbers are factors of the number $2 \times 3 \times 5 \times 7 \times 11 \times 13$?
9. How many different selections can be made from the letters a_1, a_2, a_3, a_4, a_5, b, c, d?
10. In question 9, how many selections are possible if the suffixes are removed from the a's?
11. How many different selections can be made from the letters of the word HEELED?
12. How many different selections can be made from the letters of the word INABILITY?
13. Find how many different selections can be made from the letters of the word POSSESS.
14. Obtain a formula for the number of different selections that can be made of one or more letters taken from n letters, p of one type, q of another and the remainder being different.
15. Find the number of ways that 9 children can be divided into
    (a) a group of 5 and a group of 4 children,
    (b) three groups of 3 children.
16. Find in how many ways 11 people can be divided into three groups containing 3, 4 and 4 people.

17. In how many ways can 15 people be divided into three groups of 5 people?

18. A group of 15 boys are to be divided into two teams of 7 and the remaining boy to act as referee. In how many ways can this be done?

19. Find how many different selections can be made from the letters of the word BANANA.

Exercise 7E   Examination questions

1. How many different ways are there of selecting an executive committee of three from a general committee of eighteen members?   (S.U.J.B.)

2. To mark a cub leader's retirement, ex-cubs were asked to send three photographs for inclusion in a scrapbook to be presented to the leader. How many different ways could an ex-cub who had 13 suitable photographs select 3?   (S.U.J.B.)

3. When the scrapbook in question 2 was nearly complete the person compiling it had 9 photographs left but only room to include n of them. He calculated that there were 84 different ways of selecting the n from the 9. Find n, given that it is an even number.   (S.U.J.B.)

4. Find how many four digit numbers can be formed from the six digits 2, 3, 5, 7, 8 and 9, without repeating any digit.
Also find how many of these numbers
(a) are less than 7000,   (b) are odd.   (London)

5. How many different 6 digit numbers greater than 500 000 can be formed by using the digits 1, 5, 7, 7, 7, 8?   (Cambridge)

6. An athlete owns six pairs of running shoes. How many different ways are there of selecting
(a) 4 shoes from the 12;   (b) 2 left shoes and 2 right shoes;   (c) 2 pairs of shoes?   (S.U.J.B.)

7. Seven students are eligible for selection to a delegation of four students from a school to attend a conference. Two of them will not attend together but each is prepared to attend in the absence of the other. In how many different ways can the delegation be chosen?   (London)

8. Calculate the number of different 7-letter arrangements which can be made with the letters of the word MAXIMUM.
In how many of these do the 4 consonants all appear next to one another?   (Cambridge)

9. How many different words of five letters can be formed from 7 different consonants and 4 different vowels if no two consonants or vowels can come together and no repetitions are allowed? How many can be formed if each letter could be repeated any number of times?   (S.U.J.B.)

10. Given that a triangle can be formed by joining three non-collinear points, find the number of different triangles that can be formed using the points A, B, C, D, E, F, G, H, I, J K, L, if ABCDE and EFGHIJKL are two straight lines.   (London)

11. Find the number of different permutations of the 8 letters of the word SYLLABUS.

    Find the number of different selections of 5 letters which can be made from the letters of the word SYLLABUS.   (London)

12. It is given that the number of different ways in which n people can be seated at a round table is $(n - 1)!$.

    Two of the n people ($n > 3$) are to be kept separate in the seating arrangement. Find the number of different ways in which this can be done.   (Cambridge)

13. I have six balls, three red, one blue, one brown and one green. In how many ways can I arrange them in a line?

    In how many of these arrangements will no two red balls come together?   (S.U.J.B.)

14. A small holiday hotel advertises for a manager and 7 other members of staff. There are 4 applicants for the position of manager and 10 other people apply for the other jobs at the hotel. Find the number of different ways of selecting a group of people for the 8 jobs.

    The hotel has 4 single rooms, 6 double rooms and 5 family rooms. For a particular week, 4 individuals book single rooms, 3 couples book double rooms and 3 families book family rooms. Given that all the rooms are available for that week, find the number of different possible arrangements of bookings amongst the rooms.

    One afternoon, 12 guests organise a game requiring 2 teams of 6. Find the number of different ways of selecting the teams.

    Given that the 12 guests consist of 6 adults and 6 children and that each team must contain at least 2 adults, find the number of different ways of selecting the teams.   (J.M.B.)

15. (a) A book club offers a choice of 20 books of which a member chooses six. Find the number of different ways in which a member may make his choice.

    Given that 12 of the 20 books on offer are novels and that the other 8 are biographies, find the number of different ways in which a member chooses 6 so that

    (i) he has 3 novels and 3 biographies,
    (ii) he has at least 4 biographies.

    (b) A group of 6 boys and 5 girls are to be photographed together. The girls are to sit on 5 chairs placed in a row and the boys are to stand in a line behind them. Find the number of different possible arrangements.

    For a second photograph, the boys and girls are to be arranged with 3 boys and 3 girls standing whilst 3 boys and 2 girls are seated on the chairs in front of them, and in each row the boys and girls are to occupy alternate places.

    Find the number of different possible arrangements.   (J.M.B.)

8
Series and the binomial theorem

A set of numbers, stated in a definite order, such that each number can be obtained from the previous number according to some rule, is a **sequence**.
Each number of the sequence is called a **term**.
Consider the following   3, 5, 7, 9, 11, ...
$\qquad\qquad\qquad\quad$ 1, 4, 9, 16, 25, ...
$\qquad\qquad\qquad\quad$ 1, 2, 4, 8, 16, ...
Each of these is a sequence.
The '...' at the end of each sequence show that each one could go on indefinitely, i.e. the sequence is **infinite**. However, if we wished to restrict our attention to a limited number of terms of a sequence, we could write 3, 5, 7, 9, 11, ... 47. The final single full stop shows that the sequence ends when the number 47 is reached. Such a sequence is said to be **finite**.

An expression for the nth term (written u_n) of a sequence is useful since any specific term of the sequence can be obtained from it. The nth term of the sequence 1, 4, 9, 16, 25, ... is n^2.
Thus, for $n = 1$ we obtain the first term, written $u_1 = 1^2 = 1$
$\qquad$ for $n = 2$ we obtain the second term $\qquad u_2 = 2^2 = 4$
$\qquad$ for $n = 3$ we obtain the third term $\qquad u_3 = 3^2 = 9$ etc.
The sum of the terms of a sequence is called a **series**.
Thus $\quad 3 + 5 + 7 + 9 + 11 + ...$
$\qquad\quad 1 + 4 + 9 + 16 + 25 + ...$
and $\quad 1 + 2 + 4 + 8 + 16 + ...$ are all examples of series.

8.1 Arithmetic progressions

Consider the series $3 + 5 + 7 + 9 + 11 + ...$. The first term of the series is 3 and each subsequent term is obtained by adding a constant, 2, to the previous term. We say that the terms progress arithmetically and a series of this type is called an Arithmetic Progression or A.P. In the general case, if the first term is denoted by a and the constant (called the **common difference**) is d, the general arithmetic progression would be:
$\qquad a + (a + d) + (a + 2d) + (a + 3d) + ... + [a + (n - 1)d] + ...$
where $[a + (n - 1)d]$ is the expression for the nth term of the series.
It should be noted that the common difference d may be negative, in which case the terms of the progression would decrease, e.g. 15, $13\frac{1}{2}$, 12, $10\frac{1}{2}$, ...

Sum of the first n terms of an A.P.

Writing the sum of the first n terms of an A.P. as S_n, it follows that
$\qquad S_n = a + (a + d) + (a + 2d) + ... + [a + (n - 2)d] + [a + (n - 1)d]$ $\qquad$ [1]
Rewriting these terms in the reverse order
$\qquad S_n = [a + (n - 1)d] + [a + (n - 2)d] ... + (a + d) \quad + a$ $\qquad$ [2]

Adding [1] and [2]

$$2S_n = [2a + (n - 1)d] + [2a + (n - 1)d] + \ldots + [2a + (n - 1)d] + [2a + (n - 1)d]$$
$$= n[2a + (n - 1)d]$$

$$\therefore \quad S_n = \frac{n}{2}[2a + (n - 1)d]$$

The nth term, $[a + (n - 1)d]$, is the last term involved in the sum to n terms. Writing this last term as l, we can write

$$S_n = \frac{n}{2}(a + l) \text{ as an alternative form for } S_n \text{ in which } l = [a + (n - 1)d].$$

Summary

For the Arithmetic Progression $\quad a + (a + d) + (a + 2d) + (a + 3d) + \ldots$
the nth term is given by $\qquad u_n = a + (n - 1)d$

the sum to n terms is given by $\quad S_n = \frac{n}{2}[2a + (n - 1)d] \quad$ or $\quad \frac{n}{2}(a + l)$

Example 1

Find u_{15} and S_8 of (a) the A.P. $2 + 5 + 8 + 11 + \ldots$ (b) the A.P. in which the first term is 37 and the common difference is -4.

(a) $a = 2 \quad d = 5 - 2 = 3$

using $\quad u_n = a + (n - 1)d$

$u_{15} = 2 + (15 - 1)3$

$\qquad = 44$

using $\quad S_n = \frac{n}{2}[2a + (n - 1)d]$

$S_8 = \frac{8}{2}[2(2) + (8 - 1)3]$

$\qquad = 100$

(b) $a = 37 \quad d = -4$

using $\quad u_n = a + (n - 1)d$

$u_{15} = 37 + (15 - 1)(-4)$

$\qquad = -19$

using $\quad S_n = \frac{n}{2}[2a + (n - 1)d]$

$S_8 = \frac{8}{2}[2(37) + (8 - 1)(-4)]$

$\qquad = 184$

Example 2

Find the sum of the Arithmetical Progression $8 \cdot 5 + 12 + 15 \cdot 5 + 19 + \ldots + 103$.

The nth term is 103 $\quad \therefore \qquad a + (n - 1)d = 103$

$\qquad$ but $\quad a = 8 \cdot 5$ and $d = 3 \cdot 5$

$\qquad \therefore \quad 8 \cdot 5 + (n - 1)(3 \cdot 5) = 103$

$\qquad$ giving $\qquad\qquad n = 28$

using $\quad S_n = \frac{n}{2}(a + l)$

$S_{28} = \frac{28}{2}(8 \cdot 5 + 103)$

$\qquad = 1561$

or using $\quad S_n = \frac{n}{2}[2a + (n - 1)d]$

$S_{28} = \frac{28}{2}[2(8 \cdot 5) + (28 - 1)(3 \cdot 5)]$

$\qquad = 1561$

If sufficient information about a progression is given, it is possible to determine the first term a and the common difference d.

Example 3

The fifth term of an A.P. is 23 and the twelfth term is 37. Find the first term, the common difference and the sum of the first eleven terms.

$$u_5 = 23 \qquad\qquad\qquad u_{12} = 37$$
$$\therefore \quad a + 4d = 23 \quad \dots [1] \qquad\qquad \therefore \quad a + 11d = 37 \quad \dots [2]$$

Solving equations [1] and [2] simultaneously gives $a = 15$ and $d = 2$.

Using $\quad S_n = \dfrac{n}{2}[2a + (n - 1)d]$

$$S_{11} = \frac{11}{2}[2(15) + (11 - 1)2] = 275$$

The first term is 15, the common difference is 2 and the sum of the first 11 terms is 275.

Example 4

In an A.P. $u_{10} = 3$ and $S_6 = 76\cdot5$; find a, d and the smallest value of n such that $S_n < 0$.

Using $\quad u_n = a + (n - 1)d$ $\qquad$ Using $\quad S_n = \dfrac{n}{2}[2a + (n - 1)d]$

$$u_{10} = a + 9d \qquad\qquad\qquad S_6 = \frac{6}{2}[2a + 5d]$$

Thus $\quad 3 = a + 9d$ $\qquad\qquad$ and $\quad 76\cdot5 = 6a + 15d$

Solving these equations simultaneously gives $a = 16\cdot5$ and $d = -1\cdot5$.

Now suppose $S_n = 0$ i.e. $\qquad \dfrac{n}{2}[2a + (n - 1)d] = 0$

then $\qquad \dfrac{n}{2}[33 + (n - 1)(-1\cdot5)] = 0$

$$\frac{n}{2}(34\cdot5 - 1\cdot5n) = 0$$

or $\qquad\qquad\qquad \dfrac{3n}{4}(23 - n) = 0$

Thus for $S_n = 0$, either $n = 0$ (clearly not the required solution in this case)
$\qquad\qquad\qquad$ or $n = 23$

Thus for $S_n < 0$ we require $n > 23$.

The first term is $16\cdot5$, the common difference is $-1\cdot5$ and the smallest value of n for which S_n is negative is 24.

Exercise 8A

1. Write down the next two terms in each of the following sequences:
  (a) $-8, -5, -2, 1, 4, 7,$ ____, ____. $\qquad$ (b) $1, 4, 9, 16, 25, 36,$ ____, ____.
  (c) $0\cdot1, 0\cdot01, 0\cdot001, 0\cdot0001,$ ____, ____. $\qquad$ (d) $\frac{3}{4}, 1\frac{1}{2}, 2\frac{1}{4}, 3, 3\frac{3}{4}, 4\frac{1}{2},$ ____, ____.

2. Write down the first term and the common difference for each of the following A.P.s.
  (a) $8 + 11 + 14 + 17 + \dots$ $\qquad\qquad$ (b) $23 + 25 + 27 + 29 + \dots$
  (c) $19 + 16 + 13 + 10 + \dots$ $\qquad\qquad$ (d) $13\frac{1}{2} + 15 + 16\frac{1}{2} + 18 + \dots$
  (e) $-11\cdot5 - 9 - 6\cdot5 - 4 - \dots$ $\qquad\qquad$ (f) $6\frac{1}{4} + 6\frac{3}{4} + 7\frac{1}{4} + 7\frac{3}{4} + \dots$
  (g) $-8 - 7 - 6 - 5 - \dots$ $\qquad\qquad$ (h) $6 + 3 + 0 - 3 - \dots$

3. Find the number of terms in each of the following A.P.s.
   (a) $5 + 8 + 11 + 14 + \ldots + 59 + 62$.
   (b) $1 + 6 + 11 + 16 + \ldots + 501 + 506$.
   (c) $-193 - 189 - 185 - \ldots - 21 - 17$.
   (d) $2\frac{1}{4} + 2\frac{17}{20} + 3\frac{9}{20} + \ldots + 20\frac{1}{4} + 20\frac{17}{20}$.
4. Find the 18th term of a series that has an nth term given by $(2 + 3n)$.
5. Find the 31st term of a series that has an nth term given by $\frac{1}{3}(10 + 2n)$.
6. Find the 50th term of a series that has an nth term given by $\frac{1}{2}(32 - n)$.
7. Find the 6th and 7th terms of a series that has an nth term given by $(-1)^n(2n + 1)$.
8. Find an expression for the nth term of each of the following A.P.s. and use your answer to write down the 100th term of each series.
   (a) $5 + 8 + 11 + 14 + \ldots$
   (b) $5 + 2 - 1 - 4 - \ldots$
   (c) $12\frac{1}{2} + 16 + 19\frac{1}{2} + 23 + \ldots$
9. For each of the following A.P.s, state which is the first term to exceed 1000.
   (a) $7 + 12 + 17 + 22 + 27 + \ldots$ (b) $-24 - 21{\cdot}5 - 19 - 16{\cdot}5 - 14 - \ldots$
10. For each of the following A.P.s, state which is the first term to be negative.
    (a) $843 + 836 + 829 + 822 + \ldots$ (b) $56{\cdot}3 + 55{\cdot}4 + 54{\cdot}5 + 53{\cdot}6 + \ldots$
11. State the values of a and d in each of the following A.P.s, and find S_n as indicated.
    (a) $2 + 6 + 10 + 14 + \ldots, S_{12}$ (b) $10 + 8 + 6 + 4 + \ldots, S_{15}$
    (c) $4{\cdot}5 + 6 + 7{\cdot}5 + 9 + \ldots, S_{19}$ (d) $15 + 13 + 11 + 9 + \ldots, S_{16}$
    (e) $7 + 3 - 1 - 5 - \ldots, S_{20}$ (f) $-6\frac{1}{2} - 5 - 3\frac{1}{2} - 2 - \ldots, S_{12}$
    (g) $-9 - 7 - 5 - 3 - \ldots, S_{16}$
12. Find the sum of each of the following A.P.s.
    (a) $2 + 4 + 6 + 8 + 10 + \ldots + 146$.
    (b) $100 + 95 + 90 + 85 + 80 + \ldots - 20$.
    (c) $4 + 10 + 16 + 22 + 28 + \ldots + 334$.
    (d) $5\frac{1}{4} + 4\frac{1}{2} + 3\frac{3}{4} + \ldots - 3$.
13. In an A.P. $u_5 = 8$ and $u_9 = 14$; find a, d and S_{10}.
14. In an A.P. $u_3 = 7{\cdot}5$ and $u_{10} = 11$; find a, d and S_8.
15. In an A.P. $u_4 = 15$ and $u_8 = 7$; find a, d and S_{22}.
16. In an A.P. $u_3 = -4$ and $u_7 = 8$; find a, d and S_9.
17. In an A.P. $u_2 = -12$ and $S_{12} = 18$; find a, d and u_6.
18. In an A.P. $u_5 = -0{\cdot}5$ and $S_7 = 21$; find a, d and u_9.
19. In an A.P. $u_{15} = 7$ and $S_9 = 18$; find a, d and u_{20}.
20. The sum of the first ten terms of an A.P. is 120 and the sum of the first twenty terms is 840. Find the sum of the first thirty terms.
21. An A.P. has a common difference d. If the sum to twenty terms is twenty-five times the first term, find in terms of d the sum to thirty terms.
22. In an A.P. $a = -23$ and $d = 2{\cdot}5$; find the least value of n such that $S_n > 0$.
23. In an A.P. $a = -61$ and $d = 4$; find the least value of n such that $S_n > 0$.
24. An A.P. has first term 10 and common difference 0·25. Find the least number of terms the A.P. can have, given that the sum of the terms exceeds 300.
25. An A.P. has first term -5 and common difference 1·5. Find the greatest number of terms the A.P. can have, given that the sum of the terms does not exceed 450.

26. The sum to n terms of a particular series is given by $S_n = 17n - 3n^2$.
    (a) Find an expression for the sum to $(n - 1)$ terms,
    (b) find an expression for the nth term of the series,
    (c) show that the series is an Arithmetic Progression and find the first term and the common difference.

27. Three consecutive terms of an A.P. have a sum of 36 and a product of 1428. Find the three terms.

28. A particular A.P. has a positive common difference and is such that for any three adjacent terms, three times the sum of their squares exceeds the square of their sum by 37·5. Find the common difference.

29. Find the common difference, the nth term and the sum to n terms of the following A.P.
$$\log_e 3 + \log_e(3^2) + \log_e(3^3) + \log_e(3^4) + \dots$$

30. Find expressions for the nth term and the sum to n terms of the A.P.
$$\log_e(ab) + \log_e(ab^2) + \log_e(ab^3) + \log_e(ab^4) + \dots$$

8.2 Geometric progressions

Consider the series $2 + 6 + 18 + 54 + 162 + \dots$
Each term of the series can be obtained by multiplying the previous term by 3. In the general case, if the first term is denoted by a and the ratio of one term to the previous term is r (called the **common ratio**) the series can then be written

$$a + ar + ar^2 + ar^3 + \dots$$

It follows that the nth term u_n is ar^{n-1}.

A series of this type is called a Geometric Series or Geometric Progression, abbreviated to G.P.

It should be noted that the common ratio r may be positive, negative and/or fractional.

Sum of the first n terms of a G.P.

Writing the sum of the first n terms of a G.P. as S_n, it follows that
$$S_n = a + ar + ar^2 + \quad \dots \quad + ar^{n-2} + ar^{n-1}$$
hence $\qquad rS_n = \qquad ar + ar^2 + ar^3 + \dots \qquad\qquad + ar^{n-1} + ar^n$
subtracting
$$S_n - rS_n = a - ar^n$$
or $\qquad S_n(1 - r) = a(1 - r^n)$

i.e. $\qquad S_n = \dfrac{a(1 - r^n)}{(1 - r)}$ and this is the formula for the sum to n terms of a G.P. with first term a and common ratio r.

If the common ratio r is greater than 1, it is more convenient to use this result in the alternative form: $S_n = \dfrac{a(r^n - 1)}{(r - 1)}$.

Example 5

Find u_5 and S_5 of (a) the G.P. $\frac{12}{25} + \frac{6}{5} + 3 + \dots$
(b) the G.P. in which $a = 27$ and $r = \frac{2}{3}$.

(a) $a = \frac{12}{25}$ $r = \frac{6}{5} \div \frac{12}{25}$

                  $= \frac{5}{2}$

  using $u_n = ar^{n-1}$

          $u_5 = \frac{12}{25}(\frac{5}{2})^4$

   $\therefore$ $u_5 = 18\frac{3}{4}$

  using $S_n = \dfrac{a(r^n - 1)}{(r - 1)}$

        $S_5 = \frac{12}{25}\left[\dfrac{(\frac{5}{2})^5 - 1}{\frac{5}{2} - 1}\right]$

   $\therefore$ $S_5 = 30\cdot93$

  Thus $u_5 = 18\frac{3}{4}$ and $S_5 = 30\cdot93$.

(b) $a = 27$ $r = \frac{2}{3}$

  using $u_n = ar^{n-1}$

          $u_5 = 27(\frac{2}{3})^4$

   $\therefore$ $u_5 = 5\frac{1}{3}$

  using $S_n = \dfrac{a(1 - r^n)}{(1 - r)}$

        $S_5 = 27\left[\dfrac{1 - (\frac{2}{3})^5}{1 - \frac{2}{3}}\right]$

   $\therefore$ $S_5 = 70\frac{1}{3}$

  Thus $u_5 = 5\frac{1}{3}$ and $S_5 = 70\frac{1}{3}$.

Example 6

A geometric series has first term 27 and common ratio $\frac{4}{3}$. Find the least number of terms the series can have if its sum exceeds 550.

$$a = 27 \quad r = \tfrac{4}{3}$$

Now suppose that $S_n = 550$ i.e. $\dfrac{a(r^n - 1)}{r - 1} = 550$

                 then $27\dfrac{(\frac{4}{3})^n - 1}{\frac{4}{3} - 1} = 550$

                           $(\tfrac{4}{3})^n - 1 = \tfrac{550}{81}$

                             $(\tfrac{4}{3})^n = \tfrac{631}{81}$

     taking logs $n \log_e(\tfrac{4}{3}) = \log_e(\tfrac{631}{81})$

     hence $n = \log_e(\tfrac{631}{81}) \div \log_e(\tfrac{4}{3})$

                            $= 7\cdot136$

Thus for $S_n > 550$, we require $n > 7\cdot136$, i.e. $n = 8$.

If sufficient information about a geometric progression is given, it is possible to determine the first term a and the common ratio r of the progression.

Example 7

In a G.P. $u_3 = 32$ and $u_6 = 4$; find a, r and the sum of the first eight terms of the G.P.

      $u_3 = ar^2 = 32$ and $u_6 = ar^5 = 4$

  $\therefore$ $\dfrac{ar^5}{ar^2} = \dfrac{4}{32}$ i.e. $r^3 = \dfrac{1}{8}$ giving $r = \dfrac{1}{2}$

substituting for r in u_3 gives $a = 128$

using $S_n = \dfrac{a(1 - r^n)}{1 - r}$

      $S_8 = 128\dfrac{(1 - (\frac{1}{2})^8)}{1 - \frac{1}{2}}$

        $= 255$

Thus $a = 128$, $r = \frac{1}{2}$ and the sum of the first eight terms is 255.

Arithmetic mean and geometric mean

If some number x is inserted between a pair of numbers a and b such that a, x, b are in arithmetic progression, then x is said to be the **arithmetic mean** of a and b.

It follows that $\quad x - a = b - x$

$$\text{or} \qquad x = \frac{a + b}{2}.$$

Thus for any three consecutive terms of an A.P. the middle term is the arithmetic mean of the other two and is equal to half their sum.

If some number y is inserted between a pair of numbers p and q such that p, y, q are in geometric progression, then y is said to be the **geometric mean** of p and q.

It follows that $\qquad \dfrac{y}{p} = \dfrac{q}{y}$

$$\text{or} \qquad y = \sqrt{(pq)}$$

Thus for any three consecutive terms of a G.P., the middle term is the geometric mean of the other two and is equal to the square root of their product.

Example 8

For the numbers 4 and 9, find (a) the arithmetic mean, (b) the geometric mean.

(a) Arithmetic mean $= \dfrac{4 + 9}{2}$

$\qquad\qquad\qquad = 6\frac{1}{2}$

(i.e. 4, $6\frac{1}{2}$, 9 are in A.P.)

(b) Geometric mean $= \sqrt{(4 \times 9)}$

$\qquad\qquad\qquad = 6$

(i.e. 4, 6, 9 are in G.P.)

Sum to infinity of a G.P.

Consider the infinite geometric progression
$$18 + 1{\cdot}8 + 0{\cdot}18 + 0{\cdot}018 + 0{\cdot}0018 + 0{\cdot}00018 + \ldots$$
The first term a is 18 and the common ratio r is $\frac{1}{10}$.
For this G.P.
$$S_2 = 19{\cdot}8$$
$$S_3 = 19{\cdot}98$$
$$S_4 = 19{\cdot}998$$
$$S_5 = 19{\cdot}9998$$
$$S_6 = 19{\cdot}99998$$
$$S_7 = 19{\cdot}999998$$
$$S_8 = 19{\cdot}9999998$$
Clearly the sum S_n approaches the value 20. By taking a sufficiently large value of n, we can make S_n as near to 20 as we wish, i.e. we can make $(20 - S_n)$ as small as we wish. We say that S_n tends towards a **limiting** value of 20 as n approaches infinity and this is written
$$S_n \to 20 \quad \text{as} \quad n \to \infty \quad \text{or} \quad \lim_{n \to \infty} S_n = 20$$
Consider the general infinite G.P.
$$a + ar + ar^2 + ar^3 + ar^4 + \ldots$$
and suppose also that $-1 < r < 1$, i.e. $|r| < 1$.

Now $S_n = \dfrac{a(1 - r^n)}{1 - r}$ but with $|r| < 1$, r^n will approximate to zero for large values of n, i.e. $r^n \to 0$ as $n \to \infty$.

Thus $\qquad S_n \to \dfrac{a(1)}{1 - r} \quad$ as $\quad n \to \infty$

or $\qquad \lim_{n \to \infty} S_n = \dfrac{a}{1 - r}$

The expression $\dfrac{a}{1 - r}$ gives the **sum to infinity** of a geometric series with first term a and common ratio r. It must be remembered that S_∞ only exists for infinite geometric series in which $|r| < 1$. The terms of the series will decrease in magnitude and the series is said to be **convergent**.
When $|r| \geqslant 1$ we have various possibilities:

  if $r > 1$ then, as $n \to \infty$, $r^n \to \infty$ and $S_n \to +\infty$ or $-\infty$ dependent on the sign of a,

  if $r < -1$ then, as $n \to \infty$, S_n alternates between being large and positive and being large and negative,

  if $r = 1$, the series becomes $a + a + a + a + \ldots$ and $S_n = na$. Hence, as $n \to \infty$, $S_n \to +\infty$ or $-\infty$ dependent on the sign of a,

  if $r = -1$, the series becomes $a - a + a - a + a - \ldots$ and $S_n = 0$ for even n, and $S_n = a$ for odd n.

Any series whose sum does not approach some finite limit as $n \to \infty$ is said to be **divergent**. Thus any geometric series for which $|r| \geqslant 1$ is divergent.

Example 9

Find the sum to infinity of the geometric series $16 + 12 + 9 + \ldots$

$$a = 16, \quad r = \tfrac{12}{16} = \tfrac{3}{4}$$
Thus $|r| < 1$; hence the series is convergent and S_∞ will exist.

$$\therefore \quad S_\infty = \dfrac{a}{1 - r} = \dfrac{16}{1 - \tfrac{3}{4}}$$
$$= 64$$
The sum to infinity of the series is 64.

Example 10

Express the recurring decimal $0\cdot2\dot{3}\dot{5}$ as a fraction in its lowest terms.

$$0\cdot2\dot{3}\dot{5} = 0\cdot2353535\ldots = 0\cdot2 + 0\cdot035 + 0\cdot00035 + 0\cdot0000035 + \ldots$$
$$= 0\cdot2 + \text{an infinite geometric series with } a = 0\cdot035 \text{ and } r = \tfrac{1}{100}$$
$$= 0\cdot2 + S_\infty$$

using $\quad S_\infty = \dfrac{a}{1 - r}$

$$S_\infty = \dfrac{0\cdot035}{1 - \tfrac{1}{100}}$$
$$= \dfrac{35}{1000} \times \dfrac{100}{99} = \dfrac{7}{198}$$
$$\therefore \quad 0\cdot2\dot{3}\dot{5} = 0\cdot2 + \tfrac{7}{198}$$
$$= \tfrac{1}{5} + \tfrac{7}{198} = \tfrac{233}{990}.$$

Summary

For the geometric series $a + ar + ar^2 + ar^3 + \ldots$

the nth term is given by $u_n = ar^{n-1}$,

the sum to n terms is given by $S_n = \dfrac{a(1 - r^n)}{1 - r}$ or $S_n = \dfrac{a(r^n - 1)}{r - 1}$,

and for $|r| < 1$ the series is convergent with $S_\infty = \dfrac{a}{1 - r}$.

Exercise 8B

1. For each of the following G.P.s, state the common ratio and the next two terms:
   (a) $4 + 20 + 100 + 500 + \ldots$
   (b) $24 + 12 + 6 + \ldots$
   (c) $45 + 15 + 5 + \ldots$

2. Using the formula $S_n = \dfrac{a(r^n - 1)}{r - 1}$, find S_{10} for the G.P. 2, 6, 18, 54, $\ldots$

3. Using the formula $S_n = \dfrac{a(1 - r^n)}{1 - r}$, find S_5 and S_6 for
   the G.P. 18, -9, $4\frac{1}{2}$, $\ldots$ and hence deduce the value of u_6.

4. Using the formula $S_n = \dfrac{a(r^n - 1)}{r - 1}$, find S_7 for the G.P. in which
   (a) $a = \frac{3}{8}$ and $r = 4$,   (b) $a = \frac{3}{8}$ and $r = -4$.

5. Find the value of the common ratio of the G.P. that has third term equal to 6 and eighth term equal to 1458.

6. Find the value of the common ratio of the G.P. that has second term equal to 4 and fifth term equal to $-\frac{1}{16}$.

7. In a G.P. the seventh term equals 8 and the ninth term equals 18; find the possible values of the common ratio.

8. In a G.P. the fourth term equals 6 and the eighth term equals 96; find the possible values of the common ratio and the corresponding values of the first term.

9. Find the sum of the first ten terms of a G.P. that has a sixth term of $\frac{32}{33}$ and a seventh term of $1\frac{31}{33}$.

10. Find the sum of the first seven terms of a G.P. that has an eighth term of $\frac{2}{3}$ and a fifth term of 18.

11. Find the arithmetic mean of each of the following pairs of numbers:
    (a) 8, 20   (b) -5, -1 (c) -2, 10   (d) 3, 12.

12. Find the geometric mean of each of the following pairs of numbers:
    (a) 4, 16   (b) 8, 18   (c) $1\frac{1}{4}$, 5   (d) $1\frac{3}{4}$, 7.

13. The geometric mean of two numbers a and $b (b > a)$ is equal to four-fifths of the arithmetic mean of the two numbers. If $a = 6$, find the value of b.

14. For each of the following series state whether they are convergent or divergent and, for those that are convergent, find the sum to infinity:
    (a) $1 + 4 + 7 + 10 + 13 + \ldots$ (b) $16 + 8 + 4 + 2 + 1 + \ldots$
    (c) $84 - 42 + 21 - 10\frac{1}{2} + 5\frac{1}{4} - \ldots$ (d) $125 + 25 + 5 + 1 + \ldots$
    (e) $8 + 12 + 18 + 27 + 40\frac{1}{2} + \ldots$ (f) $64 - 16 + 4 - 1 + \frac{1}{4} - \ldots$

15. For each of the following geometric series, find the range of values of x for which the sum to infinity of the series exists:

    (a) $x + x^2 + x^3 + x^4 + \ldots$ (b) $1 + \dfrac{x}{3} + \dfrac{x^2}{9} + \dfrac{x^3}{27} + \ldots$

    (c) $1 + \dfrac{1}{x} + \dfrac{1}{x^2} + \dfrac{1}{x^3} + \ldots$ (d) $(x + \frac{1}{2}) + (x + \frac{1}{2})^2 + (x + \frac{1}{2})^3 + \ldots$

16. Find the sum to infinity of the G.P. having a second term of -9 and a fifth term of $\frac{1}{3}$.

17. The fourth, eighth and fourteenth terms of an A.P., common difference 0·5, are in geometric progression. Find the first term of the A.P. and the common ratio of the G.P.

18. The fourth, seventh and sixteenth terms of an A.P. are in geometric progression. If the first six terms of the A.P. have a sum of 12, find the common difference of the A.P. and the common ratio of the G.P.

19. The third, fifth and seventeenth terms of an A.P. are in geometric progression. Find the common ratio of the G.P.

20. A mathematical child negotiates a new pocket money deal with her unsuspecting father in which she receives 1p on the first day of the month, 2p on the second day, 4p on the third day, 8p on the fourth day, 16p on the fifth day, ... until the end of the month. How much would the child receive during the course of a month of 30 days? (Give your answer to the nearest million pounds.)

21. Find the common ratio of a G.P. that has a first term of 5 and a sum to infinity of 15.

22. Find the first term of a G.P. that has a common ratio of $\frac{2}{3}$ and a sum to infinity of 40.

23. Find the third term of a G.P. that has a common ratio of $-\frac{1}{4}$ and a sum to infinity of 8.

24. The sum of the first two terms of a G.P. is 9 and the sum to infinity of the G.P. is 25. If the G.P. has a positive common ratio r, find r and the first term.

25. A geometric series has first term 2 and common ratio $\frac{3}{2}$. Find the greatest number of terms the series can have without its sum exceeding 125.

26. A geometric series has first term 5 and common ratio 3. Find the least number of terms the series can have if its sum exceeds 2000.

27. Express the following recurring decimals as fractions in their lowest terms:
    (a) $0\cdot\dot{7}$ (b) $0\cdot\dot{5}\dot{0}$ (c) $0\cdot4\dot{5}\dot{2}$ (d) $0\cdot3\dot{6}\dot{0}$

28. Prove that the arithmetic mean of two different positive numbers exceeds the geometric mean of the same two numbers.

8.3 Summation of other series and proof by induction

The sums of certain finite series can be proved by the method of induction. This method can only be used to prove that a *given* expression is the required sum because the method does not produce the expression itself. The following example illustrates its use.

Example 11

Prove that the sum of the series $1^2 + 2^2 + 3^2 + \ldots + n^2$

is $\dfrac{n}{6}(n + 1)(2n + 1)$.

To prove this result by induction, we proceed as follows.

(a) Show that the result is true for a particular value of n, say $n = 1$.
  If $n = 1$, $1^2 = 1$ and $\frac{1}{6}(1 + 1)(2 + 1) = 1$
  Hence the result is true for $n = 1$.

(b) Prove that, *if* the result is true for some general value of n, say $n = k$, it
  must also be true for the next value of n, i.e. $n = k + 1$.
  Assume the result is true for $n = k$,

  i.e. $1^2 + 2^2 + 3^2 + \ldots + k^2 = \dfrac{k}{6}(k + 1)(2k + 1)$

From this it follows that

$$1^2 + 2^2 + 3^2 + \ldots + k^2 + (k + 1)^2 = \dfrac{k}{6}(k + 1)(2k + 1) + (k + 1)^2$$

$$= \dfrac{(k + 1)}{6}[k(2k + 1) + 6(k + 1)]$$

$$= \dfrac{(k + 1)}{6}(2k^2 + 7k + 6)$$

$$\therefore \quad 1^2 + 2^2 + 3^2 + \ldots + (k + 1)^2 = \dfrac{(k + 1)}{6}(k + 2)(2k + 3)$$

$$= \dfrac{n}{6}(n + 1)(2n + 1) \quad \text{with} \quad n = k + 1.$$

Thus *if* the result is true for $n = k$, it must also be true for $n = k + 1$.

From (a) $1^2 + 2^2 + 3^2 + \ldots + n^2 = \dfrac{n}{6}(n + 1)(2n + 1)$ *is* true for $n = 1$

and using (b), if it is true for $n = 1$, it must be true for $n = 2$. Using (b)
again, if it is true for $n = 2$, it must be true for $n = 3$, and so on. Hence the
result is true for all positive integral values of n, i.e. for every finite number
of terms.

$$\text{Thus} \quad 1^2 + 2^2 + 3^2 + \ldots + n^2 = \dfrac{n}{6}(n + 1)(2n + 1).$$

A more concise way of writing the series $1^2 + 2^2 + 3^2 + \ldots + n^2$ is to
use the symbol Σ (a Greek letter pronounced 'sigma').

$\displaystyle\sum_{r=1}^{n} f(r)$ means the series with first term $f(1)$, second term $f(2)$, $\ldots$ last term $f(n)$.

Thus, $1^2 + 2^2 + 3^2 + \ldots + n^2 = \displaystyle\sum_{r=1}^{n} r^2$ and the result proved above can

then be written:

$$\sum_{r=1}^{n} r^2 = \dfrac{n}{6}(n + 1)(2n + 1).$$

Example 12

Write down the series $\displaystyle\sum_{r=5}^{10} r(r + 1)$ in full.

For $r = 5$ $r(r + 1) = 5 \times 6 = 30$,
for $r = 6$ $r(r + 1) = 6 \times 7 = 42$ and so on to $r = 10$.

Thus $\displaystyle\sum_{r=5}^{10} r(r + 1) = 5 \times 6 + 6 \times 7 + 7 \times 8 + 8 \times 9 + 9 \times 10 + 10 \times 11$
$$= 30 + 42 + 56 + 72 + 90 + 110.$$

Example 13

Express the series $1 + 4 + 7 + 10 + 13 + \ldots + 298$ in the form $\displaystyle\sum_{r=1}^{n} f(r)$.

In $\displaystyle\sum_{r=1}^{n} f(r)$, $f(r)$ is the expression for the rth term of the series.

By inspection, $1 + 4 + 7 + 10 + 13 + \ldots + 298$ has rth term $(3r - 2)$.
Alternatively, had we not been able to state this by inspection, we could say:
$1 + 4 + 7 + 10 + \ldots$ is an arithmetic progression with $a = 1$ and $d = 3$.
Thus the rth term is $a + (r - 1)d$
$$= 1 + (r - 1)3$$
$$= 3r - 2$$
As the final term is 298, the greatest value of r in this series is that for which
$3r - 2 = 298$, i.e. $r = 100$.

$$\therefore \quad 1 + 4 + 7 + 10 + 13 + \ldots + 298 = \sum_{r=1}^{100} (3r - 2).$$

Example 14

Use the method of induction to prove that $\displaystyle\sum_{r=1}^{n} 2^{r-1} = 2^n - 1$.

If $n = 1$ L.H.S. $= \displaystyle\sum_{r=1}^{1} 2^{r-1} = 2^{1-1}$ R.H.S. $= 2^1 - 1$
$$= 1 \qquad\qquad\qquad\qquad = 1$$
Thus the statement is true for $n = 1$.

Assume the statement is true for $n = k$, i.e. $\displaystyle\sum_{r=1}^{k} 2^{r-1} = 2^k - 1$

then $\displaystyle\sum_{r=1}^{k+1} 2^{r-1} = \sum_{r=1}^{k} 2^{r-1} + 2^{k+1-1}$
$$= 2^k - 1 + 2^k$$
$$= 2(2^k) - 1$$

$$\therefore \quad \sum_{r=1}^{k+1} 2^{r-1} = 2^{k+1} - 1$$
$$= 2^n - 1 \quad \text{with} \quad n = k + 1$$

Thus if $\displaystyle\sum_{r=1}^{n} 2^{r-1} = 2^n - 1$ is true for $n = k$, it is also true for $n = k + 1$.

However it is true for $n = 1$, and so it must be true for $n = 2$ and also for $n = 3, 4, 5 \ldots$

Thus $\displaystyle\sum_{r=1}^{n} 2^{r-1} = 2^n - 1$ for all positive integer values of n.

The following results hold for the sigma notation:

$$\sum_{r=1}^{n} af(r) = a \sum_{r=1}^{n} f(r) \qquad \text{and} \qquad \sum_{r=1}^{n} [f(r) + g(r)] = \sum_{r=1}^{n} f(r) + \sum_{r=1}^{n} g(r)$$

Proof: $\displaystyle\sum_{r=1}^{n} af(r)$ Proof: $\displaystyle\sum_{r=1}^{n} [f(r) + g(r)]$

$$= af(1) + af(2) + \ldots + af(n)$$
$$= a[f(1) + f(2) + \ldots + f(n)]$$
$$= a \sum_{r=1}^{n} f(r)$$

$$= f(1) + g(1) + \ldots f(n) + g(n)$$
$$= [f(1) + \ldots f(n)] + [g(1) + \ldots g(n)]$$
$$= \sum_{r=1}^{n} f(r) + \sum_{r=1}^{n} g(r)$$

Example 15

(a) Using the standard result $\displaystyle\sum_{r=1}^{n} r^2 = \frac{n}{6}(n + 1)(2n + 1)$, obtain an expression for $\displaystyle\sum_{r=1}^{n} (2r^2 - 1)$, simplifying your answer as far as is possible.

(b) Use your answer for part (a) to sum the series $1 + 7 + 17 + 31 + \ldots + 799$.

(a) $\displaystyle\sum_{r=1}^{n} (2r^2 - 1) = 2\sum_{r=1}^{n} r^2 - \sum_{r=1}^{n} 1$

$$= \frac{n}{3}(n + 1)(2n + 1) - n \qquad \text{[Note: } \sum_{r=1}^{n} 1 = 1 + 1 + 1 \ldots + 1 = n]$$

$$= \frac{n}{3}[(n + 1)(2n + 1) - 3]$$

$$= \frac{n}{3}(n + 2)(2n - 1)$$

(b) $1 + 7 + 17 + 31 + \ldots + 799 = \displaystyle\sum_{r=1}^{20} (2r^2 - 1)$ since $u_{20} = 799$

$$= \frac{20}{3}(20 + 2)(40 - 1)$$

$$= 5720$$

Exercise 8C

1. Write each of the following series in full:

(a) $\displaystyle\sum_{r=1}^{5} r$ (b) $\displaystyle\sum_{r=1}^{5} r^2$ (c) $\displaystyle\sum_{r=7}^{9} (2r - 1)$ (d) $\displaystyle\sum_{r=5}^{10} \frac{2520}{r}$ (e) $\displaystyle\sum_{r=1}^{5} (-1)^r r$ (f) $\displaystyle\sum_{r=1}^{5} (-1)^{r+1} r$

2. Express each of the following series in the form $\displaystyle\sum_{r=1}^{n} f(r)$

(a) $1 + 2 + 3 + 4 + \ldots + 50$,   (b) $1^3 + 2^3 + 3^3 + 4^3 + \ldots + 50^3$,
(c) $5 + 10 + 15 + 20 + 25 + 30$,   (d) $3 + 5 + 7 + 9 + 11 + 13$,
(e) $3 - 6 + 9 - 12 + 15 - 18$,   (f) $-1 + 4 - 9 + 16 - 25 + 36 - \ldots - 169$.

3. Prove each of the following statements by the method of induction.

(a) $1 + 2 + 3 + 4 + \ldots + n = \dfrac{n}{2}(n + 1)$

(b) $1^3 + 2^3 + 3^3 + \ldots + n^3 = \dfrac{n^2}{4}(n + 1)^2$

(c) $1 \times 3 + 2 \times 4 + 3 \times 5 + \ldots + n(n + 2) = \frac{n}{6}(n + 1)(2n + 7)$.

(d) $\frac{1}{1 \times 2} + \frac{1}{2 \times 3} + \frac{1}{3 \times 4} + \ldots + \frac{1}{n(n + 1)} = \frac{n}{n + 1}$

In the remaining questions of this exercise, the following standard results may be quoted and used without proof:

$$\sum_{r=1}^{n} r = \frac{n}{2}(n + 1); \quad \sum_{r=1}^{n} r^2 = \frac{n}{6}(n + 1)(2n + 1); \quad \sum_{r=1}^{n} r^3 = \frac{n^2}{4}(n + 1)^2$$

4. Write each of the following series in the form $\sum_{r=1}^{n} f(r)$ and, using the standard results for $\sum r$, $\sum r^2$ and $\sum r^3$, evaluate each summation.

(a) $1 + 2 + 3 + 4 + \ldots + 100$,

(b) $1 + 4 + 9 + 16 + 25 + \ldots + 484$,

(c) $1 + 8 + 27 + 64 + 125 + \ldots + 3375$,

(d) $1 + 3 + 5 + 7 + 9 + \ldots + 101$,

(e) $1 \times 3 + 2 \times 4 + 3 \times 5 + 4 \times 6 + \ldots + 20 \times 22$,

(f) $1 \times 1 + 2 \times 3 + 3 \times 5 + 4 \times 7 + 5 \times 9 + \ldots + 13 \times 25$.

5. Using the standard result for $\sum_{r=1}^{n} r^3$ evaluate (a) $\sum_{r=1}^{24} r^3$, (b) $\sum_{r=1}^{15} r^3$, (c) $\sum_{r=16}^{24} r^3$.

6. (a) Using the standard results for $\sum r^2$ and $\sum r$, obtain an expression for $\sum_{r=1}^{n} r(r + 1)$ and simplify your answer as far as possible.

(b) Use your answer from part (a) to sum the series
$1 \times 2 + 2 \times 3 + 3 \times 4 + 4 \times 5 + \ldots + 28 \times 29$.

7. (a) Using the standard results for $\sum r^3$, $\sum r^2$ and $\sum r$, obtain an expression for $\sum_{r=1}^{n} r(r + 1)(r + 2)$ and simplify your answer as far as possible.

(b) Use your answer from part (a) to sum the series
$1 \times 2 \times 3 + 2 \times 3 \times 4 + 3 \times 4 \times 5 + \ldots + 20 \times 21 \times 22$.

8. (a) Using the standard results for $\sum r^2$ and $\sum r$, obtain an expression for $\sum_{r=1}^{n} (r + 1)(2r + 1)$ and simplify your answer as far as possible.

(b) Use your answer from part (a) to sum the series
$2 \times 3 + 3 \times 5 + 4 \times 7 + \ldots + 21 \times 41$.

9. Prove that $\sum_{r=1}^{n} r^2(r + 1) = \frac{n}{12}(n + 1)(n + 2)(3n + 1)$ by using

(a) the method of induction,   (b) the standard results for $\sum r^3$ and $\sum r^2$.

10. Find expressions in terms of n for the sums of the following series, to the number of terms stated, simplifying your answers as far as is possible.

(a) $1 + 2 + 3 + 4 + \ldots$ to $2n$ terms,

(b) $1^2 + 2^2 + 3^2 + 4^2 + \ldots$ to $(n + 1)$ terms,

(c) $1^3 + 2^3 + 3^3 + 4^3 + \ldots$ to $2n$ terms,

(d) $7 + 9 + 11 + 13 + 15 + \ldots$ to $2n$ terms,

(e) $2^3 + 4^3 + 6^3 + 8^3 + \ldots$ to n terms,

(f) $2^3 + 4^3 + 6^3 + 8^3 + \ldots$ to $2n$ terms,

(g) $1 \times 4 + 2 \times 5 + 3 \times 6 + 4 \times 7 + \ldots$ to n terms,

(h) $1 \times 2^2 + 2 \times 3^2 + 3 \times 4^2 + 4 \times 5^2 + \ldots$ to n terms,

(i) $2 \times 1 + 4 \times 3 + 6 \times 5 + 8 \times 7 + \ldots$ to n terms,

(j) $1^2 \times 1 + 2^2 \times 3 + 3^2 \times 5 + 4^2 \times 7 + \ldots$ to n terms.

11. Using the standard result for $\displaystyle\sum_{r=1}^{n} r^3$ show that $\displaystyle\sum_{r=n+1}^{2n} r^3 = \frac{n^2}{4}(3n + 1)(5n + 3)$.

12. Prove each of the following statements by the method of induction.

(a) $\displaystyle\sum_{r=1}^{n} [a + (r - 1)d] = \frac{n}{2}[2a + (n - 1)d]$ (b) $\displaystyle\sum_{r=1}^{n} ap^{r-1} = \frac{a(p^n - 1)}{p - 1}$.

13. Use the method of induction to prove that $6^n - 1$ is divisible by 5 for all positive integral values of n (Hint: show that it is true for $n = 1$ and show that, if $6^k - 1 = 5A$ where A is a positive integer, then $6^{k+1} - 1 = 5B$, where B is a positive integer).

Prove also that $8^n - 7n + 6$ is divisible by 7 for all positive integral values of n.

8.4 The binomial theorem for a positive integral index

By expanding, it can be shown that $(a + x)^1 = a + x$

$$(a + x)^2 = a^2 + 2ax + x^2$$
$$(a + x)^3 = a^3 + 3a^2x + 3ax^2 + x^3$$
$$(a + x)^4 = a^4 + 4a^3x + 6a^2x^2 + 4ax^3 + a^4.$$

Extracting the coefficients of a and x we obtain Pascal's triangle:

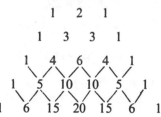

and the next two lines would be

Pascal's triangle can be used to expand expressions of the type $(a + x)^n$.

Example 16

Expand $(2 - x)^5$.

Selecting from Pascal's triangle, the line that commences 1 5 ... gives:

1   5   10   10   5   1

Inserting decreasing powers of 2, from 2^5 to 2^0, and increasing powers of $(-x)$, from $(-x)^0$ to $(-x)^5$

$$(2-x)^5 = 1 \times 2^5(-x)^0 + 5 \times 2^4(-x)^1 + 10 \times 2^3(-x)^2 + 10 \times 2^2(-x)^3 + 5 \times 2^1(-x)^4 + 1 \times 2^0(-x)^5$$
$$= 32 - 80x + 80x^2 - 40x^3 + 10x^4 - x^5$$

Pascal's triangle can be used to obtain the expansion of $(a + x)^n$ for any positive integral values of n. However, for large values of n the task of obtaining the relevant line of the triangle would be tedious. Indeed, as we

shall see later, we may only require the first few terms of the expansion, or perhaps just one particular term and in such cases it is better to use the **binomial theorem**.
According to this theorem

$$(1+x)^n = \sum_{r=0}^{n} {}^nC_r x^r \quad \left(\text{remember that } {}^nC_r = \frac{n!}{(n-r)!\,r!}\right)$$

$$= {}^nC_0 x^0 + {}^nC_1 x^1 + {}^nC_2 x^2 + {}^nC_3 x^3 + \ldots + {}^nC_n x^n$$

$$\therefore \quad (1+x)^n = 1 + {}^nC_1 x + {}^nC_2 x^2 + {}^nC_3 x^3 + \ldots + x^n$$

$$\text{or} \quad (1+x)^n = 1 + \frac{nx}{1!} + \frac{n(n-1)x^2}{2!} + \frac{n(n-1)(n-2)x^3}{3!} + \ldots + x^n$$

These are two forms of the binomial expansion of $(1+x)^n$ for positive integral values of n. The R.H.S. gives the series expansion of $(1+x)^n$ in ascending powers of x.

Proof of the binomial expansion of $(1 + x)^n$

We know that $(1+x)^n = (1+x)(1+x)(1+x)\ldots(1+x)$

n brackets

When expanding the R.H.S. the constant term is obtained by multiplying the *n* 1's together and it is therefore 1.

The term involving x is obtained by multiplying $(n-1)$ 1's and one x together. The x can be chosen from any one of the n brackets, so the coefficient of x is ${}^nC_1$.

The term involving x^2 is obtained by multiplying $(n-2)$ 1's and two x's together. The x's can be chosen from any two of the n brackets, so the coefficient of x^2 is ${}^nC_2$. We continue this process until eventually we multiply the n x's together to get x^n.

Thus $(1+x)^n = 1 + {}^nC_1 x + {}^nC_2 x^2 + \ldots + x^n$

The binomial theorem for $(1+x)^n$ can also be proved by using the method of induction. The proof uses the fact that ${}^nC_r + {}^nC_{r-1} = {}^{n+1}C_r$, proved below.

$${}^nC_r + {}^nC_{r-1} = \frac{n!}{(n-r)!\,r!} + \frac{n!}{(n-r+1)!\,(r-1)!}$$

$$= \frac{(n-r+1)\,n!}{(n-r+1)!\,r!} + \frac{n!\,r}{(n-r+1)!\,r!}$$

$$= \frac{n!}{(n-r+1)!\,r!}(n-r+1+r)$$

$$= \frac{(n+1)!}{(n+1-r)!\,r!} = {}^{n+1}C_r$$

Proof of the binomial expansion by induction

To prove $(1 + x)^n = {}^nC_0 + {}^nC_1x + {}^nC_2x^2 + {}^nC_3x^3 + \ldots + {}^nC_{n-1}x^{n-1} + {}^nC_nx^n$.

If $n = 1$, L.H.S. $= (1 + x)^1$　　　　R.H.S. $= {}^1C_0 + {}^1C_1x$
　　　　　　　　$= 1 + x$　　　　　　　　　　$= 1 + x$

Hence the result is true for $n = 1$.

Assume that the result is correct for $n = k$,

i.e.　　　　$(1+x)^k = {}^kC_0 + {}^kC_1x + {}^kC_2x^2 + {}^kC_3x^3 + \ldots + {}^kC_{k-1}x^{k-1} + {}^kC_kx^k$

then　　$(1+x)^{k+1} = (1+x)({}^kC_0 + {}^kC_1x + {}^kC_2x^2 + {}^kC_3x^3 + \ldots + {}^kC_{k-1}x^{k-1} + {}^kC_kx^k)$

$\qquad = {}^kC_0 + {}^kC_1x + {}^kC_2x^2 + {}^kC_3x^3 + \ldots + {}^kC_{k-1}x^{k-1} + {}^kC_kx^k +$

$\qquad\qquad {}^kC_0x + {}^kC_1x^2 + {}^kC_2x^3 + {}^kC_3x^4 + \ldots + {}^kC_{k-1}x^k + {}^kC_kx^{k+1}$

$\qquad = 1 + ({}^kC_1 + {}^kC_0)x + ({}^kC_2 + {}^kC_1)x^2 + ({}^kC_3 + {}^kC_2)x^3 + \ldots + ({}^kC_k + {}^kC_{k-1})x^k + {}^kC_kx^{k+1}$

$\qquad = 1 + {}^{k+1}C_1x + {}^{k+1}C_2x^2 + {}^{k+1}C_3x^3 + \ldots + {}^{k+1}C_kx^k + x^{k+1}$

$\qquad = 1 + {}^nC_1x + {}^nC_2x^2 + {}^nC_3x^3 + \ldots + x^n$　with　$n = k + 1$

Thus if the result is true for $n = k$, it is also true for $n = k + 1$. However, the result is true for $n = 1$ and is therefore true for $n = 2, 3, 4, \ldots$ and so on. Thus the binomial expansion of $(1 + x)^n$ is proved for positive integral values of n.

Using　$(1 + x)^n = 1 + {}^nC_1x + {}^nC_2x^2 + {}^nC_3x^3 + \ldots + x^n$

$$(a+x)^n = \left[a\left(1 + \frac{x}{a}\right)\right]^n$$

$$= a^n\left(1 + \frac{x}{a}\right)^n = a^n\left[1 + {}^nC_1\left(\frac{x}{a}\right) + {}^nC_2\left(\frac{x}{a}\right)^2 + {}^nC_3\left(\frac{x}{a}\right)^3 + \ldots + \left(\frac{x}{a}\right)^n\right]$$

Thus　$(a + x)^n = a^n + {}^nC_1a^{n-1}x + {}^nC_2a^{n-2}x^2 + {}^nC_3a^{n-3}x^3 + \ldots + x^n$

or　　$(a + x)^n = a^n + \dfrac{n}{1!}a^{n-1}x + \dfrac{n(n-1)}{2!}a^{n-2}x^2 + \dfrac{n(n-1)(n-2)}{3!}a^{n-3}x^3 + \ldots + x^n$

This is the binomial expansion of $(a + x)^n$ for positive integral values of n.

Example 17

Use the binomial theorem to expand $(2 + 3x)^5$.

Using the binomial expansion of $(a + x)^n$ with $a = 2, n = 5$ and $3x$ substituted for x:

$(2 + 3x)^5 = 2^5 + {}^5C_12^4(3x)^1 + {}^5C_22^3(3x)^2 + {}^5C_32^2(3x)^3 + {}^5C_42^1(3x)^4 + (3x)^5$
$\qquad\quad = 32 + \qquad\quad 240x + \qquad\quad 720x^2 + \qquad\quad 1080x^3 + \qquad\quad 810x^4 + 243x^5$

Example 18

Use the binomial expansion to expand $(2 - x)^{10}$ in ascending powers of x up to and including the term in x^3.

Using the expansion of $(a + x)^n$ with $a = 2$, $n = 10$ and $(-x)$ substituted for x:

$$(2 - x)^{10} = 2^{10} + \frac{10}{1!}2^9(-x) + \frac{10 \times 9}{2!}2^8(-x)^2 + \frac{10 \times 9 \times 8}{3!}2^7(-x)^3$$
$$+ \text{ terms of the order } x^4 \text{ and higher}$$

$\therefore$ $(2 - x)^{10} = 1024 - 5120x + 11\,520x^2 - 15\,360x^3 \ldots$ is the required expansion.

Example 19

Expand $(1 + x + x^2)(1 - x)^8$ in ascending powers of x up to and including the term in x^3.

Using the expansion of $(1 + x)^n$ with $n = 8$ and $(-x)$ substituted for x:
$$(1 - x)^8 = 1 - 8x + 28x^2 - 56x^3 + \ldots$$
$\therefore$ $(1 + x + x^2)(1 - x)^8 = (1 + x + x^2)(1 - 8x + 28x^2 - 56x^3 \ldots)$
Expanding and ignoring terms of order x^4 and above gives
$$(1 + x + x^2)(1 - x)^8 = 1 - 8x + 28x^2 - 56x^3 \ldots$$
$$+ \quad x - \quad 8x^2 + 28x^3 \ldots$$
$$+ \quad x^2 - \quad 8x^3 \ldots$$
$\therefore$ $(1 + x + x^2)(1 - x)^8 = 1 - 7x + 21x^2 - 36x^3 \ldots$

Example 20

Expand $(1 + x - x^2)^7$ in ascending powers of x up to and including the term in x^3.

Using the expansion of $(1 + x)^n$ with $n = 7$ and $(x - x^2)$ substituted for x:

$$(1 + x - x^2)^7 = 1 + 7 \times (x - x^2) + \frac{7 \times 6}{2!}(x - x^2)^2 + \frac{7 \times 6 \times 5}{3!}(x - x^2)^3 \ldots$$

Expanding and ignoring terms of order x^4 and above gives
$$(1 + x - x^2)^7 = 1 + 7(x - x^2) + 21(x^2 - 2x^3) + 35(x^3) \ldots$$
$$= 1 + 7x + 14x^2 - 7x^3 \ldots$$

Approximate values

Example 21

Expand $(2 + x)^4$ and use your expansion to find (a) $(2 \cdot 1)^4$, (b) $(1 \cdot 9)^4$.

By using the binomial theorem, or Pascal's triangle, we obtain
$$(2 + x)^4 = 16 + 32x + 24x^2 + 8x^3 + x^4$$

(a) $(2 \cdot 1)^4 = (2 + 0 \cdot 1)^4 = 16 + 32(0 \cdot 1) + 24(0 \cdot 1)^2 + 8(0 \cdot 1)^3 + (0 \cdot 1)^4$
$$= 16 + 3 \cdot 2 + 0 \cdot 24 + 0 \cdot 008 + 0 \cdot 0001$$
$\therefore$ $(2 \cdot 1)^4 = 19 \cdot 4481$

(b) $(1 \cdot 9)^4 = (2 - 0 \cdot 1)^4 = 16 + 32(-0 \cdot 1) + 24(-0 \cdot 1)^2 + 8(-0 \cdot 1)^3 + (-0 \cdot 1)^4$
$$= 13 \cdot 0321$$
$\therefore$ $(1 \cdot 9)^4 = 13 \cdot 0321$

A similar procedure can be adopted for evaluating higher powers of numbers but this could involve a large number of terms. However, if we ensure that x is sufficiently small, the terms involving the higher powers of x will approximate to zero and can be neglected. The following example demonstrates this technique.

Example 22

Expand $(1 + 3x)^{10}$ in ascending powers of x, up to and including the term in x^3. Hence evaluate $(1 \cdot 003)^{10}$ correct to five places of decimals.

Using the binomial expansion of $(1 + x)^n$ with $n = 10$ and $3x$ substituted for x, we obtain

$$(1 + 3x)^{10} = 1 + 30x + 405x^2 + 3240x^3 + \ldots$$
as the required expansion.

$$(1 \cdot 003)^{10} = [1 + 3(\tfrac{1}{1000})]^{10} \approx 1 + 30(\tfrac{1}{1000}) + 405(\tfrac{1}{1000})^2 + 3240(\tfrac{1}{1000})^3$$
$$= 1 + 0 \cdot 03 + 0 \cdot 000\,405 + 0 \cdot 000\,003\,240$$
$$= 1 \cdot 030\,41 \quad \text{correct to five decimal places.}$$

Notice that to obtain an answer correct to five decimal places we need to know whether the figure in the sixth decimal place is less than 5 or not. In the last example the first four terms give a total of $1 \cdot 030\,408\,24$ which becomes $1 \cdot 030\,41$ when corrected to 5 places of decimals. From these first four terms we can see that the terms in x^4 and higher powers will not be large enough to alter the result and so our answer is indeed correct to five decimal places.

Example 23

Use the binomial expansion of $(1 + x)^8$ to find $(1 \cdot 01)^8$ correct to five significant figures.

In this example we do not know how many terms of the expansion are required to give the necessary accuracy. In such cases, start by finding the first four terms and be prepared to find more if they are required later.

The binomial expansion gives $\quad (1 + x)^8 = 1 + 8x + 28x^2 + 56x^3 + \ldots$

$$(1 \cdot 01)^8 = (1 + \tfrac{1}{100})^8 = 1 + 8(\tfrac{1}{100}) + 28(\tfrac{1}{100})^2 + 56(\tfrac{1}{100})^3 + \ldots$$
$$= 1 + 0 \cdot 08 + 0 \cdot 002\,8 + 0 \cdot 000\,056 + \ldots$$

We need to know whether the sixth significant figure is less than 5 or not. As the answer is clearly between 1 and 10, the sixth significant figure will be in the fifth decimal place. The first four terms give a total of $1 \cdot 082\,856$ which becomes $1 \cdot 0829$ when corrected to 5 significant figures. Clearly the terms in x^4 and higher will not alter this answer. Thus we have sufficient terms for the required accuracy and

$$(1 \cdot 01)^8 \approx 1 \cdot 082\,856$$
$$= 1 \cdot 082\,9 \quad \text{correct to five significant figures.}$$

Exercise 8D

1. Expand the following using Pascal's triangle:
   (a) $(3 + x)^3$,   (b) $(5 + 2x)^3$,   (c) $(2 + x)^4$,   (d) $(2 - x)^4$,

   (e) $(2y + x)^5$,   (f) $(2x - 3y)^5$,   (g) $\left(x - \dfrac{1}{x}\right)^4$,   (h) $\left(x - \dfrac{2}{x}\right)^5$.

2. Expand the following using the binomial theorem:

   (a) $(1 + 3x)^4$,   (b) $(2x + y)^4$,   (c) $(2 - 3x)^6$,   (d) $\left(2x + \dfrac{3}{x}\right)^5$.

3. Use the binomial theorem to expand $(1 + x)^{12}$ in ascending powers of x up to and including the term in x^3.

4. Use the binomial theorem to expand $(a + x)^{12}$ in ascending powers of x up to and including the term in x^3.

5. Find the first four terms in the series expansion of $(a - 3x)^{10}$ in ascending powers of x.

6. If x is such that terms involving x^5 and higher powers can be neglected, find an approximate expansion of $\left(1 + \dfrac{x}{2}\right)^{20}$.

7. Expand $(2 + x)^5$ and use your expansion to find (a) $(2 \cdot 1)^5$, (b) $(1 \cdot 9)^5$.

8. Expand each of the following expressions in ascending powers of x, up to and including the term in x^3:
   (a) $(1 + 2x)(1 - x)^{10}$, (b) $(1 - 3x)(1 + x)^6$,
   (c) $(1 + x^2)(1 + 2x)^8$, (d) $(1 - x + x^2)(1 + x)^4$,
   (e) $(1 + 2x - x^2)(1 + 2x)^7$, (f) $(1 + x + x^2)(1 - 3x)^5$.

9. Expand each of the following expressions in ascending powers of x, up to and including the term in x^3:
   (a) $(1 + x + x^2)^6$, (b) $(1 + 2x - x^2)^5$, (c) $(1 - 2x + 4x^2)^8$.

10. Expand $(1 - x)^{11}$ in ascending powers of x up to and including the term in x^4. Hence evaluate $(0 \cdot 98)^{11}$ correct to three places of decimals.

11. Expand $(1 + x)^{15}$ in ascending powers of x up to and including the term in x^3. Hence evaluate $(1 \cdot 01)^{15}$ correct to three places of decimals.

12. Expand $(1 + 2x)^{12}$ in ascending powers of x up to and including the term in x^2. Hence evaluate $(1 \cdot 02)^{12}$ correct to two significant figures.

13. Expand $(2 + x)^7$ in ascending powers of x up to and including the term in x^3. Hence evaluate $(2 \cdot 01)^7$ correct to six significant figures.

14. Use binomial expansions to evaluate the following to the stated degree of accuracy:
    (a) $(1 \cdot 01)^9$ correct to four decimal places,
    (b) $(0 \cdot 998)^7$ correct to seven decimal places,
    (c) $(0 \cdot 99)^{10}$ correct to four decimal places,
    (d) $(1 \cdot 99)^{10}$ correct to four significant figures.

15. When $(1 + ax)^{10}$ is expanded in ascending powers of x, the series expansion is $A + Bx + Cx^2 + 15x^3 + \ldots$; find the values of a, A, B and C.

16. When $(1 + ax)^n$ is expanded in ascending powers of x, the series expansion is $1 + 2x + \tfrac{15}{8}x^2 + \ldots$; find the values of n and a.

17. If x is sufficiently small to allow any terms in x^3 or higher powers of x to be neglected, show that
    $$(1 + x)^7(1 + 2x)^4 \approx 1 + 15x + 101x^2.$$
    Obtain a similar approximation for $(1 + x)^7(1 - 2x)^4$, again neglecting terms in x^3 or higher powers of x.

18. If x is sufficiently small to allow any terms in x^5 or higher powers of x to be neglected, show that
    $$(1 + x)^6(1 - 2x^3)^{10} \approx 1 + 6x + 15x^2 - 105x^4.$$

Particular terms of a binomial expansion

By considering the general term of a binomial expansion, a term involving a particular power of x may be found.

Example 24

Find the term involving x^{12} in the expansion of $(3 + 2x)^{15}$.

Using the expansion $(a + x)^n = a^n + {}^nC_1a^{n-1}x + \ldots + {}^nC_ra^{n-r}x^r + \ldots + x^n$, the general term u_{r+1} is given by ${}^nC_ra^{n-r}x^r$.

Substituting $a = 3$, $2x$ for x and $n = 15$, we have ${}^{15}C_r3^{15-r}(2x)^r$.

To find the term involving x^{12} we need $r = 12$ (i.e. the 13th term).

Thus the required term is ${}^{15}C_{12}3^3(2x)^{12} = \dfrac{15!}{12! \, 3!}3^32^{12}x^{12}$

$$= \frac{15!}{12!}3^22^{11}x^{12}$$

Example 25

Find the term independent of x in the binomial expansion of $\left(3x - \dfrac{2}{x^2}\right)^{18}$.

The general term u_{r+1} in the expansion of $(a + x)^n$ is given by ${}^nC_ra^{n-r}x^r$.

Substituting $3x$ for a, $\left(\dfrac{-2}{x^2}\right)$ for x and $n = 18$, we have

$$u_{r+1} = {}^{18}C_r(3x)^{18-r}\left(\frac{-2}{x^2}\right)^r$$
$$= {}^{18}C_r3^{18-r}(-2)^rx^{18-3r}$$

This term is independent of x if $18 - 3r = 0$, i.e. $r = 6$

Thus the required term is $u_7 = {}^{18}C_63^{12}(-2)^6$
$$= {}^{18}C_63^{12}2^6$$

Exercise 8E

1. Find the term involving x^8 in the expansion of $(4 + x)^{12}$.
2. Find the term involving x^5 in the expansion of $(3 + 2x)^8$.
3. Find the 4th term when $(2 - x)^9$ is expanded in ascending powers of x.
4. Find the coefficient of x^3 in the expansion of   (a) $(5 + 3x)^8$,   (b) $(7 - 2x)^7$.
5. Find the term involving x^{10} in the expansion of $(3 + 2x^2)^7$.
6. Find the term independent of x in the expansion of each of the following:

   (a) $\left(3x + \dfrac{1}{x}\right)^{10}$,   (b) $\left(2x^2 + \dfrac{4}{x}\right)^{12}$,   (c) $\left(\dfrac{3}{x^2} - 2x\right)^6$,   (d) $(1 + x^2)\left(2x + \dfrac{1}{x}\right)^{10}$

8.5 The binomial theorem for any rational index

It can be shown that the binomial expansion of $(1 + x)^n$ for positive integral values of n (stated on page 223) is also true for negative and/or fractional values of n. However, for these values of n the term ${}^nC_r$ is meaningless as the factorial notation has no meaning when applied to negative numbers and fractions. However, from our definition

$$^nC_r = \frac{n!}{(n-r)!\, r!} \qquad \ldots [1]$$

$$= \frac{n(n-1)(n-2) \ldots (n-r+1)}{r!} \qquad \ldots [2]$$

The second form does have meaning for negative and fractional values of n.

Thus we use the notation $\binom{n}{r}$ to mean $\dfrac{n(n-1)(n-2) \ldots (n-r+1)}{r!}$

Thus $(1+x)^n = 1 + \binom{n}{1}x + \binom{n}{2}x^2 + \binom{n}{3}x^3 + \ldots$

$$= 1 + nx + \frac{n(n-1)}{2!}x^2 + \frac{n(n-1)(n-2)}{3!}x^3 + \ldots \quad \text{for all } n \in \mathbb{R}$$

$$\text{provided } -1 < x < 1, \text{ i.c. } |x| < 1$$

Note that the condition $|x| < 1$ is necessary because with n fractional or negative, the series is infinite and the condition $|x| < 1$ ensures that this infinite series converges.

Example 26

Expand $(1 + 3x)^{-5}$ in ascending powers of x stating the first four terms and the range of values of x for which the expansion is valid.

Using the expansion

$$(1+x)^n = 1 + \frac{nx}{1!} + \frac{n(n-1)}{2!}x^2 + \frac{n(n-1)(n-2)}{3!}x^3 + \ldots$$

we obtain

$$(1+3x)^{-5} = 1 + \frac{(-5)}{1!}(3x) + \frac{(-5)(-6)}{2!}(3x)^2 + \frac{(-5)(-6)(-7)}{3!}(3x)^3 + \ldots$$

$$= 1 - 15x + 135x^2 - 945x^3 + \ldots$$

The expansion is valid for $|3x| < 1$, i.e. $|x| < \frac{1}{3}$.

Example 27

Expand $\dfrac{(1+2x)^2}{(2-x)}$ in ascending powers of x up to and including the term in x^3, and state the values of x for which the expansion is valid.

$\dfrac{(1+2x)^2}{(2-x)} = (1+2x)^2(2-x)^{-1}$. However, we cannot expand $(2-x)^{-1}$ in this form as the binomial expansion for negative and/or fractional n only applies to $(1+x)^n$, not $(a+x)^n$.

However, $(2-x)^{-1} = \left[2\left(1 - \frac{x}{2}\right)\right]^{-1}$

$$= 2^{-1}\left(1 - \frac{x}{2}\right)^{-1}$$

$$= \frac{1}{2}\left(1 - \frac{x}{2}\right)^{-1}$$

$$\therefore \quad \frac{(1 + 2x)^2}{(2 - x)} = (1 + 2x)^2 \times \frac{1}{2}\left(1 - \frac{x}{2}\right)^{-1}$$

$$= (1 + 4x + 4x^2)\frac{1}{2}\left[1 + (-1)\left(-\frac{x}{2}\right) + \frac{(-1)(-2)}{2!}\left(-\frac{x}{2}\right)^2 + \frac{(-1)(-2)(-3)}{3!}\left(-\frac{x}{3}\right)^3 + \dots\right]$$

$$= (1 + 4x + 4x^2)\left(\frac{1}{2} + \frac{x}{4} + \frac{x^2}{8} + \frac{x^3}{16} + \dots\right)$$

$$= \frac{1}{2} + \frac{9x}{4} + \frac{25x^2}{8} + \frac{25x^3}{16} + \dots$$

The expansion is valid for $\left|-\frac{x}{2}\right| < 1$ i.e. $|x| < 2$.

Example 28

Using the binomial theorem, find $\sqrt{9 \cdot 18}$ correct to five places of decimals.

$$\sqrt{9 \cdot 18} = \sqrt{9(1 + 0 \cdot 02)} = 3(1 + \tfrac{2}{100})^{1/2}$$

The expression $(1 + \tfrac{2}{100})^{1/2}$ can be expanded using the binomial theorem as $|\tfrac{2}{100}| < 1$ as required.

$$\therefore \quad \sqrt{9 \cdot 18} = 3\left[1 + (\tfrac{1}{2})(\tfrac{2}{100}) + \frac{(\tfrac{1}{2})(-\tfrac{1}{2})(\tfrac{2}{100})^2}{2!} + \frac{(\tfrac{1}{2})(-\tfrac{1}{2})(-\tfrac{3}{2})(\tfrac{2}{100})^3}{3!} + \dots\right]$$

$$= \quad 3 + \quad 0 \cdot 03 \quad - \quad 0 \cdot 000\,15 \quad + \quad 0 \cdot 000\,001\,5 \quad \dots$$

$$\sqrt{9 \cdot 18} \approx 3 \cdot 029\,85$$

If x is such that $|x| > 1$, the expression $(1 + x)^n$ can be expanded by the binomial theorem using the rearrangement $(1 + x)^n = x^n\left(1 + \frac{1}{x}\right)^n$ as $\left|\frac{1}{x}\right| < 1$ if $|x| > 1$.

The expansion obtained will then be in ascending powers of $\frac{1}{x}$.

Example 29

Expand $(3 + x)^{-2}$ in ascending powers of $\frac{1}{x}$, stating the first four terms only and the values of x for which the expansion is valid.

$$(3 + x)^{-2} = \left[x\left(\frac{3}{x} + 1\right)\right]^{-2}$$

$$= \frac{1}{x^2}\left(1 + \frac{3}{x}\right)^{-2}$$

$$= \frac{1}{x^2}\left[1 + (-2)\left(\frac{3}{x}\right) + \frac{(-2)(-3)}{2!}\left(\frac{3}{x}\right)^2 + \frac{(-2)(-3)(-4)}{3!}\left(\frac{3}{x}\right)^3 + \dots\right]$$

$$= \frac{1}{x^2}\left[1 - \frac{6}{x} + \frac{27}{x^2} - \frac{108}{x^3} + \dots\right]$$

$$= \frac{1}{x^2} - \frac{6}{x^3} + \frac{27}{x^4} - \frac{108}{x^5} + \dots$$

The expansion is valid for $\left|\dfrac{3}{x}\right| < 1$ i.e. $|x| > 3$.

Exercise 8F

1. For each of the following, state the values of x for which a binomial expansion of the given expression in ascending powers of x, is valid.

   (a) $(1 + 2x)^{-3}$ (b) $(1 - 2x)^{12}$ (c) $\dfrac{1}{(1 - 2x)}$ (d) $\dfrac{1}{\sqrt{(1 + \frac{1}{2}x)}}$

2. Expand each of the following in ascending powers of x, up to and including the term in x^3, and state the values of x for which the expansion is valid.

   (a) $\dfrac{1}{1 + x}$ (b) $\sqrt[3]{(1 + x)}$ (c) $\dfrac{1}{\sqrt{(1 + 2x)}}$ (d) $\dfrac{1}{(1 - 2x)^2}$

   (e) $\dfrac{1}{(1 + 3x)^2}$ (f) $\sqrt{[(1 + 2x)^3]}$ (g) $\sqrt{(4 + x)}$ (h) $\dfrac{1}{\sqrt{(100 - 50x)}}$

3. Expand $\dfrac{1 - x}{(1 + 2x)^3}$ in ascending powers of x, up to and including the term in x^4, and state the values of x for which the expansion is valid.

4. When the terms in x^4 and higher powers of x are neglected, the series expansion of $\dfrac{2 + 3x - x^2}{(1 + 2x)^3}$ in ascending powers of x gives $a + bx + cx^2 + dx^3$. Find the values of a, b, c and d.

5. Expand the expression $(1 + x - 6x^2)^{-1}$ in ascending powers of x up to and including the term in x^3.

6. Expand $(1 + x)^{1/2}$ in ascending powers of x, up to and including the term in x^3, and state the values of x for which the expansion is valid. Hence calculate $\sqrt{1\cdot08}$ correct to four places of decimals.

7. Expand each of the following in ascending powers of $\dfrac{1}{x}$, stating the first 4 non-zero terms and the value of x for which the expansion is valid.

   (a) $(2 - x)^{-3}$ (b) $(1 + 2x)^{-1}$ (c) $(4 - 3x)^{-2}$

8. Use binomial expansions to evaluate the following to the stated degree of accuracy:

   (a) $\sqrt{0\cdot96}$ correct to five decimal places,
   (b) $\sqrt{104}$ correct to six significant figures,
   (c) $\sqrt{4\cdot08}$ correct to four decimal places,
   (d) $\sqrt[4]{1\cdot08}$ correct to five places of decimals,
   (e) $\sqrt[3]{8\cdot72}$ correct to five places of decimals.

9. Expand $\sqrt{\left(\dfrac{1 - x}{1 + 2x}\right)}$ in ascending powers of x, up to and including the term in x^3. State the values of x for which the expansion is valid.

10. Expand $\sqrt{\left(\dfrac{1 + x}{1 - x}\right)}$ in ascending powers of x, up to and including the term in x^3. State the values of x for which the expansion is valid and by substituting $x = 0\cdot2$, find an approximation for $\sqrt{1\cdot5}$.

11. When $(1 + ax)^{3/2}$ is expanded in ascending powers of x, the first four terms are $A + Bx + \frac{27}{8}x^2 + Cx^3$. Find the values of a, A, B and C.

12. When $(1 + bx)^n$ is expanded in ascending powers of x, the first three terms of the expansion are $1 - \frac{3}{2}x - \frac{27}{100}x^2$. Find the values of n and b.

Exercise 8G   Examination questions

1. The sum of the first six terms of an arithmetic progression is 72 and the second term is seven times the fifth term.
Find the first term and the common difference.
Find the sum of the first ten terms.   (A.E.B)

2. The sum of the first two terms of a geometric progression is 9, and the sum of the first four terms is 45. Calculate the two possible values of the fifth term in the progression.   (A.E.B.)

3. (i) Find a number x which, when added to each of the numbers, 21, 27 and 29, produces three numbers whose squares are in arithmetic progression.
  (ii) Find a number y which, when added to each of the numbers, 21, 27 and 29, produces three numbers which are in geometric progression.
                                                                        (Oxford)

4. (i) The first term of an arithmetic progression is 7, the last term is 43 and the sum of the terms of the progression is 250. Find the number of terms and the common difference.
  (ii) The sum of the first and second terms of a geometric progression is 12, and the sum of the third and fourth terms is 48. Find the two possible values of the common ratio and the corresponding values of the first term. Find also the sum to ten terms of each of the series.
                                                                        (Oxford)

5. (i) The first term of an arithmetic progression is 2 and the common difference is 3. Find the sum to n terms of this progression.
Find the value of n for which the sum of the progression is 610.
Find, also, the least value of n for which the sum exceeds 1000.

  (ii) Find $S_n = \sum_{r=0}^{n-1} 3^r$.

Find also the least value of n such that $S_n > 10\,000$.   (London)

6. (a) The first three terms of an arithmetic series are $-3\frac{1}{8}$, $-1\frac{7}{8}$, $-\frac{5}{8}$. Find the sum of the first 70 terms of this series.
  (b) The first three terms of a geometric series are 2, $-\frac{1}{2}$, $\frac{1}{8}$. Find the sum to infinity of this series.
Find also the least value of n for which the sum of the first n terms of this series differs from the sum to infinity by less than 10^{-5}.
                                                                        (Cambridge)

7. The population of a colony of insects increases in such a way that if it is N at the beginning of a week, then at the end of the week it is $a + bN$, where a and b are constants and $0 < b < 1$.
Starting from the beginning of the week when the population is N, write down an expression for the population at the end of 2 weeks. Show that at the end of c consecutive weeks the population is

$$a\left(\frac{1 - b^c}{1 - b}\right) + b^c N.$$

When $a = 2000$ and $b = 0.2$, it is known that the population takes about 4 weeks to increase from N to $2N$. Estimate a value for N from this information. (A.E.B.)

8. Prove that $\sum_{r=1}^{n} r(r+1) = \frac{1}{3}n(n+1)(n+2)$.

Evaluate $\sum_{r=1}^{20} r(r-1)$. (London)

9. (a) For all positive integral values of n, the sum of the first n terms of a series is $3n^2 + 2n$. Find the nth term in its simplest form.
   (b) Show that, for all positive integral values of n, $7^n + 2^{2n+1}$ is divisible by 3.
   (c) Prove, by induction or otherwise, that
   $$(1)(4)(7) + (2)(5)(8) + \ldots + n(n+3)(n+6)$$
   $$= \frac{1}{4}n(n+1)(n+6)(n+7).$$ (J.M.B.)

10. Expand $(2-x)^5$ in ascending powers of x up to and including the term in x^2. Use these terms to find the value of $2 \cdot 01^5$, giving your answer to three decimal places. (S.U.J.B.)

11. Find, in ascending powers of x, the first three terms in the expansion of $(2-3x)^8$. Use the expansion to find the value of $(1 \cdot 997)^8$ correct to the nearest whole number. (Cambridge)

12. Obtain the term independent of x in the expansion of $\left(2x^3 - \dfrac{1}{x}\right)^{20}$

You may leave the answer in terms of factorials. (S.U.J.B.)

13. Find the non-zero value of b if the coefficient of x^2 in the expansion of $(b+2x)^6$ is equal to the coefficient of x^5 in the expansion of $(2+bx)^8$. (A.E.B.)

14. The first four terms in the expansion of $(1-x)^n$ are $1 - 6x + px^2 + qx^3$. Show that $p = 15$ and obtain the value of q. (S.U.J.B.)

15. (a) Find the first four terms in the expansion of $(1+3x)^{1/3}$ in ascending powers of x and state the range of values of x for which the expansion is valid. Use the expansion to determine $\sqrt[3]{8 \cdot 24}$ to five decimal places.
   (b) The coefficient of x^2 in the expansion of $(1-2x)^n$ is 24. Calculate the two possible values of n. (S.U.J.B.)

16. Obtain the binomial expansion of $(1+2x)^{-1/2}$ in ascending powers of x as far as the term in x^3, assuming the value of x to be such that the expansion is valid. Simplify the coefficients. Hence, or otherwise, write down the first four terms in the binomial expansion of $(1-2x)^{-1/2}$.

Use your expansion to evaluate $\dfrac{1}{\sqrt{0 \cdot 998}}$ to 7 decimal places. (A.E.B.)

17. (a) In the expansion of $(1+ax)^n$, the first three terms are
   $$1 - \tfrac{5}{2}x + \tfrac{75}{8}x^2.$$
   Find n and a, and state the range of values of x for which the expansion is valid.
   (b) Expand $(1+x)^{1/2}$ in ascending powers of x as far as the term in x^2, and hence find an approximation for $\sqrt{1 \cdot 08}$. Deduce that $\sqrt{12} \approx 3 \cdot 464$. (S.U.J.B.)

18. Find the values of the constants a and b for which the expansions, in ascending powers of x, of the two expressions $(1 + 2x)^{1/2}$ and $\dfrac{1 + ax}{1 + bx}$, up to and including the term in x^2, are the same.

With these values of a and b, use the result $(1 + 2x)^{1/2} \approx \dfrac{1 + ax}{1 + bx}$, with $x = -\frac{1}{100}$, to obtain an approximate value for $\sqrt{2}$ in the form p/q, where p and q are positive integers.   (Cambridge)

9
Probability

9.1 Simple probability

Suppose an unbiased coin is tossed. The result is equally likely to be a head as a tail. The total number of possible outcomes of a single toss is 2, i.e. a head or a tail. We say that the probability of obtaining a head is $\frac{1}{2}$, or more briefly P(head) = $\frac{1}{2}$.

The probability of a particular outcome of a trial (i.e. an attempt at something) is expressed as a fraction $\frac{a}{b}$, where a is the number of ways in which the particular outcome can occur and b is the total number of possible outcomes of the trial.

Suppose a six-sided die is thrown. There are six possible outcomes of this trial (or throw) and they are all equally likely. Only one of these outcomes is that of obtaining a three.

The probability of obtaining a three is therefore $\frac{1}{6}$. Thus P(3) = $\frac{1}{6}$.

In a similar way the probability of obtaining a number greater than 4 with one throw of a die is $\frac{2}{6}$ i.e. $\frac{1}{3}$, because there are a total of 6 possible outcomes of which two (5 and 6) satisfy our requirement.

Example 1

A six-sided die is to be thrown. Calculate the probability that the result will be (a) an odd number,   (b) a number less than 3,   (c) a multiple of 3.

(a) For one trial, total number of outcomes = 6
              number of odd outcomes = 3 (i.e. 1, 3 or 5)
                    P(odd number) = $\frac{3}{6}$ = $\frac{1}{2}$

(b) For one trial, total number of outcomes = 6
              number of scores less than 3 = 2 (i.e. 1 or 2)
                    P(less than 3) = $\frac{2}{6}$ = $\frac{1}{3}$

(c) For one trial, total number of outcomes = 6
              number of multiples of 3 = 2 (i.e. 3 or 6)
                    P(multiple of 3) = $\frac{2}{6}$ = $\frac{1}{3}$

Example 2

One card is to be drawn from a pack of 52 cards. Calculate the probability that the card will be (a) a diamond, (b) a court card, (c) a black card, (i.e. a club or a spade).

A pack of cards consists of 52 cards; 4 suits of 13 cards with each suit having 3 court cards (king, queen and jack).

Number of possible outcomes = 52
(a)   number of diamonds = 13
                     P(diamond) = $\frac{13}{52}$ = $\frac{1}{4}$
(b)   number of court cards = 4 × 3 = 12
                     P(court card) = $\frac{12}{52}$ = $\frac{3}{13}$
(c)   number of black cards = 26
                     P(black card) = $\frac{26}{52}$ = $\frac{1}{2}$

Example 3

A bag contains 8 discs of which 4 are red, 3 are blue and 1 is yellow.
Calculate the probability that when one disc is drawn from the bag it will be
(a) red, (b) yellow, (c) blue, (d) not blue.

There are 8 discs altogether so the total number of possible outcomes is 8.
              (a)   number of red discs = 4
                                P(red) = $\frac{4}{8}$ = $\frac{1}{2}$
              (b)   number of yellow discs = 1
                                P(yellow) = $\frac{1}{8}$
              (c)   number of blue discs = 3
                                P(blue) = $\frac{3}{8}$
              (d) number of discs not blue = 5
                                P(not blue) = $\frac{5}{8}$

A probability of 1 means that the event is certain to occur and a probability
of zero means that the event cannot occur.
In example 3, the probability of obtaining a disc that is either red, blue or
yellow would be 1 because all eight discs meet the requirements i.e.
              P(either red, blue or yellow) = $\frac{8}{8}$ = 1
On the other hand the probability of obtaining a green disc would be zero as
there are no discs that meet this requirement i.e.
              P(green) = $\frac{0}{8}$ = 0
Notice also that in example 3, P(not blue) = $\frac{5}{8}$
              and P(blue) = $\frac{3}{8}$.
This is a particular illustration of the rule: If the probability of an event
occurring is p, then the probability of it not occurring is $(1 - p)$.

Set notation

Set notation can be useful when considering a number of different events
that may occur in a given trial. Consider one number selected from the set
of numbers {1, 2, 3, 4, 5, 6, 7, 8, 9}. Considering this set as our universal set,
$\mathscr{E}$, it clearly contains a number of subsets.

For example, the subset of odd numbers = {1, 3, 5, 7, 9}
              the subset of multiples of three = {3, 6, 9}

If A is the event of the selected number being odd and B is the event of the
selected number being a multiple of three, we can show the sets of outcomes
as a Venn diagram:

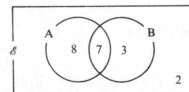

Using the notation $n(\text{X})$ for the number of elements in set X, and X′ as the complement of set X, we see that

$$P(A) = \frac{n(A)}{n(\mathscr{E})} = \frac{5}{9} \qquad \text{i.e.} \quad P(\text{odd number}) = \frac{5}{9}$$

$$P(B) = \frac{n(B)}{n(\mathscr{E})} = \frac{3}{9} = \frac{1}{3} \quad \text{i.e.} \quad P(\text{multiple of 3}) = \frac{1}{3}$$

$$P(A \cap B) = \frac{n(A \cap B)}{n(\mathscr{E})} = \frac{2}{9} \qquad \text{i.e.} \quad P(\text{odd number } and \text{ multiple of 3}) = \frac{2}{9}$$

$$P(B') = \frac{n(B')}{n(\mathscr{E})} = \frac{6}{9} = \frac{2}{3} \quad \text{i.e.} \quad P(not \text{ a multiple of 3}) = \frac{2}{3}$$

$$P(A \cup B) = \frac{n(A \cup B)}{n(\mathscr{E})} = \frac{6}{9} = \frac{2}{3} \quad \text{i.e.} \quad P(\text{multiple of 3 or odd number or both}) = \frac{2}{3}$$

Example 4

One element is randomly selected from a universal set of 20 elements. Sets A and B are subsets of the universal set and $n(A) = 15$, $n(B) = 10$ and $n(A \cap B) = 7$. If P(A) is the probability of the selected element belonging to set A, find (a) P(A), (b) P(A ∩ B), (c) P(A′), (d) P(A ∪ B).

First draw a Venn diagram, showing the number of elements in each subset:
(hint: consider $n(A \cap B)$ first)
From the Venn diagram:

(a) P(A) $= \frac{15}{20} = \frac{3}{4}$
(b) P(A ∩ B) $= \frac{7}{20}$
(c) P(A′) $= \frac{5}{20} = \frac{1}{4}$
(d) P(A ∪ B) $= \frac{18}{20} = \frac{9}{10}$

Exercise 9A

1. What is the probability of obtaining a tail when a fair coin is tossed once?
2. A six-sided die is thrown once. What is the probability that the result will be
   (a) an even number,   (b) a number less than 6,
   (c) a number greater than 2?
3. A bag contains ten coloured discs of which 4 are blue, 3 are green, 2 red and 1 yellow. Find the probability that when one disc is randomly selected from the bag it will be
   (a) red,   (b) blue,   (c) yellow,   (d) not red.
4. Twenty cards are numbered from 1 to 20. One card is then taken at random. What is the probability that the number on the card will be
   (a) a multiple of 4,   (b) even,   (c) greater than 15,   (d) divisible by 5,
   (e) not a multiple of 6?

5. A letter is chosen at random from the letters a, b, c, d, e, f, g, h. Calculate the probability that the letter chosen will be
(a) a consonant,   (b) a vowel,   (c) not f, g or h.

6. One girl is chosen at random, from 4 girls Anne, Bonny, Linda and Dawn. Calculate the probability that the girl chosen will be
(a) Anne,   (b) either Linda or Dawn.

7. Of 12 beads, 7 are coloured green, 3 are orange and 2 are black. Calculate the probability that when one bead is selected at random, it will
(a) be orange,   (b) not be green.

8. What is the probability of drawing a court card from the 26 black cards of a pack?

9. One card is drawn from a pack of 52 cards. Find the probability that the card drawn will   (a) be a club,   (b) be an ace,   (c) be a red card,   (d) not be a spade.

10. A universal set has 24 elements and A and B are subsets of the universal set such that $n(A) = 14$, $n(B) = 9$ and $n(A \cap B) = 6$. If P(A) is the probability of selecting an element belonging to set A, calculate
(a) P(A),   (b) P(A $\cap$ B),   (c) P(B'),   (d) P(A $\cup$ B)'.

11. One element is selected at random from a universal set of 15 elements. R and Q are subsets of the universal set and $n(R) = 6$, $n(Q) = 8$ and $n(R \cap Q) = 2$. Calculate
(a) P(Q),   (b) P(R $\cup$ Q),   (c) P(R'),   (d) P(R $\cap$ Q).

12. A and B are subsets of the universal set and $n(A) = 25$, $n(B) = 20$, $[n(A \cup B)'] = 20$ and there are 50 elements in the universal set. When one element is selected at random, calculate
(a) P(A),   (b) P(A $\cup$ B),   (c) P(B'),   (d) P(A $\cap$ B).

9.2 Sum and product laws

Possibility space

Suppose that a trial T_1 has possible outcomes a, b, c, d or e. We say that the set a, b, c, d, e is the possibility space, or sample space, for the trial T_1. Now suppose that two trials are carried out, trial T_1 possibility space {a, b, c, d, e} and trial T_2 possibility space {f, g, h} The pairs of outcomes for the two trials can be shown as points on a possibility space diagram. Each dot represents a pair of outcomes.

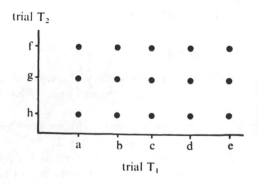

Example 5

Suppose that a coin is tossed and a die is thrown. Draw a possibility space diagram and use it to find the probability of obtaining both a head on the coin and an even number on the die.

The possibility space diagram is:
There are twelve possible outcomes of which only three (circled in the diagram) satisfy the requirements.
Thus P(head and an even number) = $\frac{3}{12}$ = $\frac{1}{4}$

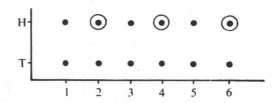

Independent events

Two events are said to be independent if the outcome of one event has no bearing upon the outcome of the other. In the last example the appearance of say, a head, on the coin does not affect the result of throwing the die. Thus the events are independent.

For two independent events A and B

P(A and B) = P(A) × P(B) or, in set notation,

P(A ∩ B) = P(A) × P(B)

$$\left(\begin{array}{c} \text{Conversely, we can say that two events A and B are independent if} \\ \text{P(A ∩ B) = P(A) × P(B)} \end{array} \right)$$

Thus in example 5, P(head) = $\frac{1}{2}$ P(even number) = $\frac{3}{6}$ = $\frac{1}{2}$
∴   P(head and even number) = $\frac{1}{2}$ × $\frac{1}{2}$ = $\frac{1}{4}$ as required.

Example 6

A card is to be selected from a pack of cards and a die is to be thrown. Find the probability that the outcome will be (a) a spade and an even number on the die, (b) a red card and a number less than 3 on the die.

The selection of the card and the throwing of the die are independent events.

(a)   P(a spade) = $\frac{13}{52}$
                P(even number on die) = $\frac{3}{6}$ (either 2, 4 or 6)
    P(a spade and an even number) = P(a spade) × P(an even number on die)
                                = $\frac{13}{52}$ × $\frac{3}{6}$
                                = $\frac{1}{8}$
(b)   P(a red card) = $\frac{26}{52}$
                P(1 or 2 on die) = $\frac{2}{6}$
    P(a red card and a 1 or 2 on die) = P(a red card) × P(a 1 or 2 on die)
                                = $\frac{26}{52}$ × $\frac{2}{6}$
                                = $\frac{1}{6}$

Mutually exclusive events

As was seen on page 191 two events are said to be mutually exclusive if the occurrence of one automatically excludes the possibility of the other occurring. In tossing a coin, if the outcome is a head this excludes the possibility of obtaining a tail on that trial, so the appearance of a head and the appearance of a tail are mutually exclusive events.

Note that, if two events are mutually exclusive, the set of events A and the set of events B would be shown as disjoint sets on a Venn diagram.

Either – Or

If the probability of an event A occurring is p_1 and the probability of a second mutually exclusive event B occurring is p_2, then the probability of either A or B occurring will be $(p_1 + p_2)$

i.e. $P(A \text{ or } B) = P(A) + P(B)$

or, using set notation $P(A \cup B) = P(A) + P(B)$ for mutually exclusive events.
Conversely, we can say that two events are mutually exclusive

if $P(A \cup B) = P(A) + P(B)$.

Example 7

One card is selected from a pack of cards. What is the probability that the card selected will be either a diamond or a black ace?

Number of cards in pack $= 52$

$$P(\text{a diamond}) = \tfrac{13}{52}$$
$$P(\text{a black ace}) = \tfrac{2}{52}$$

But these are mutually exclusive events since the occurrence of one excludes the other (i.e. the card selected cannot be a black ace and a diamond);
thus $P(\text{a diamond or a black ace}) = P(\text{a diamond}) + P(\text{a black ace})$

$$= \tfrac{13}{52} + \tfrac{2}{52} = \tfrac{15}{52}$$

Events that are not mutually exclusive

Suppose we want the probability of one card selected from a pack being either a heart or a queen. These events are not mutually exclusive as the card could be both a heart and a queen, i.e. it could be the queen of hearts. We consider the possibility-space diagram:

	A	2	3	4	5	6	7	8	9	10	J	Q	K
♣	x	x	x	x	x	x	x	x	x	x	x	⊗	x
♦	x	x	x	x	x	x	x	x	x	x	x	⊗	x
♥	⊗	⊗	⊗	⊗	⊗	⊗	⊗	⊗	⊗	⊗	⊗	⊗	⊗
♠	x	x	x	x	x	x	x	x	x	x	x	⊗	x

Thus $P(\heartsuit) = \tfrac{13}{52} = \tfrac{1}{4}$, $P(Q) = \tfrac{4}{52} = \tfrac{1}{13}$, and counting the circled entries in the diagram $P(\heartsuit \text{ or } Q) = \tfrac{16}{52} = \tfrac{4}{13}$.
Therefore $P(\heartsuit \text{ or } Q) \neq P(\heartsuit) + P(Q)$ because, as we can see from the diagram, the right-hand side of such a calculation would include the queen of hearts twice.
In general, for two events A and B

$$P(A \text{ or } B) = P(A) + P(B) - P(A \text{ and } B)$$

or using set notation

$$P(A \cup B) = P(A) + P(B) - P(A \cap B)$$

The reader should notice that for mutually exclusive events $P(A \cap B) = 0$ and the above rule reduces to $P(A \cup B) = P(A) + P(B)$ as required.

Example 8

A coin is tossed and a die is thrown. Find the probability that the outcome will be a head or a number greater than 4, or both.

These events are clearly not mutually exclusive.
Also, $P(\text{head}) = \frac{1}{2}$ and $P(>4) = \frac{2}{6}$
thus, $P(\text{head and a number greater than 4}) = P(\text{head}) \times P(\text{number} > 4)$
$$= \frac{1}{2} \times \frac{2}{6} = \frac{1}{6}$$

Using the result $P(A \text{ or } B) = P(A) + P(B) - P(A \text{ and } B)$
thus, $P(\text{Head or number} > 4) = \frac{1}{2} + \frac{2}{6} - \frac{1}{6}$
$$= \frac{2}{3}$$

Alternatively, we could say
$P(\text{neither a head nor a number} > 4) = P(\text{tail}) \times P(\text{number} \leqslant 4)$
$$= \frac{1}{2} \times \frac{2}{3} = \frac{1}{3}$$
$\therefore$ $P(\text{head or number} > 4) = 1 - \frac{1}{3}$
$$= \frac{2}{3}$$

Example 9

If A and B are two events such that $P(A) = \frac{1}{4}$, $P(B) = \frac{1}{2}$ and $P(A \cap B) = \frac{1}{8}$ find (a) $P(A \cup B)$, (b) $P[(A \cup B)']$.

METHOD 1 (by calculation)
(a) $P(A \cup B) = P(A) + P(B) - P(A \cap B)$
$$= \frac{1}{4} + \frac{1}{2} - \frac{1}{8}$$
$$= \frac{5}{8}$$
(b) $P[(A \cup B)'] = 1 - P(A \cup B)$
$$= 1 - \frac{5}{8}$$
$$= \frac{3}{8}$$

METHOD 2 (by Venn diagram)
Show the probabilities on a Venn diagram, (fill in $P(A \cap B)$ first):

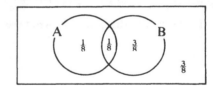

(a) $P(A \cup B)$ $= \frac{3}{8} + \frac{1}{8} + \frac{1}{8} = \frac{5}{8}$
(b) $P[(A \cup B)'] = \frac{3}{8}$.

Successive trials

Suppose that A and B are outcomes of a particular trial. If the trial is repeated and the outcome of the first trial does not affect the outcome of the second trial (i.e. the outcomes of trial 2 are independent of the outcomes of trial 1) then
$P(A \text{ on 1st trial followed by } B \text{ on 2nd trial}) = P(A) \times P(B)$
i.e. $P(A, B) = P(A) \times P(B)$

However, in some cases the outcome of trial 1 will affect the outcome of trial 2, i.e. the outcomes are *dependent* or *conditional* events. For example, if the successive trials involve the selection of one object from a group of objects the card, coloured disc or number etc. that is drawn on the first trial may, or may not, be replaced before the second trial takes place. If it is not replaced, then the probabilities of the various outcomes of the second trial will clearly be different from those of the first trial. It is usual therefore to refer to *replacement* or *non-replacement* problems.

Tree diagrams

A tree diagram can be used to show the probabilities of certain outcomes occurring when two or more trials take place in succession. The outcome is written at the end of the branch and the fraction on the branch gives the probability of the outcome occurring.

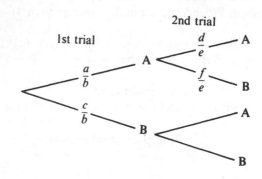

For each trial the number of branches is equal to the number of possible outcomes of that trial. In the diagram there are two possible outcomes, A and B, of each trial.

Example 10

A bag contains 8 marbles of which 3 are red and 5 are blue. One marble is drawn at random, its colour noted and the marble replaced in the bag. A marble is again drawn from the bag and its colour noted. Find the probability that the marbles drawn will be
(a) red followed by blue, (b) red and blue in any order, (c) of the same colour.

Draw a tree diagram showing the probabilities of each outcome of the two trials.

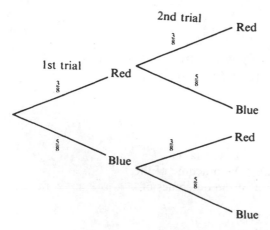

P(red followed by red) $= \frac{3}{8} \times \frac{3}{8} = \frac{9}{64}$

P(red followed by blue) $= \frac{3}{8} \times \frac{5}{8} = \frac{15}{64}$

P(blue followed by red) $= \frac{5}{8} \times \frac{3}{8} = \frac{15}{64}$

P(blue followed by blue) $= \frac{5}{8} \times \frac{5}{8} = \frac{25}{64}$

(a) P(red followed by blue) $= \frac{15}{64}$
(b) P(red and blue in any order) $= \frac{15}{64} + \frac{15}{64} = \frac{15}{32}$
(c) P(both of same colour) $= \frac{9}{64} + \frac{25}{64} = \frac{17}{32}$

Alternatively, these results can be calculated without the use of a tree diagram.
The two trials are independent events, and for each trial P(red) = $\frac{3}{8}$ and P(blue) = $\frac{5}{8}$.

(a) P(red followed by blue) = P(red) × P(blue)

$$= \tfrac{3}{8} \times \tfrac{5}{8} = \tfrac{15}{64}$$

(b) P(red and blue in any order) = P(red then blue) + P(blue then red)

$$= \text{P(red)} \times \text{P(blue)} + \text{P(blue)} \times \text{P(red)}$$
$$= \tfrac{3}{8} \times \tfrac{5}{8} + \tfrac{5}{8} \times \tfrac{3}{8}$$
$$= \tfrac{15}{32}$$

(c) P(both of same colour) = P(red then red) + P(blue then blue)

$$= \text{P(red)} \times \text{P(red)} + \text{P(blue)} \times \text{P(blue)}$$
$$= \tfrac{3}{8} \times \tfrac{3}{8} + \tfrac{5}{8} \times \tfrac{5}{8}$$
$$= \tfrac{17}{32}$$

Example 11

A bag contains 7 discs, 2 of which are red and 5 are green. Two discs are
removed at random and their colours noted. The first disc is not replaced
before the second is selected. Find the probability that the discs will be
(a) both red, (b) of different colours, (c) the same colour.

METHOD 1 (by calculation)
After the first disc is removed only 6 discs
remain in the bag and the number of each colour
will depend upon the result of the first trial.

(a) 1st trial P(R) = $\frac{2}{7}$
    2nd trial P(R) = $\frac{1}{6}$
    thus P(R, R) = $\frac{2}{7} \times \frac{1}{6} = \frac{1}{21}$

(b) P(different) = P(G, R) + P(R, G)
$$= \tfrac{5}{7} \times \tfrac{2}{6} + \tfrac{2}{7} \times \tfrac{5}{6}$$
$$= \tfrac{10}{21}$$

(c) P(same colour) = 1 − P(different)
$$= \tfrac{11}{21}$$

METHOD 2 (by tree diagram)

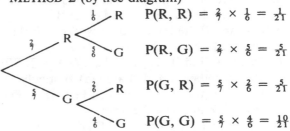

P(R, R) = $\frac{2}{7} \times \frac{1}{6} = \frac{1}{21}$

P(R, G) = $\frac{2}{7} \times \frac{5}{6} = \frac{5}{21}$

P(G, R) = $\frac{5}{7} \times \frac{2}{6} = \frac{5}{21}$

P(G, G) = $\frac{5}{7} \times \frac{4}{6} = \frac{10}{21}$

(a) P(R, R) = $\frac{1}{21}$

(b) P(different) = P(R, G) + P(G, R)
$$= \tfrac{5}{21} + \tfrac{5}{21}$$
$$= \tfrac{10}{21}$$

(c) P(same) = P(R, R) + P(G, G)
$$= \tfrac{1}{21} + \tfrac{10}{21}$$
$$= \tfrac{11}{21}$$

Example 11 shows the use of the general rule that, for two dependent events,
A and B,

$$\text{P(A and B)} = \text{P(A)} \times \text{P(B|A)} \text{ or } \text{P(A} \cap \text{B)} = \text{P(A)} \times \text{P(B|A)}$$

where P(B|A) represents the probability of event B occurring given that
event A has occurred.

Summary

The sum and product laws of probability are:

i $P(A \cup B) = P(A) + P(B) - P(A \cap B)$ which, for mutually exclusive events reduces to
$P(A \cup B) = P(A) + P(B)$.

ii $P(A \cap B) = P(A) \times P(B)$ for independent events
and $P(A \cap B) = P(A) \times P(B|A)$ for dependent (or conditional) events.

Exercise 9B

1. If $P(A) = \frac{2}{3}$ and $P(B) = \frac{1}{3}$, find
   (a) $P(A \cup B)$ if A and B are mutually exclusive events,
   (b) $P(A \cap B)$ if A and B are independent events.
2. Given that $P(A) = \frac{2}{3}$, $P(B) = \frac{1}{2}$, $P(A \cap B) = \frac{1}{3}$ and $P(A \cup B) = \frac{5}{6}$ state whether each of the following are true or false:
   (a) A and B are mutually exclusive events,
   (b) A and B are independent events.
3. If $P(A) = \frac{1}{2}$, $P(B) = \frac{1}{4}$ and $P(C) = \frac{1}{3}$ find
   (a) $P(A \cap B)$, if A and B are independent events,
   (b) $P(A \cup C)$, if A and C are mutually exclusive events.
4. If A and B are independent events such that $P(A) = \frac{1}{2}$ and $P(B) = \frac{1}{4}$, find   (a) $P(A \cap B)$,   (b) $P(A \cup B)$.
5. If A and B are independent events such that $P(A) = \frac{2}{3}$ and $P(A \cap B) = \frac{1}{3}$, find   (a) $P(B)$,   (b) $P(A \cup B)$.
6. Events A and B are such that $P(A) = \frac{3}{4}$, $P(B) = \frac{1}{2}$ and $P(A \cap B) = \frac{3}{8}$.
   State whether A and B are   (a) mutually exclusive,   (b) independent.
7. Events A and B are such that $P(A) = \frac{1}{2}$, $P(B) = \frac{7}{12}$ and $P[(A \cup B)'] = \frac{1}{4}$.
   State whether A and B are
   (a) mutually exclusive,   (b) independent.
8. Find the probability of drawing either a black ace or a red court card in one draw from a pack of cards.
9. The numbers 1, 2, 4, 6, 10, 12, 13, 16, 17 and 20 are written on ten pieces of card and one is drawn at random. Find the probability that the number on the card drawn will be either an odd number or a multiple of 3.
10. Find the probability that when a die is thrown and a card is drawn from a pack, the outcome will be a black card and an odd number on the die.
11. A card is drawn from a pack and then replaced. A second card is then drawn. Find the probability that both cards will be diamonds.
12. A coin is tossed twice. Find the probability that the outcome of the two trials will be   (a) two heads,   (b) two tails.
13. A die is thrown and a card is drawn from a pack. Find the probability that the outcome is an even number on the die and a black court card.
14. A card is drawn from a pack and a die is thrown. Find the probability that the card is not a club and the number on the die is greater than 4.
15. To start a game a player has to throw a 6 with a die. Find the probability that a player starts at
    (a) his first throw,   (b) his second throw,   (c) either his first or his second throw.
16. Two dice are thrown and the scores added together. Illustrate the

possibility-space diagrammatically. Find the probability that the total is
(a) 4,   (b) greater than 4.

17. A coin is tossed and a die is thrown. Draw the possibility-space diagram
and find the probability that the outcome of the trial is
(a) a tail and an odd number,   (b) a tail and a multiple of 3.

18. Two cards are drawn from a pack of cards, the first being replaced
before the second is drawn. Find the probability that the cards will be
(a) a diamond followed by a spade,   (b) a king followed by a black card,
(c) a red card followed by a club,   (d) a red card and a club in any order,
(e) two aces.

19. A number is selected at random from the set {2, 4, 6, 8} and another
number is selected at random from the set {1, 3, 5, 7}. The two numbers
so obtained are multiplied together. Draw a possibility-space diagram
and find the probability that the product formed will be
(a) even,   (b) more than ten.

20. Two numbers are selected at random, one from each of the sets
{1, 3, 5, 7} and {2, 4, 6, 8}. The numbers obtained are added together.
Draw a possibility-space diagram and find the probability that the sum
obtained is   (a) less than eight,   (b) greater than ten.

21. Three cards are drawn in succession from a pack of cards with
replacement taking place. Find the probability that the cards will be
(a) all red,   (b) a heart followed by two spades,
(c) a heart and two spades in any order.

22. Two cards are drawn from a pack of cards without replacement. Find
the probability that the two cards will be
(a) a club followed by a spade,   (b) a red ace followed by a spade,
(c) a black 7 followed by a red card,   (d) two hearts.

23. A bag contains 9 discs, 2 of which are green and 7 yellow. Two discs are
removed at random in succession, without replacement. Find the
probability that the discs will
(a) both be green,   (b) be of the same colour,   (c) be of different colours.

24. Two different pupils are chosen at random from a group of 3 boys and 5
girls. Find the probability that the two pupils chosen will be
(a) the two youngest pupils,   (b) two boys.

25. A coin is tossed and a die is thrown. Find the probability that the result
will be a head on the coin or a six on the die (or both).

26. A card is drawn from a pack of 52 cards and a die is thrown. Find the
probability of obtaining an ace card or a three on the die (or both).

27. A coin is tossed and a card selected from a pack of 52 cards. Find the
probability of obtaining a head on the coin or a court card (or both).

28. A die is biased so that the probability of throwing a six is $\frac{1}{10}$. Two
players each throw the die once. Find the probability that one or both
players throw a six. ·

29. Four different numbers are chosen at random from the numbers 1, 2, 3,
4, 5, 6, 7, 8, 9. Find the probability that the numbers chosen will be
(a) all odd,   (b) two odd and two even numbers.

30. Three discs are chosen at random, and without replacement, from a bag
containing 3 red, 8 blue and 7 white discs. Find the probability that the
discs chosen will be
(a) all red,   (b) all blue,   (c) one of each colour.

9.3 Further conditional probability

In the last section we saw that, for conditional events A and B,

$$P(A \cap B) = P(A) \times P(B|A)$$

Some questions require $P(B|A)$ to be found and the rearranged form
$P(B|A) = \dfrac{P(A \cap B)}{P(A)}$ is useful.

Example 12

A bag contains five discs, three of which are red. A box contains six discs, four of which are red. A card is selected at random from a normal pack of 52 cards, if the card is a club a disc is removed from the bag and if the card is not a club a disc is removed from the box. Find (a) the probability that the disc removed will be red, (b) the probability that, if the removed disc is red it came from the bag.

Such questions are best solved by means of a tree diagram. Let C be the event of the card being a club and R the event of the disc being red. Draw the tree diagram:
(R′ is the event of the disc *not* being red.)

$$P(C \cap R) = \tfrac{1}{4} \times \tfrac{3}{5} = \tfrac{3}{20} \quad \text{(a) } P(R) = P(C \cap R) + P(C' \cap R)$$
$$= \tfrac{3}{20} + \tfrac{1}{2}$$
$$= \tfrac{13}{20}$$

$$P(C \cap R') = \tfrac{1}{4} \times \tfrac{2}{5} = \tfrac{1}{10}$$

$$P(C' \cap R) = \tfrac{3}{4} \times \tfrac{2}{3} = \tfrac{1}{2} \quad \text{(b) We now require } P(C|R)$$
$$P(C|R) = \frac{P(R \cap C)}{P(R)}$$

$$P(C' \cap R') = \tfrac{3}{4} \times \tfrac{1}{3} = \tfrac{1}{4} \qquad = \frac{P(C \cap R)}{P(R)}$$

$$= \frac{\tfrac{3}{20}}{\tfrac{13}{20}} = \frac{3}{13}$$

Exercise 9C

1. If A and B are dependent events such that $P(A) = \tfrac{5}{8}$ and $P(B|A) = \tfrac{3}{7}$, find $P(A \cap B)$.
2. If A and B are dependent events such that $P(A) = \tfrac{2}{3}$ and $P(A \cap B) = \tfrac{3}{10}$, find $P(B|A)$.
3. Suppose that two different cards are removed from a pack of 52 cards. If event A is that of the first card being a Jack and event B is that of the second card being a court card (J, Q, K), find (a) $P(B|A)$, (b) $P(A \cap B)$.
4. A bag contains 3 yellow and 1 blue disc, whereas a box contains 2 yellow and 3 blue discs. A card is drawn from a pack of cards and if this card is a court card, a disc is drawn from the bag, otherwise a disc is drawn from the box.
   (a) Find the probability that the disc drawn is yellow,
   (b) given that the disc drawn is blue, find the probability that it came from the bag.
5. A bag contains 4 golf balls and 2 other balls, whereas a box contains 2 golf balls and 3 other balls. A die is thrown and if a number less than 3

results a ball is drawn from the bag, otherwise a ball is drawn from the box. Find the probability that
(a) the ball drawn is a golf ball,
(b) if the ball drawn is a golf ball, it came from the box.

6. A school is divided into two parts: Upper school, 400 boys and 200 girls, Lower school, 400 girls and 300 boys. A first pupil is chosen at random from the school. If this pupil comes from the Lower school, a second pupil is chosen from the Upper school; if the first pupil comes from the Upper school, the second pupil is chosen from the Lower school. Find the probability that
(a) the second pupil chosen will be a girl,
(b) if the second pupil chosen is a boy, he is a member of the Upper school.

7. A bag contains 20 discs of which one quarter are white; a similar box contains 15 discs of which one third are white. A card is drawn from a pack, and if the card drawn is a 7, 8 or 9 a disc is drawn from the bag, otherwise a disc is drawn from the box. Find the probability that
(a) the disc drawn is white,
(b) if the disc drawn is white, it came from the box.

8. A hospital diagnoses that a patient has contracted a virus X but it is not known which one of the three strains of the virus X_1, X_2 or X_3 the patient has. For a patient having virus X, the probability of it being X_1, X_2 or X_3 is $\frac{1}{2}$, $\frac{3}{8}$ or $\frac{1}{8}$ respectively. The probability of a recovery (event R) is $\frac{1}{2}$ for X_1, $\frac{2}{3}$ for X_2 and $\frac{1}{4}$ for X_3. Find
(a) the probability that the patient will recover,
(b) the probability that, if the patient recovers, he had virus X_3.

9. A box contains 6 discs of which 2 are blue; a similar bag contains 5 discs of which 3 are blue. A coin, biased so that a head is twice as likely as a tail, is tossed. If the outcome is a head, a disc is removed from the box, otherwise a disc is removed from the bag.
(a) Find the probability that the disc will not be blue,
(b) given that the disc is blue, find the probability that it came from the box.

10. Use the box and bag of question 9. When the weighted coin shows a head, two discs are removed from the box without replacement and, if not a head, two discs are removed from the bag, without replacement. Find the probability that
(a) both discs are blue,   (b) if both discs are blue, they came from the bag.

9.4 Probability involving permutations and combinations

Example 13

A team of five children is randomly selected from Alec, Bob, Charles, David, Emma, Frances, Gina and Helen. What is the probability that both Gina and Helen are in the team?

There are 8 children from whom to choose the team.
Number of ways of choosing a team of 5 from 8 = $^8C_5$
If Gina and Helen are to be included, the other 3 to complete the team have

to be chosen from the other 6 (A, B, C, D, E, F)
Number of ways of choosing a team including Gina and Helen = $^6C_3$

$$P(\text{G, H included}) = \frac{\text{number of teams including G and H}}{\text{total number of possible teams}}$$

$$= \frac{^6C_3}{^8C_5} = \frac{6!}{3!\,3!} \times \frac{5!\,3!}{8!}$$

$$= \frac{5}{14}$$

Example 14

Find the probability that when a hand of 7 cards is dealt from a shuffled pack of 52 cards it contains (a) all 4 aces, (b) exactly 3 aces, (c) at least 3 aces.

(a) Total number of possible hands = $^{52}C_7$
   number of hands with 4 aces = $^{48}C_3$ (since the other 3 cards must be chosen from the other 48 cards)

$$P(\text{hand has 4 aces}) = \frac{^{48}C_3}{^{52}C_7} = \frac{1}{7735}$$

(b) number of hands with 3 aces and 4 cards not aces = $^4C_3 \times {}^{48}C_4$

$$P(\text{exactly 3 aces}) = 4 \times \frac{^{48}C_4}{^{52}C_7} = \frac{9}{1547}$$

(c) P(at least 3 aces) = P(exactly 3 aces) + P(4 aces)

$$= \frac{1}{7735} + \frac{9}{1547}$$

$$= \frac{46}{7735}$$

Binomial Probability

Suppose that a trial is repeated a number of times, say 3, and that in each trial there are two possible outcomes A and B. If $P(A) = a$ and $P(B) = b$, and the outcome of each trial is independent of the previous trials, the tree diagram would be:

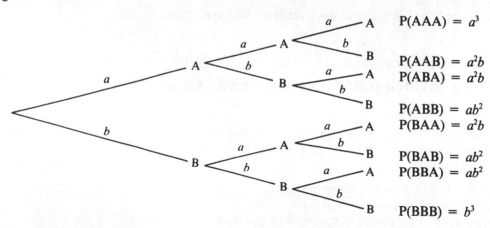

$$P(AAA) = a^3$$
$$P(AAB) = a^2b$$
$$P(ABA) = a^2b$$
$$P(ABB) = ab^2$$
$$P(BAA) = a^2b$$
$$P(BAB) = ab^2$$
$$P(BBA) = ab^2$$
$$P(BBB) = b^3$$

The probability of obtaining outcome B on each occasion = P(BBB) = b^3

or of obtaining outcome A exactly twice = P(AAB) + P(ABA) + P(BAA)

$$= 3a^2b$$

Clearly for more trials, say 5, the tree diagram would become very large and so an alternative method is needed. If the trial is carried out five times the various outcomes would be:

5 A's   4 A's, 1 B   3 A'S, 2 B's   2 A'S, 3 B's   1 A, 4 B'S   5 B's

The 5 A's occur in five trials in $^5C_5 = 1$ way (AAAAA)

The 4 A's occur in five trials in $^5C_4 = 5$ ways (BAAAA, ABAAA, AABAA, AAABA, AAAAB)

The 3 A's occur in five trials in $^5C_3 = 10$ ways etc.

Thus the probabilities of the various combinations of A's and B's would be

$^5C_5a^5$ $^5C_4a^4b^1$ $^5C_3a^3b^2$ $^5C_2a^2b^3$ $^5C_1ab^4$ $^5C_0b^5$

or $1a^5$ $5a^4b^1$ $10a^3b^2$ $10a^2b^3$ $5ab^4$ $1b^5$

Notice that these numbers could also be obtained from the sixth row of Pascal's Triangle:

Similarly the seventh row would give the coefficients for 6 trials and so on.

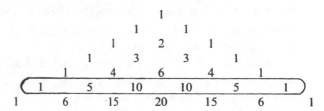

Example 15

A coin is biased so that P(head) = $\frac{2}{3}$ and P(tail) = $\frac{1}{3}$. If the coin is tossed 6 times find the probability of obtaining

(a) 6 heads, (b) exactly 5 heads, (c) at least 5 heads, (d) at least 1 tail,
(e) 3 heads and 3 tails with the heads occurring on successive tosses of the coin.

First list the various combinations of probabilities with h = P(head) = $\frac{2}{3}$ and t = P(tail) = $\frac{1}{3}$ These are:

$^6C_6h^6$ $^6C_5h^5t^1$ $^6C_4h^4t^2$ $^6C_3h^3t^3$ $^6C_2h^2t^4$ $^6C_1h^1t^5$ $^6C_0t^6$

or $1h^6$ $6h^5t^1$ $15h^4t^2$ $20h^3t^3$ $15h^2t^4$ $6h^1t^5$ $1t^6$

(a) P(6 heads) = $1h^6 = (\frac{2}{3})^6$
$$= \tfrac{64}{729}$$

(b) P(5 heads) = $6h^5t^1 = 6(\frac{2}{3})^5(\frac{1}{3})$
$$= \tfrac{64}{243}$$

(c) P(at least 5 heads) = P(exactly 5 heads) + P(6 heads)
$$= \tfrac{64}{243} + \tfrac{64}{729}$$
$$= \tfrac{256}{729}$$

(d) P(at least 1 tail) = $1 -$ P(no tails) = $1 -$ P(6 heads)
$$= 1 - \tfrac{64}{729}$$
$$= \tfrac{665}{729}$$

(e) P(HHHTTT) = $\left(\dfrac{2}{3}\right)^3\left(\dfrac{1}{3}\right)^3 = \dfrac{2^3}{3^6}$

However, there are $\dfrac{4!}{3!}$ ways of arranging the 3 heads and 3 tails with the 3 heads kept together:

HHHTTT, THHHTT, TTHHHT, TTTHHH.

Thus the required probability is $4 \times \dfrac{2^3}{3^6} = \dfrac{32}{729}$

Exercise 9D

1. There are only 2 boys in a group of 6 pupils. A group of 5 pupils is to be selected. Find the probability that both boys are in the group selected.

2. A group of 7 pupils are designated T, U, V, W, X, Y and Z. Find the probability that when a group of 4 pupils are selected, at random, it will include X, Y and Z.

3. A bag contains 10 discs of which 3 are red and 7 are blue. If 6 discs are selected at random, find the probability that all the red discs are selected.

4. Four letters are chosen at random from the letters a, b, c, d, e, f. Find the probability that both vowels are in the group chosen.

5. There are 5 green, 4 yellow and 3 blue discs in a bag, from which 4 discs are chosen at random. Find the probability that the 4 discs selected will contain
   (a) exactly 3 blue discs,   (b) exactly 3 yellow discs,   (c) at least one green disc.

6. From a well shuffled pack of 52 cards a hand of 6 cards is dealt. Find the probability that the hand will contain
   (a) 4 queens,   (b) exactly 3 queens,   (c) at least 3 queens.

7. A hand of 7 cards is dealt from a shuffled pack of 52 cards. Find the probability that the hand will contain
   (a) exactly 6 black cards,   (b) all black cards,   (c) at least 6 black cards.

8. An unbiased coin is tossed 7 times. Find the probability of obtaining
   (a) 7 tails,   (b) exactly 6 tails,   (c) at least one head.

9. A fair coin is tossed ten times. Find the probability of obtaining
   (a) exactly 9 heads,   (b) no tails,   (c) more than 8 heads.

10. A fair coin is tossed 9 times. Find the probability of obtaining
    (a) 5 heads and 4 tails,   (b) 7 tails and 2 heads.

11. A coin is biased so that P(head) = $\frac{1}{4}$ and P(tail) = $\frac{3}{4}$. If the coin is tossed 5 times find the probability of obtaining
    (a) no tails,   (b) exactly 1 tail,   (c) 3 tails and 2 heads,
    (d) at least 4 heads,   (e) 2 heads and 3 tails with the heads occurring in succession.

12. A fair coin is tossed 12 times. Find the probability of obtaining more than 9 tails.

13. A marksman fires at a target and the probability of hitting the bull with any one shot is $\frac{2}{3}$. If he fires 6 shots find the probability that he obtains
    (a) 6 bulls,   (b) no bulls,   (c) at least 3 bulls.

14. Ten unbiased dice are thrown. Find the probability of obtaining exactly seven sixes.

15. A fair die is thrown 6 times. Find the probability of obtaining
    (a) 6 odd numbers,   (b) only 1 even number,   (c) at least 4 odd numbers.

16. A firm manufactures radios. The probability that any radio fails the quality check is $0 \cdot 1$. Find the probability that in a case of 15 radios
    (a) all pass the quality check,
    (b) exactly 1 fails the quality check,
    (c) at least 13 pass the quality check.

17. A die is biased so that P(odd number) = $\frac{1}{3}$ and P(even number) = $\frac{2}{3}$. If this die is thrown 8 times, find the probability of obtaining

(a) an even number 7 times,   (b) no odd numbers,

(c) an equal number of odd and even numbers,

(d) 5 odd numbers and 3 even numbers with the even numbers occurring in succession,

(e) alternate odd and even numbers.

Exercise 9E   *Examination questions*

1. From a pack of 52 playing cards two cards are drawn without replacement. Calculate, as a fraction in its lowest terms, the probability that

(a) both are Hearts,

(b) both are red,

(c) neither is a Club.   (A.E.B)

2. A card is drawn at random from an ordinary pack of 52 playing cards and it is not returned to the pack. A second card is drawn at random. Calculate, showing details of your working, the probabilities:

(i)   that the first card is a spade and the second card is the king of spades;

(ii) that the first card is a king and the second card is the queen of the same suit;

(iii) that the two cards are numerically the same or have the same rank (e.g. are both queens);

(iv) that one card is a heart and the other a spade.   (Oxford)

3. Box A contains 5 pieces of paper numbered 1, 3, 5, 7, 9.
Box B contains 3 pieces of paper numbered 1, 4, 9.
One piece of paper is removed at random from each box.
Calculate the probability that the two numbers obtained have

(i)   the same value,

(ii) a sum greater than 3,

(iii) a product that is exactly divisible by 3.   (Cambridge)

4. Two fair cubical dice are thrown. Calculate the probability that the sum of the scores is

(a) exactly 5,

(b) less than 5,

(c) at least 6.   (A.E.B.)

5. In a game, a player rolls two balls down an inclined plane so that each ball finally settles in one of five slots and scores the number of points allotted to that slot as shown in the diagram on the right.
It is possible for both balls to settle in one slot and it may be assumed that each slot is equally likely to accept either ball.
The player's score is the sum of the points scored by each ball.
Draw up a table showing all the possible scores and the probability of each.
If the player pays 10p for each game and receives back a number of pence equal to his score, calculate the player's expected gain or loss per 50 games.   (Cambridge)

2 | 4 | 7 | 4 | 2

6. (a) Two dice are thrown together, and the scores added. What is the probability that (i) the total score exceeds 8? (ii) the total score is 9, or the individual scores differ by 1, or both?

(b) A bag contains 3 red balls and 4 black ones. 3 balls are picked out, one at a time and not replaced. What is the probability that there will be 2 red and 1 black in the sample?   (S.U.J.B.)

7. A, B and C fire one shot each at a target.
The probability that A will hit the target is $\frac{1}{2}$.
The probability that B will hit the target is $\frac{1}{4}$.
The probability that C will hit the target is $\frac{1}{3}$.
If they fire together, calculate the probability that
(i)   all three shots hit the target,   (ii) C's shot only hits the target,
(iii) at least one shot hits the target,   (iv) exactly two shots hit the target.
    (Cambridge)

8. The birthdays of Jack and Jill are in the first seven days in January.
Find the probability that next year
(a) both have their birthday on Monday,
(b) Jack and Jill have their birthdays on the same day,
(c) they have their birthdays on different days,
(d) Monday is the birthday of one or both.   (A.E.B.)

9. Calculate
(i) the number of arrangements of 8 different books on a shelf,
(ii) the probability that, in any one of these arrangements, 3 particular books are together.   (Cambridge)

10. Three dice are to be rolled. Find the probability of scoring a double but not a triple.   (London)

11. The probability that it will rain on a given morning is $\frac{1}{4}$.
If it rains the probability that Mr X misses his train is $\frac{2}{3}$.
If it does not rain the probability that Mr X catches his train is $\frac{5}{6}$.
If he catches his train the probability that he is early for work is $\frac{4}{5}$.
If he misses his train the probability that he is late for work is $\frac{3}{5}$.
Calculate the probability that, on a given morning,
(i) it rains and Mr X is late for work,
(ii) it does not rain and he is early for work.   (Cambridge)

12. Two rugby teams, A and B, play a series of three matches. The probability that team A wins any given match is $\frac{1}{2}$ while the probability that team B wins any given match is $\frac{2}{5}$.
Calculate the probability that
(i)   all three matches are drawn,
(ii) the teams are level after two matches,
(iii) the series is drawn.   (Cambridge)

13. In order to start in a game of chance a player throws a fair cubical die until he obtains a six. He then records whatever scores he obtains on subsequent throws.
For example:
throws of 2, 4, 3, 6, 4, 6, 2, 5 give recorded scores of 4, 6, 2, 5.
Calculate the probability that
(a) the first score recorded is that of the player's fourth throw,
(b) the player does not record a score in his first five throws.
The player has seven throws. Calculate the probability that he will have recorded
(c) exactly three fives and a three,
(d) a total score of three.   (A.E.B.)

14. Two events A and B are such that
$P(A) = 0.2$, $P(A' \cap B) = 0.22$, $P(A \cap B) = 0.18$.
Evaluate (i) $P(A \cap B')$, (ii) $P(A|B)$.   (J.M.B.)

15. A card is drawn at random from a pack of 52 cards and is then replaced. Let A denote the event 'an ace is drawn', and let R denote the event 'a red card is drawn'. Calculate the values of $P(A)$, $P(R)$, $P(A \cap R)$ and $P(A \cup R)$.
If this experiment is performed four times, find the probability
(a) that an ace will be drawn for the first time at the third attempt,
(b) that a red card will be drawn exactly twice.
(c) that a red card will be drawn at least once.
Find also how many times the experiment must be performed to ensure that the probability of a red card being drawn at least once exceeds 0.99.   (A.E.B.)

16. Two players, X and Y, play a game in which X throws 6 coins and Y throws an unbiased die. Player X wins if the number of heads is greater than the number on the die.
(i) If X throws 5 heads find the probability that he wins.
(ii) If Y throws the number 3, show that the probability of X winning is $\frac{11}{32}$.   (Cambridge)

17. Previous experience indicates that, of the students entering upon a particular diploma course, 90% will successfully complete it.
One year, 15 students commence the course. Calculate, correct to 3 decimal places, the probability that
(i) all 15 successfully complete the course,
(ii) only 1 student fails,
(iii) no more than 2 students fail,
(iv) at least 2 students fail.   (Cambridge)

18. The probability that a marksman will hit a target is $\frac{4}{5}$. He fires 10 shots. Calculate, correct to 3 decimal places, the probability that he will hit the target
(i) at least 8 times,
(ii) no more than 7 times.
If he hits the target exactly 7 times, calculate the probability that the 3 misses are with successive shots.   (Cambridge)

19. The probability that a certain football team, playing at home, will win a match is $\frac{2}{3}$.

Calculate the probability that in their next 6 home matches, the team will win
(i)   exactly 4 matches,
(ii) at least 4 matches,
(iii) exactly 4 successive matches.   (Cambridge)

20. Out of a large population it may be assumed that 5% of unmarried men of forty years of age will marry within five years. Calculate the probabilities that out of a sample of 8 unmarried men who are 40 years old, selected at random,
(i)   none will get married within 5 years;
(ii) just three will get married within 5 years;
(iii) at least four will get married within 5 years.
If the size of the sample increases by 4, calculate the probability that just 6 of the 12 men are still unmarried after five years.   (Oxford)

21. A player can play on either of two gambling machines, A and B. He chooses one of the two machines at random, and plays two games. The probability of winning a game on A is $\frac{1}{3}$, and the probability of winning a game on B is $\frac{1}{4}$. If he loses both of these two games, he plays a third game on the other machine; otherwise he plays the third game on the same machine. Find the probability that he
(i)   wins the first game;
(ii) changes machines after the second game;
(iii) plays the third game on A;
(iv) wins the third game.   (Oxford)

22. The events A and B are such that
$$P(A) = \tfrac{1}{2},$$
$$P(A|B') = \tfrac{2}{3},$$
$$P(A|B) = \tfrac{3}{7},$$
where B′ is the event 'B does not occur'. Find $P(A \cap B)$, $P(A \cup B)$, $P(B)$, $P(B|A)$.
State, with reasons, whether A and B are
(i)   independent,
(ii) mutually exclusive.   (Cambridge)

23. Three people independently each think of an integer in the set $\{1, 2, 3, 4, 5, 6, 7\}$. Find, in fractional form, the probability that
(a) all three of the integers selected are greater than 4,
(b) all three of the integers selected are greater than 5,
(c) the least integer selected is 5,
(d) the three integers selected are different given that the least integer selected is 5,
(e) the sum of the three integers selected is more than 15.   (A.E.B.)

10

Calculus I: Differentiation

10.1 The gradient of a curve

In chapter 3 we saw that the gradient of a straight line is the same at all points on the line. With a curve however the gradient, or steepness, will depend upon where we are on the curve.

The gradient at a point P on a curve is defined as the gradient of the *tangent* drawn to the curve at the point P, i.e. the gradient of the line that just touches the curve at the point P.

If we wished to find the gradient of a curve at some particular point, we could accurately draw the curve, draw the tangent to the curve at the required point and then measure the gradient of this tangent.

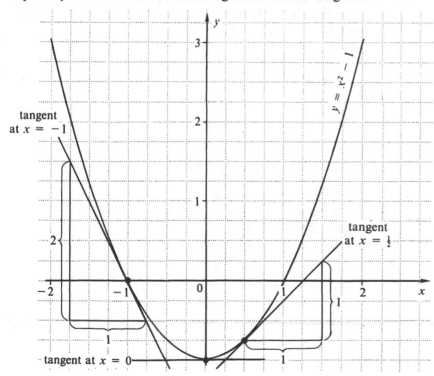

The graph shown is that of $y = x^2 - 1$ for $-2 \leqslant x \leqslant 2$ with the tangents drawn at $x = -1$, $x = 0$ and at $x = \frac{1}{2}$. From the graph we can see that the gradient of $y = x^2 - 1$ is

$-\frac{2}{1} = -2$ at $x = -1$ (the negative sign is because y *decreases* by 2 units as x *increases* by 1 unit)

0 at $x = 0$,

and $\frac{1}{1} = 1$ at $x = \frac{1}{2}$.

Clearly it is not easy to draw these tangents with any high degree of accuracy and so we need an alternative method for finding accurately the gradient at particular points on a curve.

Suppose that we wish to find the gradient of a
curve $y = f(x)$ at a point P(x, y) on the curve.
Consider a second point, Q, lying on the
curve, near to P and with x-coordinate given
by $x + \delta x$ where δx is used to denote a small
increment of length in the direction of the x-axis.

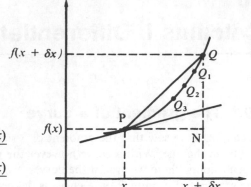

Thus the gradient of the chord PQ $= \dfrac{\text{QN}}{\text{PN}}$

$$= \frac{f(x + \delta x) - f(x)}{x + \delta x - x}$$

$$= \frac{f(x + \delta x) - f(x)}{\delta x}$$

Now if we move Q nearer to P, say to points Q_1, Q_2, ... then the gradient
of the chords PQ, PQ_1, PQ_2, ... will give better and better approximations
for the gradient of the tangent at P and therefore, the gradient of the curve at P.
Thus the gradient of the curve $y = f(x)$ at some point P(x, y) on the curve is given by

$$\lim_{\delta x \to 0} \left[\frac{f(x + \delta x) - f(x)}{\delta x} \right]$$

If we say that P has coordinates (x, y) and Q has coordinates $(x + \delta x, y + \delta y)$

then the gradient $= \lim\limits_{\delta x \to 0} \left[\dfrac{y + \delta y - y}{x + \delta x - x} \right]$

$$= \lim_{\delta x \to 0} \left(\frac{\delta y}{\delta x} \right)$$

We write $\dfrac{dy}{dx}$ (pronounced 'dee y by dee x') for $\lim\limits_{\delta x \to 0} \left(\dfrac{\delta y}{\delta x} \right)$ and we call $\dfrac{dy}{dx}$

the gradient function, derived function or differential coefficient of y with
respect to x. The gradient function gives a formula by which the gradient at
any point on the line can be determined. The process of finding the
differential coefficient of a function is called *differentiation*. A shorthand
notation that is sometimes used is that if $y = f(x)$ we write $\dfrac{dy}{dx}$ as $f'(x)$ or
simply y'.

$$\boxed{\begin{aligned} \text{Thus} \quad \frac{dy}{dx} &= \lim_{\delta x \to 0} \left(\frac{\delta y}{\delta x} \right) \\[2mm] &= \lim_{\delta x \to 0} \left[\frac{f(x + \delta x) - f(x)}{\delta x} \right] \end{aligned}}$$

Note   Though we choose to use a fractional form of representation, $\dfrac{dy}{dx}$ is

a limit and is not a fraction, i.e. $\dfrac{dy}{dx}$ does not mean $dy \div dx$. $\dfrac{dy}{dx}$ means y

differentiated with respect to x. Thus, $\dfrac{dp}{dx}$ means p differentiated with respect

to x. The '$\dfrac{d}{dx}$' is the 'operator', operating on some function of x.

Example 1

Find, from first principles, the differential coefficient of y with respect to x if $y = x^2$.

Here $f(x) = x^2$. Using
$$\frac{dy}{dx} = \lim_{\delta x \to 0} \left[\frac{f(x + \delta x) - f(x)}{\delta x} \right]$$

$$\frac{dy}{dx} = \lim_{\delta x \to 0} \left[\frac{(x + \delta x)^2 - x^2}{\delta x} \right]$$

$$= \lim_{\delta x \to 0} \left[\frac{2x\delta x + (\delta x)^2}{\delta x} \right]$$

$$= \lim_{\delta x \to 0} (2x + \delta x) = 2x$$

Thus if $y = x^2$, $\dfrac{dy}{dx} = 2x$.

Alternatively, this could be set out as follows:
$$y = x^2 \qquad \qquad \ldots [1]$$

Let x change by a small amount δx and the consequent change in y be δy, then
$$y + \delta y = (x + \delta x)^2 \qquad \qquad \ldots [2]$$

Subtracting equation [1] from equation [2]
$$y + \delta y - y = (x + \delta x)^2 - x^2$$
or
$$\delta y = (x + \delta x)^2 - x^2$$
$$\therefore \qquad \frac{\delta y}{\delta x} = \frac{(x + \delta x)^2 - x^2}{\delta x}$$

By definition
$$\frac{dy}{dx} = \lim_{\delta x \to 0} \left(\frac{\delta y}{\delta x} \right) = \lim_{\delta x \to 0} \left[\frac{(x + \delta x)^2 - x^2}{\delta x} \right]$$

$$= \lim_{\delta x \to 0} \left[\frac{2x\delta x + (\delta x)^2}{\delta x} \right]$$

$$= \lim_{\delta x \to 0} [2x + \delta x]$$

$$= 2x$$

Example 2

If $f(x) = 4x + 2x^2$, find $f'(x)$ from first principles and hence calculate $f'(2)$ and $f'(-2)$.

$$f'(x) = \frac{dy}{dx} = \lim_{\delta x \to 0} \left[\frac{f(x + \delta x) - f(x)}{\delta x} \right]$$

$$= \lim_{\delta x \to 0} \left[\frac{4(x + \delta x) + 2(x + \delta x)^2 - 4x - 2x^2}{\delta x} \right]$$

$$= \lim_{\delta x \to 0} \left[\frac{4\delta x + 4x\delta x + 2(\delta x)^2}{\delta x} \right]$$

$$= \lim_{\delta x \to 0} [4 + 4x + 2\delta x]$$

$$\therefore \qquad \qquad f'(x) = 4 + 4x$$
Since $\qquad \qquad f'(x) = 4 + 4x$
then $\qquad \qquad f'(2) = 4 + 4(2)$
$$= 12$$
and $\qquad f'(-2) = 4 + 4(-2)$
$$= -4$$

Thus if $f(x) = 4x + 2x^2$, then $f'(x) = 4 + 4x$, $f'(2) = 12$ and $f'(-2) = -4$.

Exercise 10A

1. Draw the graph of $y = x^2 + x - 6$ for $-5 \leqslant x \leqslant 6$. Draw the tangents to this curve at $x = 3$, $x = 1$ and $x = -2$, and hence find a value for the gradient of the curve at each of these points.

2. Draw the graph of $y = \dfrac{x^2 - 4x}{4}$ for $0 \leqslant x \leqslant 6$. Draw the tangents to the curve at $x = 4$, $x = 3$ and $x = 2$ and hence find a value for the gradient of the curve at each of these points.

3. Differentiate each of the following from first principles to find $\dfrac{dy}{dx}$.

   (a) $y = 5x$ (b) $y = 9x + 5$ (c) $y = 3x^2$

   (d) $y = x^3$ (e) $y = x^2 + 3x$ (f) $y = 5x - x^2 + 7$

   (g) $y = \dfrac{1}{x}$ (h) $y = \dfrac{1}{x^2}$

4. If $f(x) = 3x - 2x^2$ find $f'(x)$ from first principles and hence evaluate $f'(4)$ and $f'(-1)$.

5. If $f(x) = 2x^2 + 5x - 3$ find $f'(x)$ from first principles and hence evaluate $f'(-1)$ and $f'(-2)$.

6. If $f(x) = x^3 - 2x$ find $f'(x)$ from first principles and hence evaluate $f'(1)$, $f'(0)$ and $f'(-1)$.

10.2 Differentiation of ax^n

In order to avoid differentiating functions from first principles, as we have done in 10.1, we can establish certain rules.

Suppose we wish to find $\dfrac{dy}{dx}$ given that $y = x^n$.

In this case $f(x) = x^n$, and
$$\frac{dy}{dx} = \lim_{\delta x \to 0} \left[\frac{f(x + \delta x) - f(x)}{\delta x} \right]$$

$$= \lim_{\delta x \to 0} \left[\frac{(x + \delta x)^n - x^n}{\delta x} \right] \qquad \ldots [1]$$

Now for positive integer values of n, $(x + \delta x)^n$ can be expanded by the binomial theorem to give a finite series:

$$(x + \delta x)^n = x^n + {}^nC_1 x^{n-1} \delta x + {}^nC_2 x^{n-2} (\delta x)^2 + {}^nC_3 x^{n-3} (\delta x)^3 + \ldots + (\delta x)^n$$

substituting this in [1], gives

$$\frac{dy}{dx} = \lim_{\delta x \to 0} [nx^{n-1} + {}^nC_2 x^{n-2} \delta x + {}^nC_3 x^{n-3} (\delta x)^2 + \ldots + (\delta x)^{n-1}]$$

i.e. $\dfrac{dy}{dx} = nx^{n-1}$

Thus the differential coefficient of x^n is nx^{n-1} and although the above proof is for positive integer values of n, the result applies for all rational values of n. This result, together with the three rules stated below, enable us to differentiate many functions.

(i)   If $y = f(x) + g(x) - h(x)$ then

$$\frac{dy}{dx} = \frac{d}{dx}(f(x)) + \frac{d}{dx}(g(x)) - \frac{d}{dx}(h(x))$$
$$= f'(x) + g'(x) - h'(x)$$

(ii) If $y = af(x)$ where a is a constant then
$$\frac{dy}{dx} = a\frac{d}{dx}(f(x))$$
$$= af'(x)$$

(iii) If $y = a$ where a is a constant, we can write this as
$$y = ax^0$$
then $\frac{dy}{dx} = (0)ax^{-1} = 0$

Thus the differential coefficient of a constant is zero.

[Rules for differentiating $f(x) \times g(x)$ and $\frac{f(x)}{g(x)}$ will be obtained later.]

Note: The rule for differentiating $y = ax^n$, i.e. $\frac{dy}{dx} = anx^{n-1}$, can be remembered in words as:
'multiply by the power and decrease the power by one'.

Example 3

Differentiate the following with respect to x.
(a) $y = x^8$ (b) $y = 6$ (c) $y = 3x^5$ (d) $y = 8\sqrt{x}$
(e) $y = 3x^2 - 6x + \frac{2}{x^2}$ (f) $y = (2x + 3)(x + 1)$

(a) If $y = x^8$
$\frac{dy}{dx} = 8x^7$
(Multiply by the power and decrease the power by one.)

(b) If $y = 6$
$\frac{dy}{dx} = 0$
(Differentiation of a constant term gives zero.)

(c) If $y = 3x^5$
$\frac{dy}{dx} = 15x^4$
(Multiply by the power and decrease the power by one.)

(d) If $y = 8\sqrt{x}$
$= 8x^{1/2}$
$\frac{dy}{dx} = \frac{1}{2} \times 8x^{-1/2}$
$= \frac{4}{\sqrt{x}}$

(e) If $y = 3x^2 - 6x + \frac{2}{x^2}$
$= 3x^2 - 6x + 2x^{-2}$
$\frac{dy}{dx} = 6x - 6 - 4x^{-3}$
$= 6x - 6 - \frac{4}{x^3}$

(f) If $y = (2x + 3)(x + 1)$
$= 2x^2 + 5x + 3$
$\frac{dy}{dx} = 4x + 5$

Example 4

Find the gradient of the curve $y = x^2 + 7x - 2$ at the point (2, 16).
$$y = x^2 + 7x - 2$$
$$\frac{dy}{dx} = 2x + 7$$

Thus at (2, 16), $\frac{dy}{dx} = 2(2) + 7 = 11$

At the point (2, 16) the curve $y = x^2 + 7x - 2$ has a gradient of 11.

Example 5

Find the points on the curve $y = x^3 + 3x^2 - 6x - 10$ where the gradient is 3.

$$y = x^3 + 3x^2 - 6x - 10$$
$$\frac{dy}{dx} = 3x^2 + 6x - 6$$

If the gradient is 3, then $\frac{dy}{dx} = 3$

i.e. $3x^2 + 6x - 6 = 3$
$$x^2 + 2x - 3 = 0$$

giving $x = -3$ or 1.   If $x = -3$ $y = -27 + 27 + 18 - 10$
$$= 8$$
and if $x = 1$ $y = 1 + 3 - 6 - 10$
$$= -12$$

The curve $y = x^3 + 3x^2 - 6x - 10$ has a gradient of 3 at $(-3, 8)$ and at $(1, -12)$.

Tangents and normals

Suppose that some point P lies on a curve $y = f(x)$. The line passing through P, perpendicular to the tangent to the curve at P, is said to be the *normal* to the curve at P. To find the equation of the tangent or normal to a curve at some point $P(x_1, y_1)$ on the curve we use $y - y_1 = m(x - x_1)$ [see page 87] with the gradient m determined by differentiation.

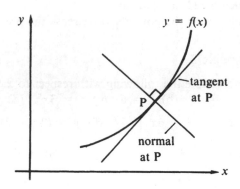

Example 6

Find the equation of the tangent and normal to the curve $y = x^2 - 4x + 1$ at the point $(-2, 13)$.

$$y = x^2 - 4x + 1$$
$$\therefore \quad \frac{dy}{dx} = 2x - 4$$

at $(-2, 13)$, $\frac{dy}{dx} = 2(-2) - 4 = -8$

Thus the tangent at $(-2, 13)$ has gradient -8 and the normal has gradient $\frac{1}{8}$.

Using $y - y_1 = m(x - x_1)$,   equation of tangent is: equation of normal is:
$$y - 13 = -8(x - -2) \qquad\qquad y - 13 = \tfrac{1}{8}(x - -2)$$
i.e. $\qquad\qquad y = -8x - 3$ $\qquad\qquad$ i.e. $\qquad 8y = x + 106$

Higher derivatives

We can repeat the differentiation process to find the differential coefficient of the differential coefficient of y with respect to x. This is called the second

differential of y with respect to x and is written $\frac{d^2y}{dx^2}$ or, if $y = f(x)$ it is written as $f''(x)$.

Thus if $y = 3x^3 - 6x + 4$

$$\frac{dy}{dx} = 9x^2 - 6$$

and $\frac{d^2y}{dx^2} = 18x$.

Exercise 10B

Differentiate the following functions with respect to x.

1. x^5 **2.** x^3 **3.** $12x^2$ **4.** $5x^4$ **5.** $16x$
6. $3x^2$ **7.** $9x^3$ **8.** $3x^4$ **9.** 7 **10.** $2x$
11. $5x^7$ **12.** 6 **13.** $x^{5/3}$ **14.** $x^{3/4}$ **15.** $x^{2/5}$
16. $x^{3/5}$ **17.** $6x^{2/3}$ **18.** $8x^{1/4}$ **19.** $\sqrt{x}$ **20.** $\sqrt{x^3}$
21. $4\sqrt{x}$ **22.** $\frac{5}{x}$ **23.** $\frac{3}{x^2}$ **24.** $\frac{8}{\sqrt{x}}$ **25.** $\frac{2}{x^3}$

Find the gradient function $\frac{dy}{dx}$ for each of the following:

26. $y = x^2 + 7x - 4$
27. $y = x - 7x^2$
28. $y = 6x^2 - 7x + 8$
29. $y = 3 + x$
30. $y = x^3 + 7x^2 - 2$
31. $y = 3x^6 - 2x^2 + 6x - 8$
32. $y = 3x^2 + 7x - 4 + \frac{1}{x}$

33. $y = 3x - \frac{5}{x} + \frac{6}{x^2}$

34. $y = 5x - \frac{3}{\sqrt{x}}$

35. $y = (x + 3)(x + 1)$
36. $y = (x + 4)(x - 2)$
37. $y = (2x + 3)(x + 2)$

Find the gradients of the following lines at the points indicated:

38. $y = x^2 + 4x$ at $(0, 0)$
39. $y = 5x - x^2$ at $(1, 4)$
40. $y = 3x^3 - 2x$ at $(2, 20)$
41. $y = 5x + x^3$ at $(-1, -6)$
42. $y = (x + 1)(2x + 3)$ at $(2, 21)$
43. $y = 2x^3 - x^2 - 6$ at $(2, 6)$

44. $y = 3x + \frac{1}{x}$ at $(1, 4)$

45. $y = 2x^2 - x + \frac{4}{x}$ at $(2, 8)$

Find the coordinates of any points on the following lines where the gradient is as stated.

46. $y = x^2$, gradient 8
47. $y = x^2$, gradient -8
48. $y = x^2 - 4x + 5$, gradient 2
49. $y = 5x - x^2$, gradient 3
50. $y = x^4 + 2$, gradient -4
51. $y = x^3 + 3x^2 - 5x - 10$, gradient 4
52. $y = x^3 + x^2 - x + 1$, gradient zero.

53. $y = \frac{12}{x}$, gradient -3

54. If $f(x) = x^3 + 4x$ find
   (a) $f(1)$,   (b) $f'(x)$,   (c) $f'(1)$,
   (d) $f''(x)$,   (e) $f''(1)$.

55. If $f(x) = 3x^2 + \frac{24}{x}$ find

   (a) $f(2)$,   (b) $f'(x)$,   (c) $f'(2)$,
   (d) $f''(x)$,   (e) $f''(2)$.

56. If $f(x) = 3x^2 - 4x$ find the value of a given that $f'(a) = 5$
57. If $f(x) = x^3 + 1$ find the value of a given that $f''(a) = 24$
58. Find the second differential of y with respect to x for each of the following:
   (a) $y = 6x^2 + 7$
   (b) $y = 5x^3 + 6x - 5$

   (c) $y = 2 + \frac{3}{x}$.

59. If $y = 3x^2 - x$ show that

$$y\frac{d^2y}{dx^2} + \frac{dy}{dx} - 6y + 1 = 6x.$$

60. Find the gradient of the curve $y = x^2 + 6x - 4$ at the point where the curve cuts the y-axis.
61. Find the coordinates of the points where the curve $y = x^2 - x - 12$ cuts the x-axis and determine the gradient of $y = x^2 - x - 12$ at these points.

62. The line $y = 4x - 5$ cuts the curve $y = x^2 - 2x$ at two points. Find the gradient of $y = x^2 - 2x$ at these two points.

63. The gradient of $y = x^2 - 4x + 6$ at $(3, 3)$ is the same as the gradient of $y = 8x - 3x^2$ at (a, b). Find the values of a and b.

64. Find the coordinates of the point on the curve $y = x^2 - 6x + 3$ where the tangent is parallel to the line $y = 2x + 3$.

65. Find the coordinates of the point on the curve $y = x - x^2$ where the tangent is parallel to the line $2y + x - 3 = 0$.

66. Find the coordinates of the point on the curve $y = 3x^2 - 9x + 10$ where the normal is parallel to the line $3y - x + 4 = 0$.

67. Find the equations of the tangent and the normal to the following curves at the points indicated:
    (a) $y = x^2$ at $(3, 9)$
    (b) $y = 5 - 2x^2$ at $(-1, 3)$
    (c) $y = x^2 - 5x - 4$ at the point on the curve with an x-coordinate of 5,
    (d) $y = 4 + x - 2x^2$ at the point on the curve with an x-coordinate of 1.

68. T is the tangent to the curve $y = x^2 + 6x - 4$ at $(1, 3)$ and N is the normal to the curve $y = x^2 - 6x + 18$ at $(4, 10)$. Find the coordinates of the point of intersection of T and N.

69. The tangent to the curve $y = ax^2 + bx + 2$ at $(1, \frac{1}{2})$ is parallel to the normal to the curve $y = x^2 + 6x + 10$ at $(-2, 2)$. Find the values of a and b.

10.3 Maximum, minimum and points of inflexion

Figure 1 shows the sketch of a function of x, say $y = f(x)$.

At the points A, B and C the tangent to the curve is parallel to the x-axis and therefore the gradient of the curve at A, B and C is zero,

i.e. $\dfrac{dy}{dx} = 0$ at A, B and C.

The point A is said to be a maximum point on the curve since the value of y at A is clearly greater than the values of y at points on the curve close to A.

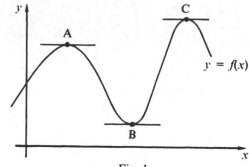

Fig. 1

It is important to realise that at a maximum point such as A, the value of y may not be the greatest value of y on the entire curve. The important fact is that the value of y at A is greater than at points close to A. Thus A is really a *local* maximum but we usually refer to such points simply as maximum points.

In a similar way C is another (local) maximum point on the curve.

The value of y at B is less than the values of y at points on the curve close to B and so B is said to be a (local) minimum point on the curve.

Such maximum and minimum points are said to be *turning* points on the curve. To locate such points without having to draw the graph we have only to find the points at which $\dfrac{dy}{dx} = 0$. However there may be some point on a curve at which $\dfrac{dy}{dx} = 0$, but the point is neither a maximum nor a minimum.

In Figure 2, D is such a point because the tangent is parallel to the x-axis but D is neither a maximum point nor a minimum point. D is called a point of *inflexion*.

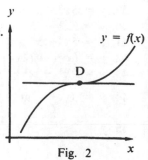

Fig. 2

A point on a curve at which $\frac{dy}{dx} = 0$, i.e. maximum points like A or C in

Figure 1, minimum points like B in Figure 1 or points of inflexion such as D in Figure 2 are called *stationary* points.

Having located these stationary points, we can distinguish between them i.e. determine whether they are maximum points, minimum points or points of inflexion, by considering the sign of the gradient at points close to, and on either side of these points.

For increasing values of x, as we approach and pass through a maximum point such as A, the gradient of the curve changes from positive ('uphill' for increasing x) to zero, at A, to negative ('downhill' for increasing x).

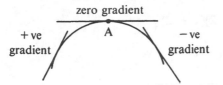

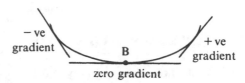

On the other hand, as we approach and pass through a minimum point such as B, the gradient changes from negative through zero to positive.

For points of inflexion, such as D and E, the gradient of the curve does not change sign as we pass through the point.

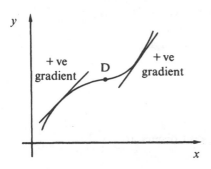

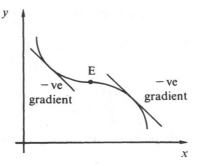

Thus, to find the stationary points on a curve $y = f(x)$ and to distinguish between them, we proceed as follows:

1. Find the gradient function $\frac{dy}{dx}$ of the curve.

2. Equate to zero the expression for $\frac{dy}{dx}$.

3. Find the values of x (i.e. $x_1, x_2, x_3 \ldots$) which satisfy this equation.

4. Consider the sign of $\frac{dy}{dx}$ on either side of these points.

5. Find the values $y_1, y_2, y_3 \ldots$ which correspond to $x_1, x_2, x_3 \ldots$

Example 7

Find the coordinates of any stationary points on the curve
$y = x^3 - 2x^2 - 4x$ and distinguish between these points.

$$y = x^3 - 2x^2 - 4x$$
$$\Rightarrow \frac{dy}{dx} = 3x^2 - 4x - 4$$

At stationary points, $\frac{dy}{dx} = 0$ i.e. $3x^2 - 4x - 4 = 0$

$$(3x + 2)(x - 2) = 0$$
$$\text{thus}\quad x = -\tfrac{2}{3}\ \text{or}\ 2$$

We must now consider the sign of $\frac{dy}{dx}$ as x increases through these points,

$$\text{now,}\quad \frac{dy}{dx} = (3x + 2)(x - 2)$$

$\therefore$ If x is 'just' less than $-\tfrac{2}{3}$ (say $-0\cdot8$) $\frac{dy}{dx} = (-\text{ve})(-\text{ve}) = +\text{ve}$

and   if x is 'just' more than $-\tfrac{2}{3}$ (say $-0\cdot5$) $\frac{dy}{dx} = (+\text{ve})(-\text{ve}) = -\text{ve}$

Thus the gradient changes from $+$ve ($/$) to $-$ve ($\backslash$) as we pass through
$x = -\tfrac{2}{3}$, i.e. a maximum point.

If x is 'just' less than 2 (say $1\cdot5$) $\frac{dy}{dx} = (+\text{ve})(-\text{ve}) = -\text{ve}$

and   if x is 'just' more than 2 (say $2\cdot5$) $\frac{dy}{dx} = (+\text{ve})(+\text{ve}) = +\text{ve}$

Thus the gradient changes from $-$ve ($\backslash$) to $+$ve ($/$) as we pass through
$x = 2$, i.e. a minimum point.

When $x = -\tfrac{2}{3}$, $y = (-\tfrac{2}{3})^3 - 2(-\tfrac{2}{3})^2 - 4(-\tfrac{2}{3})$
$$= 1\tfrac{13}{27}$$
when $x = 2$, $y = (2)^3 - 2(2)^2 - 4(2)$
$$= -8$$

Thus a maximum point occurs at $(-\tfrac{2}{3}, 1\tfrac{13}{27})$ and a minimum point occurs at
$(2, -8)$.

Notice that we did not need to evaluate $\frac{dy}{dx}$ at points on either side of the
stationary points, we had only to determine the *sign* of the gradient function.
For this reason the factorised form of $\frac{dy}{dx}$ was useful.

Example 8

Find the coordinates of any stationary points on the curve $y = x^4 + 2x^3$
and distinguish between them. Hence sketch the curve.

$$y = x^4 + 2x^3$$
$$\frac{dy}{dx} = 4x^3 + 6x^2 = 2x^2(2x + 3)$$

At stationary points $\frac{dy}{dx} = 0$ i.e. $x = 0$ or $-\frac{3}{2}$

Using $\dfrac{dy}{dx} = 2x^2(2x + 3)$

For x just less than 0 (say $-0\cdot1$) $\dfrac{dy}{dx} = (+\text{ve})(+\text{ve}) = +\text{ve}$ (/)

For x just more than 0 (say $0\cdot1$) $\dfrac{dy}{dx} = (+\text{ve})(+\text{ve}) = +\text{ve}$ (/)

Thus at $x = 0$ there is a point of inflexion.

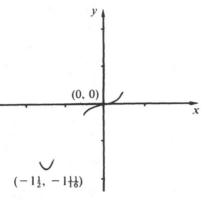

For x just less than $-\frac{3}{2}$ (say $-1\cdot6$) $\dfrac{dy}{dx} = (+\text{ve})(-\text{ve}) = -\text{ve}$ (\)

For x just more than $-\frac{3}{2}$ (say -1) $\dfrac{dy}{dx} = (+\text{ve})(+\text{ve}) = +\text{ve}$ (/)

Thus at $x = -\frac{3}{2}$ there is a minimum point.

If $x = 0$, $y = 0$.   If $x = -\frac{3}{2}$, $y = -1\frac{11}{16}$
Thus a minimum point occurs at $(-1\frac{1}{2}, -1\frac{11}{16})$ and a point of
inflexion occurs at $(0, 0)$.
These facts can be used to start the sketch as shown.

When sketching, it is also useful to know where the curve cuts the axes.

Clearly when $x = 0$, $y = 0$
but also, when $y = 0$, $x^4 + 2x^3 = 0$
       i.e. $x^3(x + 2) = 0$
       giving $x = 0$ or -2.
Thus we can complete the sketch as shown on the right.

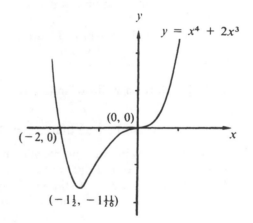

Note
When sketching the graph of $y = f(x)$, it is useful to
examine the behaviour of $f(x)$ as $x \to \pm\infty$.

In the above example $y = x^4 + 2x^3$ and so
as $x \to \pm\infty$ the x^4 term will dominate.
Thus as $x \to +\infty$, $y \to +\infty$
and as $x \to -\infty$, $y \to +\infty$ which agrees with our sketch.

Curve sketching is considered in greater detail in the next chapter.

Exercise 10C

For questions **1** to **8**, find the coordinates of any stationary points on the
given curves and distinguish between them.
1. $y = 2x^2 - 8x$
2. $y = 18x - 20 - 3x^2$
3. $y = x^3 - x^2 - x + 7$
4. $y = x^3 + 3x^2 - 9x - 5$
5. $y = 1 - 3x + x^3$
6. $y = x^3 - 3x^2 + 3x - 1$
7. $y = (x - 1)(x^2 - 6x + 2)$
8. $y = x^3 + 6x^2 + 12x + 12$
For questions **9** to **14**, sketch the curves of the given equations, clearly
indicating on your sketch the coordinates of any stationary points and of
any points where the lines cut the axes.
9. $y = (1 - x)(x - 5)$
10. $y = x^2 + 2x - 3$
11. $y = x^2 - 8x - 20$
12. $y = x^3 - 4x^2 + 4x$
13. $y = x(x - 3)(x + 5)$
14. $y = 3x^4 - 4x^3$

Use of the second differential

Figure 1 shows the graph of some function $y = f(x)$. A is a minimum point, B is a point of inflexion and C is a maximum point. Figure 2 shows the corresponding graph of $f'(x)$ plotted against x.

The gradient of the curve shown in Figure 2 will be given by $\frac{d}{dx}[f'(x)]$, i.e. $f''(x)$ or $\frac{d^2y}{dx^2}$.

Notice that the minimum point A, in Figure 1, corresponds to a point at which there is a +ve gradient in Figure 2,

i.e. $\frac{d^2y}{dx^2} > 0$ for a minimum point.

The maximum point C, in Figure 1, corresponds to a point at which there is a −ve gradient in Figure 2,

i.e. $\frac{d^2y}{dx^2} < 0$ for a maximum point.

The point of inflexion B, in Figure 1, corresponds to a point where the gradient is zero in Figure 2,

i.e. $\frac{d^2y}{dx^2} = 0$ for a point of inflexion.

In fact it can also be true that $\frac{d^2y}{dx^2} = 0$ at maximum and minimum points and so, if we wish to use the second differential as an aid to determining the nature of stationary points,

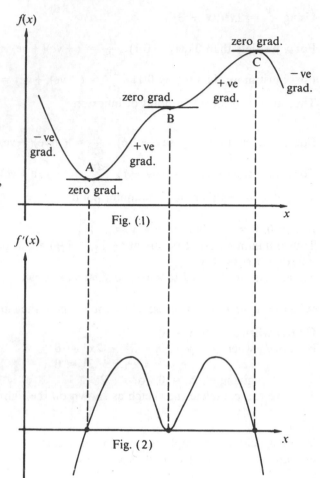

Fig. (1)

Fig. (2)

we can say: For any point (x_1, y_1) on the curve $y = f(x)$ for which $f'(x_1) = 0$ then,

(a) if $f''(x_1) > 0$ then the point is a minimum point,
(b) if $f''(x_1) < 0$ then the point is a maximum point,
(c) if $f''(x_1) = 0$ then the point is either a maximum point, a minimum point or a point of inflexion and, to determine which it is, we consider the sign of $f'(x)$ on either side of the point.

Example 9

Find the coordinates of any stationary points on the curve $y = 5x^6 - 12x^5$ and distinguish between them. Hence sketch the curve.

$$
\begin{aligned}
y &= 5x^6 - 12x^5 &&= x^5(5x - 12) \\
y' &= 30x^5 - 60x^4 &&= 30x^4(x - 2) \\
y'' &= 150x^4 - 240x^3 &&= 30x^3(5x - 8)
\end{aligned}
$$

At stationary points $y' = 0$ i.e. $30x^4(x - 2) = 0$ giving $x = 0$ or $x = 2$

when $x = 2$, $y'' = 30(2)^3(10 - 8)$ i.e. positive $\Rightarrow$ minimum,

when $x = 0$, $y'' = 0 \Rightarrow$ further investigation of y' needed either side of $x = 0$,

when x is just less than zero, say -0.1, y' is negative (\)
when x is just more than zero, say $+0.1$, y' is negative (\) } $\Rightarrow$ point of inflexion

when $x = 0$, $y = 0$ and when $x = 2$, $y = -64$.

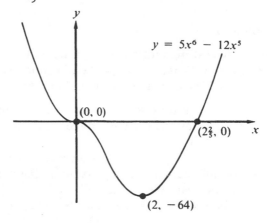

Thus $(0, 0)$ is a point of inflexion
and $(2, -64)$ is a minimum point.

From $y = x^5(5x - 12)$ we see that
the axes are cut at $(0, 0)$ and $(2\frac{2}{5}, 0)$ and
a sketch of the curve can be made as shown.

Note also that as $x \to +\infty$, $y \to +\infty$
and as $x \to -\infty$, $y \to +\infty$

Example 10

Find (a) the coordinates and nature of any turning points on the curve $y = x^3 + 3x^2 - 9x + 6$
  (b) the maximum value of $x^3 + 3x^2 - 9x + 6$ in the range
     (i) $-4 \leqslant x \leqslant 2$ (ii) $-4 \leqslant x \leqslant 4$

(a) $y = x^3 + 3x^2 - 9x + 6$
  $y' = 3x^2 + 6x - 9 \qquad = 3(x - 1)(x + 3)$
  $y'' = 6x + 6$
  At stationary points $y' = 0$ i.e. $(x - 1)(x + 3) = 0$
      giving $x = 1$ or $x = -3$
  when $x = 1$ $y'' = 12$ i.e. positive $\Rightarrow$ minimum
  when $x = -3$ $y'' = -12$ i.e. negative $\Rightarrow$ maximum
  when $x = 1$ $y = 1$ and when $x = -3$ $y = 33$
  Thus $(1, 1)$ is a minimum point and $(-3, 33)$ is a maximum point.

(b) From (a) we know that the graph of $y = x^3 + 3x^2 - 9x + 6$ must look
  something like that shown on the right.
  Thus, remembering that $(-3, 33)$ is a *local* maximum the expression
  $x^3 + 3x^2 - 9x + 6$ will exceed the value 33 for sufficiently large x.
  (i) for $-4 \leqslant x \leqslant 2$:
     if $x = 2$, $x^3 + 3x^2 - 9x + 6 = 8 + 12 - 18 + 6$
             $= 8$
     $\therefore$ in the range $-4 \leqslant x \leqslant 2$,
       $x^3 + 3x^2 - 9x + 6$ has a maximum value of 33
  (ii) for $-4 \leqslant x \leqslant 4$:
     if $x = 4$, $x^3 + 3x^2 - 9x + 6 = 64 + 48 - 36 + 6$
             $= 82$
     $\therefore$ in the range $-4 \leqslant x \leqslant 4$,
       $x^3 + 3x^2 - 9x + 6$ has a maximum value of 82.

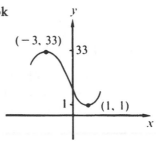

The above examples show how the stationary points of a curve may be
located when the equation of the curve is given. However, in some instances
the equation may not be given. This is likely to be the case when the
problem relates to a real life or practical situation and then the first task is
to express the given information as an equation relating two variables. The

stationary points can then be determined and their practical significance interpreted. The following example illustrates such a case.

Example 11

A company that manufactures dog food wishes to pack the food in closed cylindrical tins. What should be the dimensions of each tin if each is to have a volume of 128π cm³ and the minimum possible surface area?

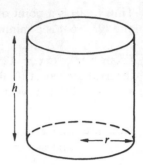

Suppose that each tin has base radius r and perpendicular height h. We require the dimensions for the surface area S to be a minimum and so we need an expression for S.

$$S = 2\pi r^2 + 2\pi rh$$

We cannot differentiate this expression, in this form, as S is given as a function of *two* variables r and h.

However the volume $V = \pi r^2 h$ and this volume is to be 128π cm³.

$$\therefore \quad 128\pi = \pi r^2 h \quad \text{or} \quad h = \frac{128}{r^2}$$

$$\therefore \qquad S = 2\pi r^2 + \frac{256\pi}{r}$$

and $\quad \dfrac{dS}{dr} = 4\pi r - \dfrac{256\pi}{r^2}$

When $\quad \dfrac{dS}{dr} = 0, \quad$ then $\quad 4\pi r - \dfrac{256\pi}{r^2} = 0$

$$\text{giving} \quad r = 4 \text{ cm}$$

$$\text{and} \quad h = \frac{128}{4^2} = 8 \text{ cm}$$

$\dfrac{d^2S}{dr^2} = 4\pi + \dfrac{512\pi}{r^3}$, which is positive for $r = 4 \Rightarrow$ minimum.

Thus the surface area S has a minimum value when $r = 4$ cm and $h = 8$ cm. Each tin should have a base radius of 4 cm and perpendicular height 8 cm.

Exercise 10D

For questions **1** to **8** find the coordinates of any stationary points on the given curves and distinguish between them.

1. $y = x^2 + 10x + 10$
2. $y = \frac{1}{4} + 9x - 3x^2$
3. $y = x^3 - 3x^2 - 24x$
4. $y = 3x^2 + 45x - 75 - x^3$
5. $y = 2x^3 - 24x$
6. $y = x^4 + 3$
7. $y = x^6 - 6x^4$
8. $y = 15x^3 - x^5$

For questions **9** to **14**, sketch the curves of the given equations clearly indicating on your sketch the coordinates of any stationary points and of any points where the lines cut the axes.

9. $y = 3x - x^2$
10. $y = x^2 - 4$
11. $y = (x - 2)^2$
12. $y = x^4 - 4x^3$
13. $y = (x - 2)^2(x - 1)$
14. $y = x^5 - 5x^4$

15. Find (a) the coordinates and nature of any turning points on the curve
$$y = x^3 - \tfrac{3}{2}x^2 - 6x + 12,$$
     (b) the maximum value of $x^3 - \tfrac{3}{2}x^2 - 6x + 12$ in the range
$$-3 \leqslant x \leqslant 3,$$
     (c) the minimum value of $x^3 - \tfrac{3}{2}x^2 - 6x + 12$ in the range
$$-3 \leqslant x \leqslant 3.$$

16. Find (a) the coordinates and nature of any turning points on the curve
$$y = 36x - 3x^2 - 2x^3,$$
     (b) the minimum and maximum values of $36x - 3x^2 - 2x^3$ in the
range $-5 \leqslant x \leqslant 5$.

17. If $S = 4r^2 - 10r + 7$, find the minimum value of S and the value of r which gives this minimum value.

18. If $V = 30t - 6t^2$, find the maximum value of V and the value of t for which it occurs.

19. If $A = xy$ and $2x + 5y = 100$, find the maximum value of A and the values of x and y which give this maximum value.

20. If $V = 4rx + 2r^2$ and $3r + x = 5$, find the maximum value of V and the values of r and x that give this maximum value.

21. A rectangular enclosure is formed by using 1200 m of fencing. Find the greatest possible area that can be enclosed in this way and the corresponding dimensions of the rectangle.

22. An open metal tank with a square base is made from 12 m² of sheet metal. Find the length of the side of the base for the volume of the tank to be a maximum and find this maximum volume.

23. A piece of wire of length 60 cm is cut into two pieces. Each piece is then bent to form the perimeter of a rectangle which is twice as long as it is wide. Find the lengths of the two pieces of wire if the sum of the areas of the rectangles is to be a minimum.

24. A cylindrical can is made so that the sum of its height and the circumference of its base is 45π cm. Find the radius of the base of the cylinder if the volume of the can is a maximum.

25. A ship is to make a voyage of 200 km at a constant speed. When the speed of the ship is v km/h the cost is $£(v^2 + 4000/v)$ per hour. Find the speed at which the ship should travel so that the cost of the voyage is a minimum.

10.4 Small changes

In section 10.1 we defined $\dfrac{dy}{dx}$ as $\lim\limits_{\delta x \to 0}\left(\dfrac{\delta y}{\delta x}\right)$. Thus, if δx is small then

$$\frac{\delta y}{\delta x} \approx \frac{dy}{dx} \quad \text{or} \quad \delta y \approx \frac{dy}{dx}\delta x.$$

Thus if y is given as a function of x we can determine the change in y corresponding to some given small change in x.

Example 12

If $y = 2x^2 - 3x$ find the approximate change in y when x increases from 6 to 6·02.

$$y = 2x^2 - 3x$$

$$\frac{dy}{dx} = 4x - 3$$

using $\delta y \approx \frac{dy}{dx} \delta x$

$\delta y \approx (4x - 3)\delta x$ and if $x = 6$ and $\delta x = 0·02$

$\delta y \approx (24 - 3)(0·02) = 0·42$

When x changes from 6 to 6·02 the value of y changes by approximately 0·42.

Example 13

In calculating the area of a circle it is known that an error of $\pm 3\%$ could have been made in the measurement of the radius. Find the possible percentage error in the area.

We are told that $\dfrac{\delta r}{r} = \pm \dfrac{3}{100}$ and we require $\dfrac{\delta A}{A}$.

For small δr $\dfrac{\delta A}{\delta r} \approx \dfrac{dA}{dr}$

but $A = \pi r^2$ i.e. $\dfrac{dA}{dr} = 2\pi r$

$\therefore$ $\dfrac{\delta A}{\delta r} \approx 2\pi r$

$\delta A \approx 2\pi r \delta r$

$\therefore$ $\dfrac{\delta A}{A} \approx \dfrac{2\pi r \delta r}{\pi r^2} = \dfrac{2\delta r}{r}$

$$= \pm \frac{6}{100}$$

Thus the possible percentage error in the area will be $\pm 6\%$.

Example 14

Find an approximate value for $\sqrt{(16·08)}$

For this we use $y = \sqrt{x}$ with $(x, y) = (16, 4)$ and $(x + \delta x, y + \delta y) = (16 + 0·08, 4 + \delta y)$

$$\frac{dy}{dx} = \frac{1}{2\sqrt{x}}$$

For small δx, $\dfrac{\delta y}{\delta x} \approx \dfrac{dy}{dx}$ i.e. $\delta y \approx \dfrac{dy}{dx} \delta x$

$\therefore$ $\delta y \approx \dfrac{1}{2\sqrt{x}} \delta x$

$$= \frac{1}{2\sqrt{16}}(0·08)$$

$$= 0·01$$

$\therefore$ $y + \delta y \approx 4 + 0·01$ and so $\sqrt{(16·08)} \approx 4·01$

Exercise 10E

1. If $y = x^2 + 2x$ find the approximate increase in y when x increases from 9 to 9·01.

2. If $y = x + \dfrac{1}{x}$ find the approximate increase in y when x increases from 2 to 2·04.

3. It is noticed that when x increases slightly from an initial value of 7, the value of y increases by 0·7. Find the approximate increase in x that caused this increase in y given that x and y are related according to the rule $y = (2x + 3)(x + 2)$.

4. If $y = 5x^4$ and x is increased by 2% of its original value, find the corresponding percentage increase in y.

5. Find an approximate value for the increase in the volume of a sphere when the radius of the sphere increases from 10 cm to 10·1 cm. (Leave π in your answer).

6. Find the percentage increase in the volume of a cube when all the edges of the cube are increased in length by 2%.

7. Find an approximate value for the increase in the radius of a sphere given that this increase causes the surface area to increase from 100π cm² to $100\cdot4\pi$ cm².

8. A rectangle is known to be twice as long as it is wide. If the width is measured as 20 cm $\pm$ 0·2 cm find the area in the form $A \pm b$.

9. The time period, T, of a pendulum of length l is given by $T = 2\pi\sqrt{\dfrac{l}{g}}$

where π and g are constants. Find the approximate percentage increase in T when the length of the pendulum increases by 4%.

10. A solid circular cylinder has base radius 5 cm and height 12·5 cm. Find the approximate increase in the surface area of the cylinder when the radius of the base increases to 5·04 cm. (Leave π in your answer).

11. Find an approximate value for $(5\cdot02)^3$.

12. Find an approximate value for $\sqrt[3]{(64\cdot96)}$.

10.5 More about points of inflexion

We saw on page 266 that when examining points on the curve $y = f(x)$ for which $f'(x) = 0$, if $f''(x) < 0$ we have a maximum point, if $f''(x) > 0$ we have a minimum point and if $f''(x) = 0$ we *could* have a point of inflexion. Indeed at all points of inflexion $f''(x) = 0$ but it is possible to have a point of inflexion (a, b) i.e. $f''(a) = 0$, for which $f'(a) \neq 0$. A point of inflexion is said to occur at any point P on a curve at which the tangent to the curve at P crosses the curve at P. Thus in the diagrams shown below A, B and C are all points of inflexion, but only A is a stationary point on the curve as the tangent at A is parallel to the x-axis, (i.e. $f'(x) = 0$ at A). The second set of diagrams show that the gradient of $f'(x)$, i.e. $f''(x)$, is zero at A, B and C.

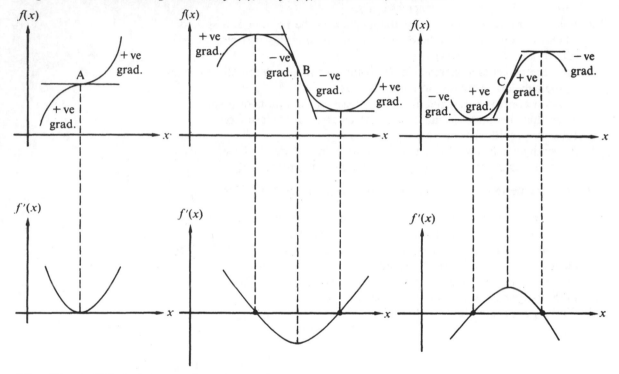

Thus if $y = f(x)$ stationary points occur where $f'(x) = 0$. These stationary points may be turning points (maximum or minimum) or points of inflexion. Points of inflexion that are not turning points may also occur if $f'(x) \neq 0$ but $f''(x) = 0$.

Example 15

Find the coordinates of any points of inflexion on the curve $y = x^3(x - 1)$ and determine whether they are horizontal points of inflexion.

$$y = x^3(x - 1)$$
$$= x^4 - x^3$$
$$y' = 4x^3 - 3x^2 = x^2(4x - 3)$$
$$y'' = 12x^2 - 6x = 6x(2x - 1)$$

At points of inflexion $y'' = 0$ i.e. $6x(2x - 1) = 0$

$$\text{giving } x = 0 \quad \text{or} \quad x = \tfrac{1}{2}$$

At $x = 0$, $y' = 0$ and $y = 0$
At $x = \frac{1}{2}$, $y' \neq 0$ and $y = -\frac{1}{16}$
Thus at $(\frac{1}{2}, -\frac{1}{16})$ $y'' = 0$ but $y' \neq 0 \Rightarrow$ Point of inflexion but not a horizontal one
     at $(0, 0)$ $y'' = 0$ and $y' = 0 \Rightarrow$ Stationary value, i.e. maximum, minimum
                        or point of inflexion.

$$\text{Since } y' = x^2(4x - 3)$$
if x is just less than 0 (say -0.1) $y' = (+\text{ve})(-\text{ve}) = -\text{ve} \ (\backslash)$
if x is just more than 0 (say 0.1) $y' = (+\text{ve})(-\text{ve}) = -\text{ve}(\backslash)$
$\therefore$ $(0, 0)$ is a horizontal point of inflexion.
Points of inflexion occur at $(0, 0)$ and $(\frac{1}{2}, -\frac{1}{16})$. The point of inflexion at $(0, 0)$ is horizontal.

Exercise 10F

Find the coordinates of any points of inflexion that occur on the following curves, indicating clearly any points that are horizontal points of inflexion.
1. $y = x^4 - 2x^3$ **2.** $y = x(15 - 4x^2)$
3. $y = 2x^3 - 9x^2 + 12x + 1$ **4.** $y = x^3 - x^2 - 3x + 10$
5. $y = x^3 - 3x^2 + 3x - 4$ **6.** $y = (x - 1)^3$
7. $y = x^3 - 12x^2 + 130$ **8.** $y = x^5 - 5$
9. $y = x^3 - x^2 - x + 1$

Exercise 10G Examination questions

1. The curve $y = ax^3 - 2x^2 - x + 7$ has a gradient of 3 at the point where $x = 2$. Determine the value of a.   (S.U.J.B.)

2. Find the equation of the tangent to the curve $y = 2x^2 - 3$ at the point $(2, 5)$. Find the co-ordinates of the point where this tangent meets the x-axis.   (S.U.J.B.)

3. The normal to the curve $y = x^2 - 4x$ at the point $(3, -3)$ cuts the x-axis at A and the y-axis at B. Find the equation of the normal and the coordinates of A and B.   (A.E.B.)

4. (a) Differentiate $(3x + \sqrt{x})^2$ with respect to x.
   (b) Find the x co-ordinates of the points on the curve $y = \dfrac{x^2 - 1}{x}$ at which the gradient of the curve is 5.   (Cambridge)

5. The tangent to the curve $y = 2x^2 + ax + b$ at the point $(-2, 11)$ is perpendicular to the line $2y = x + 7$. Find the value of a and of b.   (Cambridge)

6. The curve $y = ax^2 + 12x + 1$ has a turning point when $x = 2$. Calculate a. Is the point a maximum or a minimum?   (S.U.J.B.)

7. The curve $y = ax^2 + bx + c$ has a maximum point at $(2, 18)$ and passes through the point $(0, 10)$. Evaluate a, b and c.   (Cambridge)

8. Calculate the co-ordinates of the turning points of the curve
$$y = x^3 - 6x^2 + 9x.$$
   Sketch the curve.   (Cambridge)

9. (i) A rectangular block has a base x centimetres square. Its total surface area is 150 cm². Prove that the volume of the block is $\frac{1}{2}(75x - x^3)$ cm³.
   Calculate (a) the dimensions of the block when its volume is maximum; (b) this maximum volume. Show that your answer is a maximum rather than a minimum.

   (ii) Write down the gradient of the function $4x^2 + \dfrac{27}{x}$. Hence find the value of x for which the function has a maximum or minimum value, and state which it is, giving reasons.   (Oxford)

10. A solid right circular cylinder of height h and radius r has a *total* external surface area of 600 cm².
    Show that $h = \dfrac{300}{\pi r} - r$ and hence express the volume, V, in terms of r.

    If h and r can vary, find $\dfrac{dV}{dr}$ and $\dfrac{d^2V}{dr^2}$ in terms of r. Show that V has a maximum and find the corresponding value of r in terms of π.
    Calculate the ratio $h:r$ in this maximum case.   (A.E.B.)

11. A cylinder of volume V is to be cut from a solid sphere of radius R.
    Prove that the maximum value of V is $\dfrac{4\pi R^3}{3\sqrt{3}}$.   (A.E.B.)

12. The equation of a curve is $y = 2x^3 - 7x^2 + 15$. Write down an expression for $\dfrac{dy}{dx}$ and hence find

    (i) the equation of the tangent to the curve at (2, 3),
    (ii) the approximate change in y as x increases from 2 to 2·03, stating whether this is an increase or a decrease.   (Cambridge)

13. Given that $y = x^{-1/3}$ use the calculus to determine an approximate value for $\dfrac{1}{\sqrt[3]{0·9}}$.   (Cambridge)

14. (i)

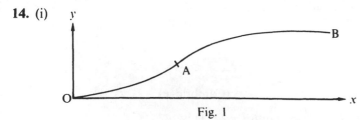

Fig. 1

Fig. 1 represents part of a switch-back ride at a fair. The section O to A is part of the curve whose equation is $y = x^2$ and the section from A to B is part of the curve whose equation is $3y = -20 + 20x - 2x^2$. Find the coordinates of A. Show that the gradients for each section are equal at A, and find this gradient.

(ii) A solid metal cylinder has a height of 10 cm. It is machined until its radius is reduced by 0·5%, its height remaining at 10 cm. Use the method of small increments to find the percentage decrease in volume. No credit will be given for any other method.   (Oxford)

11
Sketching functions I

11.1 General methods

For the function $y = f(x)$, we can plot values of x against the corresponding values of y and obtain an accurate graph of the function. A less accurate representation, which we call a sketch, is adequate for many purposes provided that the sketch still shows the salient and noteworthy features of the function.

In earlier chapters we have already encountered the idea of sketching straight lines, quadratic curves, cubic curves and various other functions. Thus if asked to sketch the graph of an equation of the type $y = mx + c$, we know it will be a straight line and need only find two points on the line. If asked to sketch the graph of a function of the type $f(x) = ax^2 + bx + c$, we expect a quadratic curve which will be either a 'hill' $\wedge$ (if $a < 0$) or a 'valley' $\smile$ (if $a > 0$). By finding (a) where the curve cuts the axes and (b) the coordinates and nature of any turning points, a good sketch can be made.

If asked to sketch the graph of a function of the type $f(x) = ax^3 + bx^2 + cx + d$, we expect a cubic curve which will have both a 'hill' and a 'valley': $\sim$ (if $a > 0$) or $\sim$ (if $a < 0$). [These 'hills' and 'valleys' may merge into a point of inflexion: $\sim$ or $\sim$]. Again, by finding intercepts with the axes and the coordinates and nature of any turning points, the curve can be sketched.

However, not all functions are of one of the above types and so we need a general approach to sketching. The following steps will give information from which the graphs of many types of functions, $y = f(x)$, can be sketched.

1. Is there any obvious symmetry?
   If the equation is unchanged when $(-x)$ is substituted for x, the graph will be symmetrical about the y-axis (i.e. an even function, $f(-x) = f(x)$).
   If the equation is unchanged when $(-y)$ is substituted for y, the graph will be symmetrical about the x-axis. [Applies to graphs of the type $y^2 = f(x)$]
   If $f(-x) = -f(x)$ the function is an odd function and the graph is symmetrical for a 180° turn about the origin.

2. Find where the line cuts the x and y axes.

3. Examine the behaviour of the function as $x \to \pm\infty$.

4. Investigate any places where the function is undefined (e.g. $f(x) = 1/x$ is not defined for $x = 0$).

5. If the above steps indicate the presence of a turning point, find its location and nature.

Important notes

(a) For simple functions, e.g. $f(x) = x$, $g(x) = x^2$, $h(x) = 1/x$, point **1**, the symmetry aspect is worthy of consideration but for more complicated functions it is not always easy to consider this aspect. Thus only consider **1** if symmetry is obvious.

(b) For most functions, it is not necessary to follow all five steps; as soon as sufficient information has been gained to enable a sketch of the function to be made, there is no need to search for more. Indeed, point **5** can be difficult unless the function is easy to differentiate, and even when turning points are present a reasonably accurate sketch can often be made without having to locate them precisely.

(c) The behaviour of a function as $x \rightarrow \pm\infty$ (point **3**), and values of x for which a function is undefined (point **4**), are illustrated in the following example.

Example 1

For each of the following functions, (i) examine the behaviour of the function as $x \rightarrow \pm\infty$, (ii) find any values of x for which the function is undefined and investigate the function on either side of this value of x.

(a) $f(x) = \dfrac{3}{x}$, (b) $f(x) = 3 + \dfrac{2}{x}$, (c) $f(x) = \dfrac{5x - 1}{x}$, (d) $f(x) = 3 - \dfrac{2}{x^2}$.

(a) $f(x) = \dfrac{3}{x}$

(i) For x, a large +ve number, $f(x)$ is small and positive we write this: as $x \rightarrow +\infty$, $f(x) \rightarrow 0^+$. (i.e. $f(x)$ is 'just' greater than zero)

For x, a large −ve number, $f(x)$ is small and negative we write this: as $x \rightarrow -\infty$, $f(x) \rightarrow 0^-$. (i.e. $f(x)$ is 'just' less than zero)

(ii) $f(x)$ is undefined for $x = 0$

For x *just* greater than zero, $f(x)$ is a large +ve number.

For x *just* less than zero, $f(x)$ is a large −ve number.

Thus: as $x \rightarrow 0^+$, $f(x) \rightarrow +\infty$

as $x \rightarrow 0^-$, $f(x) \rightarrow -\infty$.

(b) $f(x) = 3 + \dfrac{2}{x}$

(i) For x, a large +ve number, $\dfrac{2}{x}$ is small and positive.

Thus: as $x \rightarrow +\infty$, $f(x) \rightarrow 3^+$. (i.e. $f(x)$ is 'just' greater than 3)

For x, a large −ve number, $\dfrac{2}{x}$ is small and negative.

Thus: as $x \rightarrow -\infty$, $f(x) \rightarrow 3^-$. (i.e. $f(x)$ is 'just' less than 3)

(ii) $f(x)$ is undefined for $x = 0$

For x *just* greater than zero, $f(x)$ is a large +ve number.

For x *just* less than zero, $f(x)$ is a large −ve number.

Thus: as $x \rightarrow 0^+$, $f(x) \rightarrow +\infty$.

as $x \rightarrow 0^-$, $f(x) \rightarrow -\infty$.

(c) $f(x) = \dfrac{5x - 1}{x}$

$\quad = 5 - \dfrac{1}{x}$ (i) As $x \to +\infty, f(x) \to 5^-$.
                 As $x \to -\infty, f(x) \to 5^+$.
          (ii) $f(x)$ is undefined for $x = 0$
                 As $x \to 0^+, f(x) \to -\infty$.
                 As $x \to 0^-, f(x) \to +\infty$.

(d) $f(x) = 3 - \dfrac{2}{x^2}$ (i) As $x \to +\infty, f(x) \to 3^-$.
                  As $x \to -\infty, f(x) \to 3^-$.
           (ii) $f(x)$ is undefined for $x = 0$
                  As $x \to 0^+, f(x) \to -\infty$.
                  As $x \to 0^-, f(x) \to -\infty$.

Example 2

Sketch the graph of the function
(a) $f(x) = 5 + 4x - x^2$,

(b) $g(x) = \begin{cases} 5 & \text{for } x \leqslant 0 \\ 5 + 4x - x^2 & \text{for } 0 \leqslant x \leqslant 5 \\ 0 & \text{for } x \geqslant 5. \end{cases}$

(a) Let $y = 5 + 4x - x^2$

   x-axis   When $y = 0, 5 + 4x - x^2 = 0$
                  i.e. $(5 - x)(1 + x) = 0$
                  cuts x-axis at $(5, 0)$ and $(-1, 0)$
   y-axis   When $x = 0, y = 5$
                  cuts y-axis at $(0, 5)$
   $x \to \pm\infty$ For large x the x^2 term dominates
                  Thus as $x \to +\infty, y \to -\infty$
                      as $x \to -\infty, y \to -\infty$
   f(x) undefined   There is no value of x for
                  which $f(x)$ is undefined.
   Max/min $y = 5 + 4x - x^2$
                  $y' = 4 - 2x$
                  $y'' = -2$
                  Thus a maximum exists at $(2, 9)$

The sketch can then be completed.

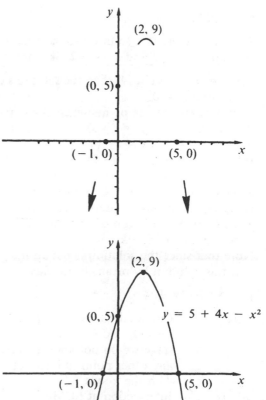

(b) The sketch of $g(x)$ will be the same as that
of $f(x)$ for $0 \leqslant x \leqslant 5$, but will equal 5 for
$x \leqslant 0$ and 0 for $x \geqslant 5$.
Thus the sketch will be:

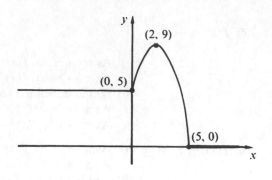

Notice that, although the nature of $g(x)$
changes noticeably at $x = 0$ and at $x = 5$,
the line itself does not 'break'; $g(x)$ is said
to be a *continuous* function (as indeed was
$f(x)$).

An example of a *discontinuous* function
would be:

$$h(x) = \begin{cases} 5 & \text{for} \quad x \leqslant 0 \\ 5 + 4x - x^2 & \text{for} \quad 0 \leqslant x \leqslant 5 \\ 5 & \text{for} \quad x > 5 \end{cases}$$

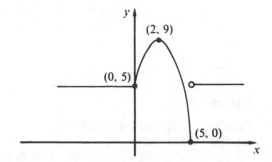

Example 3

Show that there is a solution to the equation $x^3 - 3x + 4 = 0$, that lies
between $x = -3$ and $x = -2$. Sketch the curve given by $y = x^3 - 3x + 4$.

Let $y = x^3 - 3x + 4$, then the solutions of $x^3 - 3x + 4 = 0$ are obtained
by putting $y = 0$.
Now y is zero at the points where the curve intersects the x-axis.
When $x = -3$, $y = (-3)^3 - 3(-3) + 4$
or $y = -14$
When $x = -2$, $y = (-2)^3 - 3(-2) + 4$
or $y = +2$
Since for $x = -3$, $y < 0$ and for $x = -2$, $y > 0$ there must exist some
value of x between -3 and -2 for which $y = 0$.
Thus there is a solution of the equation $x^3 - 3x + 4 = 0$ lying between
$x = -3$ and $x = -2$.
Note that since the solution is not an integer value, it would not be easy to
find this solution by means of the factor theorem as used in section 5.5.

$y = x^3 - 3x + 4$

x-axis	Curve cuts x-axis between $(-3, 0)$ and $(-2, 0)$. At this stage we do not know if there is any other point of intersection with the x-axis.
y-axis	Intersection at $(0, 4)$
$x \to \pm\infty$	For large x, x^3 dominates. Thus as $x \to +\infty$, $y \to +\infty$ and as $x \to -\infty$, $y \to -\infty$.

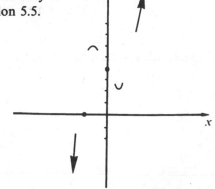

y undefined	There is no value of x for which y is undefined.
Max/min	$y = x^3 - 3x + 4$
	$y' = 3x^2 - 3$
	$y'' = 6x$
	Thus there exists a max at $(-1, 6)$ and a min at $(1, 2)$.

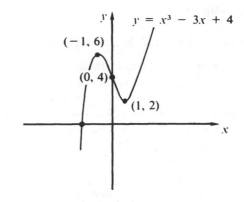

The sketch can then be completed.

Example 4

Make a sketch of the curve given by $y = \dfrac{x + 5}{x}$.

$y = \dfrac{x + 5}{x}$ can be written as $y = 1 + \dfrac{5}{x}$.

x-axis	$y = 0$ when $x = -5$
	$\therefore$ cuts x-axis at $(-5, 0)$
y-axis	No y-axis intercept because y is not defined for $x = 0$
$x \to \pm\infty$	As $x \to +\infty$, $y \to 1^+$
	As $x \to -\infty$, $y \to 1^-$

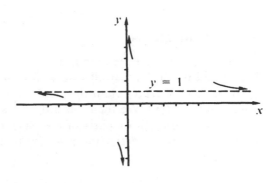

The line $y = 1$ is called an *asymptote* to the curve, meaning that it is a straight line that the curve gets nearer and nearer to without actually touching. The line $y = 1$ is a horizontal asymptote and is shown as a broken line on the graph.

y undefined	y is undefined for $x = 0$.
	As $x \to 0^+$, $y \to +\infty$.
	As $x \to 0^-$, $y \to -\infty$.

The line $x = 0$, i.e. the y-axis, is a vertical asymptote to the curve.

Max/min	$y' = \dfrac{-5}{x^2}$. Thus there is no value of x for which $y' = 0$ and therefore no turning points.

Note, also, that the gradient is always negative.

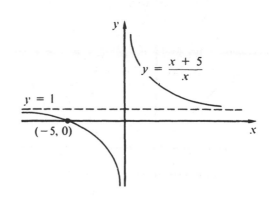

The sketch can then be completed.

Example 5

Make a sketch of the curve given by $y = 1 + \dfrac{1}{x^2}$.

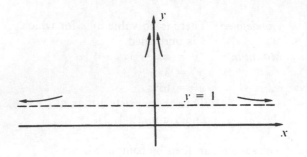

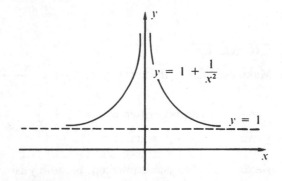

Symmetry   Equation unchanged if x replaced
by $(-x)$; hence symmetrical about
y-axis.

x-axis   No value of x for which $y = 0$.

y-axis   No y-axis intersection as y is not
defined for $x = 0$.

$x \to \pm\infty$ As $x \to +\infty$, $y \to 1^+$.
As $x \to -\infty$, $y \to 1^+$.

y undefined y is undefined for $x = 0$
As $x \to 0^+$, $y \to +\infty$.
As $x \to 0^-$, $y \to +\infty$.
$x = 0$ is a vertical asymptote.

Max/min $y' = \dfrac{-2}{x^3}$. Thus there is no value

of x for which $y' = 0$ and
therefore no turning points.

The sketch can then be completed.

Modulus functions

We first encountered the modulus function, i.e. $y = |f(x)|$, in section 0.2.
We should recall that for all values of x, the modulus of $f(x)$ i.e. $|f(x)|$, is a
positive quantity. To obtain the graph of $y = |f(x)|$, we first make a light
sketch of $y = f(x)$. We then retain those parts of $y = f(x)$ for which y is
positive and draw those parts of $y = f(x)$ for which y is negative as positive
values i.e. any part of the line $y = f(x)$ that lies below the x-axis is reflected
in the x-axis.

The graphs of $y = |x|$, $y = |3 + 2x - x^2|$ and $y = \left|\dfrac{x + 5}{x}\right|$ are shown

below. $\left(\text{See example 4 for } y = \dfrac{x + 5}{x}.\right)$

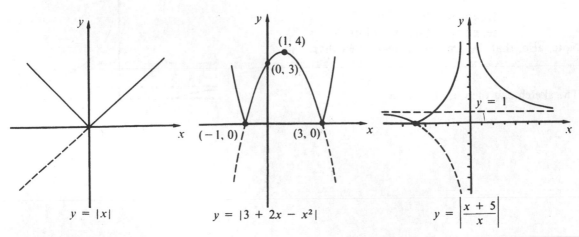

$y = |x|$ $y = |3 + 2x - x^2|$ $y = \left|\dfrac{x + 5}{x}\right|$

Exercise 11A

1. If each of the following were sketched, state which would give straight lines.
   (a) $y = 2x + 3$ (b) $y = x^2 + 4x$ (c) $2y + 3x = 8$
   (d) $y = x^3$ (e) $\dfrac{x}{3} + \dfrac{y}{4} = 1$ (f) $y = 2 + \dfrac{3}{x}$

2. Find the coordinates of the points A to J shown in the sketches below.

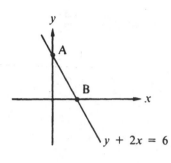

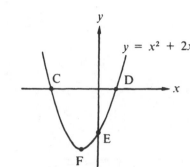

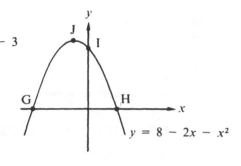

3. Find the coordinates of the points A to D and the values of the constants a to c shown in the sketches below.

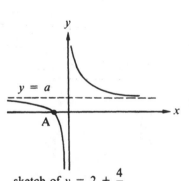

sketch of $y = 2 + \dfrac{4}{x}$

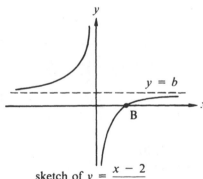

sketch of $y = \dfrac{x - 2}{x}$

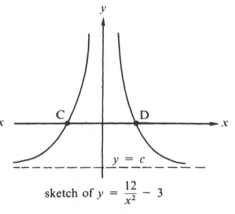

sketch of $y = \dfrac{12}{x^2} - 3$

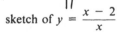

4. Make sketch graphs of the following lines giving the coordinates of any maximum points, minimum points and any intersections with the axes.
   (a) $5y = 15 - 3x$
   (b) $y = |2x + 4|$
   (c) $y = 16 - x^2$
   (d) $y = |2 - x - x^2|$

5. Show that there is a solution to the equation $x^3 - 6x^2 + 9x + 1 = 0$ between $x = -1$ and $x = 0$. Sketch the curve given by $y = x^3 - 6x^2 + 9x + 1$.

6. Sketch the graphs of the functions:
   (a) $f(x) = \begin{cases} 8 & \text{for} \quad x \leqslant -3 \\ x^2 - 1 & \text{for} \quad -3 \leqslant x \leqslant 2 \\ 3 & \text{for} \quad x \geqslant 2 \end{cases}$

   (b) $g(x) = \begin{cases} 2 & \text{for} \quad x < -2 \\ x^2 + 2x & \text{for} \quad -2 \leqslant x \leqslant 1 \\ 4 - x & \text{for} \quad x \geqslant 1 \end{cases}$

   (c) $h(x) = \begin{cases} x^2 + 2x - 3 & \text{for} \quad x \leqslant 1 \\ 6x - 5 - x^2 & \text{for} \quad x \geqslant 1 \end{cases}$

7. Make a sketch of $y = f(x)$ given that $f(x)$ is an odd function and
$$f(x) = \begin{cases} x^2 & \text{for } 0 \leqslant x \leqslant 2 \\ 4 & \text{for } x > 2 \end{cases}$$

8. Make a sketch of $y = f(x)$ given that $f(x)$ is an even function and
$$f(x) = \begin{cases} x^2 - 1 & \text{for } 0 \leqslant x \leqslant 3 \\ 8 & \text{for } x > 3 \end{cases}$$

9. Make a sketch of $y = f(x)$ for $-6 \leqslant x < 6$ given that
$$f(x) = \begin{cases} x(3 - x) & \text{for } 0 \leqslant x \leqslant 3 \\ x - 3 & \text{for } 3 \leqslant x < 6 \end{cases}$$
and $f(x)$ is periodic with period 6.

Make sketch graphs of the following functions, labelling asymptotes, and giving the coordinates of any turning points and intersections with axes.

10. $f(x) = 1 + \dfrac{3}{x}$

11. $f(x) = \dfrac{2}{x} - 2$

12. $f(x) = \left| 2 - \dfrac{8}{x} \right|$

13. $f(x) = \dfrac{2 - x}{x}$

14. $f(x) = \dfrac{2x - 3}{x}$

15. $f(x) = \dfrac{4}{x^2} - 1$

16. $f(x) = \left| 1 - \dfrac{4}{x^2} \right|$

11.2 Further considerations

When sketching functions of the type $y = ax \pm \dfrac{b}{cx}, y = ax \pm \dfrac{b}{cx^2}$

or $y = ax^2 \pm \dfrac{b}{x}$ the techniques of section 11.1 are still applicable. However, with these functions it should be noted that as $x \to \pm\infty$, $y \to ax$, ax or ax^2 respectively.

Example 6

Make a sketch of the function $f(x) = x - \dfrac{1}{x}$.

Let $y = x - \dfrac{1}{x}$.

x-axis $y = 0$ when $x - \dfrac{1}{x} = 0$
i.e. $x^2 = 1$ or $x = \pm 1$
Cuts x-axis at $(-1, 0)$ and $(1, 0)$

y-axis No y-axis intersection as y is not defined for $x = 0$.

$x \to \pm\infty$ As $x \to \pm\infty$, $y \to x$ i.e. $y = x$ is an asymptote (called an oblique asymptote).

In fact, for $y = x - \dfrac{1}{x}$
as $x \to +\infty$, $y \to x^-$ and
as $x \to -\infty$, $y \to x^+$

y undefined y is undefined for $x = 0$.
As $x \to 0^+$, $y \to -\infty$.
As $x \to 0^-$, $y \to +\infty$.
$x = 0$ is a vertical asymptote.

Max/min $y' = 1 + \dfrac{1}{x^2}$ which will never equal zero. $\therefore$ no turning points.

Notice also that the gradient is always positive.

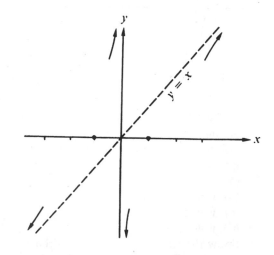

The sketch can then be completed.

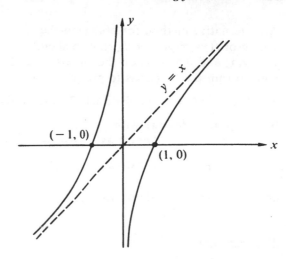

Example 7

Make a sketch of the curve given by $y = x^2 + \dfrac{2}{x}$.

x-axis When $y = 0$, $x^2 + \dfrac{2}{x} = 0$

i.e. $x = \sqrt[3]{(-2)}$
Cuts x-axis at $(-\sqrt[3]{2}, 0)$.

y-axis No y-axis intersection as y is not defined for $x = 0$.

$x \to \pm\infty$ $y \to x^2$, i.e. the curve approximates to that of $y = x^2$.

In fact, for $y = x^2 + \dfrac{2}{x}$,

as $x \to +\infty$, $y \to (x^2)^+$.
as $x \to -\infty$, $y \to (x^2)^-$.

y undefined y is undefined for $x = 0$.
Thus $x = 0$ is a vertical asymptote.
As $x \to 0^+$, $y \to +\infty$.
As $x \to 0^-$, $y \to -\infty$.

Max/min $y' = 2x - \dfrac{2}{x^2}$, $y'' = 2 + \dfrac{4}{x^3}$

Thus min at $(1, 3)$.

The sketch can then be completed.

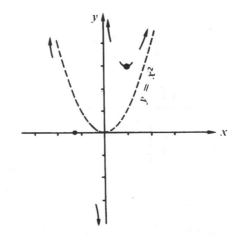

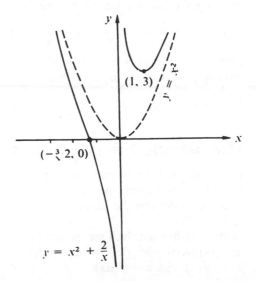

An alternative method for obtaining the
sketch of $f(x) = g(x) + h(x)$, is to sketch
$y = g(x)$ and $y = h(x)$ on the same axes and
then to sum the functions to give $y = f(x)$.

For $f(x) = x^2 + \dfrac{2}{x}$, i.e. Example 7, this would

then appear as follows with:

$\qquad y = x^2 \qquad$ shown as $\quad$ -------------------

$\qquad y = \dfrac{2}{x}, \qquad$ shown as $\quad$ - - - - - - - - -

and $\quad y = x^2 + \dfrac{2}{x} \quad$ shown as $\quad$ _____

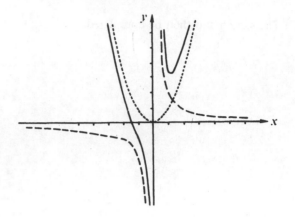

Exercise 11B

1. Find the coordinates of the points A to D and the values of the constants
a to c in the sketches shown below.

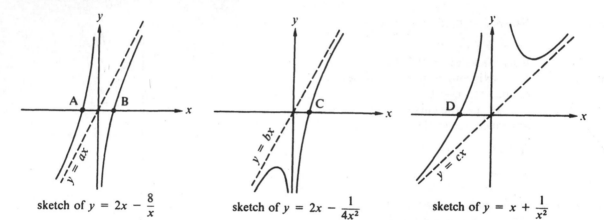

sketch of $y = 2x - \dfrac{8}{x}$ $\qquad$ sketch of $y = 2x - \dfrac{1}{4x^2}$ $\qquad$ sketch of $y = x + \dfrac{1}{x^2}$

Make sketch graphs of each of the following, labelling any asymptotes and
showing the coordinates of any turning points and the intersections with the axes.

2. $y = x + \dfrac{1}{x}$ $\quad$ **3.** $y = x - \dfrac{9}{x}$ $\quad$ **4.** $y = \dfrac{9}{x} - x$ $\quad$ **5.** $y = \left| x + \dfrac{4}{x} \right|$

Make sketch graphs of each of the following. The precise location of any
turning points need not be given.

6. $y = x - \dfrac{1}{x^2}$ $\quad$ **7.** $y = \dfrac{8}{x^2} - x$ $\quad$ **8.** $y = \left| x^2 - \dfrac{1}{x} \right|$ $\quad$ **9.** $y = \left| x^2 + \dfrac{1}{x} \right|$

11.3 Simple transformations

The reader will be familiar with the simple transformations, i.e. translations,
reflections, rotations and enlargements. From the graph of $y = f(x)$ it is
possible to deduce the graphs of other functions which are transformations
of $y = f(x)$.

In this section we shall see how the graph of $y = f(x)$ can help us to draw
the graphs of $y = f(x) + a$, $y = f(x - a)$, $y = -f(x)$, $y = f(-x)$,
$y = af(x)$ and $y = f(ax)$.

$y = f(x) + a$

Consider a point (x_1, y_1) on the graph of $y = f(x)$, i.e. $y_1 = f(x_1)$. The point on $y = f(x) + a$ with an x-coordinate of x_1, will have a y-coordinate of $(y_1 + a)$. Thus for every point (x_1, y_1) on $y = f(x)$, there exists a point $(x_1, y_1 + a)$ on $y = f(x) + a$. Therefore, if we translate the graph of $y = f(x)$ by a units parallel to the y-axis, we obtain the graph of $y = f(x) + a$.

e.g.

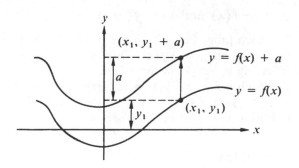

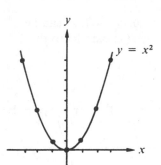
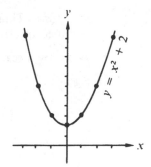
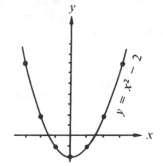

$y = f(x - a)$

Consider a point (x_1, y_1) on the graph of $y = f(x)$, i.e. $y_1 = f(x_1)$. The point on $y = f(x - a)$ with a y-coordinate of y_1 will have an x coordinate of $(x_1 + a)$. Thus for every point (x_1, y_1) on $y = f(x)$, there exists a point $(x_1 + a, y_1)$ on $y = f(x - a)$. Therefore, if we translate the graph of $y = f(x)$ by a units parallel to the x-axis, we obtain the graph of $y = f(x - a)$.

e.g.

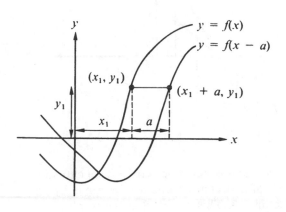

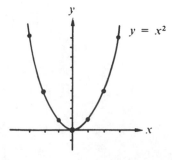
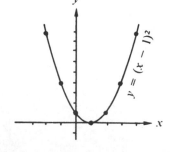
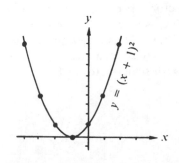

$y = -f(x)$ and $y = f(-x)$

For each point (x_1, y_1) on the graph of $y = f(x)$ there exists a point
$(x_1, -y_1)$ on $y = -f(x)$. Thus the graph of $y = -f(x)$ can be obtained by
reflecting $y = f(x)$ in the x-axis.
For each point (x_1, y_1) on the graph of $y = f(x)$, there exists a point
$(-x_1, y_1)$ on $y = f(-x)$. Thus the graph of $y = f(-x)$ can be obtained by
reflecting $y = f(x)$ in the y-axis.

$y = af(x)$

For each point (x_1, y_1) on the graph of $y = f(x)$
there exists a point (x_1, ay_1) on the graph of
$y = af(x)$.
Thus the graph of $y = af(x)$ can be obtained by
stretching $y = f(x)$ parallel to the y-axis by a
scale factor a.

e.g.

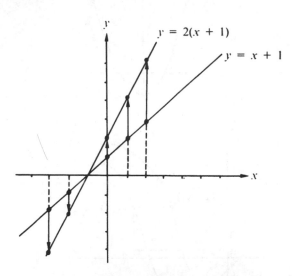

$y = f(ax)$

For each point (x_1, y_1) on the graph of $y = f(x)$,
there exists a point $\left(\dfrac{x_1}{a}, y_1\right)$ on the graph of
$y = f(ax)$. Thus the graph of $y = f(ax)$ can be
obtained by stretching $y = f(x)$ parallel to the
x-axis by a scale factor $\dfrac{1}{a}$.

e.g.

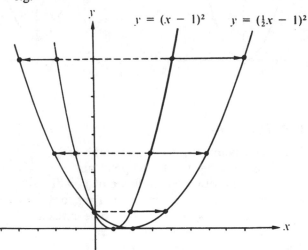

Note If a is negative, say -3, we can still
talk of a stretch parallel to the y-axis by scale
factor -3. This means that the graph is
reflected in the x-axis and then stretched by a
scale factor 3 because, for negative a,
$y = af(x)$ is a combination of $y = -f(x)$ and
$y = |a|f(x)$. Thus a reflection in the x-axis
could be viewed as a stretch parallel to the
y-axis with scale factor -1.

Example 8

Find the equation of the curve obtained when the graph of
$y = 2x^2 + 3x - 1$ is first reflected in the y-axis and then translated
$+2$ units in the direction Ox.

A reflection in the y-axis transforms $y = f(x)$ to $y = f(-x)$
Thus $y = 2x^2 + 3x - 1$ is transformed to $y = 2(-x)^2 + 3(-x) - 1$
$$= 2x^2 - 3x - 1$$
A transformation of $+ 2$ units in the direction $0x$ transforms $y = f(x)$ to
$y = f(x - 2)$
Thus $y = 2x^2 - 3x - 1$ transforms to $y = 2(x - 2)^2 - 3(x - 2) - 1$
$$= 2x^2 - 11x + 13$$
Thus under a reflection in the y-axis followed by a translation of $+2$ units in
the direction $0x$, the curve $y = 2x^2 + 3x - 1$ transforms to $y = 2x^2 - 11x + 13$.
Alternatively the same result could be obtained using the methods of
chapter 6, as follows.
Any point on the line $y = 2x^2 + 3x - 1$ will have coordinates of the form
$(k, 2k^2 + 3k - 1)$.
Thus if (x', y') lies on the required image line

$$\begin{pmatrix} x' \\ y' \end{pmatrix} = \begin{pmatrix} -1 & 0 \\ 0 & 1 \end{pmatrix}\begin{pmatrix} k \\ 2k^2 + 3k - 1 \end{pmatrix} + \begin{pmatrix} 2 \\ 0 \end{pmatrix}$$
$$= \begin{pmatrix} -k + 2 \\ 2k^2 + 3k - 1 \end{pmatrix}$$

i.e. $x' = -k + 2$ and $y' = 2k^2 + 3k - 1$
Eliminating k from these equations gives $y' = 2(x')^2 - 11x' + 13$.
The required image line has equation $y = 2x^2 - 11x + 13$, as before.

Exercise 11C

1. (a) Draw x and y axes for $-5 \leqslant x \leqslant 5$ and $-6 \leqslant y \leqslant 6$.
   (b) By using integer values of x plot the graph of $y = 3 + 2x - x^2$ for
      $-2 \leqslant x \leqslant 4$.
   (c) If $f(x) = 3 + 2x - x^2$ draw the graphs of the following functions on the
      same pair of axes used for (b),
      $y = f(x) + 2, \quad y = f(x - 1), \quad y = f(-x)$
      and give the equation of each function.
2. (a) Draw x- and y-axes for $-6 \leqslant x \leqslant 6$ and $-8 \leqslant y \leqslant 8$.
   (b) By using integer values of x, plot the graph of $y = x^2 - 2x$ for
      $-2 \leqslant x \leqslant 4$.
   (c) If $f(x) = x^2 - 2x$ draw the graphs of the following functions on the same
      pair of axes used for (b),
      $y = f(x) - 3, \quad y = -f(x), \quad y = f(x + 4)$
      and give the equation of each function.
3. Draw x- and y-axes for $-4 \leqslant x \leqslant 10$ and $-4 \leqslant y \leqslant 10$.
   By using integer values of x, plot the graph of $y = x^2$ for $-3 \leqslant x \leqslant 3$ and, on
   the same axes, draw $y = (x - 2)^2, \quad y = (\tfrac{1}{2}x - 2)^2, \quad y = (\tfrac{1}{2}x - 2)^2 - 4$.

4. Find the equation of the curve obtained when the graph of
   $y = x^2 + x + 1$ is:
   (a) Translated $+4$ units in the direction of Oy, (O being the origin).
   (b) Translated -3 units in the direction of Oy.
   (c) Reflected in the y-axis.
   (d) Reflected in the x-axis.
   (e) Translated $+2$ units in the direction of Ox.
   (f) Translated -2 units in the direction of Ox.
   (g) Stretched parallel to the y-axis, scale factor 2, followed by a
   translation of $+2$ in the direction of Oy.
   (h) Translated $+2$ units in the direction of Oy followed by a stretch
   parallel to Oy, scale factor 2.

5. State the transformation that must be given to the graph of $y = x^n$ to
   obtain the graph of:
   (a) $y = x^n + 2$ (b) $y = (-x)^n$ (c) $y = (x - 3)^n$

   (d) $y = 2x^n$ (e) $y = \left(\dfrac{x}{2}\right)^n$ (f) $y = (x + 2)^n - 3$

   (g) $y = (2x)^n - 3$ (h) $y = 2(x - 1)^n + 5$

6. Prove that the curve obtained when $y = \dfrac{5x}{2} - \dfrac{x^2}{4} - 6$ is stretched

   parallel to Ox by a scale factor $\frac{1}{2}$ and then reflected in the x-axis is the
   same as that obtained when $y = x^2 + x$ is translated by $+3$ units
   parallel to Ox. Find the equation of this curve.

11.4 $\dfrac{1}{f(x)}$

A sketch of the graph $y = f(x)$ can be used to obtain a sketch of $y = \dfrac{1}{f(x)}$.
It is necessary to bear in mind the following important points.

1. For values of x for which $f(x) > 0$, then $\dfrac{1}{f(x)} > 0$.

2. For values of x for which $f(x) < 0$, then $\dfrac{1}{f(x)} < 0$.

3. If $f(x) = 0$ when $x = a$, i.e. $f(a) = 0$, then $\dfrac{1}{f(x)}$ is not defined for this

   value of x and $x = a$ is a vertical asymptote of $y = \dfrac{1}{f(x)}$.

4. If $f(x)$ cuts the y-axis at the point $(0, b)$, then $\dfrac{1}{f(x)}$ will cut the y-axis at

   the point $\left(0, \dfrac{1}{b}\right)$.

5. As $f(x) \rightarrow \pm\infty$, then $\dfrac{1}{f(x)} \rightarrow 0$.

6. If there is a maximum/minimum at $x = x_1$ on $y = f(x)$ then there is a

   minimum/maximum at $x = x_1$ on $y = \dfrac{1}{f(x)}$.

Example 9

Make a sketch of $y = x^2 - 2x - 8$ and hence sketch $y = \dfrac{1}{x^2 - 2x - 8}$.

$y = x^2 - 2x - 8$
$\quad = (x - 4)(x + 2)$

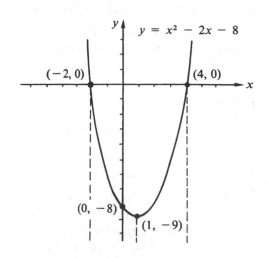

x-axis	Cuts x-axis at $(-2, 0)$ and at $(4, 0)$.
y-axis	Cuts y-axis at $(0, -8)$.
$x \to \pm\infty$	As $x \to +\infty,\, y \to +\infty$. As $x \to -\infty,\, y \to +\infty$.
y undefined	There is no value of x for which y is undefined.
Max/min	$y' = 2x - 2$ and $y'' = 2$ $\therefore$ min at $(1, -9)$

$y = \dfrac{1}{x^2 - 2x - 8}$

$\quad = \dfrac{1}{f(x)}$

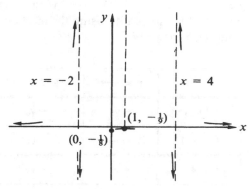

Vertical asymptote where $f(x) = 0$, i.e.
$x = -2$ and $x = 4$. $\dfrac{1}{f(x)}$ has the same sign
as $f(x)$ on either side of these asymptotes.

As $f(x) \to \infty$, $\dfrac{1}{f(x)} \to 0$, i.e. when $x \to \pm\infty$.

$f(x)$ cuts y-axis at $(0, -8) \Rightarrow \dfrac{1}{f(x)}$ cuts y-axis at $(0, -\tfrac{1}{8})$

$f(x)$ has min at $(1, -9) \Rightarrow \dfrac{1}{f(x)}$ has max at $(1, -\tfrac{1}{9})$

The sketch can then be completed.

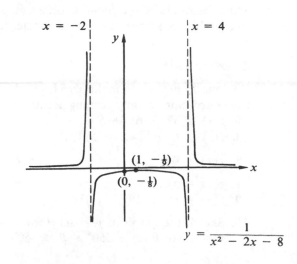

Notice from the graph of $y = \dfrac{1}{(x - 4)(x + 2)}$ that the ranges of values y can take for real x are $y \leqslant -\frac{1}{8}, y > 0$.

We say that the range of the function $f(x) = \dfrac{1}{(x - 4)(x + 2)}$ for the domain $\{x \in \mathbb{R}: x \neq 4, x \neq -2\}$ is $\{y \in \mathbb{R}: y \leqslant -\frac{1}{8} \text{ or } y > 0\}$. This range of values of y can be determined algebraically using the method shown in Example 10.

Example 10

Calculate the range of the function $f(x) = \dfrac{1}{(x - 4)(x + 2)}$ for real x.

Let $y = \dfrac{1}{(x - 4)(x + 2)}$ $\therefore y = \dfrac{1}{x^2 - 2x - 8}$ ... [1]

From [1] $yx^2 - 2xy - 8y = 1$

i.e. $yx^2 - 2xy - 8y - 1 = 0$

For $y \neq 0$ this is a quadratic in x.   For $y = 0$ the result $-1 = 0$

Thus, for real x, $4y^2 - 4(y)(-8y - 1) \geqslant 0$ is obtained which is not possible.

i.e. $y(9y + 1) \geqslant 0$ $\therefore y \neq 0$

Solving this inequality using the methods of chapter 5, page 143:

	$y < -\frac{1}{8}$	$-\frac{1}{8} < y < 0$	$y > 0$
y	$-$ve	$-$ve	$+$ve
$9y + 1$	$-$ve	$+$ve	$+$ve
$y(9y + 1)$	$+$ve	$-$ve	$+$ve

Thus the range of the function is $\{y \in \mathbb{R}: y \leqslant -\frac{1}{8} \text{ or } y > 0\}$.

(Note: this technique is practised further in Chapter 14.)

Exercise 11D

For questions **1** to **9**, make sketch graphs of $f(x)$ and $1/f(x)$ showing clearly the coordinates of any turning points and of any intersections with the axes.

1. $f(x) = x^2 - 6x + 5$ **2.** $f(x) = 1 - x^2$

3. $f(x) = (x - 2)(x - 5)$ **4.** $f(x) = (3 - x)(x + 1)$

5. $f(x) = 4x - x^2 - 3$ **6.** $f(x) = \dfrac{x^2 - 4x - 12}{8}$

7. $f(x) = x(x - 4)$ **8.** $f(x) = (x - 2)^2$

9. $f(x) = (x + 2)(x - 1)^2$

10. Sketch the graph of $y = \sin\theta$ for $-360° \leqslant \theta \leqslant 360°$. Hence sketch $y = \csc\theta$ for $-360° \leqslant \theta \leqslant 360°$.

11. Sketch the graph of $y = \cos \theta$ for $-360° \leqslant \theta \leqslant 360°$. Hence sketch $y = \sec \theta$ for $-360° \leqslant \theta \leqslant 360°$.
12. Sketch the graph of $y = \tan \theta$ for $-360° \leqslant \theta \leqslant 360°$. Hence sketch $y = \cot \theta$ for $-360° \leqslant \theta \leqslant 360°$.
13. Calculate the range of the following functions for real x.

(a) $\dfrac{1}{x^2 - x - 6}$ (b) $\dfrac{1}{2x^2 - 7x + 3}$ (c) $\dfrac{1}{4x - 3 - x^2}$ (d) $\dfrac{6}{x^2 + 3x - 4}$

Exercise 11E Examination questions

1. On two separate diagrams sketch the graphs of
   (i) $y = x^2 + 2$ for $-6 \leqslant x \leqslant 4$,
   (ii) $y = |8 - 3x|$ for $1 \leqslant x \leqslant 5$.
   In each case state the range of values of y.   (Cambridge)

2. Using a *separate* diagram for each, *sketch* the curves
   (a) $y = x(x - 1)$,
   (b) $y = x^2(x - 1)$,
   (c) $y = x^2(x - 1)^2$.   (A.E.B.)

3. Calculate the range of values of m for which the line $y = mx$ intersects the curve $y = x^2 - 4x + 9$ in two distinct points. Sketch the curve and indicate on it the significance of the values of m you have found.
          (S.U.J.B.)

4. The curve with equation
   $$y = ax^3 + bx^2 + cx + d,$$
   where a, b, c and d are constants, passes through the points $(0, 3)$ and $(1, 0)$. At these points the curve has gradients -7 and 0, respectively.
   (i)   Find the values of a, b, c and d.
   (ii) Show that the curve crosses the x-axis at the point $(3, 0)$.
   (iii) Find the x-coordinate of the maximum point on the curve.
   (iv) Sketch the curve.   (Cambridge)

5. On separate diagrams, **sketch** the graphs of
   (i) $y = (x - 3)^2 + 2$,
   (ii) $y = 2 \sin \left(x - \dfrac{\pi}{2} \right)$ for $-\dfrac{\pi}{2} \leqslant x \leqslant \dfrac{3\pi}{2}$.
   In each case state the range of values of y.   (Cambridge)

6. Sketch the curve $y = \cos x + 2$ for the interval $0 \leqslant x \leqslant 2\pi$.
   On the same diagram sketch the curve $y = \dfrac{1}{\cos x + 2}$ for the same interval.   (Cambridge)

7. Find $\dfrac{dy}{dx}$ when $y = x + \dfrac{9}{x}$. Find also the maxima and minima of y, distinguishing carefully between them.
   Show that $\dfrac{dy}{dx} < 1$ for all values of x.
   Draw a sketch of the curve $y = x + \dfrac{9}{x}$.   (Oxford)

8. On separate diagrams, sketch the graphs of
   (a) $y = x(x - 5)^2$ (b) $y^2 = x(x - 5)^2$

9. (a) Sketch the graph of $y = (x + 1)(x - 2)$ and hence sketch the graph
   of $z = 1/\{(x + 1)(x - 2)\}$. By calculation find the range of values
   which z can take.   (S.U.J.B.)

10. Prove that, for all real x,

$$0 < \frac{1}{x^2 + 6x + 10} \leqslant 1.$$

   Sketch the curve

$$y = \frac{1}{x^2 + 6x + 10}.$$ (Oxford)

11. Sketch separately the graphs of
    (i) $y = 2(x + 2)(x - 1)(x - 2)$;   (ii) $y = (x - 1)(x - 3)^2$
    showing the coordinates of the points where they cross or touch the axes.
    Find a cubic function $f(x)$ such that the graph of $y = f(x)$ crosses the
    x-axis at $x = 1$, touches the x-axis at $x = 3$ and crosses the y-axis at
    $y = 18$. From consideration of the sketch graph of $y = f(x)$ show that if
    the equation $f(x) = k$ has three real roots then k must be negative. Give
    the sign of each root in this case. Show, also, that if $k > 0$ then the
    equation $f(x) = k$ has just one real root and give the range of values
    of k for which this root is negative.   (S.U.J.B.)

12. Write down the condition for the equation $ax^2 + bx + c = 0$ $(a \neq 0)$
    to have no real roots. Sketch the graph of $y = x^2 + 3x + 7$ and hence
    sketch the graph of $y = 1/(x^2 + 3x + 7)$.
    Sketch the graph of $y = (x - 1)(x + 5)$ and hence sketch the graph of
    $z = 1/((x - 1)(x + 5))$.
    By calculation find the range of values which z can take.   (S.U.J.B.)

12
Calculus II: Integration

12.1 The reverse of differentiation

In chapter 10 we saw that if we know the equation of a curve, say $y = f(x)$ then, by differentiation, we can find the gradient function $\frac{dy}{dx}$.

If, instead, we are given the gradient function $\frac{dy}{dx}$, can we obtain the equation of the curve?

This reverse process is called **integration**.

Suppose $\frac{dy}{dx} = 2x$. To find y, we require a function that differentiates to give $2x$. Clearly, $y = x^2$ is such a function, but there are many others, e.g. $y = x^2 - 3$, $y = x^2 + 1$, $y = x^2 + 7$, etc.

We say that $y = x^2 + c$, where c is a constant, gives $2x$ when differentiated. Thus if we **integrate** $2x$, we obtain $x^2 + c$. Given the gradient function of a curve, integration gives the 'family' of curves to which that gradient function applies. Given more information about the curve, we may be able to determine the value of c, the constant of integration, and hence the equation of a particular curve.

In simple cases this process of integration can be carried out by inspection.

$$\text{Thus, given} \quad \frac{dy}{dx} = 2x \quad \text{then} \quad y = x^2 + c$$

$$\text{given} \quad \frac{dy}{dx} = 4x^3 \quad \text{then} \quad y = x^4 + c$$

$$\text{given} \quad \frac{dy}{dx} = x^4 \quad \text{then} \quad y = \frac{1}{5}x^5 + c$$

In the general case, given $\frac{dy}{dx} = ax^n$ then $y = \frac{ax^{n+1}}{n+1} + c.$

This general statement can be verified by differentiation:

$$\text{If} \quad y = \frac{ax^{n+1}}{n+1} + c$$

$$\frac{dy}{dx} = \frac{a(n+1)x^{n+1-1}}{n+1}$$

$$= ax^n \quad \text{as required.}$$

Note that the statement 'If $\frac{dy}{dx} = ax^n$, then $y = \frac{ax^{n+1}}{n+1} + c$' does not apply for $n = -1$, because in such cases we have $y = \frac{ax^0}{0} + c$, which is meaningless.

We can now state the rule for integrating ax^n:

$$\boxed{\text{If } \frac{dy}{dx} = ax^n \text{ then } y = \frac{ax^{n+1}}{n+1} + c \text{ provided } n \neq -1}$$

This result can be readily remembered in words as 'increase the power of x by one, and divide by the new power.'

In chapter 10 we saw that, if $y = af(x) + bg(x) - ch(x)$, then

$$\frac{dy}{dx} = af'(x) + bg'(x) - ch'(x).$$

It follows therefore that, given $\frac{dy}{dx} = af'(x) + bg'(x) - ch'(x)$, we can obtain an expression for y by integrating each of the functions $f'(x)$, $g'(x)$ and $h'(x)$ in turn so as to get $y = af(x) + bg(x) - ch(x) + d$, where d is the constant of integration.

Note that, although the three functions $f'(x)$, $g'(x)$ and $h'(x)$ have been integrated, we need only one constant of integration since this can incorporate the constants from all three integrations.

Example 1

Find an expression for y if $\frac{dy}{dx}$ is given by

(a) $3x^2$ (b) $2x^5$ (c) $6x^2 - 4x + 2$ (d) $x^2(2x + 1)$

(a)　　$\dfrac{dy}{dx} = 3x^2$　　　　　　　　(b)　　$\dfrac{dy}{dx} = 2x^5$

　　　　$y = \dfrac{3x^3}{3} + c$　　　　　　　　　　$y = \dfrac{2x^6}{6} + c$

　i.e.　　$y = x^3 + c$　　　　　　　i.e.　　$y = \dfrac{x^6}{3} + c$

(c)　　$\dfrac{dy}{dx} = 6x^2 - 4x + 2$　　　(d)　　$\dfrac{dy}{dx} = 2x^3 + x^2$

　　　$y = \dfrac{6x^3}{3} - \dfrac{4x^2}{2} + \dfrac{2x^1}{1} + c$　　　　　$y = \dfrac{2x^4}{4} + \dfrac{x^3}{3} + c$

　i.e.　$y = 2x^3 - 2x^2 + 2x + c$　　i.e.　$y = \dfrac{x^4}{2} + \dfrac{x^3}{3} + c$

Example 2

A curve passes through the point $(1, 7)$ and the gradient of the curve at the point (x, y) is given by $(5 - 2x)$. Find the equation of the curve.

Gradient of the curve is　$\dfrac{dy}{dx} = 5 - 2x$

$$\therefore \quad y = 5x - \frac{2x^2}{2} + c$$

　　　　i.e.　$y = 5x - x^2 + c$

But the curve passes through the point $(1, 7)$, so substituting these coordinates in the equation of the curve

$$7 = 5(1) - 1^2 + c \quad \text{or} \quad c = 3$$

Thus the equation of the curve is $y = 5x - x^2 + 3$.

The integral sign

We have seen that, if $\dfrac{dy}{dx} = ax^n$ then, by integration $y = \dfrac{ax^{n+1}}{n+1} + c$

(provided $n \neq -1$).

Using the integral sign: $\displaystyle\int$ we write this as

$$\int ax^n dx = \frac{ax^{n+1}}{n+1} + c \quad \text{for} \quad n \neq -1$$

with the 'dx' signifying that the integration is carried out with respect to the variable x.

$\displaystyle\int g(x)dx$ is read as 'the integral of $g(x)$ with respect to x' and we

use the extended S symbol because, as we shall see, integration can be considered as a summation.
From the rules given in chapter 10 for differentiation, it follows that

$$\int af(x)dx = a\int f(x)dx$$

and $\displaystyle\int [af(x) + bg(x) - ch(x)]dx = a\int f(x)dx + b\int g(x)dx - c\int h(x)dx$

Example 3

Find (a) $\displaystyle\int 4x^5 dx$ (b) $\displaystyle\int \frac{6}{x^3}dx$ (c) $\displaystyle\int \left(5 - x^2 + \frac{18}{x^4}\right)dx$

(a) $\displaystyle\int 4x^5 dx$ (b) $\displaystyle\int \frac{6}{x^3}dx$ (c) $\displaystyle\int \left(5 - x^2 + \frac{18}{x^4}\right)dx$

$= \dfrac{4x^6}{6} + c$ $= \displaystyle\int 6x^{-3}dx$ $= \displaystyle\int (5 - x^2 + 18x^{-4})dx$

$= \dfrac{2x^6}{3} + c$ $= \dfrac{6x^{-2}}{-2} + c$ $= 5x - \dfrac{x^3}{3} + \dfrac{18x^{-3}}{-3} + c$

$\qquad\qquad\qquad = -\dfrac{3}{x^2} + c$ $= 5x - \dfrac{x^3}{3} - \dfrac{6}{x^3} + c$

Since there is an arbitrary constant, c, in each of these solutions, we say that these are *indefinite* integrals.

Second Order Differentials

Remembering that the differential of $\dfrac{dy}{dx}$ with respect to x is $\dfrac{d^2y}{dx^2}$, it follows that

$$\int \frac{d^2y}{dx^2}dx = \frac{dy}{dx} + c$$

Example 4

Find y as a function of x, given that $\dfrac{d^2y}{dx^2} = 15x - 2$ and that when $x = 2$,

$\dfrac{dy}{dx} = 25$ and $y = 20$.

Given $\dfrac{d^2y}{dx^2} = 15x - 2$

Integrating both sides of this equation with respect to x,

$$\frac{dy}{dx} = \int (15x - 2)dx$$

$$= \frac{15x^2}{2} - 2x + c$$

But $\dfrac{dy}{dx} = 25$ when $x = 2$,

$\therefore \qquad 25 = \dfrac{15(4)}{2} - 2(2) + c \quad$ i.e. $c = -1$

$\therefore \qquad \dfrac{dy}{dx} = \dfrac{15x^2}{2} - 2x - 1$

Integrating both sides of this equation with respect to x,

$$y = \frac{5x^3}{2} - x^2 - x + d$$

But $y = 20$ when $x = 2$, giving $d = 6$.

The required equation is $y = \dfrac{5x^3}{2} - x^2 - x + 6$.

Exercise 12A

1. Find an expression for y if $\dfrac{dy}{dx}$ is given by

   (a) $3x^2$ (b) $2x$ (c) x^3 (d) $2x^4$ (e) 5

   (f) $3x^5$ (g) $\sqrt{x}$ (h) $\dfrac{6}{x^2}$ (i) $-\dfrac{4}{x^3}$ (j) $\dfrac{1}{\sqrt{x}}$

   (k) $2x^3 + 3x^2$ (l) $5x + 1$ (m) $2x + 9x^2$

   (n) $5x^4 - 6x$ (o) $8x^3 - 12x^2$ (p) $x(4 - 3x)$

   (q) $3x(x - 2)$ (r) $2x(x^3 - 4)$ (s) $(3x - 1)(x + 1)$

   (t) $(x - 6)(x - 2)$

2. Integrate the following functions with respect to x.

   (a) $8x^3$ (b) $12x$ (c) $5x^2$ (d) 7 (e) $7 - 2x$

   (f) $\dfrac{6}{x^3}$ (g) $-\dfrac{12}{x^5}$ (h) $\dfrac{3x}{\sqrt{x}}$ (i) $\dfrac{5x^2}{\sqrt{x}}$ (j) $\dfrac{3x^4 + 6}{x^2}$

   (k) $4x^3 + 3x^2 + 2x + 1$ (l) $2x^2(3 - 4x)$

   (m) $x^4 + x^2 + \dfrac{1}{x^2} + \dfrac{1}{x^4}$.

3. Find

(a) $\displaystyle\int 12x\,dx$

(b) $\displaystyle\int (x^3 + x)\,dx$

(c) $\displaystyle\int x(x + 1)\,dx$

(d) $\displaystyle\int (x + 6)(x - 4)\,dx$

(e) $\displaystyle\int \frac{5}{x^4}\,dx$

(f) $\displaystyle\int \left(10x^4 + 8x^3 - \frac{6}{x^2}\right)dx$

(g) $\displaystyle\int \frac{x^4 + 1}{x^2}\,dx$

(h) $\displaystyle\int \frac{(1 - 3x)}{\sqrt{x}}\,dx$

(i) $\displaystyle\int \left(x + \frac{1}{x}\right)\left(x - \frac{1}{x}\right)dx$

4. The gradient of a curve at the point (x, y) on the curve is given by $6x$. If the curve passes through the point $(1, 4)$, find the equation of the curve.

5. Find the equation of the curve passing through the point $(-2, 6)$ and having gradient function $(3x^2 - 2)$.

6. The gradient of a curve at the point (x, y) on the curve is given by $2(1 - x)$ and the curve passes through the point $(-1, 5)$. Find the equation of the curve.

7. Find S as a function of t given that $\dfrac{dS}{dt} = 6t^2 + 12t + 1$ and when $t = -2$, $S = 5$.

8. Find V as a function of h given that $\dfrac{dV}{dh} = 2(7h - 2)$ and when $h = 2$, $V = 21$.

9. Find A as a function of p given that $\dfrac{dA}{dp} = 5 - 4p$ and when $p = 3$, $A = -2$.

10. The gradient of a curve at the point (x, y) on the curve is given by $(3x^2 + 8)$. If the curve and the line $2x - y - 1 = 0$ cut the y-axis at the same point, find the equation of the curve.

11. The gradient function of a curve is given by $(2x - 3)$ and the curve cuts the x-axis at two points: $A(5, 0)$ and B. Find the equation of the curve and the coordinates of B.

12. The gradient of a curve at the point (x, y) on the curve is given by $(2x - 4)$. If the minimum value of y is 3, find the equation of the curve.

13. Find y as a function of x given that $\dfrac{d^2y}{dx^2} = 4 - 6x$ and that when $x = 2$, $\dfrac{dy}{dx} = -4$ and $y = 7$.

14. Find y as a function of x given that $\dfrac{d^2y}{dx^2} = 6x - 4$, $y = 4$ when $x = 1$ and $y = 2$ when $x = -1$.

15. Find y as a function of x given that $\dfrac{d^2y}{dx^2} = 30x$, $y = 32$ when $x = 2$ and $y = 5$ when $x = -1$.

12.2 The area under a curve

Suppose A_1 is the area bounded by the curve $y = f(x)$, the x-axis and the lines $x = a$ and $x = b$ (see Figure 1). We say that A_1 is the area 'under' the curve from $x = a$ to $x = b$. One way to estimate this area would be to divide it into strips. Since each of these strips approximates to a rectangle (Figure 2), we can then sum the areas of these rectangles. This would give an approximate value for A_1; the more rectangles we use, the greater is the accuracy. Consider one such rectangle, width δx (Figure 3).

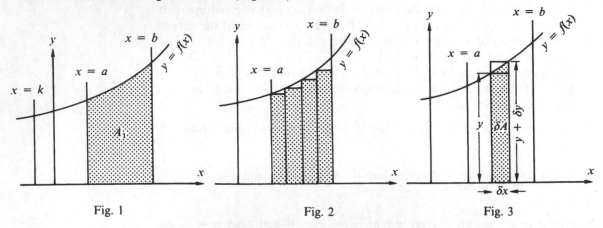

Fig. 1   Fig. 2   Fig. 3

In Figure 3, if δA is the shaded area, $y\,\delta x < \delta A < (y + \delta y)\delta x$

Thus $\quad y < \dfrac{\delta A}{\delta x} < y + \delta y$

Now as $\delta x \to 0$ (i.e. we increase the number of rectangles)

$\dfrac{\delta A}{\delta x} \to \dfrac{dA}{dx}$ and $\delta y \to 0$

Thus $\quad \dfrac{dA}{dx} = y$ or $A = \displaystyle\int y\,dx$

This integration will give an area function $A(x)$ and will involve a constant of integration c. As we substitute a value for x into the function $A(x)$, say $x = b$, we will obtain an answer for the area under the curve from a right-hand boundary of $x = b$ to some left-hand boundary. The position of the left-hand boundary will determine the value of c, the constant of integration. Suppose we take $x = k$ as the left-hand boundary, then

$\qquad A(a) = $ Area from $x = k$ to $x = a$.
$\qquad A(b) = $ Area from $x = k$ to $x = b$.
$\therefore\quad A(b) - A(a) = $ Area from $x = a$ to $x = b$.

As $A = \displaystyle\int y\,dx$, we write $A(b) - A(a)$ as $\displaystyle\int_a^b y\,dx$.

$\therefore\quad A_1 = \displaystyle\int_a^b y\,dx$.

$y\,\delta x < \delta A < y\,\delta x + \delta y \delta x$
As $\delta x \to 0$ $\delta y \to 0$ and so $\delta y \delta x$ becomes negligible compared with $y\,\delta x$.
Thus as $\delta x \to 0$, $\delta A \to y\,\delta x$.

But $\quad A_1 = \displaystyle\sum_{x=a}^{x=b} \delta A$

$\therefore \quad A_1 = \displaystyle\lim_{\delta x \to 0} \sum_{x=a}^{x=b} y\,\delta x$

The area under the curve can therefore be found as the limit of a sum or by integration. Thus integration is a process of summation and

$$A_1 = \lim_{\delta x \to 0} \sum_{x=a}^{x=b} y\, \delta x = \int_a^b y\, dx, \quad \text{where} \quad y = f(x).$$

Definite integrals

Note that $\int_a^b f(x)\, dx$ is known as a *definite* integral because the limits of integration, i.e. $x = a$ and $x = b$, are known.

Suppose $\quad\displaystyle\int f(x)\, dx = F(x) + c$

then $\quad\displaystyle\int_{x=a}^{x=b} f(x)\, dx = (F(b) + c) - (F(a) + c)$

$$= F(b) - F(a)$$

We usually write this: $\quad\displaystyle\int_{x=a}^{x=b} f(x)\, dx = \left[F(x) \right]_a^b$

$$= F(b) - F(a)$$

We see that the constants of integration cancel out so that in the case of a definite integral there is no need to give an arbitrary constant in the result.

Example 5

Evaluate the following definite integrals: (a) $\displaystyle\int_{-1}^{1} (2x - 3)\, dx$ (b) $\displaystyle\int_{1/4}^{1/2} \frac{1}{x^3}\, dx$

(a) $\displaystyle\int_{-1}^{1} (2x - 3)\, dx$

$= \left[x^2 - 3x \right]_{-1}^{1}$

$= [1^2 - 3(1)] - [(-1)^2 - 3(-1)]$

$= -2 - 4$

$= -6$

(b) $\displaystyle\int_{1/4}^{1/2} \frac{1}{x^3}\, dx = \int_{1/4}^{1/2} x^{-3}\, dx$

$= \left[\dfrac{x^{-2}}{-2} \right]_{1/4}^{1/2}$

$= \left[-\dfrac{1}{2x^2} \right]_{1/4}^{1/2}$

$= \left(-\dfrac{1}{2(\frac{1}{2})^2} \right) - \left(-\dfrac{1}{2(\frac{1}{4})^2} \right)$

$= -2 + 8$

$= +6$

Calculation of the area under a curve

When we calculate the area under a curve, the important first step is to make a sketch of the curve. We must then remember that an area lying 'above' the x-axis will have a positive value, whereas areas lying 'below' the x-axis will be negative. In some cases the required area may lie both 'above' and 'below' the x-axis and particular care is needed in these situations.

Example 6

Find the area between the curve $y = x(x - 3)$ and the x-axis.

First, make a sketch of the curve $y = x(x - 3)$
In this case the required area lies 'below' the x-axis.

Using $A = \displaystyle\int y\,dx$ and substituting for y from the equation

of the curve, as we cannot integrate y with respect to x.

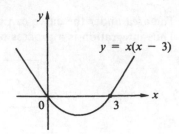

$$\therefore \quad A = \int_0^3 x(x - 3)\,dx$$

$$= \left[\frac{x^3}{3} - \frac{3x^2}{2} \right]_0^3$$

$$= \left(9 - \frac{27}{2} \right) - (0) = -4\tfrac{1}{2}$$

The area has a negative sign, as was anticipated, and the numerical value is $4\tfrac{1}{2}$ sq. units.

Example 7

Find the area between the curve $y = x(4 - x)$ and the x-axis from $x = 0$ to $x = 5$.
First, make a sketch of the curve $y = x(4 - x)$
The sketch shows that the required area is in
two parts; one part lies above the x-axis and
therefore has a positive area, the other part
lies below the x-axis and has a negative area.

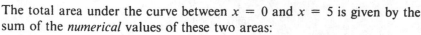

Using $A = \displaystyle\int y\,dx$ and calculating the two areas separately,

$$A_1 = \int_0^4 x(4 - x)\,dx \qquad\qquad A_2 = \int_4^5 x(4 - x)\,dx$$

$$= \left[2x^2 - \frac{x^3}{3} \right]_0^4 \qquad\qquad = \left[2x^2 - \frac{x^3}{3} \right]_4^5$$

$$= \left(32 - \frac{64}{3} \right) - (0) \qquad\qquad = \left(50 - \frac{125}{3} \right) - \left(32 - \frac{64}{3} \right)$$

$$= +\frac{32}{3} = +10\tfrac{2}{3} \qquad\qquad = -2\tfrac{1}{3}$$

The total area under the curve between $x = 0$ and $x = 5$ is given by the
sum of the *numerical* values of these two areas:

$$\text{required area} = 10\tfrac{2}{3} + 2\tfrac{1}{3} = 13 \text{ sq. units.}$$

Note In the last example, it is possible to calculate $\displaystyle\int_0^5 x(4 - x)\,dx$, but this

would not give the correct answer for the required area. Instead we would
obtain an answer of $10\tfrac{2}{3} - 2\tfrac{1}{3}$ i.e. $8\tfrac{1}{3}$, as the following working shows:

$$\int_0^5 x(4 - x)\,dx = \left[2x^2 - \frac{x^3}{3} \right]_0^5$$

$$= \left(50 - \frac{125}{3} \right) - (0) = 8\tfrac{1}{3}$$

Discontinuous functions

Although we may be able to evaluate $\int_{x=a}^{x=b} f(x)\,dx$, this does not mean that
the value obtained has any geometrical significance. In order that the definite
integral has a meaning we must ensure that $f(x)$ is defined and continuous
for this range of values of x, $a \leqslant x \leqslant b$.
The following examples illustrate this point:

(i) $\int_{-1}^{+1} \frac{1}{x}\,dx$ has no meaning since $\frac{1}{x}$ is not defined for $x = 0$.

(ii) $\int_{0}^{3} \frac{1}{x-2}\,dx$ has no meaning since $\frac{1}{x-2}$ is not defined for $x = 2$.

(iii) $\int_{0}^{2a} \frac{1}{x(x-a)}\,dx$ has no meaning since $\frac{1}{x(x-a)}$ is not defined for $x = 0$
or for $x = a$.

(iv) $\int_{-2}^{+2} \sqrt{(x+1)}\,dx$ has no meaning since $\sqrt{(x+1)}$ is not defined for
$-2 \leqslant x < -1$.

Area defined by two curves

An area can be defined by two curves and in this case it is essential to make
a sketch and to determine the points of intersection of the two curves.
Suppose the curves $y = f(x)$ and $y = g(x)$
intersect at the points where $x = a$ and
$x = b$.
The area between the curve $y = f(x)$ and the
x-axis from $x = a$ to $x = b$ is given by

$$\int_{a}^{b} f(x)\,dx.$$

The area between the curve $y = g(x)$ and the
x-axis from $x = a$ to $x = b$ is given by

$$\int_{a}^{b} g(x)\,dx.$$

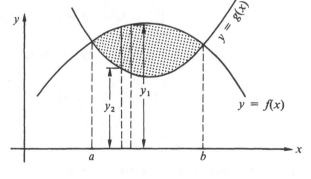

The shaded area between the two curve is then $\int_{a}^{b} f(x)\,dx - \int_{a}^{b} g(x)\,dx$ and

this may be written as $\int_{a}^{b} [f(x) - g(x)]\,dx.$

Alternatively, the second form of this solution can be obtained directly by
considering a strip of width δx, drawn parallel to the y-axis. The length of
this strip is $y_1 - y_2$, where $y_1 = f(x)$ and $y_2 = g(x)$, and the area of the
strip is then $[f(x) - g(x)]\delta x$ and the result follows.

Example 8

Find the area enclosed between the curves $y = 2x^2 + 3$ and $y = 10x - x^2$.

First, make a sketch of the two curves, noting that the curves will intersect at the points where

$$2x^2 + 3 = 10x - x^2$$
$$3x^2 - 10x + 3 = 0$$
$$(3x - 1)(x - 3) = 0 \quad \text{i.e. at} \quad x = \tfrac{1}{3} \quad \text{and} \quad x = 3.$$

The curve $y = 2x^2 + 3$ intersects the y-axis at $(0, 3)$ and does not cut the x-axis.
The curve $y = 10x - x^2$ intersects the axes at $(0, 0)$ and at $(10, 0)$.
The information is sufficient for a sketch to be made.

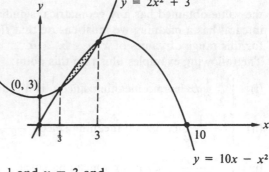

The area enclosed by $y = 10x - x^2$, the ordinates $x = \tfrac{1}{3}$ and $x = 3$ and the x-axis is

$$\int_{1/3}^{3} (10x - x^2)\,dx.$$

The area enclosed by $y = 2x^2 + 3$, the ordinates $x = \tfrac{1}{3}$ and $x = 3$ and the x-axis is

$$\int_{1/3}^{3} (2x^2 + 3)\,dx.$$

The shaded area is
$$\int_{1/3}^{3} (10x - x^2)\,dx - \int_{1/3}^{3} (2x^2 + 3)\,dx$$
$$= \int_{1/3}^{3} (10x - x^2 - 2x^2 - 3)\,dx$$
$$= \int_{1/3}^{3} (10x - 3x^2 - 3)\,dx$$
$$= \left[5x^2 - x^3 - 3x \right]_{1/3}^{3}$$
$$= (45 - 27 - 9) - (\tfrac{5}{9} - \tfrac{1}{27} - 1)$$
$$= 9\tfrac{13}{27} \text{ sq. units.}$$

Area between a curve and the y-axis

Suppose we wish to find the area between some curve $y = f(x)$ and the y-axis, from $y = a$ to $y = b$.
Considering a strip of length x and width δy, drawn parallel to the x-axis, we see that

$$A = \lim_{\delta y \to 0} \sum_{y=a}^{y=b} x\,\delta y$$

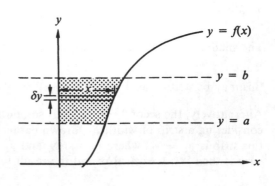

Example 9

Find the area enclosed between the curve $y^2 = 9 - x$ and the y-axis.

First we make a sketch of the curve $y^2 = 9 - x$.

Symmetry   The equation is unchanged if y is replaced by $(-y)$. Hence the curve is symmetrical about the x-axis.

y-axis   Cuts y-axis at $(0, 3)$ and at $(0, -3)$.

x-axis   Cuts x-axis at $(9, 0)$.

$x \to \pm\infty$ $y = \pm\sqrt{(9 - x)}$
∴ as $x \to +\infty$, y is undefined.
As $x \to -\infty$, $y = \pm\sqrt{(9 + \infty)}$
i.e. as $x \to -\infty$, $y \to \pm\infty$
slowly by comparison with x.

y undefined y is undefined for $x > 9$ because $(9 - x)$ will be negative. Thus the sketch can be completed and the required area shown shaded:

$$\text{Required area} = \int_{y=-3}^{y=3} x \, dy$$

Now we cannot integrate x with respect to y, so we substitute for x,

$$\therefore \quad A = \int_{y=-3}^{y=3} (9 - y^2) \, dy$$

which gives $A = 36$ sq. units

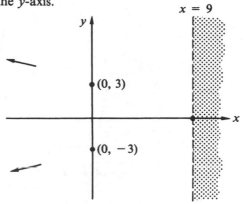

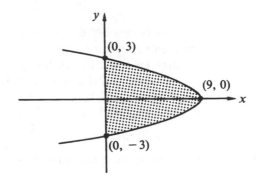

Exercise 12B

1. Evaluate the following definite integrals.

(a) $\int_1^5 2x \, dx$

(b) $\int_0^2 3x^2 \, dx$

(c) $\int_{-1}^4 (6 - 2x) \, dx$

(d) $\int_{-1}^1 (1 + x) \, dx$

(e) $\int_{-1}^3 (3x - 2) \, dx$

(f) $\int_{-4}^0 (x^2 + x + 1) \, dx$

(g) $\int_1^2 \frac{1}{x^2} \, dx$

(h) $\int_4^9 \frac{1}{\sqrt{x}} \, dx$

(i) $\int_0^4 (x^3 - 2x - 3\sqrt{x}) \, dx$

(j) $\int_1^4 \left(\frac{x^4 - x^3 + \sqrt{x} - 1}{x^2} \right) dx$

2. For each of the following, find the coordinate of points A, B and C and find the shaded area.

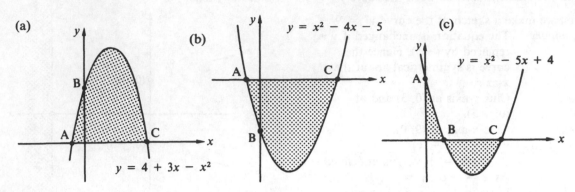

(a)

$y = 4 + 3x - x^2$

(b)

$y = x^2 - 4x - 5$

(c)

$y = x^2 - 5x + 4$

3. State why each of the following has no meaning

(a) $\int_{-3}^{4} \frac{1}{x} dx$ (b) $\int_{0}^{4} \frac{1}{x^2} dx$ (c) $\int_{-3}^{3} \frac{1}{x-1} dx$ (d) $\int_{-5}^{0} \frac{1}{x^2-1} dx$ (e) $\int_{-2}^{2} \sqrt{x} \, dx$.

4. Find the area between the line $y = 2x + 3$ and the x-axis from $x = 4$ to $x = 6$.

5. Find the area between the curve $y = x^3$ and the x-axis from $x = 1$ to $x = 2$.

6. Find the area enclosed by the lines $y = x^2 + 2$, the x-axis, $x = 1$ and $x = 3$.

7. Find the area between the curve $y = 10 + 3x - x^2$ and the x-axis from $x = -1$ to $x = 2$.

8. Find the area enclosed by $y = 3 + 2x - x^2$ and the x-axis.

9. Find the area enclosed by $y = x^2 - 6x$ and the x-axis.

10. Find the area between $y = 1 + \dfrac{4}{x^2}$ and the x-axis from $x = 1$ to $x = 2$.

11. Find the area between the curve $y = x^2 - 6x + 5$ and the x-axis from $x = 0$ to $x = 5$.

12. Find the area between the curve $y = 4 - x^2$ and the x-axis from $x = 0$ to $x = 3$.

13. Find the total area enclosed between $y = (x^2 - 1)(x - 3)$ and the x-axis.

14. Find the total area between the curve $y = \dfrac{4}{x^2} - 1$ and the x-axis from $x = 1$ to $x = 3$.

15. Using area $= \displaystyle\int_{y=a}^{y=b} x \, dy$, find the following shaded areas:

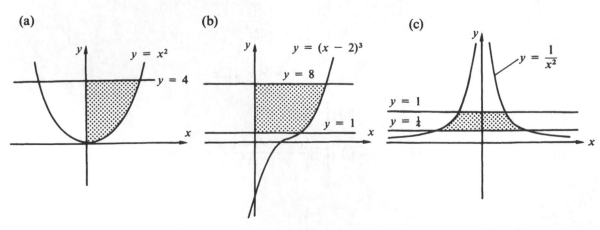

(a)

$y = x^2$

$y = 4$

(b)

$y = (x - 2)^3$

$y = 8$

$y = 1$

(c)

$y = \dfrac{1}{x^2}$

$y = 1$

$y = \frac{1}{4}$

16. Find the area enclosed between the curve $y^2 = 4 - x$ and the y-axis.
17. Find the area enclosed by the curve $(y - 1)^2 = x$, the y-axis and the line $y = 3$.
18. Find the area enclosed between the curve $y = x^2$ and the straight line $y = x$.
19. The line $y = x + 8$ cuts the curve $y = 12 + x - x^2$ at two points A and B (B being in the first quadrant). Find the coordinates of A and B and find the area enclosed between the curve $y = 12 + x - x^2$ and the straight line AB.
20. The sketch graph shows the lines $y = 7x + 7$ and $y = x^3 + 3x^2 - 6x - 8$. Find the coordinates of the points A, B, C, D, E and F and find the shaded area.

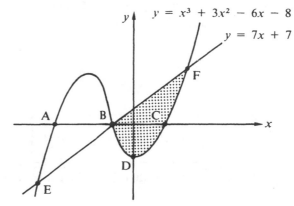

21. Find the area enclosed between the curves $y = x^2 + 6$ and $y = 12 + 4x - x^2$.
22. Find the area enclosed between the curves $y = x^2 - 4x$ and $y = 6x - x^2$.

12.3 Volume of revolution

Suppose that the area enclosed by the line $y = \frac{3}{4}x$, the x-axis and the line $x = 8$ is rotated about the x-axis through one revolution. The solid formed is referred to as a solid of revolution and, in this case, will be a cone. In this particular case the volume of the cone can be readily calculated since its base radius is 6 units and the perpendicular height of the cone is 8 units:

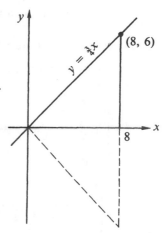

$$\text{volume of solid of revolution (i.e. the cone)}$$
$$= \tfrac{1}{3} \times \pi r^2 \times h$$
$$= \tfrac{1}{3} \times \pi 6^2 \times 8$$
$$= 96\pi \text{ cubic units.}$$

If instead of a straight line, a curve with equation $y = f(x)$ is rotated in the same way, the volume of the solid of revolution can be calculated by the use of calculus.

Suppose the area bounded by the curve $y = f(x)$, the x-axis and the ordinates $x = a$ and $x = b$ is rotated about the x-axis through one revolution.

Consider the elementary strip PQ, thickness δx, length y, drawn parallel to the y-axis. The result of rotating this elementary strip through one revolution about the x-axis is to produce an elementary disc of radius y and thickness δx. Let δV be the volume of this disc, then

$$\delta V = \pi y^2 \delta x.$$

Thus, as we allow $\delta x \to 0$, the sum of all these volumes δV will tend towards V, the volume of the solid of revolution.

$$\therefore \qquad V = \lim_{\delta x \to 0} \sum_{x=a}^{x=b} \pi y^2 \, \delta x$$

$$= \int_a^b \pi y^2 \, dx$$

Thus $\boxed{V = \int_a^b \pi y^2 \, dx \quad \text{where} \quad y = f(x)}$

This definite integral can then be evaluated to find V, the volume of revolution.

Example 10

Find the volume of the solid of revolution formed by rotating the area enclosed by the curve $y = x + x^2$, the x-axis and the ordinates $x = 2$ and $x = 3$ through one revolution about the x-axis.

First sketch the situation.

The volume of revolution V is given by

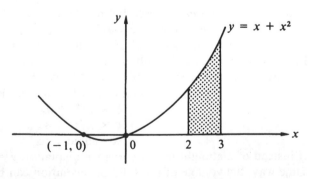

$$V = \int_a^b \pi y^2 \, dx$$

$$= \int_2^3 \pi (x + x^2)^2 \, dx$$

$$= \pi \int_2^3 (x^2 + 2x^3 + x^4) \, dx$$

$$= \pi \left[\frac{x^3}{3} + \frac{x^4}{2} + \frac{x^5}{5} \right]_2^3$$

$$= \pi \left[\left(9 + \frac{81}{2} + \frac{243}{5} \right) - \left(\frac{8}{3} + 8 + \frac{32}{5} \right) \right] = 81\tfrac{1}{30}\pi \text{ cubic units}$$

The volume of revolution is $81\tfrac{1}{30}\pi$ cubic units.

Rotation about the *y*-axis

The volume of the solid of revolution formed by rotating an area through one revolution about the *y*-axis can be found in a similar way.

The elementary strip will be drawn parallel to the *x*-axis. In this case, the elementary discs will have radius *x* and thickness δy.

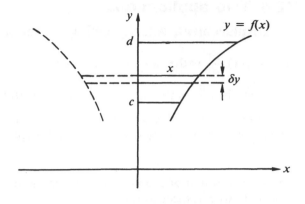

Thus $V = \lim\limits_{\delta y \to 0} \sum \pi x^2 \, \delta y$

$$V = \int_{y=c}^{y=d} \pi x^2 \, dy$$

Note that the integration will be with respect to *y* and hence the limits of the integration will be values of *y*.

Exercise 12C

Find the volumes of the solids formed when each of the areas of questions **1** to **10** perform one revolution about the *x*-axis.

1. The area enclosed by the curve $y = x^2$, the *x*-axis and the line $x = 2$.
2. The area between the line $y = x + 1$ and the *x*-axis from $x = 1$ to $x = 3$.
3. The area between the line $y = 3x + 2$ and the *x*-axis from $x = 0$ to $x = 1$.
4. The area enclosed by the curve $y = x^3$, the *x*-axis and the line $x = 2$.
5. The area enclosed by the curve $y = x^2 + 1$, the *x*-axis, $x = -1$ and $x = 1$.
6. The area enclosed by the curve $y = x^3 + 1$, the *x*-axis and the line $x = 1$.
7. The area enclosed by the curve $y = 4x - x^2$ and the *x*-axis.
8. The area enclosed by the curve $y = x^2 - x^3$ and the *x*-axis.
9. The area enclosed by the curve $y = x - \dfrac{1}{x}$ the *x*-axis and the line $x = 2$.
10. The area enclosed between the curve $y = 4x - x^2$ and the line $y = 2x$.
11. Find the equation of the straight line joining the origin to the point with coordinates (h, r). Hence find a formula for the volume of a right circular cone of base radius *r* and height *h*.

Find the volumes of the solids formed when each of the areas of questions **12** to **14** perform one revolution about the *y*-axis.

12. The area lying in the first quadrant and bounded by the curve $y = x^2$, the *y*-axis and the line $y = 4$.
13. The area lying in the first quadrant and bounded by the curve $y = 2x^2 + 1$, the *y*-axis and the lines $y = 2$ and $y = 5$.
14. The area lying in the first quadrant and bounded by the *y*-axis, the curve $y = x^3$ and the line $y = 3x + 2$.

12.4 Two applications

A. Displacement, velocity and acceleration

If $y = f(x)$, $\dfrac{dy}{dx}$ is the rate of change of y with respect to x. Similarly, if

$u = f(v)$, then $\dfrac{du}{dv}$ is the rate of change of u with respect to v.

Now the velocity v of a body is defined as the rate of change of the displacement s of a body from some fixed origin, with respect to time,

i.e. $v = \dfrac{ds}{dt}$

The acceleration a of a body is defined as the rate of change of the velocity of a body with respect to time,

i.e. $a = \dfrac{dv}{dt}$

$ = \dfrac{d}{dt}\left(\dfrac{ds}{dt}\right)$

$ = \dfrac{d^2 s}{dt^2}$

So displacement, velocity and acceleration are linked together by the process of differentiation with respect to time.

In the reverse order, a, v and s are linked together by integration, for if we integrate an expression for the acceleration of a body, with respect to time, we obtain an expression for the velocity of the body at time t,

i.e. from $\qquad\qquad a = \dfrac{dv}{dt}$ it follows that $v = \displaystyle\int a\,dt$

and similarly, from $\quad v = \dfrac{ds}{dt}$ it follows that $s = \displaystyle\int v\,dt$

This can be summarised diagrammatically:

```
┌─────────────────────────────────────────────┐
│   Differentiate with respect to time          ╲
└─────────────────────────────────────────────  ╲
```

$$\text{displacement} - \text{velocity} - \text{acceleration}$$
$$s \qquad\qquad v \qquad\qquad a$$

```
╱─────────────────────────────────────────────┐
╱   Integrate with respect to time              │
  └─────────────────────────────────────────────┘
```

Example 11

The displacement s metres of a body from an origin O at time t seconds is given by $s = 2t^2 - 3t + 6$. Find (a) the displacement, (b) the velocity and (c) the acceleration of the body when $t = 1\frac{1}{2}$.

Given $\qquad\qquad s = 2t^2 - 3t + 6$

(a) When $t = 1\frac{1}{2}$, $s = 2(1\frac{1}{2})^2 - 3(1\frac{1}{2}) + 6$

i.e. $\qquad\qquad s = 6$ m

(b) Since $\qquad\qquad v = \dfrac{ds}{dt}$

$\qquad\qquad\qquad v = 4t - 3$

when $t = 1\frac{1}{2}$, $v = 4(1\frac{1}{2}) - 3$

i.e. $v = 3$ m/s

(c) Using $a = \dfrac{dv}{dt}$

$a = 4$ m/s²

The displacement is 6 m, the velocity is 3 m/s and the acceleration is constant and is 4 m/s².

Example 12

A body moves in a straight line. At time t seconds its acceleration is given by $a = 6t + 1$. When $t = 0$, the velocity v of the body is 2 m/s and its displacement s from the origin O is 1 metre. Find expressions for v and s in terms of t.

Given that $a = \dfrac{dv}{dt} = 6t + 1$

$$v = \int (6t + 1)\,dt$$

$$v = 3t^2 + t + c$$

but when $t = 0$, $v = 2$ thus $2 = 0 + c$ ∴ $c = 2$

Substituting for c, $v = 3t^2 + t + 2$

Using $v = \dfrac{ds}{dt} = 3t^2 + t + 2$

$$s = \int (3t^2 + t + 2)\,dt$$

i.e. $s = t^3 + \dfrac{t^2}{2} + 2t + d$

but when $t = 0$, $s = 1$ thus $1 = 0 + d$ ∴ $d = 1$

Substituting for d, $s = t^3 + \dfrac{t^2}{2} + 2t + 1$

Thus at time t, $v = 3t^2 + t + 2$ and $s = t^3 + \dfrac{t^2}{2} + 2t + 1$.

B. Determination of centroids

The centroid of an area is that point about which the area is evenly spread. Thus if the area possesses a line of symmetry, the centroid will lie on that line.

If $(\bar{x}, \bar{y})$ is the centroid of an area A, then $A\bar{x}$ is called the first moment of area about the y-axis and $A\bar{y}$ is called the first moment of area about the x-axis.

Considering A as the sum of a number of smaller areas of which δA, centroid (x, y), is typical,

then $A\bar{x} = \lim\limits_{\delta A \to 0} \sum (\delta A \times x)$

and $A\bar{y} = \lim\limits_{\delta A \to 0} \sum (\delta A \times y)$

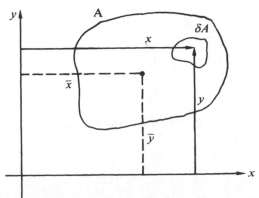

Suppose we wish to find the coordinates of the centroid of an area under a curve $y = f(x)$ from $x = a$ to $x = b$.

Consider the total area A, centroid $(\bar{x}, \bar{y})$, divided into elementary strips like PQ, parallel to the y-axis and at a distance x from it. For small values of δx, the centroid of PQ can be considered to be at $G\left(x, \dfrac{y}{2}\right)$ and if the area of the strip is δA, it follows that $\delta A \approx y\,\delta x$.

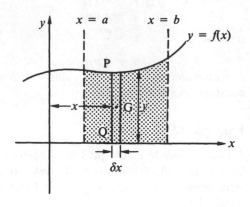

Considering the moment of area about the y-axis

$$A\bar{x} = \lim_{\delta A \to 0} \sum_{x=a}^{x=b} x\,\delta A$$

$$= \lim_{\delta x \to 0} \sum_{x=a}^{x=b} xy\,\delta x \quad \text{because as} \quad \delta A \to 0,\ \delta x \to 0$$

$$= \int_{x=a}^{x=b} xy\,dx$$

Similarly, considering the moment of area about the x-axis

$$A\bar{y} = \lim_{\delta A \to 0} \sum_{x=a}^{x=b} \frac{y}{2}\,\delta A$$

$$= \lim_{\delta x \to 0} \sum_{x=a}^{x=b} \frac{y^2}{2}\,\delta x$$

$$= \int_{x=a}^{x=b} \frac{y^2}{2}\,dx$$

Thus
$$\boxed{A\bar{x} = \int_{x=a}^{x=b} xy\,dx \quad \text{and} \quad A\bar{y} = \int_{x=a}^{x=b} \frac{y^2}{2}\,dx}$$

Example 13

Find the coordinates of the centroid of the area lying in the first quadrant and enclosed by the curve $y^2 = 16x$, the x-axis and the lines $x = 9$ and $x = 1$.

Thus
$$A\bar{x} = \int_1^9 xy\,dx \qquad \ldots [1]$$

and
$$A = \int_1^9 y\,dx \qquad \ldots [2]$$

From [2]
$$A = \int_1^9 4x^{1/2}\,dx$$

$$= \frac{8}{3}(27 - 1) = \frac{8 \times 26}{3}$$

Substituting this into [1] gives

$$\frac{8 \times 26}{3}\bar{x} = \int_1^9 4x^{3/2}\,dx$$

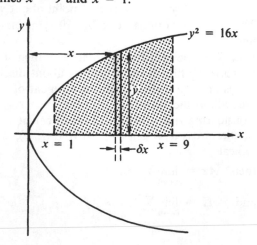

$$= \frac{242 \times 8}{5}$$

$\therefore \qquad \bar{x} = \frac{363}{65}$ or 5·58 correct to 2 decimal places

Considering moment of area about the *x*-axis

$$A\bar{y} = \int_{x=1}^{x=9} \frac{y^2}{2} dx$$

$$\frac{8 \times 26}{3}\bar{y} = \int_{1}^{9} 8x \, dx \quad \text{(substituting for } y \text{ from } y^2 = 16x)$$

$$= 320$$

$\therefore \qquad \bar{y} = \frac{60}{13}$ or 4·62 correct to 2 decimal places

Notes: (i) In the last example the area considered had no lines of symmetry and so both coordinates $\bar{x}$ and $\bar{y}$ had to be calculated using integration.

(ii) For a body of uniform density the centroid of the body will coincide with its centre of gravity.

A similar technique may be used to determine the centroid of a solid of revolution.

Example 14

Find the centroid of the volume of revolution formed by rotating, through one revolution about the *x*-axis, the area in the first quadrant bounded by the curve $y = 3x^2$, the *x*-axis and the line $x = 2$.

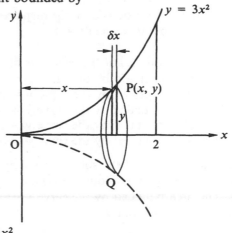

The centroid of the volume of revolution will lie on the axis of symmetry, i.e. the *x*-axis. Consider an elementary disc PQ formed by rotating an elementary strip, thickness δx and radius y, through one revolution about the *x*-axis.

Volume of elementary disc PQ $= \pi y^2 \, \delta x$

moment of PQ about *y*-axis $= x\pi y^2 \, \delta x$

Taking moments about the *y*-axis:

$$V\bar{x} = \int_{0}^{2} \pi x y^2 \, dx \quad \text{where} \quad V = \int_{0}^{2} \pi y^2 \, dx$$

$$\bar{x}\int_{0}^{2} \pi 9x^4 dx = \int_{0}^{2} 9\pi x^5 \, dx \quad \text{substituting for } y \text{ from } y = 3x^2$$

giving $\qquad \bar{x} = \frac{5}{3}$

The centroid of the volume of revolution is at the point $(1\frac{2}{3}, 0)$.

Exercise 12D

Displacement, velocity and acceleration

In questions **1** to **6**, s metres, v m/s and a m/s² represent respectively the displacement, velocity and acceleration of a body, relative to an origin O, at time t seconds.

1. If $s = 5t^3 - t$ find expressions for v and a in terms of t.
2. If $s = 4t^2 - t^3$ find the displacement, velocity and acceleration when
   (a) $t = 0$,   (b) $t = 2$.
3. If $v = 3t^2 - 8t$ and $s = 3$ when $t = 0$, find expressions for a and s in terms of t.
4. If $v = t^2 - 4t + 3$ and $s = 4$ when $t = 3$, find
   (a) the values of t when the body is at rest,
   (b) the acceleration when $t = 5$,
   (c) the displacement when $t = 1$.
5. If $a = 1 - t$ and, when $t = 2$, $v = 1$ and $s = 4\frac{2}{3}$ find expressions for v and s in terms of t.
6. If $a = 6t - 12$ and, when $t = 0$, $v = 9$ and $s = 6$ find the values of t when the body is at rest and the displacement of the body from O at these times.
7. A body starts from rest at an origin O and its acceleration at any time t seconds later is given by $a = (3 - 2t)$ m/s². Find the displacement of the body from O when it is next at rest.
8. A body starts from rest at an origin O and its acceleration at time t seconds later is given by
   $a = t + 3$ for $0 \leqslant t \leqslant 6$ and $a = 3t/2$ for $t \geqslant 6$.
   Find an expression for s, the displacement from O at time t for $t \geqslant 6$ and hence find s when $t = 8$.

Centroids

Find the coordinates of the centroids of the following areas.

9. The area enclosed by the curve $y = 4 - x^2$ and the x-axis.
10. The area enclosed by the curve $y = 3x - x^2$ and the x-axis.
11. The area bounded by the curve $y = x^2 + 2$, $x = -2$, $x = 2$ and the x-axis.
12. The area bounded by the curve $y = x^2$, $x = 3$ and the x-axis.
13. The area bounded by the curve $y = x^3$, $x = 3$ and the x-axis.

Find the centroid of the solid of revolution formed by rotating the following areas one revolution about the x-axis.

14. The area bounded by the curve $y = x^2$, the x-axis and the line $x = 4$.
15. The area between $y = x^3 + 1$ and the x-axis from $x = 0$ to $x = 2$.

12.5 Mean value of a function

Consider the area under a curve $y = f(x)$ from $x = a$ to $x = b$.

If this area is A, then $A = \displaystyle\int_{x=a}^{x=b} y\,dx$.

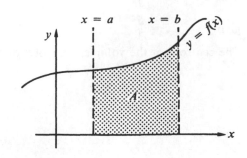

Now consider a rectangle of the same area A and of the same base $(b - a)$. The height of this rectangle would equal the mean value of the function $f(x)$ in the range $a \leqslant x \leqslant b$ i.e. (mean value of $f(x)$ in $a \leqslant x \leqslant b$) $\times (b - a) = A$

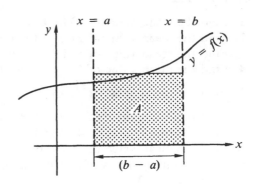

Thus we define the mean value of a function $y = f(x)$

in the range $a \leqslant x \leqslant b$ as $\dfrac{1}{(b - a)} \displaystyle\int_a^b f(x)\, dx.$

Example 15

Find the mean value with respect to x of the function $(5x^2 - 4x)$ for $1 \leqslant x \leqslant 3$.

$$\text{Given:} \quad f(x) = 5x^2 - 4x$$

By definition, the mean value is $\quad \dfrac{1}{(b - a)} \displaystyle\int_a^b f(x)\, dx$

$$= \frac{1}{(3 - 1)} \int_1^3 (5x^2 - 4x)\, dx$$

$$= \tfrac{1}{2}\left[\tfrac{5}{3}x^3 - 2x^2 \right]_1^3$$

$$= \tfrac{1}{2}(\tfrac{5}{3} \times 27 - 2 \times 9 - \tfrac{5}{3} \times 1 + 2 \times 1)$$
$$= 27\tfrac{1}{3}$$

The mean value of $(5x^2 - 4x)$ over the range $1 \leqslant x \leqslant 3$ is $27\tfrac{1}{3}$.

Example 16

A thin rod of length $2l$ is of variable density. Calculate the mean value of the density of the rod, if at a distance x from one end, the density of the rod is cx^2, where c is a constant.

Given that the density is a function of x such that
$$f(x) = cx^2 \quad \text{for } 0 \leqslant x \leqslant 2l$$

$$\text{mean value of density} = \frac{1}{(b - a)} \int_a^b f(x)\, dx$$

$$= \frac{1}{(2l - 0)} \int_0^{2l} cx^2\, dx$$

$$= \frac{1}{2l}\left[\frac{cx^3}{3} \right]_0^{2l}$$

$$= \frac{1}{2l}\left(\frac{8l^3 c}{3} \right)$$

$$= \frac{4l^2 c}{3}$$

The mean value of the density of the rod is $\dfrac{4l^2 c}{3}$.

Exercise 12E

1. Find the mean value of x^2 for $0 \leqslant x \leqslant 2$.
2. Find the mean value of $2x + 3$ for $1 \leqslant x \leqslant 7$.
3. Find the mean value of x^3 for $0 \leqslant x \leqslant 4$.
4. Find the mean value of $3x^2 + 2x$ for $0 \leqslant x \leqslant 2$.
5. Find the mean value of $\dfrac{1}{x^2}$ for $1 \leqslant x \leqslant 5$.
6. Find the mean value of $x^2 + 4$ for $-2 \leqslant x \leqslant 3$.
7. Find the mean value of $4 - x^2$ for $-2 \leqslant x \leqslant 2$.
8. Find the mean value of $4x^3 - 6x^2 + 2x - 1$ for $-1 \leqslant x \leqslant 3$.
9. The velocity v in m/s of a body t seconds after timing commences is given by $v = 3t^2 - 4$.
   (a) Find the mean velocity during the interval $t = 1$ to $t = 3$.
   (b) Find the mean acceleration during the interval $t = 1$ to $t = 3$.
10. The tension T newtons in a particular spring depends on the extension x metres of the spring from its natural length in accordance with the rule $T = 30x$. Find the mean tension in the spring as x increases from $0{\cdot}1$ m to $0{\cdot}2$ m.
11. The kinetic energy k joules of a 10 kg body depends on the velocity v m/s in accordance with the rule $k = 5\,v^2$. Find the mean kinetic energy possessed by the body as v increases from 1 m/s to 7 m/s.

Exercise 12F   *Examination questions*

1. Integrate with respect to x:
   (i) $\dfrac{1}{x^3}$ (ii) $\sqrt{x} + \dfrac{1}{\sqrt{x}}$.   (S.U.J.B.)

2. Evaluate $\displaystyle\int_1^2 \frac{x^3 + 1}{x^2}\,dx$.   (S.U.J.B.)

3. (a) Integrate with respect to x,
       $(\sqrt{x} - x)^2$
   (b) Evaluate
   $$\int_1^2 \left(3x + \frac{1}{x^2} - \frac{1}{x^4}\right) dx$$ (A.E.B.)

4. At any point (x, y) on a certain curve,
   $$\frac{dy}{dx} = (3x - 2)(x + 2).$$
   Given that it passes through $(-1, 1)$ find the equation of the curve.
   Find, and distinguish between, the turning points of the curve.   (A.E.B.)

5. A particle P moves in a straight line and passes a fixed point O with a velocity of V m/s. Its acceleration, a m/s^2, is given by $a = 16 - 4t$ for $0 \leqslant t \leqslant 3$ and $a = t + 1$ for $t \geqslant 3$, where t is the time in seconds after passing O. Given that the velocity of P when $t = 3$ is 38 m/s, find
   (i) the value of V,
   (ii) the velocity of P when $t = 4$.   (Cambridge)

6. The figure shows part of the curve
   $y = 6x - x^2$. Calculate the shaded
   area.

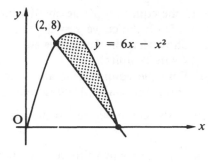

(Cambridge)

7. The figure shows part of the curve
   $$y = x^2 + \frac{1}{x}.$$
   Find
   (i) the co-ordinates of the point A,
   (ii) the volume generated when the
        shaded region is rotated through 360°
        about the x-axis.

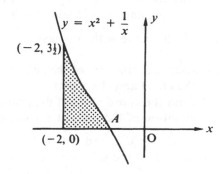

(Cambridge)

8. Sketch the curve $y = 4 - \dfrac{1}{x^2}$, indicating the asymptotes and the

   co-ordinates of the points of intersection of the curve and the x-axis.
   Find the area in the first quadrant bounded by the curve, the line $x = 2$
   and the x-axis.   (S.U.J.B.)

9. The points P(0, 2) and Q(3, 2) lie on the curve $y = x^2 - 3x + 2$. Find
   the equation of the normal to the curve at each of the points P and Q.
   Find the coordinates of R, the point where the normals meet.
   The curve cuts the x-axis at the points A and B, A being the point
   nearer to the origin. Obtain the coordinates of A and B, and sketch the
   curve for the range of values $0 \leqslant x \leqslant 3$.
   Show that the area of the region enclosed by the arc of the curve
   between P and Q, and the lines PR and QR is $5\frac{1}{4}$.   (J.M.B.)

10. The area bounded by the curve $y = x^2 + 1$, the x-axis and the
    ordinates $x = -1$ and $x = 1$ is rotated through four right-angles about
    the x-axis to form a solid of revolution. Calculate the volume of this
    solid.
    The same area is now rotated through two right-angles about the y-axis
    to form a new solid. Find the volume of this new solid.   (Oxford)

11. Find the equation of the chord which joins the points A(-2, 3) and
    B(0, 15) on the curve $y = 15 - 3x^2$.
    (a) Show that the finite area enclosed by the curve and the chord AB is
        4 square units.
    (b) Find the volume generated when this area is rotated through 360°
        about the x-axis, leaving your answer in terms of π.   (A.E.B.)

12. Given the curve whose equation is
    $$y = x^{-1/2},$$
    find (i) the mean value of $\dfrac{1}{y}$, with respect to x, in the interval
    $1 \leqslant x \leqslant 4$;

    (ii) the area of the region R bounded by the curve, the x-axis and
         the lines $x = 1$ and $x = 4$;   (J.M.B.)

13. Sketch the curve with equation
    $$y = (x^2 - 1)(x + 2).$$
    (a) Calculate the area of the finite region above the x-axis bounded by
        the curve and the x-axis.
    (b) Find the coordinates of the point of inflexion on the curve and the
        equation of the tangent at this point.   (London)

14. The points P(3, 2) and Q(0, 1) lie on the curve $y^2 = x + 1$. Calculate
    the volume of the solid generated when the region bounded by the lines
    $y = 0$, $x = 0$, $x = 3$ and the arc PQ of the curve is rotated completely
    about the x-axis. Give your answer as a multiple of π.
    S is the region bounded by the lines $y = 1$, $x = 3$ and the arc PQ of the
    curve. Show that when S is rotated completely about the x-axis the
    volume of the solid generated is $\dfrac{9\pi}{2}$.

    When S is rotated completely about the line $x = 3$ show that the volume
    of the solid generated is
    $$\pi \int_1^2 (4 - y^2)^2 \, dy.$$
    Calculate this volume in terms of π.   (J.M.B.)

13

Calculus III: Further techniques

13.1 Function of a function

When $y = x^2 + 4$, we say that y is a function of x.
If $y = (x^2 + 4)^5$, we say that y is a function (the fifth power) of a function $(x^2 + 4)$ of x.
Suppose that $y = u^5$ and that $u = x^2 + 4$, then this again leads to $y = (x^2 + 4)^5$. In this case we have used another variable u which links together the variables y and x.
Suppose $y = f(u)$ and $u = g(x)$ and let x change by a small amount δx, and the consequent change in u be δu and the change in y be δy.

Then $\dfrac{\delta y}{\delta x} = \dfrac{\delta y}{\delta u} \times \dfrac{\delta u}{\delta x}$

As $\delta x \to 0$, so also $\delta u \to 0$ and $\delta y \to 0$.
So taking limits as $\delta x \to 0$ we have

$$\boxed{\dfrac{dy}{dx} = \dfrac{dy}{du} \times \dfrac{du}{dx}}$$

This is also referred to as the **chain rule**.
The introduction of a third variable may be used to enable us to differentiate a function of a function.

Example 1

Find $\dfrac{dy}{dx}$ if (a) $y = (3x^2 - 2)^4$, (b) $y = \dfrac{1}{\sqrt{(x^2 - 2)}}$

(a)
$$y = (3x^2 - 2)^4$$
Let $u = 3x^2 - 2$ then $y = u^4$
$$\therefore \quad \frac{du}{dx} = 6x \quad \text{and} \quad \frac{dy}{du} = 4u^3$$
Using $\dfrac{dy}{dx} = \dfrac{dy}{du} \times \dfrac{du}{dx}$
$$\frac{dy}{dx} = 4u^3 \times 6x$$
$$\frac{dy}{dx} = 24x(3x^2 - 2)^3$$

(b)
$$y = \frac{1}{\sqrt{(x^2 - 2)}} = (x^2 - 2)^{-1/2}$$
Let $u = x^2 - 2$ then $y = u^{-1/2}$
$$\therefore \quad \frac{du}{dx} = 2x \quad \text{and} \quad \frac{dy}{du} = -\frac{1}{2}u^{-3/2}$$
$$= -\frac{1}{2u^{3/2}}$$
Using $\dfrac{dy}{dx} = \dfrac{dy}{du} \times \dfrac{du}{dx}$
$$\frac{dy}{dx} = -\frac{1}{2u^{3/2}} \times 2x$$
$$\frac{dy}{dx} = -\frac{x}{(x^2 - 2)^{3/2}}$$

Note: The final answer should always be given in terms of the variable used in the question and not the third variable, u, which was of our invention.

In general if $y = [f(x)]^n$, letting $u = f(x)$ then $y = u^n$

$$\text{then} \quad \frac{du}{dx} = f'(x) \text{ and } \frac{dy}{du} = nu^{n-1}$$

$$\boxed{\text{Thus if } \quad y = [f(x)]^n, \quad \frac{dy}{dx} = n[f(x)]^{n-1}f'(x)}$$

With some practice it is quite possible, and permissible, to write down the answer to differentiations of this type without showing the substitution and the introduction of the third variable.

Example 2

Find $\frac{dy}{dx}$ if (a) $y = (4x^3 - 7x)^6$, (b) $y = \sqrt{(5x - 2x^2)}$, (c) $y = \frac{1}{3x^3 - 4x}$.

(a) $y = (4x^3 - 7x)^6$

$\frac{dy}{dx} = 6(4x^3 - 7x)^5(12x^2 - 7)$

(b) $y = \sqrt{(5x - 2x^2)}$

$\quad = (5x - 2x^2)^{1/2}$

$\frac{dy}{dx} = \frac{1}{2}(5x - 2x^2)^{-1/2} \times (5 - 4x)$

i.e. $\frac{dy}{dx} = \frac{5 - 4x}{2\sqrt{(5x - 2x^2)}}$

(c) $\qquad y = \frac{1}{3x^3 - 4x} = (3x^3 - 4x)^{-1}$

$\frac{dy}{dx} = (-1)(3x^3 - 4x)^{-2} \times (9x^2 - 4)$

i.e. $\frac{dy}{dx} = \frac{4 - 9x^2}{(3x^3 - 4x)^2}$

Example 3

Find the equation of (a) the tangent and (b) the normal to the curve $y = \frac{5}{x^2 - 3}$ at the point (2, 5).

Note   Substituting $x = 2$ in the equation of the curve gives $y = \frac{5}{2^2 - 3} = 5$.

So the point (2, 5) does lie on the curve.

(a) $\qquad y = \frac{5}{x^2 - 3} = 5(x^2 - 3)^{-1}$

$\therefore \quad \frac{dy}{dx} = 5(-1)(x^2 - 3)^{-2} \times (2x)$

$\qquad = \frac{-10x}{(x^2 - 3)^2}$

Gradient of tangent at the point (2, 5) is $\frac{-10(2)}{(2^2 - 3)^2} = -20$.

Using $y - y_1 = m(x - x_1)$, equation of tangent at (2, 5) is $\quad y - 5 = -20(x - 2)$

or $y + 20x = 45$

(b) Gradient of normal at the point (2, 5) is $\dfrac{-1}{\text{(gradient of tangent)}} = \dfrac{1}{20}$.

Equation of normal at (2, 5) is $y - 5 = \dfrac{1}{20}(x - 2)$

or $20y = x + 98$.

The equation of the tangent is $y + 20x = 45$ and the equation of the normal is $20y = x + 98$.

Example 4

Find the coordinates of any stationary points on the curve $y = (3x - 1)^4$ and state the nature of any such points.

$$y = (3x - 1)^4$$

Differentiating this as a function of a function,

$$\frac{dy}{dx} = 4(3x - 1)^3 \times (3)$$
$$= 12(3x - 1)^3$$

If $\dfrac{dy}{dx} = 0$ then $12(3x - 1)^3 = 0$, i.e. $x = \tfrac{1}{3}$.

If x is 'just' less than $\tfrac{1}{3}$ (say 0), $\dfrac{dy}{dx} = 12(-1)^3$ i.e. negative

If x is 'just' more than $\tfrac{1}{3}$ (say $\tfrac{1}{2}$), $\dfrac{dy}{dx} = 12(1\tfrac{1}{2} - 1)^3$ i.e. positive

Thus the gradient changes from negative to positive as we pass through $x = \tfrac{1}{3}$, i.e. a minimum point.
When $x = \tfrac{1}{3}$, $y = (3 \times \tfrac{1}{3} - 1)^4 = 0$
A minimum point occurs at the point $(\tfrac{1}{3}, 0)$.

Exercise 13A

Differentiate the following with respect to x:

1. $(3x + 2)^4$
2. $(2x - 3)^5$
3. $(1 - 4x)^7$
4. $(2 + 9x)^5$
5. $(3 + x^2)^5$
6. $(1 - x^3)^8$
7. $(2x^4 + 1)^6$
8. $(x^2 + 5)^{10}$
9. $(1 + 3x)^{-3}$
10. $(1 - 4x^2)^{-1}$
11. $(5x - 3)^{1/2}$
12. $(x^2 + 1)^{-3/2}$
13. $\dfrac{1}{1 - 2x}$
14. $\dfrac{1}{x^2 + 3}$
15. $\sqrt{(x^2 + 4x + 1)}$
16. $\dfrac{1}{\sqrt{(x^2 - 5)}}$
17. $\dfrac{1}{1 + \sqrt{x}}$
18. $(1 + (1 + x^2)^5)^4$
19. $\sqrt{(2 + \sqrt{x})}$
20. $\left(\dfrac{1}{\sqrt{(3x^2 + 1)}}\right)^3$

Find the equation of (a) the tangent and (b) the normal to the following curves at the given points.

21. $y = (3x - 5)^4$ at the point (2, 1)
22. $y = \dfrac{1}{3 - x^2}$ at the point $(2, -1)$
23. $y = \dfrac{1}{3 + x^2}$ at the point $(-1, \tfrac{1}{4})$
24. $y = \sqrt{(1 + 2x)}$ at the point (4, 3)

25. Find the coordinates of any stationary points on the following curves and state the nature of each.

(a) $y = \dfrac{1}{1 + x^2}$ (b) $y = (2x - 5)^4$

(c) $y = (2x - 5)^3$ (d) $y = \dfrac{1}{x^2 + 4x}$

26. The displacement, s metres, of a body from an origin O at time t seconds is given by $s = \sqrt{(1 + 2t)}$. When $t = 12$ find
(a) the displacement from O   (b) the velocity and
(c) the acceleration of the body.

13.2 Integration of $f'(x)[f(x)]^n$

Although we have established a rule for integrating a power of x, i.e.

$\displaystyle\int x^n dx = \dfrac{x^{n+1}}{n + 1} + c$, there are few such rules and we are dependent

upon our ability to recognize the type of expression to be integrated.

Consider $\displaystyle\int 24x^3(x^4 + 7)^5 dx$.

If we go back to our idea of integration being the reverse of differentiation, we can see that this integrand, i.e. the expression which is to be integrated, may have come from differentiating $(x^4 + 7)^6$.

The expression $(x^4 + 7)^6$ is a function of a function and as we saw in section 13.1, the differential of this expression is $6(x^4 + 7)^5 \times (4x^3)$ or $24x^3(x^4 + 7)^5$.

Thus the integrand is exactly the differential of $(x^4 + 7)^6$ with respect to x, and hence

$$\int 24x^3(x^4 + 7)^5 \, dx = (x^4 + 7)^6 + c.$$

This is rather an artificial example in that the numbers have been carefully chosen to produce the desired result. The method would still work had a number other than 24 appeared in the integrand.

In this type of integration we are expecting to see an integrand which has come from differentiating a function of a function of x. The key to the method lies in recognizing that the integrand is a function to some power multiplied by the differential of that function (or some scalar multiple of it). We can state this in general terms:

$$\int f'(x)[f(x)]^n \, dx = \frac{1}{n + 1}[f(x)]^{n+1} + c.$$

It is better *not* to remember such a statement, but rather to understand how the integrand has been built up by differentiating a function of a function.

Example 5

Find (a) $\int 20(7 + 5x)^3 \, dx$ (b) $\int (2 - 3x)^6 \, dx$.

(a) $\int 20(7 + 5x)^3 \, dx$

We suspect that this may have come from differentiating $(7 + 5x)^4$ as a function of a function. This would give $4(7 + 5x)^3 \times (5)$ which is exactly the integrand,

$\therefore \quad \int 20(7 + 5x)^3 \, dx = (7 + 5x)^4 + c$

(b) $\int (2 - 3x)^6 \, dx$

We suspect that this may have come from differentiating $(2 - 3x)^7$ as a function of a function. This would give $7(2 - 3x)^6 \times (-3)$ which is $-21(2 - 3x)^6$ and this is a scalar multiple of the integrand,

$\therefore \quad \int (2 - 3x)^6 \, dx = -\frac{1}{21}(2 - 3x)^7 + c$

Example 6

Find (a) $\int x^4(1 + 2x^5)^3 \, dx$ (b) $\int (4x - 2)(x^2 - x + 4)^5 \, dx$.

(a) $\int x^4(1 + 2x^5)^3 \, dx$

If this has come from differentiating $(1 + 2x^5)^4$, then we expect to see the differential of $(1 + 2x^5)^4$, or a multiple of it, in the integrand.

Now $\frac{d}{dx}(1 + 2x^5)^4$ is $40x^4(1 + 2x^5)^3$,

$\therefore \quad \int x^4(1 + 2x^5)^3 \, dx = \frac{1}{40}(1 + 2x^5)^4 + c$

(b) $\int (4x - 2)(x^2 - x + 4)^5 \, dx$

We suspect this may have come from differentiating $(x^2 - x + 4)^6$.

Now $\frac{d}{dx}(x^2 - x + 4)^6$

$= 6(x^2 - x + 4)^5 \times (2x - 1)$

$\therefore \quad \int (4x - 2)(x^2 - x + 4)^5 \, dx$

$= \frac{1}{3}(x^2 - x + 4)^6 + c$

Note   Integrations of the type shown in Examples 5 and 6 can also be performed by the method of substitution which is explained in more detail in chapter 20.

Exercise 13B

Find

1. $\int 24(4 + 3x)^7 \, dx$

2. $\int 42(2 + 7x)^5 \, dx$

3. $-\int 14(3 - 2x)^6 \, dx$

4. $\int 10x(x^2 + 4)^4 \, dx$

5. $\int (2 + 3x)^5 \, dx$

6. $\int (1 + 2x)^4 \, dx$

7. $\int (1 - 6x)^3 \, dx$

8. $\int 2(3 + 4x)^4 \, dx$

9. $\int 6(1 - 2x)^4 \, dx$

10. $\int 2x(1 + x^2)^3 \, dx$

11. $\int x^2(1 - x^3)^4 \, dx$

12. $\int x^3(x^4 + 6)^3 \, dx$

13. $\int (2x + 1)(x^2 + x + 3)^4 \, dx$

14. $\int (3x^2 - 1)(x^3 - x + 4)^4 \, dx$

15. $\displaystyle\int (1 + 2x)(4 + x + x^2)^5\, dx$ **16.** $\displaystyle\int (x - 1)(x^2 - 2x + 4)^7\, dx$

Evaluate the following definite integrals

17. $\displaystyle\int_0^1 16x(x^2 + 5)^3\, dx$ **18.** $\displaystyle\int_1^7 (x - 7)^5\, dx$

19. $\displaystyle\int_0^1 x(x^2 + 4)^4\, dx$ **20.** $\displaystyle\int_{-1}^1 (3 + 2x)^5\, dx$

21. $\displaystyle\int_1^2 (2x - 1)(x^2 - x + 1)^5\, dx$ **22.** $\displaystyle\int_0^2 (1 - 2x)(x^2 - x - 3)^3\, dx$

23. $\displaystyle\int_1^4 \frac{(1 + \sqrt{x})^5}{\sqrt{x}}\, dx$

13.3 Related rates of change

Suppose that y is expressed as a function of x, say $y = f(x)$. If we know the rate at which x changes with respect to some other variable, say time t

$\left(\text{i.e. we know } \dfrac{dx}{dt}\right)$, then we can use the chain rule to find the rate of change

of y with respect to this third variable t, $\left(\text{i.e. we can find } \dfrac{dy}{dt}\right)$.

i.e. Given $y = f(x)$ and $\dfrac{dx}{dt}$, we can find $\dfrac{dy}{dt}$ by using $\dfrac{dy}{dt} = \dfrac{dy}{dx} \times \dfrac{dx}{dt}$

$$= f'(x)\frac{dx}{dt}.$$

Example 7

If the radius r of a sphere is increasing at 2 cm/s, find the rate at which the volume of the sphere is increasing when the radius is 3 cm (leave your answer in terms of π).

We know that the volume of a sphere is given by $V = \frac{4}{3}\pi r^3$ and we are given that $\dfrac{dr}{dt} = 2$ cm/s

We require $\dfrac{dV}{dt}$; using $\dfrac{dV}{dt} = \dfrac{dV}{dr} \times \dfrac{dr}{dt}$

$$\frac{dV}{dt} = \frac{d}{dr}\left(\frac{4}{3}\pi r^3\right) \times \frac{dr}{dt}$$

$$= \frac{4}{3}\pi \times 3r^2 \times 2$$

$$= 8\pi r^2 \text{ cm}^3/\text{s}$$

Thus when $r = 3$ cm, $\dfrac{dV}{dt} = 72\pi$ cm^3/s

When the radius of the sphere is 3 cm, the volume of the sphere is increasing at 72π cm^3/s.

Example 8

Water is pumped into an empty trough which
is 200 cm long, at the rate of 33 000 cm³/s.
The uniform cross-section of the trough is an
isosceles trapezium with the dimensions
shown. Find the rate at which the depth of the
water is increasing at the instant when this
depth is 20 cm.

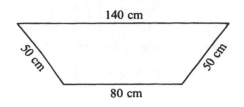

We require $\dfrac{dh}{dt}$; thus, given $\dfrac{dV}{dt}$, we want a
formula linking V and h so that we can use
the chain rule.
The total depth of the trough is 40 cm.

When the depth of water is h, $\dfrac{x}{30} = \dfrac{h}{40}$

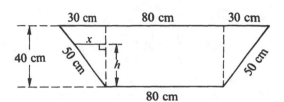

$$\text{or} \qquad x = \tfrac{3}{4}h$$

The volume V of water in the trough when the
depth of water is h, is given by

$$V = \tfrac{1}{2}(80 + 80 + 2x) \times h \times 200 \text{ cm}^3$$
$$= (160 + \tfrac{3}{2}h) \times 100h$$

Using $\dfrac{dV}{dt} = \dfrac{dV}{dh} \times \dfrac{dh}{dt}$

$$\dfrac{dV}{dt} = \dfrac{d}{dh}(16\,000h + 150h^2) \times \dfrac{dh}{dt}$$

$$= (16\,000 + 300h) \times \dfrac{dh}{dt}$$

But $\dfrac{dV}{dt} = 33\,000$ cm³/s, so when $h = 20$ cm,

$$33\,000 = (16\,000 + 300(20)) \times \dfrac{dh}{dt}$$

i.e. $\dfrac{dh}{dt} = 1\tfrac{1}{2}$ cm/s

When the depth of water is 20 cm, the depth is increasing at the rate of
$1\tfrac{1}{2}$ cm/s.

Note that in the last example we could have used $\dfrac{dh}{dt} = \dfrac{dh}{dV} \times \dfrac{dV}{dt}$, where

$$\dfrac{dV}{dt} = 33\,000 \quad \text{and} \quad \dfrac{dh}{dV} = 1 \bigg/ \dfrac{dV}{dh} = \dfrac{1}{16\,000 + 300h}$$

$$\dfrac{dh}{dt} = \dfrac{33\,000}{16\,000 + 300h}$$

$$= 1\tfrac{1}{2} \text{ cm/s} \quad \text{when} \quad h = 20 \text{ cm.}$$

The justification that $\dfrac{dh}{dV} = 1 \bigg/ \dfrac{dV}{dh}$, or that $\dfrac{dy}{dx} = 1 \bigg/ \dfrac{dx}{dy}$, is as follows.

By definition, $\dfrac{dy}{dx} = \lim\limits_{\delta x \to 0} \left(\dfrac{\delta y}{\delta x} \right)$

$$\therefore \qquad \frac{dy}{dx} = \lim_{\delta x \to 0} \left[\frac{1}{\left(\frac{\delta x}{\delta y} \right)} \right] \qquad \qquad \text{... [1]}$$

but as $\delta x \to 0$, then so also $\delta y \to 0$. Thus [1] can be written

$$\frac{dy}{dx} = \lim_{\delta y \to 0} \left[1 \Big/ \frac{\delta x}{\delta y} \right]$$

Thus
$$\boxed{\frac{dy}{dx} = \frac{1}{\left(\frac{dx}{dy} \right)}} \quad \text{as required.}$$

Exercise 13C

1. If $T = 5p^2 + \dfrac{3}{p}$, find $\dfrac{dT}{dq}$ when $p = 2$ given that for that value of p,
   $\dfrac{dp}{dq} = 4$.

2. If $A = (2t + 3)^5$ and $x = 2t^2 + 6t$, find an expression for $\dfrac{dA}{dx}$ in terms
   of t.

3. If r, the radius of a circle, increases at the rate of 2 cm/s, find an
   expression in terms of r for the rate at which the area of the circle is
   increasing.

4. If the radius of a sphere is increasing at 1 cm/s, find the rate at which
   the surface area is increasing when the radius is 5 cm. (Leave π in your
   answer.)

5. Air is being pumped into a spherical balloon at a rate of 54 cm³/s. Find
   the rate at which the radius is increasing when the volume of the balloon
   is 36π cm³.

6. If the volume of a sphere increases at the rate of 6 cm³/s, find the rate of
   increase in the surface area of the sphere at the instant when its radius is
   4 cm.

7. Oil is dripping onto a surface at the rate of $\frac{1}{10}\pi$ cm³/s and forms a
   circular film which may be considered to have a uniform depth of
   0·1 cm. Find the rate at which the radius of the circular film is increasing
   when this radius is 5 cm.

8. A closed right-circular cylinder has base radius r cm and height $3r$ cm. If
   r is increased at a rate of 1 millimetre per second, find expressions in
   terms of r for the rate of increase of
   (a) the total external surface area and   (b) the volume of the cylinder.

9. A hollow cone of base radius a and height $3a$ is held vertex downwards.
   The cone is initially empty and liquid is poured into it at a rate of 4π
   cm³/s. Find the rate at which the depth of the liquid in the vessel is
   increasing 16 seconds after the pouring commenced.

10. A container is in the shape of a cone of semi-vertical angle 30°, with its
    vertex downwards. Liquid flows into the container at the rate of
    $\dfrac{\sqrt{3}\pi}{4}$ cm³/s. At the instant when the radius of the circular surface of

the liquid is 5 cm, find the rate of increase in
- (a) the radius of the circular surface of the liquid,
- (b) the area of the circular surface of the liquid.
11. If the velocity of a body is given by $v = s^2 - 2$, where s is the displacement of the body from an origin O at time t, find an expression for the acceleration of the body in terms of s.

13.4 Parameters

The cartesian equation of a line is a law which links the x-coordinate and the y-coordinate of the general point (x, y) on the line. It can sometimes be useful to involve a third variable, say t, and to express x and y each as a function of this third variable. These two equations, $x = f(t)$ and $y = g(t)$, are said to be the **parametric equations** of the line and t is the **parameter**. If we use $x = f(t)$ and $y = g(t)$ to eliminate t, we return to the cartesian equation of the line.

Example 9

Find the cartesian equation for each of the following parametric forms:

(a) $x = 2t - 1, y = 1 - t^2$, (b) $x = \dfrac{1}{1 + t}, y = t^2 + 4$.

(a) $x = 2t - 1, y = 1 - t^2$;

hence $t = \dfrac{x + 1}{2}$.

Substituting this value of t in $y = 1 - t^2$,

$$y = 1 - \left(\frac{x + 1}{2}\right)^2$$

i.e. $4y = 3 - x^2 - 2x$
The cartesian equation is
$4y = 3 - x^2 - 2x$.

(b) $x = \dfrac{1}{1 + t}, y = t^2 + 4$;

hence $t = \dfrac{1}{x} - 1$.

Substituting this value of t in $y = t^2 + 4$,

$$y = \left(\frac{1}{x} - 1\right)^2 + 4$$

i.e. $y = \dfrac{1}{x^2} - \dfrac{2}{x} + 5$
The cartesian equation is

$$y = \frac{1}{x^2} - \frac{2}{x} + 5.$$

$\dfrac{dy}{dx}$ from parametric equations

We have seen that it is usually possible to obtain the cartesian equation from the two parametric equations by eliminating the parameter t. If an expression for $\dfrac{dy}{dx}$ is required, it is not usually necessary (or desirable) to obtain the cartesian equation.
Suppose $x = f(t)$ and $y = g(t)$,

then using the chain rule, $\dfrac{dy}{dx} = \dfrac{dy}{dt} \times \dfrac{dt}{dx}$, we can determine $\dfrac{dy}{dx}$.

Example 10

Find $\dfrac{dy}{dx}$ in terms of the parameter t if: (a) $x = 2t^3$, $y = 4t^2 + 1$ (b) $x = \dfrac{3}{t}$, $y = \sqrt{(1 + t^2)}$

(a) $x = 2t^3$, $y = 4t^2 + 1$

$\therefore \quad \dfrac{dx}{dt} = 6t^2, \dfrac{dy}{dt} = 8t$

Using $\dfrac{dy}{dx} = \dfrac{dy}{dt} \times \dfrac{dt}{dx}$

$\dfrac{dy}{dx} = 8t \times \dfrac{1}{6t^2}$

i.e. $\dfrac{dy}{dx} = \dfrac{4}{3t}$

(b) $x = \dfrac{3}{t} = 3t^{-1}$, $y = \sqrt{(1 + t^2)} = (1 + t^2)^{1/2}$

$\therefore \quad \dfrac{dx}{dt} = \dfrac{-3}{t^2}, \qquad \dfrac{dy}{dt} = \dfrac{1}{2}(1 + t^2)^{-1/2} \times 2t$

Using $\dfrac{dy}{dx} = \dfrac{dy}{dt} \times \dfrac{dt}{dx}$

$\dfrac{dy}{dx} = \dfrac{1}{2}(1 + t^2)^{-1/2} \times 2t \times \left(\dfrac{t^2}{-3}\right)$

i.e. $\dfrac{dy}{dx} = \dfrac{-t^3}{3\sqrt{(1 + t^2)}}$

The last example made use of the fact that $\dfrac{dt}{dx} = \dfrac{1}{\left(\dfrac{dx}{dt}\right)}$, which was proved on page 324.

Second differential

Particular care is needed when finding the second differential from the parametric equations.
One method is to find the cartesian equation by eliminating the parameter and then to find $\dfrac{dy}{dx}$ and $\dfrac{d^2y}{dx^2}$ in the usual way.
Alternatively, it is often better to work in terms of the parameter throughout as the following example illustrates.

Example 11

Find $\dfrac{dy}{dx}$ and $\dfrac{d^2y}{dx^2}$ in terms of t given that $x = \dfrac{1}{t}$, $y = 3t^2 + 2$.

$\qquad x = \dfrac{1}{t} = t^{-1} \qquad y = 3t^2 + 2$

$\qquad \dfrac{dx}{dt} = \dfrac{-1}{t^2} \qquad \dfrac{dy}{dt} = 6t$

Using $\dfrac{dy}{dx} = \dfrac{dy}{dt} \times \dfrac{dt}{dx}$

$\qquad \dfrac{dy}{dx} = 6t \times (-t^2) \qquad\qquad \left(\dfrac{dx}{dt} = \dfrac{-1}{t^2}, \text{ hence } \dfrac{dt}{dx} = -t^2\right)$

$\qquad\quad = -6t^3$

In order to find $\dfrac{d^2y}{dx^2}$, we need to differentiate $\dfrac{dy}{dx}$ with respect to x, so we must therefore differentiate $(-6t^3)$ with respect to x.

$\qquad \dfrac{d^2y}{dx^2} = \dfrac{d}{dx}(-6t^3)$

$\qquad\qquad = \dfrac{d}{dt}(-6t^3) \times \dfrac{dt}{dx}$ 　　　(Chain rule used to enable $-6t^3$ to be differentiated with respect to t)

$$= (-18t^2) \times (-t^2)$$
$$= 18t^4$$

Thus $\dfrac{dy}{dx} = -6t^3$ and $\dfrac{d^2y}{dx^2} = 18t^4$

Exercise 13D

Find the cartesian equation for each of the following parametric forms
1. $x = \sqrt{t}, y = t^2 + 4$
2. $x = 4 - t, y = t^2 - 10$
3. $x = 2 + 3t, y = \dfrac{1}{t}$
4. $x = 2t^2 + 3, y = 4 - t^2$
5. $x = 3 + 2t, y = 4t^2 - 9$
6. $x = 3t + 2, y = (t + 3)(t - 2)$
7. $x = \dfrac{1}{t - 1}, y = \dfrac{1}{t + 1}$
8. $x = \dfrac{t}{2t - 1}, y = \dfrac{t}{t + 3}$

For questions 9 to 11, remember that $\sin^2 \theta + \cos^2 \theta = 1$.
9. $x = 2 \sin \theta, y = \cos \theta$
10. $x = \sin 2\theta, y = \cos \theta$
11. $x = \sin \theta + 2 \cos \theta,$
    $y = 2 \sin \theta + \cos \theta$

In questions 12 to 19, x and y are given in terms of the parameter t. For each question find $\dfrac{dy}{dx}$ in terms of t.
12. $x = 3t^2, y = t^3$
13. $x = 4t^2, y = 8t$
14. $x = 3t^2 + 4t + 1, y = t^3 + t^2$
15. $x = t^2 + 4, y = \sqrt{t}$
16. $x = 2 + 3t, y = \dfrac{1}{t}$
17. $x = t^2 + 3t, y = 3t + 4$
18. $x = \dfrac{1}{t - 1}, y = \dfrac{1}{t + 1}$
19. $x = \dfrac{1}{t + 1}, y = t^2 + 4$

In questions 20 to 23, the equation of a curve is given parametrically with parameter t. Find the equation of (a) the tangent and (b) the normal at the given point on the curve.
20. $x = t^2 - 1, y = 2t^3$ at the point where $t = 2$
21. $x = \dfrac{1}{t + 1}, y = t^2$ at the point where $t = 1$
22. $x = t^2 + 5t, y = t^3$ at the point $(-4, -1)$
23. $x = \sqrt{(t^2 + 3)}, y = 3t + 4$ at the point $(2, 7)$
24. Find the coordinates of the turning point on the curve $x = (t^2 - 2)^2, y = t^2 - 6t$.
25. By finding $\dfrac{dy}{dt}$ and $\dfrac{dx}{dt}$ find $\dfrac{d^2y}{dx^2}$ in terms of t given that
    (a) $x = t + 3, y = t^2 + 4$
    (b) $x = 3 - 2t^2, y = \dfrac{1}{t}$
    (c) $x = t^2 + 2t, y = t^3 - 3t$
26. Copy and complete the following table for $x = 2t^2, y = 4t$

t	-4	-3	-2	-1	0	1	2	3	4
x	32		8						
y	-16		-8						

Draw x- and y-axes and hence plot the graph of the curve $x = 2t^2, y = 4t$ for $-4 \leqslant t \leqslant 4$.

27. Copy and complete the following table for $x = 4 \sin \theta$, $y = \cos \theta$
(where θ is in radians).

θ	0	$\dfrac{\pi}{6}$	$\dfrac{\pi}{3}$	$\dfrac{\pi}{2}$	$\dfrac{2\pi}{3}$	$\dfrac{5\pi}{6}$	π	$\dfrac{7\pi}{6}$	$\dfrac{4\pi}{3}$	$\dfrac{3\pi}{2}$	$\dfrac{5\pi}{3}$	$\dfrac{11\pi}{6}$	2π
x			3·46		3·46				−3·46		−3·46		
y		0·87				−0·87		−0·87				0·87	

Draw x- and y-axes and hence plot the graph of the curve $x = 4 \sin \theta$,
$y = \cos \theta$ for $0 \leqslant \theta \leqslant 2\pi$.

28. Copy and complete the following table for $x = 2t(t^2 - 1)$, $y = 4t^2$

t	-2	$-1\frac{1}{2}$	-1	$-\frac{1}{2}$	0	$\frac{1}{2}$	1	$1\frac{1}{2}$	2
x									
y									

Draw x- and y- axes and hence plot the graph of the curve $x = 2t(t^2 - 1)$,
$y = 4t^2$ for $-2 \leqslant t \leqslant 2$.

13.5 Product rule

Suppose u and v are functions of x and

$$y = uv \qquad \qquad \ldots [1]$$

Let x change by a small amount δx and let the consequent change in u be
δu, in v be δv and in y be δy.

Then $\qquad\qquad y + \delta y = (u + \delta u)(v + \delta v) \qquad \qquad \ldots [2]$

Subtracting equation [1] from equation [2],

$$y + \delta y - y = (u + \delta u)(v + \delta v) - uv$$

i.e. $\qquad\qquad \delta y = u\delta v + v\delta u + \delta u \delta v$

$$\frac{\delta y}{\delta x} = u\frac{\delta v}{\delta x} + v\frac{\delta u}{\delta x} + \delta u \frac{\delta v}{\delta x}$$

By definition, as $\delta x \to 0$, $\dfrac{\delta y}{\delta x} \to \dfrac{dy}{dx}$, $\dfrac{\delta v}{\delta x} \to \dfrac{dv}{dx}$ and $\dfrac{\delta u}{\delta x} \to \dfrac{du}{dx}$;

also as $\delta x \to 0$, both δu and δv approach 0.

$$\therefore \quad \frac{dy}{dx} = u\frac{dv}{dx} + v\frac{du}{dx} + 0 \times \frac{dv}{dx}$$

$$\boxed{\therefore \quad \frac{dy}{dx} = u\frac{dv}{dx} + v\frac{du}{dx}}$$

This important result should be memorised.

Example 12

Differentiate the following with respect to x (a) $y = x(x + 3)^4$ (b) $y = (2x + 1)^3(x - 1)^4$

(a) $y = x(x + 3)^4$

This is of the form $y = uv$ where
$u = x$ and $v = (x + 3)^4$
Using the product rule
$$\frac{dy}{dx} = x \times 4(x + 3)^3 + (x + 3)^4 \times 1$$
$$= (x + 3)^3[4x + x + 3]$$
$$= (x + 3)^3(5x + 3)$$

(b) $y = (2x + 1)^3(x - 1)^4$

This is of the form $y = uv$ where
$u = (2x + 1)^3$ and $v = (x - 1)^4$
Using the product rule
$$\frac{dy}{dx} = (2x + 1)^3 \times 4(x - 1)^3 + (x - 1)^4 \times 3(2x + 1)^2 \times 2$$
$$= 2(2x + 1)^2(x - 1)^3[2(2x + 1) + 3(x - 1)]$$
$$= 2(2x + 1)^2(x - 1)^3(7x - 1)$$

13.6 Quotient rule

Suppose u and v are functions of x and that $y = \dfrac{u}{v}$.

Writing this as $y = uv^{-1}$, we can use the product rule to give
$$\frac{dy}{dx} = u\frac{d}{dx}(v^{-1}) + v^{-1}\frac{du}{dx}$$
$$= u \times -1(v^{-2})\frac{dv}{dx} + \frac{1}{v} \times \frac{du}{dx} \qquad \left[\text{by chain rule} \quad \frac{d}{dx}(v^{-1}) = \frac{d}{dv}(v^{-1}) \times \frac{dv}{dx}\right]$$
$$= \frac{-u\dfrac{dv}{dx} + v\dfrac{du}{dx}}{v^2}$$

$$\therefore \quad \boxed{\frac{dy}{dx} = \frac{v\dfrac{du}{dx} - u\dfrac{dv}{dx}}{v^2}}$$

This important result should be memorised.

Example 13

Differentiate the following with respect to x:

(a) $y = \dfrac{2x + 3}{1 - 5x}$ (b) $y = \dfrac{(3x + 1)^4}{(5x - 2)^3}$

(a) $y = \dfrac{2x + 3}{1 - 5x}$

This is of the form $y = \dfrac{u}{v}$ where

$u = 2x + 3$ and $v = 1 - 5x$.

Using the quotient rule,
$$\frac{dy}{dx} = \frac{(1 - 5x)2 - (2x + 3)(-5)}{(1 - 5x)^2}$$
$$= \frac{2 - 10x + 10x + 15}{(1 - 5x)^2}$$
$$\therefore \quad \frac{dy}{dx} = \frac{17}{(1 - 5x)^2}$$

(b) $y = \dfrac{(3x + 1)^4}{(5x - 2)^3}$

This is of the form $y = \dfrac{u}{v}$ where

$u = (3x + 1)^4$ and $v = (5x - 2)^3$.

Using the quotient rule,
$$\frac{dy}{dx} = \frac{(5x - 2)^3 4(3x + 1)^3(3) - (3x + 1)^4 3(5x - 2)^2(5)}{(5x - 2)^6}$$
$$= \frac{3(5x - 2)^2(3x + 1)^3}{(5x - 2)^6}[4(5x - 2) - 5(3x + 1)]$$
$$\therefore \quad \frac{dy}{dx} = \frac{3(3x + 1)^3}{(5x - 2)^4}(5x - 13).$$

Note In the differentiation of products and quotients, particular care over the algebraic simplification is necessary.

Exercise 13E

Differentiate the following with respect to x, simplifying your answers where possible.

1. $y = x(2x + 5)^7$ **2.** $y = x^2(2x + 5)^7$

3. $y = x^4(x + 3)^5$ **4.** $y = x^4(x^2 + 3)^5$

5. $y = x^3(3 - x^4)^5$ **6.** $y = (x^2 - 7)(x^4 + 1)^2$

7. $y = (2x^2 + 1)^3(3x - 1)^7$ **8.** $y = (3x + 5)^9(2x - 1)^6$

9. $y = (4 - x)^5(2x + 1)^3$ **10.** $y = x\sqrt{(x + 3)}$

11. $y = x^2\sqrt{(x + 3)}$ **12.** $y = (x + 3)^4\sqrt{x}$

13. $y = (x + 3)^2\sqrt{(2x^2 + 1)}$ **14.** $y = \dfrac{2x}{x + 3}$

15. $y = \dfrac{3x}{x - 5}$ **16.** $y = \dfrac{3x - 1}{2x + 3}$

17. $y = \dfrac{1 - 5x}{3x + 2}$ **18.** $y = \dfrac{2x}{x^2 + 4}$

19. $y = \dfrac{1 - x}{x^2 + 3}$ **20.** $y = \dfrac{3 - x}{x^2 + 16}$

21. $y = \dfrac{x^5}{(x + 1)^3}$ **22.** $y = \dfrac{3x}{(x - 2)^4}$

23. $y = \dfrac{(6x + 5)^2}{(2x - 3)^3}$ **24.** $y = \dfrac{4x + 3}{\sqrt{(2x - 1)}}$

25. If $y = x(2x + 3)^5$, find $\dfrac{d^2y}{dx^2}$, simplifying your answer as far as is possible.

26. If $f(x) = \dfrac{x}{x + 1}$, find   (a) $f'(x)$ (b) $f'(-2)$ (c) $f''(x)$ (d) $f''(0)$

27. If $x = \dfrac{t}{t + 6}$ and $y = \dfrac{3t}{2 - t}$, find $\dfrac{dy}{dx}$ in terms of t.

28. If $x = \dfrac{t^2}{t + 1}$ and $y = \dfrac{t - 1}{t + 2}$, find $\dfrac{dy}{dx}$ in terms of t.

For questions **29** to **32**, find the equation of (a) the tangent and (b) the normal to the following curves at the given point on the curve.

29. $y = x^2(x - 1)^5$ at the point $(2, 4)$

30. $y = x\sqrt{(x^2 - 3)}$ at the point $(2, 2)$

31. $y = \dfrac{x + 3}{x^2 - 4x + 1}$ at the point $(1, -2)$

32. $y = \dfrac{3x^2}{x + 1}$ at the point $(2, 4)$

Find the coordinates of any stationary points on the following curves and state the nature of these stationary points.

33. $y = x(x - 5)^4$ **34.** $y = x^2(x - 3)^4$

35. $y = \dfrac{1 - x}{x^2 + 8}$ **36.** $y = \dfrac{x^5}{(x - 2)^3}$

37. If $x = \dfrac{1}{t + 1}$ and $y = \dfrac{1}{t - 1}$, find $\dfrac{d^2y}{dx^2}$ in terms of t.

38. If $x = \dfrac{3t}{t + 3}$ and $y = \dfrac{4t + 1}{t - 2}$, find $\dfrac{d^2y}{dx^2}$ in terms of t.

13.7 Implicit functions

An implicit function is one in which a relationship between, say, two variables x and y is given without having y as an explicit or clearly defined function of x.

Thus in $y = 3x^2 - 7x + 1$, y is given as an explicit function of x, whereas in $y^2 - 3yx = x^2$, y is not given explicitly as a function of x.

An implicit function involving y and x can be differentiated with respect to x as it stands.

Suppose $\qquad y^2 + 4x = 6x^2$,

Differentiating each term with respect to x,

$$\frac{d}{dx}(y^2) + \frac{d}{dx}(4x) = \frac{d}{dx}(6x^2)$$

i.e. $\quad \frac{d}{dy}(y^2)\frac{dy}{dx} + \frac{d}{dx}(4x) = \frac{d}{dx}(6x^2)$

Thus $\qquad 2y\frac{dy}{dx} + 4 = 12x$

It should be noted that we had to rearrange $\frac{d}{dx}(y^2)$ using the chain rule,

$\frac{d}{dx}(y^2) = \frac{d}{dy}(y^2)\frac{dy}{dx}$ to enable y^2 to be differentiated with respect to y.

Example 14

Find $\frac{dy}{dx}$ in terms of x and y if: (a) $y^3 + 6x = x^2$ (b) $3y^2 + 2y + xy = x^3$.

(a) $y^3 + 6x = x^2$

Differentiating each term with respect to x,

$$3y^2\frac{dy}{dx} + 6 = 2x$$

i.e. $\qquad \frac{dy}{dx} = \frac{2x - 6}{3y^2}$

(b) $3y^2 + 2y + xy = x^3$

Differentiating each term with respect to x,

$$6y\frac{dy}{dx} + 2\frac{dy}{dx} + \left(x\frac{dy}{dx} + y\right) = 3x^2$$

$\therefore \qquad \frac{dy}{dx}(6y + 2 + x) = 3x^2 - y$

i.e. $\qquad \frac{dy}{dx} = \frac{3x^2 - y}{x + 6y + 2}.$

Example 15

Find the equation of (a) the tangent and (b) the normal to the curve $3x^2 - xy - 2y^2 + 12 = 0$ at the point $(2, 3)$.

$3x^2 - xy - 2y^2 + 12 = 0$

[Note that since
$$3(2)^2 - 2(3) - 2(3)^2 + 12$$
$$= 12 - 6 - 18 + 12$$
$$= 0, \text{ the point } (2, 3) \text{ does lie on the curve.}]$$

(a) Differentiating the equation of the curve with respect to x,

$$6x - x\frac{dy}{dx} - y - 4y\frac{dy}{dx} = 0$$

thus
$$\frac{dy}{dx} = \frac{6x - y}{x + 4y}$$

Gradient at the point $(2, 3)$ $= \dfrac{6(2) - 3}{2 + 4(3)}$

$$= \frac{9}{14}$$

Equation of tangent at $(2, 3)$ is $y - 3 = \dfrac{9}{14}(x - 2)$

or $14y = 9x + 24$.

(b) Gradient of normal is $\dfrac{-1}{(\text{gradient of tangent})} = -\dfrac{14}{9}$ at the point $(2, 3)$.

Equation of normal at $(2, 3)$ is $y - 3 = -\dfrac{14}{9}(x - 2)$

or $9y + 14x = 55$

The equation of the tangent is $14y = 9x + 24$ and the equation of the normal is $9y + 14x = 55$.

Exercise 13F

In questions **1** to **9**, find $\dfrac{dy}{dx}$ in terms of x and y.

1. $x^2 + y^2 = 10$ **2.** $2x^2 + y^2 = 4x$

3. $6x^2 + 2y^3 = 8x + 4y$ **4.** $2x^3 - 5y^2 + 6x - 10y = 6$

5. $2x^2 + 2y^2 + 3x = 10 + 7y$ **6.** $x^2 + xy + y^2 = 0$

7. $x^3 + 3xy - y^2 = 6$ **8.** $2x^3 + 3xy^2 - y^3 = 0$

9. $3x^2 + 2y^2 - 5x + xy + 6y = 8$

10. Find the gradient of the curve $x^2 + 6y^2 = 10$ at the point $(2, -1)$.

11. Find the gradient of the curve $x^3 + 4xy = 15 + y^2$ at the point $(2, 1)$.

For questions **12** to **14**, find the equation of (a) the tangent and (b) the normal to the curve at the given point on the curve.

12. $5y^2 - 3x^2 - x + y = 0$ at the point $(1, -1)$

13. $3x^2 - 2xy + y^2 = 9$ at the point $(-2, -3)$

14. $x^3 + 3x^2y = 2y^2$ at the point $(-1, 1)$

Exercise 13G Examination questions

1. Find the exact value of

$$\int_0^1 x\sqrt{(4 - 3x^2)}\,dx. \qquad \text{(Cambridge)}$$

2. (a) Find the co-ordinates of the stationary points of the curve

$$y = \frac{x^2}{x - 2}.$$

(b) Given that $y = x\sqrt{(x + 1)}$, show that $\dfrac{dy}{dx} = \dfrac{3x + 2}{2\sqrt{(x + 1)}}$.

Hence, or otherwise, evaluate $\displaystyle\int_3^8 \frac{(3x + 2)}{\sqrt{(x + 1)}}\,dx.$ \qquad (Cambridge)

3. Differentiate the function $\dfrac{x + 3}{\sqrt{(1 + x^2)}}$ with respect to x. Find the value of x at which the function has a maximum or minimum, and determine which of these it is. \qquad (Oxford)

4. Given that $y = \dfrac{x^2 - 1}{2x^2 + 1}$, find $\dfrac{dy}{dx}$ and state the set of values of x for which $\dfrac{dy}{dx}$ is positive.

Find the greatest and least values of y for $0 \leqslant x \leqslant 1$. \qquad (London)

5. The radius of a circular disc is increasing at a constant rate of 0·003 cm/s. Find the rate at which the area is increasing when the radius is 20 cm. \qquad (Cambridge)

6. The volume, V, of a sphere of radius r, is $\dfrac{4\pi r^3}{3}$ and the surface area, A, is $4\pi r^2$.

The volume is increasing at the steady rate of 3 cm³/s.

Find $\dfrac{dr}{dt}$, where t is the time in seconds.

Calculate the value of $\dfrac{dA}{dt}$ in cm²/s at the instant when the radius is 12 cm. \qquad (A.E.B.)

7. The radius r centimetres of a circular blot on a piece of blotting paper t seconds after it was first observed is given by the formula

$$r = \frac{1 + 2t}{1 + t}.$$

Calculate
(i) the radius of the blot when it was first observed;
(ii) the time at which the radius of the blot was $1\frac{1}{2}$ cm;
(iii) the rate of increase of the area of the blot when the radius was $1\frac{1}{2}$ cm.
By considering the expression for $2 - r$ in terms of t or otherwise, show that the radius of the blot never reaches 2 cm. \qquad (Oxford)

8. Find the cartesian equations of the curves which are defined parametrically by (i) $x = 2\sin\theta, y = \cos^2\theta,$
(ii) $x = t(t - 1), y = 1 + t.$ \qquad (Cambridge)

9. A curve is defined parametrically by the equations

$$x = t^3 - 6t + 4, \ y = t - 3 + \frac{2}{t}.$$

Find

(i) the equations of the normals to the curve at the points where the curve meets the x-axis,

(ii) the coordinates of their point of intersection.   (Cambridge)

10. A curve is given by the parametric equations:

$$x = \frac{(1 - t)}{(1 + t)}, \ y = (1 - t)(1 + t)^2.$$

Find dy/dx and d^2y/dx^2 in terms of t. Find also the equation of the tangent to the curve at the point where $t = 2$.

(S.U.J.B.)

11. Find the equations of the tangents to the curve $y^2 + 3xy + 4x^2 = 14$ at the points where $x = 1$.   (S.U.J.B.)

12. Given that $y^2 - 5xy + 8x^2 = 2$, prove that $\dfrac{dy}{dx} = \dfrac{5y - 16x}{2y - 5x}$.

The distinct points P and Q on the curve $y^2 - 5xy + 8x^2 = 2$ each have x-coordinate 1. The normals to the curve at P and Q meet at the point N. Calculate the coordinates of N.   (A.E.B.)

13. Sketch the curve given parametrically by

$$x = t^2, \ y = t^3.$$

Show that an equation of the normal to the curve at the point A(4, 8) is $x + 3y - 28 = 0$.

This normal meets the x-axis at the point N. Find the area of the region enclosed by the arc OA of the curve, the line segment AN and the x-axis.   (London)

14
Sketching functions II

14.1 Rational functions

In this section we consider functions $f(x)$ of the type $f(x) = \dfrac{g(x)}{h(x)}$. The five
basic investigations for curve sketching used in chapter 11 are still applicable
for functions of this type, i.e.
(i)   symmetry if obvious,
(ii) intersection with axes,
(iii) behaviour as $x \rightarrow \pm\infty$
(iv) $f(x)$ undefined,
(v)   maximum/minimum.
Examples 1 to 7 show the ways in which these investigations give
information from which a sketch of $\dfrac{g(x)}{h(x)}$ can be made. We restrict our
attention to functions $\dfrac{g(x)}{h(x)}$ for which $h(x)$ is a polynomial of order 2 or less.

The reader should note that in the examples of this chapter, the behaviour of
the function on either side of the vertical asymptotes is not investigated.
These investigations are not always easy and the information gained from
the other investigations usually allows the behaviour of the function near
these asymptotes to be determined.

Type I $f(x) = \dfrac{a}{bx + c}$

Example 1
Make a sketch of the curve given by $y = \dfrac{5}{x + 2}$

x-axis	No value of x for which $y = 0$ $\therefore$ no intercept with x-axis.
y-axis	Cuts y-axis at $(0, 2\tfrac{1}{2})$
$x \rightarrow \pm\infty$	As $x \rightarrow \pm\infty,\ y \rightarrow \dfrac{5}{x}$

Thus for x a large positive number, y is small
and positive, thus as $x. \rightarrow +\infty,\ y \rightarrow 0^+$. For
x a large negative number, y is small and
negative, thus as $x \rightarrow -\infty,\ y \rightarrow 0^-$.

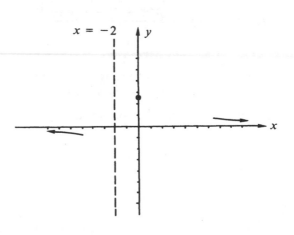

y undefined y is undefined for $x = -2$.
Thus $x = -2$ is a vertical asymptote.

Max/min $y' = \dfrac{-5}{(x + 2)^2}$

Thus no turning points and the gradient is always negative.

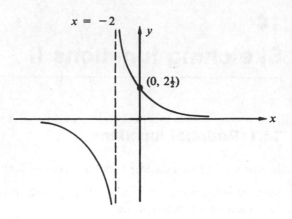

The sketch can then be completed.

Type II $f(x) = \dfrac{ax + b}{cx + d}$

Example 2

Make a sketch of the curve given by $y = \dfrac{x + 3}{x - 1}$

x-axis   Cuts x-axis at $(-3, 0)$

y-axis   Cuts y-axis at $(0, -3)$

$x \to \pm\infty$ By writing $y = \dfrac{1 + 3/x}{1 - 1/x}$

As $x \to \pm\infty$, $y \to 1$
$\therefore$ $y = 1$ is a horizontal asymptote. (It is not necessary to consider $x \to +\infty$ and $x \to -\infty$ separately for functions of this type.)

y undefined y is undefined for $x = 1$.
Thus $x = 1$ is a vertical asymptote.

Max/min $y' = \dfrac{1(x - 1) - 1(x + 3)}{(x - 1)^2}$

$= \dfrac{-4}{(x - 1)^2}$

Thus no turning points and the gradient is always negative.

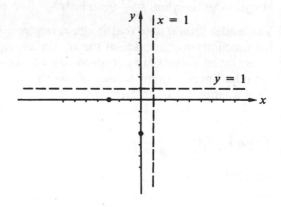

The sketch can then be completed.

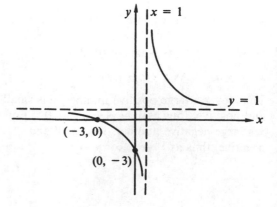

Type III $f(x) = \dfrac{g(x)}{h(x)}$ for $h(x)$ a quadratic function.

Some functions of this type have a range that is restricted in some way for $x \in \mathbb{R}$. It is useful to examine the range of $f(x)$ first to determine whether such restrictions exist. In addition, such examinations will indicate whether any maximum or minimum points exist.

Note   In the following examples, when $f(x)$ is an improper fraction, i.e. order of $g(x) \geqslant$ order of $h(x)$, then $f(x)$ is rearranged to eliminate these improper fractions when considering $x \to \pm\infty$.

Example 3

Sketch the curve given by $y = \dfrac{3x + 3}{x(3 - x)}$

If $y = \dfrac{3x + 3}{x(3 - x)}$ ... [1]   then $y = \dfrac{3x + 3}{3x - x^2}$ ... [2]

Range of values of y   From [2] $3xy - x^2y = 3x + 3$

                                i.e. $x^2y + 3x(1 - y) + 3 = 0$

For $y \neq 0$, this is a quadratic in x.   For $y = 0$, $3x + 3 = 0$,

Thus, for real x, $[3(1 - y)]^2 - 4y(3) \geqslant 0$ $x = -1$

i.e. $(3y - 1)(y - 3) \geqslant 0$ $\therefore$ y can equal zero.

We solve this inequality using the methods of chapter 5, page 143.

	$y < \tfrac{1}{3}$	$\tfrac{1}{3} < y < 3$	$y > 3$
$3y - 1$	$-$ve	$+$ve	$+$ve
$y - 3$	$-$ve	$-$ve	$+$ve
$(3y - 1)(y - 3)$	$+$ve	$-$ve	$+$ve

Thus the ranges of values y can take for real x are $y \leqslant \tfrac{1}{3}$, $y \geqslant 3$.
We can therefore shade a region on our sketch where the curve cannot exist. Note also that for each value of y in the allowed ranges, there will exist two distinct values of x, except at $y = 3$ and $y = \tfrac{1}{3}$ where a repeated root will occur, and at $y = 0$ where a single root occurs. From $y \leqslant \tfrac{1}{3}$ we expect a (local) maximum at $y = \tfrac{1}{3}$, and from $y \geqslant 3$ we expect a (local) minimum at $y = 3$.

x-axis   Cuts x-axis at $(-1, 0)$.
y-axis   No y-axis intercept as y not
                  defined for $x = 0$.

$x \to \pm\infty$ As $x \to \pm\infty$, $y \to \dfrac{3x}{-x^2}$.

                Thus as $x \to +\infty$, $y \to 0^-$; as $x \to -\infty$, $y \to 0^+$

y undefined $x = 0$ and $x = 3$ are vertical asymptotes.

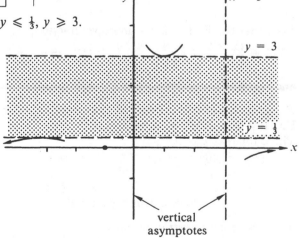

Max/min   When $y = \frac{1}{3}$, $x = -3$
max at $(-3, \frac{1}{3})$;
when $y = 3$, $x = 1$
min at $(1, 3)$
The sketch can then be completed.

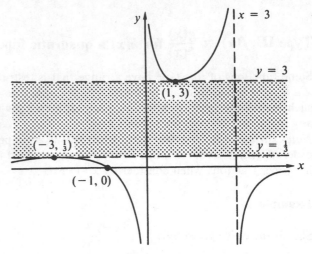

Note For some sketches it can also be useful to determine the sign of the function throughout its domain by constructing a table. For the last function, remembering that $f(x)$ can only change sign where the curve cuts the x-axis (i.e. $x = -1$) and at vertical asymptotes (i.e. $x = 0$ and $x = 3$), the table would be as shown on the right.

The reader should verify that these results agree with the sketch.

	$x < -1$	$-1 < x < 0$	$0 < x < 3$	$x > 3$
$3x + 3$	$-$ve	$+$ve	$+$ve	$+$ve
x	$-$ve	$-$ve	$+$ve	$+$ve
$3 - x$	$+$ve	$+$ve	$+$ve	$-$ve
$\dfrac{3x + 3}{x(3 - x)}$	$+$ve	$-$ve	$+$ve	$-$ve

Example 4

Sketch the curve given by $y = \dfrac{(x - 5)(x - 1)}{(x + 1)(x - 3)}$

Note that the R.H.S. is an improper fraction.
If $y = \dfrac{(x - 5)(x - 1)}{(x + 1)(x - 3)}$ ... [1]   then $y = \dfrac{x^2 - 6x + 5}{x^2 - 2x - 3}$ ... [2]

and $y = 1 + \dfrac{8 - 4x}{x^2 - 2x - 3}$ ... [3]

Range of values of y From [2], $(y - 1)x^2 + 2x(3 - y) - 3y - 5 = 0$
For $y \neq 1$, this is a quadratic in x.
Thus, for real x,

    $[2(3 - y)]^2 - 4(y - 1)(-3y - 5) \geqslant 0$

i.e. $y^2 - y + 1 \geqslant 0$

or $(y - \frac{1}{2})^2 + \frac{3}{4} \geqslant 0$

           which is true for all real y

For $y = 1$, $4x - 8 = 0$,
          $x = 2$
$\therefore$ y can equal 1.

Thus there is no restriction on y.

Note also that, for each value of y, there will exist two real distinct values of x, except where $y = 1$ for which there is one value for x, i.e. $x = 2$.

x-axis	Cuts *x*-axis at $(1, 0)$ and $(5, 0)$ (from [1])	
y-axis	Cuts *y*-axis at $(0, -1\frac{2}{3})$ (from [2])	
$x \to \pm\infty$	As $x \to \pm\infty$, $y \to 1 - \dfrac{4x}{x^2}$ (from [3])	

$\therefore$ as $x \to +\infty$ $y \to 1^-$

as $x \to -\infty$ $y \to 1^+$

(see note below)

y undefined	$x = -1$ and $x = 3$ are vertical asymptotes (from [1])
Max/min	Range of values suggest no max/min.

The sketch can then be completed:

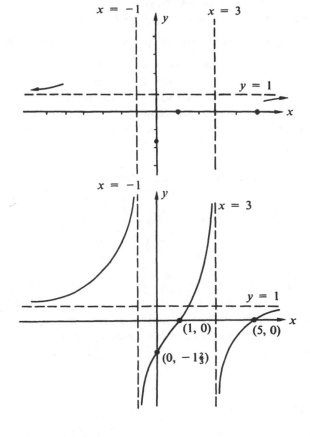

Note Alternatively, the horizontal asymptote could be determined by writing
$y = \dfrac{1 - 6/x + 5/x^2}{1 - 2/x - 3/x^2}$, then as $x \to \pm\infty$, $y \to 1$

Example 5

Sketch the curve given by $y = \dfrac{12}{x^2 + 2x - 3}$

If $y = \dfrac{12}{x^2 + 2x - 3}$ ... [1]   then $y = \dfrac{12}{(x + 3)(x - 1)}$ ... [2]

Range of values of y From [1], $x^2y + 2xy - 3y - 12 = 0$

For $y \neq 0$, this is a quadratic in x.   For $y = 0$, we obtain $-12 = 0$

Thus, for real x, $4y^2 - 4y(-3y - 12) \geq 0$ which is impossible.

i.e. $y(y + 3) \geq 0$ $\therefore$ $y \neq 0$.

	$y < -3$	$-3 < y < 0$	$y > 0$
y	$-$ve	$-$ve	$+$ve
$y + 3$	$-$ve	$+$ve	$+$ve
$y(y + 3)$	$+$ve	$-$ve	$+$ve

i.e. $y \leq -3, y \geq 0$

$\therefore$ the ranges of values y can take for real x are: $y \leq -3, y > 0$.

$y = -3$ will be a local maximum and for the remainder of the range there will be two distinct real values of x for each value of y.

x-axis	No x-axis intercept
y-axis	Cuts y-axis at $(0, -4)$
$x \to \pm\infty$	As $x \to \pm\infty$, $y \to \dfrac{12}{x^2}$ (from [1])
	$\therefore$ as $x \to +\infty$, $y \to 0^+$
	as $x \to -\infty$, $y \to 0^+$
y undefined	$x = -3$ and $x = 1$ are vertical asymptotes (from [2])
Max/min	When $y = -3$, $x = -1$. Max at $(-1, -3)$

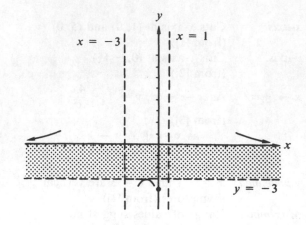

The sketch can then be completed.

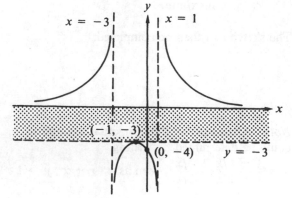

Note   Alternatively the method of section 11.4 page 288 could have been used for this

Example 6

Sketch the curve given by $y = \dfrac{2 - 3x}{x^2 + 3x + 3}$

In this case the denominator does not factorise.

Range of values of y   From $y = \dfrac{2 - 3x}{x^2 + 3x + 3}$, we have $x^2 y + 3x(y + 1) + 3y - 2 = 0$

For $y \neq 0$, this is a quadratic in x.
Thus, for real x, $[3(y + 1)]^2 - 4y(3y - 2) \geqslant 0$
i.e. $(3y + 1)(y - 9) \leqslant 0$

For $y = 0$, $3x = 2$
$x = \tfrac{2}{3}$
$\therefore$ y can equal zero.

	$y < -\tfrac{1}{3}$	$-\tfrac{1}{3} < y < 9$	$y > 9$
$3y + 1$	$-$ve	$+$ve	$+$ve
$y - 9$	$-$ve	$-$ve	$+$ve
$(3y + 1)(y - 9)$	$+$ve	$-$ve	$+$ve

i.e. $-\tfrac{1}{3} \leqslant y \leqslant 9$

$\therefore$ the range of values y can take for real x is $-\tfrac{1}{3} \leqslant y \leqslant 9$.
$y = -\tfrac{1}{3}$ will be a minimum and $y = 9$ a maximum. For every other value of y in the permitted range there will correspond two distinct values for x except $y = 0$ for which there is one value of x, i.e. $x = \tfrac{2}{3}$.

x-axis	Cuts *x*-axis at $(\tfrac{2}{3}, 0)$
y-axis	Cuts *y*-axis at $(0, \tfrac{2}{3})$

$x \to \pm\infty$ As $x \to \pm\infty$, $y \to \dfrac{-3x}{x^2}$

$\therefore$ as $x \to +\infty$, $y \to 0^-$
        as $x \to -\infty$, $y \to 0^+$

y undefined   No values for which
        $x^2 + 3x + 3 = 0$

Max/min   When $y = -\tfrac{1}{3}$, $x = 3$
        Min at $(3, -\tfrac{1}{3})$
        when $y = 9$, $x = -1\tfrac{2}{3}$
        Max at $(-1\tfrac{2}{3}, 9)$

The sketch can then be completed.

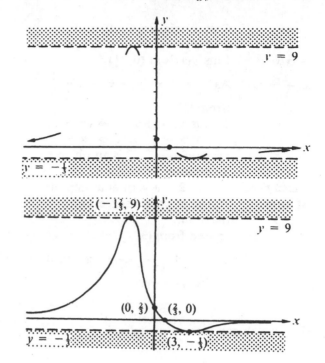

Example 7

Sketch the curve given by $y = \dfrac{x^2 + 4x + 3}{x + 2}$

Note that in this case the R.H.S. is an improper fraction.

If $y = \dfrac{x^2 + 4x + 3}{x + 2}$ ... [1]

then $y = \dfrac{(x + 3)(x + 1)}{x + 2}$ ... [2]

and $y = x + 2 - \dfrac{1}{x + 2}$ ... [3]

Range of values of y
From [1] $x^2 + x(4 - y) + 3 - 2y = 0$
Thus, for real x $(4 - y)^2 - 4(1)(3 - 2y) \geqslant 0$
giving $y^2 + 4 \geqslant 0$
which is true for all y.
Thus there is no restriction on y; for each
value of y, there exist two distinct values
for x.

x-axis	Cuts x-axis at $(-3, 0)$ and $(-1, 0)$
y-axis	Cuts y-axis at $(0, 1\frac{1}{2})$
$x \to \pm\infty$	As $x \to \pm\infty$, $y \to x + 2 - \dfrac{1}{x}$

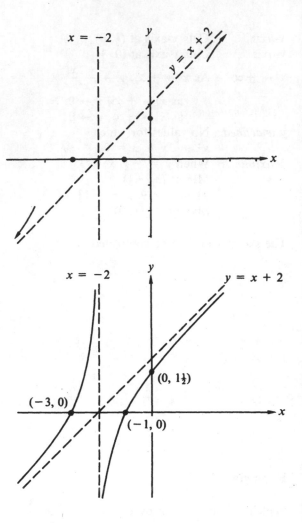

(from [3])

$$\therefore \quad \text{as } x \to +\infty, \; y \to (x + 2)^-$$
$$\text{as } x \to -\infty, \; y \to (x + 2)^+$$

hence $y = x + 2$ is an oblique asymptote

y undefined $x = -2$ is a vertical asymptote.

Max/min Range of values suggest no max/min.

Indeed from [3],

$$y' = 1 + \frac{1}{(x + 2)^2} \text{ so gradient}$$

is always $+$ve.

The sketch can then be completed:

Exercise 14A

1. Make sketch graphs of the following functions.

(a) $f(x) = \dfrac{4}{x - 2}$ (b) $f(x) = \dfrac{4}{2 - x}$ (c) $f(x) = \left|\dfrac{5}{x - 1}\right|$

2. Make sketch graphs of the following functions.

(a) $f(x) = \dfrac{x + 4}{x - 2}$ (b) $f(x) = \dfrac{3 - 2x}{x - 1}$ (c) $f(x) = \dfrac{x - 5}{x - 1}$

(d) $f(x) = \dfrac{2x + 5}{1 - x}$ (e) $f(x) = \left|\dfrac{3x}{2 - x}\right|$ (f) $f(x) = \left|\dfrac{x - 1}{3 - x}\right|$

3. Show that for real x the function $f(x) = \dfrac{3x - 1}{(x + 3)(x - 1)}$ can take all

real values. Sketch $y = \dfrac{3x - 1}{(x + 3)(x - 1)}$.

4. Find the range of values the function $f(x) = \dfrac{12x}{x^2 + 2x + 4}$ can take for

real x. Sketch $y = \dfrac{12x}{x^2 + 2x + 4}$.

5. Show that for x real $\quad 0 < \dfrac{4}{x^2 + 2x + 2} \leqslant 4$. Sketch $y = \dfrac{4}{x^2 + 2x + 2}$.

6. Show that for real x the function $f(x) = \dfrac{x + 2}{(x + 3)(x - 1)}$ can take all

real values. Sketch $y = \dfrac{x + 2}{(x + 3)(x - 1)}$.

7. Show that for real x the function $f(x) = \dfrac{(x + 1)(x - 6)}{(x + 3)(x - 2)}$ can take all

real values. Sketch $y = \dfrac{(x + 1)(x - 6)}{(x + 3)(x - 2)}$.

8. Show that for real x the function $f(x) = \dfrac{x^2 - x - 6}{x - 1}$ can take all real

values. Sketch $y = \dfrac{x^2 - x - 6}{x - 1}$.

For questions **9** to **20**, make sketch graphs of the given functions and in each case state the range of values the function can take for real x.

9. $f(x) = \dfrac{12}{x^2 - 2x - 3}$

10. $f(x) = \dfrac{8x}{(x + 1)^2}$

11. $f(x) = \dfrac{2x + 5}{x^2 + 5x + 4}$

12. $f(x) = \dfrac{4x - 5}{x^2 - 1}$

13. $f(x) = \dfrac{x}{(x + 5)(x - 1)}$

14. $f(x) = \dfrac{4(x + 1)}{x^2 + 2x + 2}$

15. $f(x) = \dfrac{3x - 2}{x^2 - 3x + 2}$

16. $f(x) = \dfrac{2x - 1}{(x - 1)^2}$

17. $f(x) = \dfrac{2x^2 - 5x}{x^2 - 1}$

18. $f(x) = \dfrac{(x - 3)(x + 1)}{(x - 1)^2}$

19. $f(x) = \dfrac{x^2 + 3}{x - 1}$

20. $f(x) = \left| \dfrac{4x - 3}{(x - 1)^2} \right|$

21. For real x, $f(x) = \dfrac{x^2 - k}{x - 2}$ can take any real value. Find the range of

values k can take and sketch $f(x)$ for $k = 9$.

22. Find the range of values k can take given that, for real x, $f(x) = \dfrac{x^2 + kx}{x + 2}$

can take any real value.

23. Find the range of values k can take given that, for real x, $f(x) = \dfrac{x^2 + 3x}{x + k}$

can take any real value.
Find the range of values $f(x)$ can take when $k = -1$ and x is real.
Sketch $f(x)$ for $k = -1$.

14.2 Inequalities

We saw in chapter 5 that, for a quadratic inequality, say $x^2 - 3x - 4 < 0$, the range of values that x can take can be found by sketching, by completing the square, or by considering the signs of the factors:

Sketching

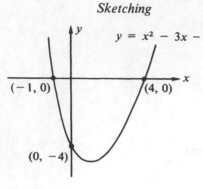

Completing the square

$$x^2 - 3x - 4 < 0$$
$$(x - \tfrac{3}{2})^2 - \tfrac{9}{4} - 4 < 0$$
$$(x - \tfrac{3}{2})^2 < \tfrac{25}{4}$$
$$-\tfrac{5}{2} < (x - \tfrac{3}{2}) < \tfrac{5}{2}$$
giving $\quad -1 < x < 4$

Signs of the factors

$$x^2 - 3x - 4 < 0$$
$$\therefore \quad (x - 4)(x + 1) < 0$$

	$x < -1$	$-1 < x < 4$	$x > 4$
$x - 4$	$-$ve	$-$ve	$+$ve
$x + 1$	$-$ve	$+$ve	$+$ve
$(x-4)(x+1)$	$+$ve	$-$ve	$+$ve

$$\therefore \quad -1 < x < 4$$

Thus for $\quad x^2 - 3x - 4 < 0$,
$$-1 < x < 4$$

Similar techniques can be used to find the range of possible values x can take in more complicated inequalities, as the following examples show.

Example 8

Find the range (or ranges) of values that x can take if $x - 2 < \dfrac{8}{x}$.

By sketching

In this method we sketch $y = x - 2$ and $y = \dfrac{8}{x}$ on the same graph.

Important points will be where these lines meet.

At points of intersection $x - 2 = \dfrac{8}{x}$ which,

provided $x \neq 0$, gives $x^2 - 2x - 8 = 0$
giving $x = -2$ or $x = 4$.

For $x - 2 < \dfrac{8}{x}$, we look for x values for

which the line $y = x - 2$ is 'lower' than the

curve $y = \dfrac{8}{x}$. (The relevant parts of the x-axis

are shown in heavy type on the sketch.)

Thus for $x - 2 < \dfrac{8}{x}$ we must have $x < -2$ or $0 < x < 4$.

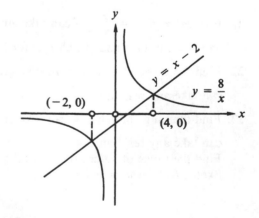

By calculation $x - 2 < \dfrac{8}{x}$

i.e. $x - 2 - \dfrac{8}{x} < 0$ or $\dfrac{(x - 4)(x + 2)}{x} < 0.$

Now the function $y = f(x)$ may change sign as the graph of $y = f(x)$ cuts the x-axis and at vertical asymptotes. Thus for $y = \dfrac{(x - 4)(x + 2)}{x}$, the critical values of x are 4, -2 and 0. We can then construct a table from which we can see that $\dfrac{(x - 4)(x + 2)}{x} < 0$ for $x < -2$ and for $0 < x < 4$.

	$x < -2$	$-2 < x < 0$	$0 < x < 4$	$x > 4$
$x - 4$	$-$ve	$-$ve	$-$ve	$+$ve
$x + 2$	$-$ve	$+$ve	$+$ve	$+$ve
x	$-$ve	$-$ve	$+$ve	$+$ve
y	$-$ve	$+$ve	$-$ve	$+$ve

Example 9

Find the solution set of the inequality $\dfrac{x + 4}{x + 1} < \dfrac{x - 2}{x - 4}.$

By sketching:

The graphical solution is obtained by sketching $y = \dfrac{x + 4}{x + 1}$ and $y = \dfrac{x - 2}{x - 4}$ on the same axes.

The sketch can be made using the methods developed earlier in this chapter. Notice that the two curves will intersect where

$x^2 - 16 = (x - 2)(x + 1),$

i.e. $x = 14.$

For $\dfrac{x + 4}{x + 1} < \dfrac{x - 2}{x - 4}$, we look for values of x for which the curve $y = \dfrac{x + 4}{x + 1}$ is 'lower' than the curve $y = \dfrac{x - 2}{x - 4}.$

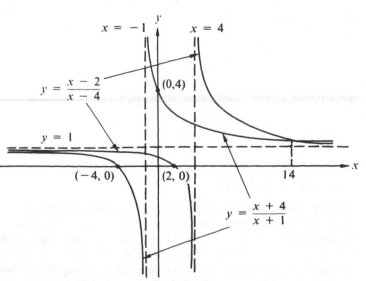

Thus for $\dfrac{x + 4}{x + 1} < \dfrac{x - 2}{x - 4}$ we must have

$x < -1$ or $4 < x < 14,$

i.e. the solution set is $\{x \in \mathbb{R}: x < -1 \text{ or } 4 < x < 14\}.$

By calculation: $\dfrac{x + 4}{x + 1} < \dfrac{x - 2}{x - 4}$ i.e. $\dfrac{x + 4}{x + 1} - \dfrac{x - 2}{x - 4} < 0$

or $\dfrac{x - 14}{(x + 1)(x - 4)} < 0$

	$x < -1$	$-1 < x < 4$	$4 < x < 14$	$x > 14$
$x - 14$	$-$ve	$-$ve	$-$ve	$+$ve
$x + 1$	$-$ve	$+$ve	$+$ve	$+$ve
$x - 4$	$-$ve	$-$ve	$+$ve	$+$ve
$\dfrac{x - 14}{(x + 1)(x - 4)}$	$-$ve	$+$ve	$-$ve	$+$ve

Thus for $\dfrac{x + 4}{x + 1} < \dfrac{x - 2}{x - 4}$, we must have $x < -1$ or $4 < x < 14$, or in set notation $\{x \in \mathbb{R}: x < -1 \text{ or } 4 < x < 14\}$.

Example 10

Find the range (or ranges) of values x can take if $\dfrac{x - 2}{x^2 - x + 1} > 0$.

Since $x^2 - x + 1 = (x - \frac{1}{2})^2 + \frac{3}{4}$, the denominator of $\dfrac{x - 2}{x^2 - x + 1}$

is always positive. Thus the sign of $\dfrac{x - 2}{x^2 - x + 1}$ depends on the sign of

the numerator, $x - 2$.

Now $\qquad x - 2 > 0$ if $x > 2$, hence

$\dfrac{x - 2}{x^2 - x + 1} > 0$ if $x > 2$.

This result, as the reader can verify, can also be obtained by sketching.

Modulus inequalities

Simple modulus inequalities have been solved on page 4.

For example, we know that $|3x + 1| > 8$

means that $\qquad\qquad\qquad 3x + 1 < -8 \quad$ or $\quad 3x + 1 > 8$

i.e. $\qquad\qquad\qquad\qquad x < -3 \quad$ or $\qquad\quad x > 2\frac{1}{3}$

In order to solve more complicated modulus inequalities, we need other techniques as shown in the following examples.

Example 11

Find the values of x such that $|2x - 3| > |x + 3|$.

By sketching:
In this method, we sketch $y = |2x - 3|$ and
$y = |x + 3|$ on the same axes.
Important points will be where the lines meet.
At points of intersection:

$$2x - 3 = x + 3 \quad \text{or} \quad 2x - 3 = -(x + 3)$$

i.e. $\qquad x = 6 \qquad\qquad$ i.e. $\quad x = 0$

For $|2x - 3| > |x + 3|$, we look for values
of x for which the line $y = |x + 3|$ is 'lower'
than the line $y = |2x - 3|$.
Thus for $|2x - 3| > |x + 3|$ we must have

$$x < 0 \text{ or } x > 6.$$

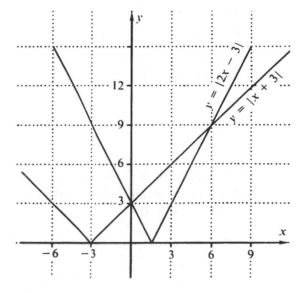

By calculation: $\qquad\qquad |2x - 3| > |x + 3|$

Since both sides of the inequality are positive (or zero) for all real values of
x, we can square both sides of the inequality.

i.e. $\quad (2x - 3)^2 > (x + 3)^2$
$\qquad 3x^2 - 18x > 0$
or $\quad 3x(x - 6) > 0$
giving $\; x < 0 \;$ or $\; x > 6.$
(See table.)

	$x < 0$	$0 < x < 6$	$x > 6$
x	$-$ve	$+$ve	$+$ve
$x - 6$	$-$ve	$-$ve	$+$ve
$x(x - 6)$	$+$ve	$-$ve	$+$ve

Thus for $|2x - 3| > |x + 3|$ we must have $x < 0 \quad$ or $\quad x > 6.$

Example 12

Find the range (or ranges) of values x can take for $x + 6 > |2x + 3|$.

For $x + 6 > |2x + 3|$, since the R.H.S. is necessarily positive (or zero), it
follows that the L.H.S. must be positive; hence $x > -6$.
For $x > -6$, both sides of the inequality are positive,
hence squaring gives

$\qquad (x + 6)^2 > (2x + 3)^2$
or $\qquad 0 > 3x^2 - 27$
i.e. $\qquad 9 > x^2$

Thus for $x + 6 > |2x + 3|$ we must have $-3 < x < 3$.

Exercise 14B

1. The graph shows the lines $y = 2x + 1$, $y = 3$ and $y = \dfrac{x + 4}{x - 2}$.

Use the graph to find the solution sets of the following inequalities.

(a) $\dfrac{x + 4}{x - 2} < 3$,

(b) $2x + 1 > \dfrac{x + 4}{x - 2}$.

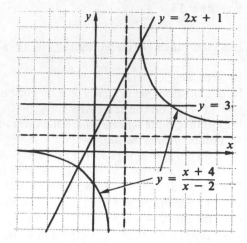

For each of the inequalities in questions **2** to **27**, find the range (or ranges) of values x can take for the inequality to be true.

2. $(2x - 3)^2 > 1$

3. $(x - 1)^2 < 7 - x$

4. $x + 2 > \dfrac{15}{x}$

5. $13 > 2x + \dfrac{15}{x}$

6. $\dfrac{7}{4 - x} > 2$

7. $x < \dfrac{8}{x - 2}$

8. $x^4 - 10x^2 + 9 > 0$

9. $4x^4 - 17x^2 + 4 < 0$

10. $\dfrac{3x + 2}{x - 2} > 1$

11. $\dfrac{3x - 10}{x - 4} < 2$

12. $\dfrac{3x + 1}{2x - 5} < 1$

13. $x > \dfrac{4 - x}{x - 1}$

14. $x < \dfrac{4x}{x + 1}$

15. $\dfrac{x^2 + 3}{x - 1} > 6$

16. $\dfrac{x + 1}{x^2 + x + 1} > 0$

17. $x - 2 < \dfrac{6 - x}{x}$

18. $\dfrac{(x - 6)(x + 1)}{x} > -4$

19. $x - 2 < \dfrac{6x - 9}{x + 2}$

20. $\dfrac{3 - x}{x - 8} < \dfrac{1}{x}$

21. $\dfrac{x - 2}{x + 1} > \dfrac{x - 6}{x - 2}$

22. $\dfrac{x + 1}{x - 1} < \dfrac{x + 3}{x + 2}$

23. $\dfrac{2x + 3}{x + 4} > \dfrac{x}{x - 2}$

24. $\dfrac{3x - 8}{2x - 1} > \dfrac{x - 4}{x + 1}$

25. $\dfrac{3x}{x - 8} < \dfrac{2x - 1}{x - 5}$

26. $\dfrac{x}{4x - 8} < \dfrac{1}{x}$

27. $\dfrac{x - 1}{x} > \dfrac{2}{3 - x}$

28. The graph shows the lines $y = |2x + 3|$, $y = 2x$, $y = |x + 3|$ and $y = 4$. Use the graph to find the solution sets of the following inequalities.

(a) $|x + 3| < 4$,

(b) $2x > |x + 3|$,

(c) $|2x + 3| > |x + 3|$.

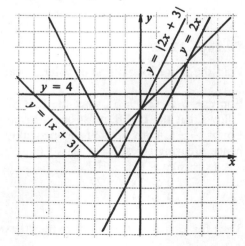

For each of the inequalities in questions **29** to **43**, find the range (or ranges) of values x can take for the inequality to be true.

29. $|2x - 3| > 5$ **30.** $|4x + 2| < x + 8$ **31.** $|2x + 1| > x + 5$

32. $|x| > |2x + 3|$ **33.** $|2x + 5| > |x + 1|$ **34.** $|2x - 3| > 4|x|$

35. $|x + 1| > |x - 3|$ **36.** $6 - x > |3x - 2|$ **37.** $3|x + 5| > |x + 3|$

38. $3|x - 1| < |x - 3|$ **39.** $2|x - 3| > |x|$ **40.** $|x| > 2|x + 1|$

41. $\left| \dfrac{x}{x - 3} \right| < 2$ **42.** $\left| \dfrac{x}{x - 3} \right| < x$ **43.** $\left| \dfrac{2x - 4}{x + 1} \right| < 4$

Exercise 14C   *Examination questions*

1. Sketch the curve $y = \dfrac{1}{1 + x}$.

Determine (i) the equation of the tangent to the curve at the point where $x = 0$;

           (ii) the equation of the other tangent to the curve which is parallel to the tangent in part (i).   (S.U.J.B.)

2. Sketch the curve $y = \dfrac{x}{x - 1}$. Find the ranges of values of x for which

$$\frac{x}{x - 1} > -1.$$ (Oxford)

3. Sketch the curve

$$y = \frac{x}{x - 2}$$

and write down the equation of its mirror image in the y-axis. (London)

4. State the equations of the asymptotes of the graph of the function

$$f: \quad x \rightarrow \frac{3x}{x - 2},$$

where $x \in \mathbb{R}$ and $x \neq 2$.

Sketch the graph of the function showing clearly the asymptotes.

        (London)

5. Prove that the function $y = \dfrac{9}{x + 2} - \dfrac{1}{x}$ has a maximum value at $x = 1$.

Find the value of x for which y has a minimum value.

SKETCH the graph of the function. (Graph paper is not required—a sketch showing the main relevant features will suffice.)

Prove that the x-coordinate of any point of inflexion on the graph will be a solution of the equation $(x + 2)^3 = 9x^3$. (You are NOT asked to solve this equation.)   (S.U.J.B.)

6. Show that, for real x,

$$-\frac{1}{4} \leqslant \frac{x}{x^2 + 4} \leqslant \frac{1}{4}.$$

Sketch the curve $y = \dfrac{x}{x^2 + 4}$, showing the coordinates of the turning points.

        (Cambridge)

7. Given that $y(x - 1) = x^2 + 3$, where x is real, show that y cannot take any value between -2 and 6.
   Find the equations of the asymptotes of the curve
   $$y = \frac{x^2 + 3}{x - 1}$$
   and sketch the curve, showing the coordinates of the turning points.
   (Cambridge)

8. In each of the following cases, determine the range of values of x for which
   (i) $1 - 2x > 0$,
   (ii) $(x + 1)(x - 2) > 0$,
   (iii) $\dfrac{(x + 1)(x - 2)}{1 - 2x} > 0$. (Cambridge)

9. Find the set of real values of x for which
   $|x - 2| > 2|x + 1|$. (London)

10. Find the set of values of x for which
    $|x - 1| - |2x + 1| > 0$. (London)

11. Given that $y = \dfrac{3x + k}{x^2 - 1}$, where x is real and k is a constant, show that y can take all real values if $|k| < 3$. (London)

12. Find the ranges of values of x which satisfy the inequalities
    (a) $x^2 - 3x < 0$,
    (b) $\dfrac{2}{x - 1} < x$. (A.E.B.)

13. Obtain the set of values of x for which
    $$\frac{1}{x - 2} > \frac{1}{x + 2},$$ (London)

14. Find, in each case, the set of values of x for which
    (i) $x(x - 2) > x + 4$,
    (ii) $x - 2 > \dfrac{x + 4}{x}$, (Cambridge)

15. Find the equation of the tangent to the curve $y = x(x - 3)$ at the point $P(2, -2)$. Using the same axes sketch the curve and the tangent at P. Using sketches, or otherwise, find the values of x for which
    (a) $x(x - 3) \leqslant x - 4$,
    (b) $x - 3 < \dfrac{x - 4}{x}$,
    (c) $\dfrac{1}{x - 3} < \dfrac{x}{x - 4}$. (London)

16. Given that $y(x^2 - 4) = 2x + 5$ where x is real, prove that y cannot take any value in the range $-1 < y < -\tfrac{1}{4}$.
    Sketch the graph of the curve with equation $y = \dfrac{2x + 5}{x^2 - 4}$, indicating clearly the asymptotes.
    Calculate the ranges of values of x for which $\left|\dfrac{2x + 5}{x^2 - 4}\right| \geqslant \tfrac{1}{4}$. (Cambridge)

15
Trigonometry II

15.1 Sum and product formulae

We saw in section 4.7, that

$$\sin (A + B) = \sin A \cos B + \cos A \sin B \qquad \ldots [1]$$
and
$$\sin (A - B) = \sin A \cos B - \cos A \sin B \qquad \ldots [2]$$
adding these we get
$$\sin (A + B) + \sin (A - B) = 2 \sin A \cos B \qquad \ldots [3]$$
and subtracting we get
$$\sin (A + B) - \sin (A - B) = 2 \cos A \sin B. \qquad \ldots [4]$$

If we write $C = A + B$ and $D = A - B$, then $A = \dfrac{C + D}{2}$ and $B = \dfrac{C - D}{2}$

and we then have
$$\sin C + \sin D = 2 \sin \frac{C + D}{2} \cos \frac{C - D}{2}$$

and
$$\sin C - \sin D = 2 \cos \frac{C + D}{2} \sin \frac{C - D}{2}$$

We can, in a similar way, use the expansions of $\cos (A \pm B)$,
i.e.
$$\cos (A + B) = \cos A \cos B - \sin A \sin B \qquad \ldots [5]$$
and
$$\cos (A - B) = \cos A \cos B + \sin A \sin B, \qquad \ldots [6]$$

to give
$$\cos C + \cos D = 2 \cos \frac{C + D}{2} \cos \frac{C - D}{2}$$

and
$$\cos C - \cos D = -2 \sin \frac{C + D}{2} \sin \frac{C - D}{2}$$

These four results should be memorised:

$$\sin C + \sin D = 2 \sin \frac{C + D}{2} \cos \frac{C - D}{2}$$

$$\sin C - \sin D = 2 \cos \frac{C + D}{2} \sin \frac{C - D}{2}$$

$$\cos C + \cos D = 2 \cos \frac{C + D}{2} \cos \frac{C - D}{2}$$

$$\cos C - \cos D = -2 \sin \frac{C + D}{2} \sin \frac{C - D}{2}$$

Note carefully the pattern which runs through these identities and also the negative sign which appears on the right-hand side of the last one.

Example 1

Prove that $\dfrac{\sin 5A - \sin 3A}{\cos 3A + \cos 5A} = \tan A$.

Applying the above results: $\dfrac{\sin 5A - \sin 3A}{\cos 3A + \cos 5A} = \dfrac{2 \cos \dfrac{5A + 3A}{2} \sin \dfrac{5A - 3A}{2}}{2 \cos \dfrac{3A + 5A}{2} \cos \dfrac{3A - 5A}{2}}$

$$= \frac{2 \cos 4A \sin A}{2 \cos 4A \cos (-A)}$$

$$= \frac{\sin A}{\cos A} \quad \text{since} \quad \cos (-A) = \cos A$$

$$\therefore \quad \frac{\sin 5A - \sin 3A}{\cos 3A + \cos 5A} = \tan A \quad \text{as required.}$$

Note   It is also useful to be able to use these four standard results the other way round, i.e. to express products as sums or differences.

From equations [3], [4], [5] and [6], we can write:

$$\begin{array}{l}
2 \sin A \cos B = \sin (A + B) + \sin (A - B) \\[4pt]
2 \cos A \sin B = \sin (A + B) - \sin (A - B) \\[4pt]
2 \cos A \cos B = \cos (A + B) + \cos (A - B) \\[4pt]
2 \sin A \sin B = \cos (A - B) - \cos (A + B)
\end{array}$$

Note very carefully the similarities, and also the differences, between these four relations and the previous ones.

Example 2

Prove $\dfrac{\sin 8\theta \cos \theta - \sin 6\theta \cos 3\theta}{\cos 2\theta \cos \theta - \sin 3\theta \sin 4\theta} = \tan 2\theta$.

$$\frac{\sin 8\theta \cos \theta - \sin 6\theta \cos 3\theta}{\cos 2\theta \cos \theta - \sin 3\theta \sin 4\theta} = \frac{\frac{1}{2}(\sin 9\theta + \sin 7\theta) - \frac{1}{2}(\sin 9\theta + \sin 3\theta)}{\frac{1}{2}(\cos 3\theta + \cos \theta) - \frac{1}{2}(\cos \theta - \cos 7\theta)}$$

$$= \frac{\sin 7\theta - \sin 3\theta}{\cos 3\theta + \cos 7\theta}$$

$$= \frac{2 \cos 5\theta \sin 2\theta}{2 \cos 5\theta \cos 2\theta}$$

$$= \frac{\sin 2\theta}{\cos 2\theta}$$

$$\therefore \quad \frac{\sin 8\theta \cos \theta - \sin 6\theta \cos 3\theta}{\cos 2\theta \cos \theta - \sin 3\theta \sin 4\theta} = \tan 2\theta$$

Example 3

Solve the equation $\sin x + \sin 5x = \sin 3x$ for $0° \leqslant x \leqslant 180°$.

$$\sin x + \sin 5x = \sin 3x$$

$$\therefore \qquad\qquad 2 \sin 3x \cos 2x = \sin 3x$$

hence $\qquad \sin 3x(2 \cos 2x - 1) = 0$

$\therefore \quad \sin 3x = 0$, which gives $\quad 3x = 0°, 180°, 360°, 540°, \ldots$

i.e. $\qquad\qquad\qquad\qquad\qquad x = 0°, 60°, 120°, 180°, \ldots$

or $\cos 2x = \frac{1}{2}$, which gives $\quad 2x = 60°, 300°, 420°, \ldots$

i.e. $\qquad\qquad\qquad\qquad\qquad x = 30°, 150°, 210°, \ldots$

Thus, solutions in the range $0° \leqslant x \leqslant 180°$ are $x = 0°, 30°, 60°, 120°, 150°, 180°$.

Exercise 15A

1. Express each of the following as the product of two trigonometrical functions.
   (a) $\sin 5\theta + \sin 3\theta$ (b) $\cos 5\theta + \cos \theta$ (c) $\sin 7\theta - \sin \theta$
   (d) $\cos 5\theta - \cos \theta$ (e) $\cos 7\theta + \cos 3\theta$ (f) $\sin \theta + \sin 4\theta$
   (g) $\cos \theta - \cos 3\theta$ (h) $\sin 4\theta - \sin 6\theta$

2. Express each of the following as the sum or difference of two trigonometrical functions
   (a) $2 \sin 4\theta \cos 2\theta$ (b) $2 \cos 4\theta \sin \theta$ (c) $-2 \sin 4\theta \sin 3\theta$
   (d) $2 \sin 5\theta \cos 2\theta$ (e) $2 \cos 5\theta \sin 2\theta$ (f) $2 \cos 4\theta \cos 3\theta$
   (g) $-2 \sin \dfrac{5\theta}{2} \sin \dfrac{3\theta}{2}$ (h) $2 \sin \theta \sin 6\theta$

3. Evaluate the following without using a calculator or mathematical tables
   (a) $\sin 75° + \sin 15°$ (b) $\sin 105° - \sin 15°$ (c) $\cos 105° - \cos 15°$
   (d) $\sin 37\frac{1}{2}° \cos 7\frac{1}{2}°$

4. Prove the following identities
   (a) $\dfrac{\sin \theta - \sin \alpha}{\cos \theta + \cos \alpha} = \tan \left(\dfrac{\theta - \alpha}{2} \right)$ (b) $\sin 3\theta + \sin \theta = 4 \sin \theta - 4 \sin^3 \theta$

   (c) $\cos 3\theta + \cos \theta = 4 \cos^3 \theta - 2 \cos \theta$ (d) $\cos 3\theta - \cos 7\theta = 4 \sin 5\theta \sin \theta \cos \theta$

   (e) $\dfrac{\sin x + \sin 5x}{\sin 2x + \sin 4x} = 2 \cos x - \sec x$ (f) $\dfrac{\cos 3x - \cos 5x}{2 \sin 2x \cos 2x} = 2 \sin x$

   (g) $\sin x + \sin 3x + \sin 5x + \sin 7x = 4 \sin 4x \cos 2x \cos x$

   (h) $\dfrac{\sin x + \sin 2x + \sin 3x}{\cos x + \cos 2x + \cos 3x} = \tan 2x$

5. Solve the following equations for $0 \leqslant x \leqslant 180°$
   (a) $\sin 3x + \sin x = 0$ (b) $\sin 5x - \sin x = 0$
   (c) $\sin 5x - \sin x = \cos 3x$ (d) $\cos 4x - \cos 2x + \sin 3x = 0$
   (e) $\cos 5x + \cos x = \sin 5x + \sin x$ (f) $\sin 3x - \sin 2x = \sin 6x + \sin x$

6. Solve the following equations for $-180° \leqslant x \leqslant 180°$
   (a) $\cos 4x + \cos 2x = 0$ (b) $\cos 5x - \cos x = 0$
   (c) $\sin 5x + \sin x = \sin 3x$ (d) $\sin 5x - \sin x + \sqrt{3} \cos 3x = 0$

15.2 $a \cos x \pm b \sin x$

The expression $a \cos x \pm b \sin x$ can be expressed in the form $R \cos (x \mp \alpha)$, where α is an acute angle.

For, $a \cos x + b \sin x = \sqrt{(a^2 + b^2)} \left[\dfrac{a}{\sqrt{(a^2 + b^2)}} \cos x + \dfrac{b}{\sqrt{(a^2 + b^2)}} \sin x \right]$

Now if $\cos \alpha = \dfrac{a}{\sqrt{(a^2 + b^2)}}$ then $\sin \alpha = \dfrac{b}{\sqrt{(a^2 + b^2)}}$:

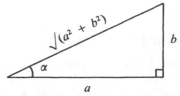

thus $a \cos x + b \sin x = \sqrt{(a^2 + b^2)}[\cos x \cos \alpha + \sin x \sin \alpha]$

i.e. $a \cos x + b \sin x = \sqrt{(a^2 + b^2)}\cos (x - \alpha)$ as required.

Also $a \cos x - b \sin x = \sqrt{(a^2 + b^2)}\left[\dfrac{a}{\sqrt{(a^2 + b^2)}} \cos x - \dfrac{b}{\sqrt{(a^2 + b^2)}} \sin x\right]$

$\qquad\qquad\qquad\qquad = \sqrt{(a^2 + b^2)}[\cos x \cos \alpha - \sin x \sin \alpha]$

$\qquad\qquad\qquad\qquad = \sqrt{(a^2 + b^2)}\cos (x + \alpha)$ where $\tan \alpha = \dfrac{b}{a}$.

Example 4

Express (a) $4 \cos x - 5 \sin x$ in the form $R \cos (x + \alpha)$,
(b) $2 \sin x + 5 \cos x$ in the form $R \sin (x + \alpha)$.

(a) $4 \cos x - 5 \sin x$

Now $\sqrt{(4^2 + 5^2)} = \sqrt{41}$, and so we write

$\qquad 4 \cos x - 5 \sin x = \sqrt{41}\left[\dfrac{4}{\sqrt{41}} \cos x - \dfrac{5}{\sqrt{41}} \sin x\right]$

$\qquad\qquad\qquad\qquad = \sqrt{41}(\cos x \cos \alpha - \sin x \sin \alpha)$

$\qquad\qquad\qquad\qquad\qquad$ where α is given by:

$\qquad\qquad\qquad\qquad = \sqrt{41} \cos (x + \alpha)$ where $\tan \alpha = \tfrac{5}{4}$

thus $4 \cos x - 5 \sin x = \sqrt{41} \cos (x + 51\cdot34°)$

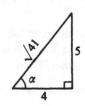

(b) $2 \sin x + 5 \cos x$

Now $\sqrt{(2^2 + 5^2)} = \sqrt{29}$, and so we write

$\qquad 2 \sin x + 5 \cos x = \sqrt{29}\left[\dfrac{2}{\sqrt{29}} \sin x + \dfrac{5}{\sqrt{29}} \cos x\right]$

$\qquad\qquad\qquad\qquad = \sqrt{29}(\sin x \cos \alpha + \cos x \sin \alpha)$

$\qquad\qquad\qquad\qquad\qquad$ where α is given by:

$\qquad\qquad\qquad\qquad = \sqrt{29} \sin (x + \alpha)$ where $\tan \alpha = \tfrac{5}{2}$

thus $2 \sin x + 5 \cos x = \sqrt{29} \sin (x + 68\cdot20°)$

Example 5

Find the maximum value of $24 \sin \theta - 7 \cos \theta$ and the smallest positive value of θ that gives this maximum value.

Now $\sqrt{(24^2 + 7^2)} = 25$ and so we write

$\qquad 24 \sin \theta - 7 \cos \theta = 25\left[\dfrac{24}{25} \sin \theta - \dfrac{7}{25} \cos \theta\right]$

$\qquad\qquad\qquad\qquad = 25(\sin \theta \cos \alpha - \cos \theta \sin \alpha)$

$\qquad\qquad\qquad\qquad\qquad$ where α is given by:

$\qquad\qquad\qquad\qquad = 25 \sin (\theta - 16\cdot26°)$

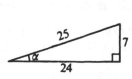

Hence the maximum value of $24 \sin \theta - 7 \cos \theta$ is 25 and this occurs when $(\theta - 16\cdot26°) = 90°$,   i.e.   when $\theta = 106\cdot26°$

Example 6

Solve $5 \cos x - 2 \sin x = 2$ for $-180° \leqslant x \leqslant 180°$.

Now $\qquad\qquad \sqrt{(5^2 + 2^2)} = \sqrt{29}$ and so we rearrange the equation to:

$$\frac{5}{\sqrt{29}} \cos x - \frac{2}{\sqrt{29}} \sin x = \frac{2}{\sqrt{29}}$$

$\therefore \qquad \cos x \cos \alpha - \sin x \sin \alpha = \dfrac{2}{\sqrt{29}} \quad$ where α is given by:

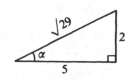

i.e. $\qquad \cos (x + 21\!\cdot\!80°) = \dfrac{2}{\sqrt{29}}$

Now $\qquad \cos^{-1}\left(\dfrac{2}{\sqrt{29}}\right) = 68\!\cdot\!20° \quad$ and a positive cosine gives:

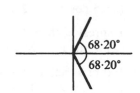

Thus $\qquad x + 21\!\cdot\!80° = 68\!\cdot\!20° \quad$ or $\quad x + 21\!\cdot\!80° = -68\!\cdot\!20°$

i.e. $\qquad\qquad\qquad x = 46\!\cdot\!40° \quad$ or $\quad x = -90°.$

The *t*-substitution

Equations of the type encountered in Example 6 (i.e. $a \cos x + b \sin x = c$) can also be solved by using the substitution $t = \tan \dfrac{x}{2}$:

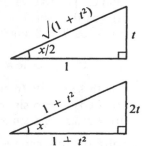

Then since $\quad \tan 2A = \dfrac{2 \tan A}{1 - \tan^2 A},$

it follows that $\tan x = \dfrac{2t}{1 - t^2} \quad$ i.e.

Thus, when $t = \tan \dfrac{x}{2},$ $\sin x = \dfrac{2t}{1 + t^2}$ and $\cos x = \dfrac{1 - t^2}{1 + t^2}.$

Example 7

Solve the equation $5 \cos x - 2 \sin x = 2$ for $-180° \leqslant x \leqslant 180°$ using the substitution $t = \tan \dfrac{x}{2}.$

Substituting $t = \tan \dfrac{x}{2},$ i.e. $\cos x = \dfrac{1 - t^2}{1 + t^2}$ and $\sin x = \dfrac{2t}{1 + t^2}$

$$5\left(\frac{1 - t^2}{1 + t^2}\right) - 2\left(\frac{2t}{1 + t^2}\right) = 2$$

$$5 - 5t^2 - 4t = 2 + 2t^2$$

i.e. $\qquad 7t^2 + 4t - 3 = 0$

giving $\qquad t = -1 \quad$ or $\quad t = \tfrac{3}{7}$

hence $\quad \tan \dfrac{x}{2} = -1$

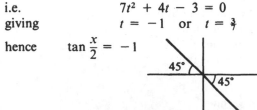

$\qquad\qquad\qquad$ or $\quad \tan \dfrac{x}{2} = \tfrac{3}{7}$

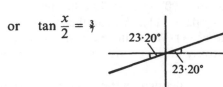

i.e. $\qquad \dfrac{x}{2} = -45°, 135°, \ldots \qquad\qquad$ i.e. $\qquad \dfrac{x}{2} = -156\!\cdot\!80°, 23\!\cdot\!20°, \ldots$

$\qquad\qquad x = -90°, 270°, \ldots \qquad\qquad\qquad\qquad x = -313\!\cdot\!60°, 46\!\cdot\!40°, \ldots$

The solutions in the range $-180° \leqslant x \leqslant 180°$ are $x = -90°, 46\!\cdot\!40°$

Exercise 15B

1. Express each of the following in the form $R \cos (x + \alpha)$ with $0° \leqslant \alpha \leqslant 90°$.
   (a) $3 \cos x - 4 \sin x$ (b) $2 \cos x - 5 \sin x$ (c) $3 \cos x - 2 \sin x$

2. Express each of the following in the form $R \cos (x - \alpha)$ with $0° \leqslant \alpha \leqslant 90°$.
   (a) $5 \cos x + 12 \sin x$ (b) $\cos x + 2 \sin x$ (c) $3 \cos x + 2 \sin x$

3. Express each of the following in the form $R \sin (x + \alpha)$ with $0° \leqslant \alpha \leqslant 90°$.
   (a) $7 \sin x + 24 \cos x$ (b) $3 \sin x + \cos x$ (c) $5 \sin x + 3 \cos x$

4. Express each of the following in the form $R \sin (x - \alpha)$ with $0° \leqslant \alpha \leqslant 90°$.
   (a) $3 \sin x - 4 \cos x$ (b) $3 \sin x - 7 \cos x$ (c) $5 \sin x - 7 \cos x$

5. Find the maximum value of each of the following expressions and the smallest positive value of θ that gives this maximum value.
   (a) $\cos \theta + \sin \theta$ (b) $3 \cos \theta + 4 \sin \theta$ (c) $8 \sin \theta + 15 \cos \theta$
   (d) $4 \cos \theta - 3 \sin \theta$ (e) $3 \sin \theta - 2 \cos \theta$ (f) $5 \cos \theta + 2 \sin \theta$

6. If $t = \tan (\tfrac{1}{2}x)$ show that $\tan x = \dfrac{2t}{1 - t^2}$, $\sin x = \dfrac{2t}{1 + t^2}$, $\cos x = \dfrac{1 - t^2}{1 + t^2}$.

   Use these results to solve
   (a) $7 \cos x + 6 \sin x = 2$ for $0 \leqslant x \leqslant 360°$
   (b) $9 \cos x - 8 \sin x = 12$ for $-360° \leqslant x \leqslant 360°$

7. Solve the following equations for $0° \leqslant x \leqslant 360°$.
   (a) $8 \cos x - 15 \sin x = 5$ (b) $8 \cos x - 15 \sin x = 10$
   (c) $2 \cos x - \sin x = 1$ (d) $5 \cos x + 2 \sin x = 3$
   (e) $3 \sin x - 5 \cos x = -4$ (f) $2 \cos 2x - 4 \sin x \cos x = \sqrt{6}$

8. Solve the following equations for $-180° \leqslant \theta \leqslant 180°$.
   (a) $\sin \theta + 2 \cos \theta = 0\cdot5$ (b) $2 \cos \theta + 3 \sin \theta = 1$
   (c) $\cos \theta - 7 \sin \theta = -2$ (d) $5 \sin \theta - 3 \cos \theta = 2$
   (e) $7 \sin \theta - 9 \cos \theta = 5$ (f) $3 + 2 \sin 2\theta = 2 \sin \theta + 3 \cos^2 \theta$

15.3 Inverse trigonometric functions

For the domain $x \in \mathbb{R}$, the functions $f(x) = \sin x$, $g(x) = \cos x$ and $h(x) = \tan x$ are many-to-one and therefore have no inverse functions. However, if we restrict the sine function to the domain $-90° \leqslant x \leqslant 90°$, the cosine function to the domain $0° \leqslant x \leqslant 180°$ and the tangent function to the domain $-90° < x < 90°$, the functions are one-to-one for these domains (see page 111). We can then consider the inverse functions $f^{-1}(x)$, $g^{-1}(x)$ and $h^{-1}(x)$. We write these functions as $\sin^{-1} x$, $\cos^{-1} x$ and $\tan^{-1} x$ (or alternatively arcsin x, arccos x and arctan x).

Thus, $\sin^{-1} x$ is defined as the angle θ such that $-90° \leqslant \theta \leqslant 90°$
   (or in radians, $-\pi/2 \leqslant \theta \leqslant \pi/2$) and $\sin \theta = x$,
   $\cos^{-1} x$ is defined as the angle θ such that $0° \leqslant \theta \leqslant 180°$
   (or in radians, $0 \leqslant \theta \leqslant \pi$) and $\cos \theta = x$,
   $\tan^{-1} x$ is defined as the angle θ such that $-90° < \theta < 90°$
   (or in radians, $-\pi/2 < \theta < \pi/2$) and $\tan \theta = x$.

Notes   1. $\sin^{-1} x$ must not be confused with $1/\sin x$ which is $(\sin x)^{-1}$ or
            cosec x.
       2. When solving the equation $\sin \theta = x$, we can obtain an infinite
          number of solutions. The solution that lies in the range of $\sin^{-1} x$,
          i.e. $-90° \leqslant \theta \leqslant 90°$, is sometimes referred to as the **principal**
          value of θ. Similarly for $\cos \alpha = x$, the principal value of α lies in
          the range $0° \leqslant \alpha \leqslant 180°$ and for $\tan \beta = x$ the principal value of
          β lies in the range $-90° < \beta < 90°$.

Example 8

Evaluate the following, giving your answers in radians and leaving π in your
answers.

(a) $\sin^{-1} \frac{1}{2}$ (b) $\cos^{-1}(-1)$ (c) $\tan^{-1} 0$ (d) $\tan^{-1}(-1/\sqrt{3})$

(a) We require the value of x in radians such that $\sin x = \frac{1}{2}$ and
    $-\pi/2 \leqslant x \leqslant \pi/2$; therefore $x = 30°$ or $\pi/6$ rad.
(b) We require the value of x in radians such that $\cos x = -1$ and
    $0 \leqslant x \leqslant \pi$; therefore $x = 180°$ or π rad.
(c) We require the value of x in radians such that $\tan x = 0$ and
    $-\pi/2 < x < \pi/2$; therefore $x = 0°$ or 0 rad.
(d) We require the value of x in radians such that $\tan x = -1/\sqrt{3}$ and
    $-\pi/2 < x < \pi/2$; therefore $x = -30°$ or $-\pi/6$ rad.

Graphs of the inverse functions

From chapter 1, page 44, we know that the graph of $f^{-1}(x)$ can be
obtained by reflecting the graph of $f(x)$ in the line $y = x$. The graphs of
$y = \sin^{-1} x$, $y = \cos^{-1} x$ and $y = \tan^{-1} x$ can be found in this way.

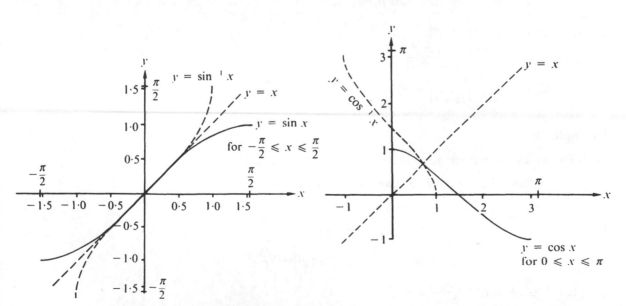

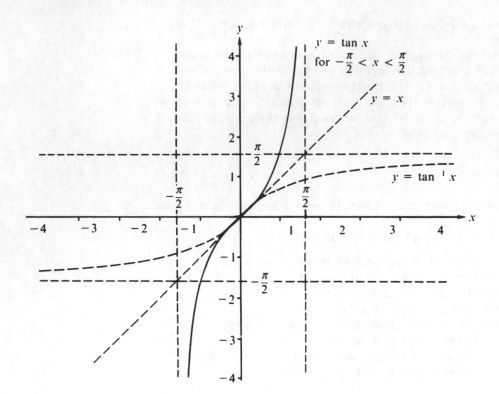

Example 9

Without the use of mathematical tables or a calculator, find $\tan \theta$ if
$\theta = \tan^{-1} \frac{5}{12} + \tan^{-1} \frac{7}{24}$.

Let $\qquad \alpha = \tan^{-1} \frac{5}{12}$ i.e. $\tan \alpha = \frac{5}{12}$
and $\qquad \beta = \tan^{-1} \frac{7}{24}$ i.e. $\tan \beta = \frac{7}{24}$
Thus $\qquad \theta = \alpha + \beta$
and so $\tan \theta = \tan (\alpha + \beta)$

$$= \frac{\tan \alpha + \tan \beta}{1 - \tan \alpha \tan \beta}$$

$\therefore \qquad \tan \theta = \dfrac{\frac{5}{12} + \frac{7}{24}}{1 - (\frac{5}{12})(\frac{7}{24})} = \frac{204}{253}.$

Example 10

Without the use of mathematical tables, or a calculator, evaluate
$\tan^{-1} \frac{\sqrt{3}}{2} + \tan^{-1} \frac{\sqrt{3}}{5}$, leaving your answer in terms of π.

Let $\qquad\qquad\qquad \alpha = \tan^{-1} \frac{\sqrt{3}}{2}$ i.e. $\tan \alpha = \frac{\sqrt{3}}{2}$

and $\qquad\qquad\qquad \beta = \tan^{-1} \frac{\sqrt{3}}{5}$ i.e. $\tan \beta = \frac{\sqrt{3}}{5}$

$\therefore \quad \tan^{-1} \frac{\sqrt{3}}{2} + \tan^{-1} \frac{\sqrt{3}}{5} = \alpha + \beta$

Note also that α and β must each be between 0 and $\pi/4$ because they are inverse tangents of positive numbers less than one. Thus we expect $(\alpha + \beta)$ to be between 0 and $\pi/2$.

$$\text{Now} \qquad \tan(\alpha + \beta) = \frac{\tan \alpha + \tan \beta}{1 - \tan \alpha \tan \beta}$$

$$= \frac{\sqrt{3}/2 + \sqrt{3}/5}{1 - (\sqrt{3}/2)(\sqrt{3}/5)}$$

$$= \sqrt{3}$$

$$\therefore \qquad \alpha + \beta = \frac{\pi}{3}$$

$$\therefore \quad \tan^{-1} \frac{\sqrt{3}}{2} + \tan^{-1} \frac{\sqrt{3}}{5} = \frac{\pi}{3}.$$

Example 11

Find a positive value of x that satisfies the equation

$\tan^{-1} 3x + \tan^{-1} x = \frac{\pi}{4}$, giving your answer correct to 3 decimal places.

Let $\qquad\qquad \alpha = \tan^{-1} 3x$ i.c. $\tan \alpha = 3x$

and $\qquad\qquad \beta = \tan^{-1} x$ i.e. $\tan \beta = x$

Thus $\qquad \alpha + \beta = \frac{\pi}{4}$

$$\therefore \qquad \tan(\alpha + \beta) = \tan \frac{\pi}{4} = 1 \qquad \ldots [1]$$

$$\frac{\tan \alpha + \tan \beta}{1 - \tan \alpha \tan \beta} = 1$$

$$\frac{3x + x}{1 - (3x)(x)} = 1$$

i.e. $\quad 3x^2 + 4x - 1 = 0$

Solving this quadratic by the formula and taking the positive root gives $x = 0\cdot215$.

Note: The negative solution of the equation $3x^2 + 4x - 1 = 0$, i.e. $x = -1\cdot55$, does not satisfy the equation $\tan^{-1} 3x + \tan^{-1} x = \pi/4$, as can be verified using a calculator. This 'other solution' arises because, for $\alpha + \beta = \pi/4$, we have said $\tan(\alpha + \beta) = \tan \pi/4 = 1$, see [1] above. However $\tan(\alpha + \beta)$ equals 1 for values of $(\alpha + \beta)$ other than the $\pi/4$ value we require it to be. Hence the 'solution' $-1\cdot55$ arises but must be discarded as it does not satisfy the orginal equation we are asked to solve. When solving equations of this type it is advisable to check that the solutions obtained do satisfy the equation.

Example 12

Solve the equation $\cos^{-1} x + \cos^{-1}(x\sqrt{3}) = \dfrac{\pi}{2}$.

Let $\alpha = \cos^{-1} x$ i.e. $\cos \alpha = x$ and $\beta = \cos^{-1}(x\sqrt{3})$ i.e. $\cos \beta = x\sqrt{3}$

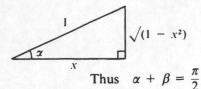
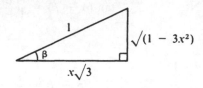

Thus $\alpha + \beta = \dfrac{\pi}{2}$

Taking the cosine of both sides of this equation, (we choose the cosine as $\cos \pi/2 = 0$)

$$\text{gives} \quad \cos(\alpha + \beta) = \cos\frac{\pi}{2} = 0$$

i.e.
$$\cos \alpha \cos \beta - \sin \alpha \sin \beta = 0$$
$$x \times x\sqrt{3} - \sqrt{[(1 - x^2)(1 - 3x^2)]} = 0$$
$$x^2\sqrt{3} = \sqrt{(1 - 4x^2 + 3x^4)}$$

Squaring both sides of this equation:
$$3x^4 = 1 - 4x^2 + 3x^4$$
$$\therefore \quad 4x^2 = 1 \quad \text{giving} \quad x = \pm\tfrac{1}{2}.$$

By substitution in the original equation it can be shown that $x = -\tfrac{1}{2}$ is not a possible solution for $\cos^{-1} x$ having a range $0 \leqslant x \leqslant \pi$.
Thus the only solution of the equation is $x = \tfrac{1}{2}$.

Exercise 15C

1. Evaluate the following giving your answers in degrees (do not use a calculator).

(a) $\sin^{-1} 0$ (b) $\sin^{-1}\left(\dfrac{\sqrt{3}}{2}\right)$ (c) $\cos^{-1}\left(\dfrac{\sqrt{3}}{2}\right)$ (d) $\sin^{-1}\left(-\dfrac{1}{2}\right)$

(e) $\cos^{-1}\left(-\dfrac{1}{2}\right)$ (f) $\sin^{-1}\left(-\dfrac{\sqrt{3}}{2}\right)$ (g) $\tan^{-1}(\sqrt{3})$ (h) $\tan^{-1}(-\sqrt{3})$

2. Evaluate the following giving your answers in radians (leave π in your answers and do not use a calculator).

(a) $\sin^{-1} 1$ (b) $\sin^{-1}(-1)$ (c) $\cos^{-1}(\tfrac{1}{2})$ (d) $\cos^{-1}\left(-\dfrac{\sqrt{3}}{2}\right)$

(e) $\sin^{-1}\left(\dfrac{1}{\sqrt{2}}\right)$ (f) $\tan^{-1}\left(\dfrac{1}{\sqrt{3}}\right)$ (g) $\sin^{-1}\left(-\dfrac{1}{\sqrt{2}}\right)$ (h) $\cos^{-1}\left(-\dfrac{1}{\sqrt{2}}\right)$

3. (a) Draw x- and y-axes using a scale of 2 cm to 1 unit and such that $-1 \leqslant x \leqslant 4$ and $-1 \leqslant y \leqslant 4$.

(b) Copy and complete the following table of values for $y = \cos x$ with $0 \leqslant x \leqslant \pi$. (Use a calculator to obtain the values of y.)

x rad	0	0·25	0·5	0·75	1	1·25	1·5	1·75	2	2·25	2·5	2·75	3	3·14 (i.e. π)
y (correct to 2 d.p's)	1	0·97	0·88									−0·92	−0·99	−1

Hence draw the graph of $y = \cos x$ for $0 \leqslant x \leqslant \pi$.

(c) By plotting (1, 0), (0·97, 0·25), (0·88, 0·5) etc. draw the graph of $y = \cos^{-1} x$ on the same pair of axes used for (b).

4. Without using a calculator find
   (a) $\sin \theta$ if $\theta = \sin^{-1}\left(\frac{3}{5}\right) + \sin^{-1}\left(\frac{5}{13}\right)$
   (b) $\cos \theta$ if $\theta = \cos^{-1}\left(\frac{5}{13}\right) - \cos^{-1}\left(\frac{8}{17}\right)$
   (c) $\tan \theta$ if $\theta = \tan^{-1}\left(\frac{3}{4}\right) + \tan^{-1}\left(\frac{7}{24}\right)$
   (d) $\sin \theta$ if $\theta = \sin^{-1}\left(\frac{4}{5}\right) - \cos^{-1}\left(\frac{8}{17}\right)$
   (e) $\cos \theta$ if $\theta = \cos^{-1}\left(\frac{24}{25}\right) - \sin^{-1}\left(\frac{15}{17}\right)$

5. Evaluate the following. (Give your answers in terms of π and do not use a calculator).
   (a) $\sin^{-1}\left(\frac{1}{2}\right) + \cos^{-1}\left(\frac{1}{2}\right)$
   (b) $2 \tan^{-1}\left(\frac{1}{3}\right) + \tan^{-1}\left(\frac{1}{7}\right)$

6. Prove that
   $$\sin\left(2 \sin^{-1} x + \cos^{-1} x\right) = \sqrt{(1 - x^2)}$$

7. Without the use of a calculator show that
   $$2 \sin^{-1}\left(\frac{3}{5}\right) - \cos^{-1}\left(\frac{35}{37}\right) = \sin^{-1}\left(\frac{756}{925}\right)$$

8. Find a positive value of x that satisfies the equation
   $$\tan^{-1}(3x) + \tan^{-1}(2x) = \frac{\pi}{4}$$

9. Find, correct to three decimal places, a positive value of x that satisfies the equation
   $$\tan^{-1} x + \tan^{-1}(2x) = \tan^{-1} 2$$

10. Solve the equation
    $$\cos^{-1} x + \cos^{-1}(x\sqrt{8}) = \frac{\pi}{2}$$

11. Solve the equation
    $$2 \sin^{-1}(x\sqrt{6}) + \sin^{-1}(4x) = \frac{\pi}{2}$$

12. Solve the equation
    $$2 \sin^{-1}\left(\frac{x}{2}\right) + \sin^{-1}(x\sqrt{2}) = \frac{\pi}{2}$$

15.4 General solutions

In all the trigonometric equations encountered so far in this book, the required solutions have been restricted to a given interval, say $-\pi$ to π or 0 to 2π, etc. Without such a restriction, there would be an infinite number of solutions. Nevertheless, it is useful to be able to give an expression for the general solution in terms of some letter, usually n. Then, by allowing n to take the integer values (i.e. $n \in \mathbb{Z}$), any particular solution of the equation can be obtained. We now consider three basic types of equation.

Type 1 $\cos \theta = \cos \alpha$

Example 13

Find the general solution of the equation $\cos \theta = \cos \pi/6$.

Now $\cos \pi/6 = \sqrt{3}/2$, and so we require all values of θ for which $\cos \theta = \sqrt{3}/2$.
We can show two solutions in the range $-\pi \leqslant \theta \leqslant \pi$ in a diagram:

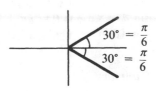

Other solutions will occur as we add or subtract multiples of 2π to these two solutions.
Thus the general solution of $\cos \theta = \cos \pi/6$ is $\theta = 2n\pi \pm \pi/6$ for $n \in \mathbb{Z}$.

Extending this idea, we can say that the general solution of the equation $\cos \theta = \cos \alpha$ is $\theta = 2n\pi \pm \alpha$ for $n \in \mathbb{Z}$ where α is in radians (or $360n° \pm \alpha$ for α in degrees).

Type 2   tan θ = tan α

Example 14

Find the general solution of the equation $\tan \theta = \tan \pi/3$.

Now $\tan \pi/3 = \sqrt{3}$ and so we require all values of θ for which $\tan \theta = \sqrt{3}$.
We can show two solutions in the range $-\pi \leqslant \theta \leqslant \pi$ in a diagram:
Other solutions will occur as we add or subtract multiples of 2π to
these two solutions.
Thus the general solution of $\tan \theta = \tan \pi/3$ is $\theta = n\pi + \pi/3$ for $n \in \mathbb{Z}$.

Again we can extend this idea to say that the general solution of the
equation $\tan \theta = \tan \alpha$ is $\theta = n\pi + \alpha$ for $n \in \mathbb{Z}$ where α is in radians (or
$180n° + \alpha$ for α in degrees).

Type 3   sin θ = sin α

Example 15

Find the general solution of the equation $\sin \theta = \sin \pi/4$.

Now $\sin \pi/4 = 1/\sqrt{2}$ and so we require all values of θ for which $\sin \theta = 1/\sqrt{2}$.
We can show two solutions in the range $-\pi \leqslant \theta \leqslant \pi$ in a diagram:
Other solutions will occur as we add or subtract multiples of 2π to
these two solutions.
Thus the general solution of $\sin \theta = \sin \pi/4$ is $\theta = 2n\pi + \pi/4, 2n\pi + \pi - \pi/4$.

$$\text{i.e.}\quad \theta = \begin{cases} 2n\pi + \pi/4 \\ (2n + 1)\pi - \pi/4 \end{cases}$$

$$\text{Alternatively we could write } \theta = \begin{cases} n\pi + \pi/4 & \text{for even } n \\ n\pi - \pi/4 & \text{for odd } n \end{cases}$$

$$\text{i.e.}\quad \theta = n\pi + (-1)^n \pi/4$$

Again we can extend this idea to say that the general solution of the equation
$\sin \theta = \sin \alpha$ is $\theta = n\pi + (-1)^n \alpha$ for $n \in \mathbb{Z}$ where α is in radians
(or $180n° + (-1)^n \alpha$ for α in degrees).

Summary

> For $\cos \theta = \cos \alpha$, the general solution is $\theta = 2n\pi \pm \alpha$ for $n \in \mathbb{Z}$
>
> For $\tan \theta = \tan \alpha$, the general solution is $\theta = n\pi + \alpha$ for $n \in \mathbb{Z}$
>
> For $\sin \theta = \sin \alpha$, the general solution is $\theta = n\pi + (-1)^n \alpha$ for $n \in \mathbb{Z}$
>
> $$\text{i.e.}\quad \theta = \begin{cases} 2n\pi + \alpha \\ (2n + 1)\pi - \alpha \end{cases}$$

To determine the general solution to a trigonometrical equation, it is best to use the first principles method shown in the previous examples, i.e. determine the general solution from a diagram showing the solutions in the range $-\pi$ to π. However, for some types of equation the ability to quote and use the above standard results can lead to a very concise method of solution. The 'alternative solutions' given for examples 18 to 20 demonstrate this technique.

Example 16

Find the general solution of the equation $\sin 2\theta = -\frac{1}{2}$.

For $\sin \theta = -\frac{1}{2}$, we have two solutions in the range $-\pi \leqslant \theta \leqslant \pi$ as shown:

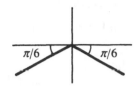

Other solutions will occur as we add or subtract multiples of 2π to these. Thus for $\sin 2\theta = -\frac{1}{2}$

$$2\theta = \begin{cases} 2n\pi - \pi/6 \\ 2n\pi + \pi + \pi/6 \end{cases} \quad \text{i.e.} \quad \theta = \begin{cases} n\pi - \pi/12 \\ n\pi + 7\pi/12 \end{cases}$$

Example 17

Find the general solution to the equation $2 \cos^2 \theta + 3 \cos \theta + 1 = 0$

$$2 \cos^2 \theta + 3 \cos \theta + 1 = 0$$
$$\therefore \quad (2 \cos \theta + 1)(\cos \theta + 1) = 0$$
i.e. $\cos \theta = -\frac{1}{2}$ or $\cos \theta = -1$
For $-\pi \leqslant \theta \leqslant \pi$, solutions are as shown in the diagrams:

Other solutions will occur as we add or subtract multiples of 2π to these.

Thus, for $\cos \theta = -\frac{1}{2}$ and for $\cos \theta = -1$
$$\theta = 2n\pi \pm 2\pi/3 \qquad\qquad \theta = 2n\pi + \pi$$
Thus $\qquad \theta = 2n\pi \pm 2\pi/3, (2n + 1)\pi$

Example 18

Find the general solution of the equation $\cos 5\theta = \cos 3\theta$.

If $\qquad\qquad\qquad \cos 5\theta = \cos 3\theta$
then $\qquad\qquad\quad \cos 5\theta - \cos 3\theta = 0$
$$\therefore \quad -2 \sin \frac{5\theta + 3\theta}{2} \sin \frac{5\theta - 3\theta}{2} = 0$$
$$-2 \sin 4\theta \sin \theta = 0$$
Thus, either $\sin 4\theta = 0 \qquad$ or $\quad \sin \theta = 0$
For $-\pi \leqslant \theta \leqslant \pi$, solutions for $\sin \theta = 0$ are as shown in the diagram:

Thus for $\quad \sin 4\theta = 0 \qquad$ and for $\quad \sin \theta = 0$
$$4\theta = n\pi$$
$$\theta = n\pi/4 \qquad\qquad\qquad \theta = n\pi$$
Now, as n takes the integer values, the solutions given by $n\pi/4$ will contain all those given by $n\pi$.

Thus the general solution of $\cos 5\theta = \cos 3\theta$ is $\theta = n\pi/4$.

Alternative solution for $\cos 5\theta = \cos 3\theta$
From the summary on page 362

$\qquad\cos \theta = \cos \alpha$ has the general solution $\qquad \theta = 2n\pi \pm \alpha$

Thus $\quad \cos 5\theta = \cos 3\theta$ has the general solution $\quad 5\theta = 2n\pi \pm 3\theta$

$\qquad\qquad\qquad\qquad\text{and either}\quad 5\theta = 2n\pi + 3\theta \quad \text{or} \quad 5\theta = 2n\pi - 3\theta$

$\qquad\qquad\qquad\qquad\qquad\qquad \theta = n\pi \qquad\qquad \text{or} \qquad \theta = n\pi/4$

Thus, as before, the general solution is $\theta = n\pi/4$

Example 19

Find the general solution of the equation $\sin 5\theta = \cos 3\theta$.

If $\qquad\qquad\qquad\qquad\qquad \sin 5\theta = \cos 3\theta$

then $\qquad\qquad\qquad\qquad\quad \sin 5\theta = \sin (\pi/2 - 3\theta)$

$\qquad\qquad \sin 5\theta - \sin (\pi/2 - 3\theta) = 0$

$\therefore \qquad 2 \cos (\theta + \pi/4) \sin (4\theta - \pi/4) = 0$

Thus either $\quad \cos (\theta + \pi/4) = 0 \qquad\qquad \text{or} \qquad \sin (4\theta - \pi/4) = 0$

Solution for $\qquad\qquad \cos \theta = 0 \qquad\qquad \text{and} \qquad\qquad \sin \theta = 0 \quad$ are shown:

Thus for $\qquad \cos (\theta + \pi/4) = 0 \qquad \text{and for} \quad \sin (4\theta - \pi/4) = 0$

$\qquad\qquad\qquad \theta + \pi/4 = n\pi + \pi/2 \qquad\qquad\qquad 4\theta - \pi/4 = n\pi$

$\qquad\qquad\qquad \theta = n\pi + \pi/4 \quad \text{or} \quad \theta = n\pi/4 + \pi/16$

$\cos \theta = 0$

$\sin \theta = 0$

Alternative solution for $\sin 5\theta = \cos 3\theta$

$\qquad\qquad \sin 5\theta = \cos 3\theta \quad$ can be written $\quad \sin 5\theta = \sin (\pi/2 - 3\theta)$

From the summary on page 362

$\qquad\qquad \sin \theta = \sin \alpha \quad$ has the general solution $\quad \theta = \begin{cases} 2n\pi + \alpha \\ (2n + 1)\pi - \alpha \end{cases}$

Thus $\qquad\quad \sin 5\theta = \sin (\pi/2 - 3\theta)$ has the general solution

$\qquad\qquad 5\theta = \begin{cases} 2n\pi + \pi/2 - 3\theta \\ (2n + 1)\pi - \pi/2 + 3\theta \end{cases}$

and either $\qquad 5\theta = 2n\pi + \pi/2 - 3\theta \quad$ giving $\quad \theta = n\pi/4 + \pi/16$

or $\qquad\qquad 5\theta = 2n\pi + \pi/2 + 3\theta \quad$ giving $\quad \theta = n\pi \quad + \pi/4$

Example 20

Find the general solution of the equation $\tan 5\theta = \tan (2\theta + \pi/6)$

$\tan 5\theta = \tan (2\theta + \pi/6) \qquad$ gives $\qquad \dfrac{\sin 5\theta}{\cos 5\theta} = \dfrac{\sin (2\theta + \pi/6)}{\cos (2\theta + \pi/6)}$

$\therefore \quad \sin 5\theta \cos (2\theta + \pi/6) - \cos 5\theta \sin (2\theta + \pi/6) = 0$

giving $\qquad\qquad\qquad \sin [5\theta - (2\theta + \pi/6)] = 0$

$\qquad\qquad\qquad \text{i.e.} \quad \sin (3\theta - \pi/6) = 0$

Solutions for $\sin \theta = 0$ are as shown in the diagram:

Thus for $\quad \sin (3\theta - \pi/6) = 0$

$\qquad\qquad 3\theta - \pi/6 = n\pi \quad$ giving $\quad \theta = n\pi/3 + \pi/18$

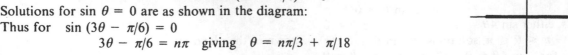

Alternative solution for $\tan 5\theta = \tan (2\theta + \pi/6)$
From the summary on page 362

$\qquad\qquad \tan \theta = \tan \alpha \quad$ has the general solution $\quad \theta = n\pi + \alpha$

Thus $\quad \tan 5\theta = \tan (2\theta + \pi/6) \quad$ has the general solution $\quad 5\theta = n\pi + 2\theta + \pi/6$

$\qquad\qquad\qquad\qquad\qquad\qquad\qquad \text{i.e.} \quad \theta = n\pi/3 + \pi/18 \quad \text{as before.}$

Exercise 15D

Find the general solutions to the following equations, giving your answers in radians.

1. $\cos \theta = \frac{1}{2}$

2. $\tan \theta = -1$

3. $\sin \theta = -\frac{\sqrt{3}}{2}$

4. $\cos 3\theta = 1$

5. $\tan 2\theta = -\frac{1}{\sqrt{3}}$

6. $\sin 5\theta = \frac{1}{2}$

7. $\cos \theta \sin 2\theta = \cos \theta$

8. $2 \sin^2 \theta - \sin \theta = 1$

9. $\cos 7\theta = \cos 3\theta$

10. $\cos 7\theta = \cos 5\theta$

11. $\sin 5\theta = \sin 3\theta$

12. $\tan 4\theta = \tan 2\theta$

13. $\cos \left(5\theta + \frac{\pi}{2} \right) = \cos 2\theta$

14. $\sin 4\theta = \cos 3\theta$

15. $\cos \left(4\theta + \frac{\pi}{2} \right) = \cos \left(2\theta - \frac{\pi}{6} \right)$

16. $\tan 7\theta = \tan \left(3\theta + \frac{\pi}{4} \right)$

Find the general solutions of the following equations, giving your answers in degrees.

17. $\sin \theta = 0{\cdot}1$

18. $2 \cos 2\theta - 1{\cdot}2 = 0$

19. $6 \cos^2 \theta - \cos \theta = 2$

20. $2 \sin 2\theta = \sin \theta$

21. $3 \sin \theta + 4 \cos \theta = 1$

22. $5 \cos \theta - 2 \sin \theta = 2$

15.5 Small angles

We can obtain approximate values for the trigonometric ratios of small angles.

Suppose θ is the angle in radians subtended by the arc PA of the circle radius r.
The tangent to the circle at P meets OA produced at the point T.
Now considering areas

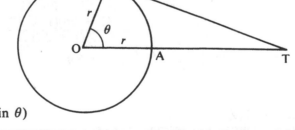

$$\triangle OPT > \text{sector } OPA > \triangle OPA$$

$\therefore \quad \frac{1}{2}r^2 \tan \theta > \quad \frac{1}{2}r^2\theta \quad > \frac{1}{2}r^2 \sin \theta$

i.e. $\quad \tan \theta > \quad \theta \quad > \sin \theta$

$$\frac{1}{\cos \theta} > \quad \frac{\theta}{\sin \theta} \quad > 1 \quad \text{(dividing by } \sin \theta)$$

But for small angles θ, $\frac{1}{\cos \theta}$ approaches 1, i.e. $\lim_{\theta \to 0} \left(\frac{1}{\cos \theta} \right) = 1$

hence $\quad \lim_{\theta \to 0} \left(\frac{\theta}{\sin \theta} \right) = 1$.

$\therefore \quad$ as $\quad \theta \to 0, \sin \theta \to \theta$

or $\qquad\qquad \sin \theta \approx \theta \quad$ for small values of θ.

Again, from $\quad \tan \theta > \quad \theta \quad > \sin \theta$

$$1 > \frac{\theta}{\tan \theta} > \cos \theta \quad \text{(dividing by } \tan \theta)$$

hence $\quad \lim_{\theta \to 0} \left(\frac{\theta}{\tan \theta} \right) = 1 \quad \therefore \quad$ as $\theta \to 0, \tan \theta \to \theta \quad$ or $\quad \tan \theta \approx \theta$ for small values of θ.

Since $\cos \theta = \sqrt{(1 - \sin^2 \theta)}$
$ \approx \sqrt{(1 - \theta^2)}$ for small values of θ
$ \approx 1 - \tfrac{1}{2}\theta^2$ expanding by the binomial theorem.

Thus we have, $\boxed{\text{for small } \theta: \sin \theta \approx \theta;\ \cos \theta \approx 1 - \tfrac{1}{2}\theta^2;\ \tan \theta \approx \theta}$

Note that these approximations depend upon θ being measured in radians.

Example 21

Find   (a) $\lim\limits_{\theta \to 0}\left(\dfrac{\cos 2\theta - 1}{\theta \sin 5\theta}\right)$ (b) $\lim\limits_{\theta \to 0}\left(\dfrac{\sin 3\theta + \tan 5\theta}{2\theta}\right)$

(a) $\lim\limits_{\theta \to 0}\left(\dfrac{\cos 2\theta - 1}{\theta \sin 5\theta}\right)$

$= \dfrac{1 - \tfrac{1}{2}(2\theta)^2 - 1}{\theta(5\theta)}$

$= \dfrac{-2\theta^2}{5\theta^2} = -\dfrac{2}{5}$

(b) $\lim\limits_{\theta \to 0}\left(\dfrac{\sin 3\theta + \tan 5\theta}{2\theta}\right)$

$= \dfrac{3\theta + 5\theta}{2\theta}$

$= 4.$

Example 22

Find an expression, involving θ, that $\dfrac{21 + 7\tan \theta - 20\cos \theta}{1 + \sin 2\theta}$ approximates to for small values of θ.

Using the approximations for $\sin \theta$, $\cos \theta$ and $\tan \theta$,

$\dfrac{21 + 7\tan \theta - 20\cos \theta}{1 + \sin 2\theta} \approx \dfrac{21 + 7\theta - 20(1 - \tfrac{1}{2}\theta^2)}{1 + 2\theta}$

$\phantom{\dfrac{21 + 7\tan \theta - 20\cos \theta}{1 + \sin 2\theta}} \approx \dfrac{1 + 7\theta + 10\theta^2}{1 + 2\theta}$

$\phantom{\dfrac{21 + 7\tan \theta - 20\cos \theta}{1 + \sin 2\theta}} \approx \dfrac{(1 + 2\theta)(1 + 5\theta)}{(1 + 2\theta)}$

i.e. $\dfrac{21 + 7\tan \theta - 20\cos \theta}{1 + \sin 2\theta} \approx 1 + 5\theta.$

Exercise 15E

1. If θ is a small angle measured in radians approximate values for $\sin \theta$, $\cos \theta$ and $\tan \theta$ can be obtained using $\sin \theta \approx \theta$, $\tan \theta \approx \theta$ and $\cos \theta \approx 1 - \tfrac{1}{2}\theta^2$. More accurate values can be obtained using a calculator. Using these facts, copy and complete the following tables:

$\theta = 0{\cdot}1$ rad	$\sin \theta$	$\cos \theta$	$\tan \theta$
Approx. value			
Calculator value (correct to 4 decimal places)			

$\theta = 0{\cdot}02$ rad	$\sin \theta$	$\cos \theta$	$\tan \theta$
Approx. value			
Calculator value (correct to 4 decimal places)			

2. Find (a) $\lim\limits_{\theta \to 0} \dfrac{\sin 2\theta}{\theta}$ (b) $\lim\limits_{\theta \to 0} \dfrac{\sin 2\theta}{2\theta}$ (c) $\lim\limits_{\theta \to 0} \dfrac{4\theta}{\sin \theta}$ (d) $\lim\limits_{\theta \to 0} \dfrac{\tan 2\theta}{\sin \theta}$

3. Find the expressions involving θ that the following approximate to for small values of θ. (Do *not* ignore terms in θ^2).

(a) $\dfrac{\cos \theta - 1}{\sin \theta}$
(b) $\dfrac{\sin \theta}{1 - \cos 2\theta}$
(c) $\dfrac{\sin 3\theta + \tan \theta}{\cos 2\theta}$

(d) $\sin (\theta + 45°)$
(e) $\cos (\theta + 30°)$
(f) $\dfrac{1 + \sin \theta}{5 + 3 \tan \theta - 4 \cos \theta}$

4. Given that $1° \approx 0.0175$ radians and $(0.0175)^2 \approx 0.000\,306$ find, without the use of a calculator, approximate values for (a) $\sin 2°$ (b) $\tan 12'$ (c) $\cos 1°$

15.6 Differentiation of trigonometric functions

Suppose $\qquad y = \sin x \qquad\qquad$ ... [1] where x is measured in radians.
Let x change by a small amount δx and the consequent change in y be δy, then
$$y + \delta y = \sin (x + \delta x) \qquad\qquad ... [2]$$

Subtracting equation [1] from equation [2]
$$\delta y = \sin (x + \delta x) - \sin x$$
$$= 2 \cos (x + \tfrac{1}{2}\delta x) \sin (\tfrac{1}{2}\delta x)$$
i.e. $\qquad\qquad \dfrac{\delta y}{\delta x} = \cos (x + \tfrac{1}{2}\delta x)\dfrac{\sin (\tfrac{1}{2}\delta x)}{(\tfrac{1}{2}\delta x)}$

By definition, $\qquad \dfrac{dy}{dx} = \lim\limits_{\delta x \to 0}\left(\dfrac{\delta y}{\delta x}\right)$

$$= \lim\limits_{\delta x \to 0}\left\{\cos (x + \tfrac{1}{2}\delta x)\dfrac{\sin (\tfrac{1}{2}\delta x)}{(\tfrac{1}{2}\delta x)}\right\}$$

Now from section 15.5 $\lim\limits_{\theta \to 0}\left(\dfrac{\sin \theta}{\theta}\right) = 1$ $\qquad\qquad \therefore \dfrac{dy}{dx} = \cos x.$

In a similar way if $y = \cos x \qquad\qquad$ ... [1]
Let x change by a small amount δx and the consequent change in y be δy,
then $\qquad\qquad y + \delta y = \cos (x + \delta x) \qquad\qquad ... [2]$

Subtracting equation [1] from equation [2]
$$\delta y = \cos (x + \delta x) - \cos x$$
$$= -2 \sin (x + \tfrac{1}{2}\delta x) \sin (\tfrac{1}{2}\delta x)$$
$$\dfrac{\delta y}{\delta x} = -\sin (x + \tfrac{1}{2}\delta x)\dfrac{\sin (\tfrac{1}{2}\delta x)}{(\tfrac{1}{2}\delta x)}$$

By definition, $\qquad \dfrac{dy}{dx} = \lim\limits_{\delta x \to 0}\left(\dfrac{\delta y}{\delta x}\right)$

$$= \lim\limits_{\delta x \to 0}\left\{-\sin (x + \tfrac{1}{2}\delta x)\dfrac{\sin (\tfrac{1}{2}\delta x)}{(\tfrac{1}{2}\delta x)}\right\} \qquad \therefore \dfrac{dy}{dx} = -\sin x$$

From these two results: $\begin{cases} y = \sin x, \\ \dfrac{dy}{dx} = \cos x \end{cases}$ and $\begin{cases} y = \cos x, \\ \dfrac{dy}{dx} = -\sin x \end{cases}$

we can obtain the differential coefficients of the other trigonometric ratios.

Differential of tan x and cot x

Since $y = \tan x$ and $y = \cot x$

$$= \frac{\sin x}{\cos x} \qquad\qquad\qquad = \frac{\cos x}{\sin x}$$

$$\frac{dy}{dx} = \frac{\cos x(\cos x) - \sin x(-\sin x)}{\cos^2 x} \qquad \frac{dy}{dx} = \frac{\sin x(-\sin x) - \cos x(\cos x)}{\sin^2 x}$$

$$= \frac{1}{\cos^2 x} \qquad\qquad\qquad = \frac{-1}{\sin^2 x}$$

$$= \sec^2 x \qquad\qquad\qquad = -\operatorname{cosec}^2 x$$

Differential of sec x and cosec x

Since $y = \sec x$ and $y = \operatorname{cosec} x$

$$= \frac{1}{\cos x} = (\cos x)^{-1} \qquad\qquad = \frac{1}{\sin x} = (\sin x)^{-1}$$

$$\frac{dy}{dx} = (-1)(\cos x)^{-2}(-\sin x) \qquad \frac{dy}{dx} = (-1)(\sin x)^{-2}(\cos x)$$

$$= \frac{\sin x}{\cos^2 x} \qquad\qquad\qquad = \frac{-\cos x}{\sin^2 x}$$

$$\therefore \quad \frac{dy}{dx} = \sec x \tan x \qquad\qquad \therefore \quad \frac{dy}{dx} = -\operatorname{cosec} x \cot x$$

To summarise these results, for the angle x in radians,

$y = \sin x$	$y = \cos x$	$y = \tan x$
$y' = \cos x$	$y' = -\sin x$	$y' = \sec^2 x$
$y = \operatorname{cosec} x$	$y = \sec x$	$y = \cot x$
$y' = -\operatorname{cosec} x \cot x$	$y' = \sec x \tan x$	$y' = -\operatorname{cosec}^2 x$

(Note that all of the 'co-' functions give a minus sign on differentiation)

If $y = \sin f(x)$ this can be written as $y = \sin u$, where $u = f(x)$ and then $\frac{dy}{dx}$ can be found from $\frac{dy}{dx} = \frac{dy}{du} \times \frac{du}{dx}$. Thus expressions such as $\sin f(x)$ can be treated as a function of a function as in section 13.1. Similarly, if $y = \tan^n x$, this may be written as $y = u^n$ where $u = \tan x$ and then $\frac{dy}{dx}$ is again $\frac{dy}{du} \times \frac{du}{dx}$.

The following examples illustrate this method.

Example 23

Find $\dfrac{dy}{dx}$ if   (a) $y = \sin 4x$,   (b) $y = \cos (x^3 - 1)$.

(a) $y = \sin 4x$

letting $u = 4x$ then $y = \sin u$

$$\frac{dy}{dx} = \frac{dy}{du} \times \frac{du}{dx}$$

$$= (\cos u)4$$

$$= 4 \cos 4x$$

(b) $y = \cos (x^3 - 1)$

letting $u = x^3 - 1$ then $y = \cos u$

$$\frac{dy}{dx} = \frac{dy}{du} \times \frac{du}{dx}$$

$$= (-\sin u)3x^2$$

$$= -3x^2 \sin (x^3 - 1)$$

Example 24

Find $\dfrac{dy}{dx}$ if   (a) $y = \tan^5 x$,   (b) $y = \sec^3 x$.

(a) $y = \tan^5 x$

letting $u = \tan x$ then $y = u^5$

$$\frac{dy}{dx} = \frac{dy}{du} \times \frac{du}{dx}$$

$$= 5u^4 \sec^2 x$$

$$= 5 \tan^4 x \sec^2 x$$

(b) $y = \sec^3 x$

letting $u = \sec x$ then $y = u^3$

$$\frac{dy}{dx} = \frac{dy}{du} \times \frac{du}{dx}$$

$$= 3u^2 \sec x \tan x$$

$$= 3 \sec^3 x \tan x$$

With practice the reader will soon be able to differentiate such functions directly,

e.g.   if $y = \cos 4x$, $\dfrac{dy}{dx} = -\sin 4x \dfrac{d}{dx}(4x) = -4 \sin 4x$

if $y = \sin 7x$, $\dfrac{dy}{dx} = \cos 7x \dfrac{d}{dx}(7x) \quad = 7 \cos 7x$

if $y = \tan 6x$, $\dfrac{dy}{dx} = 6 \sec^2 6x$

if $y = \sin (x^3 + 3)$, $\dfrac{dy}{dx} = 3x^2 \cos (x^3 + 3)$

if $y = \sin^4 x$, $\dfrac{dy}{dx} = 4 \sin^3 x \cos x$

if $y = \sin^5 6x$ $\dfrac{dy}{dx} = 5 \sin^4 6x \times 6 \cos 6x = 30 \sin^4 6x \cos 6x$

However, for more complicated functions, it is advisable to proceed more steadily and to show the use of the chain rule.

Example 25

Find $\dfrac{dy}{dx}$ if $y = \cot^4 (3x^2 + 2x - 1)$.

Let $u = 3x^2 + 2x - 1$ then $y = \cot^4 u$ and $\dfrac{dy}{du} = -4 \cot^3 u \operatorname{cosec}^2 u$

Now, $\dfrac{dy}{dx} = \dfrac{dy}{du} \times \dfrac{du}{dx}$

$$= -4 \cot^3 u \operatorname{cosec}^2 u (6x + 2)$$

$$= -8(3x + 1) \cot^3 u \operatorname{cosec}^2 u$$

∴ $\dfrac{dy}{dx} = -8(3x + 1) \cot^3 (3x^2 + 2x - 1) \operatorname{cosec}^2 (3x^2 + 2x - 1).$

Example 26

Find $\dfrac{dy}{dx}$ if (a) $y = \sin^2 x \cos 3x$,   (b) $y = \dfrac{\sin^4 3x}{6x}$.

(a) $y = \sin^2 x \cos 3x$

$\quad \dfrac{dy}{dx} = \sin^2 x(-3 \sin 3x) + \cos 3x(2 \sin x \cos x)$

$\qquad = \sin x(2 \cos 3x \cos x - 3 \sin x \sin 3x).$

(b) $y = \dfrac{\sin^4 3x}{6x}$

$\quad \dfrac{dy}{dx} = \dfrac{6x(4 \sin^3 3x \times 3 \cos 3x) - (\sin^4 3x \times 6)}{36x^2}$

$\qquad = \dfrac{\sin^3 3x}{6x^2}(12x \cos 3x - \sin 3x).$

Exercise 15F

Differentiate the following with respect to x.

1. $\sin x$
2. $\cos x$
3. $\tan x$
4. $\operatorname{cosec} x$
5. $\sec x$
6. $\cot x$
7. $\sin 2x$
8. $\cos 5x$
9. $\cos 7x$
10. $\sin 9x$
11. $\tan 3x$
12. $\sec 5x$
13. $\operatorname{cosec} 6x$
14. $\cot 2x$
15. $\cos x°$
16. $\tan 2x°$
17. $\cos (x^2 + 1)$
18. $\sin (2x^2 + 1)$
19. $\sin (x^3)$
20. $\sec (4x^2 + 1)$
21. $\cot (2x^2 + 1)$
22. $\cot (3x + 4)$
23. $\sin^2 x$
24. $\cos^3 x$
25. $\tan^4 x$
26. $\sin^7 x$
27. $\cos^4 x$
28. $\sec^6 x$
29. $\sin^3 2x$
30. $\cos^4 3x$
31. $\sin^2 (6x + 1)$
32. $\cos^3 (3 - 2x)$
33. $\tan^3 5x$
34. $\sec^7 5x$
35. $x^2 \sin x$
36. $x^2 \cos 3x$
37. $x^3 \tan 3x$
38. $\sin 3x \cos x$
39. $\cos 4x \sin x$
40. $\cos^4 x \sin 4x$
41. $\sin^3 x \cos 3x$
42. $\cos^4 x \cos 4x$
43. $\sin^3 x \sin 3x$
44. $\dfrac{\sin 2x}{x^2}$
45. $\dfrac{\cos^4 x}{x^2}$
46. $\dfrac{\cos^3 5x}{6x}$

47. For each of the following find the gradient of the curve at the given point on the curve
    (a) $y = 3 \cos 2x$ at $(\pi/6, \frac{3}{2})$,   (b) $y = 2 \cos x + 3 \sin x$ at $(0, 2)$,
    (c) $y = x \sin x$ at $(\pi/2, \pi/2)$,   (d) $y = \sin^6 x$ at $(\pi/4, \frac{1}{8})$.

48. For each of the following find the coordinates of any points on the curve where the gradient is as stated and for which $0 \leqslant x \leqslant \pi$.
    (a) $y = \sin 2x$, gradient 1,
    (b) $y = 3 \sin x - \sin^3 x$, gradient $\frac{3}{8}$.

49. Find the equation of the tangent and of the normal to the curve $y = 3 + x \cos x$ at the point $(0, 3)$ on the curve.

50. Find the coordinates and nature of any stationary point on the following curves restricting your solutions to those for which $0 \leqslant x \leqslant 2\pi$.
    (a) $y = 8 \sin (\frac{3}{4}x)$ (b) $y = x \sin x + \cos x$ (c) $y = \sin x + \cos x$

51. Find $\dfrac{dy}{dx}$ and $\dfrac{d^2y}{dx^2}$ in terms of θ for each of the following
    (a) $x = \cos \theta, y = \sin \theta$,   (b) $x = \cos 2\theta, y = \cos \theta$.

15.7 Integration of trigonometric functions

It follows from the differentiation of $\sin x$ and $\cos x$ that

$$\int \sin x \, dx = -\cos x + c \quad \text{and} \quad \int \cos x \, dx = \sin x + c,$$

and since if　$y = \sin ax$　　　　and if　　$y = \cos ax$

$$\frac{dy}{dx} = a \cos ax \qquad\qquad \frac{dy}{dx} = -a \sin ax$$

it follows that

$$\int \sin ax \, dx = -\frac{1}{a} \cos ax + c \quad \text{and} \quad \int \cos ax \, dx = \frac{1}{a} \sin ax + c$$

In some cases, trigonometric functions can be integrated directly as the reverse of differentiation, as we can see what function would differentiate to give the integrand.

Example 27

Find the following indefinite integrals　(a) $\displaystyle\int \cos 7x \, dx$,　(b) $\displaystyle\int 4 \operatorname{cosec} x \cot x \, dx$,　(c) $\displaystyle\int \sec^2 6x \, dx$.

(a) $\displaystyle\int \cos 7x \, dx$

Now　$\dfrac{d}{dx}(\sin 7x) = 7 \cos 7x$

$\therefore \displaystyle\int \cos 7x \, dx = \tfrac{1}{7} \sin 7x + c$

(b) $\displaystyle\int 4 \operatorname{cosec} x \cot x \, dx$

Now　$\dfrac{d}{dx}(\operatorname{cosec} x) = -\operatorname{cosec} x \cot x$

$\therefore \displaystyle\int 4 \operatorname{cosec} x \cot x \, dx = -4 \operatorname{cosec} x + c$

(c) $\displaystyle\int \sec^2 6x \, dx$

Now　$\dfrac{d}{dx}(\tan 6x) = 6 \sec^2 6x$

$\therefore \displaystyle\int \sec^2 6x \, dx = \tfrac{1}{6} \tan 6x + c$

Example 28

Find the following indefinite integrals　(a) $\displaystyle\int 6 \cos x \sin^2 x \, dx$,　(b) $\displaystyle\int \sin^4 2x \cos 2x \, dx$.

(a) $\displaystyle\int 6 \cos x \sin^2 x \, dx$

Now　$\dfrac{d}{dx}(\sin^3 x) = 3 \sin^2 x \cos x$

$\therefore \displaystyle\int 6 \cos x \sin^2 x \, dx = 2 \sin^3 x + c$

(b) $\displaystyle\int \sin^4 2x \cos 2x \, dx$

Now　$\dfrac{d}{dx}(\sin^5 2x) = 5 \sin^4 2x \, 2 \cos 2x$

$= 10 \sin^4 2x \cos 2x$

$\therefore \displaystyle\int \sin^4 2x \cos 2x \, dx = \tfrac{1}{10} \sin^5 2x + c$

Example 29

Find the following indefinite integrals   (a) $\displaystyle\int \cot^5 x \csc^2 x \, dx$,   (b) $\displaystyle\int \sec^4 x \tan x \, dx$.

(a) $\displaystyle\int \cot^5 x \csc^2 x \, dx$

   Now $\dfrac{d}{dx}(\cot^6 x) = -6 \cot^5 x \csc^2 x$

   $\therefore \quad \displaystyle\int \cot^5 x \csc^2 x \, dx = -\tfrac{1}{6} \cot^6 x + c$

(b) $\displaystyle\int \sec^4 x \tan x \, dx = \int \sec^3 x(\sec x \tan x) \, dx$

   Now $\dfrac{d}{dx}(\sec^4 x) = 4 \sec^3 x \sec x \tan x$

   $\therefore \quad \displaystyle\int \sec^4 x \tan x \, dx = \tfrac{1}{4} \sec^4 x + c.$

Example 30

Find $\displaystyle\int 8 \sin 4x \sin x \, dx$.

Now $8 \sin 4x \sin x = 4(\cos 3x - \cos 5x)$ (using the result of section 15.1)

$\therefore \quad \displaystyle\int 8 \sin 4x \sin x \, dx = \tfrac{4}{3} \sin 3x - \tfrac{4}{5} \sin 5x + c.$

$\displaystyle\int \sin^n x \, dx, \int \cos^n x \, dx$

Example 31 shows the method to use for these integrations when n is odd and Example 32 shows the method to use when n is even.

Example 31

Find $\displaystyle\int \cos^7 x \, dx$

$\displaystyle\int \cos^7 x \, dx = \int \cos x(\cos^6 x) \, dx$

$\qquad\qquad = \displaystyle\int \cos x(1 - \sin^2 x)^3 \, dx$

$\qquad\qquad = \displaystyle\int \cos x(1 - 3 \sin^2 x + 3 \sin^4 x - \sin^6 x) \, dx$

$\qquad\qquad = \displaystyle\int (\cos x - 3 \sin^2 x \cos x + 3 \sin^4 x \cos x - \sin^6 x \cos x) \, dx$

$\therefore \quad \displaystyle\int \cos^7 x \, dx = \sin x - \sin^3 x + \tfrac{3}{5} \sin^5 x - \tfrac{1}{7} \sin^7 x + c$

Example 32

Find $\displaystyle\int \sin^4 x \, dx$.

$$\int \sin^4 dx = \int (\sin^2 x)^2 \, dx$$

$$= \int [\tfrac{1}{2}(1 - \cos 2x)]^2 \, dx \quad \text{using } \sin^2 A = \tfrac{1}{2}(1 - \cos 2A)$$

$$= \tfrac{1}{4} \int (1 - 2 \cos 2x + \cos^2 2x) \, dx$$

$$= \tfrac{1}{4} \int (1 - 2 \cos 2x + \tfrac{1}{2} + \tfrac{1}{2} \cos 4x) \, dx, \quad \text{using } \cos^2 A = \tfrac{1}{2}(1 + \cos 2A)$$

$$= \tfrac{1}{4}\left(x - \sin 2x + \frac{x}{2} + \frac{1}{8} \sin 4x \right) + c$$

$$= \frac{3x}{8} - \frac{1}{4} \sin 2x + \frac{1}{32} \sin 4x + c.$$

Exercise 15G

Find the following indefinite integrals

1. $\displaystyle\int \sin x \, dx$ 2. $\displaystyle\int \sec x \tan x \, dx$ 3. $\displaystyle\int \sin 3x \, dx$

4. $\displaystyle\int \cos 2x \, dx$ 5. $\displaystyle\int \sec^2 2x \, dx$ 6. $\displaystyle\int \operatorname{cosec}^2 4x \, dx$

7. $\displaystyle\int \operatorname{cosec} 2x \cot 2x \, dx$ 8. $\displaystyle\int \cos 4x \, dx$ 9. $\displaystyle\int 2 \sin 4x \, dx$

10. $\displaystyle\int 4 \sin 2x \, dx$ 11. $\displaystyle -\int 6 \sin (2x + 1) dx$ 12. $\displaystyle\int 6 \cos (3x - 4) dx$

13. $\displaystyle\int (1 + \cos 2x) \, dx$ 14. $\displaystyle\int (3 - 8 \sin 4x) \, dx$ 15. $\displaystyle\int 8 \sin x \cos^3 x \, dx$

16. $\displaystyle\int \cos^5 x \sin x \, dx$ 17. $\displaystyle\int 10 \cos x \sin^4 x \, dx$ 18. $\displaystyle\int 12 \cos^5 2x \sin 2x \, dx$

19. $\displaystyle\int \cos 2x \sin^4 2x \, dx$ 20. $\displaystyle\int 50 \sin^4 5x \cos 5x \, dx$ 21. $\displaystyle\int \sec^6 x \tan x \, dx$

22. $\displaystyle\int \tan^6 x \sec^2 x \, dx$ 23. $\displaystyle\int \sin^3 x \, dx$ 24. $\displaystyle\int \sin^5 x \, dx$

25. $\displaystyle\int \cos^5 x \, dx$ 26. $\displaystyle\int \cos^3 4x \, dx$ 27. $\displaystyle\int \cos^2 x \, dx$

28. $\displaystyle\int \sin^2 x \, dx$ 29. $\displaystyle\int \cos^4 x \, dx$ 30. $\displaystyle\int \sin^4 3x \, dx$

31. $\displaystyle\int 2 \cos 3x \sin x \, dx$ 32. $\displaystyle\int 2 \sin 5x \cos x \, dx$ 33. $\displaystyle\int \cos 3x \cos 5x \, dx$

34. $\displaystyle\int 6 \sin 3x \sin x \, dx$

35. Find the following indefinite integrals (hint: simplify each expression first).

(a) $\displaystyle\int \sin x \cot x \, dx$ (b) $\displaystyle\int (\sin^2 x + \cos^2 x) \, dx$ (c) $\displaystyle\int (\cos^2 x - \sin^2 x) \, dx$

(d) $\displaystyle\int \frac{1}{1 - \sin^2 x} \, dx$ (e) $\displaystyle\int \cos x (\sec x + \tan x) \, dx$ (f) $\displaystyle\int \frac{2 \tan x}{\sin 2x} \, dx$

Evaluate the following definite integrals

36. $\displaystyle\int_0^{\pi/2} (1 - \sin x) \, dx$ **37.** $\displaystyle\int_0^{\pi/4} \sin 2x \, dx$ **38.** $\displaystyle\int_0^{\pi} \cos\left(3x + \frac{\pi}{2}\right) dx$

39. $\displaystyle\int_0^{\pi/4} \sin^3 x \cos x \, dx$ **40.** $\displaystyle\int_0^{\pi/4} \sec^2 x \tan^3 x \, dx$ **41.** $\displaystyle\int_{\pi/8}^{\pi/4} 2 \cos^2 2x \, dx$

42 $\displaystyle\int_{-\pi/2}^{\pi/2} \cos^3 x \, dx$ **43.** $\displaystyle\int_0^{\pi/2} \sin 3x \cos 5x \, dx$

44. Find the area between the curve $y = \sin x$ and the x-axis from $x = 0$ to $x = \pi$.

45. Find the area between the curve $y = 3 \cos x$ and the x-axis from $x = 0$ to $x = \pi/2$.

46. Find the area between the curve $y = \sin x + 3 \cos x$ and the x-axis from $x = 0$ to $x = \pi/2$.

47. Find the volume of the solid of revolution formed by rotating about the x-axis the area between the curve $y = \sin x$ and the x-axis, from $x = 0$ to $x = \pi$.

48. Find the volume of the solid of revolution formed by rotating about the x-axis the area between the curve $y = \sin x + \cos x$ and the x-axis, from $x = 0$ to $x = \pi/2$.

15.8 Differentiation of inverse trigonometric functions

Suppose $y = \sin^{-1} x$
then $x = \sin y$

and $\dfrac{dx}{dy} = \cos y \quad \dots [1]$

Using $\dfrac{dy}{dx} = 1/(dx/dy)$, (see page 324) and $\sin^2 y + \cos^2 y = 1$, we can

rearrange [1]   to give $\dfrac{dy}{dx} = \dfrac{1}{\sqrt{(1 - \sin^2 y)}}$

$= \dfrac{1}{\sqrt{(1 - x^2)}}$

> Thus, if $y = \sin^{-1} x$, then $\dfrac{dy}{dx} = \dfrac{1}{\sqrt{(1 - x^2)}}$

In a similar way, it can be shown that

> if $y = \cos^{-1} x$, $\dfrac{dy}{dx} = \dfrac{-1}{\sqrt{(1 - x^2)}}$
>
> and if $y = \tan^{-1} x$, $\dfrac{dy}{dx} = \dfrac{1}{1 + x^2}.$

Example 33

Find $\dfrac{dy}{dx}$ if $y = \sin^{-1}\dfrac{x}{a}$ where a is a constant.

Let $\quad u = \dfrac{x}{a}, \quad$ then $\quad y = \sin^{-1} u$

thus $\quad \dfrac{du}{dx} = \dfrac{1}{a} \quad$ and $\quad \dfrac{dy}{du} = \dfrac{1}{\sqrt{(1 - u^2)}}$

Using $\quad \dfrac{dy}{dx} = \dfrac{dy}{du} \times \dfrac{du}{dx}$

$\qquad \dfrac{dy}{dx} = \dfrac{1}{\sqrt{(1 - u^2)}} \times \dfrac{1}{a}$

$\qquad\quad = \dfrac{1}{\sqrt{(1 - x^2/a^2)}} \times \dfrac{1}{a}$

$\qquad \dfrac{dy}{dx} = \dfrac{1}{\sqrt{(a^2 - x^2)}}.$

Example 34

Find $\dfrac{dy}{dx}$ if (a) $y = \cos^{-1}\dfrac{x}{5}$, (b) $y = \cos^{-1}(x^2 - 1)$.

(a) $\qquad y = \cos^{-1}\dfrac{x}{5}$

Let $\quad u = \dfrac{x}{5}, \quad$ then $\quad y = \cos^{-1} u$

$\qquad \dfrac{du}{dx} = \dfrac{1}{5} \quad$ and $\quad \dfrac{dy}{du} = \dfrac{-1}{\sqrt{(1 - u^2)}}$

Using $\quad \dfrac{dy}{dx} = \dfrac{dy}{du} \times \dfrac{du}{dx}$

$\qquad = \dfrac{-1}{\sqrt{(1 - u^2)}} \times \dfrac{1}{5}$

$\qquad = \dfrac{-1}{\sqrt{\left(1 - \dfrac{x^2}{25}\right)}} \times \dfrac{1}{5}$

$\qquad \dfrac{dy}{dx} = \dfrac{-1}{\sqrt{(25 - x^2)}}$

(b) $\qquad y = \cos^{-1}(x^2 - 1)$

Let $\quad u = x^2 - 1, \quad$ then $\quad y = \cos^{-1} u$

$\qquad \dfrac{du}{dx} = 2x \quad$ and $\quad \dfrac{dy}{du} = \dfrac{-1}{\sqrt{(1 - u^2)}}$

Using $\quad \dfrac{dy}{dx} = \dfrac{dy}{du} \times \dfrac{du}{dx}$

$\qquad = \dfrac{-1}{\sqrt{(1 - u^2)}} \times 2x$

$\qquad = \dfrac{-2x}{\sqrt{[1 - (x^2 - 1)^2]}}$

$\qquad = \dfrac{-2x}{\sqrt{(2x^2 - x^4)}}$

$\therefore \quad \dfrac{dy}{dx} = \dfrac{-2}{\sqrt{(2 - x^2)}}$

$$\int \dfrac{1}{\sqrt{(a^2 - x^2)}}\, dx \text{ and } \int \dfrac{1}{(a^2 + x^2)}\, dx$$

From $\dfrac{d}{dx}(\sin^{-1} x) = \dfrac{1}{\sqrt{(1 - x^2)}}$, it can be shown that $\dfrac{d}{dx}\left(\sin^{-1}\dfrac{x}{a}\right) = \dfrac{1}{\sqrt{(a^2 - x^2)}}$

and from $\dfrac{d}{dx}(\tan^{-1} x) = \dfrac{1}{(1 + x^2)}$, it can be shown that $\dfrac{d}{dx}\left(\tan^{-1}\dfrac{x}{a}\right) = \dfrac{a}{(a^2 + x^2)}.$

Thus to integrate expressions involving $\dfrac{1}{\sqrt{(a^2 - x^2)}}$ we look for a solution of the form $\sin^{-1}\dfrac{x}{a} + c$

and to integrate an expression of the form $\dfrac{1}{(a^2 + x^2)}$ we look for a solution of the form $\tan^{-1}\dfrac{x}{a}.$

Example 35

Find   (a) $\int \dfrac{4}{x^2 + 16}\,dx$ (b) $\int \dfrac{2}{36 + x^2}\,dx$.

(a) $\int \dfrac{4}{x^2 + 16}\,dx$

Since this involves $\dfrac{1}{x^2 + 16}$ i.e. of the

form $\dfrac{1}{x^2 + a^2}$, we suspect it has come

from the differentiation of $\tan^{-1}\dfrac{x}{4}$ which

is $\dfrac{4}{x^2 + 16}$.

Thus $\int \dfrac{4}{x^2 + 16}\,dx = \tan^{-1}\dfrac{x}{4} + c$.

(b) $\int \dfrac{2}{36 + x^2}\,dx$

The differential of $\tan^{-1}\dfrac{x}{6}$ is $\dfrac{6}{x^2 + 36}$.

Thus $\int \dfrac{2}{36 + x^2}\,dx = \dfrac{1}{3}\int \dfrac{6}{36 + x^2}\,dx$

$\qquad\qquad\qquad = \dfrac{1}{3}\tan^{-1}\dfrac{x}{6} + c$.

Example 36

Evaluate $\displaystyle\int_{-1\cdot5}^{0} \dfrac{1}{\sqrt{(9 - x^2)}}\,dx$.

This is of the form $\displaystyle\int \dfrac{1}{\sqrt{(a^2 - x^2)}}\,dx$ and we can therefore quote the solution:

$$\int_{-1\cdot5}^{0} \dfrac{1}{\sqrt{(9 - x^2)}}\,dx = \left[\sin^{-1}\dfrac{x}{3}\right]_{-1\cdot5}^{0}$$

$$= \sin^{-1}(0) - \sin^{-1}\left(-\dfrac{1\cdot5}{3}\right)$$

$$= 0 - \left(-\dfrac{\pi}{6}\right) \qquad\qquad \therefore \int_{-1\cdot5}^{0} \dfrac{1}{\sqrt{(9 - x^2)}}\,dx = \dfrac{\pi}{6}$$

The reader should note that in order to integrate more complicated functions

of this type, e.g. $\displaystyle\int \dfrac{1}{\sqrt{(a^2 - Ax^2)}}\,dx$, for $A \neq 1$ or $\displaystyle\int \dfrac{1}{a^2 + Ax^2}\,dx$, for $A \neq 1$,

we use the method of 'changing the variable' which is explained in chapter 20.

Exercise 15H

Differentiate the following with respect to x

1. $\sin^{-1} x$
2. $\tan^{-1}\dfrac{x}{a}$
3. $\sin^{-1}\dfrac{x}{4}$
4. $\cos^{-1} 3x$
5. $\tan^{-1} 4x$
6. $\sin^{-1} 6x$
7. $\sin^{-1}(2x - 1)$
8. $\tan^{-1}(1 - 3x)$
9. $\sin^{-1}(x^2 - 1)$
10. $x \sin^{-1} x$
11. $x \tan^{-1} x$
12. $(x^2 + 1)\tan^{-1} x$

Find the following indefinite integrals

13. $\displaystyle\int \dfrac{1}{\sqrt{(4 - x^2)}}\,dx$
14. $\displaystyle\int \dfrac{1}{\sqrt{(16 - x^2)}}\,dx$
15. $\displaystyle\int \dfrac{3}{9 + x^2}\,dx$

16. $\displaystyle\int \frac{1}{25 + x^2}\, dx$ **17.** $\displaystyle\int \frac{1}{\sqrt{(49 - x^2)}}\, dx$ **18.** $\displaystyle\int \frac{1}{49 + x^2}\, dx$

19. $\displaystyle\int \frac{1}{\sqrt{(25 - x^2)}}\, dx$ **20.** $\displaystyle\int \frac{2}{100 + x^2}\, dx$

Evaluate the following definite integrals (leave π in your answers).

21. $\displaystyle\int_{-3}^{3} \frac{1}{\sqrt{(36 - x^2)}}\, dx$ **22.** $\displaystyle\int_{-2}^{2} \frac{1}{4 + x^2}\, dx$ **23.** $\displaystyle\int_{-\sqrt{3}}^{3} \frac{\sqrt{3}}{x^2 + 3}\, dx$

24. $\displaystyle\int_{0}^{3\cdot 2} \frac{1}{\sqrt{(3 - x^2)}}\, dx$

Exercise 151 *Examination questions*

1. Find the limit, as $\theta \to 0$, of
$$\frac{\sin (x + \theta) - \sin (x - \theta)}{2\theta}.$$
(London)

2. Leaving all answers as multiples of π, find the values of x for which $0 < x < \pi$ and
$$\cos x + \cos 3x = 0.$$
(London)

3. Given that $y = \dfrac{\sin \theta - 2 \sin 2\theta + \sin 3\theta}{\sin \theta + 2 \sin 2\theta + \sin 3\theta}$, prove that $y = -\tan^2 \dfrac{\theta}{2}$.

Find (a) the exact value of $\tan^2 15°$ in the form $p + q\sqrt{r}$, where p, q and r are integers,

  (b) the values of θ between $0°$ and $360°$ for which $2y + \sec^2 \dfrac{\theta}{2} = 0$.
(A.E.B.)

4. (a) Express $1\cdot 2 \cos x + 1\cdot 6 \sin x$ in the form $R \cos (x - \alpha)$, where R is a positive constant and $0° < \alpha < 90°$. Hence, or otherwise,
  (i) find the maximum and minimum values of $1\cdot 2 \cos x + 1\cdot 6 \sin x$,
  (ii) solve the equation $1\cdot 2 \cos x + 1\cdot 6 \sin x = 1\cdot 5$, giving the values of x between $0°$ and $180°$.
 (b) Prove that, for all values of x,
$$\sin (x + 30°) - \sin (x - 30°) = \cos x.$$
 (c) Solve the equation $\sin 4\theta + \sin 2\theta = \cos \theta$, giving the values of θ in the range $0° < \theta < 360°$.
(J.M.B.)

5. (a) If $t = \tan \dfrac{\theta}{2}$, show that $\sin \theta = \dfrac{2t}{1 + t^2}$ and derive an expression for $\cos \theta$ in terms of t.
  Hence, or otherwise, solve the equation
$$3 \sin \theta + \cos \theta = 2$$
  for values of θ in the range $0° \leqslant \theta \leqslant 180°$.
 (b) Prove the identity
$$\frac{\sin 4\theta + \sin 2\theta}{\cos 4\theta + \cos 2\theta} = \tan 3\theta.$$
(S.U.J.B.)

6. Given that
$$3 \sin x - \cos x \equiv R \sin (x - \alpha),$$
where $R > 0$ and $0° < \alpha < 90°$, find the values of R and α correct to one decimal place.

Hence find one value of x between $0°$ and $360°$ for which the curve $y = 3 \sin x - \cos x$ has a turning point.   (London)

7. Find the general solution of the equation
$$\tan \theta - \sin \theta = 1 - \cos \theta.$$ (Cambridge)

8. (i) Find the general solution, in radians, of the equation
$$\sin 3x = -\tfrac{1}{2}.$$
(ii) By putting $\tan (\theta/2) = t$, or otherwise, find the general solution of the equation
$$2 \cos \theta - \sin \theta = 1,$$
giving your answers to the nearest tenth of a degree.   (London)

9. (a) Find the general solution of the equation
$$\tan x + \sec x = 3 \cos x.$$
(b) Express $5 \sin^2 x - 3 \sin x \cos x + \cos^2 x$ in the form
$a + b \cos (2x - \alpha)$ where a, b, α are independent of x. ·
Hence, or otherwise, find the maximum and minimum values of
$$5 \sin^2 x - 3 \sin x \cos x + \cos^2 x$$
as x varies.   (Cambridge)

10. If $x = t^2 \sin 3t$ and $y = t^2 \cos 3t$, find $\dfrac{dy}{dx}$ in terms of t, and show that the curve defined by these parametric equations is parallel to the x-axis at points where $\tan 3t = \dfrac{2}{3t}$.   (S.U.J.B.)

11. Find
$$\int \sin x(1 + \cos^2 x)\, dx.$$ (Cambridge)

12. Prove from first principles that the derivative of $\sin x$ is $\cos x$.
$$\left[\text{You may use } \lim_{\theta \to 0} \frac{\sin \theta}{\theta} = 1 \text{ without proof.}\right]$$
When x increases from π to $\pi + \varepsilon$, where ε is small, the increment in
$$\frac{\sin x}{x}$$
is approximately equal to $p\varepsilon$. Find p in terms of π.   (J.M.B.)

13. (a) Differentiate with respect to x
   (i) $\tan 2x$,   (ii) $\sin^3 3x$,   (iii) $(1 + \sin x)^5$.
(b) If $y = \dfrac{\sin x}{x}$, prove that
$$x\frac{d^2y}{dx^2} + 2\frac{dy}{dx} + xy = 0.$$ (S.U.J.B.)

14. Sketch the curve $y = 1 + \sin x$ for $0 \leqslant x \leqslant \frac{3}{2}\pi$.

Show that $(1 + \sin x)^2 = \frac{3}{2} + 2 \sin x - \frac{1}{2} \cos 2x$.

Hence show that the volume of the solid of revolution formed when the region $(0 \leqslant x \leqslant \frac{3}{2}\pi)$ bounded by the curve $y = 1 + \sin x$, the y-axis and the x-axis is rotated through one revolution about the x-axis is
$$\frac{1}{4}\pi(9\pi + 8).$$
(Cambridge)

15. (a) Differentiate with respect to x:

(i) $x \sin 2x$;   (ii) $\tan\left(3x + \dfrac{\pi}{4}\right)$;   (iii) $(1 + \cos x)^2$.

(b) Evaluate the following integrals:

(i) $\displaystyle\int_0^{\pi/3} \sin\left(2x + \frac{\pi}{3}\right) dx$;   (ii) $\displaystyle\int_{\pi/6}^{\pi/3} \cos 3x\, dx$.   (S.U.J.B.)

16. Given that $\sin^{-1} x$, $\cos^{-1} x$ and $\sin^{-1}(1 - x)$ are acute angles,
(i) prove that $\sin[\sin^{-1} x - \cos^{-1} x] = 2x^2 - 1$,
(ii) solve the equation $\sin^{-1} x - \cos^{-1} x = \sin^{-1}(1 - x)$.   (A.E.B.)

17. Differentiate $x \sin^{-1} mx$, where m is a constant.
Hence, or otherwise, integrate $\sin^{-1} mx$.   (J.M.B.)

18. (i) For the curve $y = \sin x \cos^3 x$, where $0 \leqslant x \leqslant \pi$, find the x- and y-coordinates of the points at which $\dfrac{dy}{dx} = 0$.

Sketch the curve.

(ii) For the curve $y^2 = \sin x \cos^3 x$, where $0 \leqslant x \leqslant \frac{1}{2}\pi$, show that
$$\left(\frac{dy}{dx}\right)^2 = \frac{1}{4}\cot x(\cos^2 x - 3 \sin^2 x)^2,$$
provided that $x \neq 0$.
Sketch the curve.   (Cambridge)

16
Coordinate geometry II

16.1 Angle between two straight lines

Consider the two lines $y = m_1 x + c_1$ and $y = m_2 x + c_2$ and suppose that they make angles of θ_1 and θ_2 respectively with the positive x-axis. Let θ be the acute angle between the two lines, then

$$\theta = \theta_1 - \theta_2$$

i.e. $\tan \theta = \tan (\theta_1 - \theta_2)$

$$= \frac{\tan \theta_1 - \tan \theta_2}{1 + \tan \theta_1 \tan \theta_2}$$

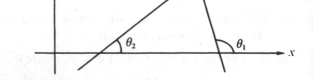

But $\tan \theta_1$ and $\tan \theta_2$ are the gradients of these lines and hence $m_1 = \tan \theta_1$ and $m_2 = \tan \theta_2$

$$\therefore \quad \tan \theta = \frac{m_1 - m_2}{1 + m_1 m_2}.$$

Note that the angle θ between the lines depends only on the gradients of the lines; the constants c_1 and c_2 do not affect the angle θ.

Example 1

Find the tangent of the acute angle between the pair of lines whose equations are $3y = x - 7$ and $2y = 3 - 4x$.

	$3y = x - 7$		$2y = 3 - 4x$
i.e.	$y = \frac{1}{3}x - \frac{7}{3}$	i.e.	$y = \frac{3}{2} - 2x$
	gradient $= \frac{1}{3} = m_1$		gradient $= -2 = m_2$

If the angle between the lines is θ, then

$$\tan \theta = \frac{m_1 - m_2}{1 + m_1 m_2}$$

$$= \frac{\frac{1}{3} - (-2)}{1 + (\frac{1}{3})(-2)}$$

i.e. $\tan \theta = 7$.   The tangent of the acute angle between the two lines is 7.

Angle between two curves

The angle between two curves, at the point P where they intersect, is defined as the angle between the tangents to the curves at that point P. Thus, to find the angle between two curves, we need to find the gradients of the curves at the point at which they intersect.

Example 2

The two curves $y = 2x^2 - 3$ and $y = x^2 - 5x + 3$ intersect at two points, one of which, P, is in the fourth quadrant. Find the tangent of the acute angle between these curves at the point P.

$$\left. \begin{array}{l} y = 2x^2 - 3 \\ y = x^2 - 5x + 3 \end{array} \right\} \quad \text{These curves intersect where } 2x^2 - 3 = x^2 - 5x + 3$$
$$x^2 + 5x - 6 = 0$$

i.e. $\qquad\qquad\qquad x = -6 \quad \text{or} \quad x = 1$

The points of intersection are $(-6, 69)$ and $(1, -1)$ and P is $(1, -1)$.

$y = 2x^2 - 3$	$y = x^2 - 5x + 3$
gradient $= \dfrac{dy}{dx} = 4x$	gradient $= \dfrac{dy}{dx} = 2x - 5$
At P, $x = 1$, $\dfrac{dy}{dx} = 4 = m_1$	At P, $x = 1$, $\dfrac{dy}{dx} = 2 - 5$
	$= -3 = m_2$

If θ is the angle between the curves at the point P, then

$$\tan \theta = \frac{m_1 - m_2}{1 + m_1 m_2}.$$
$$= \frac{4 - (-3)}{1 + (4)(-3)}$$

i.e. $\qquad\qquad \tan \theta = -\dfrac{7}{11}$

Since $\tan \theta$ is negative, θ is an obtuse angle, and $\tan (180° - \theta) = \dfrac{7}{11}$

The tangent of the acute angle between the curves at the point of intersection in the fourth quadrant is $\frac{7}{11}$.

Alternatively we could use vector methods to determine the angle between two straight lines, see question 14, Exercise 2E page 68.

Exercise 16A

1. Find the tangent of the acute angle between the following pairs of lines:
   (a) $y = 2x - 3$ (b) $y = 2x + 4$ (c) $4y = 3x - 4$
       $y = x + 4$ $2y = x - 10$ $y + 2x = 3$
2. Calculate the angles of the triangle ABC where A, B and C are the points $(-2, 2)$, $(2, 4)$ and $(7, -1)$ respectively. (Hint: sketch the triangle first to determine whether any of the angles are obtuse).
3. The straight line $y = x + 2$ cuts the curve $y = x^2$ at two places. Find the tangent of the acute angle between the line and the curve at each of these two points.
4. The curve $y = x^2 + 4x - 5$ cuts the curve $y = 1 - x^2$ at two points A and B with A lying on the x-axis. Find the acute angle between the curves at the point B, giving your answer to the nearest degree.
5. The curves $y = 2x^2 - 5x - 2$ and $y = 1 + 3x - x^2$ intersect at two points A and B with A lying in the first quadrant. Find the acute angle between the curves at the point A giving your answer to the nearest degree.
6. Find the equations of the two straight lines that make an angle of $45°$ with the line $y = 3x - 2$ and that pass through the point $(6, 4)$.

16.2 Distance of a point from a line

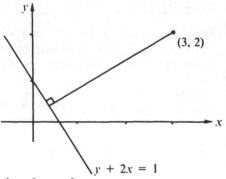

Suppose we want to find the shortest distance
from the point (3, 2) to the line $y + 2x = 1$.
We could proceed as follows:

gradient of $y + 2x = 1$ is -2

∴   gradient of line perpendicular to $y + 2x = 1$ is $\frac{1}{2}$.
Line through the point (3, 2) perpendicular to
$y + 2x = 1$ has equation $y - 2 = \frac{1}{2}(x - 3)$

i.e. $2y = x + 1$

Solving $2y = x + 1$ and $y + 2x = 1$ simultaneously gives $x = \frac{1}{5}$ and $y = \frac{3}{5}$.
Thus the lines $2y = x + 1$ and $y + 2x = 1$ intersect at $(\frac{1}{5}, \frac{3}{5})$.
Length of the line joining the points (3, 2) and $(\frac{1}{5}, \frac{3}{5})$ is given by

$$\sqrt{[(3 - \tfrac{1}{5})^2 + (2 - \tfrac{3}{5})^2]} = \tfrac{7}{5}\sqrt{5} \text{ units.}$$

Alternatively we could express the equation $y + 2x = 1$ in vector form and
proceed as in chapter 2 (see page 67 example 22). However, at times it is more
convenient to quote and use the formula stated and proved below.

The perpendicular distance from the point (x_1, y_1) to the line $ax + by + c = 0$

is given by $\dfrac{ax_1 + by_1 + c}{\sqrt{(a^2 + b^2)}}$,

i.e. we substitute the coordinates of the point into the equation of the line and divide
by $\sqrt{(a^2 + b^2)}$.

For the point (3, 2) and the line $y + 2x = 1$ (i.e. $2x + y - 1 = 0$) this formula
gives the perpendicular distance as

$$\frac{2(3) + 2 - 1}{\sqrt{(2^2 + 1^2)}} = \frac{7}{\sqrt{5}} = \frac{7}{5}\sqrt{5} \quad \text{as required.}$$

The formula quoted above may be proved as follows:

Let P be the point (x_1, y_1) and AB the line
$ax + by + c = 0$. PL is the perpendicular
from P to the line AB.
The line through P, parallel to the y-axis, cuts the
x-axis at M, AB at H and the line through the
origin parallel to AB at Q.

Gradient of $ax + by + c = 0$ is $\dfrac{-a}{b} = \tan \theta$

where θ is the angle made with the x-axis, and

the y-intercept, i.e. OB, is $-\dfrac{c}{b}$.

Then   PL sec θ = PH

$= $ PM $-$ MQ $-$ QH

$= y_1 - $ OM tan θ $-$ OB

$= y_1 - x_1\left(-\dfrac{a}{b}\right) - \left(-\dfrac{c}{b}\right)$

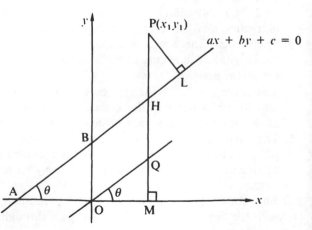

$$PL \sqrt{(1 + \tan^2\theta)} = \frac{ax_1 + by_1 + c}{b}$$

$$PL = \frac{ax_1 + by_1 + c}{b\sqrt{(1 + a^2/b^2)}} = \frac{ax_1 + by_1 + c}{\sqrt{(a^2 + b^2)}} \quad \text{as required.}$$

Example 3

Find the perpendicular distance from the line $3y = 4x - 1$ to the points (a) $(1, 3)$, (b) $(1, -2)$.

$$3y = 4x - 1$$

or

$$0 = 4x - 3y - 1$$

(a) Using $\dfrac{ax_1 + by_1 + c}{\sqrt{(a^2 + b^2)}}$

Required distance $= \dfrac{4(1) - 3(3) - 1}{\sqrt{(4^2 + 3^2)}}$

$$= -\frac{6}{5}$$

(b) Using $\dfrac{ax_1 + by_1 + c}{\sqrt{(a^2 + b^2)}}$

Required distance $= \dfrac{4(1) - 3(-2) - 1}{\sqrt{(4^2 + 3^2)}}$

$$= \frac{9}{5}$$

Notice that one result is negative and the other positive. This is due to the fact that the two points lie on opposite sides of the line.
Thus the points $(1, 3)$ and $(1, -2)$ are respectively $1\frac{1}{5}$ and $1\frac{4}{5}$ units from the line $3y = 4x - 1$, and they lie on different sides of the line.

Example 4

Find the equations of the lines that bisect the angles between the lines $3x - 4y + 13 = 0$ and $12x + 5y - 32 = 0$.

Any point on a line that bisects one of the angles between the line $3x - 4y + 13 = 0$ and the line $12x + 5y - 32 = 0$ will be equidistant from these two lines. Thus we require the locus of all the points which are equidistant from $3x - 4y + 13 = 0$ and $12x + 5y - 32 = 0$. Suppose that (x, y) is the general point equidistant from the two lines, then using $\dfrac{ax_1 + by_1 + c}{\sqrt{(a^2 + b^2)}}$ it follows that

$$\frac{3x - 4y + 13}{\sqrt{(3^2 + 4^2)}} = \pm\frac{12x + 5y - 32}{\sqrt{(12^2 + 5^2)}}$$

i.e. $13(3x - 4y + 13) = \pm 5(12x + 5y - 32)$
which gives $3x + 11y - 47 = 0$ and $11x - 3y + 1 = 0$.

The equations of the bisectors of the angles between the lines $3x - 4y + 13 = 0$ and $12x + 5y - 32 = 0$ are $3x + 11y - 47 = 0$ and $11x - 3y + 1 = 0$.

Note: The two bisectors will be at right angles to each other (i.e. in the example above the gradients of the bisectors are $-\frac{3}{11}$ and $\frac{11}{3}$).

Exercise 16B

1. In each of the following find the perpendicular distance from the given point to the given line.
   (a) $4y + 3x + 6 = 0$, $(2, -1)$ (b) $12y = 5x - 1$, $(-3, -2)$

(c) $x + y + 6 = 0$, $(1, 3)$ (d) $2y = x + 2$, $(1, 4)$
(e) $4y = 2x + 1$, $(1, -3)$

2. Find which two of the points A(1, 1), B(0, −1) and C(−1, 2) lie on the same side of the line $2x + 3y = 1$ and find how far the third point is from the line.

3. Find the possible values of k given that the point $(4, k)$ is the same distance from $9x + 8y + 1 = 0$ as $(2, 5)$ is from $y = 12x + 2$.

4. Points A, B and C have coordinates $(7, 9)$, $(-1, 2)$ and $(2, 6)$ respectively.
   Find (a) the equation of the straight line through B and C,
         (b) the length of BC,
         (c) the perpendicular distance from A to BC,
         (d) the area of the triangle ABC.

5. Find the locus of the points that are equidistant from the two lines
   $6y = 7x + 1$, $9x + 2y + 3 = 0$.

6. Find the equations of the lines that bisect the angles between the following pairs of lines
   (a) $3x + y + 3 = 0$ (b) $y = x + 1$ (c) $3x = 4y + 2$
       $3y = x - 1$ $7x + y + 3 = 0$ $12y = 5x + 2$

Conic sections

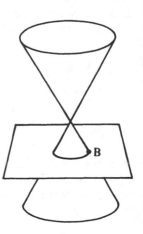

Let us consider a hollow double cone placed so that its circular bases are horizontal and, therefore, the axis of the double cone is vertical. Let B be a point on the surface of the cone and suppose that a plane, which contains the point B, intersects this double cone.

The intersection of the plane and the cone may be thought of as a 'slice' made by cutting the double cone through the point B.

If the plane is horizontal the intersection with the cone will be a circle; if the angle of the plane is then gradually increased the shape of the intersection will change to an ellipse, a parabola and a hyperbola. These four situations are shown below and they are referred to as conic sections:

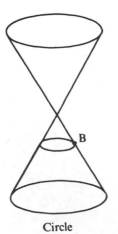

Circle

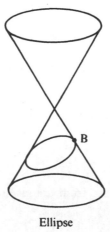

Ellipse

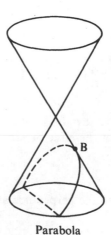

Parabola

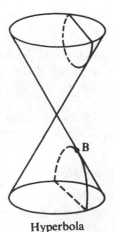

Hyperbola

We now consider the four types of curve in more detail and although we shall define each curve using the idea of a locus, it can be shown that such definitions are consistent with the conic section idea explained above.

16.3 The circle

The circle is defined as the locus of all points, P(x, y), which are equidistant from some given point C, (a, b). Suppose that the distance of the points P, from the given point C(a, b) is r, then

$$CP^2 = (x - a)^2 + (y - b)^2$$
$$= r^2$$

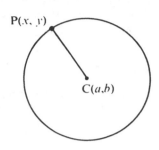

Thus the required locus is $(x - a)^2 + (y - b)^2 = r^2$, where r is the radius of the circle and the point (a, b) is its centre.

If the coordinates of the point C are (0, 0), i.e. C is at the origin, then the equation becomes $x^2 + y^2 = r^2$, thus

Circle centre (0, 0) and radius r has equation	$x^2 + y^2 = r^2$
Circle centre (a, b) and radius r has equation	$(x - a)^2 + (y - b)^2 = r^2$

Example 5

Find the centre and the radius of the circles with equations:
(a) $(x - 2)^2 + y^2 = 25$ (b) $x^2 + y^2 - 4x - 2y = 4$

(a) $(x - 2)^2 + y^2 = 25$
Comparing this equation with
$(x - a)^2 + (y - b)^2 = r^2$
the centre is (2, 0) and the radius is 5 units.

(b) $x^2 + y^2 - 4x - 2y = 4$
which can be rearranged as
$(x^2 - 4x + 4) + (y^2 - 2y + 1) = 4 + 4 + 1$
i.e. $(x - 2)^2 + (y - 1)^2 = 3^2$
Comparing this equation with
$(x - a)^2 + (y - b)^2 = r^2$
the centre is (2, 1) and the radius is 3 units

Note: the general equation of the circle, $(x - a)^2 + (y - b)^2 = r^2$, may be written as $x^2 + y^2 - 2ax - 2by + (a^2 + b^2 - r^2) = 0$. Notice that the coefficients of x^2 and y^2 are equal in this second order equation and that there is no xy term. This will be true for all equations of circles.

Intersection of a line and a circle

Consider a straight line $y = mx + c$ and a circle $(x - a)^2 + (y - b)^2 = r^2$.

There are three possible situations:

1. the line cuts the circle in two distinct places, i.e. part of the line is a chord of the circle,

2. the line *touches* the circle, i.e. the line is a tangent to the circle,

3. the line neither cuts nor touches the circle.

Given the equations of a line and a circle we can determine which situation applies by solving the two equations simultaneously. If the resulting quadratic equation has two distinct roots, situation 1 applies; a repeated root indicates that situation 2 applies; no real roots indicates that situation 3 applies.

Alternatively, we can find the centre (a, b) and radius, r, of the circle and determine the perpendicular distance from the line to the point (a, b). If this distance is less than r then situation 1 applies; if the distance is equal to r then situation 2 applies; if the distance exceeds r then situation 3 applies.

Example 6

Show that part of the line $3y = x + 5$ is a chord of the circle $x^2 + y^2 - 6x - 2y - 15 = 0$ and find the length of this chord.

METHOD 1
Substituting $x = 3y - 5$ into $x^2 + y^2 - 6x - 2y - 15 = 0$ gives, on simplification

$$y^2 - 5y + 4 = 0$$

i.e. $(y - 4)(y - 1) = 0$
giving $y = 4$ or $y = 1$.
When $y = 4, x = 7$
and when $y = 1, x = -2$
Thus the line cuts the circle in 2 distinct points $(7, 4)$ and $(-2, 1)$.
Thus part of the line is a chord of the circle.
The length of the chord is the distance of the point $(-2, 1)$ from the point $(7, 4)$.
i.e.   length of the chord
$$= \sqrt{\{(7 - -2)^2 + (4 - 1)^2\}}$$
$$= \sqrt{(81 + 9)}$$
$$= 3\sqrt{10} \text{ units}$$

METHOD 2
Rearranging $x^2 + y^2 - 6x - 2y - 15 = 0$
gives $(x - 3)^2 + (y - 1)^2 = 25$
$\therefore$ the circle has radius 5 units and centre $(3, 1)$.
Perpendicular distance of $(3, 1)$ from the line $3y = x + 5$ is given by

$$\frac{3 - 3(1) + 5}{\sqrt{(1^2 + 3^2)}}$$
$$= \frac{5}{\sqrt{10}} = \tfrac{1}{2}\sqrt{10}$$
$$= 1 \cdot 58 \quad \text{(to two decimal places)}.$$

Thus, as the perpendicular distance from $3y = x + 5$ to the centre of the circle is less than the radius of the circle, the line $3y = x + 5$ will cut the circle in two points Thus part of the line is a chord of the circle.

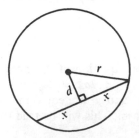

From the diagram, the length of the chord
$$= 2x$$
$$= 2\sqrt{(r^2 - d^2)}$$
$$= 2\sqrt{\{25 - (\tfrac{1}{2}\sqrt{10})^2\}}$$
$$= 3\sqrt{10} \text{ units}$$

It should be realised that, in the above example, method 1 would not have been quite so easy to use had the quadratic equation not factorised, but had instead required the use of the formula to solve it.

Gradient at a given point on a curve

The gradient at a given point is the gradient of the tangent to that curve at the given point, and we can find this directly from the equation of the curve by implicit differentiation.

Example 7

Find the equation of (i) the tangent and (ii) the normal to the circle with equation $x^2 + y^2 - 6x - 2y - 3 = 0$ at the point (5, 4) on the circle.

$$x^2 + y^2 - 6x - 2y - 3 = 0$$

differentiating with respect to x,

$$2x + 2y\frac{dy}{dx} - 6 - 2\frac{dy}{dx} = 0$$

i.e.

$$\frac{dy}{dx} = \frac{3 - x}{y - 1}$$

at the point (5, 4)

$$\frac{dy}{dx} = \frac{3 - 5}{4 - 1} = -\frac{2}{3}$$

(i) gradient of tangent is $-\frac{2}{3}$

Using $y - y_1 = m(x - x_1)$,
the equation of the tangent is

$$y - 4 = -\tfrac{2}{3}(x - 5)$$

i.e. $3y + 2x = 22$

(ii) gradient of normal is $\frac{3}{2}$

Using $y - y_1 = m(x - x_1)$,
the equation of the normal is

$$y - 4 = \tfrac{3}{2}(x - 5)$$

i.e. $2y - 3x + 7 = 0$

At the point (5, 4) on the circle $x^2 + y^2 - 6x - 2y - 3 = 0$, the equations of the tangent and of the normal are respectively $3y + 2x = 22$ and $2y - 3x + 7 = 0$.

Circle through three given points

Three non-collinear points define a circle, i.e. there is one, and only one circle which can be drawn through three non-collinear points. The equation of any circle may be written as $x^2 + y^2 + 2gx + 2fy + c = 0$, where g, f and c are constants.

Thus, if we are given the coordinates of three points on the circumference of a circle, we can substitute these values of x and y into the equation of the circle and obtain three equations which can be solved simultaneously to find the constants g, f and c.

Example 8

Find the equation of the circle passing through the points (0, 1), (4, 3) and (1, −1).

Suppose the equation of the circle is

$$x^2 + y^2 + 2gx + 2fy + c = 0.$$

Substituting the coordinates of each of the three points into this equation gives:

$$
\begin{aligned}
1 + 2f + c &= 0 && \dots [1] \\
25 + 8g + 6f + c &= 0 && \dots [2] \\
2 + 2g - 2f + c &= 0 && \dots [3]
\end{aligned}
$$

Multiplying equation [3] by 4 and then subtracting from equation [2], gives
$$17 + 14f - 3c = 0 \qquad \ldots [4]$$
Multiplying equation [1] by 3 and adding to equation [4], gives
$$20 + 20f = 0 \text{ or } f = -1.$$
Then from equation [1] $\qquad\qquad\qquad c = 1$
and from equation [3] $\qquad\qquad\qquad g = -2\tfrac{1}{2}.$
The equation of the circle which passes through $(0, 1)$, $(4, 3)$ and $(1, -1)$ is
$$x^2 + y^2 - 5x - 2y + 1 = 0.$$

Note: In the above example, the coordinates of three points on the
circumference were given. The data may sometimes be given in a different
form and it is often necessary to use some other geometrical fact concerning
the circle, e.g. the perpendicular bisector of a chord passes through the
centre of the circle, in order to find the equation of the circle.

Intersecting circles

Consider two circles, of radius r_1 and r_2 $(r_1 > r_2)$ with their centres distance d apart.
There are a number of possible situations:

(i) Circles *touch* externally

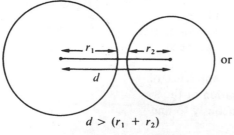

$$d = r_1 + r_2$$

(ii) Circles *touch* internally

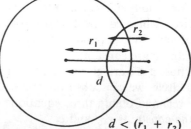

$$d = r_1 - r_2$$

(iii) Circles do not intersect

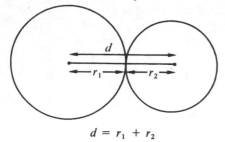 or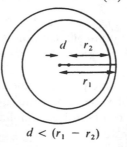

$$d > (r_1 + r_2) \qquad\qquad d < (r_1 - r_2)$$

(iv) Circles intersect at two distinct points

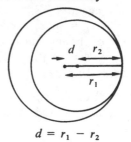

$$d < (r_1 + r_2)$$

Note also that:

(a) Circles having the same centre
are said to be *concentric*

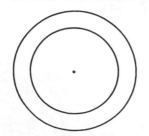

(b) Two circles that cut at right-angles
are said to be *orthogonal*

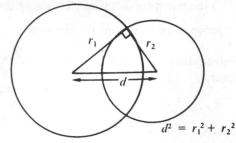

$$d^2 = r_1{}^2 + r_2{}^2$$

Example 9

Prove that the circles $x^2 + y^2 + 2x - 8y + 5 = 0$ and
$x^2 + y^2 - 4x - 4y + 7 = 0$ are orthogonal.

$$x^2 + y^2 + 2x - 8y + 5 = 0$$
i.e. $(x + 1)^2 + (y - 4)^2 = 12$
centre $(-1, 4)$ and radius is $\sqrt{12}$ units

$$x^2 + y^2 - 4x - 4y + 7 = 0$$
i.e. $(x - 2)^2 + (y - 2)^2 = 1$
centre $(2, 2)$ and radius is 1 unit

Now $r_1^2 + r_2^2 = (\sqrt{12})^2 + 1^2$
$$= 13$$
and the (distance between their centres)$^2 = (2 - (-1))^2 + (2 - 4)^2$
$$= 9 + 4 = 13.$$
Thus, since $d^2 = r_1^2 + r_2^2$, the circles $x^2 + y^2 + 2x - 8y + 5 = 0$ and
$x^2 + y^2 - 4x - 4y + 7 = 0$ are orthogonal.

Example 10

Show that the circles $x^2 + y^2 - 2y - 4 = 0$ and $x^2 + y^2 - x + y - 12 = 0$
intersect in two distinct points and find the equation of the common chord.

The circles have equations $\qquad x^2 + y^2 - 2y - 4 = 0$ ... [1]
and $\qquad\qquad\qquad\qquad x^2 + y^2 - x + y - 12 = 0$ ... [2]
Consider the equation
$$x^2 + y^2 - 2y - 4 - (x^2 + y^2 - x + y - 12) = 0 \quad \text{... [3]}$$
This simplifies to $x - 3y + 8 = 0$, i.e. the equation of a straight line.
If the point, P(a, b), satisfies both equations [1] and [2], then it will also
satisfy $x - 3y + 8 = 0$.
If some other point, Q(c, d), satisfies equations [1] and [2], then it also will
satisfy $x - 3y + 8 = 0$.
But if this is so, then PQ is the common chord of the two circles.
Thus $x - 3y + 8 = 0$ is the equation of the common chord of the two circles.

Solving the equation of the common chord, $\quad x - 3y + 8 = 0$
with one of the circles $\qquad\qquad\qquad x^2 + y^2 - 2y - 4 = 0,$
i.e. $\qquad\qquad\qquad\qquad (3y - 8)^2 + y^2 - 2y - 4 = 0$
giving $\qquad\qquad\qquad\qquad\qquad y^2 - 5y + 6 = 0$
i.e. $\qquad\qquad\qquad\qquad\qquad y = 2,$ or $y = 3,$ and then
substituting in the equation of the common chord, the distinct points of
intersection are $(-2, 2)$ and $(1, 3)$.

Note: In this example we were able to find the common chord without
having first to find the points of intersection.
If two circles are such that they do not intersect, then the equation
analogous to equation [3] above, would, when solved with one of the circle
equations, not yield any *real* points.

Parametric forms

It is sometimes useful to express the equation
of a circle using a parameter θ. For a circle
with centre at the origin and radius r,
i.e. $x^2 + y^2 = r^2$, we can write the
parametric equations:

$$x = r \cos \theta \text{ and } y = r \sin \theta.$$

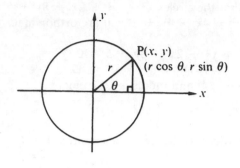

These equations can also be thought of as
giving the coordinates of a point $P(x, y)$ on
the circle (and the equations together give the
equation of the circle).
In a similar way the circle $(x - a)^2 + (y - b)^2 = r^2$ can be expressed
using the parameter θ:

$$x = a + r \cos \theta,$$
$$y = b + r \sin \theta.$$

Example 11

Find the equation of the tangent to the circle $x^2 + y^2 = 4$, at the general
point $(2 \cos \theta, 2 \sin \theta)$. Hence find the equation of the tangent at the point
$(\sqrt{2}, \sqrt{2})$.

$$x = 2 \cos\theta \qquad\qquad\qquad y = 2 \sin \theta$$
$$\therefore \qquad \frac{dx}{d\theta} = -2 \sin \theta \qquad\qquad \frac{dy}{d\theta} = 2 \cos \theta$$

Thus using $\dfrac{dy}{dx} = \dfrac{dy}{d\theta} \times \dfrac{d\theta}{dx}$

$$\frac{dy}{dx} = 2 \cos \theta \left(\frac{1}{-2 \sin \theta}\right) = -\cot \theta.$$

Thus the tangent at the point $(2 \cos \theta, 2 \sin \theta)$ has gradient $-\cot \theta$.
Using $y - y_1 = m(x - x_1)$, the equation of the tangent at $(2 \cos \theta, 2 \sin \theta)$ is
$$y - 2 \sin \theta = -\cot \theta \, (x - 2 \cos\theta)$$
$$\therefore \quad y \sin \theta - 2 \sin^2 \theta = -x\cos\theta + 2\cos^2\theta$$
i.e. $\quad y \sin \theta + x \cos \theta = 2 \qquad\qquad\qquad \dots [1]$

At the point $(\sqrt{2}, \sqrt{2})$, $\theta = \dfrac{\pi}{4}$ and equation [1] becomes $\dfrac{y}{\sqrt{2}} + \dfrac{x}{\sqrt{2}} = 2$
$$\text{i.e.} \quad x + y = 2\sqrt{2}.$$

The reader should verify that the same answer is obtained by differentiating
$x^2 + y^2 = 4$ implicitly to determine the gradient at the point $(\sqrt{2}, \sqrt{2})$ and
then using the form $y - y_1 = m(x - x_1)$ as before.

Exercise 16C

1. State which of the following are equations of circles.
   (a) $x^2 + 2xy + y^2 = 4$ $\qquad\qquad$ (b) $x^2 + y^2 = 25$
   (c) $x^2 + y^2 = 19$ $\qquad\qquad\qquad$ (d) $2x^2 + 3x - y^2 + 2y = 16$
   (e) $x^2 + 3x - y^2 = 7$ $\qquad\qquad$ (f) $x^2 + y^2 + 2x - 8y = 1$

2. Find the centre and radius of each of the following circles.
   (a) $x^2 + y^2 = 10$ (b) $(x - 3)^2 + y^2 = 25$
   (c) $(x - 2)^2 + (y + 1)^2 = 18$ (d) $x^2 + y^2 + 2x - 2y = 2$
   (e) $x^2 + y^2 - 6x + 2y = 6$ (f) $x^2 + y^2 + 4x = 6$
   (g) $4x^2 + 4y^2 + 4x = 99$ (h) $2x^2 + 2y^2 - 2x + 2y = 1$

3. Prove that the circles $x^2 + y^2 - 4x - 6y = 12$ and
   $x^2 + y^2 - 4x - 6y = 3$ are concentric and find the coordinates of the
   common centre.

4. Prove that the line $y = 3x - 1$ neither cuts nor touches the circle
   $(x - 4)^2 + (y - 1)^2 = 9$.

5. Prove that the line $y = x + 3$ neither cuts nor touches the circle
   $x^2 + y^2 - 6x + 8y = 7$.

6. Prove that the line $y = 2x - 3$ is a tangent to the circle
   $(x - 5)^2 + (y - 2)^2 = 5$.

7. Prove that the line $y = 3x + 1$ is a tangent to the circle
   $x^2 + y^2 - 14x - 4y + 13 = 0$.

8. Prove that part of the line $y = 2x + 3$ forms a chord to the circle
   $(x - 4)^2 + (y - 1)^2 = 28$ and find the length of this chord.

9. Prove that part of the line $y = 3x + 5$ forms a chord to the circle
   $x^2 + y^2 - 2x - 6y + 5 = 0$ and find the length of this chord.

10. Find the equation of (i) the tangent and (ii) the normal to each of the
    following circles at the given point on the circle
    (a) $x^2 + y^2 = 2x - 6y$, $(2, -6)$,
    (b) $x^2 + y^2 + 4x - 2y = 21$, $(3, 2)$,
    (c) $4x^2 - 4x + 4y^2 + 8y = 15$, $(1\frac{1}{2}, 1)$.

11. Find the equation of the circle passing through the given three points.
    (a) $(1, 0)$, $(0, 1)$, $(3, 4)$,
    (b) $(2, -1)$, $(1, 3)$, $(1, -4)$,
    (c) $(2, 2)$, $(-2, 1)$, $(2, 3)$.

12. Prove that the circles $x^2 + y^2 - 6x - 12y + 40 = 0$ and
    $x^2 + y^2 - 4y = 16$ are orthogonal.

13. Prove that the following pairs of circles *touch* each other and state
    whether the contact is internal or external.
    (a) $x^2 + y^2 - 2x = 0$, $x^2 + y^2 - 8x + 12 = 0$
    (b) $x^2 + y^2 - 2x - 2y = 18$, $x^2 + y^2 - 14x - 8y + 60 = 0$
    (c) $x^2 + y^2 - 12x - 2y = 12$, $x^2 + y^2 - 4x + 4y + 4 = 0$
    (d) $x^2 + y^2 - 4x + 2y = 8$, $x^2 + y^2 + 6x - 13y + 22 = 0$

14. Prove that the circle $x^2 + y^2 - 2x - 6y + 1 = 0$ cuts the circle
    $x^2 + y^2 - 8x - 8y + 31 = 0$ in two distinct places and find the
    equation of the common chord.

15. Prove that the circle $x^2 + y^2 - 2x - 6y = 15$ cuts the circle
    $x^2 + y^2 - 16x - 4y + 59 = 0$ in two distinct places and find the
    equation of the common chord.

16. Find the equation of the tangent to the circle $(x - 2)^2 + (y - 3)^2 = 16$
    at the general point $((2 + 4 \cos \theta), (3 + 4 \sin \theta))$.
    Hence find the equation of the tangent at the point $(4, 3 + 2\sqrt{3})$.

17. Find the equation of the circle having AB as diameter where A is the
    point $(1, 8)$ and B is the point $(3, 14)$.

18. Points A$(0, 2)$ and B$(4, -2)$ lie on the circumference of a given circle.
    Points C$(-3, -3)$ and D$(7, 2)$ lie outside the circle but the centre of the

circle lies on CD. Find the equation of the circle.

19. Find the equation of the circle having AB as a chord, where A is the point (3, 4), B is the point (6, 1) and the tangent to the circle at point A is the line $2y = x + 5$.

16.4 The parabola

The parabola is defined as the locus of those points equidistant from a fixed point and a fixed straight line. The fixed point is called the *focus* and the fixed straight line is called the *directrix*.

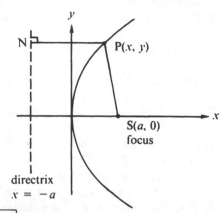

Suppose the line $x = -a$ is the directrix and the point S(a, 0) is taken as the focus, then if P(x, y) is a point on the parabola,

by definition $\qquad\qquad$ PS = PN
hence $\qquad\qquad\qquad$ PS^2 = PN^2
i.e. $\qquad (x - a)^2 + (y - 0)^2 = (x + a)^2$
which gives $\qquad\qquad\qquad y^2 = 4ax$
as the equation of the parabola.

> parabola $y^2 = 4ax$, focus (a, 0), directrix $x = -a$.

Thus,

Notes:

The origin is clearly a point on this parabola [S is (a, 0), i.e. a units from the origin, and the origin is a units from the directrix].

The equation $y^2 = 4ax$ is unchanged by substituting $(-y)$ for y, so the parabola has the x-axis as a line of symmetry.

Since $y^2 \geqslant 0$, then for $a > 0$, $x \geqslant 0$, i.e. the parabola lies in the first and fourth quadrants.

Example 12

Sketch the graph of $(y - 2)^2 = 12(x - 1)$ showing clearly the focus and the directrix of the parabola.

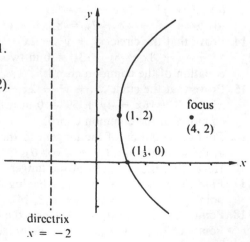

$(y - 2)^2 = 12(x - 1)$
This equation can be written
$\qquad Y^2 = 12\,X \quad$ where $\quad Y = y - 2 \quad$ and $\quad X = x - 1$.
Thus the curve $(y - 2)^2 = 12(x - 1)$ is a parabola with $a = 3$ and origin at $X = 0$, $Y = 0$, i.e. the point (1, 2).
The parabola $y^2 = 4ax$ has origin (0, 0), focus (a, 0) and directrix $x = -a$.
Thus with $a = 3$ and origin at (1, 2) the focus is at (4, 2) and the directrix is $x = -2$.
When $x = 0$, $(y - 2)^2 = -12$, thus the curve does not cut the y-axis.
When $y = 0$, $(-2)^2 = 12(x - 1)$, giving $x = 1\frac{1}{3}$.
Thus the curve cuts the x-axis at $(1\frac{1}{3}, 0)$.

Notice that the graph of $(y - 2)^2 = 12(x - 1)$ is that of $y^2 = 12x$ translated by 1 unit parallel to the x-axis and 2 units parallel to the y-axis which is consistent with the ideas of section 11.3, page 284.

Example 13

Show that the line $y = 3x + 1$ *touches* the parabola $y^2 = 12x$.

If the given line *touches* the parabola, i.e. is a tangent to the parabola, solving the two equations $y = 3x + 1$ and $y^2 = 12x$ simultaneously should give a quadratic equation with a repeated root.

Substituting $y = 3x + 1$ into $y^2 = 12x$ gives $(3x + 1)^2 = 12x$

$$\therefore \quad 9x^2 + 6x + 1 = 12x$$
$$9x^2 - 6x + 1 = 0$$
$$(3x - 1)(3x - 1) = 0$$
$$\therefore \quad x = \tfrac{1}{3} \text{ or } \tfrac{1}{3}, \text{ i.e. a repeated root.}$$

Thus the line $y = 3x + 1$ touches the parabola $y^2 = 12x$ at the point $(\tfrac{1}{3}, 2)$.

Gradient at a particular point

As with the circle, the gradient at a particular point on a parabola can be determined by implicit differentiation.

Example 14

Show that the point A$(2, -4)$ lies on the parabola $y^2 = 8x$ and find the equation of the normal to the parabola at the point A.

Substituting $x = 2$, $y = -4$ in the equation $y^2 = 8x$, gives $(-4)^2 = 8(2)$, so the point A does lie on the parabola $y^2 = 8x$.
Differentiating implicitly, with respect to x,

$$2y\frac{dy}{dx} = 8$$

i.e. $\dfrac{dy}{dx} = \dfrac{4}{y}.$

$\therefore$ the gradient at the point A$(2, -4)$ is $\dfrac{4}{-4} = -1$

Thus the gradient of the normal at A is $+1$ and, using $y - y_1 = m(x - x_1)$, the equation of the normal at A is $(y - -4) = 1(x - 2)$.

Thus the equation of the normal at the point $(2, -4)$ on the parabola $y^2 = 8x$ is $y = x - 6$.

Example 15

Find the equations of the tangents drawn from the point (1, 3) to the parabola $y^2 = -16x$.

Suppose the tangents have equations of the form $y = mx + c$. These tangents pass through the point (1, 3), hence

$$3 = m(1) + c$$
$$\therefore \quad 3 = m + c \qquad \ldots [1]$$

Substituting for y from $y = mx + c$ into $y^2 = -16x$ gives

$$m^2x^2 + 2x(mc + 8) + c^2 = 0.$$

Now as $y = mx + c$ is a tangent to the curve $y^2 = -16x$, this quadratic must have a repeated root, (i.e. $b^2 - 4ac = 0$).

$$\therefore \qquad \qquad 4(mc + 8)^2 - 4m^2c^2 = 0$$

giving

$$m = \frac{-4}{c}$$

substituting this value of m in equation [1] gives

$$3 = \frac{-4}{c} + c$$

i.e. $c^2 - 3c - 4 = 0$

giving $\begin{cases} c = 4 \\ m = -1 \end{cases}$ or $\begin{cases} c = -1 \\ m = 4. \end{cases}$

Thus the required tangents have equations $y = -x + 4$ and $y = 4x - 1$.

Parametric form

The general equation of the parabola, $y^2 = 4ax$, may be expressed in parametric form as $x = at^2$, $y = 2at$ where t is the parameter.
As in the case of the circle, these two equations may be used for the parametric coordinates of a point on the parabola. Since, if we eliminate 't'

between these equations, we get $x = a\left(\dfrac{y}{2a}\right)^2$ or $y^2 = 4ax$, it follows that

the point $(at^2, 2at)$ lies on the parabola for all values of t.
Also, therefore, for each value of t there is one and only one point on the parabola.

Example 16

Find the equation of the normal to the parabola $y^2 = 4ax$ at the point $(at^2, 2at)$.

With $x = at^2$ and $y = 2at$

then $\dfrac{dx}{dt} = 2at$ and $\dfrac{dy}{dt} = 2a$ Using $\dfrac{dy}{dx} = \dfrac{dy}{dt} \times \dfrac{dt}{dx}$

$$\dfrac{dy}{dx} = 2a\dfrac{1}{2at} = \dfrac{1}{t}.$$

Thus the gradient of the tangent at $(at^2, 2at)$ is $\dfrac{1}{t}$ and therefore the gradient of the normal at this point is $-t$.

Using $y - y_1 = m(x - x_1)$, the equation of the normal at $(at^2, 2at)$ is

$$y - 2at = -t(x - at^2)$$

i.e. $y + xt = 2at + at^3$ is the equation of the normal at the point $(at^2, 2at)$.

Example 17

The point T $(at^2, 2at)$ lies on the parabola $y^2 = 4ax$ and L is the point $(-a, 2a)$. M is the mid-point of TL. Find the equation of the locus of M as T moves on the parabola.

M will have coordinates $\left(\dfrac{at^2 - a}{2}, \dfrac{2at + 2a}{2} \right)$

$$= \left(\frac{a}{2}(t^2 - 1), a(t + 1) \right)$$

As T moves on the parabola the parameter t varies and the parametric

equations of the locus of M will be: $\quad x = \dfrac{a}{2}(t^2 - 1), \quad y = a(t + 1)$.

Eliminating t from these equations gives $y^2 = 4(x + y)$, the cartesian equation of the locus of M.

Exercise 16D

1. Given that the parabola $y^2 = 4ax$ has focus at $(a, 0)$ write down the coordinates of the foci of the following parabolas.
   (a) $y^2 = 4x$ (b) $y^2 = -8x$ (c) $y^2 = 20x$ (d) $y^2 = 9x$.
2. Given that the parabola $y^2 = 4ax$ has directrix $x = -a$ write down the equation of the directrix of each of the following parabolas.
   (a) $y^2 = 12x$ (b) $y^2 = -12x$ (c) $y^2 = 20x$ (d) $y^2 = -2x$.
3. Sketch the following parabolas showing clearly the focus and directrix of each one.
   (a) $(y - 2)^2 = 4(x - 3)$ (b) $(y + 2)^2 = 8(x - 1)$
   (c) $y^2 + 8y = 4x - 12$.
4. Prove that the line $y = 2x + 2$ *touches* the parabola $y^2 = 16x$ and find the coordinates of the point where this occurs.
5. Prove that the line $2y + 1 = 4x$ *touches* the parabola $y^2 + 4x = 0$ and find the coordinates of the point where this occurs.
6. Prove that the line $y = x + 6$ cuts the parabola $y^2 = 32x$ at two distinct points and find the coordinates of these two points.
7. Prove that the line $y = x - 3$ cuts the parabola $y^2 = 4x$ at two distinct points. Find the coordinates of these two points and the equations of the tangents to the parabola at these points.
8. Find the equations of the tangents drawn from $(-2, 3)$ to $y^2 = 8x$.
9. Find the equations of the tangents drawn from $(-4, 4)$ to $y^2 = 12x$.
10. Find the equation of the tangent to the parabola $y^2 = 4ax$ at the point $(at^2, 2at)$. If this tangent passes through the point $(-a, 0)$ show that $t = \pm 1$.
11. Find the equation of the normal to the parabola $y^2 = 4ax$ at the point $(at^2, 2at)$. If this normal passes through the point $(6a, 0)$ find the possible values of t.
12. Find the coordinates of the point where the normal to the parabola $y^2 = 4ax$ at the point $(ap^2, 2ap)$ cuts
    (a) the x-axis   (b) the y-axis.
13. Verify that the point $T(t^2, 2t)$ lies on the parabola $y^2 = 4x$.
    The line drawn from T, parallel to the y-axis, meets the line $y = 1$ at L. Find the equation of the locus of M, the mid-point of TL, as t varies.

14. Points P(ap^2, $2ap$) and Q(aq^2, $2aq$) lie on the parabola $y^2 = 4ax$. Find the equation of the chord PQ.
    If this chord passes through the focus of the parabola show that $pq = -1$.
    (Note: Such chords are called focal chords).

15. Find the possible values of t given that the tangent to the parabola $y^2 = 4ax$ at the point (at^2, $2at$) passes through the point $(-3a, 2a)$.

16. Prove that the tangents to the parabola $y^2 = 4ax$ at the points P(ap^2, $2ap$) and Q(aq^2, $2aq$) intersect at the point R(apq, $a(p + q)$).

17. If the normal to the parabola $y^2 = 4ax$ at the point P(ap^2, $2ap$) cuts the parabola again at the point Q(aq^2, $2aq$) prove that $p^2 + pq + 2 = 0$.

18. The tangent to the parabola $y^2 = 4ax$ at the point P(ap^2, $2ap$) meets the y-axis at L. Show that as P moves on the parabola the locus of M, the mid-point of PL has equation $2y^2 = 9ax$.

19. Points P(ap^2, $2ap$) and Q(aq^2, $2aq$) lie on the parabola $y^2 = 4ax$. Find the locus of the mid-point of the chord PQ given that P and Q are those points for which $pq = 2a$.

16.5 The ellipse and the hyperbola

The parabola, in section 16.4 was defined as being the locus of points equidistant from a fixed point, the focus, and a fixed line, the directrix, i.e. those points P(x, y) for which PS = PB.

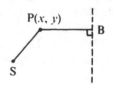

If instead of PS and PB being equal, we have the two lengths in a constant ratio, say $e : 1$, the locus of all such points P(x, y) will give the conic sections.

If $\dfrac{PS}{PB} = e$, called the eccentricity, then for $e = 1$, the locus of P is a parabola,
for $0 < e < 1$ the locus of P is an ellipse,
for $e > 1$, the locus of P is a hyperbola.

The ellipse

Suppose that S is the focus and the line JK the directrix. Take the x-axis as passing through S, perpendicular to JK. Suppose that JK cuts the x-axis at C (see Figure 1).

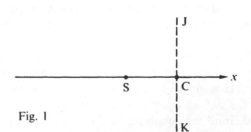

Fig. 1

We require all points P(x, y) that are such that

(distance from P to S):(distance from P to JK) = $e:1$.

Two such points P′ and P″ will lie on the x-axis with P′ dividing SC internally in the ratio $e:1$ and P″ dividing SC externally in the ratio $e:1$.

Taking the origin as the mid-point of P′P″, we say P′ has coordinates (a, 0) and P″($-a$, 0). S is the point (s, 0) and JK is the line $x = k$ (see Fig 2).

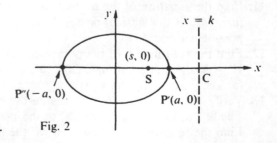

Fig. 2

Thus $SP' = e \times P'C$ gives $a - s = e(k - a)$
and $SP'' = e \times P''C$ gives $a + s = e(k + a)$

Adding these equations gives $k = \dfrac{a}{e}$,

and subtracting these equations gives $s = ae$ (see Figure 3).

Taking P (x, y) as the general point on the ellipse
$$PS = ePB$$
$\therefore$
$$PS^2 = e^2PB^2$$

$\therefore$ $(x - ae)^2 + (y - 0)^2 = e^2\left(\dfrac{a}{e} - x\right)^2$

      $x^2 - 2xae + a^2e^2 + y^2 = a^2 - 2aex + x^2e^2$

i.e. $(1 - e^2)x^2 + y^2 = a^2(1 - e^2)$

thus $\dfrac{x^2}{a^2} + \dfrac{y^2}{a^2(1 - e^2)} = 1$

writing $b^2 = a^2(1 - e^2)$, this gives
$$\dfrac{x^2}{a^2} + \dfrac{y^2}{b^2} = 1$$

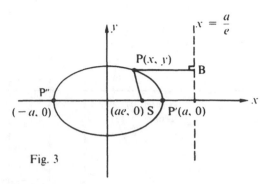

Fig. 3

Note that writing $(-x)$ for x and $(-y)$ for y in this equation leaves it unaltered. Thus the x and y axes are both lines of symmetry of the ellipse. There is therefore a symmetrically placed second focus and directrix, as shown in the diagram below:

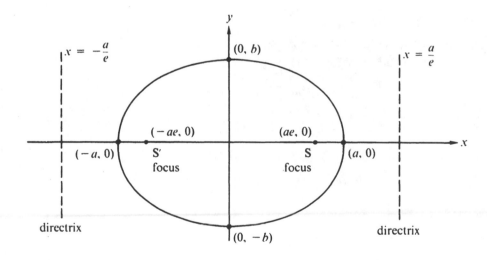

Summary

For the ellipse $\dfrac{x^2}{a^2} + \dfrac{y^2}{b^2} = 1$, $b^2 = a^2(1 - e^2)$, $e < 1$,

foci at $(\pm ae, 0)$, directrices $x = \pm\dfrac{a}{e}$

The hyperbola

In a similar way we can show that the hyperbola is as below:

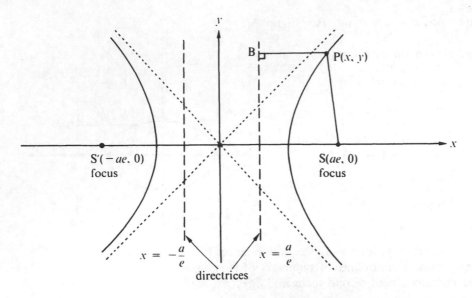

By definition $\qquad\qquad\qquad$ $SP = e\,PB,\ e > 1$

i.e. $\qquad\qquad\qquad\qquad\quad$ $SP^2 = e^2 PB^2$

thus $\qquad\qquad$ $(x - ae)^2 + (y - 0)^2 = e^2\left(x - \dfrac{a}{e}\right)^2$

$\therefore \qquad\qquad x^2 - 2aex + a^2e^2 + y^2 = e^2 x^2 - 2aex + a^2$

$\qquad\qquad\qquad\qquad (e^2 - 1)a^2 = x^2(e^2 - 1) - y^2$

or $\qquad\qquad\qquad\qquad\qquad 1 = \dfrac{x^2}{a^2} - \dfrac{y^2}{a^2(e^2 - 1)}$

writing $b^2 = a^2(e^2 - 1)$ gives $\quad \dfrac{x^2}{a^2} - \dfrac{y^2}{b^2} = 1$

Summary

For the hyperbola $\dfrac{x^2}{a^2} - \dfrac{y^2}{b^2} = 1$, $b^2 = a^2(e^2 - 1)$, $e > 1$,

foci at $(\pm ae,\ 0)$, directrices $x = \pm\dfrac{a}{e}$

Notes:

1. In the diagram of the hyperbola, the lines to which the curve tends at infinity are shown and these are the asymptotes of the hyperbola.

   Writing the equation $\dfrac{x^2}{a^2} - \dfrac{y^2}{b^2} = 1$ $\quad$ as $\quad$ $y^2 = b^2\left(\dfrac{x^2}{a^2} - 1\right)$

   $\qquad\qquad\qquad\qquad$ i.e. $\quad y = \pm\dfrac{bx}{a}\sqrt{\left(1 - \dfrac{a^2}{x^2}\right)}$

   As $x \to \infty$, $\dfrac{a^2}{x^2} \to 0$ and $y = \pm\dfrac{bx}{a}$ which are the equations of the asymptotes.

2. The special case when $e = \sqrt{2}$, gives

$$\frac{x^2}{a^2} - \frac{y^2}{b^2} = 1 \quad \text{and} \quad b^2 = a^2((\sqrt{2})^2 - 1)$$

$$\text{or} \quad b^2 = a^2$$

$$\therefore \quad x^2 - y^2 = a^2. \qquad \qquad \dots [1]$$

Furthermore, the asymptotes are $y = x$ and $y = -x$ and are therefore perpendicular to each other.

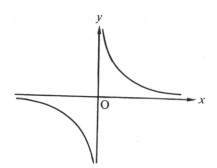

Using the asymptotes as the coordinate axes it can be shown that equation [1] becomes $xy = c^2$.

The curve is a **rectangular hyperbola** as shown on the right.

Example 18

Find the equation of the tangent to the curve $4x^2 + 9y^2 = 36$ at the point $(1, \frac{4}{3}\sqrt{2})$.

$$4x^2 + 9y^2 = 36$$

differentiating with respect to x

$$8x + 18y\frac{dy}{dx} = 0$$

i.e. $$\frac{dy}{dx} = -\frac{4x}{9y}$$

$$\therefore \quad \text{at the point } (1, \tfrac{4}{3}\sqrt{2}) \text{ the gradient} = -\frac{4(1)}{9(\frac{4}{3}\sqrt{2})}$$

$$= -\frac{1}{3\sqrt{2}}$$

Thus the gradient of the tangent at $(1, \frac{4}{3}\sqrt{2})$ is $-\dfrac{1}{3\sqrt{2}}$

Using $y - y_1 = m(x - x_1)$, the equation of the tangent is $\quad y - \frac{4}{3}\sqrt{2} = -\dfrac{1}{3\sqrt{2}}(x - 1)$

$$\text{giving} \quad 3\sqrt{2} \times y + x = 9$$

Thus the tangent to the curve $4x^2 + 9y^2 = 36$, at the point $(1, \frac{4}{3}\sqrt{2})$ is $3y\sqrt{2} + x = 9$.

Example 19

Find the coordinates of the point at which the normal to the curve $xy = 8$ at the point $(4, 2)$, cuts the tangent to the curve $16x^2 - y^2 = 64$ at the point $(2\frac{1}{2}, 6)$.

$xy = 8$	$16x^2 - y^2 = 64$
$\therefore \quad y + x\dfrac{dy}{dx} = 0$	$\therefore \quad 32x - 2y\dfrac{dy}{dx} = 0$
$\dfrac{dy}{dx} = -\dfrac{y}{x}$	$\dfrac{dy}{dx} = \dfrac{16x}{y}$

At the point $(4, 2)$ the gradient of the tangent is $-\frac{2}{4} = -\frac{1}{2}$.

$\therefore$ gradient of normal at $(4, 2)$ is 2.

Equation of normal is $\quad y - 2 = 2(x - 4)$

i.e. $\quad y = 2x - 6 \dots$ [1]

At the point $(2\frac{1}{2}, 6)$ the gradient of the tangent is $\frac{16}{6}(2\frac{1}{2}) = \frac{20}{3}$.

Equation of tangent is $\quad y - 6 = \frac{20}{3}(x - \frac{5}{2})$

i.e. $\quad 3y = 20x - 32 \dots$ [2]

Solving equations [1] and [2] simultaneously gives $x = 1$, $y = -4$.

Thus the required normal and tangent intersect at the point $(1, -4)$.

Parametric forms

For the ellipse $\frac{x^2}{a^2} + \frac{y^2}{b^2} = 1$, we use the parameter θ and, since $\cos^2 \theta + \sin^2 \theta = 1$,
we can give the equation as $x = a \cos \theta$, $y = b \sin \theta$.
Thus the general point on the ellipse has parametric coordinates $(a \cos \theta, b \sin \theta)$.

For the hyperbola $\frac{x^2}{a^2} - \frac{y^2}{b^2} = 1$, we again use the parameter θ and, since $\sec^2 \theta - \tan^2 \theta = 1$,
we can give the equation as $x = a \sec \theta$, $y = b \tan \theta$.
Thus the general point on the hyperbola has parametric coordinates $(a \sec \theta, b \tan \theta)$.

For the rectangular hyperbola $xy = c^2$, we use the parameter 't' and the
equation may then be written as $x = ct$, $y = \frac{c}{t}$.

The parametric coordinates of the general point on the rectangular hyperbola are then $\left(ct, \frac{c}{t}\right)$.

Example 20

Find the equation of the tangent to the ellipse $\frac{x^2}{a^2} + \frac{y^2}{b^2} = 1$ at the point
$(a \cos \theta, b \sin \theta)$.

$$x = a \cos \theta \qquad\qquad y = b \sin \theta$$
$$\frac{dx}{d\theta} = -a \sin \theta \qquad\qquad \frac{dy}{d\theta} = b \cos \theta$$

Using $\frac{dy}{dx} = \frac{dy}{d\theta} \times \frac{d\theta}{dx}$, $\qquad \frac{dy}{dx} = b \cos \theta \frac{-1}{a \sin \theta}$

$$= -\frac{b \cos \theta}{a \sin \theta}$$

Using $y - y_1 = m(x - x_1)$, the equation of the tangent is
$$y - b \sin \theta = -\frac{b \cos \theta}{a \sin \theta}(x - a \cos \theta)$$
$\therefore \quad ay \sin \theta - ab \sin^2 \theta = -bx \cos \theta + ab \cos^2 \theta$
$\therefore \quad ay \sin \theta + bx \cos \theta = ab$

The equation of the tangent at $(a \cos \theta, b \sin \theta)$ on the ellipse $\frac{x^2}{a^2} + \frac{y^2}{b^2} = 1$
is $ay \sin \theta + bx \cos \theta = ab$.

Exercise 16E

1. Given that the ellipse $\frac{x^2}{a^2} + \frac{y^2}{b^2} = 1$ has foci at $(\pm ae, 0)$ where e is the
   eccentricity and $b^2 = a^2(1 - e^2)$, find the coordinates of the foci and the
   value of e^2 for each of the following ellipses.
   (a) $\frac{x^2}{9} + y^2 = 1$ (b) $\frac{x^2}{25} + \frac{y^2}{16} = 1$ (c) $x^2 + 4y^2 = 16$

2. Given that the hyperbola $\frac{x^2}{a^2} - \frac{y^2}{b^2} = 1$ has foci at $(\pm ae, 0)$ where e is
   the eccentricity and $b^2 = a^2(e^2 - 1)$, find the coordinates of the foci and
   the value of e^2 for each of the following hyperbolae
   (a) $\frac{x^2}{16} - \frac{y^2}{9} = 1$ (b) $9x^2 - y^2 = 9$ (c) $16x^2 - y^2 = 64$

3. Find the equation of the tangent to each of the following curves at the given point on the curve.
   (a) $x^2 + 4y^2 = 4$ at $(\sqrt{3}, \frac{1}{2})$,
   (b) $x^2 + 4y^2 = 100$ at $(-8, 3)$,
   (c) $9x^2 - y^2 = 9$ at $(-\frac{5}{3}, 4)$,
   (d) $4x^2 - y^2 = 4$ at $(\sqrt{2}, -2)$,
   (e) $xy = 9$ at $(-3, -3)$,
   (f) $xy = 16$ at $(2, 8)$.

4. Find the equation of the tangent to the ellipse $\dfrac{x^2}{a^2} + \dfrac{y^2}{b^2} = 1$ at the point $(a \cos \theta, b \sin \theta)$ and find the coordinates of the point where this tangent cuts the y-axis.

5. (a) Find the equation of the tangent to the curve $xy = c^2$ at the point $(ct_1, c/t_1)$.
   (b) Find the equation of the normal to the curve $xy = c^2$ at the point $(ct_2, c/t_2)$.
   (c) If the tangent of (a) meets the normal of (b) on the y-axis show that $2t_2 = t_1(1 - t_2{}^4)$.

6. (a) Find the equation of the chord joining the point $(ct_1, c/t_1)$ to the point $(ct_2, c/t_2)$ on the hyperbola $xy = c^2$.
   (b) By letting $t_1 = t_2 = t$ use your answer to part (a) to obtain the tangent to the curve $xy = c^2$ at the point $(ct, c/t)$.

7. Given that $y = mx + c$ is a tangent to $xy = d^2$ prove that $m = -\dfrac{c^2}{4d^2}$.

8. Given that $y = mx + c$ is a tangent to the ellipse $\dfrac{x^2}{a^2} + \dfrac{y^2}{b^2} = 1$ prove that $c^2 = a^2m^2 + b^2$.

16.6 Polar coordinates

We have, so far, used cartesian coordinates (x, y) in order to give the position of a point in a plane relative to two axes.
We will now investigate an alternative system of coordinates in which the position of any point P can be described in terms of its distance and direction from a fixed point.
Suppose that OA is a straight line with the point O fixed.
If the distance OP $= r$ and the angle AOP $= \theta$, then the position of the point P is defined by (r, θ) and we call these the polar coordinates of the point P. The angle θ is positive when measured in an anticlockwise direction from the line OA, and is called the vectorial angle.
r is known as the radius vector.
It should be noted that different polar coordinates may be used to describe the same point, e.g. $\left(2, \dfrac{\pi}{2}\right)$ could also be written as $\left(2, \dfrac{5\pi}{2}\right), \left(2, \dfrac{9\pi}{2}\right), \left(2, -\dfrac{3\pi}{2}\right)$ etc.

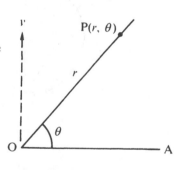

For this reason it is usual to state the polar coordinates of a point (r, θ) with $-\pi \leqslant \theta \leqslant \pi$ or $0 \leqslant \theta \leqslant 2\pi$.
Also, it is possible to give polar coordinates with a negative value of r so that $\left(-1, \dfrac{\pi}{3}\right)$ and $\left(1, -\dfrac{2\pi}{3}\right)$ in fact represent the same point. In practice it is usual to state (r, θ) with $r \geqslant 0$.
If OA in the diagram is taken as the x-axis and the y-axis be taken through the point O as indicated, then clearly the cartesian coordinates of P(x, y) can be written as $x = r \cos \theta$, and $y = r \sin \theta$.

In the same way that all the points lying on a line can be described in terms of an equation involving the cartesian coordinates x and y, so also can we describe the line in terms of a polar equation involving r and θ.

Using this polar form of the equation of a curve we can plot values of r against θ and hence represent the curve graphically. Graph paper designed for this specific purpose is obtainable, but the reader will find it possible to obtain sufficiently accurate graphs using plane paper.

Example 21

Copy and complete the following table for $r = 3 \sin \theta$.

θ	0°	30°	60°	90°	120°	150°	180°	210°	240°	270°	300°	330°	360°
r	0		2·6		2·6			−1·5	−2·6		−2·6		

Hence plot the graph of $r = 3 \sin \theta$ for $0° \leqslant \theta \leqslant 360°$.

Completing the table for $r = 3 \sin \theta$, gives

θ	0°	30°	60°	90°	120°	150°	180°	210°	240°	270°	300°	330°	360°
r	0	1·5	2·6	3·0	2·6	1·5	0	−1·5	−2·6	−3·0	−2·6	−1·5	0

and the curve can now be plotted.

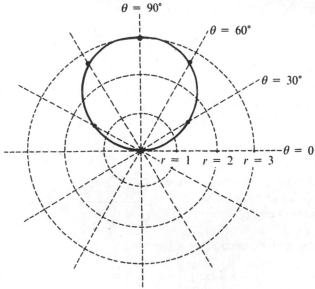

Converting between polar equations and cartesian equations

From the diagram on the right, we have
$$x = r \cos \theta, \qquad y = r \sin \theta$$
and $r^2 = x^2 + y^2, \qquad \tan \theta = \dfrac{y}{x}.$

Using these four relationships, we can convert polar equations to the equivalent cartesian equations and vice-versa.

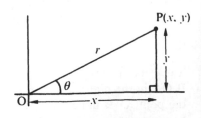

Example 22

Express (a) $x^2 + xy = 3$ in polar form,   (b) $r = 3 - \sin\theta$ in cartesian form.

(a) $x^2 + xy = 3$

     Substituting $x = r\cos\theta$ and $y = r\sin\theta$

     gives $r^2\cos^2\theta + r^2\cos\theta\sin\theta = 3$.

     i.e. $r^2\cos\theta(\cos\theta + \sin\theta) = 3$.

(b) $r = 3 - \sin\theta$

In order to be able to use $r^2 = x^2 + y^2$ and $y = r\sin\theta$, we first multiply the polar equation by r:

$$r^2 = 3r - r\sin\theta$$
$$\therefore \qquad x^2 + y^2 = 3\sqrt{(x^2 + y^2)} - y$$
i.e. $(x^2 + y^2 + y)^2 = 9(x^2 + y^2)$

which is the required cartesian form.
Note that in this case the polar form of the equation is simpler than the cartesian form.

The following are some particular types of polar equations:

(a) $r = k$, a constant   From $r^2 = x^2 + y^2$, we see that $r = k$ gives $x^2 + y^2 = k^2$, i.e. $r = k$ is the equation of a circle, centre $(0, 0)$, radius k.

(b) $\theta = k$, a constant   Now $\tan\theta = \dfrac{y}{x}$ and so $\dfrac{y}{x} = m$ where $m = \tan k$

                       i.e. $y = mx$
Thus $\theta = k$ is the equation of a straight line.
In fact, if we restrict r to positive values only, then

$\theta = k$ describes a 'half line', e.g. $\theta = \dfrac{\pi}{3}$ as shown.

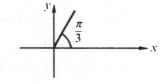

(c) $r = k\theta$ This form of polar equation will give a spiral (see Exercise 16F Question 6).

(d) $r = k\cos\theta$ Multiplying by r gives $r^2 = kr\cos\theta$

$$\therefore \qquad\qquad x^2 + y^2 = kx$$
i.e. $(x - \tfrac{1}{2}k)^2 + y^2 = \tfrac{1}{4}k^2$
Thus $r = k\cos\theta$ is the equation of a circle centre $(\tfrac{1}{2}k, 0)$ and radius $\tfrac{1}{2}k$ as shown.

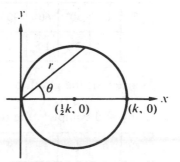

(e) $r = a\sec\theta$ If $r = a\sec\theta$ then $r = \dfrac{a}{\cos\theta}$

$$\therefore \qquad\qquad 1 = \dfrac{a}{r\cos\theta} \quad \text{or} \quad 1 = \dfrac{a}{x}.$$

thus $x = a$.
i.e. $r = a\sec\theta$ is the polar equation of the straight line $x = a$ as shown.

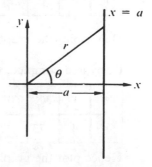

Exercise 16F

1. Write down the polar coordinates of the points A to H shown in the diagram, giving your answers in the form (r, θ) for $0 \leqslant \theta \leqslant 2\pi$ and $r \geqslant 0$

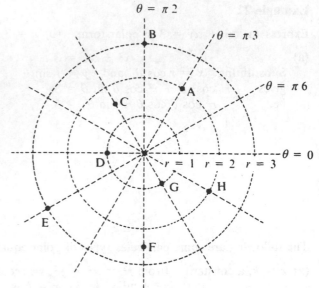

2. Find the cartesian coordinates of the points P to U having polar coordinates as follows:
P(4, 30°)   Q(5, 90°)   R(2, −90°)
S(5√2, 135°)   T(6, −120°)   U(−3, 90°).

3. Express each of the following cartesian equations in polar form, giving your answers in the form $r = f(\theta)$ or $r^2 = f(\theta)$ as appropriate.
    (a) $x = 4$ (b) $x^2 + y^2 = 4y$ (c) $2xy = 1$ (d) $y^2 = 8x$
    (e) $(x - 1)^2 + y^2 = 1$ (f) $x^2 + y^2 = 4x$ (g) $x^2 - y^2 = 8$ (h) $(x - y)^2 = 4$

4 Express each of the following polar equations in cartesian form.
    (a) $r \sin \theta = 1$ (b) $r = 3$ (c) $\theta = \dfrac{\pi}{3}$ (d) $r = \sin \theta$
    (e) $r = \sin \theta + \cos \theta$ (f) $r^2 = \sec 2\theta$ (g) $r = 3 \tan \theta$ (h) $3 = r(1 - \sin \theta)$

5. Copy and complete the following table for $r = 3 \cos \theta$.

θ	0°	30°	60°	90°	120°	150°	180°	210°	240°	270°	300°	330°	360°
r		2·6	1·5		−1·5	−2·6		−2·6				2·6	

Hence plot the graph of $r = 3 \cos \theta$ for $0° \leqslant \theta \leqslant 360°$.

6. Copy and complete the following table for $r = \dfrac{3\theta}{2\pi}$, ($\theta$ in radians).

θ	0	$\frac{1}{3}\pi$	$\frac{2}{3}\pi$	π	$\frac{4}{3}\pi$	$\frac{5}{3}\pi$	2π	$\frac{7}{3}\pi$	$\frac{8}{3}\pi$	3π	$\frac{10}{3}\pi$	$\frac{11}{3}\pi$	4π
r	0	$\frac{1}{2}$	1	$1\frac{1}{2}$									

Hence plot the graph of $r = \dfrac{3\theta}{2\pi}$ for $0 \leqslant \theta \leqslant 4\pi$.

7. Copy and complete the following table for $r = 2(1 + \cos \theta)$

θ	0°	30°	60°	90°	120°	150°	180°	210°	240°	270°	300°	330°	360°
r	4	3·73				0·27		0·27				3·73	

Hence plot the graph of $r = 2(1 + \cos \theta)$ for $0° \leqslant \theta \leqslant 360°$.

Exercise 16G   Examination questions

1. Find the coordinates of the maximum point T and the minimum point B
   of the curve

$$y = \frac{x^3}{3} - 2x^2 + 3x.$$

   Find also the point of inflexion I and show that T, I, B are collinear.
   Calculate to the nearest $0.1°$ the acute angle between TIB and the normal
   to the curve at I.   (London)

2. Obtain the equation of the normal to the curve

$$y^2 = x^3$$

   at the point (t^2, t^3). Show that the equation of the normal at the point
   where $t = \frac{1}{2}$ is

$$32x + 24y - 11 = 0.$$

   Find the perpendicular distance from the point $(-1, 2)$ to this normal.
                                                                     (J.M.B.)

3. The equation of a circle is $x^2 + y^2 - 3x - 4 = 0$.
   Find (i)   the coordinates of its centre,
        (ii) its radius,
        (iii) the coordinates of the points at which it cuts the axes.
   Show that the line whose equation is $3x + 4y = 17$ touches the circle
   and find the coordinates of its point of contact.
   Show also that this line and the tangent to the circle at the point $(3, -2)$
   intersect at a point on the x-axis and find its coordinates.   (J.M.B.)

4. Find, *by calculation*, the coordinates of the centre and the radius of the
   circle which passes through the points $(1, 1)$, $(3, 5)$ and $(-3, 1)$. (A.E.B)

5. Verify that the circle with equation

$$x^2 + y^2 - 2rx - 2ry + r^2 = 0$$

   touches both the coordinate axes.
   Find the radii of the two circles which pass through the point $(16, 2)$ and
   touch both the coordinate axes.   (Cambridge)

6. Find the values of m such that $y = mx$ is a tangent from the origin
   O $(0, 0)$ to the circle whose equation is $(x - 3)^2 + (y - 4)^2 = 1$.
   Find the cosine of the acute angle between these tangents.   (A.E.B)

7. Find the equations of the two circles which pass through the point $(2, 0)$
   and have both the y-axis and the line $y - 1 = 0$ as tangents.
   Calculate the coordinates of the second point at which the circles intersect.
                                                                     (A.E.B)

8. The fixed points A and B have coordinates $(-3a, 0)$ and $(a, 0)$ respectively.
   Find the equation of the locus of a point P which moves in the coordinate
   plane so that AP $= 3$PB. Show that the locus is a circle, S, which
   touches the axis of y and has its centre at the point $(\frac{3}{2}a, 0)$.
   A point Q moves in such a way that the perpendicular distance of Q
   from the axis of y is equal to the length of a tangent from Q to the circle
   S. Find the equation of the locus of Q.
   Show that this locus is also the locus of points which are equidistant
   from the line $4x + 3a = 0$ and the point $(\frac{3}{4}a, 0)$.   (Cambridge)

9. A point P moves in the x–y plane so that its distance from the origin, O, is twice its distance from the point with coordinates $(3a, 0)$. Show that the locus of P is a circle and obtain the coordinates of its centre and its radius. If the circle meets the x-axis in A and B, where OA $<$ OB, find the coordinates of A and B. If the tangents from O to the circle are OL and OM, find the angle LOM and the equations of OL and OM. Calculate the area of the triangle enclosed by the lines OL, OM and the tangent to the circle at A.   (S.U.J.B)

10. Find the equation of the normal to the parabola $y^2 = 4ax$ at the point $(at^2, 2at)$. The straight line $4x - 9y + 8a = 0$ meets the parabola at the points P and Q; the normals to the parabola at the points P and Q meet at R. Find the coordinates of R, and verify that it lies on the parabola.   (Oxford)

11. Prove that the equation of the tangent to the parabola $y^2 = 4ax$ at the point $P(ap^2, 2ap)$ on the curve is
$$py = x + ap^2.$$
Find the coordinates of the point of intersection, T, of the tangents at P and $Q(ap^2, 2aq)$, simplifying your answers where possible.
Given that S is the point $(a, 0)$, verify that SP . SQ $=$ ST^2.   (Cambridge)

12. The tangent to the curve $4ay = x^2$ at the point $P(2at, at^2)$ meets the x-axis at the point Q. The point S is $(0, a)$.
  (a) Prove that PQ is perpendicular to SQ.
  (b) Find a cartesian equation for the locus of the point, M, the mid-point of PS.   (A.E.B.)

13. Prove that the chord joining the points $P(ap^2, 2ap)$ and $Q(aq^2, 2aq)$ on the parabola $y^2 = 4ax$ has the equation
$$(p + q)y = 2r + 2apq.$$
A variable chord PQ of the parabola is such that the lines OP and OQ are perpendicular, where O is the origin.
  (i) Prove that the chord PQ cuts the axis of x at a fixed point, and give the x-coordinate of this point.
  (ii) Find the equation of the locus of the mid-point of PQ.   (Cambridge)

14. The point P lies on the ellipse $x^2 + 4y^2 = 1$ and N is the foot of the perpendicular from P to the line $x = 2$. Find the equation of the locus of the mid-point of PN as P moves on the ellipse.   (Cambridge)

15. Show that the x coordinates of any points of intersection of the line $y = mx + c$ and the ellipse $\dfrac{x^2}{9} + \dfrac{y^2}{4} = 1$ are given by the solutions of the quadratic equation $(4 + 9m^2)x^2 + 18mcx + (9c^2 - 36) = 0$.
If the line $y = mx + c$ is a tangent to the ellipse, prove that $c^2 = 4 + 9m^2$. The line $y = mx + c$ passes through the point $(2, 3)$. Write down a second equation connecting m and c, and hence prove that m must satisfy the equation $5m^2 + 12m - 5 = 0$.
Prove that the two tangents drawn from the point $(2, 3)$ to the ellipse are perpendicular to each other.   (S.U.J.B.)

16. Prove that the equation of the tangent to the ellipse
$$\frac{x^2}{a^2} + \frac{y^2}{b^2} = 1, (a > 0, b > 0)$$
at the point P($a \cos \theta$, $b \sin \theta$) is
$$\frac{x}{a} \cos \theta + \frac{y}{b} \sin \theta = 1.$$
The tangent at P meets the axes Ox and Oy at X and Y respectively. Find the area of triangle OXY.
The points A and B have coordinates (a, 0) and (0, b) respectively. Show that the area of triangle APB is
$$\tfrac{1}{2}ab(\cos \theta + \sin \theta - 1).$$
Prove that, as θ varies in the interval $0 < \theta < \tfrac{1}{2}\pi$, the area of triangle APB is a maximum when the tangent to the ellipse at P is parallel to AB. Prove also that triangle OXY has its minimum area when triangle APB has its maximum area. (Cambridge)

17. The ellipse
$$\frac{x^2}{a^2} + \frac{y^2}{b^2} = 1$$
intersects the positive x-axis at A and the positive y-axis at B. Determine the equation of the perpendicular bisector of AB.
(i) Given that this line intersects the x-axis at P and that M is the midpoint of AB, prove that the area of triangle PMA is
$$\frac{b(a^2 + b^2)}{8a}.$$
(ii) If $a^2 = 3b^2$, find, in terms of b, the coordinates of the points where the perpendicular bisector of AB intersects the ellipse. (J.M.B.)

18. A curve has the parametric equations $x = 3t$, $y = 3/t$. Find the equation of the tangent to the curve at the point ($3t$, $3/t$). The point P has coordinates (-5, 8) and the tangents from P to the curve touch the curve at A and B. Calculate the coordinates of A and B and the length of the chord AB. (Oxford)

19. Prove that the equation of the chord joining the points P
$$\left(cp, \frac{c}{p}\right) \text{ and } Q\left(cq, \frac{c}{q}\right) \text{ on the rectangular hyperbola } xy = c^2 \text{ is}$$
$$pqy + x = c(p + q).$$
It is given that PQ subtends a right angle at the point $R\left(cr, \frac{c}{r}\right)$ on the curve. Prove that
(i) PQ is parallel to the normal at R to the curve;
(ii) the mid-point of PQ lies on the straight line $y + r^2x = 0$. (Cambridge)

20. The tangent at the point $P\left(ct, \dfrac{c}{t}\right)$, where $t > 0$, on the rectangular

    hyperbola $xy = c^2$ meets the x-axis at A and the y-axis at B. The normal at P to the rectangular hyperbola meets the line $y = x$ at C and the line $y = -x$ at D.
    (a) Show that P is the mid-point of both AB and CD.
    (b) Prove that the points A, B, C and D form the vertices of a square.
        The normal at P meets the hyperbola again at the point Q and the mid-point of PQ is M.
    (c) Prove that, as t varies, the point M lies on the curve
        $$c^2(x^2 - y^2)^2 + 4x^3y^3 = 0. \qquad \text{(A.E.B.)}$$

21. Find a polar equation of the curve $x^2 + y^2 = 4x$ and calculate the polar coordinates of the two points P and Q where the curve intersects

    the line $r = 2\sqrt{2}\sec\left(\dfrac{\pi}{4} - \theta\right)$.

    Find the polar equations of the two half-lines from the origin which are tangents to the circle which has PQ as diameter.   (London)

17
Three-dimensional work: vectors and matrices

17.1 Three-dimensional work

From the work of earlier chapters the reader is familar with the use of coordinates to define a point in a plane in terms of its distance from two mutually perpendicular axes x and y. The position vector of such a point has been expressed in the form $a\mathbf{i} + b\mathbf{j}$, where $\mathbf{i}$ is a unit vector in the direction of the x-axis and $\mathbf{j}$ is a unit vector in the direction of the y-axis.

To consider points that are not coplanar we must introduce a third axis, the z-axis, and a corresponding third unit vector $\mathbf{k}$ in the direction of this axis.

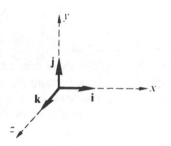

Suppose the point A shown in the diagram has coordinates (3, 4, 2).
If $\mathbf{a}$ is the position vector of point A then

$$\mathbf{a} = 3\mathbf{i} + 4\mathbf{j} + 2\mathbf{k}$$

or this may be written as $\mathbf{a} = \begin{pmatrix} 3 \\ 4 \\ 2 \end{pmatrix}$

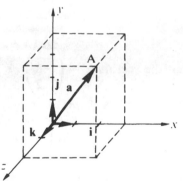

Note The convention for determining the direction of $\mathbf{k}$ is to use the 'right-handed screw rule'. For this we consider the motion of a screw at O, lying perpendicular to the $\mathbf{i} - \mathbf{j}$ plane, turning through the right angle from $\mathbf{i}$ to $\mathbf{j}$.

Thus if we draw:

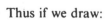

the rotation will move the screw 'out of the page':

if we draw:

the rotation will move the screw 'upwards':

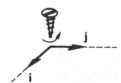

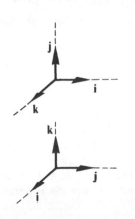

The reader will remember from chapter 2 that if $\mathbf{r} = x_1\mathbf{i} + y_1\mathbf{j}$ and
$\mathbf{s} = x_2\mathbf{i} + y_2\mathbf{j}$ then $|\mathbf{r}| = \sqrt{(x_1^2 + y_1^2)}$ and $\mathbf{r} . \mathbf{s} = x_1x_2 + y_1y_2$
$$= |\mathbf{r}||\mathbf{s}| \cos \theta \text{ where } \theta \text{ is the angle between } \mathbf{r} \text{ and } \mathbf{s}.$$

These results can be extended to three-dimensional vectors so that if
$\mathbf{r} = x_1\mathbf{i} + y_1\mathbf{j} + z_1\mathbf{k}$ and $\mathbf{s} = x_2\mathbf{i} + y_2\mathbf{j} + z_2\mathbf{k}$ then
$|\mathbf{r}| = \sqrt{(x_1^2 + y_1^2 + z_1^2)}$ and $\mathbf{r} . \mathbf{s} = x_1x_2 + y_1y_2 + z_1z_2$
$$= |\mathbf{r}||\mathbf{s}| \cos \theta$$
where θ is the angle between $\mathbf{r}$ and $\mathbf{s}$.

Example 1

Find, to the nearest degree, the angle between the vectors $\mathbf{a} = 2\mathbf{i} + \mathbf{j} - 3\mathbf{k}$ and $\mathbf{b} = \mathbf{i} + 2\mathbf{j} + \mathbf{k}$

If θ is the angle between the vectors then $\mathbf{a} . \mathbf{b} = |\mathbf{a}||\mathbf{b}| \cos \theta$

thus $2 + 2 - 3 = \sqrt{14}\sqrt{6} \cos \theta$

giving $\theta = 84°$ to the nearest degree.

As we saw in chapter 2, if three vectors are coplanar any one of the vectors
can be expressed as a combination of scalar multiples of the other two. The
converse statement is also true, i.e. if a vector can be expressed as a
combination of scalar multiples of two other vectors then the three vectors
are coplanar.

Example 2

Show that the vectors $\mathbf{a} = \begin{pmatrix} 1 \\ -3 \\ 2 \end{pmatrix}$, $\mathbf{b} = \begin{pmatrix} 6 \\ -4 \\ 2 \end{pmatrix}$ and $\mathbf{c} = \begin{pmatrix} 1 \\ -10 \\ 7 \end{pmatrix}$ are coplanar.

Suppose that $\mathbf{c} = \lambda\mathbf{a} + \mu\mathbf{b}$:
$$\begin{pmatrix} 1 \\ -10 \\ 7 \end{pmatrix} = \lambda\begin{pmatrix} 1 \\ -3 \\ 2 \end{pmatrix} + \mu\begin{pmatrix} 6 \\ -4 \\ 2 \end{pmatrix}$$
Then $1 = \lambda + 6\mu$, $-10 = -3\lambda - 4\mu$ and $7 = 2\lambda + 2\mu$.
Solving the first two of these simultaneously gives $\lambda = 4$ and $\mu = -\frac{1}{2}$ and
these values are compatible with $7 = 2\lambda + 2\mu$.
Thus we can write $\mathbf{c} = 4\mathbf{a} - \frac{1}{2}\mathbf{b}$, hence $\mathbf{a}$, $\mathbf{b}$ and $\mathbf{c}$ are coplanar.

Example 3

Find a vector that is perpendicular to $2\mathbf{i} + 3\mathbf{j} - \mathbf{k}$.

There are many vectors of the form $a\mathbf{i} + b\mathbf{j} + c\mathbf{k}$ that are perpendicular to
$2\mathbf{i} + 3\mathbf{j} - \mathbf{k}$ but for all of them
        $(a\mathbf{i} + b\mathbf{j} + c\mathbf{k}) . (2\mathbf{i} + 3\mathbf{j} - \mathbf{k}) = 0$
i.e. $2a + 3b - c = 0$
Suppose we take $a = b = 1$ then $c = 5$
Thus the vector $\mathbf{i} + \mathbf{j} + 5\mathbf{k}$ is perpendicular to $2\mathbf{i} + 3\mathbf{j} - \mathbf{k}$.

Direction cosines and direction ratios

If vector $\mathbf{r} = x\mathbf{i} + y\mathbf{j} + z\mathbf{k}$ makes angles α, β and γ with the directions of $\mathbf{i}$, $\mathbf{j}$ and $\mathbf{k}$ respectively then

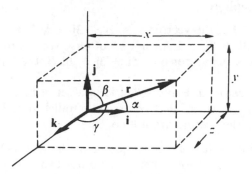

$$\cos \alpha = \frac{x}{|\mathbf{r}|} = \frac{x}{\sqrt{(x^2 + y^2 + z^2)}}$$

$$\cos \beta = \frac{y}{|\mathbf{r}|} = \frac{y}{\sqrt{(x^2 + y^2 + z^2)}}$$

$$\cos \gamma = \frac{z}{|\mathbf{r}|} = \frac{z}{\sqrt{(x^2 + y^2 + z^2)}}$$

These values $\dfrac{x}{|\mathbf{r}|}$, $\dfrac{y}{|\mathbf{r}|}$ and $\dfrac{z}{|\mathbf{r}|}$ are called the *direction cosines* of the vector $\mathbf{r} = x\mathbf{i} + y\mathbf{j} + z\mathbf{k}$

The ratios $x:y:z$ are called the *direction ratios* of the vector.

Example 4

Find the direction ratios and direction cosines of the vector $6\mathbf{i} + 3\mathbf{j} - 2\mathbf{k}$.

The direction ratios of $a\mathbf{i} + b\mathbf{j} + c\mathbf{k}$ are $a:b:c$. Thus the direction ratios of $6\mathbf{i} + 3\mathbf{j} - 2\mathbf{k}$ are $6:3:-2$.

The direction cosines of $\mathbf{r} = a\mathbf{i} + b\mathbf{j} + c\mathbf{k}$ are $\dfrac{a}{|\mathbf{r}|}$, $\dfrac{b}{|\mathbf{r}|}$ and $\dfrac{c}{|\mathbf{r}|}$. Thus the direction cosines of $6\mathbf{i} + 3\mathbf{j} - 2\mathbf{k}$ are $\frac{6}{7}$, $\frac{3}{7}$ and $-\frac{2}{7}$.

Example 5

Find the direction ratios of a vector $\mathbf{r}$ that is perpendicular to both $2\mathbf{i} + \mathbf{j} - \mathbf{k}$ and $3\mathbf{i} - \mathbf{j} - 2\mathbf{k}$.

$$
\begin{array}{llll}
\text{If} \quad \mathbf{r} = a\mathbf{i} + b\mathbf{j} + c\mathbf{k} \quad \text{then} & 2a + b - c = 0 & \dots [1] \\
\text{and} & 3a - b - 2c = 0 & \dots [2] \\
[1] + [2] \quad \text{give} & 5a - 3c = 0 \quad \text{or} \quad a = \tfrac{3}{5}c \\
\text{substitution into } [1] \quad \text{gives} & b = -\tfrac{1}{5}c
\end{array}
$$

Thus the direction ratios are $\quad a:b:c = \tfrac{3}{5}c : -\tfrac{1}{5}c : c \quad$ or $\quad 3:-1:5$

Other base vectors in three dimensions

Any vector can be expressed in the form $x\mathbf{i} + y\mathbf{j} + z\mathbf{k}$ and we say that the unit vectors $\mathbf{i}$, $\mathbf{j}$ and $\mathbf{k}$ form a set of *base vectors* for three dimensions. Whilst it is usually convenient to use $\mathbf{i}$, $\mathbf{j}$ and $\mathbf{k}$ as the base vectors, other vectors can form a set of base vectors for three dimensional space. The position vector $\mathbf{r}$ of any point R in space can be expressed as a combination of scalar multiples of the vectors $\mathbf{a}$, $\mathbf{b}$ and $\mathbf{c}$ provided $\mathbf{a}$, $\mathbf{b}$ and $\mathbf{c}$ are not coplanar and no two of the vectors $\mathbf{a}$, $\mathbf{b}$ and $\mathbf{c}$ are parallel to each other.

The vectors $\mathbf{a}$, $\mathbf{b}$ and $\mathbf{c}$ are a set of base vectors for three dimensional space and

$$\mathbf{r} = \lambda\mathbf{a} + \mu\mathbf{b} + \eta\mathbf{c}$$

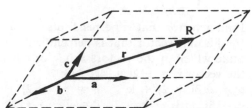

Example 6

Show that the vectors $\mathbf{a} = 2\mathbf{i} + 3\mathbf{j} + \mathbf{k}$, $\mathbf{b} = \mathbf{i} - 2\mathbf{j} + \mathbf{k}$ and $\mathbf{c} = 3\mathbf{i} + 8\mathbf{j} - \mathbf{k}$ form a set of base vectors for three dimensional space and express the vector $\mathbf{d} = 5\mathbf{i} - 3\mathbf{j} + 6\mathbf{k}$ in the form $\lambda\mathbf{a} + \mu\mathbf{b} + \eta\mathbf{c}$.

The vectors $\mathbf{a}$, $\mathbf{b}$ and $\mathbf{c}$ will form a set of base vectors provided
  (a) no two of the vectors are parallel to each other and
  (b) the vectors are not coplanar.

Condition (a) is true as no one of the vectors $\mathbf{a}$, $\mathbf{b}$ or $\mathbf{c}$ is a scalar multiple
          of either of the other two.
      (b) If $\mathbf{a}$, $\mathbf{b}$ and $\mathbf{c}$ were coplanar we could write that
          $\mathbf{c} = \alpha\mathbf{a} + \beta\mathbf{b}$ where α and β are scalars
          i.e. $3\mathbf{i} + 8\mathbf{j} - \mathbf{k} = \alpha(2\mathbf{i} + 3\mathbf{j} + \mathbf{k}) + \beta(\mathbf{i} - 2\mathbf{j} + \mathbf{k})$
Equating coefficients of $\mathbf{i}$, $\mathbf{j}$ and $\mathbf{k}$: $3 = 2\alpha + \beta$, $8 = 3\alpha - 2\beta$ and $-1 = \alpha + \beta$.
Solving the first two of these simultaneously gives $\alpha = 2$ and $\beta = -1$.
However, these values are not consistent with $-1 = \alpha + \beta$ and so we
cannot write $\mathbf{c} = \alpha\mathbf{a} + \beta\mathbf{b}$. Thus $\mathbf{a}$, $\mathbf{b}$ and $\mathbf{c}$ are not coplanar.
Therefore $\mathbf{a}$, $\mathbf{b}$ and $\mathbf{c}$ form a set of base vectors for three dimensional space.
If $\mathbf{d} = \lambda\mathbf{a} + \mu\mathbf{b} + \eta\mathbf{c}$
    $5\mathbf{i} - 3\mathbf{j} + 6\mathbf{k} = \lambda(2\mathbf{i} + 3\mathbf{j} + \mathbf{k}) + \mu(\mathbf{i} - 2\mathbf{j} + \mathbf{k}) + \eta(3\mathbf{i} + 8\mathbf{j} - \mathbf{k})$
$\therefore$ $5 = 2\lambda + \mu + 3\eta$ ... [1]
             $-3 = 3\lambda - 2\mu + 8\eta$ ... [2]
             $6 = \lambda + \mu - \eta$ ... [3]
equation [1] minus equation [3] gives: $-1 = \lambda + 4\eta$
equation [2] added to twice equation [3] gives: $9 = 5\lambda + 6\eta$
Thus $\lambda = 3$, $\mu = 2$ and $\eta = -1$
$\therefore$ $\mathbf{d} = 3\mathbf{a} + 2\mathbf{b} - \mathbf{c}$

Exercise 17A

1. If $\mathbf{a} = 9\mathbf{i} - 2\mathbf{j} - 6\mathbf{k}$, $\mathbf{b} = 2\mathbf{i} - 6\mathbf{j} + 3\mathbf{k}$ and $\mathbf{c} = 2\mathbf{i} - \mathbf{j} + 2\mathbf{k}$ find
   (a) $|\mathbf{a}|$ (b) $|\mathbf{b}|$ (c) $|\mathbf{c}|$ (d) $\mathbf{a} \cdot \mathbf{b}$ (e) $\mathbf{b} \cdot \mathbf{c}$
   (f) the angle between $\mathbf{a}$ and $\mathbf{b}$ (to the nearest degree)
   (g) the angle between $\mathbf{b}$ and $\mathbf{c}$ (to the nearest degree).

2. If $\mathbf{a} = \begin{pmatrix} 1 \\ 1 \\ 3 \end{pmatrix}$ and $\mathbf{b} = \begin{pmatrix} -2 \\ 1 \\ 1 \end{pmatrix}$ find

   (a) $|\mathbf{a}|$ (b) $|\mathbf{b}|$ (c) the angle between $\mathbf{a}$ and $\mathbf{b}$ (to the nearest degree).
3. If $\mathbf{a} = 3\mathbf{i} + 4\mathbf{j} + 12\mathbf{k}$ find $\hat{\mathbf{a}}$, a unit vector in the direction of $\mathbf{a}$.

4. If $\mathbf{b} = \begin{pmatrix} 4 \\ 4 \\ -7 \end{pmatrix}$ find $\hat{\mathbf{b}}$, a unit vector in the direction of $\mathbf{b}$.

5. Find a vector that is perpendicular to $5\mathbf{i} - \mathbf{j} + 2\mathbf{k}$.
6. Find a unit vector that is perpendicular to $\mathbf{i} + 2\mathbf{j} - 3\mathbf{k}$.
7. State which of the vectors $\mathbf{a}$, $\mathbf{b}$, $\mathbf{c}$ or $\mathbf{d}$ listed below are perpendicular to
   the vector $\mathbf{r} = -2\mathbf{i} + 4\mathbf{j} + 6\mathbf{k}$.
   $\mathbf{a} = -3\mathbf{i} + 2\mathbf{j} + \mathbf{k}$ $\mathbf{b} = -\mathbf{i} + 2\mathbf{j} + 3\mathbf{k}$
   $\mathbf{c} = 3\mathbf{i} + 3\mathbf{j} - \mathbf{k}$ $\mathbf{d} = 4\mathbf{i} - \mathbf{j} + 2\mathbf{k}$

8. Find the direction ratios and direction cosines of the vector $14\mathbf{i} - 2\mathbf{j} + 5\mathbf{k}$.

9. Vector $\mathbf{r}$ is at right angles to both $2\mathbf{i} + 3\mathbf{j} + 4\mathbf{k}$ and $2\mathbf{i} - 3\mathbf{j} - 2\mathbf{k}$. Find the direction ratios of $\mathbf{r}$.

10. (a) Find a vector that is perpendicular to both $3\mathbf{i} - 6\mathbf{j} - 4\mathbf{k}$ and $-3\mathbf{i} + 2\mathbf{j} + 2\mathbf{k}$
    (b) Find the direction cosines of the vector obtained for part (a)
    (c) Find a unit vector that is perpendicular to both $3\mathbf{i} - 6\mathbf{j} - 4\mathbf{k}$ and $-3\mathbf{i} + 2\mathbf{j} + 2\mathbf{k}$.

11. Points A, B and C have position vectors $2\mathbf{i} + 3\mathbf{j} - \mathbf{k}$, $3\mathbf{i} + 6\mathbf{j} - 3\mathbf{k}$ and $5\mathbf{i} + 12\mathbf{j} - 7\mathbf{k}$ respectively. Prove that A, B and C are collinear.

12. Points A, B and C have position vectors $\mathbf{i} + 2\mathbf{j} - 3\mathbf{k}$, $3\mathbf{i} + \mathbf{j} - 5\mathbf{k}$ and $2\mathbf{i} - \mathbf{k}$ respectively. Prove that angle $\mathrm{B\hat{A}C}$ is a right angle.

13. Prove that the vectors $\mathbf{a} = 3\mathbf{i} + \mathbf{j} - 4\mathbf{k}$, $\mathbf{b} = 5\mathbf{i} - 3\mathbf{j} - 2\mathbf{k}$ and $\mathbf{c} = 4\mathbf{i} - \mathbf{j} - 3\mathbf{k}$ are coplanar.

14. Prove that the vectors $\mathbf{a} = \begin{pmatrix} 3 \\ 4 \\ 1 \end{pmatrix}$, $\mathbf{b} = \begin{pmatrix} 3 \\ -1 \\ 2 \end{pmatrix}$ and $\mathbf{c} = \begin{pmatrix} 1 \\ -2 \\ 1 \end{pmatrix}$ are coplanar.

15. If the points A, B and C have position vectors $\mathbf{i} + 3\mathbf{j} - 5\mathbf{k}$, $3\mathbf{i} - 2\mathbf{j} + 9\mathbf{k}$ and $8\mathbf{i} + 7\mathbf{j} - \mathbf{k}$ respectively find the angles of triangle ABC giving your answers to the nearest degree.

16. Prove that the sum of the squares of the direction cosines of any vector is one.

17. Prove that the vectors $\mathbf{a} = -\mathbf{i} + 2\mathbf{j} + 3\mathbf{k}$, $\mathbf{b} = 3\mathbf{i} + \mathbf{j} - 2\mathbf{k}$ and $\mathbf{c} = \mathbf{i} + 5\mathbf{j} + 2\mathbf{k}$ form a set of base vectors for three dimensional space and express the vector $\mathbf{d} = -3\mathbf{i} - \mathbf{j} + 6\mathbf{k}$ in the form $\lambda\mathbf{a} + \mu\mathbf{b} + \eta\mathbf{c}$.

18. Prove that the vectors $\mathbf{a} = \begin{pmatrix} 2 \\ 1 \\ -3 \end{pmatrix}$, $\mathbf{b} = \begin{pmatrix} 1 \\ 0 \\ -1 \end{pmatrix}$ and $\mathbf{c} = \begin{pmatrix} 1 \\ 2 \\ 2 \end{pmatrix}$ form a set of base vectors for three dimensional space and express the vector $\mathbf{d} = \begin{pmatrix} 7 \\ 7 \\ -4 \end{pmatrix}$ in the form $\lambda\mathbf{a} + \mu\mathbf{b} + \eta\mathbf{c}$.

17.2 Differentiation and integration of vectors

Suppose that $\quad \mathbf{a} = f(x)\mathbf{i} + g(x)\mathbf{j} + h(x)\mathbf{k}$,

then $\qquad \dfrac{d\mathbf{a}}{dx} = f'(x)\mathbf{i} + g'(x)\mathbf{j} + h'(x)\mathbf{k}$.

i.e. to differentiate a vector with respect to some variable x we differentiate each component with respect to that variable.

Similarly, to integrate a vector with respect to some variable x we integrate each component with respect to that variable.

In particular, for a particle which has a displacement vector $\mathbf{s}$, velocity vector **v** and acceleration vector **a** given in terms of the time, t, we can link these vectors by the method shown on page 308 i.e.

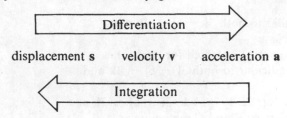

Example 7

If $\mathbf{y} = 8p^2\mathbf{i} + (8p - 4)\mathbf{j} + (p^3 + 3)\mathbf{k}$, find $\dfrac{d\mathbf{y}}{dp}$ and $\dfrac{d^2\mathbf{y}}{dp^2}$

$$\mathbf{y} = 8p^2\mathbf{i} + (8p - 4)\mathbf{j} + (p^3 + 3)\mathbf{k}$$

$$\therefore \quad \frac{d\mathbf{y}}{dp} = 16p\mathbf{i} + 8\mathbf{j} + 3p^2\mathbf{k}$$

and $\quad \dfrac{d^2\mathbf{y}}{dp^2} = 16\mathbf{i} + 6p\mathbf{k}$

Example 8

If the velocity of a body at time t is given by $\mathbf{v} = 3t^2\mathbf{i} - 2t\mathbf{j} + 4\mathbf{k}$, find expressions for the acceleration, **a**, and the displacement, **s**, of the body at time t, given that when $t = 1$, $\mathbf{s} = 3\mathbf{i} - \mathbf{j} + 2\mathbf{k}$.

$$\mathbf{v} = 3t^2\mathbf{i} - 2t\mathbf{j} + 4\mathbf{k}$$

$$\therefore \quad \mathbf{a} = \frac{d\mathbf{v}}{dt} = 6t\mathbf{i} - 2\mathbf{j}$$

also $\quad \mathbf{s} = \displaystyle\int \mathbf{v}\,dt = t^3\mathbf{i} - t^2\mathbf{j} + 4t\mathbf{k} + \mathbf{c}$

but when $t = 1$, $\quad \mathbf{s} = 3\mathbf{i} - \mathbf{j} + 2\mathbf{k}$, $\quad \therefore \quad \mathbf{c} = 2\mathbf{i} - 2\mathbf{k}$

$\therefore \qquad\qquad\qquad \mathbf{s} = (t^3 + 2)\mathbf{i} - t^2\mathbf{j} + 2(2t - 1)\mathbf{k}$

Thus at time t the displacement of the body is given by
$\mathbf{s} = (t^3 + 2)\mathbf{i} - t^2\mathbf{j} + 2(2t - 1)\mathbf{k}$ and the acceleration is given by
$\mathbf{a} = 6t\mathbf{i} - 2\mathbf{j}$.

Exercise 17B

1. Differentiate each of the following vectors with respect to t.
   (a) $3t\mathbf{i} + 4t^2\mathbf{j}$ (b) $(3t + 1)\mathbf{i} + 2t\mathbf{j} - 5t^2\mathbf{k}$
   (c) $3\mathbf{i} + 2t\mathbf{j} - 5\mathbf{k}$ (d) $(6t - 1)\mathbf{i} + 3t^3\mathbf{j} - 6t\mathbf{k}$
2. Differentiate each of the following vectors with respect to θ,
   (a) $\sin \theta\mathbf{i} + \cos \theta\mathbf{j}$ (b) $2 \cos \theta\mathbf{i} + 2\theta\mathbf{j}$
   (c) $\sin \theta\mathbf{i} - \cos \theta\mathbf{j} + \theta^2\mathbf{k}$ (d) $\sin 3\theta\mathbf{j} - \cos^3 \theta\mathbf{k}$
3. The position vector of a particle at time t seconds is given by
   $\mathbf{s} = (2t^3\mathbf{i} - 2t\mathbf{j} + \mathbf{k})$ metres.
   Find the displacement, velocity and acceleration vectors when **t** = 2.

4. The velocity vector of a particle at time t is given by $\mathbf{v} = 2t\mathbf{i} + 3t^2\mathbf{j} + 2\mathbf{k}$. Given that when $t = 0$ the particle has position vector $\mathbf{i} + \mathbf{j}$ with respect to an origin O, find the position vector of the particle with respect to O when $t = 3$.

5. At time t the displacement of a particle from an origin O is given by $\mathbf{s} = (2\sin(\pi t)\mathbf{i} + 2\cos(\pi t)\mathbf{j})$ metres. Prove that the particle is always 2 metres from O and find the velocity and speed of the particle when $\mathbf{t} = 2$ seconds.

6. The acceleration of a body at time t is given by $\mathbf{a} = 4t\mathbf{j}$. Find expressions for the velocity, $\mathbf{v}$, and the displacement, $\mathbf{s}$, from an origin O at time t given that when $t = 0$, $\mathbf{v} = 2\mathbf{i} + \mathbf{j} - 4\mathbf{k}$ and $\mathbf{s} = 3\mathbf{k}$.

7. A body moves such that its position vector when at point P is given by $\overrightarrow{OP} = 3\sin 5t\mathbf{i} + 3\cos 5t\mathbf{j}$ where O is the origin and t is the time. Prove that the velocity of the particle when at P is perpendicular to $\overrightarrow{OP}$.

8. For a body moving in a circle of radius r, at constant angular speed ω, the position vector $\mathbf{s}$, of the body at time t is given by $\mathbf{s} = \overrightarrow{OP} = r\cos\omega t\mathbf{i} + r\sin\omega t\mathbf{j}$.
Prove that the acceleration of the body has constant magnitude $r\omega^2$ and is directed along $\overrightarrow{PO}$ (see diagram).

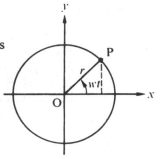

17.3 Equations of a straight line in 3-D

As we saw in section 2.5 a line that passes through the point with position vector $\mathbf{a}$ and is parallel to the vector $\mathbf{b}$ has vector equation $\mathbf{r} = \mathbf{a} + \lambda\mathbf{b}$. Thus a line passing through the point A, position vector $2\mathbf{i} + 3\mathbf{j} - \mathbf{k}$ and which is parallel to the vector $\mathbf{i} + 2\mathbf{j} + \mathbf{k}$ has vector equation $\mathbf{r} = 2\mathbf{i} + 3\mathbf{j} - \mathbf{k} + \lambda(\mathbf{i} + 2\mathbf{j} + \mathbf{k})$. This may also be written in column vector form, $\mathbf{r} = \begin{pmatrix} 2 \\ 3 \\ -1 \end{pmatrix} + \lambda\begin{pmatrix} 1 \\ 2 \\ 1 \end{pmatrix}$

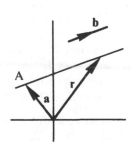

Example 9

Find the vector equation of the line which is parallel to the vector $3\mathbf{i} + 4\mathbf{j} + \mathbf{k}$ and passes through a point with position vector $2\mathbf{i} + 3\mathbf{j} - 2\mathbf{k}$.

The vector equation is $\mathbf{r} = 2\mathbf{i} + 3\mathbf{j} - 2\mathbf{k} + \lambda(3\mathbf{i} + 4\mathbf{j} + \mathbf{k})$ where $\mathbf{r}$ is the position vector of a general point on the line and λ is a scalar.

Example 10

Show that the point with position vector $\mathbf{i} - 9\mathbf{j} + \mathbf{k}$ lies on the line L, vector equation $\mathbf{r} = 3\mathbf{i} + 3\mathbf{j} - \mathbf{k} + \lambda(\mathbf{i} + 6\mathbf{j} - \mathbf{k})$.

The position vector of any point on a line must satisfy the equation of the line. Thus if $\mathbf{i} - 9\mathbf{j} + \mathbf{k}$ lies on L there must exist some valve of λ such that

$$\mathbf{i} - 9\mathbf{j} + \mathbf{k} = 3\mathbf{i} + 3\mathbf{j} - \mathbf{k} + \lambda(\mathbf{i} + 6\mathbf{j} - \mathbf{k})$$

i.e. $1 = 3 + \lambda$, $-9 = 3 + 6\lambda$ and $1 = -1 - \lambda$.
The first of these gives $\lambda = -2$ which is consistent with $-9 = 3 + 6\lambda$ and $1 = -1 - \lambda$. Thus the point with position vector $\mathbf{i} - 9\mathbf{j} + \mathbf{k}$ does lie on L.

Example 11

The vector equations of three lines are stated below:
line 1: $\mathbf{r} = 17\mathbf{i} + 2\mathbf{j} - 6\mathbf{k} + \lambda(-9\mathbf{i} + 3\mathbf{j} + 9\mathbf{k})$
line 2: $\mathbf{r} = 2\mathbf{i} - 3\mathbf{j} + 4\mathbf{k} + \mu(6\mathbf{i} + 7\mathbf{j} - \mathbf{k})$
line 3: $\mathbf{r} = 2\mathbf{i} - 12\mathbf{j} - \mathbf{k} + \eta(-3\mathbf{i} + \mathbf{j} + 3\mathbf{k})$
State which pair of lines (a) are parallel to each other,
                                     (b) intersect with each other,
                                     (c) are not parallel and do not intersect (i.e. are skew).

(a) line 1 is parallel to $-9\mathbf{i} + 3\mathbf{j} + 9\mathbf{k}$
line 2 is parallel to $6\mathbf{i} + 7\mathbf{j} - \mathbf{k}$
line 3 is parallel to $-3\mathbf{i} + \mathbf{j} + 3\mathbf{k}$
$(-9\mathbf{i} + 3\mathbf{j} + 9\mathbf{k})$ is parallel to $(-3\mathbf{i} + \mathbf{j} + 3\mathbf{k})$ as one is a scalar multiple of the other. Thus lines 1 and 3 are parallel.

(b) If two lines intersect there will exist a point whose position vector satisfies the vector equations of both lines.
If lines 1 and 2 intersect there will exist values of λ and μ such that
$17\mathbf{i} + 2\mathbf{j} - 6\mathbf{k} + \lambda(-9\mathbf{i} + 3\mathbf{j} + 9\mathbf{k}) = 2\mathbf{i} - 3\mathbf{j} + 4\mathbf{k} + \mu(6\mathbf{i} + 7\mathbf{j} - \mathbf{k})$
i.e. $17 - 9\lambda = 2 + 6\mu$, $2 + 3\lambda = -3 + 7\mu$ and $-6 + 9\lambda = 4 - \mu$
Solving $15 = 9\lambda + 6\mu$ and $10 = 9\lambda + \mu$ simultaneously gives $\mu = 1$ and $\lambda = 1$
However this is inconsistent with $5 = 7\mu - 3\lambda$
Thus lines 1 and 2 do not intersect.

If lines 2 and 3 intersect there will exist values of μ and η such that
$2\mathbf{i} - 3\mathbf{j} + 4\mathbf{k} + \mu(6\mathbf{i} + 7\mathbf{j} - \mathbf{k}) = 2\mathbf{i} - 12\mathbf{j} - \mathbf{k} + \eta(-3\mathbf{i} + \mathbf{j} + 3\mathbf{k})$
i.e. $2 + 6\mu = 2 - 3\eta$, $-3 + 7\mu = -12 + \eta$ and $4 - \mu = -1 + 3\eta$
or $\quad\quad 2\mu = -\eta$, $\quad\quad\quad\quad 9 = \eta - 7\mu$ $\quad$ and $\quad\quad\quad 5 = \mu + 3\eta$
Solving $2\mu = -\eta$ and $9 = \eta - 7\mu$ simultaneously gives $\eta = 2$ and $\mu = -1$ which is consistent with $5 = \mu + 3\eta$
Thus lines 2 and 3 intersect.

(c) From the working of parts (a) and (b) it is clear that lines 1 and 2 are not parallel and do not intersect.

Example 12

Find the perpendicular distance from the point A, position vector $\begin{pmatrix} 4 \\ -3 \\ 10 \end{pmatrix}$ to

the line L, vector equation $\mathbf{r} = \begin{pmatrix} 1 \\ 2 \\ 3 \end{pmatrix} + \lambda \begin{pmatrix} 3 \\ -1 \\ 2 \end{pmatrix}$

Let P be the point where the perpendicular from the point A meets the line L, and let P have position vector $\mathbf{p}$.

Suppose that P is the point on the line L for which $\lambda = \lambda_1$, then

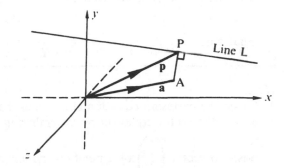

$$\mathbf{p} = \begin{pmatrix} 1 \\ 2 \\ 3 \end{pmatrix} + \lambda_1 \begin{pmatrix} 3 \\ -1 \\ 2 \end{pmatrix} = \begin{pmatrix} 1 + 3\lambda_1 \\ 2 - \lambda_1 \\ 3 + 2\lambda_1 \end{pmatrix}$$

$$\therefore \quad \overrightarrow{PA} = \mathbf{a} - \mathbf{p} = \begin{pmatrix} 4 \\ -3 \\ 10 \end{pmatrix} - \begin{pmatrix} 1 + 3\lambda_1 \\ 2 - \lambda_1 \\ 3 + 2\lambda_1 \end{pmatrix}$$

$$= \begin{pmatrix} 3 - 3\lambda_1 \\ -5 + \lambda_1 \\ 7 - 2\lambda_1 \end{pmatrix}$$

But $\overrightarrow{PA}$ is perpendicular to L and L is parallel to $\begin{pmatrix} 3 \\ -1 \\ 2 \end{pmatrix}$

Thus $\begin{pmatrix} 3 - 3\lambda_1 \\ -5 + \lambda_1 \\ 7 - 2\lambda_1 \end{pmatrix} \cdot \begin{pmatrix} 3 \\ -1 \\ 2 \end{pmatrix} = 0$ i.e. $9 - 9\lambda_1 + 5 - \lambda_1 + 14 - 4\lambda_1 = 0$

giving $\lambda_1 = 2$

Thus $\overrightarrow{PA} = \begin{pmatrix} -3 \\ -3 \\ 3 \end{pmatrix}$ and $|\overrightarrow{PA}| = \sqrt{[(-3)^2 + (-3)^2 + (3)^2]}$

$$= 3\sqrt{3} \text{ units}$$

The perpendicular distance from the point A to the line L is $3\sqrt{3}$ units.

Cartesian equation of a line in three dimensions

Consider a line that is parallel to the vector $\begin{pmatrix} p \\ q \\ r \end{pmatrix}$ and which passes through

the point A, position vector $\begin{pmatrix} a \\ b \\ c \end{pmatrix}$. if the general point on this line has

position vector $\mathbf{r} = \begin{pmatrix} x \\ y \\ z \end{pmatrix}$ then the vector equation of the line is $\begin{pmatrix} x \\ y \\ z \end{pmatrix} = \begin{pmatrix} a \\ b \\ c \end{pmatrix} + \lambda \begin{pmatrix} p \\ q \\ r \end{pmatrix}$

Thus $\left.\begin{array}{l} x = a + \lambda p \\ y = b + \lambda q \\ \text{and} \quad z = c + \lambda r \end{array}\right\}$ These are the parametric equations of the line, using the parameter λ.

Isolating λ in each equation gives $\dfrac{x - a}{p} = \dfrac{y - b}{q} = \dfrac{z - c}{r} (= \lambda)$. These are the cartesian equations of the line

---

Thus the line with vector equation $\mathbf{r} = \begin{pmatrix} a \\ b \\ c \end{pmatrix} + \lambda \begin{pmatrix} p \\ q \\ r \end{pmatrix}$ has cartesian

equations: $\qquad \dfrac{x - a}{p} = \dfrac{y - b}{q} = \dfrac{z - c}{r}$

Notes

1. Given the cartesian equation of a line in the above form it is easy to obtain the vector equation by remembering that the numerator gives the

   position vector $\begin{pmatrix} a \\ b \\ c \end{pmatrix}$ of a point on the line and the denominator gives the direction vector $\begin{pmatrix} p \\ q \\ r \end{pmatrix}$.

2. It is acceptable to give the cartesian equation of a line in the above form even when one or more of p, q and r are zero.

   For example, the line through $(0, 1, 1)$ and parallel to $y = -x$, i.e. parallel to the

   vector $\begin{pmatrix} -1 \\ 1 \\ 0 \end{pmatrix}$, has vector equation

   $\mathbf{r} = \begin{pmatrix} 0 \\ 1 \\ 1 \end{pmatrix} + \lambda \begin{pmatrix} -1 \\ 1 \\ 0 \end{pmatrix}$, (see diagram).

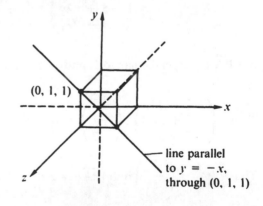

   This gives the parametric equations $\begin{cases} x = -\lambda \\ y = 1 + \lambda \\ z = 1 \end{cases}$

   The cartesian equations of the line can then be written $\dfrac{x}{-1} = \dfrac{y - 1}{1} = \dfrac{z - 1}{0}$ or simply as $\dfrac{x}{-1} = \dfrac{y - 1}{1}$, $z = 1$.

Example 13

Find the cartesian equations of the line that is parallel to the vector $2\mathbf{i} + 3\mathbf{j} + 4\mathbf{k}$ and which passes through the point A, position vector $3\mathbf{i} - \mathbf{j} + 2\mathbf{k}$.

The vector equation of the line is $\quad \mathbf{r} = 3\mathbf{i} - \mathbf{j} + 2\mathbf{k} + \lambda(2\mathbf{i} + 3\mathbf{j} + 4\mathbf{k})$
thus
$$x = 3 + 2\lambda$$
$$y = -1 + 3\lambda$$
$$z = 2 + 4\lambda$$

The cartesian equations are therefore $\quad \dfrac{x - 3}{2} = \dfrac{y + 1}{3} = \dfrac{z - 2}{4} (= \lambda)$

Example 14

Give the vector equation of the line with cartesian equations
$$\frac{x-1}{3} = \frac{y-2}{1} = \frac{z+3}{4}$$

The line is parallel to the vector $3\mathbf{i} + \mathbf{j} + 4\mathbf{k}$.
The line passes through a point with position vector $\mathbf{i} + 2\mathbf{j} - 3\mathbf{k}$.
Thus the vector equation of the line is $\mathbf{r} = \mathbf{i} + 2\mathbf{j} - 3\mathbf{k} + \lambda(3\mathbf{i} + \mathbf{j} + 4\mathbf{k})$.

Example 15

The lines L_1 and L_2 have cartesian equations
$$\frac{x-1}{3} = \frac{y+1}{2} = \frac{z-2}{1}(= \lambda) \text{ and } \frac{x-5}{1} = \frac{y-1}{1} = \frac{z}{2}(= \mu)$$
(a) State the vector equations of L_1 and L_2.
(b) Find the acute angle between L_1 and L_2 giving your answer to the nearest degree.

(a) L_1 will have vector equation $\mathbf{r} = \mathbf{i} - \mathbf{j} + 2\mathbf{k} + \lambda(3\mathbf{i} + 2\mathbf{j} + \mathbf{k})$
L_2 will have vector equation $\mathbf{r} = 5\mathbf{i} + \mathbf{j} + \mu(\mathbf{i} + \mathbf{j} + 2\mathbf{k})$

(b) The line L_1 is parallel to $3\mathbf{i} + 2\mathbf{j} + \mathbf{k}$ and the line L_2 is parallel to $\mathbf{i} + \mathbf{j} + 2\mathbf{k}$
Thus the angle between L_1 and L_2 will be the same as that
between $3\mathbf{i} + 2\mathbf{j} + \mathbf{k}$ and $\mathbf{i} + \mathbf{j} + 2\mathbf{k}$. (Note that this will be true even if the lines are skew—see page 23 for explanation of what is meant by the angle between skew lines).
If θ is the angle between L_1 and L_2 then
$(3\mathbf{i} + 2\mathbf{j} + \mathbf{k}) \cdot (\mathbf{i} + \mathbf{j} + 2\mathbf{k}) = |3\mathbf{i} + 2\mathbf{j} + \mathbf{k}||\mathbf{i} + \mathbf{j} + 2\mathbf{k}|\cos\theta$
i.e. $3 + 2 + 2 = \sqrt{14}\sqrt{6}\cos\theta$
giving $\theta = 40°$ to the nearest degree.
The acute angle between L_1 and L_2 is 40°, to the nearest degree.

Example 16

The lines L_1 and L_2 have cartesian equations
$$\frac{x}{1} = \frac{y+2}{2} = \frac{z-5}{-1} \text{ and } \frac{x-1}{-1} = \frac{y+3}{-3} = \frac{z-4}{1} \text{ respectively. Show that}$$
L_1 and L_2 intersect and find the coordinates of the point of intersection.

For the two lines to intersect there must exist some point (x_1, y_1, z_1) which lies on both lines
i.e. $$\frac{x_1}{1} = \frac{y_1+2}{2} = \frac{z_1-5}{-1} \qquad \ldots [1]$$
and $$\frac{x_1-1}{-1} = \frac{y_1+3}{-3} = \frac{z_1-4}{1} \qquad \ldots [2]$$

From equations [1] $\frac{x_1}{1} = \frac{y_1+2}{2}$ i.e. $2x_1 = y_1 + 2$

From equations [2] $\frac{x_1-1}{-1} = \frac{y_1+3}{-3}$ i.e. $-3x_1 = -y_1 - 6$

solving simultaneously gives $x_1 = 4$ and $y_1 = 6$
Using these values of x_1 and y_1 equations [1] and equations [2] both give
$z_1 = 1$.
Thus the lines L_1 and L_2 do intersect and the point of intersection has
coordinates (4, 6, 1).

Exercise 17C

1. State the vector equation of the line which is parallel to $2\mathbf{i} + 3\mathbf{j} - \mathbf{k}$
   and which passes through the point A, position vector $\mathbf{i} + \mathbf{j} + \mathbf{k}$.
2. State the vector equation of the line which passes through the point B,
   position vector $\begin{pmatrix} -1 \\ 2 \\ 1 \end{pmatrix}$ and which is parallel to the vector $\begin{pmatrix} 1 \\ 2 \\ 3 \end{pmatrix}$.
3. State a vector that is parallel to the line with vector equation
   $\mathbf{r} = 3\mathbf{i} + 4\mathbf{j} + \mathbf{k} + \lambda(2\mathbf{i} + 5\mathbf{j} + 3\mathbf{k})$.
4. Show that the point with position vector $4\mathbf{i} - \mathbf{j} + 12\mathbf{k}$ lies on the line
   with vector equation $\mathbf{r} = 2\mathbf{i} + 3\mathbf{j} + 4\mathbf{k} + \lambda(\mathbf{i} - 2\mathbf{j} + 4\mathbf{k})$.
5. Points A, B and C have position vectors $\begin{pmatrix} 5 \\ 2 \\ 3 \end{pmatrix}$, $\begin{pmatrix} -4 \\ 5 \\ -1 \end{pmatrix}$ and $\begin{pmatrix} 8 \\ 1 \\ 7 \end{pmatrix}$

   respectively. Find which of these points lie on the line with vector

   equation $\mathbf{r} = \begin{pmatrix} -1 \\ 4 \\ 1 \end{pmatrix} + \lambda \begin{pmatrix} 3 \\ -1 \\ 2 \end{pmatrix}$.
6. Points D, E and F have position vectors $\mathbf{i} - 2\mathbf{j}$, $4\mathbf{i} - \mathbf{j} + 3\mathbf{k}$ and
   $7\mathbf{i} - 8\mathbf{j} - 4\mathbf{k}$ respectively. Find which of these points lie on the line with
   vector equation $\mathbf{r} = (2\mathbf{i} - 3\mathbf{j} + \mathbf{k}) + \lambda(\mathbf{i} - \mathbf{j} - \mathbf{k})$.
7. If the point A, position vector $a\mathbf{i} + b\mathbf{j} + 3\mathbf{k}$, lies on the line L, vector
   equation $\mathbf{r} = (2\mathbf{i} + 4\mathbf{j} - \mathbf{k}) + \lambda(\mathbf{i} + \mathbf{j} + \mathbf{k})$, find the values of a and b.
8. Points A and B have position vectors $2\mathbf{i} - \mathbf{j} + \mathbf{k}$ and $5\mathbf{i} + 2\mathbf{j} - 2\mathbf{k}$
   repectively. Find
   (a) $\overrightarrow{AB}$ in the form $a\mathbf{i} + b\mathbf{j} + c\mathbf{k}$,
   (b) the vector equation of the line that passes through A and B.
9. Find the cartesian equations of the lines with vector equations
   (a) $\mathbf{r} = 2\mathbf{i} + 3\mathbf{j} - \mathbf{k} + \lambda(2\mathbf{i} + 3\mathbf{j} + \mathbf{k})$,
   (b) $\mathbf{r} = 3\mathbf{i} - \mathbf{j} + 2\mathbf{k} + \mu(3\mathbf{i} + 2\mathbf{j} - 4\mathbf{k})$,
   (c) $\mathbf{r} = 2\mathbf{i} + \mathbf{j} + \mathbf{k} + \eta(2\mathbf{i} - \mathbf{j} - \mathbf{k})$.
10. Find the vector equation of the line with parametric equations $\quad x = 2 + 3\lambda$
    $$y = 5 - 2\lambda$$
    $$z = 4 - \lambda$$
11. Find the vector equations of the lines with the following cartesian
    equations
    (a) $\dfrac{x - 2}{3} = \dfrac{y - 2}{2} = \dfrac{z + 1}{4}$ (b) $x - 3 = \dfrac{y + 2}{4} = \dfrac{z - 3}{-1}$
12. Lines L_1 and L_2 have vector equations $\mathbf{r} = 8\mathbf{i} - \mathbf{j} + 3\mathbf{k} + \lambda(-4\mathbf{i} + \mathbf{j})$
    and $\mathbf{r} = -2\mathbf{i} + 8\mathbf{j} - \mathbf{k} + \mu(\mathbf{i} + 3\mathbf{j} - 2\mathbf{k})$ respectively. Show that L_1
    and L_2 intersect and find the position vector of the point of intersection.

13. Lines L_1 and L_2 have vector equations

$$\mathbf{r} = \begin{pmatrix} 1 \\ 3 \\ -2 \end{pmatrix} + \lambda \begin{pmatrix} 4 \\ 0 \\ 1 \end{pmatrix} \text{ and } \mathbf{r} = \begin{pmatrix} 5 \\ 3 \\ 8 \end{pmatrix} + \mu \begin{pmatrix} -1 \\ 0 \\ 2 \end{pmatrix} \text{ respectively. Show}$$

that L_1 and L_2 intersect and find the position vector of the point of intersection.

14. For each of the following pairs of lines state whether the two lines intersect and, for those that do, give the coordinates of the point of intersection.

(a) $\dfrac{x-1}{2} = \dfrac{y+2}{1} = \dfrac{z-3}{-1}$; $\dfrac{x+1}{-2} = y-3 = \dfrac{z-7}{2}$.

(b) $\dfrac{x-1}{1} = \dfrac{y+1}{3} = \dfrac{z-2}{-1}$; $\dfrac{x+3}{-2} = \dfrac{y-8}{1} = \dfrac{z+2}{-1}$.

(c) $x - 2 = \dfrac{y+3}{4} = \dfrac{z-5}{2}$; $\dfrac{x-1}{-1} = \dfrac{y-8}{1} = \dfrac{z-3}{-2}$.

(d) $\dfrac{x-2}{5} = \dfrac{y-3}{-3} = \dfrac{z+1}{2}$; $\dfrac{x-9}{-3} = \dfrac{y-2}{5} = \dfrac{z-2}{-1}$.

15. For each of the pairs of lines given by the following vector equations state whether the lines are parallel lines, non-parallel coplanar lines or skew lines.

(a) $\mathbf{r} = 3\mathbf{i} + 2\mathbf{j} + 4\mathbf{k} + \lambda(\mathbf{i} + 2\mathbf{j} - \mathbf{k})$ and $\mathbf{r} = 2\mathbf{i} + 4\mathbf{j} - \mathbf{k} + \mu(3\mathbf{i} + 6\mathbf{j} - 3\mathbf{k})$.

(b) $\mathbf{r} = 2\mathbf{i} + 3\mathbf{j} + \mathbf{k} + \lambda(\mathbf{i} + 3\mathbf{j} + 2\mathbf{k})$ and $\mathbf{r} = 7\mathbf{i} + 3\mathbf{j} + 5\mathbf{k} + \mu(-\mathbf{i} + 2\mathbf{j})$.

(c) $\mathbf{r} = 2\mathbf{i} - 3\mathbf{j} - \mathbf{k} + \lambda(-\mathbf{i} + 3\mathbf{j} + 2\mathbf{k})$ and $\mathbf{r} = 3\mathbf{i} + 7\mathbf{j} + 6\mathbf{k} + \mu(3\mathbf{i} + 4\mathbf{j} + 2\mathbf{k})$.

(d) $\mathbf{r} = \mathbf{i} - 2\mathbf{j} + 4\mathbf{k} + \lambda(3\mathbf{i} + \mathbf{j} + 2\mathbf{k})$ and $\mathbf{r} = -8\mathbf{i} + 2\mathbf{j} + 3\mathbf{k} + \mu(\mathbf{i} - 2\mathbf{j} - \mathbf{k})$.

16. Find the acute angle between the lines with vector equations
$\mathbf{r} = 2\mathbf{i} + \mathbf{j} - \mathbf{k} + \lambda(2\mathbf{i} + 3\mathbf{j} + 6\mathbf{k})$ and $\mathbf{r} = \mathbf{i} + 2\mathbf{j} - 3\mathbf{k} + \mu(2\mathbf{i} - 2\mathbf{j} + \mathbf{k})$,
giving your answer to the nearest degree.

17. Find the acute angle between the lines whose equations are
$\dfrac{x-2}{-4} = \dfrac{y-3}{3} = \dfrac{z+1}{-1}$ and $\dfrac{x-3}{2} = \dfrac{y-1}{6} = \dfrac{z+1}{-5}$, giving your answer to the nearest degree.

18. The vector equations of three lines are:
line 1 $\mathbf{r} = 3\mathbf{i} - 2\mathbf{j} - \mathbf{k} + \lambda(-\mathbf{i} + 3\mathbf{j} + 4\mathbf{k})$
line 2 $\mathbf{r} = -2\mathbf{i} + 4\mathbf{j} + \mathbf{k} + \mu(-\mathbf{i} - 2\mathbf{k})$
line 3 $\mathbf{r} = -2\mathbf{i} + \mathbf{j} + \eta(2\mathbf{i} - 3\mathbf{j} + 3\mathbf{k})$
(a) Show that lines 1 and 2 intersect and find the position vector of the point of intersection.
(b) Show that lines 2 and 3 intersect and find the position vector of the point of intersection.
(c) Find the distance between these two points of intersection.

19. Two lines L_1 and L_2 lie in the x–y plane and have cartesian equations $y = m_1 x + c_1$ and $y = m_2 x + c_2$ respectively. Show that the vector equations of L_1 and L_2 can be written

$$\mathbf{r}_1 = \begin{pmatrix} x \\ y \end{pmatrix} = \begin{pmatrix} 0 \\ c_1 \end{pmatrix} + \lambda \begin{pmatrix} 1 \\ m_1 \end{pmatrix} \text{ and } \mathbf{r}_2 = \begin{pmatrix} x \\ y \end{pmatrix} = \begin{pmatrix} 0 \\ c_2 \end{pmatrix} + \mu \begin{pmatrix} 1 \\ m_2 \end{pmatrix}.$$

Use vector methods to show that if θ is the angle between L_1 and L_2

then $\tan \theta = \dfrac{m_1 - m_2}{1 + m_1 m_2}$.

(i.e. obtain the result of page 380 by vector methods).

20. For each of the following parts find the perpendicular distance from the given point to the given line,

(a) the point with position vector $\begin{pmatrix} 4 \\ 2 \\ 2 \end{pmatrix}$ and the line $\mathbf{r} = \begin{pmatrix} 3 \\ 1 \\ -1 \end{pmatrix} + \lambda \begin{pmatrix} 1 \\ -1 \\ 2 \end{pmatrix}$

(b) the point with position vector $3\mathbf{i} + \mathbf{j} - \mathbf{k}$ and the line $\mathbf{r} = \mathbf{i} - 6\mathbf{j} - 2\mathbf{k} + \lambda(\mathbf{i} + 2\mathbf{j} + 2\mathbf{k})$

(c) the point $(1, 1, 3)$ and the line $\dfrac{x + 4}{2} = \dfrac{y + 1}{3} = \dfrac{z - 1}{3}$

(d) the point $(-6, -4, -5)$ and the line $x - 5 = \dfrac{y - 6}{2} = \dfrac{z - 3}{4}$

21. Find the distance between the pairs of parallel lines listed below

(a) $\mathbf{r} = \begin{pmatrix} 2 \\ 0 \\ 3 \end{pmatrix} + \lambda \begin{pmatrix} 1 \\ -1 \\ 2 \end{pmatrix}$ and $\mathbf{r} = \begin{pmatrix} 1 \\ -1 \\ 4 \end{pmatrix} + \mu \begin{pmatrix} 1 \\ -1 \\ 2 \end{pmatrix}$

(b) $\dfrac{x - 2}{1} = \dfrac{y - 1}{-1} = \dfrac{z - 3}{2}$ and $\dfrac{x + 1}{1} = \dfrac{y - 3}{-1} = \dfrac{z - 1}{2}$

22. Find the shortest distance between the two skew lines L_1 and L_2 given

that L_1 has vector equation $\mathbf{r} = \begin{pmatrix} -1 \\ 2 \\ 3 \end{pmatrix} + \lambda \begin{pmatrix} 1 \\ 2 \\ 1 \end{pmatrix}$ and L_2 has vector

equation $\mathbf{r} = \begin{pmatrix} 0 \\ -1 \\ 1 \end{pmatrix} + \mu \begin{pmatrix} 2 \\ 1 \\ 3 \end{pmatrix}$

23. Find the shortest distance between the two skew lines given in each of the following parts.

(a) $\mathbf{r} = \begin{pmatrix} 1 \\ 2 \\ -1 \end{pmatrix} + \lambda \begin{pmatrix} 1 \\ 0 \\ 1 \end{pmatrix}$ and $\mathbf{r} = \begin{pmatrix} 2 \\ -1 \\ 0 \end{pmatrix} + \mu \begin{pmatrix} 1 \\ 1 \\ 2 \end{pmatrix}$

(b) $\dfrac{x - 2}{0} = \dfrac{y + 1}{1} = \dfrac{z}{2}$ and $\dfrac{x + 1}{1} = \dfrac{y - 1}{-3} = \dfrac{z - 1}{-2}$

24. The diagram shows the line L and the point A both lying in the x–y plane. L has cartesian equation $ax + by + c = 0$ and A is the point (x_1, y_1). A is a perpendicular distance d from L.

Show that the vector equation $\mathbf{r} = \begin{pmatrix} x \\ y \end{pmatrix} = \begin{pmatrix} 0 \\ -c/b \end{pmatrix} + \lambda \begin{pmatrix} b \\ -a \end{pmatrix}$ also

represents the line L.

If λ takes the value λ_1 at point B show that $\lambda_1 = \dfrac{b^2 x_1 - aby_1 - ac}{b(a^2 + b^2)}$

Hence show that $d = \dfrac{ax_1 + by_1 + c}{\sqrt{(a^2 + b^2)}}$

(Note: this has proved the result on page 382 by vector methods).

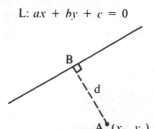

L: $ax + by + c = 0$

17.4 Equations of a plane

Vector equation of a plane

If we are given two non-parallel vectors **b** and **c** that are parallel to a particular plane and the position vector **a** of a point in the plane then the plane is uniquely defined. To obtain a vector equation of the plane we consider some general point R in the plane having position vector **r**. From the diagram **r** = **a** + $\overrightarrow{AR}$

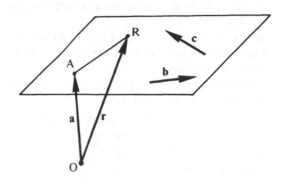

Now it is possible to move from A to R by combining a suitable number of vector **b**'s with a suitable number of vector **c**'s. Thus we can write that $\overrightarrow{AR}$ = λ**b** + μ**c** where λ and μ are scalars.
Thus   **r** = **a** + λ**b** + μ**c**

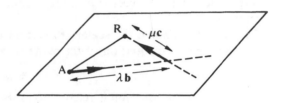

This is the vector equation for the plane that is parallel to the vectors **b** and **c** and contains the point with position vector **a**. As the equation involves the parameters λ and μ it is sometimes referred to as the parametric vector equation of the plane. The position vector **r**, of *any* point lying in the plane will satisfy this equation. (Note: The vector equation of a line **r** = **a** + λ**b** can be similarly referred to as the parametric vector equation of a line).

Example 17

Find a vector equation for the plane containing the three points A, B and C whose position vectors are 2**i** + 3**j** − **k**, 3**i** + **j** + **k** and 5**i** − 2**j** + 3**k** respectively.
As A, B and C all lie in the plane then the vectors $\overrightarrow{AB}$ and $\overrightarrow{AC}$ will lie in the plane.
But $\quad\overrightarrow{AB}$ = (3**i** + **j** + **k**) − (2**i** + 3**j** − **k**)
$\qquad\qquad$ = **i** − 2**j** + 2**k**
and $\quad\overrightarrow{AC}$ = (5**i** − 2**j** + 3**k**) − (2**i** + 3**j** − **k**)
$\qquad\qquad$ = 3**i** − 5**j** + 4**k**
Thus the vector equation of the plane is
$\qquad$ **r** = 2**i** + 3**i** − **k** + λ(**i** − 2**j** + 2**k**) + μ(3**i** − 5**j** + 4**k**)

A plane can also be uniquely defined by
stating a vector that is perpendicular to the
plane and the position vector of a point on
the plane.

Suppose that **n** is a vector perpendicular to the
plane and A is a point in the plane having
position vector **a**.

Consider some general point R in the plane
having position vector **r**.

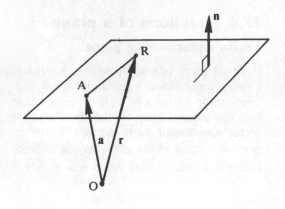

Thus $\overrightarrow{AR}$ will lie in the plane and will
therefore be perpendicular to **n**.

i.e. $\overrightarrow{AR} \cdot \mathbf{n} = 0$

thus $(-\mathbf{a} + \mathbf{r}) \cdot \mathbf{n} = 0$

or $\mathbf{r} \cdot \mathbf{n} = \mathbf{a} \cdot \mathbf{n}$

The equation $\mathbf{r} \cdot \mathbf{n} = \mathbf{a} \cdot \mathbf{n}$ is the vector equation of the plane that is
perpendicular to the vector **n** and that contains the point with position
vector **a**.

With **a** and **n** known the scalar product $\mathbf{a} \cdot \mathbf{n}$ can be evaluated. If $\mathbf{a} \cdot \mathbf{n} = p$,
a scalar, then the equation of the plane is

$$\mathbf{r} \cdot \mathbf{n} = p$$

This form of the equation is referred to as the scalar product form of the vector
equation of the plane. The position vector **r**, of *any* point in the plane will satisfy
this equation.

Distance of a plane from the origin

Suppose that **n̂** is the *unit* vector perpendicular
to the plane and d is the perpendicular distance
from the plane to the origin.

The scalar product vector equation of the plane will be

$\mathbf{r} \cdot \mathbf{\hat{n}} = \mathbf{a} \cdot \mathbf{\hat{n}}$

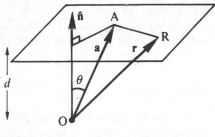

$\quad = |\mathbf{a}| \times 1 \times \cos \theta$ where θ is the angle between **a** and **n̂**

$\quad = a \cos \theta$

$\quad = d$

Thus if the equation of the plane is in the form $\mathbf{r} \cdot \mathbf{\hat{n}} = d$ where **n̂** is a unit
vector normal to the plane then d is the perpendicular distance of the plane
from the origin.

Consider the three parallel planes Π_1, Π_2 and
Π_3 with vector equations $\mathbf{r} \cdot \mathbf{\hat{n}} = 3$, $\mathbf{r} \cdot \mathbf{\hat{n}} = 1$
and $\mathbf{r} \cdot \mathbf{\hat{n}} = -2$ respectively. The
perpendicular distance of these planes from
the origin will be 3 units, 1 unit and 2 units
respectively and the significance of the
negative sign is that Π_3 is on the other side of
the origin from Π_1 and Π_2. Thus Π_1 is 2 units
from Π_2 and 5 units from Π_3.

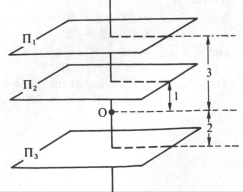

Example 18

A plane contains a point A, position vector $3\mathbf{i} + 4\mathbf{j} + 2\mathbf{k}$ and is perpendicular to the vector $\mathbf{i} + 2\mathbf{j} - 2\mathbf{k}$. Find
(a) a vector equation of the plane,
(b) the perpendicular distance of the plane from the origin,
(c) the perpendicular distance from this plane to the parallel plane
     $\mathbf{r} \cdot (\mathbf{i} + 2\mathbf{j} - 2\mathbf{k}) = -3$

(a) The scalar product form of the vector equation will be
$$\mathbf{r} \cdot (\mathbf{i} + 2\mathbf{j} - 2\mathbf{k}) = (3\mathbf{i} + 4\mathbf{j} + 2\mathbf{k}) \cdot (\mathbf{i} + 2\mathbf{j} - 2\mathbf{k})$$
$$= 3 + 8 - 4$$
$$= 7$$
The scalar product vector equation of the plane is $\mathbf{r} \cdot (\mathbf{i} + 2\mathbf{j} - 2\mathbf{k}) = 7$

(b) If we can write the vector equation in the form $\mathbf{r} \cdot \hat{\mathbf{n}} = c$ where $\hat{\mathbf{n}}$ is a unit vector normal to the plane, then c is the required distance.
Now $|\mathbf{i} + 2\mathbf{j} - 2\mathbf{k}| = 3$
From $\mathbf{r} \cdot (\mathbf{i} + 2\mathbf{j} - 2\mathbf{k}) = 7$ we can write $\mathbf{r} \cdot \dfrac{(\mathbf{i} + 2\mathbf{j} - 2\mathbf{k})}{3} = \dfrac{7}{3}$
giving the perpendicular distance of the plane from the origin as $2\frac{1}{3}$ units.

(c) The plane $\mathbf{r} \cdot (\mathbf{i} + 2\mathbf{j} - 2\mathbf{k}) = -3$ is on the other side of the origin from $\mathbf{r} \cdot (\mathbf{i} + 2\mathbf{j} - 2\mathbf{k}) = 7$ and is a perpendicular distance of
$$\dfrac{3}{\sqrt{(1^2 + 2^2 + 2^2)}} = 1 \text{ unit from the origin. Thus the perpendicular}$$
distance between the planes is $3\frac{1}{3}$ units.

Example 19

Find the position vector of the point where the line
$$\mathbf{r} = \begin{pmatrix} 5 \\ 3 \\ -1 \end{pmatrix} + \lambda \begin{pmatrix} 1 \\ -4 \\ 2 \end{pmatrix} \text{ meets the plane } \mathbf{r} \cdot \begin{pmatrix} 2 \\ 1 \\ 3 \end{pmatrix} = 12.$$

The position vector of the point of intersection will satisfy both the equation of the line and that of the plane.
If $\mathbf{r}_1$ is the position vector of the point of intersection then
$$\mathbf{r}_1 = \begin{pmatrix} 5 \\ 3 \\ -1 \end{pmatrix} + \lambda \begin{pmatrix} 1 \\ -4 \\ 2 \end{pmatrix} \quad \text{and} \quad \mathbf{r}_1 \cdot \begin{pmatrix} 2 \\ 1 \\ 3 \end{pmatrix} = 12$$
thus
$$\begin{pmatrix} 5 + \lambda \\ 3 - 4\lambda \\ -1 + 2\lambda \end{pmatrix} \cdot \begin{pmatrix} 2 \\ 1 \\ 3 \end{pmatrix} = 12$$
$$\therefore \quad 10 + 2\lambda + 3 - 4\lambda - 3 + 6\lambda = 12$$
$$\lambda = \tfrac{1}{2}$$
Thus the required position vector is $\begin{pmatrix} 5 \\ 3 \\ -1 \end{pmatrix} + \frac{1}{2}\begin{pmatrix} 1 \\ -4 \\ 2 \end{pmatrix} = \begin{pmatrix} 5\frac{1}{2} \\ 1 \\ 0 \end{pmatrix}$

Example 20

Show that the line $\mathbf{r} = 3\mathbf{i} + 3\mathbf{j} - 2\mathbf{k} + \lambda(\mathbf{i} + \mathbf{j} - \mathbf{k})$ lies in the plane
$\mathbf{r} \cdot (3\mathbf{i} - 2\mathbf{j} + \mathbf{k}) = 1$.

If the line and the plane (i) are parallel and (ii) contain a common point,
then the line must lie in the plane.

(i) The line is parallel to $\mathbf{i} + \mathbf{j} - \mathbf{k}$ and the plane is perpendicular to
$3\mathbf{i} - 2\mathbf{j} + \mathbf{k}$.
However $(\mathbf{i} + \mathbf{j} - \mathbf{k}) \cdot (3\mathbf{i} - 2\mathbf{j} + \mathbf{k}) = 3 - 2 - 1$
$= 0$
Hence the vectors $\mathbf{i} + \mathbf{j} - \mathbf{k}$ and $3\mathbf{i} - 2\mathbf{j} + \mathbf{k}$ are perpendicular to each
other.
Thus the line and plane are both perpendicular to $3\mathbf{i} - 2\mathbf{j} + \mathbf{k}$ and must
therefore be parallel to each other

(ii) The line passes through the point with position vector $3\mathbf{i} + 3\mathbf{j} - 2\mathbf{k}$.
Also $(3\mathbf{i} + 3\mathbf{j} - 2\mathbf{k}) \cdot (3\mathbf{i} - 2\mathbf{j} + \mathbf{k}) = 9 - 6 - 2$
$= 1$
i.e. $(3\mathbf{i} + 3\mathbf{j} - 2\mathbf{k})$ satisfies $\mathbf{r} \cdot (3\mathbf{i} - 2\mathbf{j} + \mathbf{k}) = 1$
Thus the point with position vector $3\mathbf{i} + 3\mathbf{j} - 2\mathbf{k}$ is common to both the line
and the plane.
With conditions (i) and (ii) proved the line must lie in the plane.

Alternatively example 20 can be solved by showing that the line and plane
contain two common points. Taking any two values for λ, for example
$\lambda = 0$ and $\lambda = 1$:
If $\lambda = 0$ $\mathbf{r} = 3\mathbf{i} + 3\mathbf{j} - 2\mathbf{k}$ Thus the point with position vector
$3\mathbf{i} + 3\mathbf{j} - 2\mathbf{k}$ lies on the given line.
If $\lambda = 1$ $\mathbf{r} = 4\mathbf{i} + 4\mathbf{j} - 3\mathbf{k}$ Thus the point with position vector
$4\mathbf{i} + 4\mathbf{j} - 3\mathbf{k}$ lies on the given line.
But $(3\mathbf{i} + 3\mathbf{j} - 2\mathbf{k}) \cdot (3\mathbf{i} - 2\mathbf{j} + \mathbf{k}) = 1$
and $(4\mathbf{i} + 4\mathbf{j} - 3\mathbf{k}) \cdot (3\mathbf{i} - 2\mathbf{j} + \mathbf{k}) = 1$
i.e.   both points also lie in the plane $\mathbf{r} \cdot (3\mathbf{i} - 2\mathbf{j} + \mathbf{k}) = 1$.
Therefore the line and plane contain two common points and so the line
must lie in the plane.

Cartesian equation of a plane

Consider a plane with equation $\mathbf{r} \cdot \mathbf{n} = p$ where $\mathbf{n} = \begin{pmatrix} a \\ b \\ c \end{pmatrix}$. Writing $\mathbf{r}$, the

position vector of a general point in the plane, as $\begin{pmatrix} x \\ y \\ z \end{pmatrix}$ gives

$$\begin{pmatrix} x \\ y \\ z \end{pmatrix} \cdot \begin{pmatrix} a \\ b \\ c \end{pmatrix} = p$$

i.e. $ax + by + cz = p$

Thus a plane perpendicular to $a\mathbf{i} + b\mathbf{j} + c\mathbf{k}$ has cartesian equation $ax + by + cz = p$.

Example 21

Find the cartesian equation of the plane with vector equation $\mathbf{r} \cdot (2\mathbf{i} + 3\mathbf{j} - 4\mathbf{k}) = 5$.
If $\mathbf{r} = x\mathbf{i} + y\mathbf{j} + z\mathbf{k}$ then $(x\mathbf{i} + y\mathbf{j} + z\mathbf{k}) \cdot (2\mathbf{i} + 3\mathbf{j} - 4\mathbf{k}) = 5$
 i.e. $2x + 3y - 4z = 5$
The cartesian equation of the plane is $2x + 3y - 4z = 5$.

The cartesian equation of a plane can also be found from its parametric
vector equation, as the following example shows.

Example 22

Find the cartesian equation of the plane with parametric vector equation

$$\mathbf{r} = \begin{pmatrix} 1 \\ 2 \\ -1 \end{pmatrix} + \lambda \begin{pmatrix} 1 \\ 1 \\ 2 \end{pmatrix} + \mu \begin{pmatrix} 0 \\ 2 \\ -1 \end{pmatrix}$$

Writing $\mathbf{r}$ as $\begin{pmatrix} x \\ y \\ z \end{pmatrix}$ gives the equations

$$x = 1 + \lambda \qquad \dots [1]$$
$$y = 2 + \lambda + 2\mu \qquad \dots [2]$$
$$\text{and} \quad z = -1 + 2\lambda - \mu \qquad \dots [3]$$

from [1] $\lambda = x - 1$
substituting this into [2] gives $\mu = \frac{1}{2}(y - 1 - x)$
substituting these expressions for λ and μ into [3] gives $5x - y - 2z = 5$
The cartesian equation of the plane is $5x - y - 2z = 5$

Example 23

The planes Π_1 and Π_2 have cartesian equations $2x + 5y - 14z = 30$ and
$2x + 5y - 14z = -15$ respectively. State the vector equations of Π_1 and
Π_2 in scalar product form, show that the planes are parallel and find the
distance between the planes.

The vector equation of Π_1 will be $\mathbf{r} \cdot (2\mathbf{i} + 5\mathbf{j} - 14\mathbf{k}) = 30$.
The vector equation of Π_2 will be $\mathbf{r} \cdot (2\mathbf{i} + 5\mathbf{j} - 14\mathbf{k}) = -15$.
As each plane is perpendicular to $2\mathbf{i} + 5\mathbf{j} - 14\mathbf{k}$ the planes Π_1 and Π_2 must
be parallel to each other.
To find the perpendicular distance of a plane from the origin we must
express the vector equation of the plane in the form $\mathbf{r} \cdot \hat{\mathbf{n}} = d$
The unit vector in the direction of $2\mathbf{i} + 5\mathbf{j} - 14\mathbf{k}$ is $\frac{1}{15}(2\mathbf{i} + 5\mathbf{j} - 14\mathbf{k})$
Thus if $\mathbf{r} \cdot (2\mathbf{i} + 5\mathbf{j} - 14\mathbf{k}) = 30$ then $\mathbf{r} \cdot \frac{1}{15}(2\mathbf{i} + 5\mathbf{j} - 14\mathbf{k}) = 2$
and if $\mathbf{r} \cdot (2\mathbf{i} + 5\mathbf{j} - 14\mathbf{k}) = -15$ then $\mathbf{r} \cdot \frac{1}{15}(2\mathbf{i} + 5\mathbf{j} - 14\mathbf{k}) = -1$
Thus the planes Π_1 and Π_2 are on opposite sides of the origin and are
respectively 2 units and 1 unit from the origin. i.e. the perpendicular distance
between the planes is 3 units.

Example 24

Find the point where the line $\dfrac{x-3}{-1} = \dfrac{y-1}{2} = \dfrac{z+3}{4}$ cuts the plane $3x - y + 2z = 8$.

The point of intersection of the line and the plane is obtained by solving the equations of the line and plane simultaneously.

From $\dfrac{x-3}{-1} = \dfrac{y-1}{2}$ $\qquad x = \dfrac{-1(y-1)}{2} + 3$ $\qquad$ i.e. $\quad x = \dfrac{7-y}{2}$ $\ \ldots$ [1]

From $\dfrac{y-1}{2} = \dfrac{z+3}{4}$ $\qquad z = \dfrac{4(y-1)}{2} - 3$ $\qquad$ i.e. $\quad z = 2y - 5$ $\ \ldots$ [2]

Substituting these values of x and z into $\quad 3x - y + 2z = 8$

$$\text{gives} \quad \frac{3(7-y)}{2} - y + 2(2y-5) = 8$$

thus $\qquad\qquad\qquad\qquad\qquad\qquad\qquad\qquad y = 5$

and from [1] and [2] $\qquad\qquad\qquad\qquad\qquad x = 1 \quad \text{and} \quad z = 5$

Thus line and plane intersect at the point $(1, 5, 5)$

Example 25

Find in both cartesian and vector forms the equation of the line of intersection of the two planes $7x - 4y + 3z = -3$ and $4x + 2y + z = 4$

The two equations represent distinct, non-parallel planes and so the intersection will be a line.

Eliminating z from the two equations gives $\quad x + 2y = 3$

$$\text{or} \qquad\qquad x = 3 - 2y$$

Similarly, eliminating y gives $\qquad\qquad 3x + z = 1$

$$\text{or} \qquad x = \frac{1-z}{3}$$

Thus the cartesian equations of the line are $\qquad x = 3 - 2y = \dfrac{1-z}{3}$

Rewriting these as $\qquad\qquad\qquad \dfrac{x-0}{1} = \dfrac{2y-3}{-1} = \dfrac{z-1}{-3}$

i.e. $\qquad \dfrac{x-0}{1} = \dfrac{y-1\frac{1}{2}}{-\frac{1}{2}} = \dfrac{z-1}{-3}(= \lambda)$

gives the vector equation $\qquad \mathbf{r} = \begin{pmatrix} x \\ y \\ z \end{pmatrix} = \begin{pmatrix} 0 \\ 1\frac{1}{2} \\ 1 \end{pmatrix} + \lambda\begin{pmatrix} 1 \\ -\frac{1}{2} \\ -3 \end{pmatrix}$

For a form of the vector equation that avoids fractions we can write $\lambda = 1 + 2\mu$ to give

$$\mathbf{r} = \begin{pmatrix} 1 \\ 1 \\ -2 \end{pmatrix} + \mu\begin{pmatrix} 2 \\ -1 \\ -6 \end{pmatrix}.$$

Example 26

Find the acute angle between the line $\dfrac{x+1}{4} = y - 2 = \dfrac{z-3}{-1}$ and the

plane $3x - 5y + 4z = 5$ giving your answer correct to the nearest degree.

The vector equation of the line is $\mathbf{r} = -\mathbf{i} + 2\mathbf{j} + 3\mathbf{k} + \lambda(4\mathbf{i} + \mathbf{j} - \mathbf{k})$
The vector equation of the plane is $\mathbf{r} \cdot (3\mathbf{i} - 5\mathbf{j} + 4\mathbf{k}) = 5$
The line is parallel to $4\mathbf{i} + \mathbf{j} - \mathbf{k}$ and the plane is perpendicular to $3\mathbf{i} - 5\mathbf{j} + 4\mathbf{k}$.
Thus, if θ is the angle between the line and the plane

$$(4\mathbf{i} + \mathbf{j} - \mathbf{k}) \cdot (3\mathbf{i} - 5\mathbf{j} + 4\mathbf{k}) = |4\mathbf{i} + \mathbf{j} - \mathbf{k}||3\mathbf{i} - 5\mathbf{j} + 4\mathbf{k}| \cos(90° - \theta)$$

i.e. $\qquad\qquad\qquad\qquad 3 = \sqrt{18}\sqrt{50} \sin\theta$

$\therefore \qquad\qquad\qquad\qquad \sin\theta = 0 \cdot 1$

giving $\qquad\qquad\qquad\qquad \theta = 6°$ (to nearest degree)
The angle between the line and the plane is $6°$, to the nearest degree.

Example 27

A plane Π has cartesian equation $5x + 2y - 4z = -22$. Find (a) the
cartesian equation of the plane parallel to Π and containing the point
A(2, 1, 1), (b) the perpendicular distance from the point A to the plane Π.

(a) Π has vector equation $\mathbf{r} \cdot (5\mathbf{i} + 2\mathbf{j} - 4\mathbf{k}) = -22$. Thus the required
plane is perpendicular to $5\mathbf{i} + 2\mathbf{j} - 4\mathbf{k}$ and passes through the point A,
position vector $2\mathbf{i} + \mathbf{j} + \mathbf{k}$.
Therefore the equation of the plane is

$$\mathbf{r} \cdot (5\mathbf{i} + 2\mathbf{j} - 4\mathbf{k}) = (2\mathbf{i} + \mathbf{j} + \mathbf{k}) \cdot (5\mathbf{i} + 2\mathbf{j} - 4\mathbf{k})$$
$$= 8$$

$\therefore$ The cartesian equation of the plane is $5x + 2y - 4z = 8$

(b) To find the distance from a plane to the origin we need the equations of
the plane to be in the form $\mathbf{r} \cdot \hat{\mathbf{n}} = d$

If $\quad \mathbf{r} \cdot (5\mathbf{i} + 2\mathbf{j} - 4\mathbf{k}) = -22 \qquad$ If $\quad \mathbf{r} \cdot (5\mathbf{i} + 2\mathbf{j} - 4\mathbf{j}) = 8$

then $\quad \mathbf{r} \cdot \dfrac{(5\mathbf{i} + 2\mathbf{j} - 4\mathbf{k})}{3\sqrt{5}} = -\dfrac{22}{3\sqrt{5}} \qquad$ then $\quad \mathbf{r} \cdot \dfrac{(5\mathbf{i} + 2\mathbf{j} - 4\mathbf{k})}{3\sqrt{5}} = \dfrac{8}{3\sqrt{5}}$

Thus the perpendicular distance from A to the plane Π is

$$\dfrac{8}{3\sqrt{5}} + \dfrac{22}{3\sqrt{5}} = \dfrac{30}{3\sqrt{5}}$$

$$= 2\sqrt{5} \text{ units}$$

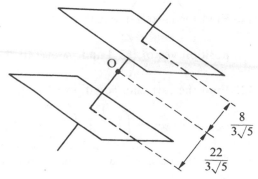

Exercise 17D

1. The plane Π has vector equation
$\mathbf{r} = 4\mathbf{i} + 3\mathbf{j} + 2\mathbf{k} + \lambda(\mathbf{i} - \mathbf{j} - \mathbf{k}) + \mu(2\mathbf{i} + 3\mathbf{j} + \mathbf{k})$. Show that the
point with position vector $7\mathbf{i} - 5\mathbf{j} - 4\mathbf{k}$ lies in the plane Π.

2. The plane Π has vector equation $\mathbf{r} = \begin{pmatrix} 2 \\ 3 \\ -1 \end{pmatrix} + \lambda \begin{pmatrix} 4 \\ 1 \\ 0 \end{pmatrix} + \mu \begin{pmatrix} 2 \\ 1 \\ 3 \end{pmatrix}$.

   Show that the point with position vector $\begin{pmatrix} 2 \\ 2 \\ -7 \end{pmatrix}$ lies in the plane Π.

3. The plane Π has vector equation $\mathbf{r} . (2\mathbf{i} - 3\mathbf{j} + \mathbf{k}) = 5$. Show that the point with position vector $\mathbf{i} - 2\mathbf{j} - 3\mathbf{k}$ lies in the plane Π.

4. Find the scalar product vector equation of the plane that is perpendicular to $\mathbf{i} - 2\mathbf{j} - \mathbf{k}$ and contains the point with position vector $2\mathbf{i} - 3\mathbf{j} + \mathbf{k}$.

5. Find the scalar product vector equation of the plane that is perpendicular to $2\mathbf{i} + 3\mathbf{j} - \mathbf{k}$ and contains the point with position vector $3\mathbf{i} - \mathbf{j} - 2\mathbf{k}$.

6. Find the vector equation of the line that passes through the point A, position vector $2\mathbf{i} + 3\mathbf{j} - \mathbf{k}$, and is perpendicular to the plane Π, vector equation $\mathbf{r} . (2\mathbf{i} - \mathbf{j} + 3\mathbf{k}) = 2$

7. State which of the lines given by the vector equations below are perpendicular to the plane $\mathbf{r} . (2\mathbf{i} + 3\mathbf{j} - \mathbf{k}) = 6$
   (a) $\mathbf{r} = 2\mathbf{i} + 3\mathbf{j} - \mathbf{k} + \lambda(2\mathbf{i} + 3\mathbf{j} - \mathbf{k})$
   (b) $\mathbf{r} = 2\mathbf{i} + 3\mathbf{j} - \mathbf{k} + \mu(\mathbf{i} + \mathbf{j} + \mathbf{k})$
   (c) $\mathbf{r} = 2\mathbf{i} - \mathbf{j} + \mathbf{k} + \eta(4\mathbf{i} + 6\mathbf{j} - 2\mathbf{k})$.

8. State which of the lines given by the vector equations below are parallel to the plane $\mathbf{r} . (\mathbf{i} - 2\mathbf{j} + 3\mathbf{k}) = 5$
   (a) $\mathbf{r} = 2\mathbf{i} + 3\mathbf{j} - \mathbf{k} + \lambda(6\mathbf{i} + 3\mathbf{j})$
   (b) $\mathbf{r} = 3\mathbf{i} + \mathbf{j} + \mu(4\mathbf{i} - \mathbf{j} - 2\mathbf{k})$
   (c) $\mathbf{r} = \mathbf{i} + \mathbf{j} + \mathbf{k} + \eta(\mathbf{i} - 2\mathbf{j} + 3\mathbf{k})$.

9. Write down the vector equations of the following planes in the form $\mathbf{r} . \mathbf{n} = p$
   (a) $4x + 2y + 3z = 4$ (b) $2x - 3y + 4z = 5$

10. Find the cartesian equation of the plane with parametric vector equation

$$\mathbf{r} = \begin{pmatrix} 3 \\ 0 \\ 1 \end{pmatrix} + \lambda \begin{pmatrix} 2 \\ -1 \\ 0 \end{pmatrix} + \mu \begin{pmatrix} 1 \\ 1 \\ 1 \end{pmatrix}.$$

11. Find the cartesian equation of the plane containing the point with position vector $\begin{pmatrix} 1 \\ 3 \\ 1 \end{pmatrix}$ and parallel to the vectors $\begin{pmatrix} 1 \\ -1 \\ 3 \end{pmatrix}$ and $\begin{pmatrix} 2 \\ 1 \\ -3 \end{pmatrix}$.

12. Find the cartesian equation of the plane containing the points with position vectors $\begin{pmatrix} 1 \\ 2 \\ -1 \end{pmatrix}, \begin{pmatrix} 2 \\ 1 \\ -2 \end{pmatrix}$ and $\begin{pmatrix} 3 \\ -3 \\ 3 \end{pmatrix}$.

13. Find the perpendicular distance from the plane $\mathbf{r} . (2\mathbf{i} - 14\mathbf{j} + 5\mathbf{k}) = 10$ to the origin.

14. Find the perpendicular distance from the plane $2x + 3y - 6z = 21$ to the origin.

15. The plane Π contains the point A, position vector $2\mathbf{i} - \mathbf{j} - 2\mathbf{k}$, and is perpendicular to the vector $4\mathbf{i} + 4\mathbf{j} - 7\mathbf{k}$. Find the perpendicular distance from the plane Π to the origin.

16. Find the position vector of the point where the line $\mathbf{r} = \begin{pmatrix} 2 \\ -1 \\ 3 \end{pmatrix} + \lambda \begin{pmatrix} 5 \\ 3 \\ 2 \end{pmatrix}$ meets

    the plane $\mathbf{r} \cdot \begin{pmatrix} 1 \\ 2 \\ -1 \end{pmatrix} = 15$.

17. Find the position vector of the point where the line
    $\mathbf{r} = -\mathbf{i} - 3\mathbf{j} + 4\mathbf{k} + \lambda(2\mathbf{i} + \mathbf{j} - 3\mathbf{k})$ cuts the plane $\mathbf{r} \cdot (\mathbf{i} - \mathbf{j} + 2\mathbf{k}) = -5$.

18. Show that the line with cartesian equations $\dfrac{x - 4}{4} = \dfrac{y - 2}{3} = z - 3$
    is parallel to the plane with cartesian equation $2x - 3y + z = 7$.

19. Show that the lines $\mathbf{r} = 2\mathbf{i} - 3\mathbf{j} + 4\mathbf{k} + \lambda(3\mathbf{i} - 2\mathbf{j} + \mathbf{k})$ and
    $\mathbf{r} = \mathbf{i} + 3\mathbf{j} + \mathbf{k} + \mu(-\mathbf{i} - 2\mathbf{j} + \mathbf{k})$ intersect and that their point of
    intersection lies on the plane $\mathbf{r} \cdot (2\mathbf{i} + \mathbf{j} + 2\mathbf{k}) = 3$

20. Find the point where the line $\dfrac{x + 2}{-1} = \dfrac{y - 1}{2} = z - 4$ cuts the plane
    $2x - y + 3z = 10$.

21. Find the point where the line $x + 1 = \dfrac{y - 2}{4} = z - 3$ cuts the plane
    $\mathbf{r} \cdot (2\mathbf{i} - \mathbf{j} + 3\mathbf{k}) = 8$.

22. Show that the line with vector equation $\mathbf{r} = \begin{pmatrix} 6 \\ -5 \\ 1 \end{pmatrix} + \lambda \begin{pmatrix} 1 \\ -2 \\ 3 \end{pmatrix}$ is

    perpendicular to the plane with vector equation $\mathbf{r} \cdot \begin{pmatrix} 1 \\ -2 \\ 3 \end{pmatrix} = -9$. Find

    the position vector of the point of intersection of the line and plane and

    the distance from the point with position vector $\begin{pmatrix} 1 \\ 1 \\ -11 \end{pmatrix}$ to this point of

    intersection.

23. (a) Find the acute angle between the line L, vector equation
        $\mathbf{r} = \mathbf{i} - 2\mathbf{j} + 3\mathbf{k} + \lambda(2\mathbf{i} + 2\mathbf{j} - \mathbf{k})$ and the normal to the plane Π,
        vector equation $\mathbf{r} \cdot (2\mathbf{i} + 3\mathbf{j} + 6\mathbf{k}) = 18$. (Give your answer to the
        nearest degree).
    (b) Find the acute angle between the line L and the plane
        $\mathbf{r} \cdot (2\mathbf{i} + 3\mathbf{j} + 6\mathbf{k}) = 18$. (Give your answer to the nearest degree).

24. Find the acute angle between each of the following pairs of planes giving
    your answers to the nearest degree.
    (a) $\mathbf{r} \cdot (\mathbf{i} - \mathbf{j} + \mathbf{k}) = 7$ and $\mathbf{r} \cdot (2\mathbf{i} - \mathbf{j} + 3\mathbf{k}) = 10$
    (b) $6x + 2y + 5z = 4$ and $x - 4y + 3z = 9$

25. For each of the parts (a) (b) and (c) state whether the given line is
    parallel to, or intersects, the given plane. If the line is parallel to the
    plane find whether or not the line lies in the plane. If the line and plane
    intersect find the position vector of the point of intersection.
    (a) line $\mathbf{r} = 2\mathbf{i} + \mathbf{j} + \mathbf{k} + \lambda(5\mathbf{i} - \mathbf{j} + 3\mathbf{k})$ plane $\mathbf{r} \cdot (2\mathbf{i} + \mathbf{j} - 3\mathbf{k}) = 7$
    (b) line $\mathbf{r} = 3\mathbf{i} + 2\mathbf{j} + 8\mathbf{k} + \lambda(2\mathbf{i} - 3\mathbf{j} - \mathbf{k})$ plane $\mathbf{r} \cdot (2\mathbf{i} + \mathbf{j} - \mathbf{k}) = 4$
    (c) line $\mathbf{r} = 2\mathbf{i} + 4\mathbf{j} - \mathbf{k} + \lambda(\mathbf{i} + 2\mathbf{j} - 3\mathbf{k})$ plane $\mathbf{r} \cdot (4\mathbf{i} + \mathbf{j} + 2\mathbf{k}) = 10$

26. A plane Π is perpendicular to the vector $2\mathbf{i} + 3\mathbf{j} + \mathbf{k}$ and contains the point A, position vector $\mathbf{i} + 3\mathbf{j} - 3\mathbf{k}$. Find the position vector of the point where the line $\mathbf{r} = 2\mathbf{i} + 3\mathbf{j} + 4\mathbf{k} + \lambda(\mathbf{i} - \mathbf{j} + 4\mathbf{k})$ meets the plane Π.

27. Points D, E and F have position vectors $\mathbf{i} + 2\mathbf{j} + 2\mathbf{k}$, $2\mathbf{i} - 3\mathbf{j} - 5\mathbf{k}$ and $-\mathbf{i} + 3\mathbf{j} - 2\mathbf{k}$ respectively. Find the equation of the plane containing D, E and F giving your answer both in parametric vector form and cartesian form.

28. Find the acute angle between the line $\dfrac{x - 6}{5} = \dfrac{y - 1}{-1} = z + 1$ and the plane $7x - y + 5z = -5$ giving your answer to the nearest degree.

29. The point A has position vector $3\mathbf{i} - \mathbf{j} - \mathbf{k}$ and the plane Π_1 has vector equation $\mathbf{r} \cdot (3\mathbf{i} - 5\mathbf{j} + 4\mathbf{k}) = 3\sqrt{2}$.
    (a) Find in scalar product form, the vector equation of a plane Π_2 that is parallel to Π_1 and contains the point A.
    (b) Find the perpendicular distance from A to the plane Π_1.

30. For each of the following find the perpendicular distance from the given point to the given plane.
    (a) The point with position vector $3\mathbf{i} + 7\mathbf{j} + \mathbf{k}$ and the plane $\mathbf{r} \cdot (\mathbf{i} + 2\mathbf{j} - 2\mathbf{k}) = 6$,
    (b) the point $(4, -1, 2)$ and the plane $2x - 2y + z = 21$.

31. The line L has vector equation $\mathbf{r} = \mathbf{i} - \mathbf{j} + 3\mathbf{k} + \lambda(\mathbf{i} + 2\mathbf{j} - 2\mathbf{k})$ and the plane Π has vector equation $\mathbf{r} \cdot (6\mathbf{i} - 2\mathbf{j} + \mathbf{k}) = -3$
    (a) Show that the line L is parallel to plane Π.
    (b) Obtain the scalar product vector equation of the plane that is parallel to Π and that contains the line L.
    (c) Find the perpendicular distance from the line L to the plane Π.

32. Show that the line $\dfrac{x - 2}{2} = \dfrac{y - 2}{-1} = \dfrac{z - 3}{3}$ is parallel to the plane $4x - y - 3z = 4$ and find the perpendicular distance from the line to the plane.

33. Find the cartesian equation of the line of intersection of the two planes $2x - 3y - z = 1$ and $3x + 4y + 2z = 3$

34. Two lines have vector equations $\mathbf{r} = \begin{pmatrix} 3 \\ -1 \\ 1 \end{pmatrix} + \lambda \begin{pmatrix} 1 \\ 2 \\ -1 \end{pmatrix}$ and

    $\mathbf{r} = \begin{pmatrix} 4 \\ 4 \\ 1 \end{pmatrix} + \mu \begin{pmatrix} -1 \\ 1 \\ 2 \end{pmatrix}$. Find the position vector of the point of

    intersection of the two lines and the cartesian equation of the plane containing the two lines.

35. The four points A, B, C and D have position vectors
    $\begin{pmatrix} 4 \\ 2 \\ -1 \end{pmatrix}$, $\begin{pmatrix} 1 \\ 2 \\ 1 \end{pmatrix}$, $\begin{pmatrix} -3 \\ 0 \\ 3 \end{pmatrix}$ and $\begin{pmatrix} 5 \\ -4 \\ 1 \end{pmatrix}$ respectively.
    The perpendicular from D to the plane containing A, B and C meets the plane at E. Find
    (a) the scalar product vector equation of the plane containing A, B and C,
    (b) the vector equation of the straight line through D and E,
    (c) the position vector of the point E.

17.5 3 × 3 Matrices

In chapter 5 we saw that a linear transformation on 2 dimensional space has an associated 2 × 2 matrix. Similarly any linear transformation on 3 dimensional space has an associated 3 × 3 matrix. If under this

transformation the points with position vectors $\begin{pmatrix} 1 \\ 0 \\ 0 \end{pmatrix}$, $\begin{pmatrix} 0 \\ 1 \\ 0 \end{pmatrix}$ and $\begin{pmatrix} 0 \\ 0 \\ 1 \end{pmatrix}$ are

transformed to the points with position vectors $\begin{pmatrix} a \\ b \\ c \end{pmatrix}$, $\begin{pmatrix} d \\ e \\ f \end{pmatrix}$ and $\begin{pmatrix} g \\ h \\ i \end{pmatrix}$ respectively

then the associated 3 × 3 matrix is $\begin{pmatrix} a & d & g \\ b & e & h \\ c & f & i \end{pmatrix}$.

Note that under a linear transformation on 3-D space the origin $(0, 0, 0)$ is

mapped onto itself: $\begin{pmatrix} a & d & g \\ b & e & h \\ c & f & i \end{pmatrix}\begin{pmatrix} 0 \\ 0 \\ 0 \end{pmatrix} = \begin{pmatrix} 0 \\ 0 \\ 0 \end{pmatrix}$

Example 28

Find the 3 × 3 matrix representing the transformation of 3-D space that reflects all points in the x–y plane (i.e. the plane $z = 0$)

Under this transformation

$$\begin{pmatrix} 1 \\ 0 \\ 0 \end{pmatrix} \rightarrow \begin{pmatrix} 1 \\ 0 \\ 0 \end{pmatrix}, \begin{pmatrix} 0 \\ 1 \\ 0 \end{pmatrix} \rightarrow \begin{pmatrix} 0 \\ 1 \\ 0 \end{pmatrix}, \begin{pmatrix} 0 \\ 0 \\ 1 \end{pmatrix} \rightarrow \begin{pmatrix} 0 \\ 0 \\ -1 \end{pmatrix}.$$

Thus the required matrix is $\begin{pmatrix} 1 & 0 & 0 \\ 0 & 1 & 0 \\ 0 & 0 & -1 \end{pmatrix}$

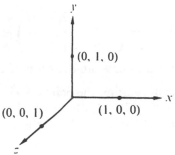

Example 29

Find the 3 × 3 matrix representing an orthogonal projection onto the x–z plane.

This projection maps any point $A(a, b, c)$ onto the point $A'(a, 0, c)$. Note that A' lies in the x–z plane and is such that AA' is perpendicular to the x–z plane—hence the term 'orthogonal' projection.

Thus $\begin{pmatrix} 1 \\ 0 \\ 0 \end{pmatrix} \rightarrow \begin{pmatrix} 1 \\ 0 \\ 0 \end{pmatrix}, \begin{pmatrix} 0 \\ 1 \\ 0 \end{pmatrix} \rightarrow \begin{pmatrix} 0 \\ 0 \\ 0 \end{pmatrix}, \begin{pmatrix} 0 \\ 0 \\ 1 \end{pmatrix} \rightarrow \begin{pmatrix} 0 \\ 0 \\ 1 \end{pmatrix}$

and the required matrix is $\begin{pmatrix} 1 & 0 & 0 \\ 0 & 0 & 0 \\ 0 & 0 & 1 \end{pmatrix}$

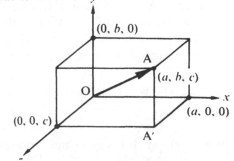

Example 30

Show that the matrix $\begin{pmatrix} -1 & 1 & 3 \\ 2 & -2 & -6 \\ -4 & 4 & 12 \end{pmatrix}$ maps all points in 3-D space

onto a line and find the cartesian equations of this line.

Under this transformation the points A(1, 0, 0), B(0, 1, 0) and C(0, 0, 1)

have images A′, B′, C′, position vectors $\begin{pmatrix} -1 \\ 2 \\ -4 \end{pmatrix}$, $\begin{pmatrix} 1 \\ -2 \\ 4 \end{pmatrix}$ and $\begin{pmatrix} 3 \\ -6 \\ 12 \end{pmatrix}$

respectively, i.e. all the images are points whose position vectors are

multiples of $\begin{pmatrix} -1 \\ 2 \\ -4 \end{pmatrix}$. Thus this transformation maps all points onto the

line passing through the origin and the point with position vector $\begin{pmatrix} -1 \\ 2 \\ -4 \end{pmatrix}$.

The vector equation of this line will be $\mathbf{r} = \begin{pmatrix} 0 \\ 0 \\ 0 \end{pmatrix} + \lambda \begin{pmatrix} -1 \\ 2 \\ -4 \end{pmatrix}$ giving the

cartesian equations $\quad -x = \dfrac{y}{2} = -\dfrac{z}{4} (= \lambda)$.

Example 31

Show that the matrix $\begin{pmatrix} 3 & -2 & 0 \\ -1 & 3 & 7 \\ 2 & -1 & 1 \end{pmatrix}$ maps all points in 3-D space onto

a plane and find the cartesian equation of this plane.

Under this transformation the points A (1, 0, 0), B (0, 1, 0) and C (0, 0, 1)
are mapped onto A′(3, −1, 2), B′(−2, 3, −1) and C′(0, 7, 1). For these
points A′, B′ and C′ to be coplanar there must exist values for λ and μ such
that

$$\lambda \begin{pmatrix} 3 \\ -1 \\ 2 \end{pmatrix} + \mu \begin{pmatrix} -2 \\ 3 \\ -1 \end{pmatrix} = \begin{pmatrix} 0 \\ 7 \\ 1 \end{pmatrix}$$

Solving $3\lambda - 2\mu = 0$ and $-\lambda + 3\mu = 7$ simultaneously gives $\lambda = 2$ and
$\mu = 3$ which is consistent with $2\lambda - \mu = 1$. Hence C′ lies in the plane
containing A′ and B′ [and the origin O (0, 0, 0)]. Thus the transformation
maps all points in 3-D space on to the plane containing the origin and the
points (3, −1, 2) and (−2, 3, −1).

As the vectors $\overrightarrow{OA}$ and $\overrightarrow{OB}$ lie in this plane the parametric vector equation

of the plane is $\mathbf{r} = \begin{pmatrix} 0 \\ 0 \\ 0 \end{pmatrix} + \lambda \begin{pmatrix} 3 \\ -1 \\ 2 \end{pmatrix} + \mu \begin{pmatrix} -2 \\ 3 \\ -1 \end{pmatrix}$.

With $\mathbf{r} = \begin{pmatrix} x \\ y \\ z \end{pmatrix}$ this gives the cartesian equation of the plane as $5x + y - 7z = 0$.

The determinant of a 3×3 matrix

Just as for linear transformations in 2-dimensional space the determinant of the associated matrix gives the area scale factor then similarly the determinant of a 3×3 matrix gives the volume scale factor for the 3-dimensional transformation defined by that matrix.

Remembering that for the 2×2 matrix $\mathbf{A} = \begin{pmatrix} a & b \\ c & d \end{pmatrix}$ we can write the

determinant of $\mathbf{A}$ as det $\mathbf{A}$, $|\mathbf{A}|$ or $\begin{vmatrix} a & b \\ c & d \end{vmatrix}$, the determinant of the 3×3

matrix $\mathbf{A} = \begin{vmatrix} a_1 & b_1 & c_1 \\ a_2 & b_2 & c_2 \\ a_3 & b_3 & c_3 \end{vmatrix}$ is defined as

$$\det \mathbf{A} = \begin{vmatrix} a_1 & b_1 & c_1 \\ a_2 & b_2 & c_2 \\ a_3 & b_3 & c_3 \end{vmatrix} = a_1 \begin{vmatrix} b_2 & c_2 \\ b_3 & c_3 \end{vmatrix} - b_1 \begin{vmatrix} a_2 & c_2 \\ a_3 & c_3 \end{vmatrix} + c_1 \begin{vmatrix} a_2 & b_2 \\ a_3 & b_3 \end{vmatrix}$$

Example 32

If a solid of volume 8 cubic units is transformed using the matrix $\mathbf{T} = \begin{pmatrix} 3 & 1 & -2 \\ 2 & 1 & 1 \\ 0 & 1 & -2 \end{pmatrix}$ find the volume of the new solid.

Volume of new body $= |\det \mathbf{T}| \times 8$

Now $\qquad \det \mathbf{T} = \begin{vmatrix} 3 & 1 & -2 \\ 2 & 1 & 1 \\ 0 & 1 & -2 \end{vmatrix}$

$$= 3 \begin{vmatrix} 1 & 1 \\ 1 & -2 \end{vmatrix} - 1 \begin{vmatrix} 2 & 1 \\ 0 & -2 \end{vmatrix} - 2 \begin{vmatrix} 2 & 1 \\ 0 & 1 \end{vmatrix}$$

$$= -9 + 4 - 4$$

$$= -9$$

Thus the volume of the new body is $9 \times 8 = 72$ cubic units.

Note: Just as a 2×2 matrix with zero determinant has no inverse then similarly a 3×3 matrix with zero determinant has no inverse and is said to be *singular*.

The inverse of a 3×3 matrix

If the 3×3 matrix $\mathbf{A}$ transforms a body B to its image B$'$, then $\mathbf{A}^{-1}$ (i.e. the inverse of $\mathbf{A}$) will transform B$'$ back to B. By definition

$$\mathbf{A}\mathbf{A}^{-1} = \mathbf{A}^{-1}\mathbf{A} = \mathbf{I} \text{ where, for } 3 \times 3 \text{ matrices, } \mathbf{I} = \begin{pmatrix} 1 & 0 & 0 \\ 0 & 1 & 0 \\ 0 & 0 & 1 \end{pmatrix} \text{ but can } \mathbf{A}^{-1}$$

be determined from $\mathbf{A}$?

One method, called the *adjoint* method, first requires an understanding of certain terms, namely the *minor* of an element of a matrix, the *cofactor* of an element of a matrix and the *adjoint* of a matrix.

Consider the matrix $\mathbf{A} = \begin{pmatrix} a_1 & b_1 & c_1 \\ a_2 & b_2 & c_2 \\ a_3 & b_3 & c_3 \end{pmatrix}$

If we delete the row and column that contains the element a_1 and find the determinant of the 2×2 matrix that we are left with we obtain the *minor* of a_1.

Thus the minor of $a_1 = \begin{vmatrix} b_2 & c_2 \\ b_3 & c_3 \end{vmatrix}$, the minor of $b_1 = \begin{vmatrix} a_2 & c_2 \\ a_3 & c_3 \end{vmatrix}$ etc.

The *cofactor* of each element is then obtained by multiplying the minor by

$$\begin{matrix} + & - & + \\ - & + & - \\ + & - & + \end{matrix}$$

± 1 according to the pattern . i.e. the cofactor is the minor with an appropriate sign attached to it.

Thus the cofactors of a_1, b_1 and c_1 are $\begin{vmatrix} b_2 & c_2 \\ b_3 & c_3 \end{vmatrix}$, $- \begin{vmatrix} a_2 & c_2 \\ a_3 & c_3 \end{vmatrix}$, $\begin{vmatrix} a_2 & b_2 \\ a_3 & b_3 \end{vmatrix}$ respectively.

If we form a new 3×3 matrix by replacing each element of $\mathbf{A}$ by its cofactor and then transpose this new matrix, we obtain the *adjoint of* $\mathbf{A}$, written adj $\mathbf{A}$. By forming the product (adj $\mathbf{A}$)$\mathbf{A}$, or $\mathbf{A}$(adj $\mathbf{A}$), we can then determine $\mathbf{A}^{-1}$ as the following example shows.

Example 33

Find the inverse of the matrix $\begin{pmatrix} 1 & -2 & 3 \\ 2 & 1 & 0 \\ 1 & -1 & 1 \end{pmatrix}$.

Let the given matrix be $\mathbf{A}$ and first check that det $\mathbf{A} \neq 0$. In this case det $\mathbf{A} = -4$.

Then find the minors of each element:
$$\begin{matrix} 1 & 2 & -3 \\ 1 & -2 & 1 \\ -3 & -6 & 5 \end{matrix}$$

and write the matrix of cofactors:
$$\begin{pmatrix} 1 & -2 & -3 \\ -1 & -2 & -1 \\ -3 & 6 & 5 \end{pmatrix}$$

Hence obtain the adjoint of $\mathbf{A}$:
$$\begin{pmatrix} 1 & -1 & -3 \\ -2 & -2 & 6 \\ -3 & -1 & 5 \end{pmatrix}$$

and determine (adj $\mathbf{A}$) $\times$ $\mathbf{A}$:
$$\begin{pmatrix} 1 & -1 & -3 \\ -2 & -2 & 6 \\ -3 & -1 & 5 \end{pmatrix}\begin{pmatrix} 1 & -2 & 3 \\ 2 & 1 & 0 \\ 1 & -1 & 1 \end{pmatrix} = \begin{pmatrix} -4 & 0 & 0 \\ 0 & -4 & 0 \\ 0 & 0 & -4 \end{pmatrix}$$

Hence   (adj $\mathbf{A}$)$\mathbf{A}$ = $-4\mathbf{I}$

or $-\tfrac{1}{4}$(adj $\mathbf{A}$)$\mathbf{A}$ = $\mathbf{I}$

Thus $\mathbf{A}^{-1}$ is given by $-\dfrac{1}{4}\begin{pmatrix} 1 & -1 & -3 \\ -2 & -2 & 6 \\ -3 & -1 & 5 \end{pmatrix}$ or $\dfrac{1}{4}\begin{pmatrix} -1 & 1 & 3 \\ 2 & 2 & -6 \\ 3 & 1 & -5 \end{pmatrix}$.

Notice that since $(\text{adj } A)A = (\det A)I$, i.e. $\dfrac{(\text{adj } A)}{\det A} \times A = I$, we could find A^{-1} by finding adj A and then dividing by $(\det A)$. However it is very easy to make an arithmetical error when determining adj A and so it is best to determine $(\text{adj } A)A$ as a means of checking for such errors.

Exercise 17E

1. If the linear transformations T_1 and T_2 are represented by the matrices
$$\begin{pmatrix} 1 & 0 & 2 \\ 2 & 1 & 0 \\ 1 & 1 & 1 \end{pmatrix} \text{ and } \begin{pmatrix} 1 & -1 & 0 \\ 2 & 0 & -3 \\ 1 & 2 & 1 \end{pmatrix} \text{ respectively find the image of the}$$
point $(2, -1, 3)$ under the transformation
(a) T_1, (b) T_2, (c) T_1 followed by T_2.

2. Find the 3×3 matrix representing the following linear transformations on 3-D space.
   (a) reflection in the y–z plane,
   (b) reflection in the plane $y = 0$,
   (c) rotation of $180°$ about the x-axis,
   (d) enlargement, scale factor 3, centre $(0, 0, 0)$,
   (e) stretch parallel to the x-axis, scale factor 2, with y–z plane fixed,
   (f) $90°$ rotation about the x-axis such that the positive y-axis maps onto the positive z-axis,
   (g) orthogonal projection onto the $x = 0$ plane,
   (h) reflection in the $y = x$ plane,
   (i) reflection in the $y + x = 0$ plane,
   (j) shear with the y–z plane fixed and $(1, 0, 0) \rightarrow (1, 0, 2)$,
   (k) orthogonal projection onto the $y = x$ plane.

3. Show that any line lying in the plane $z = 0$ and with an equation $y = ax$ is mapped onto the origin by the transformation matrix
$$\begin{pmatrix} a & -1 & 3 \\ -2a & 2 & 5 \\ -a^2 & a & a^3 \end{pmatrix}.$$

4. Give a geometrical description of the linear transformation defined
   by the matrix $A = \begin{pmatrix} 1 & 0 & 0 \\ 0 & 0 & 1 \\ 0 & 1 & 0 \end{pmatrix}$. *Hence* state the effects of the
   transformations defined by A^2 and A^3.

5. Give a geometrical description of the linear transformation defined
   by the matrix $B = \begin{pmatrix} 0 & 0 & 1 \\ 0 & 1 & 0 \\ -1 & 0 & 0 \end{pmatrix}$. Hence state the effects of the
   transformations defined by B^2, B^3 and B^4.

6. Show that each of the following matrices map any points in 3-D space onto a line and in each case find vector and cartesian equations of the line.

   (a) $\begin{pmatrix} 2 & 1 & -1 \\ 6 & 3 & -3 \\ -4 & -2 & 2 \end{pmatrix}$ (b) $\begin{pmatrix} -3 & 2 & -1 \\ -6 & 4 & -2 \\ 9 & -6 & 3 \end{pmatrix}$ (c) $\begin{pmatrix} 2 & 3 & -5 \\ -4 & -6 & 10 \\ 2 & 3 & -5 \end{pmatrix}$

7. Show that each of the following matrices map any points in 3-D space onto a plane and find the cartesian equation of the plane in each case.

(a) $\begin{pmatrix} 0 & -2 & -2 \\ 1 & 1 & 5 \\ 1 & 3 & 7 \end{pmatrix}$ (b) $\begin{pmatrix} 1 & 2 & 10 \\ 2 & -3 & -8 \\ -3 & 1 & -2 \end{pmatrix}$

8. Find the determinants of the following matrices.

(a) $\begin{pmatrix} 1 & -2 & 1 \\ 2 & -1 & 3 \\ 0 & 2 & 1 \end{pmatrix}$ (b) $\begin{pmatrix} 1 & -1 & 0 \\ -1 & 3 & 4 \\ 1 & -1 & 1 \end{pmatrix}$ (c) $\begin{pmatrix} 1 & 3 & 2 \\ 3 & -1 & 4 \\ 2 & 1 & 3 \end{pmatrix}$

(d) $\begin{pmatrix} 2 & 1 & 4 \\ -1 & 0 & 3 \\ 2 & 1 & -1 \end{pmatrix}$ (e) $\begin{pmatrix} 3 & 2 & 2 \\ 1 & 1 & -1 \\ 2 & 2 & 2 \end{pmatrix}$ (f) $\begin{pmatrix} -1 & -2 & 3 \\ 1 & -2 & -3 \\ 1 & 2 & 3 \end{pmatrix}$

9. Find, where possible, the inverses of the following matrices.

(a) $\begin{pmatrix} -1 & 1 & 0 \\ 2 & -1 & 1 \\ 1 & 1 & 1 \end{pmatrix}$ (b) $\begin{pmatrix} 3 & -1 & -4 \\ 2 & 1 & 0 \\ 0 & 1 & 2 \end{pmatrix}$ (c) $\begin{pmatrix} 4 & 3 & 1 \\ 2 & -1 & 3 \\ 1 & 0 & 1 \end{pmatrix}$

(d) $\begin{pmatrix} 2 & 1 & 2 \\ 2 & 3 & 1 \\ -1 & 0 & 2 \end{pmatrix}$ (e) $\begin{pmatrix} 2 & 1 & 1 \\ 1 & 1 & 2 \\ -1 & -2 & 3 \end{pmatrix}$ (f) $\begin{pmatrix} 2 & 5 & 1 \\ -1 & 3 & 1 \\ -1 & 2 & 1 \end{pmatrix}$

10. The transformations T_1 and T_2 are defined by the matrices

$\begin{pmatrix} 4 & 1 & 1 \\ 1 & 2 & -1 \\ 3 & 1 & 1 \end{pmatrix}$ and $\begin{pmatrix} 1 & 1 & 1 \\ 1 & 2 & -1 \\ 0 & 1 & 2 \end{pmatrix}$ respectively. If T_1 transforms

a body S, volume 10 cubic units, to its image S′ and T_2 transforms S′ to S″ find
(a) the single matrix that will transform S to S″,
(b) the single matrix that will transform S′ back to S,
(c) the volume of S′,
(d) the volume of S″.

11. Find the inverse of the matrix $\begin{pmatrix} -1 & -1 & 1 \\ 0 & -1 & 3 \\ 2 & 1 & 2 \end{pmatrix}$ and *hence* solve the

equations $\begin{cases} -x - y + z = 1 \\ -y + 3z = 8 \\ 2x + y + 2z = 8 \end{cases}$

12. Find the inverse of the matrix $\begin{pmatrix} 2 & -1 & 3 \\ 1 & 1 & -2 \\ 3 & 1 & 2 \end{pmatrix}$ and *hence* solve the

equations $\begin{cases} 2x - y + 3z = 0 \\ x + y - 2z = -1 \\ 3x + y + 2z = 5 \end{cases}$

13. (a) Find the inverse of the matrix $\begin{pmatrix} -1 & -3 & -4 \\ 2 & -1 & 0 \\ 1 & 2 & 3 \end{pmatrix}$

(b) A point (x, y, z) in three-dimensional space is transformed to its image (x', y', z') according to the rule

$$\begin{pmatrix} x' \\ y' \\ z' \end{pmatrix} = \begin{pmatrix} 0 & -3 & -4 \\ 2 & 0 & 0 \\ 1 & 2 & 4 \end{pmatrix} \begin{pmatrix} x \\ y \\ z \end{pmatrix} + \begin{pmatrix} -7 \\ -3 \\ 5 \end{pmatrix}$$

Find the coordinates of the point in three dimensional space that is invariant under the transformation.

17.6 Sets of linear equations

In chapter 6 we saw that a pair of linear equations in two unknowns could be solved simultaneously by matrix methods to find the one set of values that satisfied both equations. In questions 11 and 12 of exercise 17E we used our ability to determine the inverse of a 3 × 3 matrix to solve a set of 3 linear equations in 3 unknowns. Matrix methods are not the only method of solution since the equations could be solved by progressively eliminating the variables.

Example 34

Solve the following simultaneous equations by eliminating the variables.

$$2x + y - 2z = -4 \qquad \ldots [1]$$
$$2x + 3y + z = -2 \qquad \ldots [2]$$
$$-x + 2z = 3 \qquad \ldots [3]$$

$3 \times [1] - [2]:$ $4x - 7z = -10$ $\ldots [4]$
$4 \times [3]:$ $-4x + 8z = 12$ $\ldots [5]$
$[4] + [5]:$ $z = 2$ $\ldots [5]$

By substituting into [3] and then into [1], we obtain $x = 1$ and $y = -2$.
Thus the solution of the given set of equations is $x = 1$, $y = -2$ and $z = 2$.

The geometrical interpretation of this solution is that the point $(1, -2, 2)$ is the point that is common to the three planes
$2x + y - 2z = -4$, $2x + 3y + z = -2$ and $-x + 2z = 3$.

This method of eliminating the variables gives another method for determining the inverse of a matrix.

Suppose we wish to find the inverse of $\begin{pmatrix} -1 & 1 & 3 \\ 2 & 1 & 1 \\ 1 & 1 & 2 \end{pmatrix}$.

First let $\begin{pmatrix} -1 & 1 & 3 \\ 2 & 1 & 1 \\ 1 & 1 & 2 \end{pmatrix} \begin{pmatrix} x \\ y \\ z \end{pmatrix} = \begin{pmatrix} 1 & 0 & 0 \\ 0 & 1 & 0 \\ 0 & 0 & 1 \end{pmatrix} \begin{pmatrix} a \\ b \\ c \end{pmatrix}$

i.e. $-x + y + 3z = a$ $\ldots [1]$
$2x + y + z = b$ $\ldots [2]$
$x + y + 2z = c$ $\ldots [3]$

To solve equations [1], [2] and [3] we would attempt to eliminate the x's in two of the equations and then use these two equations to eliminate the variable y. The equivalent steps in the matrix equation are to manipulate the left-hand matrix so that two of its rows commence with a zero and then use these two rows to obtain a row commencing with two zeros.

By continuing this process until our left hand 3×3 matrix is reduced to the identity matrix and, at each stage carrying out any manipulation of the 3×3 matrix on the left hand side of the equation on the 3×3 matrix on the right hand side as well, it is possible to determine the inverse of the given matrix:

$$\begin{matrix} \text{row 1 } (r_1): \\ \text{row 2 } (r_2): \\ \text{row 3 } (r_3): \end{matrix} \quad \begin{pmatrix} -1 & 1 & 3 \\ 2 & 1 & 1 \\ 1 & 1 & 2 \end{pmatrix} \begin{pmatrix} x \\ y \\ z \end{pmatrix} = \begin{pmatrix} 1 & 0 & 0 \\ 0 & 1 & 0 \\ 0 & 0 & 1 \end{pmatrix} \begin{pmatrix} a \\ b \\ c \end{pmatrix}$$

$$\begin{matrix} r_1: \\ \text{new } r_2 = r_2 + 2r_1: \\ \text{new } r_3 = r_3 + r_1: \end{matrix} \quad \begin{pmatrix} -1 & 1 & 3 \\ 0 & 3 & 7 \\ 0 & 2 & 5 \end{pmatrix} \begin{pmatrix} x \\ y \\ z \end{pmatrix} = \begin{pmatrix} 1 & 0 & 0 \\ 2 & 1 & 0 \\ 1 & 0 & 1 \end{pmatrix} \begin{pmatrix} a \\ b \\ c \end{pmatrix}$$

$$\begin{matrix} \text{new } r_1 = 2r_1 - r_3: \\ r_2: \\ \text{new } r_3 = 3r_3 - 2r_2: \end{matrix} \quad \begin{pmatrix} -2 & 0 & 1 \\ 0 & 3 & 7 \\ 0 & 0 & 1 \end{pmatrix} \begin{pmatrix} x \\ y \\ z \end{pmatrix} = \begin{pmatrix} 1 & 0 & -1 \\ 2 & 1 & 0 \\ -1 & -2 & 3 \end{pmatrix} \begin{pmatrix} a \\ b \\ c \end{pmatrix}$$

$$\begin{matrix} \text{new } r_1 = r_1 - r_3: \\ \text{new } r_2 = r_2 - 7r_3: \\ r_3: \end{matrix} \quad \begin{pmatrix} -2 & 0 & 0 \\ 0 & 3 & 0 \\ 0 & 0 & 1 \end{pmatrix} \begin{pmatrix} x \\ y \\ z \end{pmatrix} = \begin{pmatrix} 2 & 2 & -4 \\ 9 & 15 & -21 \\ -1 & -2 & 3 \end{pmatrix} \begin{pmatrix} a \\ b \\ c \end{pmatrix}$$

$$\begin{matrix} \text{new } r_1 = -\frac{1}{2}r_1: \\ \text{new } r_2 = \frac{1}{3}r_2: \\ r_3: \end{matrix} \quad \begin{pmatrix} 1 & 0 & 0 \\ 0 & 1 & 0 \\ 0 & 0 & 1 \end{pmatrix} \begin{pmatrix} x \\ y \\ z \end{pmatrix} = \begin{pmatrix} -1 & -1 & 2 \\ 3 & 5 & -7 \\ -1 & -2 & 3 \end{pmatrix} \begin{pmatrix} a \\ b \\ c \end{pmatrix}$$

Thus $\begin{pmatrix} -1 & -1 & 2 \\ 3 & 5 & -7 \\ -1 & -2 & 3 \end{pmatrix}$ is the inverse of $\begin{pmatrix} -1 & 1 & 3 \\ 2 & 1 & 1 \\ 1 & 1 & 2 \end{pmatrix}$

This method of determining the inverse of a maxtrix is called the method of *row reduction*.

In some instances it may be necessary to interchange the rows in a matrix to avoid zeros appearing on the leading diagonal. The next example illustrates this point and that there is no need to show $\begin{pmatrix} x \\ y \\ z \end{pmatrix}$ and $\begin{pmatrix} a \\ b \\ c \end{pmatrix}$ throughout the working.

Example 35

Find the inverse of matrix $\mathbf{A} = \begin{pmatrix} -1 & 2 & 1 \\ 2 & -4 & 1 \\ 3 & -2 & -2 \end{pmatrix}$

$$\begin{matrix} r_1: \\ r_2: \\ r_3: \end{matrix} \quad \begin{pmatrix} -1 & 2 & 1 \\ 2 & -4 & 1 \\ 3 & -2 & -2 \end{pmatrix} \quad \vdots \quad \begin{pmatrix} 1 & 0 & 0 \\ 0 & 1 & 0 \\ 0 & 0 & 1 \end{pmatrix}$$

$$\begin{array}{l} r_1: \\ \text{new } r_2 = r_2 + 2r_1: \\ \text{new } r_3 = r_3 + 3r_1: \end{array} \begin{pmatrix} -1 & 2 & 1 \\ 0 & 0 & 3 \\ 0 & 4 & 1 \end{pmatrix} \quad \vdots \quad \begin{pmatrix} 1 & 0 & 0 \\ 2 & 1 & 0 \\ 3 & 0 & 1 \end{pmatrix}$$

$$\begin{array}{l} r_1: \\ \text{new } r_2 = r_3: \\ \text{new } r_3 = r_2: \end{array} \begin{pmatrix} -1 & 2 & 1 \\ 0 & 4 & 1 \\ 0 & 0 & 3 \end{pmatrix} \quad \vdots \quad \begin{pmatrix} 1 & 0 & 0 \\ 3 & 0 & 1 \\ 2 & 1 & 0 \end{pmatrix}$$

$$\begin{array}{l} \text{new } r_1 = r_1 - r_2: \\ \text{new } r_2 = 3r_2 - r_3: \\ r_3: \end{array} \begin{pmatrix} -1 & -2 & 0 \\ 0 & 12 & 0 \\ 0 & 0 & 3 \end{pmatrix} \quad \vdots \quad \begin{pmatrix} -2 & 0 & -1 \\ 7 & -1 & 3 \\ 2 & 1 & 0 \end{pmatrix}$$

$$\begin{array}{l} \text{new } r_1 = 6r_1 + r_2: \\ r_2: \\ r_3: \end{array} \begin{pmatrix} -6 & 0 & 0 \\ 0 & 12 & 0 \\ 0 & 0 & 3 \end{pmatrix} \quad \vdots \quad \begin{pmatrix} -5 & -1 & -3 \\ 7 & -1 & 3 \\ 2 & 1 & 0 \end{pmatrix}$$

$$\begin{array}{l} \text{new } r_1 = -\tfrac{1}{6}r_1: \\ \text{new } r_2 = \tfrac{1}{12}r_2: \\ \text{new } r_3 = \tfrac{1}{3}r_3: \end{array} \begin{pmatrix} 1 & 0 & 0 \\ 0 & 1 & 0 \\ 0 & 0 & 1 \end{pmatrix} \quad \vdots \quad \begin{pmatrix} \tfrac{5}{6} & \tfrac{1}{6} & \tfrac{1}{2} \\ \tfrac{7}{12} & -\tfrac{1}{12} & \tfrac{1}{4} \\ \tfrac{2}{3} & \tfrac{1}{3} & 0 \end{pmatrix}$$

Thus the inverse of matrix **A** is $\dfrac{1}{12}\begin{pmatrix} 10 & 2 & 6 \\ 7 & -1 & 3 \\ 8 & 4 & 0 \end{pmatrix}$

Note:
1. If the determinant of a matrix is zero (i.e. the matrix is singular), the row reduction method will break down because one row is reduced to all zeros.
2. An alternative method of presentation is to write **A** and **I** as a single matrix:

$$\left(\begin{array}{ccc|ccc} -1 & 2 & 1 & 1 & 0 & 0 \\ 2 & -4 & 1 & 0 & 1 & 0 \\ 3 & -2 & -2 & 0 & 0 & 1 \end{array} \right)$$ and to manipulate the rows of this matrix $(\mathbf{A}|\mathbf{I})$.

Consider the equations
$$\begin{aligned} 2x - y + z &= 5 \\ x - 3y + 2z &= 2 \\ 2x + y + 4z &= -3 \end{aligned}$$

We could determine the inverse of $\begin{pmatrix} 2 & -1 & 1 \\ 1 & -3 & 2 \\ 2 & 1 & 4 \end{pmatrix}$ by row reduction and use this inverse to solve the equations.

However it is not necessary to reduce this 3×3 matrix to the identity matrix in order to solve the equations. Instead we can stop when we have reduced it to a triangular matrix i.e. a matrix that has only zeros below the leading diagonal or only zeros above the leading diagonal.

For example $\begin{pmatrix} 2 & 1 & 3 \\ 0 & 1 & 2 \\ 0 & 0 & 3 \end{pmatrix}$ and $\begin{pmatrix} 1 & 0 & 0 \\ 1 & -1 & 0 \\ 3 & 4 & 2 \end{pmatrix}$ are triangular matrices.

The following example shows how this method can be used to solve the three equations given above.

Example 36

Solve the equations $\begin{cases} 2x - y + z = 5 \\ x - 3y + 2z = 2 \\ 2x + y + 4z = -3 \end{cases}$

Writing the equations in matrix form $\begin{pmatrix} 2 & -1 & 1 \\ 1 & -3 & 2 \\ 2 & 1 & 4 \end{pmatrix} \begin{pmatrix} x \\ y \\ z \end{pmatrix} = \begin{pmatrix} 5 \\ 2 \\ -3 \end{pmatrix}$

$$\begin{matrix} r_1: \\ \text{new } r_2 = 2r_2 - r_1: \\ \text{new } r_3 = r_3 - r_1: \end{matrix} \begin{pmatrix} 2 & -1 & 1 \\ 0 & -5 & 3 \\ 0 & 2 & 3 \end{pmatrix} \begin{pmatrix} x \\ y \\ z \end{pmatrix} = \begin{pmatrix} 5 \\ -1 \\ -8 \end{pmatrix}$$

$$\begin{matrix} r_1: \\ r_2: \\ \text{new } r_3 = 5r_3 + 2r_2: \end{matrix} \begin{pmatrix} 2 & -1 & 1 \\ 0 & -5 & 3 \\ 0 & 0 & 21 \end{pmatrix} \begin{pmatrix} x \\ y \\ z \end{pmatrix} = \begin{pmatrix} 5 \\ -1 \\ -42 \end{pmatrix}$$

Row 3 gives $z = -2$.
Using this value in row 2 gives $-5y - 6 = -1$ i.e. $y = -1$
and these values in row 1 gives $2x + 1 - 2 = 5$ i.e. $x = 3$
Thus $x = 3$, $y = -1$ and $z = -2$ is the required solution.

Given three equations $a_1x + b_1y + c_1z = p$
$a_2x + b_2y + c_2z = q$
$a_3x + b_3y + c_3z = r$
the geometrical interpretation of the solution $x = \alpha$, $y = \beta$ and $z = \gamma$ is
that the three planes defined by the three equations have a common point

(α, β, γ). However if the matrix $\begin{pmatrix} a_1 & b_1 & c_1 \\ a_2 & b_2 & c_2 \\ a_3 & b_3 & c_3 \end{pmatrix}$ is singular then the three

equations do not have a unique solution. In such a situation the equations
could either
have (a) no solutions, and the equations are said to be *inconsistent*,
or   (b) an infinite number of solutions.

The geometrical interpretation of these two possibilities is as follows:

(a) *No solutions* (i.e. inconsistent equations).
      This situation will arise when
      1. two or more of the planes are parallel
         (but not coincident). In such cases there
         is no point common to all three planes.

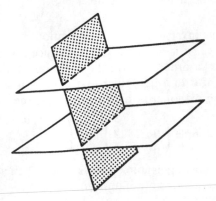

or 2. each plane is parallel to the line of
intersection of the other two. Again
there will be no point that is common
to all three planes.

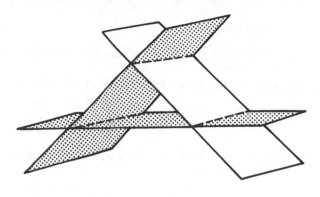

(b) *An infinite number of solutions.*
This situation will arise when
1. all three equations represent the same
plane. (i.e. all 3 planes are coincident).
Any point in the plane will provide a
solution to the equation.

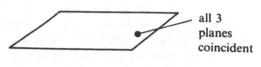

all 3
planes
coincident

or 2. two of the three planes are coincident
and the third plane is not parallel to
these two. The planes will intersect in a
line and any point on the line will
provide a solution.

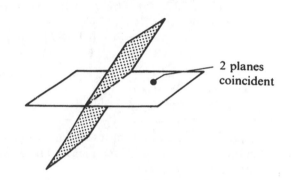

2 planes
coincident

or 3. the three planes have a common line
and any point on this line provides a
solution to the equation.

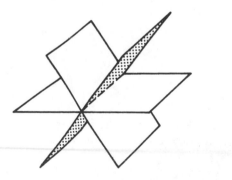

Example 37

For what value of a does the equation $\begin{pmatrix} 2 & -4 & 2 \\ 4 & -1 & 2 \\ 1 & a & 0 \end{pmatrix} \begin{pmatrix} x \\ y \\ z \end{pmatrix} = \begin{pmatrix} 5 \\ 1 \\ -3 \end{pmatrix}$ not have a unique solution?

The equation will not have a unique solution if $\begin{vmatrix} 2 & -4 & 2 \\ 4 & -1 & 2 \\ 1 & a & 0 \end{vmatrix}$ is zero.

i.e. if $-4a - 8 + 8a + 2 = 0$
giving $a = 1\tfrac{1}{2}$
The equation will not have a unique solution if $a = 1\tfrac{1}{2}$.

Example 38

Discuss the solution of the following equations when   (a) $a = 6$ and $b = 6$,
                                                       (b) $a = 5$ and $b = 0$,
giving a full geometric interpretation for each situation.

$$2x - y + 3z = 2$$
$$-4x + 2y - bz = -4$$
$$6x - 3y + 9z = a$$

(a) With $a = 6$ and $b = 6$ the equations become
$$2x - y + 3z = 2$$
$$-4x + 2y - 6z = -4 \quad \text{i.e.} \quad -2[2x - y + 3z = 2]$$
$$6x - 3y + 9z = 6 \quad \text{i.e.} \quad 3[2x - y + 3z = 2]$$
These three equations represent the same plane, $2x - y + 3z = 2$ and
so any point in the plane provides a solution to the equations. Thus
there are an infinite number of solutions. Letting $x = \lambda$ and $y = \mu$ we
can write these solutions in parametric form as $x = \lambda$, $y = \mu$,
$z = \frac{1}{3}(2 - 2\lambda + \mu)$. This gives the parametric vector equation of the
plane as
$$\mathbf{r} = \begin{pmatrix} 0 \\ 0 \\ \frac{2}{3} \end{pmatrix} + \lambda \begin{pmatrix} 1 \\ 0 \\ -\frac{2}{3} \end{pmatrix} + \mu \begin{pmatrix} 0 \\ 1 \\ \frac{1}{3} \end{pmatrix}$$

(b) With $a = 5$ and $b = 0$ the equations become
$$2x - y + 3z = 2 \qquad \qquad \text{... [1]}$$
$$-4x + 2y = -4 \qquad \qquad \text{... [2]}$$
$$6x - 3y + 9z = 5 \qquad \qquad \text{... [3]}$$
Equations [1] and [3] represent parallel planes and so there will be no
point common to all three planes. The equations have no solutions i.e.
they are inconsistent.

Example 39

Discuss the solution of the following equations when (a) $a = 5$, (b) $a = -2$,
giving a geometrical interpretation for each case.
$$-2x + y - 5z = 4$$
$$3x - y + 2z = -1$$
$$-4x + y + z = a$$

These equations do not represent coincident or parallel planes whatever the
value of a and so we attempt a solution by the diagonal matrix method:

$$\begin{matrix} r_1: \\ r_2: \\ r_3: \end{matrix} \begin{pmatrix} -2 & 1 & -5 \\ 3 & -1 & 2 \\ -4 & 1 & 1 \end{pmatrix} \begin{pmatrix} x \\ y \\ z \end{pmatrix} = \begin{pmatrix} 4 \\ -1 \\ a \end{pmatrix}$$

$$\begin{matrix} r_1: \\ \text{new } r_2 = 2r_2 + 3r_1: \\ \text{new } r_3 = r_3 - 2r_1: \end{matrix} \begin{pmatrix} -2 & 1 & -5 \\ 0 & 1 & -11 \\ 0 & -1 & 11 \end{pmatrix} \begin{pmatrix} x \\ y \\ z \end{pmatrix} = \begin{pmatrix} 4 \\ 10 \\ a-8 \end{pmatrix}$$

$$\begin{array}{l} r_1: \\ r_2: \\ \text{new } r_3 = r_3 + r_2: \end{array} \begin{pmatrix} -2 & 1 & -5 \\ 0 & 1 & -11 \\ 0 & 0 & 0 \end{pmatrix} \begin{pmatrix} x \\ y \\ z \end{pmatrix} = \begin{pmatrix} 4 \\ 10 \\ a+2 \end{pmatrix}$$

(a) for $a = 5$ row 3, $0 = a + 2$, gives a contradiction and so the equations are inconsistent. The equations have no solutions for $a = 5$ and this corresponds to the situation of each plane being parallel to the line of intersection of the other two.

(b) for $a = -2$ row 3 gives no contradiction and is true for any z. Using the parameter λ we let $z = \lambda$.

Then from row$_1$ $-2x + y - 5z = 4$
and from row$_2$ $y - 11z = 10$
     we obtain $y = 10 + 11\lambda$ and $x = 3(1 + \lambda)$.
i.e. the solutions to the equations are all of the form $x = 3 + 3\lambda$, $y = 10 + 11\lambda$, $z = \lambda$. Thus the 3 given equations represent planes

having a common line, cartesian equation $\dfrac{x - 3}{3} = \dfrac{y - 10}{11} = z$ or, in

parametric vector form $\mathbf{r} = \begin{pmatrix} 3 \\ 10 \\ 0 \end{pmatrix} + \lambda \begin{pmatrix} 3 \\ 11 \\ 1 \end{pmatrix}$.

Exercise 17F

1. Find, where possible, the inverses of the following matrices by the method of row reduction.

(a) $\begin{pmatrix} 2 & 0 & 1 \\ 4 & -1 & 2 \\ 3 & 1 & 1 \end{pmatrix}$ (b) $\begin{pmatrix} 1 & -1 & -1 \\ 3 & 2 & 1 \\ -5 & 1 & 2 \end{pmatrix}$ (c) $\begin{pmatrix} 2 & 1 & -4 \\ 1 & 2 & -1 \\ 1 & 3 & 2 \end{pmatrix}$

(d) $\begin{pmatrix} 2 & -4 & 2 \\ 3 & -2 & 1 \\ 1 & 2 & -1 \end{pmatrix}$ (e) $\begin{pmatrix} 1 & 1 & 2 \\ -1 & 3 & 1 \\ 2 & 1 & 4 \end{pmatrix}$ (f) $\begin{pmatrix} 1 & 0 & 1 \\ -1 & 3 & 2 \\ 2 & -2 & 2 \end{pmatrix}$

2. Solve the following sets of equations by eliminating the variables.

(a) $x + y = 2$
     $2y - z = -7$
     $3x + z = 16$

(b) $2x + y - 3z = -5$
     $x - 2y + 3z = -1$
     $3x + y + z = 2$

(c) $x + y - z = -4$
     $2x - 3y + 2z = 15$
     $5x + 2y + z = 1$

3. Solve the following matrix equations by reducing each 3×3 matrix to a triangular matrix. ·

(a) $\begin{pmatrix} -1 & 2 & -1 \\ 2 & 1 & 3 \\ 2 & -1 & 4 \end{pmatrix} \begin{pmatrix} x \\ y \\ z \end{pmatrix} = \begin{pmatrix} 1 \\ 7 \\ 9 \end{pmatrix}$ (b) $\begin{pmatrix} 3 & -1 & 1 \\ -2 & 2 & 1 \\ 3 & -1 & 2 \end{pmatrix} \begin{pmatrix} x \\ y \\ z \end{pmatrix} = \begin{pmatrix} 8 \\ 3 \\ 9 \end{pmatrix}$

(c) $\begin{pmatrix} 2 & 1 & 3 \\ 1 & 2 & 1 \\ -1 & 3 & 2 \end{pmatrix} \begin{pmatrix} x \\ y \\ z \end{pmatrix} = \begin{pmatrix} 13 \\ 3 \\ 4 \end{pmatrix}$

4. If matrix $\mathbf{A} = \begin{pmatrix} 1 & 3 & -1 \\ 0 & 2 & 1 \\ 1 & 2 & -1 \end{pmatrix}$ find the 3×3 matrices $\mathbf{B}$, $\mathbf{C}$ and $\mathbf{D}$ such

that

(a) $\mathbf{AB} = \begin{pmatrix} 10 & 8 & 2 \\ 8 & 5 & 2 \\ 7 & 6 & 1 \end{pmatrix}$ (b) $\mathbf{CA} = \begin{pmatrix} 1 & 1 & 0 \\ 0 & -4 & -1 \\ 2 & 2 & -2 \end{pmatrix}$ (c) $\mathbf{AD} + \mathbf{C} = \mathbf{B}$

5. Find the value of a for which the equation

$$\begin{pmatrix} -3 & -1 & 1 \\ 2 & -3 & 0 \\ a & 5 & -2 \end{pmatrix}\begin{pmatrix} x \\ y \\ z \end{pmatrix} = \begin{pmatrix} 1 \\ 2 \\ -4 \end{pmatrix}$$ does not have a unique solution.

6. For the set of equations
$$\begin{aligned} ax + by + z &= 2 \\ -x + 2y + 3z &= -2 \\ 3x - y + z &= 1 \end{aligned}$$
give a geometric interpretation for
(a) $a = 3$, $b = -1$ (b) $a = 5$, $b = -2$ (c) $a = 4$, $b = -1$.
(Give any solutions in parametric form where appropriate).

7. For the set of equations
$$\begin{aligned} -3x + ay - 6z &= b \\ x - y + 2z &= 1 \\ 2x - 2y + 4z &= 2 \end{aligned}$$
give a geometric interpretation for (a) $a = 3$, $b = 7$ (b) $a = 4$, $b - 2$.
(Give any solutions in parametric form).

8. For the set of equations
$$\begin{aligned} x + y + z &= -1 \\ 2x - 2y + z &= 5 \\ 3x - y + 2z &= a \end{aligned}$$
give a geometric interpretation for (a) $a = -1$ (b) $a = 4$.
(Give any solutions in parametric form).

9. For each of the following sets of equations find the value a must take for the equations to have solutions and find these solutions in parametric form.

(a) $\begin{aligned} x + 2y + 3z &= 2 \\ -x - 2y - 3z &= -2 \\ 2x + 4y + 6z &= a \end{aligned}$ (b) $\begin{aligned} x + 3y - z &= -3 \\ x - 2y + z &= 4 \\ -2x + 4y - 2z &= a \end{aligned}$

(c) $\begin{aligned} x + y + 3z &= -1 \\ 2x - y + 4z &= 1 \\ -x + 5y + z &= a \end{aligned}$ (d) $\begin{aligned} 2x - 2y + z &= 3 \\ -3x + 3y + 2z &= -1 \\ 2x - 2y + 3z &= a \end{aligned}$

10. For the set of equations
$$\begin{aligned} 2x + y - 3z &= 5 \\ x + 2y + 3z &= 1 \\ 2x - y + az &= b \end{aligned}$$
(a) Find the value of a for which the equations have no unique solution.
(b) With a taking this value find the value b must take for there to be an infinite number of solutions to the equations. Interpret this situation geometrically giving any relevant equations in cartesian form.

11. Show that the equations
$$\begin{aligned} x + y - z &= -1 \\ 5x + 3y + z &= 3 \\ 2x + y + z &= a \end{aligned}$$
are inconsistent for $a = 4$.
Find the value a must take for the equations to represent three planes that have a common line and find the vector equation of this line.

12. Find the inverse of the 4 × 4 matrix $\begin{pmatrix} 1 & 2 & 0 & 1 \\ 1 & 1 & 2 & 1 \\ 2 & 1 & 3 & 1 \\ 1 & 0 & 2 & -1 \end{pmatrix}$

Hence solve the equations $\begin{cases} x + 2y + t + 1 = 0 \\ x + y + 2z + t + 2 = 0 \\ 2x + y + 3z + t + 3 = 0 \\ x + 2z - t + 7 = 0 \end{cases}$

Exercise 17G Examination questions

1. The position vectors of the points A and B relative to an origin O are $3i + 2j - k$ and $2i - 3j + k$ respectively. Write down the vector $\overrightarrow{AB}$. Calculate the size of angle OAB. (A.E.B.)

2. The triangle ABC is such that $AB = 3i + 6j - 2k$ and $AC = 4i - j + 3k$. Find $B\hat{A}C$ and the area of the triangle. (Cambridge)

3. Relative to an origin O, points A and B have position vectors $2j + 9j - 6k$ and $6i + 3j + 6k$ respectively, i, j and k being orthogonal unit vectors. C is the point such that $OC = 2OA$ and D is the mid-point of AB. Find
(i)   the position vectors of C and D;
(ii) a vector equation of the line CD;
(iii) the position vector of the point of intersection of CD and OB;
(iv) the angle AOB correct to the nearest degree. (S.U.J.B)

4. Find the angle between the lines with vector equations $r = a + tc$ and $r = b + sd$, where

$$a = \begin{pmatrix} 1 \\ 2 \\ -1 \end{pmatrix} \quad b = \begin{pmatrix} 5 \\ 5 \\ -4 \end{pmatrix} \quad c = \begin{pmatrix} 0 \\ 1 \\ -1 \end{pmatrix} \quad d = \begin{pmatrix} 2 \\ 1 \\ -1 \end{pmatrix}.$$

Show that these lines intersect, and find the position vector of P, the point of intersection. Express in the vector form $r \cdot n = k$ the equation of the plane which passes through the point P, and which is perpendicular to the line joining the two points with position vectors a and b. (Oxford)

5. The points P and Q have coordinates (1, 2, 3) and (4, 6, −2) respectively and the plane π has equation $x + y - z = 24$.

(i)   Write down, in the form $r = a + tb$ where $r = \begin{pmatrix} x \\ y \\ z \end{pmatrix}$, the equation of the line PQ, and find the coordinates of the point where PQ meets π.

(ii) Calculate the cosine of the angle that PQ makes with the normal to the plane π.

(iii) Find the equation of the plane perpendicular to π containing the line PQ. (Oxford)

6. The point A has position vector $\mathbf{i} + 4\mathbf{j} - 3\mathbf{k}$ referred to the origin O. The line L has vector equation $\mathbf{r} = t\mathbf{i}$. The plane Π contains the line L and the point A. Find
   (a) a vector which is normal to the plane Π,
   (b) a vector equation for the plane Π,
   (c) the cosine of the acute angle between OA and the line L.   (London)

7. The coordinates of the points A and B are $(0, 2, 5)$ and $(-1, 3, 1)$, respectively, and the equations of the line L are
   $$\frac{x-3}{2} = \frac{y-2}{-2} = \frac{z-2}{-1}.$$
   (i)   Find the equation of the plane π which contains A and is perpendicular to L, and verify that B lies in π.
   (ii) Show that the point C in which L meets π is $(1, 4, 3)$, and find the angle between CA and CB.
   (iii) Find the coordinates of the two points P and Q on L which are such that the volume of each of the tetrahedra PABC and QABC is 9.

   (J.M.B.)

8. The lines l_1 and l_2 have the vector equations
   $$l_1: \mathbf{r} = (1 + s)\mathbf{i} + (1 - s)\mathbf{j} - 2\mathbf{k},$$
   $$l_2: \mathbf{r} = \mathbf{i} + (1 + t)\mathbf{j} - (2 - t)\mathbf{k},$$
   where $\mathbf{i}$, $\mathbf{j}$ and $\mathbf{k}$ are perpendicular unit vectors. Show that l_1 and l_2 meet in the point A with position vector $\mathbf{i} + \mathbf{j} - 2\mathbf{k}$. Show also that the lines l_3 and l_4,
   $$l_3: \mathbf{r} = (3 - u)\mathbf{i} - (1 - u)\mathbf{j} + (-2 + u)\mathbf{k},$$
   $$l_4: \mathbf{r} = (1 + v)\mathbf{i} - (1 + 3v)\mathbf{j} - v\mathbf{k},$$
   form the other two sides of a quadrilateral ABCD, and find the position vectors of B, C and D, which are the intersections of l_1 and l_3, l_3 and l_4, l_2 and l_4 respectively. Find the angle between the diagonals AC and BD. Determine whether or not the four points A, B, C and D lie in a plane, explaining your method carefully.   (Oxford)

9. The planes p and q are given by the equations $3x + 2y + z = 4$ and $2x + 3y + z = 5$ respectively. The plane π containing the point A(2, 2, 1) is perpendicular to each of the planes p and q. Find
   (a) the distance from the point A to the plane p,
   (b) the cosine of the angle between the planes p and q,
   (c) a cartesian equation for the plane π,
   (d) cartesian equations for the line of intersection, l, of the planes p and q,
   (e) the point P on line l, which is nearest to the point A.   (A.E.B.)

10. Let A be matrix
    $$\begin{pmatrix} 1 & 2 & 7 \\ 1 & 3 & 0 \\ 0 & -1 & 8 \end{pmatrix}.$$
    By performing elementary row operations on the matrix $(\mathbf{A}|\mathbf{I})$ find $\mathbf{A}^{-1}$.
    Solve the equation $\mathbf{AX} = \mathbf{K}$, when
    (i) $\mathbf{K} = \begin{pmatrix} 0 \\ 0 \\ 0 \end{pmatrix}$ (ii) $\mathbf{K} = \begin{pmatrix} -1 \\ 1 \\ 0 \end{pmatrix}$ (iii) $\mathbf{K} = \begin{pmatrix} 1 \\ 2 \\ 3 \end{pmatrix}$.   (Cambridge)

11. Find values of a, b, c, d, e, f, so that the linear transformation

$$\begin{pmatrix} x' \\ y' \\ z' \end{pmatrix} = \mathbf{M} \begin{pmatrix} x \\ y \\ z \end{pmatrix}$$

with matrix $\mathbf{M}$, where

$$\mathbf{M} = \begin{pmatrix} a & b & c \\ b & d & e \\ c & e & f \end{pmatrix},$$

maps the points with position vectors $\begin{pmatrix} 1 \\ 0 \\ 0 \end{pmatrix}, \begin{pmatrix} 1 \\ -1 \\ 0 \end{pmatrix}, \begin{pmatrix} 1 \\ -1 \\ 1 \end{pmatrix}$,

to the points with position vectors $\begin{pmatrix} 2 \\ 1 \\ 1 \end{pmatrix}, \begin{pmatrix} 1 \\ -2 \\ 1 \end{pmatrix}, \begin{pmatrix} 2 \\ -2 \\ 1 \end{pmatrix}$,

respectively.
Find the inverse matrix $\mathbf{M}^{-1}$.
Show that the transformation with matrix $\mathbf{M}^{-1}$ maps points of the line
$2x + y = 0, z = 0$ to points of the line $x = 0, 5y + 2z = 0$. (London)

12. Show that $x + a + y$ is a factor of the determinant

$$\begin{vmatrix} a & x & y \\ x & a & y \\ x & y & a \end{vmatrix}.$$

Express the determinant as a product of three factors.
Hence find all the values of θ in the range $0 \leqslant \theta \leqslant \pi$ which satisfy the
equation

$$\begin{vmatrix} 1 & \cos \theta & \cos 2\theta \\ \cos \theta & 1 & \cos 2\theta \\ \cos \theta & \cos 2\theta & 1 \end{vmatrix} = 0.$$ (J.M.B.)

13. Find the complete set of solutions of the system of equations
$$\begin{aligned} 2x - 3y + 4z &= 11, \\ 3x + 2y - z &= 9, \\ \lambda x + 12y - 11z &= \lambda \end{aligned}$$
(i) when $\lambda = 4$ (ii) when $\lambda = 5$.   (Oxford)

14. Find all the solutions of the system of equations
$$\begin{aligned} x + 2y + 2z &= -1, \\ 3x + 2y + 10z &= 5, \\ 2x - y + 9z &= 8, \end{aligned}$$
and the particular solution for which $x + y + z = 0$.
Find also the solutions (if any) of the systems obtained
(a) by replacing the first equation by
$$5x + y + 19z = 0;$$
(b) by replacing the first two equations by
$$\begin{aligned} 4x - 2y + 18z &= 16. \\ -2x + y - 9z &= -8. \end{aligned}$$ (Oxford).

15. Show that the only values of λ for which the simultaneous equations

$$x + (\lambda - 4)y + 2z = 0$$
$$2x - 6y + (\lambda + 3)z = 0$$
$$(\lambda - 5)x + 12y - 8z = 0$$

have a solution in which x, y and z are not all zero are $\lambda = 1$ and $\lambda = 4$.

Show that, when $\lambda = 4$, the three planes represented by the above equations meet in a line L. Show also that L intersects the line whose equations are

$$\frac{x - 8}{4} = y + 1 = \frac{z - 2}{10}.$$

Interpret the three equations geometrically in the case when $\lambda = 1$.

(J.M.B.)

16. Show that the simultaneous equations

$$kx + y + z = 1,$$
$$2x + ky - 2z = -1,$$
$$x - 2y + kz = -2$$

have a unique solution except for three values of k which are to be found.

Show that, when $k = 1$, the planes represented by the equations meet at a point P which lies in the plane $y = 1$, and that, when $k = -1$, the planes meet in a line L which also lies in the plane $y = 1$. Find the perpendicular distance of P from L.

(J.M.B.)

17. Find all the values of (x, y, z) which satisfy

$$\begin{bmatrix} 7 & 2 & 4 \\ 4 & 1 & 2 \\ 3 & 1 & 1 \end{bmatrix} \begin{bmatrix} x \\ y \\ z \end{bmatrix} = \begin{bmatrix} p \\ q \\ r \end{bmatrix}$$

if (p, q, r) satisfies

$$\begin{bmatrix} 2 & -7 & 5 \\ 6 & -9 & -1 \\ -4 & 5 & 2 \end{bmatrix} \begin{bmatrix} p \\ q \\ r \end{bmatrix} = \begin{bmatrix} 5 \\ 7 \\ -4 \end{bmatrix}.$$

(Oxford)

18

Algebra II

18.1 Partial fractions

We know that $\dfrac{1}{x+3} + \dfrac{2}{x-1}$ can be expressed as the single fraction

$\dfrac{3x+5}{(x+3)(x-1)}$ but, if we were given $\dfrac{3x+5}{(x+3)(x-1)}$ how would we get

back to $\dfrac{1}{x+3} + \dfrac{2}{x-1}$?

This reverse process is called expressing $\dfrac{3x+5}{(x+3)(x-1)}$ in partial fractions.

The method used is to assume that $\dfrac{3x+5}{(x+3)(x-1)} \equiv \dfrac{A}{x+3} + \dfrac{B}{x-1}$,

where A and B are constants which have to be determined. Remember that the sign $\equiv$ means that the equality holds for *all* values of x.

The right hand side (RHS) of this identity is then expressed as a single fraction

i.e. $\qquad \dfrac{A(x-1) + B(x+3)}{(x+3)(x-1)}$

This fraction and the left hand side (LHS) of the identity now have the same denominators so their numerators must be identical

i.e. $\qquad\qquad 3x + 5 \equiv A(x-1) + B(x+3)$

One, or both, of the following techniques is used to find A and B from this identity

  (i) substituting suitable values of x in both sides of the identity,
  (ii) equating the coefficients of particular powers of x.

It should be noted that the given fraction, $\dfrac{3x+5}{(x+3)(x-1)}$, is a proper

fraction and so the partial fractions assumed, $\dfrac{A}{x+3} + \dfrac{B}{x-1}$, are also

proper fractions. (Remember that an improper algebraic fraction is one for which the degree of the numerator is equal to or greater than the degree of the denominator, see page 7).

Thus, given $\dfrac{5x^2 + 13x + 1}{(x+5)(x^2-2)}$, we would assume the partial fractions to be $\dfrac{A}{x+5} + \dfrac{Bx+C}{x^2-2}$

The denominators of the algebraic fractions encountered will be of three basic types:

Type	Denominator of fraction	Example	Expression used
1.	has linear factors	$\dfrac{5}{(x-2)(x+3)}$	$\dfrac{A}{x-2}+\dfrac{B}{x+3}$
2.	has a quadratic factor which does not factorise	$\dfrac{2x+3}{(x-1)(x^2+4)}$	$\dfrac{A}{x-1}+\dfrac{Bx+C}{x^2+4}$
3.	has a repeated factor	$\dfrac{5x+3}{(x-2)(x+3)^2}$	$\dfrac{A}{x-2}+\dfrac{B}{x+3}+\dfrac{C}{(x+3)^2}$

Notes 1. When expressing an algebraic fraction in partial fractions, always factorise the denominator as far as possible first.

2. Given an improper algebraic fraction, we must first use the methods of page 7 to obtain an expression which does not contain any improper fractions (see examples 5 and 6).

3. For a fraction with a denominator having a repeated factor,

e.g. $\dfrac{5x+3}{(x-2)(x+3)^2}$, we may initially think of using $\dfrac{A}{x-2}+\dfrac{Bx+C}{(x+3)^2}$

but this is not the simplest form because

$$\frac{Bx+C}{(x+3)^2}=\frac{B(x+3)}{(x+3)^2}+\frac{C-3B}{(x+3)^2}$$

$$=\frac{B}{x+3}+\frac{C'}{(x+3)^2}\quad\text{where }C'=C-3B.$$

Example 1 (denominator has linear factors)

Express $\dfrac{1}{x^3-9x}$ in partial fractions.

The denominator factorises to $x(x-3)(x+3)$

Assume $\dfrac{1}{x(x-3)(x+3)}\equiv\dfrac{A}{x}+\dfrac{B}{x-3}+\dfrac{C}{x+3}$

R.H.S. $=\dfrac{A(x-3)(x+3)+B(x)(x+3)+C(x)(x-3)}{x(x-3)(x+3)}$

$\therefore\qquad 1\equiv A(x-3)(x+3)+B(x)(x+3)+C(x)(x-3)$

as L.H.S. and R.H.S. have the same denominators.

Put $x=0$, $1=A(-3)(+3)+B(0)+C(0)$ hence $A=-\frac{1}{9}$
Put $x=3$, $1=A(0)+B(3)(3+3)+C(0)$ hence $B=\frac{1}{18}$
Put $x=-3$, $1=A(0)+B(0)+C(-3)(-6)$ hence $C=\frac{1}{18}$

The partial fractions are $-\dfrac{1}{9x}+\dfrac{1}{18(x-3)}+\dfrac{1}{18(x+3)}.$

Example 2 (denominator has quadratic factor)

Express $\dfrac{5x^2-2x-1}{(x+1)(x^2+1)}$ in partial fractions.

Assume
$$\frac{5x^2 - 2x - 1}{(x + 1)(x^2 + 1)} \equiv \frac{A}{x + 1} + \frac{Bx + C}{x^2 + 1}$$

$$\equiv \frac{A(x^2 + 1) + (Bx + C)(x + 1)}{(x + 1)(x^2 + 1)}$$

$\therefore \qquad 5x^2 - 2x - 1 \equiv A(x^2 + 1) + (Bx + C)(x + 1)$

Put $x = -1$, $\quad 5(-1)^2 - 2(-1) - 1 = A(1 + 1) + 0$

$\therefore \qquad\qquad 5 + 2 - 1 = 2A \qquad\qquad$ hence $A = 3$

Put $x = 0$, $\qquad\qquad 0 - 0 - 1 = A(1) + C$

$\therefore \qquad\qquad\qquad -1 = 3 + C \qquad\qquad$ hence $C = -4$

Equating the coefficients of x^2 on the two sides of the identity:

$$5 = A + B \qquad\qquad \text{hence } B = 2$$

The partial fractions are $\dfrac{3}{x + 1} + \dfrac{2x - 4}{x^2 + 1}$.

Example 3 (denominator has a repeated factor)

Express $\dfrac{x + 4}{(x + 1)(x - 2)^2}$ in partial fractions.

Assume $\quad \dfrac{x + 4}{(x + 1)(x - 2)^2} \equiv \dfrac{A}{x + 1} + \dfrac{B}{x - 2} + \dfrac{C}{(x - 2)^2}$

(Note carefully the form which the partial fractions will take in this case.)

$\therefore \qquad\qquad x + 4 \equiv A(x - 2)^2 + B(x + 1)(x - 2) + C(x + 1)$

Put $x = 2$, $\qquad 2 + 4 = A(0) + B(0) + C(2 + 1) \qquad$ hence $C = 2$

Put $x = -1$, $\quad -1 + 4 = A(-1-2)^2 + B(0) + C(0) \quad$ hence $A = \frac{1}{3}$

Equating coefficients of x^2 on both sides of the identity:

$$0 = A + B \qquad\qquad \text{hence } B = -\tfrac{1}{3}$$

The partial fractions are $\dfrac{1}{3(x + 1)} - \dfrac{1}{3(x - 2)} + \dfrac{2}{(x - 2)^2}$.

Example 4 (denominator has a repeated factor)

Express $\dfrac{4x + 3}{(x - 1)^2}$ in partial fractions.

METHOD 1 as for Example 3

Assume $\quad \dfrac{4x + 3}{(x - 1)^2} \equiv \dfrac{A}{x - 1} + \dfrac{B}{(x - 1)^2}$

$\therefore \qquad 4x + 3 = A(x - 1) + B$

Put $x = 1$, $\quad 4 + 3 = A(0) + B$

$$B = 7$$

Equating coefficients of x:

$$4 = A$$

METHOD 2

$$\frac{4x + 3}{(x - 1)^2} \equiv \frac{4(x - 1) + 7}{(x - 1)^2}$$

$$= \frac{4(x - 1)}{(x - 1)^2} + \frac{7}{(x - 1)^2}$$

$$= \frac{4}{x - 1} + \frac{7}{(x - 1)^2}$$

The partial fractions are $\dfrac{4}{(x - 1)} + \dfrac{7}{(x - 1)^2}$.

Example 5

Express $\dfrac{x(x + 3)}{x^2 + x - 12}$ in partial fractions.

This expression is an improper fraction since the numerator and denominator are of the same degree.

By the methods shown on page 7

$$\frac{x^2 + 3x}{x^2 + x - 12} = 1 + \frac{2x + 12}{x^2 + x - 12}$$

$$= 1 + \frac{2x + 12}{(x + 4)(x - 3)}$$

The proper fraction $\dfrac{2x + 12}{(x + 4)(x - 3)}$ is of the 'denominator with linear

factors' type and can be expressed in partial fractions accordingly, thus

$$\frac{x^2 + 3x}{x^2 + x - 12} = 1 + \frac{18}{7(x - 3)} - \frac{4}{7(x + 4)}.$$

Example 6

Express $\dfrac{x^3}{(x + 4)(x - 1)}$ in partial fractions.

By the methods shown on page 7

$$\frac{x^3}{(x + 4)(x - 1)} = x - 3 + \frac{13x - 12}{(x + 4)(x - 1)}$$

By the methods of this chapter

$$\frac{13x - 12}{(x + 4)(x - 1)} = \frac{64}{5(x + 4)} + \frac{1}{5(x - 1)}$$

Thus $\dfrac{x^3}{(x + 4)(x - 1)} = x - 3 + \dfrac{64}{5(x + 4)} + \dfrac{1}{5(x - 1)}.$

Exercise 18A

Express the following in partial fractions.

Denominator with linear factors only.

1. $\dfrac{4x - 9}{(x - 2)(x - 3)}$

2. $\dfrac{6(5 - x)}{(x - 7)(4 - x)}$

3. $\dfrac{3 - 8x}{x(1 - x)}$

4. $\dfrac{x + 24}{x^2 - x - 12}$

5. $\dfrac{2(3x + 4)}{x^2 + 4x}$

6. $\dfrac{2x^2 + 17x + 21}{(x + 2)(x + 3)(x - 3)}$

Denominator with a quadratic factor.

7. $\dfrac{7x^2 - x + 14}{(x - 2)(x^2 + 4)}$

8. $\dfrac{8x - 1}{(x - 2)(x^2 + 1)}$

9. $\dfrac{x(3x + 2)}{(x - 3)(x^2 + 2)}$

10. $\dfrac{10x^2 - 4x + 15}{x(3 + 2x^2)}$

11. $\dfrac{3x + 5}{(x - 2)(3 + 2x^2)}$

12. $\dfrac{3x^2 + x + 9}{(x + 3)(x^2 + x + 5)}$

Denominator with a repeated factor.

13. $\dfrac{x + 3}{(x + 5)^2}$

14. $\dfrac{2x + 1}{(x - 3)^2}$

15. $\dfrac{3x^2 + 2}{x(x - 1)^2}$

16. $\dfrac{x - 3 - 2x^2}{x^2(x - 1)}$

17. $\dfrac{3}{(x + 1)(x - 2)^2}$

18. $\dfrac{5x^2 - 6x - 21}{(x - 4)^2(2x - 3)}$

Improper algebraic fractions.

19. $\dfrac{x^2 + 3x - 13}{(x + 2)(x - 1)}$

20. $\dfrac{3x^2 + 5x + 3}{x(x - 3)}$

21. $\dfrac{3x^3 + 6x^2 + 17x + 13}{(x + 3)(x^2 + 4)}$

22. $\dfrac{x^3 + 3x^2 + 10}{(x + 1)(x + 4)}$

23. $\dfrac{2x^2 + 2x + 3}{x^2 - 1}$

24. $\dfrac{x^4 - 3x^3 - 3}{x^2(x - 1)}$

Miscellaneous.

25. $\dfrac{7}{(2x - 3)(x + 2)}$

26. $\dfrac{10 + 6x - 3x^2}{(2x - 1)(x + 3)^2}$

27. $\dfrac{2x - 3}{x^2 - 1}$

28. $\dfrac{3x^2 - 1}{(x - 1)(2x - 1)^2}$

29. $\dfrac{2x^2 - 7}{(x - 3)(2x + 5)}$

30. $\dfrac{2x^3 + 11}{(x^2 + 4)(x - 3)}$

31. $\dfrac{37x - 81}{(x - 3)(x + 7)(2x - 3)}$

32. $\dfrac{3x + 7}{(1 + 2x)(x^2 - x + 2)}$

33. $\dfrac{2x^3 + 46}{(x - 1)^2(x + 3)}$

34. $\dfrac{3x^3 + 6x^2 - 11x + 1}{(1 + x)^3(x - 2)}$

35. $\dfrac{3x^2 - 10x - 24}{x(x^2 - 4)}$

36. $\dfrac{x^3 + 2x^2 + 61}{(x + 3)^2(x^2 + 4)}$

18.2 Partial fractions and series

Series expansions

Partial fractions can be used when finding the series expansions of some functions.

Example 7

Express $\dfrac{32x^2 + 17x + 18}{(2 - 3x)(1 + 2x)^2}$ in partial fractions and hence obtain its series expansion in ascending powers of x, stating the terms up to and including the term in x^3, the term in x^r and the values of x for which the expansion is valid.

Using the methods of the previous section, we find that $\dfrac{32x^2 + 17x + 18}{(2 - 3x)(1 + 2x)^2} = \dfrac{8}{2 - 3x} + \dfrac{5}{(1 + 2x)^2}$

Using the binomial expansion

$$\dfrac{8}{2 - 3x} = \dfrac{8}{2(1 - \frac{3}{2}x)}$$

$$= 4(1 - \tfrac{3}{2}x)^{-1} = 4\left[1 + (-1)(-\tfrac{3}{2}x) + \dfrac{(-1)(-2)}{1 \times 2}(-\tfrac{3}{2}x)^2 + \dfrac{(-1)(-2)(-3)}{1 \times 2 \times 3}(-\tfrac{3}{2}x)^3 + \ldots\right]$$

$$= 4[1 + \tfrac{3}{2}x + (\tfrac{3}{2}x)^2 + (\tfrac{3}{2}x)^3 + \ldots + (\tfrac{3}{2}x)^r + \ldots]$$

$$= 4 + 6x + 9x^2 + \tfrac{27}{2}x^3 + \ldots + \dfrac{3^r x^r}{2^{r-2}} + \ldots \qquad \ldots [1]$$

$$\dfrac{5}{(1 + 2x)^2} = 5(1 + 2x)^{-2} = 5\left[1 + (-2)(2x) + \dfrac{(-2)(-3)(2x)^2}{1 \times 2} + \dfrac{(-2)(-3)(-4)(2x)^3}{1 \times 2 \times 3} + \ldots\right]$$

$$= 5[1 - 2(2x) + 3(2x)^2 - 4(2x)^3 + \ldots + (-1)^r(r + 1)(2x)^r + \ldots]$$

$$= 5 - 20x + 60x^2 - 160x^3 + \ldots + (-2)^r 5(r + 1)x^r + \ldots \qquad \ldots[2]$$

Adding [1] and [2]:

$$\dfrac{8}{2 - 3x} + \dfrac{5}{(1 + 2x)^2} = 9 - 14x + 69x^2 - \tfrac{293}{2}x^3 + \ldots + [3^r 2^{2-r} + (-2)^r 5(r + 1)]x^r \ldots$$

Notice that, substituting $r = 0$ into the term in x^r gives the term in x^0, i.e. 9, substituting $r = 1$ into the term in x^r gives the term in x^1, i.e. $-14x$ etc.

The expansion of $(1 - \tfrac{3}{2}x)^{-1}$ is valid for $|\tfrac{3}{2}x| < 1$ i.e. $|x| < \tfrac{2}{3}$ and the expansion of $(1 + 2x)^{-2}$ is valid for $|2x| < 1$ i.e. $|x| < \tfrac{1}{2}$.
Thus if $|x| < \tfrac{1}{2}$, both of these conditions will be satisfied.

$$\therefore \quad \dfrac{32x^2 + 17x + 18}{(2 - 3x)(1 + 2x)^2} = 9 - 14x + 69x^2 - \tfrac{293}{2}x^3 + \ldots + [3^r 2^{2-r} + (-2)^r 5(r + 1)]x^r \ldots$$

provided $|x| < \tfrac{1}{2}$.

Exercise 18B

Express each of the given functions in partial fractions and hence obtain the series expansion of each function, stating the terms up to and including the term in x^3, the term in x^r and the values of x for which the expansion is valid.

1. $\dfrac{2 + x}{(1 - x)(1 + 2x)}$

2. $\dfrac{7x + 1}{(1 + x)(1 + 3x)}$

3. $\dfrac{8x}{(1 + 5x)(1 - 3x)}$

4. $\dfrac{2}{(1 - x)(3 - x)}$

5. $\dfrac{4x}{(3 + 2x)(2x - 1)}$

6. $\dfrac{11x + 6}{(3 + x)(3 + 4x)}$

7. $\dfrac{1 + 2x - x^2}{(1 + x)^2(1 + 2x)}$

8. $\dfrac{6 - 14x - 7x^2}{(1 - 4x)(1 - x)(2 + 3x)}$

9. $\dfrac{x^2 - 4x - 8}{(2 + x)^2(1 + x)}$

10. Express $\dfrac{11x + 3}{(1 + x^2)(1 - 5x)}$ in partial fractions and hence obtain its series expansion in ascending powers of x up to and including the term in x^3. State the range of values of x for which the expansion is valid.

11. Express $\dfrac{18 - x - x^2}{(1 + x + x^2)(2 - 3x)}$ in partial fractions and hence obtain the first four terms in the series expansion of the function in ascending powers of x.

Summation of series by the method of differences

Partial fractions enable us to sum certain series using the method of differences as the following examples show.

Example 8

(a) Express $\dfrac{2}{(2x - 1)(2x + 1)}$ in partial fractions.

(b) Find an expression for $\displaystyle\sum_{r=1}^{n} \dfrac{2}{(2r - 1)(2r + 1)}$ and state whether or not the series is convergent.

(a) Using the methods of 18.1 we obtain $\dfrac{2}{(2x - 1)(2x + 1)} = \dfrac{1}{2x - 1} - \dfrac{1}{2x + 1}$.

(b) $\displaystyle\sum_{r=1}^{n} \dfrac{2}{(2r - 1)(2r + 1)} = \sum_{r=1}^{n} \left[\dfrac{1}{2r - 1} - \dfrac{1}{2r + 1} \right] = $

$$\begin{aligned}
&\quad \dfrac{1}{1} \quad - \quad \dfrac{1}{3} \\
&+ \quad \dfrac{1}{3} \quad - \quad \dfrac{1}{5} \\
&+ \quad \dfrac{1}{5} \quad - \quad \dfrac{1}{7} \\
&\quad \vdots \\
&+ \quad \dfrac{1}{2n - 1} - \dfrac{1}{2n + 1} = \dfrac{1}{1} - \dfrac{1}{2n + 1}
\end{aligned}$$

Thus $\displaystyle\sum_{r=1}^{n} \frac{2}{(2r-1)(2r+1)} = 1 - \frac{1}{2n+1}.$

Furthermore, as $n \to \infty$, $\dfrac{1}{2n+1} \to 0$ and the summation $\to 1$. Thus S_∞ exists,

i.e. the series converges.

Thus $\displaystyle\sum_{r=1}^{n} \frac{2}{(2r-1)(2r+1)} = 1 - \frac{1}{2n+1}$ and the series is convergent with $S_\infty = 1$.

Example 9

(a) Find the sum of the series $\dfrac{5}{1 \times 2 \times 3} + \dfrac{8}{2 \times 3 \times 4} + \dfrac{11}{3 \times 4 \times 5} + \cdots + \dfrac{3n+2}{n(n+1)(n+2)}.$

(b) Deduce the value of $\displaystyle\sum_{r=1}^{\infty} \frac{3r+2}{r(r+1)(r+2)}.$

(a) In sigma notation the series can be written $\displaystyle\sum_{r=1}^{n} \frac{3r+2}{r(r+1)(r+2)}.$

Now $\dfrac{3r+2}{r(r+1)(r+2)} = \dfrac{1}{r} + \dfrac{1}{r+1} - \dfrac{2}{r+2}$

$\displaystyle\therefore \sum_{r=1}^{n} \frac{3r+2}{r(r+1)(r+2)} = \sum_{r=1}^{n} \left[\frac{1}{r} + \frac{1}{r+1} - \frac{2}{r+2} \right] = + \frac{1}{1} + \frac{1}{2} - \frac{2}{3}$

$$+ \frac{1}{2} + \frac{1}{3} - \frac{2}{4}$$
$$+ \frac{1}{3} + \frac{1}{4} - \frac{2}{5}$$
$$\vdots \qquad \vdots \qquad \vdots$$
$$+ \frac{1}{n-1} + \frac{1}{n} - \frac{2}{n+1}$$
$$+ \frac{1}{n} + \frac{1}{n+1} - \frac{2}{n+2}$$
$$= 1 + \frac{1}{2} + \frac{1}{2} - \frac{2}{n+1} + \frac{1}{n+1} - \frac{2}{n+2}$$

$\displaystyle\therefore \sum_{r=1}^{n} \frac{3r+2}{r(r+1)(r+2)} = \frac{n(2n+3)}{(n+1)(n+2)}$

(b) Considering the expression $1 + \dfrac{1}{2} + \dfrac{1}{2} - \dfrac{2}{n+1} + \dfrac{1}{n+1} - \dfrac{2}{n+2},$

we see that as $n \to \infty$, $S_n \to 2$, i.e. $\displaystyle\sum_{r=1}^{\infty} \frac{3r+2}{r(r+1)(r+2)} = 2$

This method of differences can also be used to determine $\sum f(r)$ where $f(r)$ is not a fraction and, in its simplest form, the method depends upon the ability to express $f(r)$ in the form $g(r) - g(r+a)$ where a is an integer (usually 1 or 2).

Then, if $f(r) = g(r) - g(r+1)$

$$\sum_{r=1}^{n} f(r) = \sum_{r=1}^{n} [g(r) - g(r+1)] = \quad g(1) \quad - \quad g(2)$$
$$+ \; g(2) \quad - \quad g(3)$$
$$+ \; g(3) \quad - \quad g(4)$$
$$\vdots \qquad\qquad \vdots$$
$$+ g(n-1) \quad - \quad g(n)$$
$$+ \; g(n) \quad - \quad g(n+1) = g(1) - g(n+1).$$

[Clearly Example 9 above is an extension of this basic technique and uses the relationship $f(r) = g(r) + g(r+1) - 2g(r+2)$.]

Example 10

(a) Show that $(x - 3)^2 - (x - 2)^2 \equiv 5 - 2x$.

(b) Use this identity from part (a) and the method of differences to obtain an expansion for $\sum_{r=1}^{n}(5 - 2r)$. Hence sum the series $3 + 1 - 1 - 3 - 5 \ldots - 35$.

(a) L.H.S. $= (x - 3)^2 - (x - 2)^2$
$= (x^2 - 6x + 9) - (x^2 - 4x + 4)$
$= x^2 - 6x + 9 - x^2 + 4x - 4$
$= -2x + 5 = $ R.H.S. as required.

(b) $\displaystyle\sum_{r=1}^{n} (5 - 2r) = \sum_{r=1}^{n} [(r - 3)^2 - (r - 2)^2] = \quad (-2)^2 \quad - \quad (-1)^2$
$$+ \; (-1)^2 \quad - \quad (0)^2$$
$$+ \; (0)^2 \quad - \quad (1)^2$$
$$\vdots \qquad\qquad \vdots$$
$$+ (n-4)^2 \quad - \quad (n-3)^2$$
$$+ (n-3)^2 \quad - \quad (n-2)^2$$

$$\therefore \; \sum_{r=1}^{n} (5 - 2r) = 4 - (n - 2)^2$$

$$3 + 1 - 1 - 3 - 5 \ldots - 35 = \sum_{r=1}^{20}(5 - 2r)$$
$$= 4 - (20 - 2)^2$$
$$= -320$$

Exercise 18C

1. Using the identity $2r + 1 \equiv (r + 1)^2 - r^2$ obtain an expression for $\sum_{r=1}^{n} (2r + 1)$.

2. (a) Show that $r(r + 1)(r + 2) - (r - 1)(r)(r + 1) \equiv 3r(r + 1)$

   (b) Using the identity from part (a) and the method of differences obtain an expression for $\sum_{r=1}^{n} r(r + 1)$.

3. (a) Express $\dfrac{1}{x(x + 1)}$ in partial fractions.

(b) Show that $\displaystyle\sum_{r=1}^{n} \dfrac{1}{r(r + 1)} = \dfrac{n}{n + 1}$

(c) Find the sum of the series

$$\dfrac{1}{1 \times 2} + \dfrac{1}{2 \times 3} + \dfrac{1}{3 \times 4} + \dfrac{1}{4 \times 5} + \ldots + \dfrac{1}{15 \times 16}$$

4. (a) Express $\dfrac{3}{(3x - 1)(3x + 2)}$ in partial fractions.

(b) Show that $\displaystyle\sum_{r=1}^{n} \dfrac{3}{(3r - 1)(3r + 2)} = \dfrac{1}{2} - \dfrac{1}{3n + 2}$

(c) Find the sum of the series

$$\dfrac{1}{2 \times 5} + \dfrac{1}{5 \times 8} + \dfrac{1}{8 \times 11} + \dfrac{1}{11 \times 14} + \ldots \dfrac{1}{29 \times 32} \text{ and find}$$

$$\sum_{r=1}^{\infty} \dfrac{1}{(3r - 1)(3r + 2)}$$

5. (a) Express $\dfrac{2}{x(x + 2)}$ in partial fractions.

(b) Show that $\displaystyle\sum_{x=1}^{n} \dfrac{1}{x(x + 2)} = \dfrac{3}{4} - \dfrac{2n + 3}{2(n + 1)(n + 2)}$

and determine $\displaystyle\sum_{x=1}^{\infty} \dfrac{1}{x(x + 2)}$.

(c) Find the sum of the series

$$\dfrac{1}{1 \times 3} + \dfrac{1}{2 \times 4} + \dfrac{1}{3 \times 5} + \dfrac{1}{4 \times 6} + \ldots + \dfrac{1}{9 \times 11}$$

6. (a) Show that

$$(r + 1)(2r + 1)(2r + 3) - r(2r - 1)(2r + 1) \equiv 3(2r + 1)^2$$

(b) Using the identity from part (a) and the method of differences show that

$$\sum_{r=1}^{n} (2r + 1)^2 = \dfrac{n}{3}(4n^2 + 12n + 11)$$

(c) Hence find the sum of the series $3^2 + 5^2 + 7^2 + \ldots + 21^2$.

7. Find the sums of the following series. [Express your answers in the form $a - f(n)$]

(a) $\dfrac{4}{1 \times 5} + \dfrac{4}{5 \times 9} + \dfrac{4}{9 \times 13} + \dfrac{4}{13 \times 17} + \ldots + \dfrac{4}{(4n - 3)(4n + 1)}$

(b) $\dfrac{1}{1 \times 5} + \dfrac{1}{3 \times 7} + \dfrac{1}{5 \times 9} + \dfrac{1}{7 \times 11} + \ldots + \dfrac{1}{(2n - 1)(2n + 3)}$

(c) $\dfrac{1}{1 \times 2 \times 3} + \dfrac{1}{2 \times 3 \times 4} + \dfrac{1}{3 \times 4 \times 5} + \dfrac{1}{4 \times 5 \times 6} + \ldots + \dfrac{1}{n(n + 1)(n + 2)}$

(d) $\dfrac{6}{1 \times 2 \times 4} + \dfrac{6}{2 \times 3 \times 5} + \dfrac{6}{3 \times 4 \times 6} + \dfrac{6}{4 \times 5 \times 7} + \ldots + \dfrac{1}{n(n + 1)(n + 3)}$

8. (a) Use partial fractions to show that

$$\dfrac{7}{1 \times 2 \times 3} + \dfrac{10}{2 \times 3 \times 4} + \dfrac{13}{3 \times 4 \times 5} + \ldots + \dfrac{3n + 4}{n(n + 1)(n + 2)} = \dfrac{n(5n + 9)}{2(n + 1)(n + 2)}$$

(b) State whether the series $\displaystyle\sum_{r=1}^{n} \dfrac{3r + 4}{r(r + 1)(r + 2)}$ converges as $n \to \infty$ and,

if it does, find its sum to infinity.

9. (a) Use partial fractions to show that

$$\frac{2}{1 \times 3 \times 5} + \frac{3}{3 \times 5 \times 7} + \frac{4}{5 \times 7 \times 9} + \ldots + \frac{n+1}{(2n-1)(2n+1)(2n+3)} = \frac{n(5n+7)}{6(2n+1)(2n+3)}$$

(b) State whether the series $\displaystyle\sum_{r=1}^{n} \frac{r+1}{(2r-1)(2r+1)(2r+3)}$ converges as

$n \to \infty$ and, if it does, find its sum to infinity.

18.3 Complex numbers

Page 1 discussed the extension of the basic ideas of number from positive
and negative integers to rational numbers and eventually to irrational
numbers. We can therefore solve equations like:

(i) $x^2 = 49$
    $x = +7$ or -7

(ii) $x^2 = 6\frac{1}{4} = \frac{25}{4}$
    $x = \frac{5}{2}$ or $-\frac{5}{2}$

(iii) $x^2 = 17$
    $x = \sqrt{17}$ or $-\sqrt{17}$
    i.e. $x = 4\cdot123\ldots$ or $-4\cdot123\ldots$

We now need to extend our ideas of number still further, since the equation

(iv) $x^2 = -16$

has no solution in terms of rational or irrational numbers.
We introduce, therefore, the concept of an imaginary number.
If we write i for $\sqrt{(-1)}$, then

$$x^2 = -16 = (16)(-1)$$
$$\text{and} \quad x = \pm\sqrt{(16)}\sqrt{(-1)}$$
$$\text{or} \quad x = \pm 4\text{i} \quad \text{i.e.} \quad x = 4\text{i or } x = -4\text{i}$$

Thus, by the use of the imaginary number i, to stand for $\sqrt{(-1)}$, we have
found two solutions of the equation $x^2 = -16$.

We say that bi, where $b \in \mathbb{R}$, is an *imaginary* number and that $a + i b$, where
a and $b \in \mathbb{R}$, is a *complex* number.

The following points about complex numbers should be noted:
If we write $z = a + ib$, then
    the real part of the complex number z is a, written $\text{Re}(z) = a$,
and the imaginary part of the complex number z is b, written $\text{Im}(z) = b$.
Two complex numbers are equal if, and only if, their real parts are equal
and their imaginary parts are equal.
    Thus $a + ib = c + id \Rightarrow a = c$ and $b = d$.
Complex numbers are added (or subtracted) by adding (or subtracting) their
real parts and also their imaginary parts.
The complex conjugate of $z = a + ib$ is $a - ib$ and is denoted by z^*.
It follows that the product zz^* is therefore a wholly real number, since

$$zz^* = (a + ib)(a - ib)$$
$$= a^2 - iab + iab - i^2b^2$$
$$= a^2 + b^2$$

The manipulation of complex numbers and their use is illustrated in the
following examples.

Example 11

(a) Add $(4 - 2i)$ and $(3 + 7i)$ (b) Simplify $(5 + 4i) - (3 - 2i)$ (c) Find $z + z^*$ if $z = 3 - 7i$.

(a) $(4 - 2i) + (3 + 7i)$
$= (4 + 3) + (-2i + 7i)$
$= 7 + 5i$

(b) $(5 + 4i) - (3 - 2i)$
$= 5 + 4i - 3 + 2i$
$= 2 + 6i$

(c) $z = 3 - 7i$
$\therefore \quad z^* = 3 + 7i$
$z + z^* = 3 - 7i + 3 + 7i$
$= 6$

Example 12

Multiply $(2 - 7i)$ by $(3 + 2i)$

$(2 - 7i)(3 + 2i)$
$= 6 + 4i - 21i - 14i^2$ i^2 being written for $i \times i$
$= 6 - 17i - 14(-1)$ -1 being written for i^2, since $i = \sqrt{(-1)}$
$= 20 - 17i$

Example 13

Divide $(5 + 2i)$ by $(1 - 3i)$

$$(5 + 2i) \div (1 - 3i) = \frac{5 + 2i}{1 - 3i}$$

$$= \frac{(5 + 2i)}{(1 - 3i)} \times \frac{(1 + 3i)}{(1 + 3i)} \qquad \text{The denominator } (z) \text{ is multiplied by } z^*$$
$$\text{so as to make it a real number.}$$

$$= \frac{5 + 15i + 2i + 6i^2}{1 - 9i^2}$$

$$= \frac{5 + 17i - 6}{1 - 9(-1)} \quad \text{since} \quad i^2 = -1$$

$$= \frac{1}{10}(-1 + 17i) \quad \text{or} \quad -\frac{1}{10} + \frac{17}{10}i$$

Example 14

Find the square roots of $(3 + 4i)$

Suppose the square root of $(3 + 4i)$ is $a + ib$ where a and $b \in \mathbb{R}$.
Then $3 + 4i = (a + ib)^2$
$= a^2 + 2abi + i^2b^2$
$= a^2 - b^2 + 2abi$
$\therefore \qquad 3 = a^2 - b^2$ and $4 = 2ab$
Solving simultaneously gives $a = 2$ and $b = 1$ or $a = -2$ and $b = -1$.
The square roots of $(3 + 4i)$ are $\pm(2 + i)$.

Equations

On page 11 we saw that many quadratic equations could be solved using the formula. With our extended idea of number, solutions can now be obtained for *all* quadratic equations.

Example 15

Solve $x^2 + 2x + 6 = 0$.

Using $x = \dfrac{-b \pm \sqrt{(b^2 - 4ac)}}{2a}$ with $a = 1$, $b = 2$ and $c = 6$,

$$x = \frac{-2 \pm \sqrt{(4 - 24)}}{2}$$

$$= \frac{-2 \pm \sqrt{(-20)}}{2}$$

$$= -1 \pm i\sqrt{5}$$

The solutions of the equation are $x = -1 + i\sqrt{5}$ or $x = -1 - i\sqrt{5}$.
Thus, in the cases where $b^2 - 4ac < 0$, as in the above example, we can
improve on our previous comment, in chapter 5, of 'no real roots,' by saying
that the equation has 'two complex conjugate roots', i.e. one root is the
complex conjugate of the other.

Example 16

Find the equation having roots of $(1 + 5i)$ and $(1 - 5i)$.

Sum of the roots of the equation $= 1 + 5i + 1 - 5i$
$$= 2$$
Product of the roots $= (1 + 5i)(1 - 5i)$
$$= 26$$

Thus, for $ax^2 + bx + c = 0$, $\dfrac{c}{a} = 26$ and $-\dfrac{b}{a} = 2$, so the equation can be
written as $x^2 - 2x + 26 = 0$.

Example 17

Solve $2x^3 - 12x^2 + 25x - 21 = 0$.

Consider $f(x) = 2x^3 - 12x^2 + 25x - 21$. Since $f(3) = 0$, then it follows
by the factor theorem that $f(x)$ is divisible by $(x - 3)$,
hence if $2x^3 - 12x^2 + 25x - 21 = 0$
then $\qquad (x - 3)(2x^2 - 6x + 7) = 0$ $\quad [2x^2 - 6x + 7$ obtained by long division]
$\therefore \qquad x = 3$, or $2x^2 - 6x + 7 = 0$

Solving the quadratic $\qquad x = \dfrac{6 \pm \sqrt{(36 - 56)}}{4}$

$$= \frac{3}{2} \pm \frac{i\sqrt{5}}{2}$$

The roots of the cubic equation are $x = 3$, $x = \dfrac{3}{2} + \dfrac{i\sqrt{5}}{2}$, $x = \dfrac{3}{2} - \dfrac{i\sqrt{5}}{2}$ i.e. $x = 3, \dfrac{3}{2} \pm \dfrac{i\sqrt{5}}{2}$.

Note that the complex roots of polynomial equations occur in conjugate
pairs. Thus if $a + ib$ is a root of $f(x) = 0$, then so is $a - ib$. It follows that
a polynomial equation cannot have an odd number of complex roots.

Example 18

Given that $(2 - i)$ and $(1 + 3i)$ are roots of the equation $ax^4 + bx^3 + cx^2 + dx + e = 0$, find (i) the other two roots and (ii) the sum, and the product, of the four roots of the equation.

(i)   Complex roots occur in conjugate pairs, hence
      since $x = 2 - i$ is a root of the equation, so also is $x = 2 + i$,
and since $x = 1 + 3i$ is a root of the equation, so also is $x = 1 - 3i$.

(ii)   Sum of roots $= (2 - i) + (2 + i) + (1 + 3i) + (1 - 3i)$
$$= 6$$
   Product of roots $= (2 - i)(2 + i)(1 + 3i)(1 - 3i)$
$$= (4 - i^2)(1 - 9i^2)$$
$$= 50$$
   The roots are $(2 \pm i)$, $(1 \pm 3i)$; their sum is 6 and their product 50.

Exercise 18D

1. If $z = 3 - 4i$ find   (a) Re(z),   (b) Im(z),   (c) z^*,   (d) zz^*,   (e) $(zz)^*$.
2. Simplify each of the following,
    (a) $(3 + 4i) + (2 + 3i)$ (b) $(2 - 4i) - 3(5 - 3i)$ (c) $(2i)^2$ (d) i^4
    (e) $\dfrac{1}{i^2}$ (f) $(2 + 3i)(2 - 3i)$ (g) $\dfrac{1}{(1 + i)(1 - i)}$ (h) $-\dfrac{1}{i}$
3. Simplify each of the following,
    (a) $(2 + i)(3 - i)$ (b) $(5 - 2i)(6 + i)$ (c) $(4 - 3i)(1 - i)$
    (d) $(3 + i)(2 - 5i)$ (e) $(3 + 4i)(1 - 2i)$ (f) $(2 + i)(2 - i)$
    (g) $(6 + 9i)(4 - 6i)$ (h) $(2 + i)(1 - 2i)(1 + i)$
4. Express each of the following in the form $a + ib$.
    (a) $\dfrac{20}{3 + i}$ (b) $\dfrac{4}{1 + i}$ (c) $\dfrac{2i}{1 - i}$ (d) $\dfrac{1}{1 - 2i}$
    (e) $\dfrac{5i}{1 + 2i}$ (f) $\dfrac{5}{4 - 3i}$ (g) $\dfrac{2 + 3i}{1 - i}$ (h) $\dfrac{3 - i}{1 + 2i}$
    (i) $\dfrac{3 + 2i}{3 - 2i}$ (j) $\dfrac{4i - 3}{2 + 3i}$
5. Simplify $\dfrac{a + ib}{b - ai}$
6. Solve the following equations,
    (a) $x^2 + 25 = 0$ (b) $2x^2 + 32 = 0$ (c) $4x^2 + 9 = 0$
    (d) $x^2 + 2x + 5 = 0$ (e) $x^2 - 4x + 5 = 0$ (f) $2x^2 + x + 1 = 0$.
7. Find the square roots of the following complex numbers,
    (a) $5 + 12i$ (b) $15 + 8i$ (c) $7 - 24i$
8. Find the quadratic equations having roots,
    (a) $3i, -3i$ (b) $1 + 2i, 1 - 2i$ (c) $2 + i, 2 - i$
    (d) $2 + 3i, 2 - 3i$ (e) $3 + 4i \; 3 - 4i$ (f) $3 + 5i, 3 - 5i$
9. If $a + ib$ is a root of the quadratic equation $x^2 + cx + d = 0$ show that $a^2 + b^2 = d$ and $2a + c = 0$.
10. Solve the following equations (all have at least one real root).
    (a) $x^3 - 7x^2 + 19x - 13 = 0$
    (b) $2x^3 - 2x^2 - 3x - 2 = 0$
    (c) $x^3 + 3x^2 + 5x + 3 = 0$

(d) $4x^4 - 20x^3 + 37x^2 - 31x + 10 = 0$

(e) $5x^4 + 8x^3 - 8x - 5 = 0$

11. If $5x^4 - 14x^3 + 18x^2 + 40x + 16 = (x^2 - 4x + 8)(ax^2 + bc + c)$
find a, b and c and hence find the four solutions of the equation
$5x^4 + 18x^2 + 16 = 14x^3 - 40x$.

12. If $3 - 2i$ and $1 + i$ are two of the roots of the equation
$ax^4 + bx^3 + cx^2 + dx + e = 0$ find the values of a, b, c, d and e.

13. Given that $x^3 - 1 = (x - 1)(ax^2 + bx + c)$ find the values of a, b and
c and hence find the three roots of the equation $x^3 = 1$
[These three roots can be referred to as 'the cube roots of unity' and are
often written as $1, \omega, \omega^2$].

Geometrical representation

Complex numbers can be represented geometrically using the x and y axes as
the Real (Re) and Imaginary (Im) axes. The plane of the axes is then
referred to as the complex plane and a diagram showing complex numbers is
said to be an Argand diagram.

On the Argand diagram, each complex number is represented by a line of a
certain length in a particular direction. Thus each complex number is shown
as a vector on the Argand diagram.

Thus if $P(a, b)$ is a point on the Argand diagram, the vector $\overrightarrow{OP}$ represents
the complex number $(a + ib)$; a and b are the real and imaginary
components of the complex number $(a + ib)$.

The sum and the difference of two complex numbers can be shown on an
Argand diagram in the same way as we show vectors which are added or
subtracted.

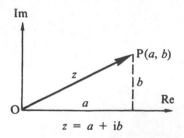
$z = a + ib$

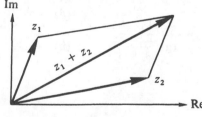

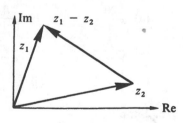

Modulus and argument of complex numbers

The modulus of a complex number, $z = a + ib$, is a measure of the
magnitude of z, and it is written as $|z|$.

Thus modulus $z = |z| = \sqrt{(a^2 + b^2)}$.

The argument of a complex number, $z = a + ib$, is the magnitude in
radians of the angle between the positive real axis and the line representing
the complex number on the Argand diagram. For the complex number z
shown in the diagram, the argument is θ where $\tan \theta = b/a$. Clearly we
could also refer to $\theta \pm 2\pi, \theta \pm 4\pi, \theta \pm 6\pi$, etc, as being the argument of z.
To avoid this complication, we say that the value of θ lying in the range
$-\pi < \theta \leqslant \pi$ is the **principal** value of the argument. We use the abbreviation
'arg' for the principal argument.

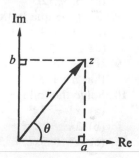

Example 19

Find the modulus and the principal argument of (a) $1 + i$, (b) $-3 + 4i$, (c) $-1 - i\sqrt{3}$
Give the arguments in radians in terms of π or correct to two decimal places as appropriate.

First show the complex numbers on an
Argand diagram.

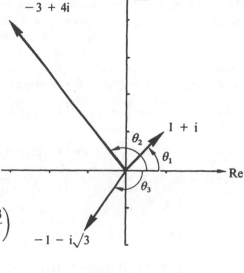

(a) $z = 1 + i$ arg $z = \theta_1$
$|z| = \sqrt{(1^2 + 1^2)}$
$= \sqrt{2}$
$\qquad\qquad\qquad\qquad = \tan^{-1}\left(\dfrac{1}{1}\right)$
$\qquad\qquad\qquad\qquad = \dfrac{\pi}{4}$

(b) $z = -3 + 4i$ arg $z = \theta_2$
$|z| = \sqrt{[(-3)^2 + (4)^2]}$
$= 5$
$\qquad\qquad\qquad\qquad = \pi - \tan^{-1}\left(\dfrac{4}{3}\right)$
$\qquad\qquad\qquad\qquad = 2 \cdot 21$

(c) $z = -1 - i\sqrt{3}$ arg $z = \theta_3$
$|z| = \sqrt{[(-1)^2 + (-\sqrt{3})^2]}$
$= 2$
$\qquad\qquad\qquad\qquad = -\pi + \tan^{-1}\left(\dfrac{\sqrt{3}}{1}\right)$
$\qquad\qquad\qquad\qquad = -\dfrac{2\pi}{3}$

Instead of writing a complex number z in the form $a + ib$, it can
be written in the form $r(\cos\theta + i \sin\theta)$ where r is the modulus
of z and θ is the argument. This form is referred to as the
modulus-argument form, or the polar coordinate form
(see page 401).

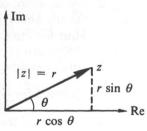

For example, from the diagram

$$i = 1\left(\cos\frac{\pi}{2} + i\sin\frac{\pi}{2}\right)$$

$$-1 - i = \sqrt{2}\left[\cos\left(-\frac{3\pi}{4}\right) + i\sin\left(-\frac{3\pi}{4}\right)\right]$$

$$2 - 2i = \sqrt{8}\left[\cos\left(-\frac{\pi}{4}\right) + i\sin\left(-\frac{\pi}{4}\right)\right]$$

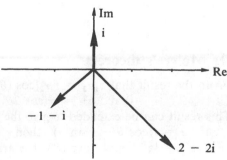

Example 20

Prove that for the complex numbers z_1 and z_2, (a) $|z_1z_2| = |z_1||z_2|$
(b) $\arg(z_1z_2) = \arg z_1 + \arg z_2$. (Assume $\pi < \arg z_1 + \arg z_2 \leqslant \pi$).

Also, if $z_1 = 3\left(\cos\frac{\pi}{3} + i\sin\frac{\pi}{3}\right)$ and $z_2 = 5\left(\cos\frac{\pi}{4} + i\sin\frac{\pi}{4}\right)$ find $|z_1z_2|$ and $\arg(z_1z_2)$.

If $z_1 = r_1(\cos\theta_1 + i\sin\theta_1)$ and $z_2 = r_2(\cos\theta_2 + i\sin\theta_2)$
then $z_1z_2 = r_1(\cos\theta_1 + i\sin\theta_1)r_2(\cos\theta_2 + i\sin\theta_2)$
$$= r_1r_2(\cos\theta_1\cos\theta_2 - \sin\theta_1\sin\theta_2 + i\sin\theta_1\cos\theta_2 + i\cos\theta_1\sin\theta_2)$$
$$= r_1r_2[\cos(\theta_1 + \theta_2) + i\sin(\theta_1 + \theta_2)]$$
thus $|z_1z_2| = r_1r_2 = |z_1||z_2|$
and $\arg(z_1z_2) = (\theta_1 + \theta_2) = \arg z_1 + \arg z_2$ as required.

For $z_1 = 3\left(\cos\frac{\pi}{3} + i\sin\frac{\pi}{3}\right)$, $|z_1| = 3$ and $\arg z_1 = \frac{\pi}{3}$

For $z_2 = 5\left(\cos\frac{\pi}{4} + i\sin\frac{\pi}{4}\right)$, $|z_2| = 5$ and $\arg z_2 = \frac{\pi}{4}$

$|z_1z_2| = |z_1||z_2|$ $\arg(z_1z_2) = \arg z_1 + \arg z_2$

$\qquad = 3 \times 5$ $\qquad = \frac{\pi}{3} + \frac{\pi}{4}$

$\qquad = 15.$ $\qquad = \frac{7\pi}{12}.$

Note: 1. The results $|z_1z_2| = |z_1||z_2|$ and $\arg(z_1z_2) = \arg z_1 + \arg z_2$ can
also be proved using $z_1 = a_1 + ib_1$ and $z_2 = a_2 + ib_2$.
2. The assumption that $\pi < \arg z_1 + \arg z_2 \leqslant \pi$ is necessary as we
use the abbreviation arg for the principal value of the argument.
Thus for $\arg(z_1z_2) = \arg z_1 + \arg z_2$ we must have
$-\pi < \arg z_1 + \arg z_2 \leqslant \pi$
3. Similar results to those obtained for $|z_1z_2|$ and $\arg(z_1z_2)$ can be

obtained for $\frac{z_1}{z_2}$, namely $\left|\frac{z_1}{z_2}\right| = \frac{|z_1|}{|z_2|}$ and $\arg\left(\frac{z_1}{z_2}\right) = \arg z_1 - \arg z_2$.
The proofs of these are left as an exercise for the reader (Exercise 18E, number 9).

De Moivre's theorem

From the result that $z_1z_2 = r_1r_2[\cos(\theta_1 + \theta_2) + i\sin(\theta_1 + \theta_2)]$ it follows
that $z^2 = r^2(\cos 2\theta + i\sin 2\theta)$.
This result can be extended to give the statement that
if $z = r(\cos\theta + i\sin\theta)$ then $z^n = r^n(\cos n\theta + i\sin n\theta)$
i.e. $|z^n| = |z|^n$ and $\arg(z^n) = n\arg z$.

For $r = 1$, we have

> $(\cos\theta + i\sin\theta)^n \equiv \cos n\theta + i\sin n\theta$
> and this is known as de Moivre's theorem.

This theorem can be proved by the method of induction and is set as an
exercise for the reader (Exercise 18E, number 11).

Example 21

Use de Moivre's theorem to prove the following identities:

(a) $\cos 4\theta = 8 \cos^4 \theta - 8 \cos^2 \theta + 1$,   (b) $\tan 4\theta = \dfrac{4 \tan \theta - 4 \tan^3 \theta}{1 - 6 \tan^2 \theta + \tan^4 \theta}$

(a) Using de Moivre's theorem with $n = 4$,

$$\cos 4\theta + i \sin 4\theta \equiv (\cos \theta + i \sin \theta)^4$$
$$= \cos^4 \theta + 4 \cos^3 \theta\, i \sin \theta + 6 \cos^2 \theta\, i^2 \sin^2 \theta + 4 \cos \theta\, i^3 \sin^3 \theta + i^4 \sin^4 \theta$$
$$= \cos^4 \theta + 4i \cos^3 \theta \sin \theta - 6 \cos^2 \theta \sin^2 \theta - 4i \cos \theta \sin^3 \theta + \sin^4 \theta$$

Equating the real parts on the two sides of the identity gives

$$\cos 4\theta = \cos^4 \theta - 6 \cos^2 \theta \sin^2 \theta + \sin^4 \theta$$
$$= \cos^4 \theta - 6 \cos^2 \theta (1 - \cos^2 \theta) + (1 - \cos^2 \theta)^2$$
$$\therefore \qquad \cos 4\theta = 8 \cos^4 \theta - 8 \cos^2 \theta + 1$$

(b) Equating the imaginary parts on the two sides of the identity for de Moivre's theorem with $n = 4$, gives

$$\sin 4\theta = 4 \cos^3 \theta \sin \theta - 4 \cos \theta \sin^3 \theta$$

but $\tan 4\theta = \dfrac{\sin 4\theta}{\cos 4\theta}$

$$= \frac{4 \cos^3 \theta \sin\theta - 4 \cos \theta \sin^3\theta}{8 \cos^4 \theta - 8 \cos^2 \theta + 1}$$

$$= \frac{4 \tan \theta - 4 \tan^3 \theta}{8 - 8 \sec^2 \theta + \sec^4 \theta} \quad \text{dividing throughout by } \cos^4 \theta$$

$$= \frac{4 \tan \theta - 4 \tan^3 \theta}{8 - 8(1 + \tan^2 \theta) + (1 + \tan^2 \theta)^2}$$

$$\tan 4\theta = \frac{4 \tan \theta - 4 \tan^3 \theta}{1 - 6 \tan^2 \theta + \tan^4 \theta}$$

Some further work on complex numbers will be found on page 535.

Exercise 18E

1. Write each of the vectors $\overrightarrow{OA}$, $\overrightarrow{OB}$, $\overrightarrow{OC}$ and $\overrightarrow{OD}$ shown on the Argand diagram, in the form $a + ib$.

2. Show the following complex numbers as position vectors on a single Argand diagram.
$3 + i$, $-4 + 2i$, $-3 - 4i$, $3 - 2i$

3. Write each of the vectors $\overrightarrow{OA}$, $\overrightarrow{OB}$, $\overrightarrow{OC}$ and $\overrightarrow{OD}$ shown on the Argand diagram, in the form $r(\cos \theta + i \sin \theta)$ with θ in radians and $-\pi < \theta \leqslant \pi$.

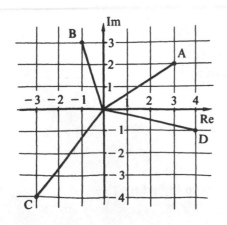

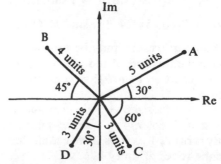

4. Find the modulus and the principal argument of each of the vectors shown in the Argand diagram.

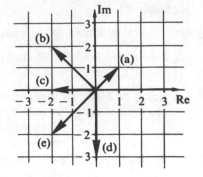

5. Find the modulus and the principal argument of the following complex numbers (give the argument in radians either in terms of π or correct to two decimal places).
(a) $5i$ (b) 7 (c) $-2i$
(d) -3 (e) $1 + \sqrt{3}i$ (f) $5\sqrt{3} - 5i$
(g) $3 - 4i$ (h) $-5 + 12i$

6. The modulus and principal arguments of the vectors z_1 to z_6 are as follows.

	z_1	z_2	z_3	z_4	z_5	z_6
modulus	5	$4\sqrt{2}$	4	12	4	$6\sqrt{2}$
argument	$\dfrac{\pi}{2}$	$-\dfrac{\pi}{4}$	π	$\dfrac{2\pi}{3}$	$-\dfrac{\pi}{3}$	$\dfrac{3\pi}{4}$

Express each of the vectors in the $a + ib$ form.

7. Find the modulus and the principal arguments of
(a) $\dfrac{1-i}{1+i}$ (b) $\dfrac{-1-7i}{4+3i}$ (c) $\dfrac{1+i}{2-i}$ (d) $\dfrac{(3+i)^2}{1-i}$.

Give the arguments in radians either in terms of π or correct to 2 decimal places.

8. (a) If $z_1 = r_1(\cos \theta_1 + i \sin \theta_1)$ and $z_2 = r_2(\cos \theta_2 + i \sin \theta_2)$ prove that $|z_1 z_2| = |z_1||z_2|$ and $\arg (z_1 z_2) = \arg z_1 + \arg z_2$. (Assume that $-\pi < \arg z_1 + \arg z_2 \leqslant \pi$).
(b) For $z_1 = 3(\cos \pi/6 + i \sin \pi/6)$ and $z_2 = 2(\cos \pi/4 + i \sin \pi/4)$ find
(i) $|z_1 z_2|$ (ii) $\arg (z_1 z_2)$ (iii) $|z_1{}^2|$ (iv) $|z_2{}^2|$ (v) $\arg (z_1{}^2)$ (vi) $\arg (z_2{}^2)$

9. (a) If $z_1 = r_1(\cos \theta_1 + i \sin \theta_1)$ and $z_2 = r_2(\cos \theta_2 + i \sin \theta_2)$ prove that $\left|\dfrac{z_1}{z_2}\right| = \dfrac{|z_1|}{|z_2|}$ and $\arg \left(\dfrac{z_1}{z_2}\right) = \arg z_1 - \arg z_2$. (Assume that $-\pi < \arg z_1 - \arg z_2 \leqslant \pi$).
(b) For $z_1 = 2(\cos 2\pi/3 + i \sin 2\pi/3)$ and $z_2 = 6[\cos (-3\pi/4) + i \sin (-3\pi/4)]$ find
(i) $\left|\dfrac{z_1}{z_2}\right|$ (ii) $\arg \left(\dfrac{z_1}{z_2}\right)$ (iii) $\left|\dfrac{z_2}{z_1}\right|$ (iv) $\arg \left(\dfrac{z_2}{z_1}\right)$

10. If $z_1 = 4(\cos 13\pi/24 + i \sin 13\pi/24)$ and
$z_2 = 2(\cos 5\pi/24 + i \sin 5\pi/24)$ find $\dfrac{z_1}{z_2}$ and $z_1 z_2$ in the form $a + ib$.
(Standard results may be used without proof).

11. Use the method of induction to prove de Moivre's theorem [i.e. $(\cos \theta + i \sin \theta)^n = \cos (n\theta) + i \sin (n\theta)$] for a positive integral value of n.

12. If $z = 2(\cos \pi/3 + i \sin \pi/3)$ find z^6.

13. If $z = 2(\cos \pi/4 + i \sin \pi/4)$ find z^6.

14. By writing 1 as $(\cos 2n\pi + i \sin 2n\pi)$ use de Moivre's theorem to find the cube roots of unity.

15. Use de Moivre's theorem to show that $\sin 3\theta \equiv 3 \sin \theta - 4 \sin^3 \theta$
   and $\cos 3\theta \equiv 4 \cos^3 \theta - 3 \cos \theta$.
   Obtain an expression for $\tan 3\theta$ in terms of $\tan \theta$.

16. Use de Moivre's theorem to show that $\sin 5\theta \equiv 16 \sin^5 \theta - 20 \sin^3 \theta + 5 \sin \theta$.

17. Use de Moivre's theorem to obtain an expression for $\tan 6\theta$ in terms of $\tan \theta$.

18. Given that $(\cos \theta + i \sin \theta)^n = \cos (n\theta) + i \sin (n\theta)$ for a positive integral value of n show that it is also true for n a negative integer. [Hint: For n negative let $n = -m$, m positive).

Further geometrical considerations

For the complex number $z = x + iy$, there is an associated point P on the Argand diagram with coordinates (x, y).
The vector $\overrightarrow{OP}$ then represents the complex number z.
Since $z = x + iy$, then z^* will be the reflection of z in the real axis.

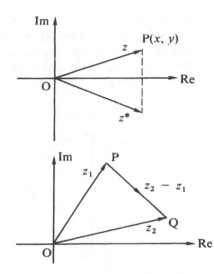

Consider the complex numbers $z_1 = x_1 + iy_1$ and $z_2 = x_2 + iy_2$ represented by the vectors $\overrightarrow{OP}$ and $\overrightarrow{OQ}$ on the Argand diagram. $|z_2 - z_1|$ is then the length of the line PQ.

On page 466 we saw that if $z_1 = r_1(\cos \theta_1 + i \sin \theta_1)$ and $z_2 = r_2(\cos \theta_2 + i \sin \theta_2)$ then
$z_1 z_2 = r_1 r_2 [\cos (\theta_1 + \theta_2) + i \sin (\theta_1 + \theta_2)]$.

Thus the effect of multiplying some complex number z_1 by another complex number z_2, modulus r_2 and argument θ_2, is to rotate z_1 anticlockwise through an angle θ_2 and multiply its length by r_2.

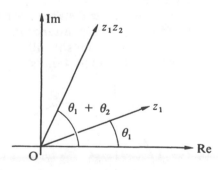

In particular, multiplying a complex number by i
($|i| = 1$, arg $i = \pi/2$) has the effect of rotating the original complex number through 90° anticlockwise.

Example 22

The diagram shows the square OABC on an Argand diagram.
If $\overrightarrow{OA} = 5 + 2i$, express the following vectors in the form $a + ib$,
(a) $\overrightarrow{CB}$, (b) $\overrightarrow{BC}$, (c) $\overrightarrow{OC}$.

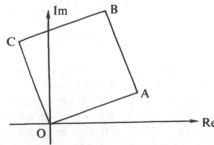

$\overrightarrow{OA} = 5 + 2i$
(a) $\overrightarrow{CB}$ is parallel to, and the same length as, $\overrightarrow{OA}$
   $\overrightarrow{CB} = \overrightarrow{OA}$
   $= 5 + 2i$.

(b) $\overrightarrow{BC} = -\overrightarrow{CB}$
   $= -5 - 2i$.

(c) $\overrightarrow{OC} = i(\overrightarrow{OA})$
   $= 5i + 2i^2$
   $= -2 + 5i$.

Locus on the Argand diagram

Consider a complex number $z = x + iy$, and the associated point P (x, y) on the Argand diagram. If we are given some condition which z must obey, then the corresponding set of all possible points P forms the locus of P.

Example 23

If the point P in the complex plane corresponds to the complex number z, find the locus of P in each of the following situations: (a) $|z| = 3$,   (b) $|z - 2| = 4$.

(a) **By geometry**
All points P such that $|z| = 3$ will be situated 3 units from the origin, i.e. the locus of P is a circle centre the origin, radius 3 units.

By algebra
Suppose $z = x + iy$, thus if $|z| = 3$, then $x^2 + y^2 = 3^2$ i.e. a circle centre the origin, radius 3 units.

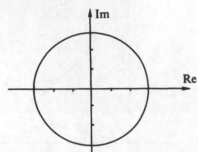

(b) **By geometry**
$|z - 2|$ gives the distance from P to the point $(2, 0)$. If this distance is 4 units, the set of all such points P will form a circle centre $(2, 0)$, radius 4 units.

By algebra
Suppose $z = x + iy$.
If $|z - 2| = 4$ then $|x + iy - 2| = 4$,
i.e. $|x - 2 + iy| = 4$
or $(x - 2)^2 + y^2 = 4^2$
i.e. a circle centre $(2, 0)$, radius 4 units.

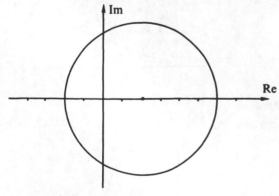

In the general case, if P (x, y) is the point corresponding to the complex number $z = x + iy$, the condition $|z - z_0| = c$, where $z_0 = a + ib$, gives the locus of P as a circle centre (a, b) radius c.
e.g. $|z - 2i| = 5$ gives a circle centre $(0, 2)$, radius 5,
    $|z - (3 + 2i)| = 4$ gives a circle centre $(3, 2)$, radius 4,
      $|z - 3 - i| = 1$ gives a circle centre $(3, 1)$, radius 1.

Example 24

If the point P in the complex plane corresponds to the complex number z, find the locus of P when

(a) $|z - 4| = |z|$,   (b) $|z + 2i| = |z - 1|$,   (c) $\arg z = \frac{\pi}{3}$,   (d) $\arg (z + 1) = \frac{3\pi}{4}$.

(a) **By geometry**

$|z - 4|$ gives the distance from P to the point (4, 0).

$|z|$ gives the distance from P to the origin. Thus the locus of $|z - 4| = |z|$ is the set of all points which are equidistant from (4, 0) and (0, 0), i.e. the perpendicular bisector of the line joining these points.

By algebra

Suppose $\qquad z = x + iy.$

If $\qquad |z - 4| = |z|$, then
$$(x - 4)^2 + y^2 = x^2 + y^2$$

i.e. $\qquad x^2 - 8x + 16 + y^2 = x^2 + y^2,$

giving the line $x = 2$ as the locus of P.

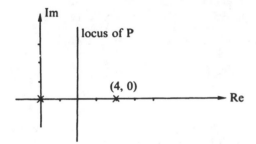

(b) **By geometry**

$|z + 2i|$, [i.e. $|z - (-2i)|$], gives the distance from P to the point (0, −2).

$|z - 1|$ gives the distance from P to the point (1, 0). Thus the locus of $|z + 2i| = |z - 1|$ is the set of all points that are equidistant from (0, −2) and (1, 0), i.e. the perpendicular bisector of the line joining these points.

By algebra

Suppose $\qquad z = x + iy.$

If $\qquad |z + 2i| = |z - 1|$
$$|x + iy + 2i| = |x + iy - 1|$$

i.e. $\qquad x^2 + (y + 2)^2 = (x - 1)^2 + y^2$

$$x^2 + y^2 + 4y + 4 = x^2 - 2x + 1 + y^2$$

giving the line $4y = -2x - 3$ as the locus of P.

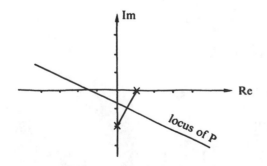

(c) **By geometry**

If arg $z = \dfrac{\pi}{3}$, then OP makes an angle of 60° with the Real axis.

Thus the locus is the set of all points such that the line joining them to the origin is at 60° to the x-axis.

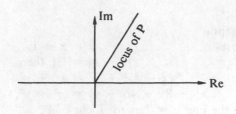

By algebra

Suppose $z = x + iy$.

As   arg $z = \dfrac{\pi}{3}$ then

$$\frac{y}{x} = \tan\frac{\pi}{3} = \sqrt{3},$$

giving the line $y = x\sqrt{3}$ as the locus of P.

In fact, it is that part of $y = x\sqrt{3}$ for which $y \geqslant 0$, as for negative y

points on the line have argument $-\dfrac{2\pi}{3}$.

(d) **By geometry**

Suppose $Z = z + 1$, then arg $Z = \dfrac{3\pi}{4}$.

This is a line making an angle of $\dfrac{3\pi}{4}$ with the Real axis, through the point where $Z = 0$,
i.e. where $z + 1 = 0$ or $z = -1$.
The locus is the set of all points on the line through the point $(-1, 0)$ making an angle of $\dfrac{3\pi}{4}$ with the Real axis.

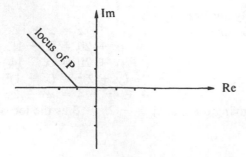

By algebra

Suppose $z = x + iy$.

As   arg $(z + 1) = \dfrac{3\pi}{4}$ then

$$\text{arg } (x + 1 + iy) = \frac{3\pi}{4},$$

i.e. $\dfrac{y}{x + 1} = \tan\dfrac{3\pi}{4} = -1,$

giving the line $y = -x - 1$ as the locus of P. As with (c) we must restrict the line to that part for which $y \geqslant 0$.

Exercise 18F

For questions **1** to **11**, find the locus of the point P corresponding to the complex number z where z obeys the given law.

For **1** to **6**, give *both* a geometrical interpretation and the cartesian equation for each locus.

1. $|z - 2i| = 4$
2. $|z - 1 - 3i| = 5$
3. $|z + 2 - 3i| = 4$
4. $|z| = |z - 2i|$
5. $|z - i| = |z - 1|$
6. $|z - 4 + i| = |z - 1 - 2i|$

For questions **7** and **8**, give the cartesian equation for each locus.
7. $z = z^*$ **8.** $z + z^* = 0$

For questions **9, 10** and **11**, show each locus on a separate Argand diagram.

9. $\arg z = \dfrac{\pi}{4}$ **10.** $\arg (z + 2) = \dfrac{\pi}{4}$ **11.** $\arg (z - i) = \dfrac{3\pi}{4}$

12. If the point P in the complex plane corresponds to the complex number $z = x + iy$ show that if $|z - 1| = 2|z + 2 - 3i|$ then the locus of P is a circle centre at $-3 + 4i$, and find the radius of the circle.

13. Find by drawing or calculation the vector $\overrightarrow{OX}$ where X is the point in the complex plane where the locus given by $|z - 3 - i| = |z - 1 - 3i|$ intersects with the locus given by $|z + 3i| = |z + 2 - i|$.

14. Find the cartesian equation for
(a) the locus given by $|z| = \sqrt{17}$,
(b) the locus given by $|z| = |z - 2|$.
Hence find the coordinates of the points in the complex plane where these two loci intersect. Show the two loci and their points of intersection on a sketch.

15. If z is the general complex number on an Argand diagram make a sketch of the Argand diagram and shade the region in which
$|z + 1 - 4i| \geqslant |z - 2 - i|$.

16. If z is the general complex number on an Argand diagram sketch the Argand diagram and shade the region in which $2 \leqslant |z - i| \leqslant 3$.

17. Show by shading on an Argand diagram the region in which both $|z| \leqslant 4$ and $|z - 3 - i| \geqslant |z - 3 - 5i|$.

Exercise 18G   Examination questions

1. Given that
$$\frac{p}{2x + 3} + \frac{q}{3x + 2} \equiv \frac{1}{(2x + 3)(3x + 2)},$$
find the values of the constants p and q.   (London)

2. Express $f(x) = \dfrac{x}{(1 - 2x)^2(1 - 3x)}$ in partial fractions.
Hence, or otherwise, find the expansion of $f(x)$ in ascending powers of x up to and including the term in x^3. For what range of values of x is the expansion valid?   (Oxford)

3. Show that $\dfrac{3}{(1 + x - 2x^2)}$ can be expressed as
$$\frac{1}{(1 - x)} + \frac{2}{(1 + 2x)}$$
Hence or otherwise expand $(1 + x - 2x^2)^{-1}$ as far as the term in x^4, the values of x being such as to make the expansion valid.
(a) By putting $x = -0.02$, use your expansion to evaluate, correct to seven decimal places, $(0.9792)^{-1}$,
(b) by giving x a suitable value, use your expansion to evaluate $(1.0098)^{-1}$, to seven decimal places.   (A.E.B.)

4. Express the function $f(x) = \dfrac{2x - 1}{(1 - x)(3 + x)}$ as the sum of partial

fractions.

Obtain the expansion of $f(x)$ in ascending powers of x, in the form
$f(x) = a + bx + cx^2 + \ldots$, and find the values of a, b and c. Show
that, when $x = 0 \cdot 1$, the error caused by using only these three terms in
evaluating $f(x)$ is slightly greater than $0 \cdot 1\%$.   (Oxford)

5. Express the function

$$f(x) = \frac{4 - 3x}{(1 - 2x)(2 + x)}$$

in partial fractions.

Find the first three terms in the expansion of $f(x)$ in ascending powers of
x, and show that the coefficient of x^n is

$$\frac{4^n + (-1)^n}{2^n}.$$ (J.M.B.)

6. Express

$$\frac{x^2 + 5x + 2}{(x + 1)^2(x - 1)}$$

in the form

$$\frac{A}{(x + 1)} + \frac{B}{(x + 1)^2} + \frac{C}{(x - 1)},$$

where A, B, C are constants to be determined.
Obtain the expansion of

$$\frac{x^2 + 5x + 2}{(x + 1)^2(x - 1)}$$

in ascending powers of $\left(\dfrac{1}{x}\right)$ up to and including the term in $\left(\dfrac{1}{x}\right)^5$

Prove that if n is even the coefficient of $\left(\dfrac{1}{x}\right)^n$ is $n + 2$. Find the

coefficient of $\left(\dfrac{1}{x}\right)^n$ if n is odd.   (Cambridge)

7. Express $\dfrac{1}{r^2 - 1}$ in partial fractions and hence evaluate

$$\sum_{r=2}^{n} \frac{1}{r^2 - 1}.$$ (Cambridge)

8. Express

$$\frac{1}{(2r - 1)(2r + 3)}$$

in partial fractions.

By multiplying by $\dfrac{1}{2r + 1}$, or otherwise, prove that

$$\sum_{r=1}^{n} \frac{1}{(2r - 1)(2r + 1)(2r + 3)} = \frac{1}{12} - \frac{1}{4(2n + 1)(2n + 3)}.$$ (J.M.B.)

9. Show that

$$\frac{1}{r!} - \frac{1}{(r + 1)!} = \frac{r}{(r + 1)!}$$

and find the corresponding expression for
$$\frac{1}{r!} - \frac{1}{(r + 1)!}.$$

S_n and T_n are the sums of the first n terms of the series whose rth terms are
$$\frac{r}{(r + 1)!} \quad \text{and} \quad \frac{(-1)^r(r + 2)}{(r + 1)!},$$

respectively. Find S_{2n} and T_{2n}, and show that
$$S_{2n} = -T_{2n}. \qquad \text{(J.M.B.)}$$

10. Given that $z_1 = 3 + 2i$ and $z_2 = 4 - 3i$,

    (i) find $z_1 z_2$ and $\frac{z_1}{z_2}$, each in the form $a + ib$;

    (ii) verify that $|z_1 z_2| = |z_1||z_2|$. (Cambridge)

11. (a) Find $(2 - i)^3$, expressing your answer in the form $a + ib$.

    (b) Verify that $2 + 3i$ is one of the square roots of $-5 + 12i$.
       Write down the other square root. (Cambridge)

12. One root of the equation
$$z^2 + az + b = 0,$$
    where a and b are real constants, is $2 + 3i$. Find the values of a and b.
    (London)

13. Given that $2 + i$ is a root of the equation
$$z^3 - 11z + 20 = 0,$$
    find the remaining roots. (London)

14. Show that $1 + i$ is a root of the equation $x^4 + 3x^2 - 6x + 10 = 0$.
    Hence write down one quadratic factor of $x^4 + 3x^2 - 6x + 10$, and
    find all the roots of the equation. (Oxford)

15. The complex number z satisfies $\frac{z}{z + 2} = 2 - i$. Find the real and
    imaginary parts of z, and the modulus and argument of z. (Oxford)

16. Expand $z = (1 + ic)^6$ in powers of c and find the five real finite values
    of c for which z is real. (J.M.B.)

17. (a) If z_1 and z_2 are complex numbers, solve the simultaneous equations
       $4z_1 + 3z_2 = 23$, $z_1 + iz_2 = 6 + 8i$, giving both answers in the
       form $x + yi$.
    (b) If $(a + bi)^2 = -5 + 12i$, find a and b given that they are both real.
       Give the two square roots of $-5 + 12i$.
    (c) In each of the following cases define the locus of the point which
       represents z in the Argand diagram. Illustrate each statement by a
       sketch.
       (i) $|z - 2| = 3$, (ii) $|z - 2| = |z - 3|$. (S.U.J.B.)

18. In the Argand diagram, the point P represents the complex number z.
    Given that
$$|z - 1 - i| = \sqrt{2},$$
    sketch the locus of P.
    Deduce the greatest and least values of $|z|$ for points P lying on the
    locus. (Cambridge)

19. Prove that the non-real cube roots of unity are

$$-\frac{1}{2} \pm i\frac{\sqrt{3}}{2}.$$

These roots are represented in an Argand diagram by the points A, B and the number $z = -2$ is represented by the point C. Show that the area of the sector of the circle with centre C through A and B which is bounded by CA, CB and the minor arc AB is $\frac{1}{2}\pi$.   (J.M.B.)

20. By using de Moivre's theorem, or otherwise, find the roots of the equation $z^4 + 4 = 0$.
Hence, or otherwise, express $z^4 + 4$ as the product of two quadratic polynomials in z with real coefficients   (Oxford)

21. State de Moivre's theorem, and hence express $\cos 5\theta$ as a polynomial in $\cos \theta$. By considering the roots of the equation $\cos 5\theta = 0$, prove that

$$\cos 5\theta = 16 \cos \theta \left(\cos \theta - \cos \frac{\pi}{10}\right)\left(\cos \theta - \cos \frac{3\pi}{10}\right)\left(\cos \theta - \cos \frac{7\pi}{10}\right)\left(\cos \theta - \cos \frac{9\pi}{10}\right).$$
                     (Oxford)

22. Prove that, when n is a positive integer,
$$(\cos \theta + i \sin \theta)^n = \cos n\theta + i \sin n\theta.$$
Given that
$$z = \cos \theta + i \sin \theta,$$
and assuming that the result above is also true for negative integers, show that
$$z^n - z^{-n} = 2i \sin n\theta.$$
Hence, or otherwise, prove that
$$16 \sin^5 \theta = \sin 5\theta - 5 \sin 3\theta + 10 \sin \theta.$$
Find all the solutions of the equation
$$4 \sin^5 \theta + \sin 5\theta = 0$$
which lie in the interval $0 \leqslant \theta \leqslant 2\pi$.   (J.M.B.)

19
Exponential and logarithmic functions

An exponential function is one in which the variable appears as an index.
For example, $f(x) = 2^x$ or more generally, $f(x) = a^x$.

19.1 Differentiating the exponential function

If we wish to find $\dfrac{d}{dx}(a^x)$ we must use the basic definition of chapter 10, i.e.:

$$\frac{d}{dx}f(x) = \lim_{\delta x \to 0}\left[\frac{f(x + \delta x) - f(x)}{\delta x}\right]$$

Thus
$$\frac{d}{dx}(a^x) = \lim_{\delta x \to 0}\left[\frac{a^{x+\delta x} - a^x}{\delta x}\right]$$

$$= \lim_{\delta x \to 0}\left[\frac{a^x(a^{\delta x} - 1)}{\delta x}\right]$$

$$= a^x \lim_{\delta x \to 0}\left(\frac{a^{\delta x} - 1}{\delta x}\right)$$

The table below shows the values of $\left(\dfrac{a^{\delta x} - 1}{\delta x}\right)$, correct to four decimal

places, for $a = 1, 2, 3$ and 4 as δx decreases.

	$\delta x = 1$	$\delta x = 0\cdot 1$	$\delta x = 0\cdot 01$	$\delta x = 0\cdot 001$	$\delta x = 0\cdot 0001$	$\delta x = 0\cdot 00001$	$\delta x = 0\cdot 000001$
$a = 1$	0	0	0	0	0	0	0
$a = 2$	1	0·7177	0·6956	0·6934	0·6932	0·6931	0·6931
$a = 3$	2	1·1612	1·1047	1·0992	1·0987	1·0986	1·0986
$a = 4$	3	1·4870	1·3959	1·3873	1·3864	1·3863	1·3862

This table suggests that there exists a value of a between 2 and 3 for which

$\lim_{\delta x \to 0}\left(\dfrac{a^{\delta x} - 1}{\delta x}\right) = 1$, i.e. for this value of a, $\dfrac{d}{dx}(a^x) = a^x$. If we call this

number e, then
$$\boxed{\frac{d}{dx}(e^x) = e^x}$$

Thus the significance of the number e (lying between 2 and 3) is that the
function e^x differentiates to give itself. It is possible to show that e is an
irrational number and, correct to 5 decimal places, its value is 2·71828. The
function $f(x) = e^x$ is known as *the* exponential function.

Example 1

Make a sketch of the curve given by $y = e^x$.

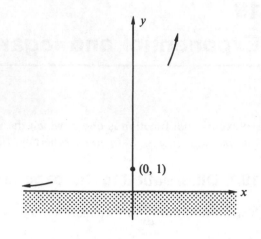

x-axis   No value of x for which $y = 0$
                  $\therefore$ does not cut the x-axis.

y-axis $x = 0 \Rightarrow y = 1$
                  $\therefore$ cuts y-axis at $(0, 1)$

$x \to \pm\infty$ As $x \to +\infty$, $y \to +\infty$ very quickly,
                  as $x \to -\infty$, $y \to 0^+$

y undefined   No value of x for which y is
                  undefined but note that, for
                  $x \in \mathbb{R}$, y cannot be negative.

max/min. $y' = e^x \Rightarrow$ no turning points;
                  gradient is always positive.

Thus the sketch can be completed:

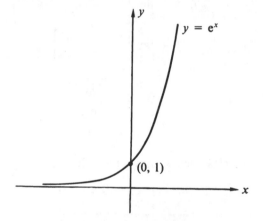

Example 2

Differentiate $y = e^{5x^2}$ with respect to x.

$$y = e^{5x^2}$$

Let $u = 5x^2$ then $y = e^u$

        $\dfrac{du}{dx} = 10x$ $\dfrac{dy}{du} = e^u$

Using $\dfrac{dy}{dx} = \dfrac{dy}{du} \times \dfrac{du}{dx}$, $\dfrac{dy}{dx} = e^u \times 10x$

                                $= 10x e^{5x^2}$

Thus, if we differentiate e^{5x^2} we obtain $10x e^{5x^2}$.

If we consider the more general case:

$$y = e^{f(x)}$$

and let $u = f(x)$ then $y = e^u$

i.e. $\dfrac{du}{dx} = f'(x)$ and $\dfrac{dy}{du} = e^u$

thus $\dfrac{dy}{dx} = e^u f'(x)$

or $\dfrac{dy}{dx} = f'(x) \times e^{f(x)}$

We can then use this result when differentiating exponential functions:

$$\text{If } \quad y = e^{f(x)}, \frac{dy}{dx} = f'(x) \times e^{f(x)}$$

Example 3

Differentiate with respect to x:

(a) $y = e^{4x}$ (b) $y = 5e^{(x^2+1)}$ (c) $y = 5x^2 + 3/e^{x^2}$ (d) $y = e^x \sin 2x$

Using the above rule:

(a) $y = e^{4x}$

$$\frac{dy}{dx} = 4e^{4x}$$

(b) $y = 5e^{(x^2+1)}$

$$\frac{dy}{dx} = 2x \times 5e^{(x^2+1)}$$

$$= 10xe^{(x^2+1)}$$

(c) $y = 5x^2 + 3/e^{x^2}$

$$= 5x^2 + 3e^{-x^2}$$

$$\frac{dy}{dx} = 10x + (-2x)3e^{-x^2}$$

$$= 10x - \frac{6x}{e^{x^2}}$$

(d) $y = e^x \sin 2x$

Using the product rule,

$$\frac{dy}{dx} = e^x(2\cos 2x) + e^x \sin 2x$$

$$= e^x(2\cos 2x + \sin 2x)$$

Integration of exponential functions

From the last section we know that if $y = e^{f(x)}$, then $\frac{dy}{dx} = f'(x)e^{f(x)}$.

Noticing that $e^{f(x)}$ features in both the expression for y and that for $\frac{dy}{dx}$, we must expect integration of $g(x)e^{f(x)}$ to feature $e^{f(x)}$. Thus when integrating $g(x)e^{f(x)}$ we first consider $\frac{d}{dx}e^{f(x)}$.

Example 4

Find the following indefinite integrals:

(a) $\displaystyle\int 10e^{5x}\,dx$ (b) $\displaystyle\int xe^{(x^2+1)}\,dx$ (c) $\displaystyle\int (\sin x + 2/e^x)\,dx$

(a) $\displaystyle\int 10e^{5x}\,dx$

Now $\dfrac{d}{dx}(e^{5x}) = 5e^{5x}$

$\therefore \displaystyle\int 10e^{5x}\,dx = 2e^{5x} + c$

(b) $\displaystyle\int xe^{(x^2+1)}\,dx$

Now $\dfrac{d}{dx}(e^{(x^2+1)}) = 2xe^{(x^2+1)}$

$\therefore \displaystyle\int xe^{(x^2+1)}\,dx = \tfrac{1}{2}e^{(x^2+1)} + c$

(c) $\int (\sin x + 2e^{-x})\,dx$

Now $\dfrac{d}{dx}(e^{-x}) = -e^{-x}$

$\therefore \int (\sin x + 2e^{-x})\,dx = -\cos x - 2e^{-x} + c$

Example 5

Evaluate $\displaystyle\int_0^1 (4xe^{x^2} + 1)\,dx$, leaving your answer in terms of e.

Now $\dfrac{d}{dx}(e^{x^2}) = 2xe^{x^2}$, $\therefore \displaystyle\int_0^1 (4xe^{x^2} + 1)\,dx = \left[2e^{x^2} + x \right]_0^1$

$= (2e + 1) - (2e^0 + 0)$

$= 2e - 1$

Thus $\displaystyle\int_0^1 (4xe^{x^2} + 1)\,dx = 2e - 1.$

Exercise 19A

Differentiate the following with respect to x.

1. e^x 2. e^{3x} 3. e^{2x} 4. e^{x^2} 5. $e^{(x^2+1)}$

6. $\dfrac{1}{e^x}$ 7. $4e^{2x}$ 8. $5e^{2x^3}$ 9. $\dfrac{3}{e^{2x}}$ 10. $e^{(5x+3)} + \dfrac{2}{x^2}$

11. $2e^{x^2} + 3e^x + \dfrac{4}{e^x}$ 12. x^2e^x 13. $3x^3e^{2x}$ 14. $\dfrac{3x^2}{e^x}$

15. $\dfrac{e^x + 4}{x^2}$

Find the following indefinite integrals

16. $\int e^x\,dx$ 17. $\int 5e^x\,dx$ 18. $\int 2e^{2x}\,dx$ 19. $\int 6e^{3x}\,dx$

20. $\int 2xe^{x^2}\,dx$ 21. $\int 4xe^{(x^2+1)}\,dx$ 22. $\int \left(-\dfrac{4}{e^{2x}}\right)dx$ 23. $\int \dfrac{e^{(x+2)} + 4}{e^x}\,dx$

Evaluate the following giving your answers in terms of e.

24. $\int_1^3 e^x\,dx$ 25. $\int_0^3 e^{-x}\,dx$ 26. $\int_1^2 2e^{(2x+1)}\,dx$ 27. $\int_{-1}^1 2e^{(1-2x)}\,dx$

28. Find the gradient of the curve $y = xe^{(x-3)}$ at the point $(3, 3)$.
29. Find the equation of the curve that has a gradient function given by $e^x + 2x$ and that passes through the point $(0, -3)$.
30. With the help of the sketch of $y = e^x$ on page 478 make sketch graphs of
    (a) $y = -e^x$ (b) $y = e^{-x}$ (c) $y = e^{x-1}$
31. Find the coordinates and nature of any turning points on the curve $y = xe^x$
32. Find the coordinates and nature of any turning points on the curve $y = x - e^x$
33. Sketch the graph of $y = x^2e^x$ indicating clearly the coordinates of any stationary points on the curve and of any points where the curve cuts the axes.

34. If $y = -e^x \cos 2x$, show that $\dfrac{d^2y}{dx^2} = 5e^x \sin(2x + \alpha)$ where $\alpha = \tan^{-1}(\frac{3}{4})$.

35. If $y = e^x \tan x$, show that $\dfrac{d^2y}{dx^2} = e^x(2 \tan^3 x + 2 \tan^2 x + 3 \tan x + 2)$

19.2 The logarithmic function

In chapter 5, we saw that if $y = \log_e x$, which may also be written as $y = \ln x$, then

$$x = e^y$$

$$\therefore \quad \frac{dx}{dy} = e^y$$

but $\dfrac{dy}{dx} = 1 \left/ \left(\dfrac{dx}{dy}\right)\right.$, (see page 324)   Thus $\dfrac{dy}{dx} = \dfrac{1}{e^y} = \dfrac{1}{x}$.

> Thus, if $y = \ln x$,   then $\dfrac{dy}{dx} = \dfrac{1}{x}$

Alternatively, this result could be obtained from $x = e^y$ by differentiating implicitly.

Example 6

Make a sketch of the curve given by $y = \ln x$.

x-axis	If $y = 0$, then $\ln x = 0$ i.e. $x = 1$. Curve cuts x-axis at $(1, 0)$
y-axis	If $x = 0$, then $y = \ln 0$, i.e. no value of y for $x = 0$ $\therefore$ curve does not cut y-axis.
$x \to \pm\infty$	$x \to +\infty$, y increases slowly $x \to -\infty$ need not be considered as x cannot be negative.
y undefined	y is undefined for negative x and for $x = 0$
max/min	$y = \ln x \Rightarrow \dfrac{dy}{dx} = \dfrac{1}{x}$. $\therefore$ no turning points. Also, as x cannot be negative the gradient is always positive and as $x \to 0^+$, the gradient $\to +\infty$.

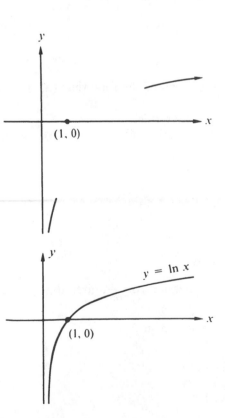

Thus the sketch can be completed.

Example 7

Differentiate $y = \ln (x^2 + 1)$ with respect to x.

$$y = \ln (x^2 + 1)$$

Let $u = x^2 + 1$ then $y = \ln u$

$$\frac{du}{dx} = 2x \qquad\qquad \frac{dy}{du} = \frac{1}{u}$$

Using $\dfrac{dy}{dx} = \dfrac{dy}{du} \times \dfrac{du}{dx}$, $\dfrac{dy}{dx} = \dfrac{1}{u} 2x$

$$\frac{dy}{dx} = \frac{2x}{x^2 + 1}.$$

Note: If we consider the more general case

$$y = \ln [f(x)]$$

and let $u = f(x)$ then $y = \ln u$

$$\frac{du}{dx} = f'(x) \qquad\qquad \frac{dy}{du} = \frac{1}{u}$$

thus $\dfrac{dy}{dx} = \dfrac{1}{u} f'(x)$

or $\dfrac{dy}{dx} = \dfrac{f'(x)}{f(x)}$

Thus we have a rule for differentiating functions of the type $\ln [f(x)]$:

$$\boxed{\text{If} \quad y = \ln [f(x)], \quad \frac{dy}{dx} = \frac{f'(x)}{f(x)}}$$

Example 8

Find $\dfrac{dy}{dx}$ for each of the following: (a) $y = \ln (5x^2 - 6)$ (b) $y = \ln (\sin 2x)$ (c) $y = \ln \left(\dfrac{x + 2}{x + 3}\right)$

Using the above rule:

(a) $y = \ln (5x^2 - 6)$

$$\frac{dy}{dx} = \frac{10x}{5x^2 - 6}$$

(b) $y = \ln (\sin 2x)$

$$\frac{dy}{dx} = \frac{2 \cos 2x}{\sin 2x}$$

$$\frac{dy}{dx} = 2 \cot 2x$$

(c) $y = \ln \left(\dfrac{x + 2}{x + 3}\right)$

$$= \ln (x + 2) - \ln (x + 3)$$

$$\frac{dy}{dx} = \frac{1}{x + 2} - \frac{1}{x + 3}$$

$$= \frac{1}{(x + 2)(x + 3)}$$

Example 9

Find an expression for $\dfrac{dy}{dx}$, given that $y = 2^x$.

If $y = 2^x$

then $\ln y = x \ln 2$ $\ldots$ [1]

giving $\dfrac{dy}{dx} = y \times \ln 2$

$\qquad\quad = 2^x \times \ln 2$

Note that in the last example the logarithm of the expression to be differentiated was taken in order to give an expression that was easier to differentiate. This method, called logarithmic differentiation, can be particularly useful when differentiating some products and quotients (see example 10) or when the variable appears as a power, as in example 9.

Example 10

Find an expression for $\dfrac{dy}{dx}$ in terms of x, given that $y = \dfrac{x^2}{\sqrt{(x-1)}}$

If $\qquad y = \dfrac{x^2}{\sqrt{(x-1)}}$

then $\quad \ln y = 2 \ln x - \tfrac{1}{2} \ln (x-1)$

differentiating with respect to x:

$\dfrac{1}{y}\dfrac{dy}{dx} = \dfrac{2}{x} - \dfrac{1}{2(x-1)}$

$\qquad\quad = \dfrac{3x-4}{2x(x-1)}$

$\therefore \quad \dfrac{dy}{dx} = \dfrac{x^2}{\sqrt{(x-1)}} \times \dfrac{3x-4}{2x(x-1)}$

$\qquad\quad = \dfrac{x(3x-4)}{2(x-1)^{3/2}}$

The reader should confirm that the same result can be obtained by use of the quotient rule.

Indefinite integrals of the form $\displaystyle\int \dfrac{f'(x)}{f(x)}dx$

As $\dfrac{d}{dx}[\ln f(x)] = \dfrac{f'(x)}{f(x)}$, it follows that $\displaystyle\int \dfrac{f'(x)}{f(x)}dx = \ln f(x) + c.$

Thus $\displaystyle\int \dfrac{1}{x}dx = \ln x + c.$

It could be that we were asked to find $\displaystyle\int \dfrac{1}{x}dx$ in situations where x could take negative values. In such cases, an answer of $\ln x + c$ has no meaning as we cannot find the logarithm of a negative number.

For $x < 0$, we proceed as follows:

$$\int \dfrac{1}{x}dx = \int \left(\dfrac{-1}{-x}\right)dx$$

$$\qquad\quad = \ln(-x) + c$$

Thus for $x > 0$, $\int \dfrac{1}{x} dx = \ln x + c$ ... [1]

and for $x < 0$, $\int \dfrac{1}{x} dx = \ln (-x) + c$ ... [2]

Combining [1] and [2], we can write the integral of $\dfrac{1}{x}$ for $\{x \in \mathbb{R}: x \neq 0\}$ as:

$$\int \frac{1}{x} dx = \ln |x| + c$$

Note: We must exclude $x = 0$, because for that value of x, $\dfrac{1}{x}$ is undefined.

Similarly,

$$\int \frac{f'(x)}{f(x)} dx = \ln |f(x)| + c$$

Example 11

Find the following indefinite integrals:

(a) $\int \dfrac{1}{x + 4} dx$ (b) $\int \dfrac{3x}{x^2 + 4} dx$.

(a) $\int \dfrac{1}{x + 4} dx$

Since the numerator is exactly the differential of the denominator, this is of the form $\int \dfrac{f'(x)}{f(x)} dx$.

Thus, $\int \dfrac{1}{x + 4} dx = \ln |x + 4| + c$

The constant c could be written as $\ln |A|$ to give an answer of $\ln |x + 4| + \ln |A|$ which is $\ln |A(x + 4)|$.

(b) $\int \dfrac{3x}{x^2 + 4} dx$

The differential of the denominator is $2x$, hence

$$\int \frac{3x}{x^2 + 4} dx = \frac{3}{2} \ln (x^2 + 4) + c$$

In this case we do not need to write $|x^2 + 4|$ since $x^2 + 4 > 0$ for all values of x.

Example 12

Find $\int \dfrac{x}{x - 4} dx$.

$$\int \frac{x}{x - 4} dx = \int \left(\frac{x - 4 + 4}{x - 4} \right) dx \quad \text{(rearranging to obtain proper fractions)}$$

$$= \int \left(1 + \frac{4}{x - 4} \right) dx$$

$$\therefore \quad \int \frac{x}{x - 4} dx = x + 4 \ln |x - 4| + c$$

Example 13

Find $\displaystyle\int \frac{2x + 7}{x^2 + 5x + 6}\,dx.$

$$\int \frac{2x + 7}{x^2 + 5x + 6}\,dx$$

The numerator of the integrand is not exactly the differential of the denominator, but the integrand can be rearranged by factorising the denominator and expressing the integrand in partial fractions.

$$\int \frac{2x + 7}{x^2 + 5x + 6}\,dx = \int \frac{2x + 7}{(x + 2)(x + 3)}\,dx$$

$$= \int \left(\frac{3}{x + 2} - \frac{1}{x + 3} \right) dx$$

$$= \int \frac{3}{x + 2}\,dx - \int \frac{1}{x + 3}\,dx$$

$$= 3 \ln |x + 2| - \ln |x + 3| + c$$

which may be written as $\ln \left| \dfrac{A(x + 2)^3}{x + 3} \right|$, where $\ln |A| = c$.

Example 14

Find $\displaystyle\int \tan x\,dx.$

Since $\tan x = \dfrac{\sin x}{\cos x}$, $\displaystyle\int \tan x\,dx = \int \frac{\sin x}{\cos x}\,dx$

$$= -\int \frac{-\sin x}{\cos x}\,dx$$

The numerator is now the differential of the denominator,

thus $\displaystyle\int \tan x\,dx = -\ln |\cos x| + c$

which may be written as $\ln |\sec x| + c$ or $\ln |A \sec x|$ where $\ln |A| = c$.

Definite integrals of the form $\displaystyle\int_a^b \frac{f'(x)}{f(x)}\,dx$

We know that $\displaystyle\int \frac{f'(x)}{f(x)}\,dx = \ln |f(x)| + c$. However, with definite integrals

of the form $\displaystyle\int_a^b \frac{f'(x)}{f(x)}\,dx$, we know the applicable range of values of x and, therefore, the corresponding sign of $f(x)$ in this range. Hence, we do not need to use the modulus sign provided that we rearrange $\dfrac{f'(x)}{f(x)}$ in those situations where $f(x)$ is negative in the range $a \leqslant x \leqslant b$. This is illustrated in method 2 of examples 15 and 16.

Example 15

Evaluate $\displaystyle\int_{3}^{5} \frac{3}{x-2}\, dx$ giving your answer correct to 3 decimal places.

METHOD 1

$$\int_{3}^{5} \frac{3}{x-2}\, dx = \left[3 \ln |x-2|\right]_{3}^{5}$$
$$= 3 \ln 3 - 3 \ln 1$$
$$= 3{\cdot}296$$

METHOD 2

$(x - 2)$ is positive for $3 \leqslant x \leqslant 5$,

$$\therefore \int_{3}^{5} \frac{3}{x-2}\, dx = \left[3 \ln (x-2)\right]_{3}^{5}$$
$$= 3 \ln 3 - 3 \ln 1 = 3{\cdot}296$$

Example 16

Evaluate $\displaystyle\int_{3}^{5} \frac{4}{1-x}\, dx$.

METHOD 1

$$\int_{3}^{5} \frac{4}{1-x}\, dx = \left[-4 \ln |1-x|\right]_{3}^{5}$$
$$= (-4 \ln 4) - (-4 \ln 2)$$
$$= 4 \ln 2 - 4 \ln 4$$
$$= 4 \ln \tfrac{1}{2}$$
$$= -2{\cdot}773$$

METHOD 2

$(1 - x)$ is negative for $3 \leqslant x \leqslant 5$,

$$\therefore \int_{3}^{5} \frac{4}{1-x}\, dx = \int_{3}^{5} \frac{-4}{x-1}\, dx$$

Now, $(x - 1)$ is positive for $3 \leqslant x \leqslant 5$,

so $$\int_{3}^{5} \frac{4}{1-x}\, dx = \left[-4 \ln (x-1)\right]_{3}^{5}$$
$$= (-4 \ln 4) - (-4 \ln 2)$$
$$= 4 \ln 2 - 4 \ln 4$$
$$= -2{\cdot}773$$

Note: (i) $\displaystyle\int_{a}^{b} \frac{f'(x)}{f(x)}\, dx$ is meaningless if $f(a)$ and $f(b)$ are of opposite sign because, in such cases, there will be a value c between a and b for which $f(c) = 0$ and so $\dfrac{f'(x)}{f(x)}$ is undefined.

For example: $\displaystyle\int_{-1}^{2} \frac{1}{x}\, dx$ is meaningless because $\dfrac{1}{x}$ is undefined for $x = 0$, which lies between -1 and 2.

$\displaystyle\int_{2}^{4} \frac{2}{2x-5}\, dx$ is meaningless because $\dfrac{2}{2x-5}$ is undefined for $x = 2\tfrac{1}{2}$ which lies between 2 and 4.

(ii) It can also be the case that $f(x) = 0$ for a value of x between a and b, even though $f(a)$ and $f(b)$ are of the same sign.

Consider $\displaystyle\int_{0}^{5} \frac{2x-3}{x^2-3x+2}\, dx$. If $f(x) = x^2 - 3x + 2$, then $f(5)$ and $f(0)$ are both positive.

But $f(x) = 0$ for $x = 1$ and $x = 2$ and both of these values lie in the range 0 to 5.

So this integral is meaningless as the integrand is undefined at $x = 1$ and $x = 2$.

Exercise 19B

Differentiate the following with respect to x.

1. $\ln x$
2. $\ln 5x$
3. $\ln 7x$
4. $-\ln 6x$
5. $3 \ln 2x$
6. $\ln (x^2 + 3)$
7. $\ln (3x - 4)$
8. $\ln (x^2 + 2x - 1)$

9. $\ln (x + 3) - \ln (x - 3)$
10. $\ln \left(\dfrac{5x}{2x - 3} \right)$
11. $\ln \left(\dfrac{6x^2}{3x - 2} \right)$

12. $\ln \left(\dfrac{x + 1}{x - 1} \right)$
13. $\ln (x + 3) + \dfrac{1}{x + 3}$
14. $\ln (x^2 - 1) + \dfrac{1}{x + 4}$

15. $\log_{10} x$
16. $\log_3 x$
17. $2e^x + \ln x$
18. $e^x \ln x$
19. $e^{x^2} \ln x$
20. $(\ln x)/e^{x^2}$

21. For each of the following use logarithmic differentiation to find $\dfrac{dy}{dx}$ in terms of x.

   (a) $y = 2x^x$ (b) $y = 2^{x^2}$ (c) $y = \dfrac{(x - 1)^2}{\sqrt{(1 + x)}}$ (d) $y = \sqrt{\left(\dfrac{1 + 2x}{1 - 2x} \right)}$.

22. If $y = \dfrac{3^x}{x + 1}$ find the value of $\dfrac{dy}{dx}$ when $x = 2$.

23. If $y = (x + 2)^3(1 - \sin 2x)^2(1 + \tan x)^3$ find the value of $\dfrac{dy}{dx}$ when $x = 0$.

24. Find $\dfrac{dy}{dx}$ in terms of x for each of the following cases

   (a) $y = 3^x$ (b) $y = 4^x$

25. Find an expression for $\dfrac{d}{dx}(a^x)$ where a is a constant. Hence find an expression for $\displaystyle\int a^x \, dx$ and evaluate $\displaystyle\int_1^2 10^x \, dx$ giving your answer correct to two decimal places.

Find the following indefinite integrals

26. $\displaystyle\int \frac{1}{x} dx$
27. $\displaystyle\int \frac{3}{x} dx$
28. $\displaystyle\int \frac{1}{2x} dx$

29. $\displaystyle\int \frac{2x}{x^2 + 1} dx$
30. $\displaystyle\int \frac{6x}{x^2 + 3} dx$
31. $\displaystyle\int \frac{6x}{x^2 - 3} dx$

32. $\displaystyle\int \frac{2x + 3}{x^2 + 3x - 1} dx$
33. $\displaystyle\int \frac{x - 2}{x^2 - 4x + 7} dx$
34. $\displaystyle\int \left(e^{2x} + \frac{3}{x} \right) dx$

35. $\displaystyle\int \left(\frac{3}{3x + 4} - \frac{2}{x + 1} \right) dx$
36. $\displaystyle\int \left(\frac{1}{x - 1} - \frac{2}{x + 3} \right) dx$.

37. $\displaystyle\int \tan 2x \, dx$
38. $\displaystyle\int \cot x \, dx$
39. $\displaystyle\int \frac{x}{x + 3} dx$

40. $\displaystyle\int \frac{2x}{x - 1} dx$
41. $\displaystyle\int \frac{2x - 4}{2x + 3} dx$

42. For each of the following explain why no meaning can be assigned to the definite integral.

(a) $\displaystyle\int_{-2}^{3} \frac{1}{x}\,dx$ (b) $\displaystyle\int_{2}^{4} \frac{1}{x-3}\,dx$ (c) $\displaystyle\int_{0}^{1} \frac{1}{2x-1}\,dx$

(d) $\displaystyle\int_{0}^{2} \frac{2(x+1)}{x^2+2x-3}\,dx$

43. Evaluate the following definite integrals, giving your answers correct to 3 decimal places.

(a) $\displaystyle\int_{3}^{7} \frac{2}{x}\,dx$ (b) $\displaystyle\int_{4}^{10} \frac{5}{x-3}\,dx$ (c) $\displaystyle\int_{1}^{3} \frac{2x}{x^2+1}\,dx$

(d) $\displaystyle\int_{2}^{5} \frac{x}{x^2-3}\,dx$ (e) $\displaystyle\int_{1}^{3} \frac{1}{x-5}\,dx$ (f) $\displaystyle\int_{3}^{5} \frac{x}{4-x^2}\,dx$

(g) $\displaystyle\int_{5}^{7} \frac{x}{x+5}\,dx$ (h) $\displaystyle\int_{-5}^{-3} \frac{x}{x-2}\,dx$

44. Using the sketch of $y = \ln x$ on page 481 make sketch graphs of
(a) $y = \ln(-x)$ (b) $y = \ln(1/x)$ (c) $y = \ln(x-3)$

45. Make sketch graphs of the following functions indicating clearly the coordinates of any stationary points on the curve and of any points where the curve cuts the axes.

(a) $y = \ln(2x+3)$ (b) $y = \dfrac{1}{x^2}\ln x$

46. Express $\dfrac{2}{1-x^2}$ in partial fractions and hence find $\displaystyle\int \frac{2}{1-x^2}\,dx$.

47. Express $\dfrac{2(x+1)}{(1+x^2)(1-x)}$ in partial fractions and hence find

$$\int \frac{2(x+1)}{(1+x^2)(1-x)}\,dx.$$

48. Express $\dfrac{3}{2x^2+5x+2}$ in partial fractions and hence evaluate

$$\int_{2}^{4} \frac{3}{2x^2+5x+2}\,dx,$$ giving your answer correct to 3 decimal places.

49. Find the equation of the curve that has a gradient function given by $\dfrac{2x}{x^2+1}$ and that passes through the point $(-1, \ln 6)$.

50. If $y = e^x \ln x$, show that $x\dfrac{d^2y}{dx^2} = (2x-1)\dfrac{dy}{dx} + (1-x)y$.

51. Using the identity $\tan^2 x = \sec^2 x - 1$ find the following indefinite integrals giving your answers in terms of $\sec x$

(a) $\displaystyle\int \tan^3 x\,dx$ (b) $\displaystyle\int \tan^5 x\,dx$

19.3 The exponential series

If we assume that the function e^x can be expressed as a series in ascending powers of x,

such that $\quad e^x = a_0 + a_1x + a_2x^2 + a_3x^3 + \ldots + a_nx^n + \ldots$

then to find the series we need to find the coefficients $a_0, a_1, a_2, \ldots$ such that

$\dfrac{d}{dx}(e^x) = e^x.$

First, substituting $x = 0$ in both sides

$$e^0 = a_0 \qquad\qquad \text{i.e.} \quad a_0 = 1$$

Differentiating the terms of the series, with respect to x:

$$\frac{d}{dx}(e^x) = e^x = a_1 + 2a_2x + 3a_3x^2 + 4a_4x^3 + \ldots + na_nx^{n-1} + \ldots$$

substituting $x = 0$ in both sides,

$$e^0 = a_1 + 0 \qquad \text{i.e.} \quad a_1 = 1$$

Differentiating again,

$$\frac{d}{dx}(e^x) = e^x = \qquad 2a_2 + 3 \times 2a_3x + 4 \times 3a_4x^2 + \ldots + n(n-1)a_nx^{n-2} + \ldots$$

substituting $x = 0$ both sides,

$$e^0 = \qquad 2a_2 + 0 \qquad\qquad \text{i.e.} \quad a_2 = \tfrac{1}{2}$$

Continuing this process we obtain $\quad a_3 = \dfrac{1}{3 \times 2}, a_4 = \dfrac{1}{4 \times 3 \times 2}$, etc.

$$\text{or} \quad a_3 = \frac{1}{3!}, \qquad a_4 = \frac{1}{4!} \quad \text{etc.}$$

$$\boxed{\text{Thus} \quad e^x = 1 + \frac{x}{1!} + \frac{x^2}{2!} + \frac{x^3}{3!} + \frac{x^4}{4!} + \ldots + \frac{x^n}{n!} + \ldots}$$

Note that we have made two assumptions, (i) that e^x can be expressed as such a series, (ii) that the sum of the series obtained by differentiating each term of a series is the differential of the sum of that series. The proof that these assumptions are true is beyond the scope of this book.

Example 17

Find a power series expansion for (a) e^{-x}, (b) $e^{x/3}$.

(a) Since $\quad e^x = 1 + \dfrac{x}{1!} + \dfrac{x^2}{2!} + \dfrac{x^3}{3!} + \dfrac{x^4}{4!} + \ldots + \dfrac{x^n}{n!} + \ldots$

substituting $-x$ for x gives:

$$e^{-x} = 1 - x + \frac{(-x)^2}{2!} + \frac{(-x)^2}{3!} + \frac{(-x)^4}{4!} + \ldots + \frac{(-x)^n}{n!} + \ldots$$

$$\therefore \quad e^{-x} = 1 - x + \frac{x^2}{2!} - \frac{x^3}{3!} + \frac{x^4}{4!} + \ldots + (-1)^n\frac{x^n}{n!} + \ldots$$

(b) Substituting $x/3$ for x in the expansion for e^x gives:

$$e^{x/3} = 1 + \frac{(x/3)}{1!} + \frac{(x/3)^2}{2!} + \frac{(x/3)^3}{3!} + \frac{(x/3)^4}{4!} + \ldots + \frac{(x/3)^n}{n!} + \ldots$$

$$\therefore \quad e^{x/3} = 1 + \frac{x}{3} + \frac{x^2}{3^2 \times 2!} + \frac{x^3}{3^3 \times 3!} + \frac{x^4}{3^4 \times 4!} + \ldots + \frac{x^n}{3^n \times n!} + \ldots$$

19.4 The logarithmic series

From the binomial expansion, we know that for $|x| < 1$

$$\frac{1}{1 + x} = (1 + x)^{-1} = 1 - x + x^2 - x^3 + \ldots$$

Integrating both sides with respect to x gives

$$\ln(1 + x) + c = \frac{x}{1} - \frac{x^2}{2} + \frac{x^3}{3} - \frac{x^4}{4} + \ldots$$

Substituting $x = 0$ in both sides gives $c = 0$

Thus for $|x| < 1$

$$\ln(1 + x) = x - \frac{x^2}{2} + \frac{x^3}{3} - \frac{x^4}{4} + \ldots + (-1)^{n+1}\frac{x^n}{n} + \ldots$$

Notes: 1. This expansion depends on the assumption that the sum of the series obtained by integrating each term of a series is the integral of the sum of that series. Again the proof of this assumption is beyond the scope of this book.

2. It can be shown that the expansion is valid for $x = 1$ as well as for $|x| < 1$.

$$\boxed{\text{Thus:} \quad \ln(1 + x) = x - \frac{x^2}{2} + \frac{x^3}{3} - \frac{x^4}{4} + \ldots + (-1)^{n+1}\frac{x^n}{n} + \ldots \quad \text{for} \quad -1 < x \leqslant 1}$$

Example 18

Find an expansion in x for $\ln(1 - x)$ and state the range of values of x for which it is valid.

$$\text{Since} \quad \ln(1 + x) = x \quad - \frac{x^2}{2} \quad + \frac{x^3}{3} \quad - \frac{x^4}{4} \quad + \ldots + (-1)^{n+1}\frac{x^n}{n} + \ldots$$

$$\text{for} \quad -1 < x \leqslant 1$$

substituting $(-x)$ for x gives

$$\ln(1 - x) = -x \quad - \frac{(-x)^2}{2} + \frac{(-x)^3}{3} - \frac{(-x)^4}{4} + \ldots + (-1)^{n+1}\frac{(-x)^n}{n} + \ldots$$

$$\text{for} \quad -1 < -x \leqslant 1$$

i.e. $\quad \ln(1 - x) = -x \quad - \frac{x^2}{2} \quad - \frac{x^3}{3} \quad - \frac{x^4}{4} \quad - \ldots - \frac{x^n}{n} - \ldots$

$$\text{for} \quad -1 \leqslant x < 1.$$

Example 19

Find an expansion, in ascending powers of x, for $\ln(2 - 5x)$ stating the terms up to and including the term in x^4 and the range of values of x for which the expansion is valid.

We know that $\quad \ln(1 + x) = x - \frac{x^2}{2} + \frac{x^3}{3} - \frac{x^4}{4} + \ldots + (-1)^{n+1}\frac{x^n}{n} + \ldots \quad \text{for} \quad -1 < x \leqslant 1$

Now $\quad \ln(2 - 5x) = \ln[2(1 - \frac{5}{2}x)]$

$$= \ln 2 + \ln(1 - \frac{5}{2}x)$$

thus $\quad \ln(2 - 5x) = \ln 2 + [-\frac{5}{2}x - \frac{1}{2}(-\frac{5}{2}x)^2 + \frac{1}{3}(-\frac{5}{2}x)^3 - \frac{1}{4}(-\frac{5}{2}x)^4 \ldots]$

$$= \ln 2 - \frac{5}{2}x - \frac{25}{8}x^2 - \frac{125}{24}x^3 - \frac{625}{64}x^4 \ldots \quad \text{for} \quad -1 < -\frac{5}{2}x \leqslant 1$$

$$\text{or} \quad -\frac{2}{5} \leqslant x < \frac{2}{5}$$

Thus $\quad \ln(2 - 5x) = \ln 2 - \frac{5}{2}x - \frac{25}{8}x^2 - \frac{125}{24}x^3 - \frac{625}{64}x^4 \ldots \quad \text{for} \quad -\frac{2}{5} \leqslant x < \frac{2}{5}$

Example 20

Find an expansion in ascending powers of x for $\ln(1 - x - 6x^2)$, giving the terms up to and including that in x^4, and state the values of x for which the expansion is valid.

$$\ln(1 - x - 6x^2) = \ln[(1 - 3x)(1 + 2x)]$$
$$= \ln(1 - 3x) + \ln(1 + 2x)$$
$$= \left(-3x - \frac{(-3x)^2}{2} + \frac{(-3x)^3}{3} - \frac{(-3x)^4}{4} \cdots\right)$$
$$+ \left(2x - \frac{(2x)^2}{2} + \frac{(2x)^3}{3} - \frac{(2x)^4}{4} + \cdots\right)$$

Thus $\ln(1 - x - 6x^2) = -x - \frac{13}{2}x^2 - \frac{19}{3}x^3 - \frac{97}{4}x^4 \cdots$
The expansion for $\ln(1 - 3x)$ is valid for $-1 < -3x \leqslant 1$, i.e. $-\frac{1}{3} \leqslant x < \frac{1}{3}$.
The expansion for $\ln(1 + 2x)$ is valid for $-1 < 2x \leqslant 1$, i.e. $-\frac{1}{2} < x \leqslant \frac{1}{2}$.
As both expansions have been used, the expansion is valid for $-\frac{1}{3} \leqslant x < \frac{1}{3}$.
Thus $\ln(1 - x - 6x^2) = -x - \frac{13}{2}x^2 - \frac{19}{3}x^3 - \frac{97}{4}x^4 \cdots$ for $-\frac{1}{3} \leqslant x < \frac{1}{3}$

Example 21

Find an expansion in ascending powers of x for $\ln\left(\frac{1 + x}{1 - x}\right)$ and state the values of x for which the expansion is valid.

$$\ln\left(\frac{1 + x}{1 - x}\right) = \ln(1 + x) - \ln(1 - x)$$

Hence $\ln\left(\frac{1 + x}{1 - x}\right) = \left(x - \frac{x^2}{2} + \frac{x^3}{3} - \frac{x^4}{4} + \frac{x^5}{5} \cdots\right) - \left(-x - \frac{x^2}{2} - \frac{x^3}{3} - \frac{x^4}{4} - \frac{x^5}{5} - \cdots\right)$

$$= 2\left(x + \frac{x^3}{3} + \frac{x^5}{5} + \cdots\right)$$

The two expansions used are valid for $-1 < x \leqslant 1$ and $-1 \leqslant x < 1$ respectively.
The combined expansion is then valid for $-1 < x < 1$.

Thus $\ln\left(\frac{1 + x}{1 - x}\right) = 2\left(x + \frac{x^3}{3} + \frac{x^5}{5} + \cdots\right)$ for $|x| < 1$.

Note: This expansion can be used to evaluate logarithms and since it converges so much more quickly than $\ln(1 + x)$, it is particularly useful for this purpose.
Using $\ln(1 + x)$, and substituting $x = 1$,
  $\ln(1 + 1) = \ln 2 = 1 - \frac{1}{2} + \frac{1}{3} - \frac{1}{4} + \frac{1}{5} - \cdots$ i.e. the fifth term is 0.2.

Using $\ln\left(\frac{1 + x}{1 - x}\right)$ and substituting $x = \frac{1}{3}$,

$$\ln\left(\frac{1 + \frac{1}{3}}{1 - \frac{1}{3}}\right) = \ln 2 = 2\left(\frac{1}{3} + \frac{(\frac{1}{3})^3}{3} + \frac{(\frac{1}{3})^5}{5} + \cdots\right)$$
$$= \frac{2}{3} + \frac{2}{81} + \frac{2}{1215} + \cdots$$ i.e. the third term is 0.0016.

Exercise 19C

1. Give the series expansion for each of the following in ascending powers of x, giving terms up to and including that in x^4.

    (a) e^{-3x} (b) e^{2x} (c) e^{x^2}

    (d) $e^{x/2}$ (e) $\dfrac{1}{\sqrt{e^x}}$ (f) $(1 + x)e^{5x}$

    (g) $(1 - 2x)e^{4x}$ (h) $(1 - x)^2 e^x$ (i) $e^{(x - x^2)}$

2. Given that $e^x = 1 + x + \dfrac{x^2}{2!} + \dfrac{x^3}{3!} + \dots \dfrac{x^r}{r!} \dots$ find expressions for

    (a) $1 + 2 + \dfrac{4}{2!} + \dfrac{8}{3!} + \dfrac{16}{4!} + \dfrac{32}{5!} + \dots$

    (b) $1 - 3 + \dfrac{9}{2!} - \dfrac{27}{3!} + \dfrac{81}{4!} - \dfrac{243}{5!} + \dots$

    (c) $-\dfrac{1}{2} + \dfrac{1}{4 \times 2!} - \dfrac{1}{8 \times 3!} + \dfrac{1}{16 \times 4!} - \dfrac{1}{32 \times 5!} + \dots$

    (d) $\dfrac{1}{2!} + \dfrac{1}{3!} + \dfrac{1}{4!} + \dfrac{1}{5!} + \dots$

    (e) $\displaystyle\lim_{n \to \infty} \sum_{r=1}^{n} \dfrac{x^r}{r!}$

    (f) $\displaystyle\lim_{n \to \infty} \sum_{r=0}^{n} \dfrac{x^{2r}}{r!}$

3. Give the series expansion for each of the following, in ascending powers of x, giving terms up to and including that in x^4 and the values of x for which the expansion is valid.

    (a) $\ln (1 + 2x)$ (b). $\ln (1 - 6x)$

    (c) $\ln (1 + \tfrac{1}{3}x)$ (d) $\ln (2 + x)$

    (e) $\ln (3 + 2x)$ (f) $\ln [(1 + x)(1 - 2x)]$

    (g) $\ln (1 + 3x + 2x^2)$ (h) $\ln (3 + 4x + x^2)$

    (i) $\ln \left(\dfrac{1 + 4x}{1 - x} \right)$ (j) $\ln \left[\dfrac{(4 - x)^2}{1 + x} \right]$

    (k) $(1 - x)^2 \ln (1 + x)$ (l) $(1 + 2x) \ln \left(\dfrac{1 + x}{1 - 5x} \right)$

4. Given that $\ln (1 + x) = x - \dfrac{x^2}{2} + \dfrac{x^3}{3} - \dfrac{x^4}{4} + \dots (-1)^{r+1}\dfrac{x^r}{r} \dots$
    for $-1 < x \leqslant 1$ express the following as logarithms:

    (a) $\dfrac{1}{2} - \left(\dfrac{1}{2}\right)^2 \dfrac{1}{2} + \left(\dfrac{1}{2}\right)^3 \dfrac{1}{3} + \left(\dfrac{1}{2}\right)^4 \dfrac{1}{4} + \dots$

    (b) $-\dfrac{1}{2} - \left(\dfrac{1}{2}\right)^2 \dfrac{1}{2} - \left(\dfrac{1}{2}\right)^3 \dfrac{1}{3} - \left(\dfrac{1}{2}\right)^4 \dfrac{1}{4} - \dots$

    (c) $\ln 6 + \dfrac{1}{3} - \dfrac{1}{9 \times 2} + \dfrac{1}{27 \times 3} - \dfrac{1}{81 \times 4} + \dots$

    (d) $\ln \left(\dfrac{3}{2}\right) + \dfrac{1}{4} + \dfrac{1}{4^2 \times 2} + \dfrac{1}{4^3 \times 3} + \dfrac{1}{4^4 \times 4} + \dots$

5. Using an appropriate number of terms of the series expansion of e^x find the value of $e^{0.1}$ and $1/e^{0.1}$ correct to four decimal places.

6. Using an appropriate number of terms of the series expansion of
ln $(1 + x)$ find the value of ln $(1 \cdot 1)$ and ln $(0 \cdot 9)$ correct to four decimal places.

7. (a) Show that for $|x| < 1$, $\ln \left(\dfrac{1 + x}{1 - x} \right) = 2 \left(x + \dfrac{x^3}{3} + \dfrac{x^5}{5} + \dfrac{x^7}{7} + \cdots \right)$

   (b) Using the result of part (a) evaluate (i) ln $\frac{5}{3}$ correct to 4 decimal places,

   (ii) ln 3 correct to 3 decimal places.

8. If $x > 1$ show that $\ln (1 + x) = \ln x + \dfrac{1}{x} - \dfrac{1}{2x^2} + \dfrac{1}{3x^3} - \dfrac{1}{4x^4} + \cdots$

9. Find the coefficient of x^r in the series expansions of the following

   (a) e^{2x} (b) $\ln \left(1 + \dfrac{x}{2} \right)$ (c) $(1 + x)e^x$ (d) $\left(2 - x - \dfrac{1}{x} \right) \ln (1 - x)$

10. Expand $e^{2x} - 3e^x - e^{-x}$ in ascending powers of x up to and including the term in x^4.

11. Obtain the series expansion of $e^x/(1 + 2x)$ up to and including the term in x^3 and state the values of x for which the expansion is valid.

12. Obtain the series expansion of $[\ln (1 + x)]/(2 + x)$ up to and including the term in x^4 and state the values of x for which the expansion is valid.

13. Find the first four non-zero terms in the series expansion of each of the following in ascending powers of x, and state the values of x for which the expansion is valid

   (a) $e^x \ln (1 + x)$ (b) $e^{-x} \ln (1 + 2x)$ (c) $e^{6x} \ln \left(1 + \dfrac{x}{2} \right)^2$

   (d) $e^x \ln \sqrt{(1 - 2x)}$

14. Find the values of the positive constants a, b and c given that when x is sufficiently small for terms in x^4, and higher powers of x, to be neglected
   then $\dfrac{e^{ax}}{2 + bx} = \dfrac{1}{2} + \dfrac{x^2}{4} - cx^3$ (assume $|bx| < 2$)

15. Find the values of a, b and c given that when x is sufficiently small for terms in x^4 and higher powers of x to be neglected
   then $e^{ax} \ln (1 + bx) = -3x - \frac{21}{2}x^2 + cx^3$ (assume $-1 < bx \leqslant 1$)

Exercise 19D   *Examination Questions*

1. Sketch the curves $y = e^x$ and $y = e^{-2x}$, using the same axes.
   The line $y = 4$ intersects the first curve at A and the second curve at B.
   Calculate the length of AB to two decimal places.   (A.E.B.)

2. Differentiate with respect to x

   (a) $\dfrac{e^{-x}}{x}$ (b) $\ln (1 + \sin x)$ (London)

3. Differentiate $x \log_e x$ with respect to x.

   Use your result to find $\displaystyle\int \log_e x \, dx$.

   Hence evaluate, to three significant figures, $\displaystyle\int_1^2 \log_e x \, dx$.   (A.E.B.)

4. Find the coordinates of the turning point on the curve
$$y = 2e^{3x} + 8e^{-3x},$$
and determine the nature of this turning point.   (Cambridge)

5. Find the equations of the tangents to the curve $y = xe^x$ at the points where $x = 1$ and $x = -1$. Find, to the nearest degree, the acute angle between these tangents.   (A.E.B.)

6. Express the function
$$f(x) = \frac{x + 2}{(x^2 + 1)(2x - 1)}$$
as the sum of partial fractions.

Hence find $\int f(x)\,dx$.   (J.M.B.)

7. Given that $y = ae^{-2x} \sin(x + \beta)$, where a and β are constants, verify that
$$\frac{d^2y}{dx^2} + 4\frac{dy}{dx} + 5y = 0.$$ (Cambridge)

8. Sketch on separate diagrams the graphs corresponding to the following equations, showing the turning points and any asymptotes parallel to the coordinate axes:

(i) $y = (x - 1)(x - 3)$, (ii) $y = \dfrac{1}{(x - 1)(x - 3)}$, $(x \neq 1, x \neq 3)$.

The line $y = -\dfrac{4}{3}$ meets the graph of $y = \dfrac{1}{(x - 1)(x - 3)}$ at the points A and B. Calculate the area of the finite region enclosed by the line AB and the arc of the graph between A and B, giving three significant figures in your answer.   (Cambridge)

9. Find the coordinates of the points of intersection of the curves
$$y = \frac{x}{x + 3} \quad \text{and} \quad y = \frac{x}{x^2 + 1}.$$
Sketch the curves on the same diagram, showing any asymptotes or turning points.
Show that the area of the finite region in the first quadrant enclosed by the two curves is
$$\tfrac{7}{2}\ln 5 - 3\ln 3 - 2.$$ (London)

10. Expand in ascending powers of x up to and including the term in x^3
(i) $(1 - x)^{1\cdot2}$, (ii) $\log_e(1 - ax)$.
Given that $(1 - x)^{1\cdot2} - \tfrac{1}{4}\log_e(1 - ax) \equiv 1 + px^2 + qx^3 + \ldots$, find the numerical values of a, p and q.   (A.E.B.)

11. Obtain the expansion in ascending powers of x of the function
$$f(x) = \ln\left\{\frac{(1 - 3x)^2}{(1 + 2x)}\right\}.$$
Give the values of the coefficients of the powers of x up to and including x^3, and give an expression for the coefficient of x^n. For what range of values of x is this expansion valid?   (Oxford)

12. Given that $\dfrac{dy}{dx} = \dfrac{2x}{1 + x^2} - 2xe^{-x^2}$ and that $y = 0$ when $x = 0$,

express y in terms of x.

Show that, for small values of $|x|$, $y \approx px^6 + qx^8$ finding the values of the constants p and q.   (A.E.B.)

13. Write down, in ascending powers of y, the first four terms in the series for e^y. By taking $e^y = 2^x$, show that

$$2^x = 1 + x \log_e 2 + \tfrac{1}{2}x^2 (\log_e 2)^2 + \tfrac{1}{6}x^3 (\log_e 2)^3 + \dots$$

Given that $2^{3x} + 5(2^x) = 6 + Ax + Bx^2 + Cx^3 + \dots$, for all real x, find the values of the constants A, B and C in terms of $\log_e 2$.   (A.E.B.)

14. Find the first three non-zero terms in the expansion, in ascending powers of x, of

$$e^{2x} + e^{-2x}.$$

Given that x is so small that its fifth and higher powers may be neglected, find the numerical values of the constants a, b and c in the approximate formula

$$\log_e (e^{2x} + e^{-2x}) \approx a + bx^2 + cx^4.$$ (J.M.B.)

15. (a) Solve the equation $\log_3 x = \log_x 9$, giving your answers correct to three significant figures.

(b) Show that $\ln (1 + x + x^2) = \ln (1 - x^3) - \ln (1 - x)$. Hence obtain the expansion of

$$\ln (1 + x + x^2)$$

in ascending powers of x up to and including the term in x^5.

Show that the coefficient of x^n in this expansion is $-\dfrac{2}{n}$ or $\dfrac{1}{n}$ according as

n is or is not a multiple of 3.

Write down the expansion of $\ln (1 - x + x^2)$ in ascending powers of x up to and including the term in x^5.   (Cambridge)

16. Given that

$$f(x) = e^{-3x} \quad \text{and} \quad g(x) = (1 - x)^{-3},$$

find the first three non-zero terms in the expansions in ascending powers of x of

(i) $f(x) - g(x)$, (ii) $\log_e f(x) - \log_e g(x)$, (iii) $\dfrac{f(x)}{g(x)}$.

In each of the expansions (i) and (ii) obtain the coefficient of x^n for $n > 1$.

(J.M.B.)

20
Calculus IV: Further integration

20.1 Change of variable

Suppose we want to integrate $x^2(2x^3 + 3)^5$ with respect to x. We could expand the bracket $(2x^3 + 3)^5$, multiply by x^2 and then integrate each term. It is, however, much better to notice that x^2 is a scalar multiple of $\frac{d}{dx}(2x^3 + 3)$ and then it follows that

$$\int x^2(2x^3 + 3)^5 \, dx = \frac{1}{36}(2x^3 + 3)^6 + c.$$

If we fail to notice this 'function of a function', an alternative method is to change the variable from x to some other suitably chosen variable. This method depends on the following theory:

Suppose $\quad y = \int f(x)\, dx \quad$ where $\quad x = g(u) \quad$ i.e. $\quad \frac{dy}{dx} = f(x) \quad$ and $\quad f(x) = f[g(u)]$

Now $\qquad \frac{dy}{du} = \frac{dy}{dx} \times \frac{dx}{du}$

$\therefore \qquad \frac{dy}{du} = f(x) \times \frac{dx}{du}$

integrating, with respect to u, gives

$$y = \int f(x)\frac{dx}{du}\, du$$

$$= \int f[g(u)]\frac{dx}{du}\, du \quad \text{but } y = \int f(x)\, dx$$

$$\boxed{\therefore \int f(x)\, dx = \int f[g(u)]\frac{dx}{du}\, du}$$

Thus, to integrate $f(x)$ with respect to x, dx is replaced by $\frac{dx}{du}du$ and the integrand is then expressed in terms of the new variable, u.

Example 1

Find $\int x^2(2x^3 + 3)^5 \, dx$.

Let $\qquad\qquad\qquad u = 2x^3 + 3 \quad$ then $\quad \frac{du}{dx} = 6x^2$

Thus $\int x^2(2x^3 + 3)^5 \, dx = \int x^2(2x^3 + 3)^5 \frac{dx}{du}\, du$

$$= \int x^2(2x^3 + 3)^5 \frac{1}{6x^2}\, du$$

$$= \int \frac{u^5}{6} du$$

$$= \frac{u^6}{36} + c$$

$$\therefore \quad \int x^2(2x^3 + 3)^5 dx = \frac{1}{36}(2x^3 + 3)^6 + c$$

As was noted at the begining of the chapter, $x^2(2x^3 + 3)^5$ could be integrated by inspection. However, there are many cases for which the method of changing the variable enables us to integrate an expression which we could not integrate by inspection.

Example 2

Find $\int 2x(x + 2)^5 dx$.

Let $u = x + 2$ then $\dfrac{du}{dx} = 1$

$$\int 2x(x + 2)^5 dx = \int 2x(x + 2)^5 \frac{dx}{du} du$$

$$= \int 2(u - 2)u^5 du$$

$$= \tfrac{2}{7}u^7 - \tfrac{2}{3}u^6 + c$$
$$= \tfrac{2}{21}u^6(3u - 7) + c$$

$$\therefore \int 2x(x + 2)^5 dx = \tfrac{2}{21}(x + 2)^6(3x - 1) + c$$

Example 3

Find $\int x\sqrt{(3x - 2)}\, dx$

Let $u = 3x - 2$ then $\dfrac{du}{dx} = 3$

$$\int x\sqrt{(3x - 2)}\, dx = \int x\sqrt{(3x - 2)} \frac{dx}{du} du$$

$$= \int \left(\frac{u + 2}{3}\right)\sqrt{u}\,\tfrac{1}{3} du$$

$$= \int \left(\frac{u^{3/2}}{9} + \frac{2u^{1/2}}{9}\right) du$$

$$= \tfrac{2}{45}u^{5/2} + \tfrac{4}{27}u^{3/2} + c$$
$$= \tfrac{2}{135}u^{3/2}(3u + 10) + c$$

$$\therefore \int x\sqrt{(3x - 2)}\, dx = \tfrac{2}{135}(3x - 2)^{3/2}(9x + 4) + c$$

Example 4

Evaluate $\displaystyle\int_0^5 \frac{x}{\sqrt{(x + 4)}}\, dx$.

Let $u = x + 4$ then $\dfrac{du}{dx} = 1$

Notice, also, that when $x = 0$, $u = 4$ and when $x = 5$, $u = 9$.

$$\int_0^5 \frac{x}{\sqrt{(x + 4)}}\, dx = \int_{u=4}^{u=9} \frac{x}{\sqrt{(x + 4)}}\frac{dx}{du} du$$

$$= \int_4^9 \frac{u - 4}{\sqrt{u}} du$$

$$= \left[\frac{2}{3}u^{3/2} - 8u^{1/2}\right]_4^9$$

$$= (\tfrac{2}{3} \times 27 - 8 \times 3) - (\tfrac{2}{3} \times 8 - 8 \times 2)$$

$$\therefore \int_0^5 \frac{x}{\sqrt{(x + 4)}}\, dx = 4\tfrac{2}{3}$$

In chapter 15 we saw that $\int \dfrac{1}{(a^2 + x^2)} dx = \dfrac{1}{a} \tan^{-1}\left(\dfrac{x}{a}\right) + c$ and

$$\int \dfrac{1}{\sqrt{(a^2 - x^2)}} dx = \sin^{-1}\left(\dfrac{x}{a}\right) + c.$$

However, as was indicated in chapter 15, for more difficult functions of this type, integration by change of variable is advisable. These more difficult

integrations will be of the form $\int \dfrac{1}{a^2 + b^2 x^2} dx, |b| \neq 1$ and

$$\int \dfrac{1}{\sqrt{(a^2 - c^2 x^2)}} dx, |c| \neq 1.$$

We can then substitute

either $u = bx$ or $x = \dfrac{a}{b} \tan u$ into $\int \dfrac{1}{a^2 + b^2 x^2} dx$

and either $u = cx$ or $x = \dfrac{a}{c} \sin u$ into $\int \dfrac{1}{\sqrt{(a^2 - c^2 x^2)}} dx.$

Example 5

Use a suitable substitution to find

(a) $\displaystyle\int \dfrac{1}{25 + 16x^2} dx$ (b) $\displaystyle\int \dfrac{1}{\sqrt{(2 - 9x^2)}} dx.$

METHOD 1

(a) Let $u = 4x$ then $\dfrac{du}{dx} = 4$

$$\int \dfrac{1}{25 + 16x^2} dx = \int \dfrac{1}{25 + u^2} \times \dfrac{1}{4} du$$

$$= \dfrac{1}{4} \int \dfrac{1}{25 + u^2} du$$

$$= \dfrac{1}{20} \tan^{-1}\left(\dfrac{u}{5}\right) + c$$

$$\therefore \int \dfrac{1}{25 + 16x^2} dx = \dfrac{1}{20} \tan^{-1}\left(\dfrac{4x}{5}\right) + c$$

(b) Let $u = 3x$ then $\dfrac{du}{dx} = 3$

$$\int \dfrac{1}{\sqrt{(2 - 9x^2)}} dx = \int \dfrac{1}{\sqrt{(2 - u^2)}} \times \dfrac{1}{3} du$$

$$= \dfrac{1}{3} \int \dfrac{1}{\sqrt{(2 - u^2)}} du$$

$$= \dfrac{1}{3} \sin^{-1}\left(\dfrac{u}{\sqrt{2}}\right) + c$$

$$\therefore \int \dfrac{1}{\sqrt{(2 - 9x^2)}} dx = \dfrac{1}{3} \sin^{-1}\left(\dfrac{3x}{\sqrt{2}}\right) + c$$

METHOD 2

Let $x = \dfrac{5}{4} \tan u$ then $\dfrac{dx}{du} = \dfrac{5}{4} \sec^2 u$

$$\int \dfrac{1}{25 + 16x^2} dx = \int \dfrac{1}{25(1 + \tan^2 u)} \times \dfrac{5}{4} \sec^2 u \, du$$

$$= \dfrac{1}{20} \int du$$

$$= \dfrac{u}{20} + c$$

$$\int \dfrac{1}{25 + 16x^2} dx = \dfrac{1}{20} \tan^{-1}\left(\dfrac{4x}{5}\right) + c$$

Let $x = \dfrac{\sqrt{2}}{3} \sin u$ then $\dfrac{dx}{du} = \dfrac{\sqrt{2}}{3} \cos u$

$$\int \dfrac{1}{\sqrt{(2 - 9x^2)}} dx = \int \dfrac{1}{\sqrt{[2(1 - \sin^2 u)]}} \times \dfrac{\sqrt{2}}{3} \cos u \, du$$

$$= \dfrac{1}{3} \int du$$

$$= \dfrac{1}{3} u + c$$

$$\int \dfrac{1}{\sqrt{(2 - 9x^2)}} dx = \dfrac{1}{3} \sin^{-1}\left(\dfrac{3x}{\sqrt{2}}\right) + c$$

Example 6

Evaluate $\displaystyle\int_0^1 \sqrt{(1 - x^2)}\, dx$.

For integrations of the type $\sqrt{(a^2 - b^2x^2)}$, $x = \dfrac{a}{b} \sin u$ is a useful substitution.

Thus, for $\displaystyle\int_0^1 \sqrt{(1 - x^2)}\, dx$, we let $x = \sin u$ then $\dfrac{dx}{du} = \cos u$.

Notice also that when $x = 0$, $u = 0$ and when $x = 1$, $u = \dfrac{\pi}{2}$.

$$\int_0^1 \sqrt{(1 - x^2)}\, dx = \int_{u=0}^{u=\pi/2} \sqrt{(1 - \sin^2 u)} \cos u\, du$$

$$= \int_0^{\pi/2} \cos^2 u\, du$$

$$= \int_0^{\pi/2} \frac{1}{2}(\cos 2u + 1)\, du$$

$$= \frac{1}{2}\left[\frac{1}{2}\sin 2u + u\right]_0^{\pi/2}$$

$$= \tfrac{1}{2}(\tfrac{1}{2} \sin \pi + \pi/2) - \tfrac{1}{2}(\tfrac{1}{2} \sin 0 + 0)$$

$$\therefore \quad \int_0^1 \sqrt{(1 - x^2)}\, dx = \frac{\pi}{4}$$

This method of substituting in order to change the variable can also be used to integrate expressions of the type $\displaystyle\int \frac{1}{ax^2 + bx + c}\, dx$ when $(ax^2 + bx + c)$ does not factorise.

Example 7

Find $\displaystyle\int \frac{1}{x^2 - 4x + 29}\, dx$.

Now $\displaystyle\int \frac{1}{x^2 - 4x + 29}\, dx = \int \frac{1}{(x - 2)^2 + 25}\, dx$

Let $u = x - 2$ then $\dfrac{du}{dx} = 1$

and $\therefore \displaystyle\int \frac{1}{x^2 - 4x + 29}\, dx = \int \frac{1}{u^2 + 25}\, du$

$$= \frac{1}{5} \tan^{-1}\left(\frac{u}{5}\right) + c$$

$$\therefore \quad \int \frac{1}{x^2 - 4x + 29}\, dx = \frac{1}{5} \tan^{-1}\left(\frac{x - 2}{5}\right) + c$$

The substitution $t = \tan \frac{1}{2}\theta$

If we let $t = \tan \frac{1}{2}\theta$ then, from $\tan 2A = \dfrac{2 \tan A}{1 - \tan^2 A}$, it follows that $\tan \theta = \dfrac{2t}{(1 - t^2)}$.

It also follows, from the right angled triangle shown,

that $\sin \theta = \dfrac{2t}{(1 + t^2)}$ and that $\cos \theta = \dfrac{(1 - t^2)}{(1 + t^2)}$.

This substitution can be useful in finding some integrals.

Example 8

Use the substitution, $t = \tan \dfrac{\theta}{2}$, to evaluate $\displaystyle\int_0^{\pi/2} \dfrac{4}{3 + 5 \sin \theta} d\theta$.

Let $t = \tan \dfrac{\theta}{2}$ then $\sin \theta = \dfrac{2t}{(1 + t^2)}$ and $\dfrac{dt}{d\theta} = \frac{1}{2} \sec^2 \dfrac{\theta}{2} = \dfrac{1 + t^2}{2}$.

Notice also that when $\theta = 0$, $t = 0$ and when $\theta = \dfrac{\pi}{2}$, $t = 1$.

Thus $\displaystyle\int_0^{\pi/2} \dfrac{4}{3 + 5 \sin \theta} d\theta = \int_{t=0}^{t=1} \dfrac{4}{[3 + 10t/(1 + t^2)]} \times \dfrac{2}{(1 + t^2)} dt$

$$= \int_0^1 \dfrac{8}{3t^2 + 10t + 3} dt$$

$$= \int_0^1 \left(\dfrac{3}{3t + 1} - \dfrac{1}{t + 3} \right) dt$$

$$= \left[\ln (3t + 1) - \ln (t + 3) \right]_0^1$$

$$= (\ln 4 - \ln 4) - (\ln 1 - \ln 3)$$

$$= \ln 3$$

$\therefore \displaystyle\int_0^{\pi/2} \dfrac{4}{3 + 5 \sin \theta} d\theta = \ln 3.$

Exercise 20A

Find the following indefinite integrals, in terms of x, using the suggested substitution

1. $\displaystyle\int 6x \sin (x^2 - 4) dx, \ u = x^2 - 4$
2. $\displaystyle\int 5x \cos (5 - x^2) dx, \ u = 5 - x^2$

3. $\displaystyle\int 3x\sqrt{(1 + x^2)} dx, \ u = 1 + x^2$
4. $\displaystyle\int 3x (x^2 + 6)^5 dx, \ u = x^2 + 6$

5. $\displaystyle\int x(x + 2)^9 dx, \ u = x + 2$
6. $\displaystyle\int 5x^2(x - 3)^8 dx, \ u = x - 3$

7. $\displaystyle\int 9x(3x + 2)^3 dx, \ u = 3x + 2$
8. $\displaystyle\int 7x(2x + 3)^5 dx, \ u = 2x + 3$

9. $\displaystyle\int \frac{3x}{\sqrt{(2x + 3)}}\,dx$, $u = 2x + 3$ 10. $\displaystyle\int \frac{1}{\sqrt{(1 - x^2)}}\,dx$, $x = \sin u$

11. $\displaystyle\int \frac{1}{4 + 9x^2}\,dx$, $x = \frac{2}{3}\tan u$ 12. $\displaystyle\int \frac{1}{\sqrt{(25 - 4x^2)}}\,dx$, $x = \frac{5}{2}\sin u$

13. $\displaystyle\int \frac{1}{\sqrt{(4 - 9x^2)}}\,dx$, $x = \frac{2}{3}\sin u$ 14. $\displaystyle\int \frac{1}{4 + 5x^2}\,dx$, $x = \frac{2}{\sqrt{5}}\tan u$

15. $\displaystyle\int \sqrt{(25 - x^2)}\,dx$, $x = 5\sin\theta$ 16. $\displaystyle\int \sqrt{(1 - 4x^2)}\,dx$, $x = \frac{1}{2}\sin\theta$

17. Given that $\displaystyle\int \frac{1}{\sqrt{(a^2 - x^2)}}\,dx = \sin^{-1}\left(\frac{x}{a}\right) + c$ and

$\displaystyle\int \frac{a}{a^2 + x^2}\,dx = \tan^{-1}\left(\frac{x}{a}\right) + c$ find the following indefinite integrals
using the suggested substitution

(a) $\displaystyle\int \frac{1}{1 + 16x^2}\,dx$, $u = 4x$ (b) $\displaystyle\int \frac{1}{\sqrt{(9 - 16x^2)}}\,dx$, $u = 4x$

(c) $\displaystyle\int \frac{1}{4 + 5x^2}\,dx$, $u = x\sqrt{5}$ (d) $\displaystyle\int \frac{1}{\sqrt{(5 - 2x^2)}}\,dx$, $u = x\sqrt{2}$

18. Use the method of example 7 to find the following integrals

(a) $\displaystyle\int \frac{3}{x^2 - 2x + 10}\,dx$ (b) $\displaystyle\int \frac{1}{x^2 + 6x + 13}\,dx$ (c) $\displaystyle\int \frac{3}{4x^2 - 12x + 13}\,dx$

Find the following indefinite integrals by using some suitable substitution.

19. $\displaystyle\int (x + 2)(2x - 3)^6\,dx$ 20. $\displaystyle\int x\sqrt{(2x + 1)}\,dx$

21. $\displaystyle\int \frac{x}{\sqrt{(2x + 1)}}\,dx$ 22. $\displaystyle\int \frac{1}{\sqrt{(25 - 4x^2)}}\,dx$

23. $\displaystyle\int \sqrt{(9 - x^2)}\,dx$ 24. $\displaystyle\int \frac{4}{x^2 + 2x + 17}\,dx$

Evaluate the following definite integrals by using some suitable substitution.

25. $\displaystyle\int_0^3 x\sqrt{(x + 1)}\,dx$ 26. $\displaystyle\int_1^2 (x + 2)(x - 1)^5\,dx$

27. $\displaystyle\int_1^2 x^2(x - 1)^5\,dx$ 28. $\displaystyle\int_1^2 x(2x - 3)^4\,dx$

29. $\displaystyle\int_0^1 4x(2x - 1)^4\,dx$ 30. $\displaystyle\int_0^{1/2} \frac{1}{1 + 4x^2}\,dx$

31. $\displaystyle\int_0^{3/2} \sqrt{(9 - 4x^2)}\,dx$

Find the following definite integrals using the substitution $t = \tan\frac{1}{2}\theta$

32. $\displaystyle\int_0^{\pi/2} \frac{3}{1 + \sin\theta}\,d\theta$ 33. $\displaystyle\int_0^{2\pi/3} \frac{3}{5 + 4\cos\theta}\,d\theta$

34. $\displaystyle\int_{-\pi/2}^{\pi/2} \frac{3}{4 + 5\cos\theta}\,d\theta$ 35. $\displaystyle\int_0^{\pi/2} \frac{5}{3\sin\theta + 4\cos\theta}\,d\theta$

20.2 Integration by parts

In section 13.5 we obtained a rule for differentiating the product of u and v, two functions of x.

From this rule we can obtain a result which enables us to integrate certain products.

If u and v are functions of x, then

$$\frac{d}{dx}(uv) = u\frac{dv}{dx} + v\frac{du}{dx}$$

integrating both sides of this equation, with respect to x

$$\int \frac{d}{dx}(uv)\,dx = \int u\frac{dv}{dx}\,dx + \int v\frac{du}{dx}\,dx$$

$$uv = \int u\frac{dv}{dx}\,dx + \int v\frac{du}{dx}\,dx$$

i.e. $$\int u\frac{dv}{dx}\,dx = uv - \int v\frac{du}{dx}\,dx$$

This is known as the formula for integration by parts.

Example 9

Find $\int x(3x - 2)^4\,dx$, using the method of integration by parts.

[Note that this integral could be found by expanding the integrand, or by means of a substitution.]

Let $u = x$ and $\dfrac{dv}{dx} = (3x - 2)^4$

then $\dfrac{du}{dx} = 1$ and $v = \dfrac{1}{15}(3x - 2)^5$

Using $\int u\dfrac{dv}{dx}\,dx = uv - \int v\dfrac{du}{dx}\,dx$,

$$\int x(3x - 2)^4\,dx = \frac{x}{15}(3x - 2)^5 - \int \frac{1}{15}(3x - 2)^5\,dx$$

$$= \frac{x}{15}(3x - 2)^5 - \frac{1}{15 \times 18}(3x - 2)^6 + c$$

$\therefore$ $$\int x(3x - 2)^4\,dx = \frac{1}{270}(3x - 2)^5(15x + 2) + c$$

Note: The choice to substitute u for x is made, since when differentiated this function becomes 1; thus the expression, $\int v\dfrac{du}{dx}\,dx$, will no longer be a product of two functions of x.

Example 10

Find $\int x \ln x \, dx$.

We know that we shall have to differentiate one, and integrate the other, function (x or $\ln x$) with respect to x. Although we would like to put $u = x$, because the differential $\frac{du}{dx} = 1$, we cannot do this since it would then mean integrating $\ln x$.

Let $\qquad u = \ln x \quad$ and $\quad \frac{dv}{dx} = x$

then $\qquad \frac{du}{dx} = \frac{1}{x} \quad$ and $\quad v = \frac{1}{2}x^2$

Using $\quad \int u\frac{dv}{dx}dx = uv - \int v\frac{du}{dx}dx,$

$\therefore \quad \int x \ln x \, dx = \ln x \left(\frac{x^2}{2}\right) - \int \frac{x^2}{2} \times \frac{1}{x}dx$

$\therefore \quad \int x \ln x \, dx = \frac{x^2}{2}\ln x - \frac{x^2}{4} + c$

Example 11

Find $\int \ln x \, dx$.

This is not seemingly a product, but we write it as one so that we can then choose to differentiate $\ln x$.

Writing $\int \ln x \, dx$ as $\int 1 \times \ln x \, dx$

Let $\qquad u = \ln x \quad$ and $\quad \frac{dv}{dx} = 1$

then $\qquad \frac{du}{dx} = \frac{1}{x} \quad$ and $\quad v = x$

Using $\quad \int u\frac{dv}{dx}dx = uv - \int v\frac{du}{dx}dx,$

$\int \ln x \, dx = (\ln x)x - \int x\frac{1}{x}dx$

$= x \ln x - \int dx$

$\therefore \quad \int \ln x \, dx = x \ln x - x + c.$

Example 12

Find   (a) $\displaystyle\int x^2 \cos x\, dx$ (b) $\displaystyle\int e^x \cos x\, dx$.

(a) $\displaystyle\int x^2 \cos x\, dx$

Let $u = x^2$ and $\dfrac{dv}{dx} = \cos x$

then $\dfrac{du}{dx} = 2x$ and $v = \sin x$

Using $\displaystyle\int u\dfrac{dv}{dx}\, dx = uv - \int v\dfrac{du}{dx}\, dx,$

$$\int x^2 \cos x\, dx = x^2 \sin x - \int (\sin x)\, 2x\, dx \qquad \ldots [1]$$

The integral on the R.H.S. is still a product, so we repeat the process:

$$\int 2x \sin x\, dx$$

Let $u = 2x$ and $\dfrac{dv}{dx} = \sin x$

then $\dfrac{du}{dx} = 2$ and $v = -\cos x$

Using $\displaystyle\int u\dfrac{dv}{dx}\, dx = uv - \int v\dfrac{du}{dx}\, dx,$

$$\int 2x \sin x\, dx = 2x(-\cos x) - \int (-\cos x)2\, dx$$
$$= -2x \cos x + 2 \sin x + c$$

Using this result in [1] above:

$$\int x^2 \cos x\, dx = x^2 \sin x - (-2x \cos x + 2 \sin x) + c$$

$\therefore$ $$\int x^2 \cos x\, dx = (x^2 - 2) \sin x + 2x \cos x + c$$

(b) $\displaystyle\int e^x \cos x\, dx$

Let $u = e^x$ and $\dfrac{dv}{dx} = \cos x$

then $\dfrac{du}{dx} = e^x$ and $v = \sin x$

Using $\displaystyle\int u\dfrac{dv}{dx}\, dx = uv - \int v\dfrac{du}{dx}\, dx,$

$$\int e^x \cos x\, dx = e^x \sin x - \int (\sin x)\, e^x\, dx \qquad \ldots [1]$$

The integral on the R.H.S. is still a product, so we repeat the process:

$$\int e^x \sin x \, dx$$

Let $\qquad u = e^x \quad$ and $\quad \dfrac{dv}{dx} = \sin x$

then $\qquad \dfrac{du}{dx} = e^x \quad$ and $\quad v = -\cos x$

Using $\qquad \displaystyle\int u \dfrac{dv}{dx} \, dx = uv - \int v \dfrac{du}{dx} \, dx,$

$$\int e^x \sin x \, dx = e^x(-\cos x) - \int (-\cos x)e^x \, dx$$

$$= -e^x \cos x + \int e^x \cos x \, dx$$

Using this result in [1] above:

$$\int e^x \cos x \, dx = e^x \sin x + e^x \cos x - \int e^x \cos x \, dx$$

$\therefore \qquad 2 \displaystyle\int e^x \cos x \, dx = e^x \sin x + e^x \cos x$

$\therefore \qquad \displaystyle\int e^x \cos x \, dx = \dfrac{1}{2}e^x(\sin x + \cos x) + c$

Exercise 20B

Use the method of integration by parts to find the following indefinite integrals

1. $\displaystyle\int x(x + 1)^4 \, dx$
2. $\displaystyle\int x\sqrt{(x - 3)} \, dx$
3. $\displaystyle\int \dfrac{x}{\sqrt{(x + 3)}} \, dx$

4. $\displaystyle\int x \sin x \, dx$
5. $\displaystyle\int x^2 \sin x \, dx$
6. $\displaystyle\int 2x \sin (3x - 1) \, dx$

7. $\displaystyle\int x^2 \ln x \, dx$
8. $\displaystyle\int (x + 1) \ln x \, dx$
9. $\displaystyle\int \dfrac{1}{x^3} \ln x \, dx$

10. $\displaystyle\int e^x \sin x \, dx$
11. $\displaystyle\int xe^x \, dx$
12. $\displaystyle\int x^2 e^{3x} \, dx$

13. $\displaystyle\int e^x \cos 2x \, dx$
14. $\displaystyle\int \ln (2x + 1) \, dx$
15. $\displaystyle\int \tan^{-1} x \, dx$

16. $\displaystyle\int \sin^{-1} x \, dx$
17. $\displaystyle\int 2x \ln (x + 1) \, dx$
18. $\displaystyle\int x \ln (x - 2) \, dx$

Evaluate the following definite integrals using the method of integration by parts.

19. $\displaystyle\int_0^1 x(x - 1)^5 \, dx$
20. $\displaystyle\int_3^6 \dfrac{x}{\sqrt{(x - 2)}} \, dx$
21. $\displaystyle\int_0^2 x(x - 2)^4 \, dx$

22. $\displaystyle\int_0^{\pi/4} x \cos 2x \, dx$
23. $\displaystyle\int_2^4 (x - 1) \ln (2x) \, dx$
24. $\displaystyle\int_1^4 \dfrac{\ln x}{x^2} \, dx$

25. $\displaystyle\int_0^2 x^3 e^x \, dx$

20.3 General methods

In order to find the value of an integral, both quickly and easily, a systematic approach is necessary. It is difficult to formulate rules for integrating, but it is possible to lay down guide lines indicating how to proceed in any given case.

Confronted with an integral, it is suggested that the reader should ask the following questions 1, 2, 3, and so on, until a positive answer is obtained and the required technique is determined.

1. Is it a standard form?

$$\int ax^n \, dx = \frac{ax^{n+1}}{n + 1} + c \qquad\qquad \int \operatorname{cosec}^2 x \, dx = -\cot x + c$$

$$\int \sin x \, dx = -\cos x + c \qquad\qquad \int e^x \, dx = e^x + c$$

$$\int \cos x \, dx = \sin x + c \qquad\qquad \int \frac{1}{x} \, dx = \ln |x| + c$$

$$\int \sec^2 x \, dx = \tan x + c \qquad\qquad \int a^x \, dx = \frac{a^x}{\ln a} + c$$

$$\int \sec x \tan x \, dx = \sec x + c \qquad\qquad \int \frac{1}{\sqrt{(a^2 - x^2)}} \, dx = \sin^{-1}\left(\frac{x}{a}\right) + c$$

$$\int \operatorname{cosec} x \cot x \, dx = -\operatorname{cosec} x + c \qquad\qquad \int \frac{1}{a^2 + x^2} \, dx = \frac{1}{a} \tan^{-1}\left(\frac{x}{a}\right) + c$$

With a knowledge of these standard results, the reader should be able to recognize integrations that are simple extensions of these standard forms:

e.g. $\displaystyle\int (x + 3)(x - 4) \, dx = \int (x^2 - x - 12) \, dx = \frac{x^3}{3} - \frac{x^2}{2} - 12x + c$

$$\int e^{2x} \, dx = \frac{1}{2}e^{2x} + c$$

$$\int \sin (2x + 3) \, dx = -\frac{1}{2} \cos (2x + 3) + c.$$

2. Is the integrand a product? If so,

    (a) is it of the form $\displaystyle\int f[g(x)]g'(x)$: Integrate by inspection or by suitable substitution.

    e.g. $\displaystyle\int 5x(x^2 + 4)^6 \, dx = \frac{5}{14}(x^2 + 4)^7 + c$ by inspection. The same result is obtained by letting $u = x^2 + 4$, and changing the variable from x to u.

    (b) If not of the above type, try integration by parts or change of variable:

    e.g. $\displaystyle\int x \ln x \, dx$ by parts gives $\frac{1}{2}x^2 \ln x - \frac{1}{4}x^2 + c.$

    $\displaystyle\int x(7 - x)^6 \, dx$ by substitution, $u = (7 - x)$, gives $\frac{1}{8}(x - 7)^7(x + 1) + c.$

3. Is the integrand a quotient? If so,

   (a) those of the type $\int \dfrac{f'(x)}{f(x)} dx$ can be integrated to give $\ln |f(x)| + c$,

   (b) those of the type $\int \dfrac{f'(x)}{[f(x)]^n} dx$ can be integrated by inspection or by change of variable:

   e.g. $\int \dfrac{2x}{(x^2 + 4)^7} dx$ is $-\dfrac{1}{6}(x^2 + 4)^{-6} + c$ by inspection or by the substitution $u = x^2 + 4$,

   (c) remember that expressing the integrand in partial fractions can enable some integrations to be carried out:

   e.g. $\displaystyle\int \dfrac{x^3 - 3x - 5}{(x + 1)(x - 2)} dx = \int \left(x + 1 + \dfrac{1}{x + 1} - \dfrac{1}{x - 2} \right) dx$

   $= \dfrac{x^2}{2} + x + \ln \left| \dfrac{x + 1}{x - 2} \right| + c,$

   (d) writing the integrand as two (or more) fractions by 'splitting' the numerator makes some integrations possible:

   e.g. $\displaystyle\int \dfrac{x + 1}{x^2 + 1} dx = \int \left(\dfrac{x}{x^2 + 1} + \dfrac{1}{x^2 + 1} \right) dx$

   $= \tfrac{1}{2} \ln (x^2 + 1) + \tan^{-1} x + c$

4. Remember the standard techniques for integrating odd and even powers of $\sin x$ and $\cos x$, and that there are seven functions for which integration is straightforward, provided that you remember the initial rearrangement.
   Note, particularly, parts (vi) and (vii) below, which have not been encountered earlier in this book:

   (i) $\displaystyle\int \tan x\, dx$: write as $\displaystyle\int \dfrac{\sin x}{\cos x} dx$, which integrates to give a log function.

   (ii) $\displaystyle\int \cot x\, dx$: write as $\displaystyle\int \dfrac{\cos x}{\sin x} dx$, which integrates to give a log function.

   (iii) $\displaystyle\int \ln x\, dx$: write as $\displaystyle\int 1 \times \ln x\, dx$ and integrate by parts.

   (iv) $\displaystyle\int \sin^{-1} x\, dx$: by a similar method to that used for $\displaystyle\int \ln x\, dx$.

   (v) $\displaystyle\int \cos^{-1} x\, dx$: by a similar method to that used for $\displaystyle\int \ln x\, dx$.

   (vi) $\displaystyle\int \sec x\, dx$: write as $\displaystyle\int \dfrac{\sec x(\sec x + \tan x)}{(\sec x + \tan x)} dx$ to give $\ln |\sec x + \tan x| + c$.

   (vii) $\displaystyle\int \operatorname{cosec} x\, dx$: write as $\displaystyle\int \dfrac{\operatorname{cosec} x(\operatorname{cosec} x - \cot x)}{(\operatorname{cosec} x - \cot x)} dx$ to give $\ln |\operatorname{cosec} x - \cot x| + c$.

5. If consideration of points 1 to 4 above do not yield a solution, try rearranging the integrand or making a suitable substitution:

    e.g. $\displaystyle\int 2 \sin 8x \cos 4x\, dx$ rearranges to $\displaystyle\int (\sin 12x + \sin 4x)\, dx$

    to give $-\dfrac{1}{12} \cos 12x - \dfrac{1}{4} \cos 4x + c.$

Exercise 20C

Find the following indefinite integrals

1. $\displaystyle\int (3x^2 - 6)\, dx$

2. $\displaystyle\int \tan x\, dx$

3. $\displaystyle\int \sin 2x\, dx$

4. $\displaystyle\int \dfrac{2}{x^2 - 1}\, dx$

5. $\displaystyle\int 3e^x\, dx$

6. $\displaystyle\int e^{3x}\, dx$

7. $\displaystyle\int \dfrac{1}{x^2}\, dx$

8. $\displaystyle\int \dfrac{1}{x}\, dx$

9. $\displaystyle\int \cos^3 2x\, dx$

10. $\displaystyle\int \dfrac{2}{(2x - 3)^3}\, dx$

11. $\displaystyle\int \dfrac{2}{2x + 3}\, dx$

12. $\displaystyle\int \dfrac{1}{\sqrt{(1 - x^2)}}\, dx$

13. $\displaystyle\int x(x - 1)^6\, dx$

14. $\displaystyle\int \dfrac{1}{4 + x^2}\, dx$

15. $\displaystyle\int \dfrac{8x}{(x^2 - 4)^5}\, dx$

16. $\displaystyle\int 2 \sin 5x \cos 4x\, dx$

17. $\displaystyle\int (x + 3)(x + 7)^5\, dx$

18. $\displaystyle\int \dfrac{1 + x}{\sqrt{(1 - x^2)}}\, dx$

19. $\displaystyle\int x \cos x\, dx$

20. $\displaystyle\int 4x(3x - 5)\, dx$

21. $\displaystyle\int \dfrac{10}{25 + x^2}\, dx$

22. $\displaystyle\int 9 \cos 3x\, dx$

23. $\displaystyle\int x^3 \ln x\, dx$

24. $\displaystyle\int 2^x\, dx$

25. $\displaystyle\int x\sqrt{(2x + 3)}\, dx$

26. $\displaystyle\int \dfrac{2x}{x^2 + 4}\, dx$

27. $\displaystyle\int \dfrac{6 - 2x}{(x + 1)(x^2 + 3)}\, dx$

28. $\displaystyle\int \ln x\, dx$

29. $\displaystyle\int 30 \cos 4x \cos x\, dx$

30. $\displaystyle\int 4 \sec^2 x\, dx$

31. $\displaystyle\int \dfrac{1}{\sqrt{(x + 5)}}\, dx$

32. $\displaystyle\int \sin^7 x\, dx$

33. $\displaystyle\int \cos^4 2x\, dx$

34. $\displaystyle\int \dfrac{\sin x}{\cos^2 x}\, dx$

35. $\displaystyle\int \dfrac{x}{x + 10}\, dx$

36. $\displaystyle\int (2x + 3)(x - 2)\, dx$

37. $\displaystyle\int xe^{-x}\, dx$

38. $\displaystyle\int \dfrac{2x^3 + 7x^2 + 2x - 10}{(x + 3)(2x - 1)}\, dx$

39. $\displaystyle\int \dfrac{1}{9 + 4x^2}\, dx$

40. $\displaystyle\int \dfrac{\sin 2x}{\cos x}\, dx$

41. $\displaystyle\int \dfrac{\sin 2x}{\cos 2x}\, dx$

42. $\displaystyle\int \dfrac{x + 2}{x^2 + 4x + 7}\, dx$

43. $\displaystyle\int \cos 7x \sin 5x\, dx$

44. $\displaystyle\int \sin^{-1}\left(\dfrac{x}{3}\right)\, dx$

45. $\displaystyle\int \sec 4x\, dx$

46. $\displaystyle\int \frac{\cos (2x + 3)}{\sin^4 (2x + 3)} dx$ **47.** $\displaystyle\int \frac{1}{x} \ln x\, dx$ **48.** $\displaystyle\int \frac{1}{\sqrt{(3 - 4x^2)}} dx$

49. $\displaystyle\int \sqrt{(16 - x^2)}\, dx$ **50.** $\displaystyle\int x \cos (x^2 + 3)\, dx$ **51.** $\displaystyle\int \tan^3 x\, dx$

52. $\displaystyle\int \frac{(1 + \cos x)}{(\sin x + x)} dx$ **53.** $\displaystyle\int \frac{x^3 + 11x + 3}{9 + x^2} dx$ **54.** $\displaystyle\int \frac{2x}{\cos^2 (x^2 + 3)} dx$

55. $\displaystyle\int \frac{2}{x^2 + 2x + 5} dx$ **56.** $\displaystyle\int \frac{1}{2x^2 - 5x + 3} dx$

20.4 Differential equations

A differential equation is an equation involving a differential coefficient

i.e. $\dfrac{dy}{dx}, \dfrac{d^2y}{dx^2}$ etc.

If the equation involves only the first of these, i.e. $\dfrac{dy}{dx}$, the equation is said to be of the first order. An equation involving the second differential coefficient,

i.e. $\dfrac{d^2y}{dx^2}$, is said to be of the second order.

In order to solve a differential equation, in the variables x and y, it is necessary to find some function $y = f(x)$ which satisfies the differential equation. Let us consider the first order differential equation, $\dfrac{dy}{dx} = 2x - 3$.

We know, by integration, that solutions to this differential equation will all be of the form $y = x^2 - 3x + c$. We say that $y = x^2 - 3x + c$ is the general solution of $\dfrac{dy}{dx} = 2x - 3$.

Indeed, if we were to draw the graphs of $y = x^2 - 3x + c$, for various values of c, we would obtain a 'family' of curves, each of which has the property that $\dfrac{dy}{dx} = 2x - 3$.

We call the complete, infinite family of curves, the solution curves of the differential equation. Given more information, say, that when $x = 1$, $y = 0$, then we can substitute these values into the general solution:

$$0 = (1)^2 - 3(1) + c \qquad c = 2$$

to give the particular solution $y = x^2 - 3x + 2$,

i.e. one particular member of the family of solution curves.

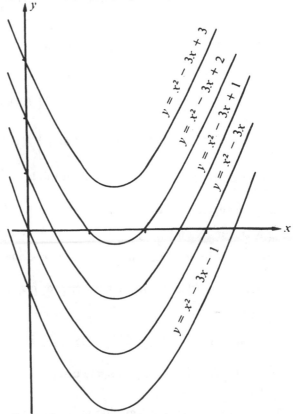

Example 13

If the general equation of a family of curves is given by $y = Ax^2$, find the corresponding differential equation for the family, giving your answer independent of the constant A.

$$y = Ax^2$$
$$\therefore \quad \frac{dy}{dx} = 2Ax$$

but $A = \dfrac{y}{x^2}$ $\therefore$ $\dfrac{dy}{dx} = \dfrac{2y}{x}$ is the required differential equation.

Solving first order differential equations

Any first order differential equation that can be expressed in the form

$$f(y)\frac{dy}{dx} = g(x),$$

can be solved by separating the variables:

Suppose $\qquad\qquad\qquad\qquad\qquad f(y)\dfrac{dy}{dx} = g(x)$

Integrating both sides with respect to x $\quad \displaystyle\int f(y)\frac{dy}{dx}dx = \int g(x)\,dx$

Earlier in this chapter we saw that the operation $\ldots \dfrac{dy}{dx}dx$ can be replaced by $\ldots dy$.

$$\therefore \qquad\qquad\qquad\qquad \int f(y)dy = \int g(x)\,dx$$

Thus, if $f(y)\dfrac{dy}{dx} = g(x)$, it follows that $\quad \displaystyle\int f(y)dy = \int g(x)dx$.

This result is easily remembered if we think of $\dfrac{dy}{dx}$ as a 'fraction' and view

the rearrangement from $f(y)\dfrac{dy}{dx} = g(x)$ to $f(y)dy = g(x)dx$ as 'putting the

x's with the dx and the y's with the dy'. It is important to realise, though,

that this is simply a useful way of remembering what to do; in fact, $\dfrac{dy}{dx}$ is not

a fraction and the method really depends on $\ldots \dfrac{dy}{dx}dx$ being equivalent to

$\ldots dy$.

The simplest case of $f(y)\dfrac{dy}{dx} = g(x)$ is for $f(y) = 1$, in which case we have

$\dfrac{dy}{dx} = g(x)$ which we can integrate directly to obtain y (see examples 14 and 15).

Remember also that integration gives rise to a constant c, and so we obtain the *general* solution of the differential equation. Only if we are given additional information, can we find a *particular* solution.

Example 14

Find the general solution of the differential equation, $\frac{dy}{dx} = 3x^2 - 4$.

$$\frac{dy}{dx} = 3x^2 - 4$$

By direct integration: or separating the variables and integrating:

$$y = x^3 - 4x + c \qquad \int dy = \int (3x^2 - 4)\, dx$$

$$y = x^3 - 4x + c$$

Thus $y = x^3 - 4x + c$ is the required general solution of the differential equation.

Example 15

Solve the differential equation, $\frac{dy}{dx} + 5 = 6x$, given that when $x = 1$, $y = 2$.

$$\frac{dy}{dx} + 5 = 6x$$

$$\therefore \qquad \frac{dy}{dx} = 6x - 5$$

$$y = 3x^2 - 5x + c$$

but when $x = 1$, $y = 2$ $\therefore$ $2 = 3(1)^2 - 5(1) + c$ giving $c = 4$

Thus $y = 3x^2 - 5x + 4$ is the required solution.

Example 16

Solve the differential equation, $\frac{dy}{dx} = \frac{2}{\sin y}$, giving y in terms of x.

$$\frac{dy}{dx} = \frac{2}{\sin y}$$

$$\therefore \quad \int \sin y\, dy = \int 2dx$$

$$-\cos y = 2x + c$$

thus $\qquad y = \cos^{-1}(-2x - c)$.

Note As c is the arbitrary constant which can take either positive or negative values, we lose no generality by writing this solution as $y = \cos^{-1}(c - 2x)$.

Example 17

Solve the equation, $\frac{dy}{dx} = \frac{1}{x^2 y}$, given that when $x = 1$, $y = 2$.

$$\frac{dy}{dx} = \frac{1}{x^2 y}$$

$$\therefore \quad \int y\, dy = \int \frac{1}{x^2}\, dx$$

$$\frac{y^2}{2} = -\frac{1}{x} + c$$

but when $x = 1$, $y = 2$

$$\therefore \qquad \frac{4}{2} = -\frac{1}{1} + c \quad \text{giving } c = 3.$$

Thus $y^2 = 6 - \frac{2}{x}$ is the required solution.

Example 18

Find the general solution of the differential equation, $(x + 2)\dfrac{dy}{dx} = 1$.

$$(x + 2)\frac{dy}{dx} = 1$$

$$\therefore \qquad \int dy = \int \frac{1}{x + 2} dx$$

$$y = \ln|x + 2| + c \qquad \qquad \dots [1]$$

If we require x in terms of y, equation [1] can be rearranged as:

$$y - c = \ln|x + 2|$$

$$\therefore \qquad |x + 2| = e^{y-c}$$

$$= e^y e^{-c}$$

$$\therefore \qquad x + 2 = Ae^y \quad \text{where} \quad |A| = e^{-c}.$$

giving $\qquad\qquad x = Ae^y - 2$

The modulus sign which appears in [1] can often be avoided in real life situations. Consider, for example, a body falling through a fluid and suppose that the rate at which the body is losing velocity at any given instant is proportional to the speed of the body at that instant,

then $\qquad \dfrac{dv}{dt} = -kv$ where k is the constant of proportionality

i.e. $\qquad \displaystyle\int \frac{1}{v} dv = -\int k\, dt$

thus $\qquad \ln v = -kt + c.$

In this particular case, we know that v cannot be negative so we do not need to write $\ln|v|$.

Example 19

Solve the differential equation $\dfrac{1}{x}\dfrac{dy}{dx} = \dfrac{\ln x}{\sin y}$,

given that when $x = e$, $y = \dfrac{\pi}{2}$.

$$\frac{1}{x}\frac{dy}{dx} = \frac{\ln x}{\sin y}$$

$$\therefore \int \sin y \, dy = \int x \ln x \, dx$$

integrating both sides, the R.H.S. being integrated by parts, we obtain

$$-\cos y = \frac{x^2}{2}\ln x - \frac{x^2}{4} + c$$

but when $x = e$, $y = \dfrac{\pi}{2}$

$$\therefore \qquad -\cos\frac{\pi}{2} = \frac{e^2}{2}\ln e - \frac{e^2}{4} + c$$

giving $\quad c = -\dfrac{e^2}{4}$

Thus $\cos y = \dfrac{x^2}{4} - \dfrac{x^2}{2}\ln x + \dfrac{e^2}{4}$ is the required solution.

Example 20

Find the general solution of the differential equation, $\dfrac{dy}{dx} = 3x^2(y + 1)$,

giving y in terms of x.

$$\frac{dy}{dx} = 3x^2(y + 1)$$

$$\therefore \quad \int \frac{1}{y + 1} dy = \int 3x^2 \, dx$$

$$\ln|y + 1| = x^3 + c$$

$$|y + 1| = e^{x^3 + c}$$

$$= e^{x^3} \times e^c$$

$$\therefore \qquad y + 1 = Ae^{x^3} \quad \text{where} \quad |A| = e^c$$

Thus $y = Ae^{x^3} - 1$ is the required general solution.

Example 21

Solve the differential equation, $(x + 2)\dfrac{dy}{dx} = (2x^2 + 4x + 1)(y - 3)$ given that when $x = 0$, $y = 7$.

$$(x + 2)\frac{dy}{dx} = (2x^2 + 4x + 1)(y - 3)$$

$$\therefore \quad \int \frac{1}{y - 3}\,dy = \int \frac{2x^2 + 4x + 1}{x + 2}\,dx$$

$$= \int \left(2x + \frac{1}{x + 2}\right)dx$$

$$\ln|y - 3| = x^2 + \ln|x + 2| + c$$

$$\therefore \quad \ln\left|\frac{y - 3}{x + 2}\right| = x^2 + c$$

$$\left|\frac{y - 3}{x + 2}\right| = e^{x^2 + c}$$

$$= e^{x^2} \times e^{c}$$

$$\therefore \quad \frac{y - 3}{x + 2} = Ae^{x^2} \quad \text{where } |A| = e^{c}$$

i.e. $\qquad y = Ae^{x^2}(x + 2) + 3$

but when $x = 0$, $y = 7$ $\quad\therefore\quad A = 2$

Thus $y = 2e^{x^2}(x + 2) + 3$ is the required solution.

Example 22

Find the general solution of the differential equation, $x\dfrac{dy}{dx} + 2\dfrac{dy}{dx} - 1 = y$ giving y in terms of x.

$$x\frac{dy}{dx} + 2\frac{dy}{dx} - 1 = y$$

$$\therefore \quad (x + 2)\frac{dy}{dx} = y + 1$$

i.e. $\displaystyle\int \frac{1}{y + 1}\,dy = \int \frac{1}{x + 2}\,dx$

$$\therefore \quad \ln|y + 1| = \ln|x + 2| + c$$

$$= \ln|A(x + 2)| \quad \text{where} \quad \ln|A| = c$$

$$\therefore \qquad y + 1 = A(x + 2)$$

Thus $y = A(x + 2) - 1$ is the required general solution.

Use of a substitution

In certain differential equations the variables may not, at first sight, appear to be separable. However, a suitable substitution may reduce the equation to a form in which the variables are separable. The substitution will, at this level, usually be suggested in any particular question.

Example 23

Use the substitution $y = vx$, where v is a function of x, to solve $\dfrac{dy}{dx} = \dfrac{x + 2y}{x}$ stating (a) the general solution, (b) the particular solution for which $y = 1\frac{1}{2}$ when $x = 3$.

(a) Now if $y = vx$, $\dfrac{dy}{dx} = v + x\dfrac{dv}{dx}$

Substituting in the given differential equation

gives $v + x\dfrac{dv}{dx} = \dfrac{x + 2vx}{x}$

$v + x\dfrac{dv}{dx} = 1 + 2v$

$\therefore \qquad\qquad x\dfrac{dv}{dx} = 1 + v$

Separating the variables

$$\int \frac{1}{v + 1}\,dv = \int \frac{1}{x}\,dx$$

$\ln |v + 1| = \ln |x| + c$

$\ln |v + 1| = \ln |Ax|$ where $\ln |A| = c$

$\therefore \qquad\qquad v + 1 = Ax$

giving $y = Ax^2 - x$

The general solution to $\dfrac{dy}{dx} = \dfrac{x + 2y}{x}$ is $y = Ax^2 - x$

(b) Substituting $y = 1\frac{1}{2}$ and $x = 3$ into the general solution gives $A = \frac{1}{2}$
Thus the required particular solution is $y = \frac{1}{2}x^2 - x$.

Note   In Example 23 the differential equation $\dfrac{dy}{dx} = \dfrac{x + 2y}{x}$ can be written as $\dfrac{dy}{dx} = 1 + 2(y/x)$ which is of the form $\dfrac{dy}{dx} = f(y/x)$. Any equation of this form is said to be *homogeneous*.

Such equations can, by using the substitution $y = vx$, be reduced to equations in which the variables are separable.

Proof: Given $\dfrac{dy}{dx} = f(y/x)$ ... [1]   and the substitution $y = vx$ ... [2]

From [2]: $\dfrac{dy}{dx} = v + x\dfrac{dv}{dx}$

Substitute into [1]: $v + x\dfrac{dv}{dx} = f(v)$

$x\dfrac{dv}{dx} = f(v) - v$

i.e.   a differential equation in which the variables are separable.

Exercise 20D

1. From each of the following equations obtain a first order differential equation that does not contain the arbitrary constant A. (For (g), (h) and (i) use implicit differentiation).

(a) $y = 3x^2 + Ax$ (b) $y = \dfrac{A}{x}$ (c) $y = 4x^2 - A$

(d) $y = Ae^{x^2}$ (e) $y = A \ln x$ (f) $y = A \cos x$

(g) $x^2 + 4y^2 = A$ (h) $y^3 = A(x^2 + 1)$ (i) $y^2 = xy + A$

Find the general solutions to the following differential equations, giving y in terms of x, in each case.

2. $\dfrac{dy}{dx} = x + 2$ **3.** $2y\dfrac{dy}{dx} = 3x^2$ **4.** $x^5\dfrac{dy}{dx} = 4y^2$

5. $\dfrac{dy}{dx} - e^x y^2 = 0$ **6.** $\dfrac{dy}{dx} = \dfrac{\cos x}{y}$ **7.** $\dfrac{dy}{dx} = \dfrac{1}{1 - x}$

8. $\dfrac{dy}{dx} + 2x \operatorname{cosec} y = 0$ **9.** $\dfrac{dy}{dx} = 2y$ **10.** $\dfrac{dy}{dx} = 3x^2(3 + y)$

11. $x\dfrac{dy}{dx} = \sec y$ **12.** $\dfrac{dy}{dx} = 2x(1 + y^2)$ **13.** $\dfrac{dy}{dx} = 4xe^{-y}$

14. $\dfrac{dy}{dx} = \dfrac{y}{x + 1}$ **15.** $(x + 1)\dfrac{dy}{dx} = x(y + 3)$ **16.** $2\dfrac{dy}{dx} = 3x^2(y^2 - 1)$

Solve the following differential equations, giving y in terms of x, in each case.

17. $\dfrac{dy}{dx} + 4 = 12x$; when $x = -2$, $y = 30$.

18. $\dfrac{dy}{dx} = y^2$; when $x = 3$, $y = -1$.

19. $3y^2\dfrac{dy}{dx} = 2x + 1$; when $x = 2$, $y = 2$.

20. $(\cos y)\dfrac{dy}{dx} = x^2 \operatorname{cosec}^2 y$; when $x = \frac{1}{2}$, $y = \dfrac{\pi}{2}$.

21. $x\dfrac{dy}{dx} = 2$; $x > 0$ and when $x = 1$, $y = -3$.

22. $x\dfrac{dy}{dx} = 2 + \dfrac{dy}{dx}$; $x > 1$ and when $x = 2$, $y = 1$.

23. $\dfrac{dy}{dx} = 4xy$; when $x = 0$, $y = 4$.

24. $\dfrac{dy}{dx} = 6x(y + 1)$; when $x = 0$, $y = 3$.

25. $\dfrac{dy}{dx} = (y - 3)(4x + 3)$; when $x = -1$, $y = 3(e^{-1} + 1)$.

26. $e^y\dfrac{dy}{dx} + \sin x = 0$; when $x = \dfrac{\pi}{2}$, $y = 1$.

27. $x\dfrac{dy}{dx} + y^2 = 0$; $x > 0$ and when $x = 1$, $y = \frac{1}{2}$.

28. $x\dfrac{dy}{dx} = y + 2$; when $x = 3$, $y = 7$.

29. $x\dfrac{dy}{dx} + 3 = y - 4\dfrac{dy}{dx}$; when $x = 1$, $y = 13$.

30. $2x\dfrac{dy}{dx} + 1 = y^2$; when $x = \frac{1}{3}$, $y = 2$.

31. $\dfrac{dy}{dx} = \dfrac{x^2y + y}{x^2 - 1}$; when $x = 0$, $y = 2$.

32. $(x^2 + 1)\dfrac{dy}{dx} = 2(y + 1)(x^2 + x + 1)$; when $x = 0$, $y = 3$.

33. Use the substitution $y = vx$, where v is a function of x, to solve

$x\dfrac{dy}{dx} = 2x - y$, stating

(a) the general solution and
(b) the particular solution for which $y = 5$ when $x = 1$.

34. Use the substitution $4x + y = z$ to solve $\dfrac{dy}{dx} = 4x + y$ given that

$y = 2$ when $x = 0$.

35. Use the substitution $z = 2x - 3y$ to solve

$(2x - 3y + 3)\dfrac{dy}{dx} = 2x - 2y + 1$ given that $y = 1$ when $x = 1$.

36. Use the substitution $y = vx$, where v is a function of x, to solve

$x^2 - y^2 + 2xy\dfrac{dy}{dx} = 0$, stating (a) the general solution and

(b) the particular solution for which $y = 4$ when $x = 2$.

37. Use the substitution $y = vx$, where v is a function of x, to find the

general solution of $(x - y)\dfrac{dy}{dx} = 2x + y$

38. Use the substitution $y = vx$, where v is a function of x, to find the

general solution of $\dfrac{dy}{dx} = \dfrac{x^2 + y^2}{x(x + y)}$

39. By substituting $x = X - 1$ and $y = Y + 3$ reduce the differential

equation $\dfrac{dy}{dx} = \dfrac{4x - y + 7}{2x + y - 1}$ to a homogeneous equation and hence find

the general solution in terms of x and y.

40. Find the general equation of the family of curves for which the gradient
at any point on the curve is the same as the y-coordinate at that point.

41. The rate at which a body loses speed at any given instant as it travels
through a resistive medium is given by kv m/s² where v is the speed of
the body at that instant and k is a positive constant. If its initial speed is
u m/s show that the time taken for the body to decrease its speed to

$\frac{1}{2}u$ m/s is $\dfrac{1}{k}\ln 2$ seconds.

42. The rate at which a body loses temperature at any instant is
proportional to the amount by which the temperature of the body, at
that instant, exceeds the temperature of its surroundings. A container of
hot liquid is placed in a room of temperature 18°C and in 6 minutes the
liquid cools from 82°C to 50°C. How long does it take for the liquid to
cool from 26°C to 20°C?

Second order differential equations

We consider three standard forms of second order differential equations.

(i) $\dfrac{d^2y}{dx^2} = f(x)$.

Differential equations of this type can be solved by integrating, with respect to x, twice. The first integration gives $\dfrac{dy}{dx}$ and the second integration gives y. Unless we are given information enabling us to determine the constants, a second order differential equation will have a general solution containing two constants.

Example 24

Solve the differential equation $\dfrac{d^2y}{dx^2} = 6x - 2$, given that when $x = 1$, $\dfrac{dy}{dx} = 2$ and $y = 3$.

$$\frac{d^2y}{dx^2} = 6x - 2$$

integrating with respect to x, $\quad \dfrac{dy}{dx} = 3x^2 - 2x + c_1$

but when $x = 1$, $\dfrac{dy}{dx} = 2$, $\therefore\quad c_1 = 1$

$$\therefore \quad \frac{dy}{dx} = 3x^2 - 2x + 1$$

integrating again, gives $\quad y = x^3 - x^2 + x + c_2$
but when $x = 1$, $y = 3$ $\quad \therefore\quad c_2 = 2$
Thus the required solution is $\quad y = x^3 - x^2 + x + 2$.

The reader should recognise this type of second order differential equation as one that is already familiar (see Example 4, chapter 12, page 296).

(ii) $a\dfrac{d^2y}{dx^2} + b\dfrac{dy}{dx} + cy = f(x)$.

At this level it is acceptable to quote and use the following results.

For $f(x) = 0$, i.e. $a\dfrac{d^2y}{dx^2} + b\dfrac{dy}{dx} + cy = 0$, we try $y = e^{px}$. On substitution into the differential equation, a quadratic in p is obtained, $ap^2 + bp + c = 0$. If this quadratic has real distinct roots p_1 and p_2, the general solution is $y = Ae^{p_1x} + Be^{p_2x}$.
If this quadratic has a repeated root p, the general solution is $y = (A + Bx)e^{px}$.
If this quadratic has unreal roots $p_1 \pm ip_2$, the general solution is $y = e^{p_1x}(A\cos p_2x + B\sin p_2x)$

For $f(x) \neq 0$, we first obtain the general solution of the differential equation with $f(x) = 0$. This is called the complementary function, and then we add a particular solution,
i.e. the general solution = complementary function + particular solution.

Example 25

Find the general solution to the differential equation $\dfrac{d^2y}{dx^2} - 4\dfrac{dy}{dx} + 3y = 0$

We try $y = e^{px}$, from which $\dfrac{dy}{dx} = pe^{px}$ and $\dfrac{d^2y}{dx^2} = p^2e^{px}$.
Substitution in the differential equation gives
$$p^2e^{px} - 4pe^{px} + 3e^{px} = 0$$
$$p^2 - 4p + 3 = 0$$
i.e. $(p - 3)(p - 1) = 0$ giving $p = 3$ or 1.
The general solution to the differential equation is $y = Ae^{3x} + Be^x$.

Example 26

Find the general solution to the differential equation $2\dfrac{d^2y}{dx^2} - \dfrac{dy}{dx} - 3y = e^{2x}$

We first find the complementary function by solving the equation
$2\dfrac{d^2y}{dx^2} - \dfrac{dy}{dx} - 3y = 0$.
Try $y = e^{px}$ then substitution in the differential equation gives
$$2p^2e^{px} - pe^{px} - 3e^{px} = 0$$
$$2p^2 - p - 3 = 0$$
i.e. $(2p - 3)(p + 1) = 0$ giving $p = \tfrac{3}{2}$ or -1
Thus the complementary function is $y = Ae^{3x/2} + Be^{-x}$.
To find a particular solution we try a solution of the form of the R.H.S. of
the differential equation i.e. $y = Ce^{2x}$
Substitution in the differential equation gives
$$2 \times 2^2Ce^{2x} - 2Ce^{2x} - 3Ce^{2x} = e^{2x}$$
giving $C = \tfrac{1}{3}$
Thus a particular solution is $y = \tfrac{1}{3}e^{2x}$ and the general solution to the
differential equation is $y = Ae^{3x/2} + Be^{-x} + \tfrac{1}{3}e^{2x}$.

Example 27

Find the general solution to the differential equation $4\dfrac{d^2y}{dx^2} + 4\dfrac{dy}{dx} + y = 3x + 4$.

To find the complementary function we solve $4\dfrac{d^2y}{dx^2} + 4\dfrac{dy}{dx} + y = 0$ by trying $y = e^{px}$
$$4p^2e^{px} + 4pe^{px} + e^{px} = 0$$
$$4p^2 + 4p + 1 = 0$$
i.e. $(2p + 1)^2 = 0$ giving $p = -\tfrac{1}{2}, -\tfrac{1}{2}$ a repeated root.
Thus the complementary function is $y = (A + Bx)e^{-x/2}$.
To find a particular solution we try a solution of the form of the R.H.S. of
the differential equation i.e. $y = Cx + D$.
Substitution in the differential equation gives $4(0) + 4C + (Cx + D) = 3x + 4$
Equating coefficients gives $C = 3$ and $D = -8$.

Thus a particular solution is $y = 3x - 8$ and the general solution to the differential equation is $y = (Ax + B)e^{-x/2} + 3x - 8$

(iii) $\dfrac{d^2y}{dx^2} \pm K^2y = $ a constant

This equation is the particular case of type (ii) obtained when $a = 1$, $b = 0$, $c = \pm K^2$ and $f(x) = $ a constant. However, this type of differential equation occurs in a number of scientific situations and it can be useful to quote the general solutions:

$$\dfrac{d^2y}{dx^2} + K^2y = c \text{ has general solution } y = A \cos Kx + B \sin Kx + \dfrac{c}{K^2}$$

$$\dfrac{d^2y}{dx^2} - K^2y = c \text{ has general solution } y = Ae^{Kx} + Be^{-Kx} - \dfrac{c}{K^2}$$

$$(A, B, K \text{ and } c \text{ are constants})$$

Exercise 20E

1. Find the general solutions to the following second order differential equations

   (a) $\dfrac{d^2y}{dx^2} = 36x^2 + 2$

   (b) $\dfrac{d^2y}{dx^2} = x + \sin x$

2. Solve the differential equation $\dfrac{d^2y}{dx^2} = 32e^{2x} + 6x$, given that when

   $x = 0$, $y = 2$ and $\dfrac{dy}{dx} = 20$.

3. Solve the differential equation $\dfrac{d^2y}{dx^2} = 10$, given that when $x = -1$, $y = 3$

   and when $x = 1$, $y = 15$.

4. Solve the differential equation $\dfrac{d^2y}{dx^2} = x \ln x$, given that when $x = 1$,

   $y = \dfrac{1}{2}$ and $\dfrac{dy}{dx} = \dfrac{1}{12}$.

5. Find the general solution to each of the following differential equations.

   (a) $\dfrac{d^2y}{dx^2} - 3\dfrac{dy}{dx} - 10y = 0$

   (b) $\dfrac{d^2y}{dx^2} - 8\dfrac{dy}{dx} + 16y = 0$

   (c) $\dfrac{d^2y}{dx^2} - 2\dfrac{dy}{dx} - 15y = e^{4x}$

   (d) $2\dfrac{d^2y}{dx^2} - 5\dfrac{dy}{dx} - 3y = 4e^{5x}$

   (e) $\dfrac{d^2y}{dx^2} - 2\dfrac{dy}{dx} + 5y = 10x + 1$

   (f) $2\dfrac{d^2y}{dx^2} + 11\dfrac{dy}{dx} + 12y = 2x^2 + 5x - 7$

6. Verify that the general solution $y = A \cos Kx + B \sin Kx + \dfrac{c}{K^2}$ satisfies

   the differential equation $\dfrac{d^2y}{dx^2} + K^2y = c$.

7. Verify that the general solution $y = Ae^{Kx} + Be^{-Kx} - \dfrac{c}{K^2}$ satisfies the

   differential equation $\dfrac{d^2y}{dx^2} - K^2y = c$.

8. Use the standard results quoted on page 519 to write down the general solutions of the following differential equations

(a) $\dfrac{d^2y}{dx^2} + 36y = 1$

(b) $\dfrac{d^2y}{dx^2} - 9y = 18$

(c) $\dfrac{d^2y}{dx^2} + 9y = 18$

(d) $4\dfrac{d^2y}{dx^2} - 9y = 36$

9. Use the standard results quoted on page 519 to solve the following differential equations

(a) $\dfrac{1}{4}\dfrac{d^2y}{dx^2} + y = 2$ given that when $x = \dfrac{\pi}{4}$, $y = 1$ and when $x = \dfrac{\pi}{2}$, $y = -1$.

(b) $\dfrac{d^2y}{dx^2} - y = 2$ given that when $x = 0$, $y = 2$ and $\dfrac{dy}{dx} = 6$.

(c) $\dfrac{d^2y}{dx^2} + \dfrac{1}{4}y = 4$ given that when $x = \pi$, $y = 19$ and $\dfrac{dy}{dx} = -1$.

(d) $4\dfrac{d^2y}{dx^2} - y = 4$ given that when $x = 0$, $\dfrac{dy}{dx} = 2$ and $\dfrac{d^2y}{dx^2} = \dfrac{3}{2}$.

Exercise 20F   Examination Questions

1. Evaluate

(a) $\displaystyle\int_0^\pi \sin\dfrac{3x}{2}\,dx$, (b) $\displaystyle\int_0^a e^{-2x}\,dx$, where $a = \log_e 2$, (c) $\displaystyle\int_1^2 \dfrac{1}{3x+2}\,dx$.

(A.E.B.)

2. Evaluate the integrals

(a) $\displaystyle\int_0^{3/4} x\sqrt{(1+x^2)}\,dx$,

(b) $\displaystyle\int_0^{\pi/4} \tan^2 x\,dx$,

(c) $\displaystyle\int_0^\pi x\sin x\,dx$,

(d) $\displaystyle\int_3^4 \dfrac{1}{x^2-3x+2}\,dx$.

Express your answer to (d) as a natural logarithm. (London)

3. Find

(a) $\displaystyle\int x\ln x\,dx$

(b) $\displaystyle\int \dfrac{x}{\sqrt{(x-2)}}\,dx$. (London)

4. Using the substitution $t = \sin x$, evaluate to two decimal places the integral

$$\int_{\pi/6}^{\pi/2} \dfrac{4\cos x}{3+\cos^2 x}\,dx.$$ (A.E.B.)

5. Evaluate the integrals

(i) $\displaystyle\int_0^1 \dfrac{2x-1}{(x-3)^3}\,dx$ (by substitution or otherwise),

(ii) $\displaystyle\int_0^{\pi/4} \sin^2 3x\,dx.$ (Oxford)

6. (a) Find the indefinite integrals of

(i) $(x - 2/x)^2$ (ii) $\cot 2x$ (iii) $\dfrac{1}{\sqrt{(1 - 2x)}}$ (iv) $x \log_e 2x$

(b) Evaluate $\displaystyle\int_0^{\pi/2} \cos^3 x \, dx$.

(c) Using the substitution $t = \log_e x$ evaluate $\displaystyle\int_e^{e^3} \dfrac{dx}{x(\log_e x)^2}$.   (S.U.J.B.)

7. Write down, or obtain, the derivative of $\tan^{-1} 2x$.

Find (i) $\displaystyle\int \dfrac{x + 2}{4x^2 + 1} dx$,   (ii) $\displaystyle\int \dfrac{4x}{4x^4 + 1} dx$.

(J.M.B.)

8. Find the volume generated when the region bounded by the curve

$$y = \left(\frac{1 - x^2}{1 + x^2}\right)^{1/2}$$

and the x-axis is rotated through four right-angles about the x-axis.

(Oxford)

9. Find y in terms of x given that

$$x\frac{dy}{dx} = y(y + 1)$$

and $y = 4$ when $x = 2$.   (London)

10. Given that

$$\frac{dy}{dx} - 2xy = 0$$

and that $y = 1$ when $x = 1$, show that $y \geqslant \dfrac{1}{e}$ for all values of x.

(Cambridge)

11. Solve the differential equation $\dfrac{dy}{dx} = \dfrac{y^2 - 1}{2 \tan x}$, given that $y = 3$ when

$x = \dfrac{\pi}{2}$.

Hence express y in terms of x.   (A.E.B.)

12. For all positive values of x the gradient of a curve at the point (x, y) is

$\dfrac{y}{x^2 + x}$. The point $A(3, 6)$ lies on this curve.

(i) Calculate the equation of the normal to the curve at A.
(ii) Find the equation of the curve in the form $y = f(x)$.   (A.E.B.)

13. Find the general solution of the differential equation

$$\frac{dy}{dx} = \frac{y^2}{(x^2 - x - 2)}$$

in the region $x > 2$. Find also the particular solution which satisfies
$y = 1$ when $x = 5$.   (Oxford)

14. (a) Evaluate $\int_1^2 x^3 \log_e x \, dx$ correct to 3 significant figures.

(b) Evaluate $\int_0^{\pi/3} \sin x \sec^2 x \, dx$.

(c) If $\dfrac{dy}{dx} = 2x + y$ and $z = 2x + y$, show that $\dfrac{dz}{dx} = 2 + z$.

Find the equation of the curve which passes through the origin and is such that $\dfrac{dy}{dx} = 2x + y$ at all points (x, y) on the curve, giving the equation in the form $y = f(x)$. (S.U.J.B.)

15. (i) Find y in terms of x, given that
$$x\frac{dy}{dx} = \cos^2 y,$$
$x > 0$ and that $y = \pi/3$ when $x = 1$.

(ii) Given that
$$xy + x^3y + Cy = 1,$$
construct a first order differential equation between x and y which is satisfied for all values of C. Deduce that the gradient at all points of the integral curves of this differential equation is never positive. (London)

16. In a certain type of chemical reaction a substance A is continuously transformed into a substance B. Throughout the reaction the sum of the masses of A and B remains constant and equal to m. The mass of B present at time t after the commencement of the reaction is denoted by x. At any instant the rate of increase of the mass of B is k times the mass of A, where k is a positive constant. Write down the differential equation relating x and t.
Solve this differential equation, given that $x = 0$ when $t = 0$.
Given also that $x = \frac{1}{2}m$ when $t = \ln 2$, determine the value of k, and show that, at time t,
$$x = m(1 - e^{-t}).$$
Hence find
(i) the value of x (in terms of m) when $t = 3 \ln 2$,
(ii) the value of t when $x = \frac{3}{4}m$. (Cambridge)

17. (i) Find the solution of the differential equation
$$\frac{dy}{dx} = \frac{\sin^2 x}{y^2}$$
which also satisfies $y = 1$ when $x = 0$.

(ii) Find the solution of the differential equation
$$\frac{d^2y}{dx^2} = 1 - 4y$$
which also satisfies $y = 1$ when $x = 0$ and $y = \frac{1}{4}$ when $x = \frac{1}{8}\pi$.
(Oxford)

18. (a) Find the solution of the differential equation

$$x\frac{dy}{dx} = 3x - 2y, \quad \text{for} \quad x > 0$$

given that $y = \frac{3}{4}$ when $x = 1$

(b) For the differential equation $\dfrac{d^2y}{dx^2} + 3\dfrac{dy}{dx} + 2y = 6e^{-3x} - 1$

find (i) the general solution

(ii) the limit to which y tends as x tends to infinity.

19. Prove that

$$\int_a^b f(x)\,dx = \int_a^b f(a + b - t)\,dt.$$

Hence prove that, if $0 < \beta < \frac{1}{2}\pi$,

$$\int_\beta^{\pi-\beta} \frac{\theta\,d\theta}{\sin\theta} = \pi \ln \cot\left(\tfrac{1}{2}\beta\right). \qquad\qquad \text{(Oxford)}$$

21
Numerical methods

21.1 The trapezium rule

As we saw in chapter 12 the area under the curve $y = f(x)$, from $x = a$ to $x = b$ is given by

$$A = \int_{x=a}^{x=b} f(x)\,dx.$$

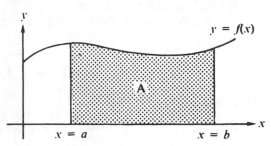

Suppose this area A is divided up into n strips each of width h. If we consider each strip to approximate to a trapezium, then

$$A \approx \tfrac{1}{2}(y_0 + y_1)h + \tfrac{1}{2}(y_1 + y_2)h + \ldots + \tfrac{1}{2}(y_{n-1} + y_n)h$$

$$\therefore \quad A \approx \tfrac{1}{2}h\left[y_0 + 2(y_1 + y_2 + \ldots + y_{n-1}) + y_n \right]$$

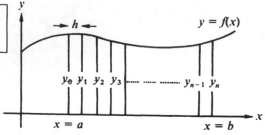

Example 1

Find an approximate value for $\int_0^1 x^2 e^x\,dx$, using the trapezium rule with 5 strips, giving your answer correct to 4 significant figures.

Using 5 strips and $0 \leqslant x \leqslant 1$ implies that $h = 0 \cdot 2$.

x	x^2	e^x	$y(= x^2 e^x)$
0	0	1·0	$y_0 = 0$
0·2	0·04	1·22140	$y_1 = 0 \cdot 04886$
0·4	0·16	1·49182	$y_2 = 0 \cdot 23869$
0·6	0·36	1·82212	$y_3 = 0 \cdot 65596$
0·8	0·64	2·22554	$y_4 = 1 \cdot 42435$
1·0	1·0	2·71828	$y_5 = 2 \cdot 71828$

Using the trapezium rule

$$\int_0^1 x^2 e^x\,dx \approx \tfrac{1}{2}(0 \cdot 2)\left[0 + 2(0 \cdot 04886 + 0 \cdot 23869 + 0 \cdot 65596 + 1 \cdot 42435) + 2 \cdot 71828 \right]$$

$$= (0 \cdot 1)(7 \cdot 4540)$$

$$\therefore \quad \int_0^1 x^2 e^x\,dx \approx 0 \cdot 7454$$

21.2 Simpson's rule

A more accurate approximation is obtained by the use of Simpson's rule which takes the top of each strip as approximating to a quadratic curve.

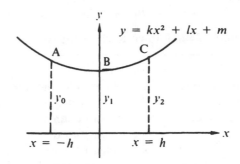

Suppose that a curve of the form $y = kx^2 + lx + m$ passes through the points A $(-h, y_0)$, B $(0, y_1)$ and C (h, y_2). The area under the curve, between the ordinates $x = -h$ and $x = h$, is given by

$$\int_{x=-h}^{x=h} (kx^2 + lx + m)\, dx = \left[\frac{kx^3}{3} + \frac{lx^2}{2} + mx\right]_{-h}^{h}$$

$$= \tfrac{2}{3}h(kh^2 + 3m) \qquad \ldots [1]$$

Since A, B and C lie on the curve,

$$y_0 = kh^2 - lh + m \qquad \ldots [2]$$
$$y_1 = m \qquad \ldots [3]$$
$$y_2 = kh^2 + lh + m \qquad \ldots [4]$$

Adding equations [2] and [4] and substituting for m from equation [3] gives

$$y_0 + y_2 - 2y_1 = 2kh^2$$

i.e. $\tfrac{1}{2}(y_0 + y_2) - y_1 = kh^2$

Substituting these values of m and kh^2 into equation [1] gives

$$\int_{x=-h}^{x=h} (kx^2 + lx + m)\, dx = \tfrac{2}{3}h\left[\tfrac{1}{2}(y_0 + y_2) - y_1 + 3y_1\right]$$

$$= \tfrac{1}{3}h(y_0 + 4y_1 + y_2)$$

Now this result is dependent only upon the values y_0, y_1 and y_2 and the distance between these ordinates. Thus

$$\int_{a}^{b} (kx^2 + lx + m)\, dx = \tfrac{1}{3}h\left[y_0 + 4y_1 + y_2\right]$$

where $f(a) = y_0$, $f(a + h) = y_1$, $f(a + 2h) = f(b) = y_2$.

Thus to obtain an approximation for $\int_{a}^{b} f(x)\, dx$, the interval from a to b is

divided into an even number of strips, say n, each of width h and we assume that the top of each strip approximates to a quadratic curve. Then if y_0, y_1, y_2, ... y_n, are the successive ordinates at the ends of these strips,

$$\int_{a}^{b} f(x)\, dx \approx \tfrac{1}{3}h\left[(y_0 + 4y_1 + y_2) + (y_2 + 4y_3 + y_4) + \ldots + (y_{n-2} + 4y_{n-1} + y_n)\right]$$

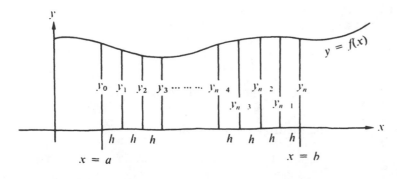

$$\therefore \quad \int_a^b f(x)\,dx \approx \tfrac{1}{3}h\left[y_0 + 4y_1 + 2y_2 + 4y_3 + 2y_4 + \ldots + 2y_{n-2} + 4y_{n-1} + y_n\right]$$

and this is known as the approximation from Simpson's rule.

Note that in order to apply this rule, the interval is divided into an *even* number of strips, i.e. there are an *odd* number of ordinates.
Numerical methods of integration can be particularly useful when the function to be integrated is one that does not integrate readily, or one for which we know the coordinates of various points on the graph of the function, but we do not know the equation of the function (as in Example 2 below).

Example 2

Given that $y = f(x)$ and that for the given values of x, the corresponding values of y are as shown in the table, use Simpson's rule with nine ordinates to find an approximate value for $\int_{-1}^7 f(x)\,dx$.

x	-1	0	1	2	3	4	5	6	7
y	17	10	5	2	1	2	5	10	17

Applying Simpson's rule, with $h = 1$, and $y_0 = 17$, $y_1 = 10$, etc.

$$\int_{-1}^7 f(x)\,dx \approx \tfrac{1}{3}\left[17 + 4(10) + 2(5) + 4(2) + 2(1) + 4(2) + 2(5) + 4(10) + 17\right]$$

$$\approx 1\tfrac{5}{3}2 = 50\cdot7 \quad \text{to one decimal place.}$$

Exercise 21A

In this exercise give your answers correct to 4 significant figures unless told otherwise.

1. Given that $y = f(x)$ and for the given values of x the corresponding values of y are as shown in the table below, use the trapezium rule to find an approximate value for $\int_1^5 f(x)\,dx$.

x	1	2	3	4	5
y	0	3	4	7	4

2. Find an approximate value for $\int_0^1 e^x\,dx$ using the trapezium rule with 5 strips.

3. Find approximate values for $\int_1^2 \dfrac{1}{x}\,dx$ using the trapezium rule with

    (a) 5 strips (b) 10 strips.

4. Given that $y = f(x)$ and for the given values of x the corresponding values of y are as shown in the table below, use Simpson's rule with eleven ordinates to find an approximate value for $\int_0^{10} f(x)\, dx$

x	0	1	2	3	4	5	6	7	8	9	10
y	2	3	5	6	7	7	8	4	3	5	7

5. Use Simpson's rule with 10 strips to find an approximation for $\int_0^1 \dfrac{1}{1 + x}\, dx$.

6. Use Simpson's rule with (a) 5 ordinates, (b) 10 ordinates to find approximations for $\int_0^2 \dfrac{1}{1 + x^2}\, dx$.

(Give your answers to 4 decimal places in each case).

7. Find an approximate value for $\int_0^1 e^{x^2}\, dx$ by (a) the trapezium rule and (b) Simpson's rule, using 8 strips in each case.

8. Find an approximate value for $\int_0^{\pi/2} \sin x\, dx$ by (a) the trapezium rule and (b) Simpson's rule, using 8 strips in each case.

9. If $A = \displaystyle\int_0^{0.6} \dfrac{1}{\sqrt{(1 - x^2)}}\, dx$

(a) find A by integration,
(b) estimate A using the trapezium rule and six strips
(c) estimate A using Simpson's rule and six strips
(d) estimate A by integrating the first four terms of the series expansion of $(1 - x^2)^{-1/2}$.

21.3 Taylor's theorem

If $f(x + h) = (x + h)^n$, we can use the binomial expansion to obtain a series expansion of $f(x + h)$ in powers of x. However, can we obtain a power series expansion when $f(x + h)$ is not of the form $(x + h)^n$? Let us suppose that such a series expansion does exist.

i.e. $\quad f(x + h) = c_0 + c_1 x + c_2 x^2 + \quad c_3 x^3 + \quad\quad c_4 x^4 + \cdots$

then $\quad f'(x + h) = \quad\quad c_1 + 2c_2 x + \quad 3c_3 x^2 + \quad\quad 4c_4 x^3 + \cdots$

$\quad f''(x + h) = \quad\quad\quad\quad 2c_2 + 3 \times 2c_3 x + \quad 4 \times 3c_4 x^2 + \cdots$

$\quad f'''(x + h) = \quad\quad\quad\quad\quad\quad 3 \times 2c_3 + 4 \times 3 \times 2c_4 x + \cdots$

$\quad f''''(x + h) = \quad\quad\quad\quad\quad\quad\quad\quad\quad 4 \times 3 \times 2c_4 + \cdots$

and so on.

Now if $x = 0$, then the above equations give:

$c_0 = f(h)$
$c_1 = f'(h)$

$2c_2 = f''(h) \quad$ i.e. $\quad c_2 = \dfrac{1}{2!} f''(h)$

$$3 \times 2c_3 = f'''(h) \quad \text{i.e.} \quad c_3 = \frac{1}{3!}f'''(h)$$

$$4 \times 3 \times 2c_4 = f''''(h) \quad \text{i.e.} \quad c_4 = \frac{1}{4!}f''''(h) \quad \text{etc.}$$

> Thus $f(x + h) = f(h) + \frac{x}{1!}f'(h) + \frac{x^2}{2!}f''(h) + \frac{x^3}{3!}f'''(h) + \frac{x^4}{4!}f''''(h) + \dots$
>
> and this result is known as Taylor's theorem.

Example 3

Use Taylor's theorem to obtain a series expansion for $\cos\left(x + \frac{\pi}{3}\right)$ stating

terms up to and including that in x^3. Hence obtain a value for cos 62° giving your answer correct to five decimal places.

In this case $f(x) = \cos x,$ $f(x + h) = \cos\left(x + \frac{\pi}{3}\right), h = \frac{\pi}{3}.$

$$f(x) = \cos x \qquad \therefore \quad f(h) = \cos\frac{\pi}{3} = \frac{1}{2}$$

$$f'(x) = -\sin x \qquad \therefore \quad f'(h) = -\sin\frac{\pi}{3} = -\frac{\sqrt{3}}{2}$$

$$f''(x) = -\cos x \qquad \therefore \quad f''(h) = -\cos\frac{\pi}{3} = -\frac{1}{2}$$

$$f'''(x) = \sin x \qquad \therefore \quad f'''(h) = \sin\frac{\pi}{3} = \frac{\sqrt{3}}{2}$$

By Taylor's theorem

$$f(x + h) = f(h) + \frac{x}{1!}f'(h) + \frac{x^2}{2!}f''(h) + \frac{x^3}{3!}f'''(h) + \dots$$

$\therefore \quad \cos\left(x + \frac{\pi}{3}\right) = \frac{1}{2} - \frac{\sqrt{3}}{2}x - \frac{1}{4}x^2 + \frac{\sqrt{3}}{12}x^3 + \dots$ which is the required expansion.

Now $\cos 62° = \cos\left(2\frac{\pi}{180} + \frac{\pi}{3}\right)$

$$= \frac{1}{2} - \frac{\sqrt{3}}{2}\left(\frac{\pi}{90}\right) - \frac{1}{4}\left(\frac{\pi}{90}\right)^2 + \frac{\sqrt{3}}{12}\left(\frac{\pi}{90}\right)^3 + \dots$$

$$= 0 \cdot 5 - 0 \cdot 030230 - 0 \cdot 000305 + 0 \cdot 000006 + \dots$$

$$= 0 \cdot 469471$$

Thus $\cos 62° = 0 \cdot 46947$ correct to five decimal places.

21.4 Maclaurin's theorem

Maclaurin's theorem is a particular case of Taylor's theorem obtained by letting $h = 0$.

Thus, from $f(x + h) = f(h) + \frac{x}{1!}f'(h) + \frac{x^2}{2!}f''(h) + \frac{x^3}{3!}f'''(h) + \dots$ with $h = 0$,

> $f(x) = f(0) + \frac{x}{1!}f'(0) + \frac{x^2}{2!}f''(0) + \frac{x^3}{3!}f'''(0) + \dots$
>
> and this result is known as Maclaurin's theorem.

Example 4

Use Maclaurin's theorem to find the first three non-zero terms in the expansion of sin x. Hence find $\sin (0.1)^{\text{rad}}$ correct to seven decimal places.

In this case
$$
\begin{aligned}
f(x) &= \sin x, & f(0) &= \sin 0 &&= 0 \\
f'(x) &= \cos x & f'(0) &= \cos 0 &&= 1 \\
f''(x) &= -\sin x & f''(0) &= -\sin 0 &&= 0 \\
f'''(x) &= -\cos x & f'''(0) &= -\cos 0 &&= -1 \\
f''''(x) &= \sin x & f''''(0) &= \sin 0 &&= 0 \\
f'''''(x) &= \cos x & f'''''(0) &= \cos 0 &&= 1
\end{aligned}
$$

By Maclaurin's theorem

$$f(x) = f(0) + \frac{x}{1!}f'(0) + \frac{x^2}{2!}f''(0) + \frac{x^3}{3!}f'''(0) + \frac{x^4}{4!}f''''(0) + \frac{x^5}{5!}f'''''(0) + \ldots$$

$$= 0 + x(1) + \frac{x^2}{2!}(0) + \frac{x^3}{3!}(-1) + \frac{x^4}{4!}(0) + \frac{x^5}{5!}(1) + \ldots$$

$$\therefore \qquad \sin x = \frac{x}{1!} - \frac{x^3}{3!} + \frac{x^5}{5!} \ldots \text{ which is the required expansion.}$$

Thus $\quad \sin (0.1 \text{ rad}) \approx \dfrac{(0.1)}{1!} - \dfrac{(0.1)^3}{3!} + \dfrac{(0.1)^5}{5!}$

$$= 0.1 - 0.00016667 + 0.00000008$$
$$= 0.09983341$$

i.e. $\quad \sin (0.1 \text{ rad}) = 0.0998334 \quad$ correct to seven decimal places.

Exercise 21B

1. Use Maclaurin's theorem to find series expansions for each of the following giving all terms up to and including that in x^5.
   (a) $\cos x$, (b) e^x, (c) $(1 + x)^n$, (d) $\ln (1 + x)$, (e) a^x, (f) $\sin^{-1} x$.
   Use your answer to (a) to determine $\cos (0.1 \text{ rads})$ correct to six decimal places. Use your answer to (f) to determine $\sin^{-1} (0.2)$ correct to four decimal places.
2. Use Maclaurin's theorem to find the series expansion of $\tan x$ giving all terms up to and including that in x^5. Hence show that if x is sufficiently small for terms in x^6 and higher powers to be neglected then
   $$\frac{\tan x}{(1 + x)} = x - x^2 + \frac{4}{3}x^3 - \frac{4}{3}x^4 + \frac{22}{15}x^5.$$
3. Using Taylor's theorem obtain a series expansion for $\sin (\theta + x)$ where θ, an acute angle, is measured in radians and is such that $\sin \theta = \frac{4}{5}$ (give the first five non-zero terms).
   Hence find $\sin (\theta + 0.1 \text{ rad})$ correct to four decimal places.
4. Using Taylor's theorem obtain a series expansion for $\tan \left(\dfrac{\pi}{4} + x\right)$, stating the terms up to and including that in x^3.
   Hence find $\tan 46°$ correct to four decimal places.

21.5 Solution of equations

In some instances it may be almost impossible to use an exact method to solve an equation. In such cases we may be able to use other techniques which give good approximations to the solution.

Two such approaches, graphical and iterative, are explained overleaf.

Graphical methods

To solve the equation $f(x) = 0$, graphically, we draw the graph of $y = f(x)$ and read from it the values of x for which $f(x) = 0$, i.e. the x-coordinates of the points where the line $y = f(x)$ cuts the x-axis.

Alternatively we could rearrange $f(x) = 0$, in the form $g(x) = h(x)$, and find the x-coordinates of the points where the lines $y = g(x)$ and $y = h(x)$ intersect.

Example 5

Copy and complete the following table for $y = x^3$ and $y = x + 1$.

x	$-1\cdot0$	$-0\cdot5$	0	$0\cdot5$	$1\cdot0$	$1\cdot5$	$2\cdot0$
$y = x^3$							

x	-1	0	1	2
$y = x + 1$				

Using one pair of axes draw the graphs of $y = x^3$ and $y = x + 1$, for $-1 \leqslant x \leqslant 2$. Use your graphs to find an approximate solution to the equation $x^3 - x - 1 = 0$ in the range $-1 \leqslant x \leqslant 2$.

$y = x^3$

x	$-1\cdot0$	$-0\cdot5$	0	$0\cdot5$	1	$1\cdot5$	2
y	$-1\cdot0$	$-0\cdot125$	0	$0\cdot125$	1	$3\cdot375$	8

$y = x + 1$

x	-1	0	1	2
y	0	1	2	3

From the intersection of the graphs, an approximate solution to the equation $x^3 = x + 1$ (i.e. $x^3 - x - 1 = 0$) is $1\cdot32$.

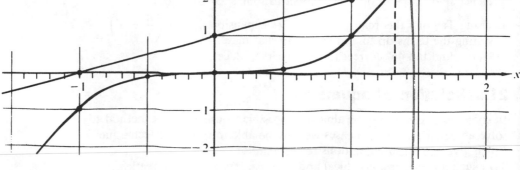

Iterative methods

A darts player will frequently use the first of his three darts as a 'marker'. He will then adjust the throwing of the second and third dart dependent upon where the first dart landed on the board. In a similar way a bowls player will use the first wood to gauge the distance of the jack and the speed of the green. Thus we can learn from one trial, or guess, so that our second attempt is an improvement upon the first. We can use similar methods to help us to obtain increasingly accurate approximations to the solutions of equations from an initial educated guess.

If we wish to solve an equation $f(x) = 0$ by an iterative method, we need to find a relationship $x_{r+1} = F(x_r)$ such that x_{r+1} is a better approximation to the solution of the equation than is x_r.

Suppose we can rearrange $f(x) = 0$ into the form $x = F(x)$, e.g. $x^2 - 3x + 1 = 0$ can be written as $x(x - 3) + 1 = 0$ to give

$x = \dfrac{1}{3 - x}$, or alternatively $x = 3 - \dfrac{1}{x}$, both

of which are rearrangements into the form $x = F(x)$.

Suppose that the graphs of $y = x$ and $y = F(x)$ are as shown, $x = X$ is the solution to $F(x) = x$, and that our initial guess is x_0. From the graph we see that $x_1 = F(x_0)$ gives a better approximation than x_0, $x_2 = F(x_1)$ gives a better approximation than x_1, and so on.

In this way we can obtain a value x_r which is as close to X as is required.

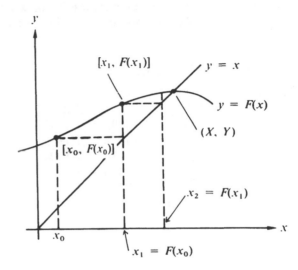

Example 6

Show that the equation $x^2 - 3x + 1 = 0$ has a root between $x = 2$ and $x = 3$. Using the rearrangement $x = 3 - \dfrac{1}{x}$, and an initial value of $x = 3$, use an iterative method to find this root correct to two decimal places.

If $f(x) = x^2 - 3x + 1$ then $f(2) = -1$
and $f(3) = 1$.

The change in sign of $f(x)$ between $x = 2$ and $x = 3$ shows that there is some value of x between 2 and 3 for which $f(x) = 0$, i.e. a solution to the equation $x^2 - 3x + 1 = 0$ lies between $x = 2$ and $x = 3$.

From the rearranged form $x = 3 - \dfrac{1}{x}$, we use the iterative formula

$x_{r+1} = 3 - \dfrac{1}{x_r}$, with $x_0 = 3$.

$$x_1 = 3 - \frac{1}{3} \quad = 2{\cdot}6667 \quad \left(\text{i.e. } x_1 = 3 - \frac{1}{x_0} \right)$$

$$x_2 = 3 - \frac{1}{2{\cdot}6667} = 2{\cdot}6250 \quad \left(\text{i.e. } x_2 = 3 - \frac{1}{x_1} \right)$$

$$x_3 = 3 - \frac{1}{2 \cdot 6250} = 2 \cdot 6190$$

$$x_4 = 3 - \frac{1}{2 \cdot 6190} = 2 \cdot 6182$$

$$x_5 = 3 - \frac{1}{2 \cdot 6182} = 2 \cdot 6181$$

Thus the required root is $x = 2 \cdot 62$, correct to two decimal places.

Note: It should not be assumed that this method will always work. In some cases the values $x_0, x_1, x_2, x_3 \ldots$ diverge. For example $x^3 + 2x - 4 = 0$ has a solution near $x = 1$. Using the rearranged form $x_{r+1} = \frac{1}{2}[4 - (x_r)^3]$, we do not obtain converging values for $x_0, x_1, x_2 \ldots$, but using $x_{r+1} = \sqrt[3]{(4 - 2x_r)}$ we obtain $x = 1 \cdot 18$, correct to two decimal places, as is shown below:

$$x_{r+1} = \sqrt[3]{(4 - 2x_r)}$$

if $x_0 = 1$,

$$\begin{aligned} x_1 &= \sqrt[3]{2} & = 1 \cdot 2599 \\ x_2 &= \sqrt[3]{[4 - 2(1 \cdot 2599)]} &= 1 \cdot 1397 \\ x_3 &= \sqrt[3]{[4 - 2(1 \cdot 1397)]} &= 1 \cdot 1983 \\ x_4 &= \sqrt[3]{[4 - 2(1 \cdot 1983)]} &= 1 \cdot 1704 \\ x_5 &= \sqrt[3]{[4 - 2(1 \cdot 1704)]} &= 1 \cdot 1839 \\ x_6 &= \sqrt[3]{[4 - 2(1 \cdot 1839)]} &= 1 \cdot 1774 \end{aligned}$$

$\therefore \quad x = 1 \cdot 18 \quad$ correct to two decimal places.

$$x_{r+1} = \frac{1}{2}[4 - (x_r)^3]$$

if $x_0 = 1$,

$$\begin{aligned} x_1 &= \frac{1}{2}(4 - 1) & = 1 \cdot 5 \\ x_2 &= \frac{1}{2}[4 - (1 \cdot 5)^3] & = 0 \cdot 3125 \\ x_3 &= \frac{1}{2}[4 - (0 \cdot 3125)^3] & = 1 \cdot 9847 \\ x_4 &= \frac{1}{2}[4 - (1 \cdot 9847)^3] & = -1 \cdot 9089 \\ x_5 &= \frac{1}{2}[4 - (-1 \cdot 9089)^3] & = 5 \cdot 4779 \end{aligned}$$

i.e. $x_0, x_1, x_2, x_3 \ldots$ do not converge.

The Newton-Raphson method

Suppose that $x \doteq x_r$ is a good approximation for the solution of $f(x) = 0$ and that the exact solution is $x = X$

$$= x_r + \varepsilon \quad \text{for some small } \varepsilon.$$

By Taylor's theorem

$$f(x_r + \varepsilon) = f(x_r) + \varepsilon f'(x_r) + \ldots$$

thus

$$f(X) \approx f(x_r) + \varepsilon f'(x_r)$$

but

$$f(X) = 0, \quad \text{so} \quad \varepsilon \approx -\frac{f(x_r)}{f'(x_r)}$$

Thus, if x_r is a good approximation for a root of $f(x) = 0$, then $x_r - \dfrac{f(x_r)}{f'(x_r)}$ is a better approximation. The Newton-Raphson iterative method is based on this statement,

i.e.

> If x_r is a good approximation for a root of $f(x) = 0$, then
> $$x_{r+1} = x_r - \frac{f(x_r)}{f'(x_r)} \text{ is a better approximation.}$$

Note that this is sometimes simply referred to as Newton's method.

Example 7

Use the Newton-Raphson method to determine the root of the equation $x^3 - x = 2$ lying in the range $-2 \leqslant x \leqslant 2$, giving your answer correct to three decimal places.

If $f(x) = x^3 - x - 2$, then
$$f(-2) = -8 + 2 - 2 = -8$$
$$f(-1) = -1 + 1 - 2 = -2$$
$$f(0) = 0 - 0 - 2 = -2$$
$$f(1) = 1 - 1 - 2 = -2$$
$$f(2) = 8 - 2 - 2 = 4$$

The change of sign of $f(x)$ between $f(1)$ and $f(2)$ indicates that $x^3 - x = 2$ has a root between $x = 1$ and $x = 2$.

If $f(x) = x^3 - x - 2, f'(x) = 3x^2 - 1$, thus using $x_{r+1} = x_r - \dfrac{f(x_r)}{f'(x_r)}$,

$$X_{r+1} = x_r - \left(\frac{x_r^3 - x_r - 2}{3x_r^2 - 1}\right)$$

if $x_0 = 1$, $\quad x_1 = 1 - \left(\dfrac{1 - 1 - 2}{3(1)^2 - 1}\right)$ $\qquad = 1 + 1 \qquad\qquad = 2$

$x_2 = 2 - \left(\dfrac{2^3 - 2 - 2}{3(2)^2 - 1}\right)$ $\qquad = 2 - \dfrac{4}{11} \qquad\qquad = 1{\cdot}63636$

$x_3 = 1{\cdot}63636 - \left[\dfrac{(1{\cdot}63636)^3 - 1{\cdot}63636 - 2}{3(1{\cdot}63636)^2 - 1}\right] = 1{\cdot}63636 - 0{\cdot}10597 = 1{\cdot}53039$

$x_4 = 1{\cdot}53039 - \left[\dfrac{(1{\cdot}53039)^3 - 1{\cdot}53039 - 2}{3(1{\cdot}53039)^2 - 1}\right] = 1{\cdot}53039 - 0{\cdot}00895 = 1{\cdot}52144$

$x_5 = 1{\cdot}52144 - \left[\dfrac{(1{\cdot}52144)^3 - 1{\cdot}52144 - 2}{3(1{\cdot}52144)^2 - 1}\right] = 1{\cdot}52144 - 0{\cdot}00006 = 1{\cdot}52138$

Thus the required root, correct to three decimal places, is $1{\cdot}521$.

Note When using the Newton-Raphson method, it can be assumed that the error obtained when finding x_{r+1} is less than the correction $\dfrac{f(x_r)}{f'(x_r)}$ used to obtain x_{r+1} from x_r.
Thus in Example 7, the error in $x_5 = 1{\cdot}52138$ is less than $0{\cdot}00006$. Therefore an answer of $1{\cdot}521$ is indeed correct to three decimal places.

In order to picture the Newton-Raphson method for solving $f(x) = 0$, consider the graph of $y = f(x)$.

Suppose that the exact solution is $x = X$ and that our initial, close estimate is x_0.
Drawing the tangent at $[x_0, f(x_0)]$, we can see that a better solution will be $x_1 = x_0 - \dfrac{f(x_0)}{\tan \theta}$.
However, $\tan \theta$ is the gradient of the curve at $[x_0, f(x_0)]$, i.e. $\tan \theta = f'(x_0)$. Thus if x_0 is our initial estimate, a better approximation is $x_1 = x_0 - \dfrac{f(x_0)}{f'(x_0)}$.

Continuing this, $x_2 = x_1 - \dfrac{f(x_1)}{f'(x_1)}, x_3 = x_2 - \dfrac{f(x_2)}{f'(x_2)}$ etc.

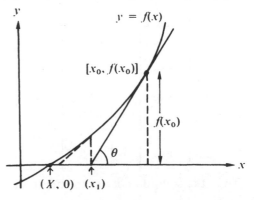

Thus we obtain the general rule $x_{r+1} = x_r - \dfrac{f(x_r)}{f'(x_r)}$ as before.

Again, it is important to realise that this method will not always work as the values $x_0, x_1, x_2, x_3 \ldots$ will not always converge to X. The following diagrams illustrate situations for which the method would fail.

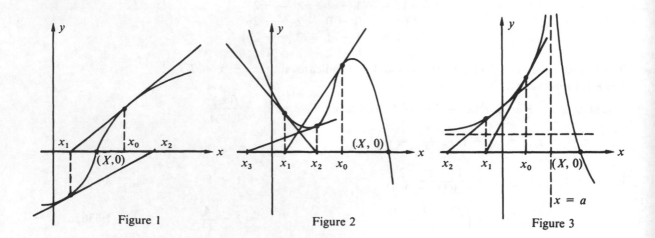

Figure 1   Figure 2   Figure 3

In Fig. 1, the values $x_0, x_1, x_2 \ldots$ diverge.

In Fig. 2, the first value x_0 was not chosen close enough to $(X, 0)$.

In Fig. 3, $y = f(x)$ has a discontinuity at $x = a$ and x_0 needs to be on the same side of $x = a$ as $(X, 0)$.

Note With the use of a calculator and particularly a programmable calculator, it is no longer so tedious to determine the root of an equation by substituting successive values of x into the equation. Again, we use the fact that if $f(a) = 0$ then, as x varies, $f(x)$ will change sign as x passes through the value $x = a$.

Thus to determine the root of $x^3 - x = 2$, lying in the range $-2 \leqslant x \leqslant 2$, (see Example 7) we could proceed as follows:

Let $f(x) = x^3 - x - 2$

x	$f(x)$	x	$f(x)$	x	$f(x)$	x	$f(x)$	x	$f(x)$
-2	$-$ve	$1\cdot1$	$-$ve	$1\cdot51$	$-$ve	$1\cdot521$	$-$ve	$1\cdot5211$	$-$ve
		$1\cdot2$	$-$ve	$1\cdot52$	$-$ve	$1\cdot522$	$+$ve	$1\cdot5215$	$+$ve
-1	$-$ve	$1\cdot3$	$-$ve	$1\cdot53$	$+$ve	$1\cdot523$		$1\cdot5219$	
		$1\cdot4$	$-$ve	$1\cdot54$		$1\cdot524$			
0	$-$ve	$1\cdot5$	$-$ve	$1\cdot55$		$1\cdot525$			
		$1\cdot6$	$+$ve	$1\cdot56$		$1\cdot526$			
1	$-$ve	$1\cdot7$		$1\cdot57$		$1\cdot527$			
		$1\cdot8$		$1\cdot58$		$1\cdot528$			
2	$+$ve	$1\cdot9$		$1\cdot59$		$1\cdot529$			

Thus the solution correct to three decimal places is $1\cdot521$.

Exercise 21C

1. Copy and complete the following table for $y = e^x$.

x	-3	-2.5	-2	-1.5	-1	-0.5	0	0.5	1	1.5	2
y (correct to 2 d.p's)	0·05	0·08	0·14								

On the same pair of axes draw the lines $y = e^x$ and $y = x + 2$ for $-3 \leqslant x \leqslant 2$ and hence find approximate values for all the solutions to the equation $e^x - x = 2$ lying in the range $-3 \leqslant x \leqslant 2$.

2. The equation $x^2 + 4x = 2$ has two roots, one near $x = 0$ and the other near $x = -4$.

 (a) Using $x_{r+1} = \dfrac{2}{x_r + 4}$ with $x_0 = 0$ find the root near $x = 0$, correct to 2 decimal places.

 (b) Why could we not use the formula $x_{r+1} = \dfrac{2}{x_r + 4}$ with $x_0 = -4$?

 (c) Using $x_{r+1} = \dfrac{2}{x_r} - 4$ with $x_0 = -4$ find the root near $x = -4$ correct to 2 decimal places.

 (d) Check your answers to (a) and (c) using the quadratic formula.

3. Show that the equation $x^3 - x + 3 = 0$ has a root in the range $-3 \leqslant x \leqslant 3$. Use the rearranged form $x = \sqrt[3]{(x - 3)}$ to obtain this root correct to two decimal places.

For questions **4** to **11**, use the Newton-Raphson iterative method to obtain a root for each equation using the given initial value, x_0, and giving your answer correct to the required degree of accuracy.

4. $x^2 = 5$; $x_0 = 2$, to 3 d.p.
5. $x^2 - 2x - 5 = 0$; $x_0 = 4$, to 4 d.p.
6. $x^3 - 5x - 2 = 0$; $x_0 = 2$, to 3 d.p.
7. $x^3 - 3x + 3 = 0$; $x_0 = -2$, to 3 d.p.
8. $\ln x = 2 - x$; $x_0 = 2$, to 3 d.p.
9. $x \ln x = 1$; $x_0 = 2$, to 4 d.p.
10. $e^x = 3 - x$; $x_0 = 1$, to 5 d.p.
11. $\sin x = x - \frac{1}{2}$, (x is in rads); $x_0 = 1$, to 4 d.p.

21.6 e^z, cos z and sin z for complex z.

Having obtained the series expansions for $\sin x$ and $\cos x$ in section 21.4,

i.e. $\sin x = x - \dfrac{x^3}{3!} + \dfrac{x^5}{5!} \ldots$ and $\cos x = 1 - \dfrac{x^2}{2!} + \dfrac{x^4}{4!} \ldots$, we will now

conclude by mentioning how these ideas can be used to assign a meaning to the exponential, sine and cosine functions of a complex number z.

If we are to assign some meaning to the idea of a complex index, it would be useful if our definition is compatible with the existing laws of indices
i.e. $e^a \times e^b = e^{a+b}$, $e^a \div e^b = e^{a-b}$, and $(e^a)^b = e^{ab}$

Now we know from earlier work that $e^x = 1 + \dfrac{x}{1!} + \dfrac{x^2}{2!} + \dfrac{x^3}{3!} + \dfrac{x^4}{4!} + \dfrac{x^5}{5!} \cdots$

If we define e^z, where z is a complex number, by a similar series

i.e. $e^z = 1 + \dfrac{z}{1!} + \dfrac{z^2}{2!} + \dfrac{z^3}{3!} + \dfrac{z^4}{4!} + \dfrac{z^5}{5!} \cdots$

it will follow that $e^{i\theta} = 1 + i\theta - \dfrac{\theta^2}{2!} - \dfrac{i\theta^3}{3!} + \dfrac{\theta^4}{4!} + \dfrac{i\theta^5}{5!} \cdots$

$$= \left(1 - \frac{\theta^2}{2!} + \frac{\theta^4}{4!} \cdots \right) + i\left(\theta - \frac{\theta^3}{3!} + \frac{\theta^5}{5!} \cdots \right)$$

$$= \cos\theta + i\sin\theta$$

Thus $e^{i\theta} = \cos\theta + i\sin\theta$ ...[1]

Similarly $e^{-i\theta} = \cos\theta - i\sin\theta$ ...[2]

However, as was mentioned earlier, we require that this definition of $e^{i\theta}$ is compatible with the existing laws of indices. The following example and question 1 of Exercise 21D show this to be the case.

Example 8

Using the definition $e^{i\theta} = \cos\theta + i\sin\theta$ prove that $e^{i\alpha}e^{i\beta} = e^{i(\alpha+\beta)}$

$e^{i\alpha} \times e^{i\beta} = (\cos\alpha + i\sin\alpha)(\cos\beta + i\sin\beta)$
$= (\cos\alpha\cos\beta - \sin\alpha\sin\beta) + i(\sin\alpha\cos\beta + \cos\alpha\sin\beta)$
$= \cos(\alpha + \beta) + i\sin(\alpha + \beta)$
$= e^{i(\alpha+\beta)}$ as required.

Let us consider equations [1] and [2]: $e^{i\theta} = \cos\theta + i\sin\theta$
$e^{-i\theta} = \cos\theta - i\sin\theta$

Adding these two equations gives $e^{i\theta} + e^{-i\theta} = 2\cos\theta$ i.e. $\cos\theta = \dfrac{e^{i\theta} + e^{-i\theta}}{2}$

and subtracting gives $e^{i\theta} - e^{-i\theta} = 2i\sin\theta$ i.e. $\sin\theta = \dfrac{e^{i\theta} - e^{-i\theta}}{2i}$

Thus we can now attribute a meaning to the functions $\sin z$ and $\cos z$ for complex values of z:

$$\cos z = \frac{e^{iz} + e^{-iz}}{2} \quad \text{and} \quad \sin z = \frac{e^{iz} - e^{-iz}}{2i}$$

Example 9 below and Exercise 21D question 2 show that, with these definitions, the existing trigonometric identities for $\sin x$ and $\cos x$ also hold for $\sin z$ and $\cos z$.

Example 9

Using the definitions $\cos z = \dfrac{e^{iz} + e^{-iz}}{2}$ and $\sin z = \dfrac{e^{iz} - e^{-iz}}{2i}$, prove that $\cos^2 z + \sin^2 z = 1$

$\cos^2 z = \left(\dfrac{e^{iz} + e^{-iz}}{2}\right)^2 = \dfrac{e^{2iz} + 2e^{iz}e^{-iz} + e^{-2iz}}{4}$ $\sin^2 z = \left(\dfrac{e^{iz} - e^{-iz}}{2i}\right)^2 = \dfrac{e^{2iz} - 2e^{iz}e^{-iz} + e^{-2iz}}{-4}$

$\qquad\qquad = \dfrac{1}{4}e^{2iz} + \dfrac{1}{2} + \dfrac{1}{4}e^{-2iz}$ $\qquad\qquad = -\dfrac{1}{4}e^{2iz} + \dfrac{1}{2} - \dfrac{1}{4}e^{-2iz}$

$\therefore \quad \cos^2 z + \sin^2 z = \dfrac{1}{2} + \dfrac{1}{2} = 1$ as required.

Exercise 21D

1. Using the definition $e^{i\theta} = \cos\theta + i\sin\theta$ prove that

   (a) $e^{i\alpha} \div e^{i\beta} = e^{i(\alpha-\beta)}$ (b) $(e^{i\theta})^n = e^{in\theta}$ (c) $\dfrac{d}{d\theta}(e^{i\theta}) = ie^{i\theta}$.

2. Using the definitions $\cos z = \dfrac{e^{iz} + e^{-iz}}{2}$ and $\sin z = \dfrac{e^{iz} - e^{-iz}}{2i}$ prove that

   (a) $\sin(-z) = -\sin z$,
   (b) $\sin w \cos z + \cos w \sin z = \sin(w + z)$,
   (c) $\cos w \cos z + \sin w \sin z = \cos(w - z)$.

3. Use the fact that $e^x \cos 2x = \mathrm{Re}\,(e^x e^{i2x})$ to find $\displaystyle\int e^x \cos 2x\, dx$.

   (Remember that $\mathrm{Re}\,(z)$ means the real part of the complex number z).
   Check your answer by performing the same integration using integration by parts.

4. Use the definition $\cos z = \dfrac{e^{iz} + e^{-iz}}{2}$ to find 2 imaginary numbers having a cosine of 4.

5. Use the definition $\sin z = \dfrac{e^{iz} - e^{-iz}}{2i}$ to find an imaginary number having a sine of i.

Exercise 21E   Examination Questions

1. Sketch the curve $y = (1 + x^2)^{1/2}$ for values of x between 0 and 1.

   Find the approximation to the integral $\displaystyle\int_0^1 (1 + x^2)^{1/2}\, dx$ given by the trapezium rule, using two trapezia, one with base from 0 to $\frac{1}{2}$, the other with base from $\frac{1}{2}$ to 1. By referring to your sketch determine whether your estimate is too large or too small, giving reasons.
   Evaluate the integral by Simpson's rule, using the same three ordinates.
   (Oxford)

2. Show that $\displaystyle\int_0^1 e^x\, dx = e - 1$. The integral is approximated by using the trapezium rule, dividing the range into n intervals each of length $h = 1/n$.
   Show that the result is $\dfrac{(e - 1)(p + 1)}{2n(p - 1)}$ where $p = e^h$.
   Hence or otherwise show that $\dfrac{h(e^h + 1)}{2(e^h - 1)}$ is very close to 1 when h is small.
   (Oxford)

3. The region enclosed by the curve $y = e^{-\frac{1}{2}x^2}$, the x-axis and the lines $x = -3$ and $x = 3$ is rotated completely about the x-axis. Showing all your working in the form of a table, use Simpson's method with six intervals to find, in terms of e and π, an estimate of the volume of the solid so formed.
   (A.E.B.)

4. Use Simpson's rule with seven equally spaced ordinates to find an estimate of the mean value of the function $\log_{10}(1 + x^3)$ between the values $x = 1$ and $x = 19$. Show your working in the form of a table and give your final answer to 3 significant figures. (A.E.B.)

5. Find an approximate value for $\int_0^{\pi/3} \sqrt{(\sec x)}\,dx$, by using Simpson's Rule with five equally spaced ordinates.

   Write down the least value of $\sec x$ for $0 \leqslant x < \dfrac{\pi}{2}$.

   Hence show that $\int_0^{\alpha} \sqrt{(\sec x)}\,dx > \alpha$ for $0 < \alpha < \dfrac{\pi}{2}$. (A.E.B.)

6. Write down the expansion of $\cos x$ in ascending powers of x, up to and including the term in x^6. Write down also the expansion of $\ln(1 + t)$ in ascending powers of t, up to and including the term in t^3. Hence, or otherwise, find the expansion of $\ln(\cos x)$ in ascending powers of x, up to and including the term in x^6. Use these terms to find an approximation to the integral
   $$\int_0^{1/2} \frac{\ln(\cos x)}{x}\,dx,$$
   giving your answer correct to 4 decimal places. (Oxford)

7. Prove that
   $$\frac{d}{dx}(\sec x \tan^n x) = \sec x\{n \tan^{n-1}x + (n + 1)\tan^{n+1} x\}$$
   for $n = 1, 2, \ldots$. Find the Maclaurin series of the function $y = \sec x$ up to and including the term in x^4. Use these terms to evaluate $\sec\tfrac{1}{10}$, to five places of decimals, showing all steps of your working. (Oxford)

8. Show that the equation $e^x - 4 \sin x = 0$ has two roots between $x = 0$ and $x = 1\!\cdot\!5$ by evaluating $e^x - 4 \sin x$ for at least three appropriate values of x.
   Using a graphical method, determine the smaller root correct to 2 decimal places. (A.E.B.)

9. Show graphically, or otherwise, that the equation $\ln x = 4 - x$ has only one real root and prove that this root lies between $2\!\cdot\!9$ and 3.
   By taking $2\!\cdot\!9$ as a first approximation to this root and applying the Newton-Raphson process once to the equation $\ln x - 4 + x = 0$, or otherwise, find a second approximation, giving your answer to 3 significant figures. (London)

10. Show, by means of a sketch graph, or otherwise, that the equation
    $$e^{2x} + 4x - 5 = 0$$
    has only one real root, and that this root lies between 0 and 1.
    Starting with the value $0\!\cdot\!5$ as a first approximation to this root, use the Newton-Raphson method to evaluate successive approximations, showing the stages of your work and ending when two successive approximations give answers which, when rounded to two decimal places, agree.
    (Cambridge)

11. Sketch the curve
$$y = 2 \cos (\tfrac{1}{2}\pi x)$$
for values of x such that $0 \leqslant x \leqslant 6$.

By drawing a suitable straight line on your diagram, show that the smallest positive root of the equation
$$4 \cos (\tfrac{1}{2}\pi x) = 6 - x$$
lies between 3 and 4.

Taking 3·5 as the first approximation to this root, use the Newton-Raphson method to find a second approximation, giving two places of decimals in your answer.   (Cambridge)

12. Find and determine the nature of the turning points on the graph of
$y = e^x - x^2 e$.

Show that the equation $(2x^2 + 1)e^{x^2} = e$ has a root near $x = 0\cdot6$ and use Newton's method once to find a second approximation to the root. Hence find, to two decimal places, the coordinates of the points of inflexion on the graph of $y = e^{x^2} - x^2 e$, and mark them clearly on a sketch of this graph.   (A.E.B.)

Answers

page 4 **Exercise A**

1. (a) F (b) F (c) T (d) T (e) F (f) T (g) T (h) F (i) T (j) F (k)T (l) F
(m) T (n) F ·(o) T (p) F

2. (a) $\{-1, 0, 1, 2, 3\}$ (b) $\{-2, -1, 0, 1, 2, 3, 4\}$ (c) $\{-3, -2, -1, 0, 1\}$
(d) $\{-2, -1, 0, 1, 2\}$ (e) $\{1, 2, 3, 4\}$ (f) $\{1, 2, 3, 4, 5\}$

3. For (a) (b) and (c) there are other possible inequality statements.
(a) $\{x \in \mathbb{Z}: -3 \leqslant x \leqslant 2\}$ (b) $\{x \in \mathbb{Z}: -5 \leqslant x \leqslant -1\}$ (c) $\{x \in \mathbb{Z}: x \leqslant 1\}$
(d) $\{x \in \mathbb{R}: x \geqslant -1\}$ (e) $\{x \in \mathbb{R}: x > -1\}$ (f) $\{x \in \mathbb{R}: x \leqslant 1\}$
(g) $\{x \in \mathbb{R}: x < -2\}$ (h) $\{x \in \mathbb{R}: -1 \leqslant x \leqslant 3\}$ (i) $\{x \in \mathbb{R}: -4 < x < 2\}$
(j) $\{x \in \mathbb{R}: -1 \leqslant x < 2\}$

4. (a) $x \leqslant 3$ (b) $x < 2$ (c) $x \geqslant 4$
(d) $x > -2$ (e) $x < -1$ (f) $x \geqslant 1\frac{1}{2}$
(g) $x \leqslant 2$ (h) $x < 5$ (i) $-1 \leqslant x \leqslant 5$
(j) $-3 \leqslant x \leqslant 1$ (k) $-3 < x < 2$ (l) $-2 < x < 3$
(m) $-5 < x \leqslant -2$ (n) $1 \leqslant x < 6$

5. (a) $\{x \in \mathbb{R}: -3 \leqslant x \leqslant 7\}$ (b) $\{x \in \mathbb{R}: -2 \leqslant x \leqslant 3\}$ (c) $\{x \in \mathbb{R}: x < -1 \text{ or } x > 6\}$
(d) $\{x \in \mathbb{R}: -3\frac{1}{2} < x < 4\}$ (e) $\{x \in \mathbb{R}: x \leqslant -\frac{3}{5} \text{ or } x \geqslant 1\}$ (f) $\{x \in \mathbb{R}: x < -1\frac{1}{3} \text{ or } x > 2\}$

page 7 **Exercise B**

1. (a) $11x + 6y$ (b) $3x - 8$ (c) $2p^2 + 8p - 3$ (d) $4x^2y + 2xy^2$
(e) $8a^2 - 2ab$

2. (a) $3x^2 - 12x$ (b) $x + 24$ (c) $13x + 3$ (d) $5xy - 6x - 4y$
(e) $x^2 - 2x - 15$ (f) $2x^2 - 5x - 12$ (g) $4x^2 - 12x + 9$ (h) $27x^3 - 27x^2 + 9x - 1$
(i) $5x^3 - 17x^2 - 32x - 18$

3. (a) $8x(2x + 3y)$ (b) $5x(3x + 2y + 4)$ (c) $(x + 5)(x - 2)$ (d) $(x + 7)(x - 2)$
(e) $(2x + 3)(x - 5)$ (f) $(3x - 2)(2x + 1)$ (g) $(2x + 5)(2x - 3)$ (h) $(x + 3)(x - 3)$
(i) $(4x + 7y)(4x - 7y)$ (j) $2x(x + 3)(x - 3)$ (k) $(a + 3)(x + 2)$ (l) $(a - 2b)(x + y)$

4. (a) $x^2 + 3x + 2$ (b) $x^2 + 6x + 7$ (c) $x^2 + 2x - 1$ (d) $3x^2 + 7x - 3$

5. (a) $13x - 13$ (b) $x - 3$ (c) $8x^2 + 11x + 10$ (d) -20

6. (a) $\dfrac{11}{12}x$ (b) $\dfrac{8x - 5}{15}$ (c) $\dfrac{x + 10}{6}$ (d) $\dfrac{x + 7}{3(x + 1)}$

(e) $\dfrac{17 - 2x}{4(2x + 3)}$ (f) $\dfrac{-3x - 5}{(x + 2)^2}$ (g) $\dfrac{x + 12}{x(x + 4)}$ (h) $\dfrac{x^2 + 9x + 26}{2(x + 4)^2}$

(i) $\dfrac{x^2 + 3x + 20}{2(x + 4)^2}$ (j) $\dfrac{14}{3}$ (k) $\dfrac{2}{5(x - 4)}$ (l) $\dfrac{5}{6x}$

(m) $\dfrac{2(2x - 3)}{x(x - 4)}$ (n) $\dfrac{3x - 5}{x(x - 1)}$ (o) $\dfrac{(4 - x)(x + 12)}{x^2}$

7. (a) $1 + \dfrac{2x - 3}{x^2 + 4x + 1}$ (b) $2 + \dfrac{3}{x^2 + 1}$ (c) $5 - \dfrac{3x + 1}{x^2 + x - 2}$

(d) $x - 3 + \dfrac{2}{x^2 - 2x + 3}$

8. (a) $1 + \dfrac{3x - 1}{x^2 + 4}$ (b) $1 - \dfrac{3}{x + 5}$ (c) $2 - \dfrac{5}{x^2 - 2x + 8}$

(d) $2 - \dfrac{5}{2x + 3}$ (e) $x + 1 + \dfrac{2}{x^2 + 3}$ (f) $x - 3 + \dfrac{2}{x + 1}$

(g) $x + 1 - \dfrac{3x}{x^2 + 2x - 4}$ (h) $x^2 + x - 2 + \dfrac{3x}{x^2 + 7}$

page 10   Exercise C

1. (a) $2\sqrt{3}$ (b) $7\sqrt{3}$ (c) $11\sqrt{2}$ (d) $8\sqrt{3}$ (e) $\sqrt{5}$ (f) 1 (g) 13 (h) -6

2. (a) $\sqrt{12}$ (b) $\sqrt{12}$ (c) $\sqrt{20}$ (d) $\sqrt{18}$ (e) $2\sqrt{5}$ (f) 36 (g) $\dfrac{1}{\sqrt{3}}$ (h) $\sqrt{12}$

(i) $\dfrac{5}{\sqrt{2}}$

3. (a) $5 + 2\sqrt{6}$ (b) $23 + 4\sqrt{15}$ (c) $7 - 2\sqrt{6}$

4. (a) $\dfrac{\sqrt{3}}{3}$ (b) $3\sqrt{3}$ (c) $\dfrac{8\sqrt{2}}{3}$ (d) $\dfrac{3 + \sqrt{2}}{7}$ (e) $\dfrac{5(3 + \sqrt{5})}{4}$

(f) $4 + \sqrt{15}$ (g) $-7 - 4\sqrt{3}$ (h) $\dfrac{11 + 4\sqrt{6}}{5}$ (i) $\dfrac{2 - 3\sqrt{2}}{7}$

page 12   Exercise D

1. (a) $-3, 4$ (b) $-8, 3$ (c) $4, 9$ (d) $-1\frac{1}{2}, 2$ (e) $-\frac{1}{2}, 4$ (f) $-3, 1\frac{1}{3}$ (g) $-6, 1$
(h) $\frac{1}{2}, 1\frac{1}{2}$ (i) $-2, 5$ (j) $-6, 4$ (k) $\frac{1}{3}, 3$ (l) $-1\frac{1}{2}, 8$

2. (a) $0{\cdot}38, 2{\cdot}62$ (b) $-1{\cdot}56, 2{\cdot}56$ (c) $-6{\cdot}19, -0{\cdot}81$
(d) $-1{\cdot}78, 0{\cdot}28$ (e) $0{\cdot}42, 1{\cdot}58$ (f) $-0{\cdot}43, 1{\cdot}18$

3. (a) $\frac{1}{2}(3 \pm \sqrt{5})$ (b) $\frac{1}{2}(3 \pm \sqrt{7})$ (c) $\frac{1}{3}(3 \pm \sqrt{6})$

4. (a) $-3 \pm \sqrt{10}$ (b) $-2 \pm \sqrt{7}$ (c) $\frac{1}{2}(-1 \pm \sqrt{5})$ (d) $\frac{1}{2}(1 \pm \sqrt{13})$
(e) $\frac{1}{2}(3 \pm \sqrt{29})$ (f) $\frac{1}{2}(3 \pm \sqrt{7})$ (g) $\frac{1}{3}(2 \pm \sqrt{10})$ (h) $\frac{1}{4}(3 \pm \sqrt{5})$
(i) $\frac{1}{3}(3 \pm 2\sqrt{3})$

5. (a) MIN (b) -7 (c) -2 **6.** (a) MIN (b) -8 (c) 3
7. (a) MAX (b) 4 (c) -1 **8.** (a) MIN (b) $-5\frac{1}{4}$ (c) $2\frac{1}{2}$
9. (a) MAX (b) 6 (c) 1 **10.** (a) MAX (b) 22 (c) 4
11. (a) MIN (b) $-\frac{1}{8}$ (c) $-\frac{3}{4}$ **12.** (a) MIN (b) $1\frac{2}{3}$ (c) $-\frac{1}{3}$
13. (a) MAX (b) $4\frac{1}{8}$ (c) $-1\frac{1}{4}$

page 16   Exercise E

1. (a) $\sin 50°$ (b) $-\cos 50°$ (c) $-\tan 50°$ (d) $\sin 40°$
(e) $-\cos 10°$ (f) $-\cos 20°$ (g) $-\tan 80°$ (h) $\sin 85°$

2. (a) $\frac{1}{2}$ (b) $\dfrac{\sqrt{3}}{2}$ (c) 1 (d) $\dfrac{1}{\sqrt{2}}$ (e) $\sqrt{3}$ (f) 1 (g) 0 (h) $\frac{1}{2}$

(i) $\dfrac{1}{\sqrt{2}}$ (j) $-\frac{1}{2}$ (k) 0 (l) -1 (m) -1 (n) $-\dfrac{\sqrt{3}}{2}$ (o) $-\sqrt{3}$ (p) $\dfrac{\sqrt{3}}{2}$

4. (a) $0{\cdot}342$ (b) $0{\cdot}342$ (c) $-0{\cdot}342$
5. (a) $0{\cdot}766$ (b) $0{\cdot}643$ (c) $0{\cdot}643$ (d) $0{\cdot}766$ (e) $-0{\cdot}766$ (f) $-0{\cdot}643$
6. $4{\cdot}9$ **7.** $3{\cdot}18$ **8.** (a) $\frac{4}{5}$ (b) $\frac{3}{4}$

9. (a) $-\frac{4}{5}$ (b) $-\frac{3}{4}$ **10.** (a) $-\frac{12}{13}$ (b) $-\frac{5}{12}$
11. (a) $40°$ (b) $60°$ (c) $20°$ (d) $20°$ (e) $30°$ (f) $5°$

page 20   Exercise F

1. (a) $6·52$ cm   (b) $5·57$ cm   (c) $3·74$'cm   (d) $5·22$ cm   (e) $8·58$ cm   (f) $6·83$ cm
2. (a) $42·2°$ (b) $78·5°$ (c) $38·9°$ (d) $40·3°$ (e) $104·5°$ (f) $81·6$ ° or $98·4°$
3. (a) 15 m²   (b) 6 cm²   (c) $14·7$ cm²   (d) 24 cm²   (e) $9·74$ cm²   (f) $17·4$ cm²
4. $\hat{B} = 70°$, $c = 5·32$ cm, $b = 7·2$ cm   **5.** $a = 4·31$ cm, $\hat{C} = 55·1°$, $\hat{B} = 79·9°$
6. Ambiguous case:- either $\hat{A} = 73·2°$, $\hat{B} = 56·8°$, $b = 8·73$ cm or $\hat{A} = 106·8°$, $\hat{B} = 23·2°$,
  $b = 4·12$ cm
7. $6·26$ km **8.** N31°E **9.** $10·9$ km, $059·4°$ **10.** $41·3$ m, $55·7$ m, $35·8$ m **11.** 6 m, 14 m, 16 m

page 25   Exercise G

1. (a) $18·4°$ (b) $22·6°$ (c) $40·6°$ **2.** (a) $13·0°$ (b) $14·0°$ (c) $64·1°$ (d) $77·0°$
3. (a) $6·78$ cm   (b) $58·0°$ (c) $66·1°$ **4.** (a) $49·1°$ (b) $20·7°$ (c) $45°$
5. (a) 4 cm   (b) $21·8°$ (c) $21·0°$ **6.** (a) $6·93$ cm   (b) $49°$ **7.** 129 m
8. (a) $7·35$ m, $67·8°$ (b) $8·66$ m, $65·2°$ **9.** $67·3$ m², $14°$
10. (a) $75·5°$ (b) $11·6$ m (c) $63·4°$ (d) $58·9°$ **11.** 71 m   **12.** N52°E, $15°$ **13.** $12°$
14. $58°$

---

page 32   Exercise 1A

1. (a) and (d)   **2.** (i) (b); (ii) (d); (iii) (c)   **3.** (a) $\{-8, -5, -2, 1, 4\}$, (b) $\{-1, 1, 7\}$
4. $\{0, 1, 2, 3, 4\}, \{1, 2, 5\}, \{0, 1, 4, 9\}$ **5.** $\{0, 1, 2\}, \{-1, 0, 1\}, \{0, 1, 2, 3\}$
6. (a) 7 (b) 1 (c) 15 (d) -3 **7.** (a) 10 (b) 10 (c) -2 (d) 3 or -2
8. (a) 18 (b) 18 (c) 6 (d) $-1\frac{1}{2}, 1$ **9.** $2, -3$ **10.** $1, -1, 17$
11. (a) $\{y: 3 \leqslant y \leqslant 7\}$ one-to-one   (b) $\{y: |y| \leqslant 2\}$ one-to-one
  (c) $\{y: 0 \leqslant y \leqslant 6\}$ one-to-one   (d) $\{y: |y| \leqslant 6\}$ one-to-one
  (e) $\{y: 0 \leqslant y \leqslant 9\}$ many-to-one   (f) $\{y: 0 \leqslant y \leqslant 5\}$ one-to-one
  (g) $\{y: 0 \leqslant y \leqslant 3\}$ many-to-one   (h) $\{y: y \geqslant 0\}$ many-to-one
  (i) $\{y: y \geqslant 0\}$ many-to-one   (j) $\{y: 0 < y \leqslant 1\}$ one-to-one
  (k) $\{y: y \geqslant 4\}$ many-to-one   (l) $\{y \in \mathbb{R}: y \neq 0\}$ one-to-one
12. (a) (i) $\mathbb{R}$ (ii) $\mathbb{R}$ (b) (i) $\mathbb{R}$ (ii) $\{y \in \mathbb{R}: y \geqslant 0\}$ (c) (i) $\{x \in \mathbb{R}: x \neq 0\}$ (ii) $\{y \in \mathbb{R}: y \neq 0\}$
  (d) (i) $\{x \in \mathbb{R}: x \neq 3\}$ (ii) $\{y \in \mathbb{R}: y \neq 0\}$
13. f is one-to-many ($f(3) = 6$ or 9) and is therefore not a function
14. g is one-to-many ($g(2) = 4$ or 6) and is therefore not a function

page 38   Exercise 1B

1. (a) 15 (b) 7 (c) 228 (d) 35 (e) 147 (f) 75
2. (a) $\{-5, -3, -1, 1, 3, 5, 7\}$ (b) $\{0, 1, 4, 9\}$ (c) $\{1, 3, 9, 19\}$ (d) $\{1, 9, 25, 49\}$
3. (a) $\{-1, 0, 3\}$ (b) $\{-3, 0, 1\}$ (c) $\{-3, 0, 1\}$ (d) $\{-1, 0, 3\}$ (e) $\{0, 1, 3\}$
4. (a) $x \to 6x - 3$ (b) $x \to 6x - 1$ (c) $x \to 2x^2 - 1$ (d) $x \to (2x - 1)^2$ (e) $x \to 3x^2$
  (f) $x \to 6x^2 - 3$ (g) $x \to 18x^2 - 1$ **5.** (a) gf (b) gg (c) fg **6.** (a) gg (b) gf (c) fg

7. (a) $x \to x^2 + 1$ (b) $x \to \dfrac{2}{x} + 1$ (c) $x \to \dfrac{4}{x^2}$ (d) $x \to \dfrac{2}{x^2}$ (e) $x \to \dfrac{2}{x^2 + 1}$

  (f) $x \to \dfrac{2}{(x + 1)^2}$

8. (a) $x \to \dfrac{x+2}{3}$ (b) $x \to 2x$ (c) $x \to 5-x$ (d) $x \to \sqrt{x}-1$ (e) $x \to \dfrac{1}{x-2}$

(f) $x \to \dfrac{1}{2-x}$ (g) $x \to \dfrac{1}{x}-2$ (h) $x \to \dfrac{1}{2x}+3$

9. (a) $x \to 2x-1$ (b) $x \to \dfrac{x+1}{2}$ (c) $x \to \dfrac{2}{x-2}$ (d) $x \to \dfrac{2}{x}+2$ (e) $x \to \dfrac{2}{2x+1}$

(f) $x \to \dfrac{1}{x}-\dfrac{1}{2}$

page 42   Exercise 1C

1. (4, 2), (6, 3)   **2.** (3, 2), (4, 2), (4, 3)   **3.** $(-2, 4), (-1, 1), (0, 0), (1, 1), (2, 4)$
4. (a) A = {2, 3}, B = {3, 4, 5}   (b) (2, 5), (3, 3), (3, 4)   (c) {(3, 3)}   **5.** (1, 2), (2, 4), (3, 6)
6. $(2, 2), (1, 5), (4, -4)$ **7.** $4, -2, -10, 4, 7$ **8.** 3   **9.** $3, -5$
10. (a) (i) $(0, -4)$ (ii) (4, 0)   (b) (i) $(0, -4)$ (ii) (2, 0)   (c) (i) (0, 12)   (ii) (6, 0)
(d) (i) (0, 3)   (ii) $(-6, 0)$ (e) (i) (0, 8)   (ii) (4, 0)   (f) (i) (0, 3)   (ii) $(\tfrac{3}{2}, 0)$
(g) (i) (0, 6)   (ii) $(-2\tfrac{2}{3}, 0)$ (h) (i) (0, 2)   (ii) (1, 0) and (2, 0)
(i) (i) $(0, -6)$ (ii) $(-3, 0)$ and (2, 0).

page 45   Exercise 1D

3. (a) even   (b) neither   (c) odd   (d) even   (e) neither
4. (a) even   (b) even   (c) neither   (d) odd.
5. (a) $y = 3x$ (b) $y = 4-x$ (c) $y = \dfrac{x+4}{2}$ (d) $y = \dfrac{1}{x}-2$

page 45   Exercise 1E

1. $0 \leqslant f(x) \leqslant 4$ **2.** (i) zero   (ii) $5, 6\tfrac{1}{2}$ (iii) $7\tfrac{3}{4}, -2$ **3.** (i) $2, -6$ (ii) 12

4. (i) $\tfrac{1}{2}(x+3)$ (ii) $4x-9$ (iii) x (iv) $\dfrac{16}{x}-3$ (v) $\dfrac{16}{x+3}$

5. (i) fg (ii) g^{-1} (iii) gg (iv) ggf (v) gf^{-1}

6. (i) $\dfrac{8}{x}-5$ (ii) x (iii) $\dfrac{4}{x}$ (iv) f^{-1} (v) gf (vi) f^2

page 50   Exercise 2A

1. 9·57 km, 111°   **2.** 471 km, 184°   **3.** 727 km, S55·3°E
4. (a) 109 km, 124°   (b) 109 km, 304°   **5.** (a) and (d)   **6.** (a), (c), (e)   **7.** (b), (c)
8. (a) $h = 2, k = 3$ (b) $h = -2, k = 3$ (c) $h = -2, k = 1$ (d) $h = -3, k = 1$
(e) $h = 2, k = 3$ (f) $h = \tfrac{1}{3}, k = \tfrac{2}{3}$ (g) $h = -1, k = 2$ (h) $h = \tfrac{1}{2}, k = \tfrac{1}{3}$
9. (a) **c**   (b) $-\mathbf{c}$ (c) $\mathbf{a}+\mathbf{c}$ (d) $\mathbf{c}+\tfrac{1}{4}\mathbf{a}$ (e) $\mathbf{c}-\tfrac{3}{4}\mathbf{a}$
10. (a) $\mathbf{b}-\mathbf{a}$ (b) $\tfrac{3}{4}(\mathbf{b}-\mathbf{a})$ (c) $\tfrac{1}{4}(\mathbf{b}-\mathbf{a})$ (d) $\tfrac{1}{4}(\mathbf{a}+3\mathbf{b})$
11. (a) $3\mathbf{a}+\mathbf{c}$ (b) $2\mathbf{a}+\mathbf{c}$ (c) $2\mathbf{a}+\tfrac{1}{2}\mathbf{c}$ (d) $2\mathbf{a}-\tfrac{1}{2}\mathbf{c}$
12. (a) $\mathbf{b}-\mathbf{a}$ (b) $\tfrac{1}{3}(\mathbf{b}-\mathbf{a})$ (c) $\tfrac{1}{3}(2\mathbf{a}-\mathbf{b})$
14. (a) $h(\mathbf{c}+\tfrac{1}{2}\mathbf{a})$ (b) $k(\mathbf{c}-\mathbf{a})$ (c) $\tfrac{2}{3}, \tfrac{2}{3}$
15. $\tfrac{1}{2}, \tfrac{2}{3}$

page 57   Exercise 2B

1. $\overrightarrow{OA} = 2i + 3j$, $\overrightarrow{OB} = 3i$, $\overrightarrow{OC} = -i + 3j$, $\overrightarrow{OD} = -3i + 2j$, $\overrightarrow{OE} = -2i - j$,
$\overrightarrow{OF} = -2i - 2j$, $\overrightarrow{OG} = -2j$, $\overrightarrow{OH} = 3i - 3j$

2. (a) (i) $4i + 3j$ (ii) 5 units   (iii) $36\cdot8°$ (b) (i) $-4i + j$ (ii) $4\cdot12$ units   (iii) $14°$
(c) (i) $2i - 3j$ (ii) $3\cdot61$ units   (iii) $56\cdot3°$

3. 4   4. $\dfrac{\sqrt{5}}{5}(i + 2j)$ 5. $\dfrac{\sqrt{10}}{10}(3i - j)$ 6. $15i + 36j$ 7. $6i - 3j$

8. $\frac{2}{3}(4i - 3j)$ 9. (a) $5i + j$ (b) $-5i - j$ 10. (a) $-2i + 8j$ (b) $2i - 8j$
11. $2i + 2j$ 12. $i + j$
13. $a + 2b$, $\frac{3}{11}a + \frac{23}{11}b$ 14. (a) $\sqrt{10}$ (b) $\sqrt{5}$ (c) $5i$ (d) 5

15. (a) $\begin{pmatrix} 4 \\ -3 \end{pmatrix}$ (b) 5 (c) $\begin{pmatrix} 7 \\ -3 \end{pmatrix}$ (d) $\sqrt{58}$ 16. $6i + j$ 17. $\begin{pmatrix} -2 \\ -1 \end{pmatrix}$

18. (a) $1\frac{5}{9}i + 2\frac{4}{9}j$ (b) $-5i + 6j$ 21. C, E and F

page 62   Exercise 2C

1. **b** and **d**   2. **g** and **h**   3. $14°$ 4. $68°$ 5. $143°$ 6. -4 or 1
7. (a) $x^2 + y^2 = 25$ (b) $2x + y = 0$ (c) $x = 2y$ (d) $4x + y = 9$ 10. 1

11. $x = 5, y = 2$; $x = -5, y = 12$ 12. (a) $\begin{pmatrix} \frac{3}{5} \\ -\frac{4}{5} \end{pmatrix}$ (b) $\pm\begin{pmatrix} \frac{4}{5} \\ \frac{3}{5} \end{pmatrix}$ 13. (a) $\begin{pmatrix} 7\sqrt{2} \\ -\sqrt{2} \end{pmatrix}$ (b) $\pm\begin{pmatrix} 3 \\ 21 \end{pmatrix}$

14. (a) $-3i + 4j$ (b) $-6i - j$ (c) $9i - 3j$ (d) $35°, 117°, 28°$ 17. (b) 23   18. $131°$
19. 4   20. $1\frac{2}{13}\sqrt{13}$ 21. $\frac{9}{5}\sqrt{5}$ 22. (b) $k\sqrt{7}, 2k\sqrt{21}$

page 64   Exercise 2D

1. (a) $a + c$ (b) $\frac{1}{2}(a + c)$ (c) $\frac{1}{2}(c - a)$ (d) $\frac{1}{2}(a + c)$
3. (a) $\frac{1}{3}(a + b)$ (b) $\frac{1}{3}(a + b)$ (c) $\frac{1}{3}(a + b)$

page 67   Exercise 2E

1. $r = 3i - j + \lambda(2i + 5j)$ 2. $r = i + 2j + \lambda(4i - j)$ 3. $r = \begin{pmatrix} 1 \\ -2 \end{pmatrix} + \lambda\begin{pmatrix} -1 \\ 3 \end{pmatrix}$

4. $r = \begin{pmatrix} -2 \\ 3 \end{pmatrix} + \lambda\begin{pmatrix} 1 \\ 5 \end{pmatrix}$ 5. $5i + 10j$ 6. $\pm(4i + 3j)$ 8. B and C

9. D and F   10. 4   11. (a) $i + 3j$ (b) $r = 3i - 2j + \lambda(i + 3j)$
12. (a) not parallel, $4i + 5j$ (b) parallel (c) parallel (d) not parallel, $9i - j$

13. $\begin{pmatrix} 6 \\ -1 \end{pmatrix}$ 14. $77°$ 15. $7a - 12b$ 16. $6i + 6j$, $6\sqrt{2}$ units

17. (a) $\sqrt{5}$ units (b) 2 units (c) $\sqrt{10}$ units

page 68   Exercise 2F

1. (a) (i) $a + b$ (ii) $-a$ (iii) $b - a$ (iv) $\frac{1}{2}(a - b)$ (v) $-\frac{1}{2}(a + b)$
(b) (i) $4i + 5j$, $\sqrt{41}$ (ii) $2\sqrt{58}$
2. $1 : 2$, $6q - 4p$, $5p - 3q$

3. $b - a$, $\frac{1}{10}(5b - 3a)$ (i) $\dfrac{n}{10}(5b - 3a)$ (ii) $-\frac{1}{2}a + (\frac{1}{2} + k)b$, $\frac{5}{3}$, $\frac{1}{3}$
4. (a) $\frac{1}{2}(3b - a)$ (b) $\frac{1}{4}(3b + c)$ (d) $1 : 1$

5. (i) $\pm\begin{pmatrix}\frac{3}{5}\\\frac{4}{5}\end{pmatrix}$ (ii) 5, 1; $-4, -\frac{2}{25}\sqrt{5}$ **6.** $-4\mathbf{i} + 7\mathbf{j}$, 45°

7. $\begin{pmatrix}1\\3\end{pmatrix}, \begin{pmatrix}4\\2\end{pmatrix}, \begin{pmatrix}3\\-1\end{pmatrix}$, 45°, 90°, right-angled isosceles, 5 sq. units.

8. (a) **If diagonals of a parallelogram cut at right angles then parallelogram is a rhombus,**
(b) **Square.**

9. (a) (i) $q^2 = rp$ (ii) $p^2 + q^2 = 2q(p + r)$ (b) $\dfrac{q(p + r)}{\sqrt{(p^2 + q^2)(q^2 + r^2)}}$

10. $\begin{pmatrix}5\\4\end{pmatrix}, \sqrt{41}$ **11.** (a) $2\mathbf{i} + 4\mathbf{j} + \lambda(4\mathbf{i} + 3\mathbf{j})$ (b) $\mu(3\mathbf{i} - 4\mathbf{j})$ (c) $-\frac{6}{5}\mathbf{i} + \frac{8}{5}\mathbf{j}$, 2 units.

page 72   Exercise 3A

(Diagrams are not drawn to full size)

1.

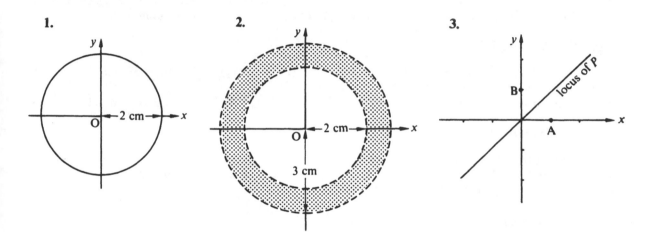

2.

3.

4.

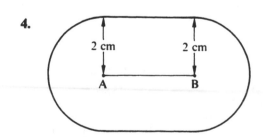

5. A pair of parallel lines, 6 cm apart, parallel to **AB** and with **AB** mid-way between them.

6.

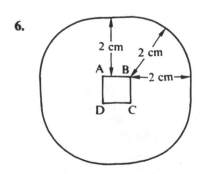

7.

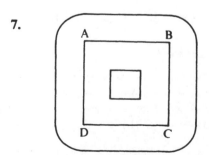

page 73   Exercise 3B

1. (a) 5   (b) 10   (c) 7   (d) 5   (e) 13   (f) $\sqrt{17}$ (g) $\sqrt{5}$ (h) $2\sqrt{5}$ (i) $2\sqrt{5}$ (j) $3\sqrt{2}$
(k) $2\sqrt{10}$ (l) $3\sqrt{5}$
2. $4\sqrt{2}$ **3.** $\sqrt{65}$ **4.** 8   **6.** 5, 10, $5\sqrt{5}$, $\hat{A}$ **7.** $2\sqrt{5}$, $3\sqrt{5}$, $\sqrt{65}$, $\hat{B}$ **8.** D   **9.** E
10. ± 4 **11.** 7 or -3 **12.** 2 or 6

page 76   Exercise 3C

1. (a) $(5, 7)$ (b) $(6, 6)$ (c) $(3, 4)$ (d) $(2\frac{1}{2}, 4)$ (e) $(2, 3)$
(f) $(4, 1)$ (g) $(-3\frac{1}{2}, 4)$ (h) $(1\frac{1}{2}, 2\frac{1}{2})$ (i) $(2, -5)$ (j) $(-5, -5)$
(k) $(-5, -5\frac{1}{2})$ (l) $(-2\frac{1}{2}, 1\frac{1}{2})$

2. (a) $\begin{pmatrix} 2 \\ 6 \end{pmatrix}$ (b) $\begin{pmatrix} -2 \\ 3 \end{pmatrix}$ (c) $\begin{pmatrix} -3 \\ 1 \end{pmatrix}$ (d) $3\mathbf{i} + \mathbf{j}$ (e) $6\mathbf{i}$ (f) $\mathbf{i} - \mathbf{j}$ **3.** $(4, 2)$ **4.** 3, 5

5. $(-2, -4)$ **6.** (a) $(3, 1)$ (b) $(6, 5)$ (c) 5   **7.** (a) $(6, 3)$ (b) $3\sqrt{5}$ **8.** $\sqrt{13}$, $\sqrt{34}$, 7
9. (a) $(7, 4)$ (b) $(7, 4)$ (c) $(8, -1)$ **10.** $(-2, 0)$ **11.** (a) $(3, 2), (2, -2), (-3, 0), (-2, 4)$
12. (a) $(8, 3\frac{4}{5})$ (b) $(-4, -1)$ **13.** (a) $(4\frac{1}{4}, 6\frac{1}{2})$ (b) $(-7, -1)$ **14.** 13 units

page 79   Exercise 3D

1. (a) 2   (b) 7   (c) 3   (d) $\frac{1}{4}$ (e) -4 (f) 3   (g) -3 (h) $-\frac{1}{4}$ (i) $\frac{1}{2}$
(j) 2   (k) $\frac{2}{3}$ (l) $-\frac{3}{4}$
2. (a) Yes   (b) No   (c) Yes   (d) Yes   (e) No   **3.** 7   **4.** 1   **5.** 2   **6.** 11, -3, -3

page 80   Exercise 3E

1. (a) $-\frac{1}{2}$ (b) $-\frac{1}{3}$ (c) $\frac{1}{2}$ (d) -2 (e) 2   (f) $-\frac{3}{2}$ (g) $-\frac{2}{3}$ (h) $-\frac{2}{3}$ (i) $-\frac{1}{4}$ (j) $\frac{5}{3}$
3. BC and AD   **4.** $\frac{5}{3}$, $-\frac{3}{5}$, $-\frac{1}{13}$, $\hat{B}$ **5.** -1 **7.** -2, 0 or 8   **8.** 5 or -3
9. (a) $-2, \frac{1}{2}, -2, \frac{1}{2}$ (b) $2\sqrt{5}, 3\sqrt{5}$ (c) a rectangle

page 85   Exercise 3F

1. (a) 3   (b) 5   (c) -3 (d) $\frac{1}{2}$ (e) $-\frac{1}{3}$ (f) -3 (g) 2   (h) 2   (i) 2   (j) 3   (k) -4 (l) $\frac{3}{2}$
(m) -3 (n) $-\frac{3}{2}$ (o) $\frac{3}{8}$
2. (a) $(0, 3)$ (b) $(0, 4)$ (c) $(0, -4)$ (d) $(0, 3)$ (e) $(0, 2)$ (f) $(0, -3)$
3. $1, -1, \frac{1}{3}\sqrt{3}, \sqrt{3}, -\frac{1}{3}\sqrt{3}$
4. (a) $(2, 0)$ (b) $(0, -2)$ **5.** (a) $(2, 0)$ (b) $(0, 3)$ **6.** (a) and (c)   **7.** (b) and (c)
8. 2, 9   **9.** $\frac{1}{2}, -3$
10. (a) $y = 4x - 10$ (b) $y = -\frac{3}{2}x + \frac{1}{2}$ (c) $y = -\frac{1}{3}x + 3$ (d) $y = \frac{3}{2}x - 11$

11. (a) $\mathbf{r} = \begin{pmatrix} 2 \\ 1 \end{pmatrix} + \lambda \begin{pmatrix} 3 \\ 4 \end{pmatrix}$ (b) $\mathbf{r} = \begin{pmatrix} 5 \\ -3 \end{pmatrix} + \lambda \begin{pmatrix} 1 \\ -4 \end{pmatrix}$ (c) $\mathbf{r} = \begin{pmatrix} 0 \\ 3 \end{pmatrix} + \lambda \begin{pmatrix} 1 \\ 2 \end{pmatrix}$ (d) $\mathbf{r} = \begin{pmatrix} 2 \\ 0 \end{pmatrix} + \lambda \begin{pmatrix} 3 \\ 2 \end{pmatrix}$

12. (a) $y = -\frac{2}{3}x - \frac{1}{3}$, $y = \frac{3}{2}x - 5$ (b) $\mathbf{r} = \begin{pmatrix} 1 \\ -1 \end{pmatrix} + \lambda \begin{pmatrix} 3 \\ -2 \end{pmatrix}$, $\mathbf{r} = \begin{pmatrix} 4 \\ 1 \end{pmatrix} + \mu \begin{pmatrix} 4 \\ 6 \end{pmatrix}$

page 89   Exercise 3G

1. (a) $y = 3x + 4$ (b) $y = 5x - 1$ (c) $y = 2x + 3$ (d) $y = \frac{1}{2}x - 1$ (e) $y = -2x + 5$
2. (a) $y = 2x + 4$ (b) $y = -x + 7$ (c) $y = 2x - 3$ (d) $y = \frac{1}{2}x - 5$ (e) $y = -3x + 1$
3. $y = -2x + 8$ **4.** $y = x - 1$ **5.** (a) $(3, 4)$ (b) -3 (c) $3y = x + 9$ **6.** $y = 2x + 1$
7. (a) $y = 3x - 3$ (b) $y = 7 - x$ **8.** $y = 2x - 2$, $x + 4y = 13$, $5x + 2y = 17$

page 92 **Exercise 3H**

1. 1·8, 32   (a) 176°F   (b) 104°F   (c) −40°F   **2.** 0·4, 10; £62   **3.** 1·5 ohm
4. 0·37, 100   (a) 100 cm³   (b) 111 cm³   **5.** 20, 0·4; 110N   **6.** 2·4, 3·6
8. 2, 5; 10   **9.** 36, 7; 11   **10.** 1, 1; 2550

page 96 **Exercise 3I**

1. (a) (4, −1) (b) (2, 1) (c) (2, 3) (d) (3, 5) (e) (1, 1), (−4, 16) (f) (2, 14), (−6, 6)
   (g) (1, 1), (5, 9) (h) ($\frac{1}{2}$, 1$\frac{1}{2}$), (2, 5) (i) (0, −1), (4, 19) (j) ($\frac{1}{3}$, $\frac{1}{9}$), (−2, 11)
   (k) (1, 2), (−4, −3) (l) (−2, −5), (4, 1)   **2.** (a) (4, 3)   **4.** 3   **5.** Parallel lines, grad. 2
6. (a) No (b) Yes (c) Yes   **7.** (1, 6)   **8.** (−1, 4), (5, 10), 6$\sqrt{2}$
9. (a) $3y + x = 21$ (b) (3, 6) (c) 2$\sqrt{10}$ **10.** (a) $3y + 4x = 28$ (b) 5 units
11. $y = 3x$ **12.** (0, −1), (6, 5), (9, 2)   **13.** −1, −4, (−1, 0)
14. (a) $x + y = 8$ (b) (4, 4) (c) 3$\sqrt{2}$

page 97 **Exercise 3J**

1. (i) $y = 3x − 6$ (ii) (2, 0)   **2.** 9   **3.** $\frac{2}{3}$, 5$\frac{1}{2}$ units, 7$\frac{1}{3}$ units, 20$\frac{1}{6}$ sq. units
4. (i) (6, −3) (ii) $4y + 5x = 18$ **5.** (−3, 5), $3y = 2x + 8$ **6.** 28°
7. (i) $8y = x + 15, 2y + x + 5 = 0$ (ii) (−7, 1) (iv) 30 sq. units (v) $\frac{6}{13}\sqrt{65}$
8. 2$\sqrt{5}$, 4$\sqrt{5}$ **9.** $y^2 − 3x^2 + 6x − 6y + 18 = 0$ **10.** (−4, −7), (7, 4)   **11.** (6, 2)
12. 140, 0·35   **13.** (i) $\frac{1}{4}$, 9 (ii) 2·7

page 104 **Exercise 4A**

1. (a) 1st (b) 3rd (c) 4th (d) 2nd (e) 3rd
2. (a) sin 20° (b) −sin 40° (c) cos 50° (d) tan 20° (e) −sin 10° (f) −tan 10°
   (g) −cos 40° (h) −sin 40° (i) cos 50° (j) tan 50° (k) −sin 10° (l) −cos 20°
   (m) −sin 50° (n) sin 40° (o) −cos 80° (p) tan 20°
3. (a) $\frac{1}{2}$ (b) $-\frac{\sqrt{3}}{2}$ (c) $\frac{1}{2}$ (d) $-\frac{1}{2}$ (e) $-\sqrt{3}$ (f) 1 (g) −1 (h) 0 (i) $-\frac{1}{2}$
   (j) 1 (k) $\frac{\sqrt{3}}{2}$ (l) $-\frac{\sqrt{3}}{2}$
4. (a) (i) 1, 90° (ii) −1, 270° (b) (i) 2, 90° (ii) −2, 270° (c) (i) 3, 0° (ii) −3, 180°
   (d) (i) 3, 90° (ii) 1, 270° (e) (i) 4, 180° (ii) 2, 0° (f) (i) 5, 90° (ii) 1, 270°
   (g) (i) 1, 45° (ii) −1, 135° (h) (i) 0, 0° (ii) −2, 90°

5. (a)

x	0	30°	45°	60°	90°	120°	135°	150°	180°	210°	225°	240°	270°	300°	315°	330°	360°
y	1	0·87	0·71	0·5	0	−0·5	−0·71	−0·87	−1	−0·87	−0·71.	−0·5	0	0·5	0·71	0·87	1

6. (a)

x	0	10°	20°	30°	40°	50°	60°	70°	80°	90°	100°	110°	120°
y	0	0·5	0·87	1	0·87	0·5	0	−0·5	−0·87	−1	−0·87	−0·5	0

   (c) (i) 12°, 48°   (ii) 72°, 108°   (iii) 15°, 45°

7. (a)

x	0	15°	30°	45°	60°	75°	90°	105°	120°	135°	150°	165°	180°
y	2	1·87	1·5	1	0·5	0·13	0	0·13	0·5	1	1·5	1·87	2

(c) (i) 36°, 144°   (ii) 51°, 129°   (iii) 67°, 113°

8. (b) 27°, 207°   **9. (b)** (i) 104°, 256° (ii) 76°, 284° (iii) 37°, 270°
10. (b) (i) 37°, 143° (ii) 17°, 163° (iii) 24°, 246°

page 108   Exercise 4B

1. (a) (v)   (b) (i)   (c) (iii)   (d) (ii)   (e) (vi)   (f) (iv)
2. (a) (ii)   (b) (iii)   ‘(c) (i)

3. (a)

(b)

(c)

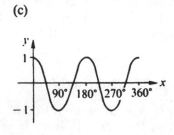

4. (a)

(b)

(c)

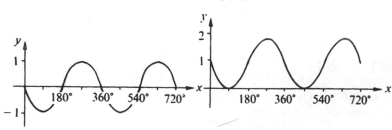

5. (a)

(b)

(c)

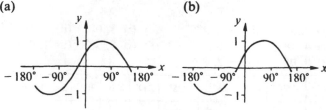

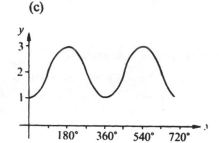

6. 2   **7.** 2   **8.** 1

page 114   Exercise 4C

1. (a) 30°, 150°   (b) 120°, 240°   (c) 45°, 225°   (d) 135°, 315° (e) 240°, 300° (f) 45°, 315°
2. (a) −30°, −150° (b) ±60°   (c) 60°, 120°   (d) −120°, 60° (e) ±135°   (f) ±150°

3. 17·5°, 162·5°   **4.** 134·4°, 225·6°   **5.** 143·1°, 323·1°   **6.** 60°, 120°, 240°, 300°

7. 63·4°, 243·4°   **8.** 56·3°, 236·3°   **9.** 120°, 150°, 300°, 330°   **10.** 30°, 150°, 210°, 330°

11. 220°, 280°   **12.** 75°, 255°   **13.** 48·2°, 311·8°   **14.** 0, 60°, 180°, 300°, 360°

15. 90°, 153·4°, 270°, 333·4°   **16.** 0, 120°, 180°, 240°, 360°   **17.** 48·6°, 90°, 131·4°, 270°

18. 90°, 104·5°, 255·5°, 270°   **19.** 0, 75·5°, 180°, 284·5°, 360°   **20.** 30°, 150°, 270°

21. 90°, 210°, 330°   **22.** 63·4°, 123·7°, 243·4°, 303·7°   **23.** 45°, 153·4°, 225°, 333·4°

24. $\pm 30°$, $\pm 150°$ **25.** $-148·3°$, $-58·3°$, $31·7°$, $121·7°$ **26.** $-115°$, $155°$

27. $\pm 90°$ **28.** 0, 33·7°, $-146·3°$, $\pm 180°$ **29.** $\pm 60°$

30. (a) $(2 \sin \theta + 1)(3 \cos \theta + 2)$ (b) $-30°$, $\pm 131·8°$, $-150°$

31. (a) $(\cos \theta - 1)(3 \sin \theta + 2)$ (b) 0, 221·8°, 318·2°, 360°

page 117   Exercise 4D

1. (a) $\sqrt{2}$ (b) 1   (c) 2   (d) 2   (e) $\sqrt{2}$ (f) -2 (g) -2 (h) -2 (i) 1
(j) 2   (k) -2 (l) $-\sqrt{2}$

2. (a) $\cos A$ (b) $\sec \theta$ (c) 1   (d) $\cos \theta$

5. (a) $x^2 + y^2 = 1$ (b) $xy = 3$ (c) $100x^2 + y^2 = 4$ (d) $x^2 + y^2 - 6x + 8 = 0$
(e) $x^2 + y^2 - 4x + 2y + 4 = 0$

6. (a) 0, 14·5°, 165·5°, $\pm 180°$ (b) $\pm 180°$ (c) 0, $\pm 48·2°$ (d) 19·5°, 160·5°   (e) $\pm 45°$, $\pm 135°$
(f) $-63·4°$, 0, 116·6°, $\pm 180°$

7. (a) 60°, 300°   (b) 55·9°, 145·9°, 235·9°, 325·9°
(c) 60°, 120°, 240°, 300°   (d) 120°, 240°   (e) 26·6°, 135°, 206·6°, 315°
(f) 68·2°, 135°, 248·2°, 315°   (g) 194·5°, 345·5°
(h) 26·6°, 45°, 206·6°, 225°   (i) 70·5°, 289·5°

page 120   Exercise 4E

1. (a) 1   (b) $\frac{1}{2}$ **2.** (a) $\sin 3A$ (b) $\sin \theta$ (c) $\cos 5\theta$ (d) 1   (e) 2

3. (a) $\frac{\sqrt{2}}{4}(\sqrt{3} + 1)$ (b) $\frac{\sqrt{2}}{4}(\sqrt{3} - 1)$ (c) $-2 - \sqrt{3}$

4. (a) (i) 1, 70°   (ii) -1, 250°   (b) (i) 1, 110°   (ii) -1, 290°   (c) (i) 1, 320°   (ii) -1, 140°

5. (a) $\frac{1}{2}$ (b) $\frac{3}{2}$ (c) $\frac{4}{7}$ **6.** (a) $\frac{33}{65}$ (b) $\frac{33}{56}$ **7.** (a) $-\frac{56}{65}$ (b) $\frac{63}{65}$

8. (a) 200°, 320°   (b) 26·4°, 253·6°   (c) 45°, 225°

page 122   Exercise 4F

1. (a) $2 \sin A$ (b) $\sin 2A$ (c) 1   **2.** $-\frac{1}{8}$ **3.** (a) $\frac{4}{5}$ (b) $\frac{3}{5}$ (c) $\frac{24}{25}$ (d) $-\frac{7}{25}$

4. (a) $\frac{4}{5}, \frac{3}{5}, \frac{4}{3}$ (b) $-\frac{4}{5}, \frac{3}{5}, -\frac{4}{3}$

5. (a) $x + 2y^2 = 0$ (b) $x = 2y^2 + 4y + 1$ (c) $2x^2 - 8x + y + 4 = 0$

6. (a) $\frac{3}{4}$ (b) $\frac{24}{25}$ (c) $\frac{\sqrt{10}}{10}$ **7.** (a) $\frac{\sqrt{5}}{4}$ (b) $\frac{\sqrt{11}}{4}$

8. (a) $3 \sin \theta - 4 \sin^3 \theta$ (b) $\frac{11}{16}$ **9.** (a) $4 \cos^3 \theta - 3 \cos \theta$ (b) $-\frac{2\sqrt{5}}{25}$

13. (a) 0, $\pm 120°$, $\pm 180°$ (b) $-135°$, 45°, $\pm 90°$ (c) $\pm 70·5°$, $\pm 120°$ (d) 0, $\pm 45°$, $\pm 135°$, $\pm 180°$

14. (a) 22·5°, 112·5°, 202·5°, 292·5°   (b) 9·6°, 90°, 170·4°, 270°   (c) 30°, 90°, 150°
(d) 14·5°, 165·5°

17. (a) 2, 1, 1   (b) -1, 2, -2 (c) -3, 1, 2

page 125   Exercise 4G

1. (a) 180°   (b) 30°   (c) 270°   (d) 120°   (e) 330°

2. (a) 2π (b) $\dfrac{\pi}{3}$ (c) $\dfrac{\pi}{4}$ (d) $\dfrac{5\pi}{6}$ (e) $\dfrac{\pi}{2}$

3. (a) 1   (b) 0   (c) 1   (d) -1 (e) -1

   (f) $\frac{1}{2}$ (g) $\frac{1}{2}$ (h) $-\dfrac{\sqrt{3}}{2}$ (i) -1 (j) $-\dfrac{1}{\sqrt{2}}$

4. (a) 1·22   (b) 4·36   **5.** (a) 172°   (b) 143°

6. (a) 0·84   (b) $-0·14$ (c) $-0·74$

7. (a) (i) 5 cm   (b) (i) 2 cm   (c) (i) 12 m
       (ii) 12·5 cm²   (ii) 4 cm²   (ii) 24 m²

8. (a) $\dfrac{\pi}{3}$ (b) $\dfrac{5\pi}{3}$ (c) π cm   (d) 5π cm

9. 1 rad   **10.** 8 cm²   **11.** 1·5 rad   **12.** 12 cm

13. (a) $\dfrac{\pi}{6}, \dfrac{5\pi}{6}$ (b) $\dfrac{\pi}{3}, \dfrac{2\pi}{3}, \dfrac{4\pi}{3}, \dfrac{5\pi}{3}$ (c) $\dfrac{3\pi}{8}, \dfrac{7\pi}{8}, \dfrac{11\pi}{8}, \dfrac{15\pi}{8}$

   (d) $0, \pi, \dfrac{7\pi}{6}, \dfrac{11\pi}{6}, 2\pi$ (e) $\dfrac{\pi}{2}$ (f) $0, \dfrac{\pi}{3}, \dfrac{5\pi}{3}, 2\pi$

14. 6 cm²   **15.** (a) 9 cm² (b) 0·42 cm²   **16.** (a) 1·70 rad (b) 13·6 cm²
18. (a) $2\frac{1}{4}$ m² (b) 0·47 m²   **19.** (a) 1·55 (b) 1·19 (c) 7·66 cm²   **20.** (c) (i) 1·7 (ii) 1·25
21. (c) 2·5 (d) 2·475

page 128   Exercise 4H

2. (a) $-146·3, 33·7$ (b) 30°, 150°   **3.** 48·6°, 131·4°, 210°, 330°

4. (a) (i) $-\dfrac{24}{7}$ (ii) $-\dfrac{33}{65}$ (iii) $\dfrac{33}{65}$ **5.** (a) $-\dfrac{1}{3}$ (b) $-\dfrac{4}{9}\sqrt{2}$ (c) $-\dfrac{7}{9}$ (d) $\dfrac{1}{\sqrt{3}}$

6. (i) 45°, 135°, 210°, 225°, 315°, 330°   (ii) 100·9°, 280·9°
7. (i) 56·3°, 135°, 236·3°, 315°   (ii) 26·6°, 63·4°, 206·6°, 243·4°
8. (a) $x = 4y^2(1 - y^2)$ (b) 26·6°, 135°
9. (a) $\frac{1}{15}(8 + 3\sqrt{5}), \frac{2}{15}(2\sqrt{5} - 3)$, acute (b) $\pm 35·3°, \pm 144·7°$

11. 1, 4, 3   **12.** (i) 123°, 303°   (ii) 32·1°, 147·9°   (iii) $\dfrac{1 + \tan\theta}{1 - \tan\theta}$, 45°, 225°

14. 4   **15.** (a) $\frac{2}{3}\pi, \frac{4}{3}\pi$ (b) $\frac{1}{3}\pi, \frac{5}{3}\pi$ (c) $\frac{3}{4}\pi, \frac{7}{4}\pi$
16. (i) 21·4 cm   (ii) 13·6 cm²   **17.** (a) $\pm\frac{1}{3}\pi, \pm 2·3$ rads   (b) 1·97 rads

page 132   Exercise 5A

1. 27   **2.** 128   **3.** 25   **4.** 49   **5.** 1   **6.** 1   **7.** 2   **8.** 4
9. 12   **10.** 3   **11.** 2   **12.** 64   **13.** $\frac{1}{25}$ **14.** $\frac{1}{7}$ **15.** $\frac{1}{100}$ **16.** $\frac{1}{2}$
17. $\frac{1}{8}$ **18.** $\frac{1}{2}$ **19.** 9   **20.** $\frac{1}{16}$ **21.** $\frac{1}{81}$ **22.** $\frac{1}{64}$ **23.** 4   **24.** $\frac{1}{4}$
25. $\frac{1}{16}$ **26.** $\frac{4}{9}$ **27.** 1   **28.** $1\frac{9}{16}$ **29.** $6\frac{1}{4}$ **30.** $\frac{3}{4}$ **31.** $1\frac{1}{3}$ **32.** $\frac{2}{3}$
33. $15\frac{5}{8}$ **34.** $\frac{8}{27}$ **35.** $\frac{4}{5}$ **36.** 2^6 **37.** 2^4 **38.** 2^6 **39.** 2^{11} **40.** 2^2
41. 2^3 **42.** 2^7 **43.** 2^6 **44.** x^7 **45.** x^6 **46.** $18x^6$ **47.** $24x^3y$ **48.** $3x^2$
49. $5x^3$ **50.** x^2 **51.** $30x^4y^3$ **52.** $12x^4y^3$ **53.** a^{-1} **54.** $a^{-1/2}$ **55.** a^{-1} **56.** a^5
57. $a^{1/2}$ **58.** $a^{-1/2}$ **59.** a^6 **60.** $a^{3/2}$ **61.** $a^{7/2}$ **62.** $a^{5/2}$ **63.** 6   **64.** 0
65. -1 **66.** -2 **67.** -2 **68.** 11   **69.** 3   **70.** 8   **71.** 4   **72.** 3
73. -2 **74.** 6   **75.** 6   **76.** 9

page 135   **Exercise 5B**

1. (a) $y = kx$ (b) $y = kx^3$ (c) $y = k\sqrt[3]{x}$ (d) $y = \dfrac{k}{x}$ (e) $y = \dfrac{k}{\sqrt{x}}$ (f) $y = kxz^2$

(g) $y = \dfrac{kx^2}{z}$ (h) $y = \dfrac{kwx}{z}$ (i) $y = k_1x + k_2z$ (j) $y = k_1x^2 + k_2z$ (k) $y = k_1\sqrt{x} + \dfrac{k_2}{z^2}$

2. (a) $y = \dfrac{5}{x^2}$ (b) 80   **3.** (a) $y = 3z^2 - \dfrac{5}{2}x$ (b) $\dfrac{1}{2}$ **4.** $F = \dfrac{50}{3}x$, 0·48 m

5. $T = 2\sqrt{l}$, 0·25 m   **6.** $F = \dfrac{3v^2}{2r}$, 12N   **7.** $C = 13 + \dfrac{u}{20}$, £35·50

8. $v = \sqrt{\dfrac{E}{\rho}}$, 5200 m/s

page 139   **Exercise 5C**

1. (a) $m^d = c$ (b) $b^q = p$ (c) $10^t = s$ (d) $e^r = z$
2. (a) 3   (b) 4   (c) 6   (d) 256   (e) $\frac{1}{100}$ (f) 2048
(g) $1\frac{1}{3}$ (h) $1\frac{1}{2}$ (i) $\frac{1}{8}$
3. (a) 5   (b) 2   (c) 4   (d) -1 (e) -2 (f) 0   (g) 1   (h) $\frac{1}{3}$
(i) $-\frac{1}{3}$ (j) $\frac{1}{3}$ (k) $\frac{2}{3}$ (l) $-\frac{2}{3}$
4. (a) $\log a + \log b + \log c$ (b) $2\log a + \log b + \log c$ (c) $3\log a + 2\log b + \log c$
(d) $\log a + \frac{1}{2}\log b$ (e) $\log a + \log b - \log c$ (f) $2\log a - \log b - 3\log c$
(g) $-\log a - \log b$ (h) $\log a + \frac{1}{2}\log b - \frac{1}{2}\log c$ (i) $1 + \log a + 2\log b$
(j) $\frac{1}{2} + \frac{1}{2}\log a - 2\log b$
5. (a) $\log_a 10$ (b) $\log_a 3$ (c) $\log_a 2$ (d) $\log_a 20$

6. (a) $\log_a x^2$ (b) $\log_a [x(x + 3)]$ (c) $\log_a\left(\dfrac{x+1}{2}\right)$ (d) $\log_a (x - 1)$

(e) $\log_a\left(\dfrac{x}{x+1}\right)$ (f) $\log_a [x^3(x+1)]$ (g) $\log_a\left(\dfrac{x^2}{1+x}\right)$ (h) $\log_a [x^2(x+1)]$

7. (a) 6   (b) 3   (c) 256
8. (a) 1·36   (b) 3·14   (c) 2·09   (d) 25·12   (e) 0·22   (f) 22·20   (g) 1·14   (h) 0·64
9. (a) 0·4   (b) 0·6   (c) 1·4   (d) 1·5   (e) $-0·3$ (f) $2\frac{1}{3}$ (g) $1\frac{3}{7}$
10. (a) 6·4   (b) 4   (c) 0·8   (d) 5·6   (e) $-1·6$ (f) $-3·2$ (g) 1·5

11. (a)

x	-3	-2	-1	0	1	2
y	0·125	0·25	0·5	1	2	4

(b)

x	0·2	0·4	0·6	0·8	1	2	3	4
y	$-2·32$	$-1·32$	$-0·74$	$-0·32$	0	1	1·58	2

12. (a) 6   (b) 12   (c) 9
13. (a) 1·6   (b) 4·1
14. 2·5, 1·4   **15.** 2·3, 6   **16.** 2, 1·5   **17.** 3, 0·5
18. (a) 2·32   (b) 2·58   (c) 1·43   (d) 1·79   (e) 5·37   (f) $-2·71$ (g) 0·326   (h) $-1·95$ (i) 0·287
19. (a) 2   (b) 1   (c) ± 1 (d) 2·16   (e) 1 or 0·585   (f) $\frac{1}{16}$ or 8
20. (a) 15   (b) 56   (c) 3   (d) 13·5   (e) 64   (f) 5 or 25
21. (a) $x = 16, y = 4$ (b) $x = \sqrt{3}, y = 9$
(c) $x = 12, y = 4$ (d) $x = \frac{1}{2}, y = \frac{1}{2}; x = 8, y = 2$
(e) $x = 3, y = 2$ (f) $x = 4, y = 4; x = 48, y = \frac{1}{3}$
(g) $x = 9, y = 6$ (h) $x = 12, y = 10$

page 144   Exercise 5D

1. (a) $(0, 4)$
(b) $(1, 0)(4, 0)$
(c) $(2\frac{1}{2}, -2\frac{1}{4})$ min

2. (a) $(0, 4)$
(b) $(-1, 0)(4, 0)$
(c) $(1\frac{1}{2}, 6\frac{1}{4})$ max

3. (a) $(0, -5)$
(b) $(-\frac{1}{2}, 0)$
$(-2\frac{1}{2}, 0)$
(c) $(-1\frac{1}{2}, 4)$ max

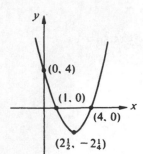
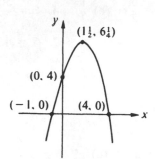
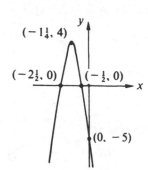

4. $a = 2, b = 2, c = 1$ **5.** $a = 1\frac{1}{2}, b = 3, c = -2$ **6.** $a = 3, b = 2, c = 2$
7. (a) $x \leqslant -3, x \geqslant 5$ (b) $3 \leqslant x \leqslant 5$ (c) $x \leqslant -1, x \geqslant 5$ (d) $x \leqslant -\frac{1}{2}, x \geqslant 3$
(e) $\frac{1}{3} \leqslant x \leqslant 6$ (f) $-2\frac{1}{2} \leqslant x \leqslant 1$
8. (a) $-6 \leqslant x \leqslant 1$ (b) $x \leqslant -4, x \geqslant -3$ (c) $5 < x < 7$ (d) $x \leqslant 2\frac{1}{3}, x \geqslant 5$
(e) $x < -1\frac{1}{2}, x > 1\frac{2}{3}$ (f) $-\frac{1}{3} \leqslant x \leqslant \frac{1}{2}$
9. (a) $x \leqslant -\sqrt{3} - 2, x \geqslant \sqrt{3} - 2$ (b) $\frac{1}{2}(5 - \sqrt{13}) \leqslant x \leqslant \frac{1}{2}(5 + \sqrt{13})$ (c) $\mathbb{R}$
(d) $\frac{1}{4}(3 - \sqrt{17}) \leqslant x \leqslant \frac{1}{4}(3 + \sqrt{17})$ (e) $\mathbb{R}$ (f) $x \leqslant -\frac{1}{3}(\sqrt{7} + 1), x \geqslant \frac{1}{3}(\sqrt{7} - 1)$
12. $k > 5$ **13.** $-1 < k < 3$

page 148   Exercise 5E

1. (a) $g(x) = 0$ (b) $f(x) = 0$ (c) $h(x) = 0$ (d) $f(x), g(x)$ (e) $h(x)$
2. (a) 129, real distinct   (b) -3, no real roots   (c) 57, real distinct   (d) 0, repeated root
(e) 37, real distinct   (f) -7, no real roots   (g) 0, repeated root   (h) -19, no real root
(i) 261, real distinct
3. (a) $\{y \in \mathbb{R}: y \geqslant -6\}$ (b) $\{y \in \mathbb{R}: y \geqslant 2\frac{3}{4}\}$ (c) $\{y \in \mathbb{R}: y \leqslant 6\}$ (d) $\{y \in \mathbb{R}: y \geqslant -5\}$
(e) $\{y \in \mathbb{R}: y \leqslant 2\frac{1}{8}\}$
4. 8   **5.** ± 20 **6.** $|b| > 12$ **7.** $k \leqslant -3, k \geqslant 1$ **8.** $k < -1, k > 15$
9. (a) $-4 \leqslant k \leqslant 1$ (b) $-\frac{1}{2} < k < 4\frac{1}{2}$ except $k = 0$ (c) $\{k \in \mathbb{R}: k \neq 0\}$
(d) $k \geqslant 3, k \leqslant \frac{1}{3}$ (e) $k < 2$ except $k = 0$ (f) $k < 0, k \geqslant 1$
10. $0 < \beta < \frac{4}{25}$ **11.** $(3, 0)$ **12.** $(1\frac{1}{2}, -3)$ **13.** (a) (ii), (b) (iii), (c) (i), (d) (iii)

page 151   Exercise 5F

1. (a) $-9, 4$ (b) $2, -5$ (c) $7, 2$ (d) $9, -3$ (e) $3\frac{1}{2}, \frac{1}{2}$ (f) $-\frac{1}{4}, -\frac{1}{4}$ (g) $-3\frac{1}{3}, -\frac{2}{3}$ (h) $-2, \frac{1}{3}$
2. (a) $x^2 + 3x - 1 = 0$ (b) $x^2 - 6x - 4 = 0$ (c) $x^2 - 7x - 5 = 0$
(d) $3x^2 + 2x - 7 = 0$ (e) $2x^2 + 5x - 4 = 0$ (f) $2x^2 + 3x - 10 = 0$
(g) $12x^2 + 3x - 4 = 0$ (h) $6x^2 + 10x + 3 = 0$
3. (a) 2   (b) -11 (c) 121   (d) 117   (e) 14   (f) -22
4. (a) $\frac{1}{2}$ (b) $2\frac{1}{2}$ (c) $6\frac{3}{4}$ (d) $3\frac{3}{4}$ (e) $2\frac{5}{8}$ (f) 5
5. (a) $-2\frac{2}{8}$ (b) $4\frac{4}{8}$ (c) 3   (d) $-2\frac{2}{3}$ (e) $-\frac{20}{24}$ (f) $4\frac{8}{27}$

7. (a) $x^2 - 8x + 14 = 0$ (b) $2x^2 - 4x + 1 = 0$ (c) $x^2 - 12x + 4 = 0$
(d) $x^2 - 8x + 8 = 0$ (e) $x^2 - 40x + 8 = 0$ (f) $4x^2 - 12x + 1 = 0$
(g) $x^2 - 12x + 34 = 0$ (h) $x^2 - 6x + 1 = 0$
8. $k = \frac{2}{9}, x = -\frac{1}{3}, -\frac{2}{3}$ **9.** $k = 1\frac{1}{4}, x = \frac{1}{2}, 2\frac{1}{2}$ **10.** (a) $-1 \leqslant k \leqslant 4$ (b) $0 < k < 3$
11. (a) $k \leqslant 1, k \geqslant 9$ (b) $0 < k < 1$

page 155   Exercise 5G

1. (a) 2 (b) -7 (c) -17 (d) 29 (e) 5 (f) 5 (g) $1\frac{1}{3}$
2. (a) $(x - 1)(x - 3)(x + 2)$ (b) $(2x - 3)(x + 4)(x + 1)$
(c) $(2x - 3)(x + 4)(x - 1)$ (d) $(3x - 1)(2x + 3)(x - 2)$
(e) $(x + 1)(x - 1)(2x - 3)(x + 5)$ (f) $(x + 1)(x + 2)(3x + 2)(2x + 3)$
3. $(x - 3)(x - 2)(x + 2); |x| < 2, x > 3$ **4.** $(x - 1)(3x - 2)(2x + 1); x < -\frac{1}{2}, \frac{2}{3} < x < 1$
5. $-1\frac{1}{2}$ **6.** -9 **7.** $-2\frac{1}{2}$ **8.** (3, 8)   **9.** (6, -1)   **10.** 12, 20; $(x - 2)(3x + 2)(2x - 5)$
12. (a) (0, 6)   **13.** (a) (0, 8)   **14.** $(x - 2)^2(2x + 3)$
(b) $(-3, 0), (-2, 0), (-1, 0)$ (b) $(-1, 0), (2, 0), (4, 0)$ (a) (0, 12)
(b) $(-1\frac{1}{2}, 0)$
(c) (2, 0)

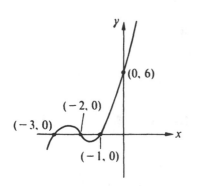

$x \leqslant -3, -2 \leqslant x \leqslant -1$

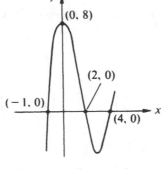

$-1 < x < 2, x > 4$

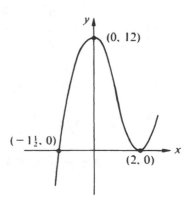

$x < -1\frac{1}{2}$

16. (a) 2 (b) -5 (c) -10 (d) $\frac{4}{3}$ (e) -6 (f) 12
17. (a) $\frac{1}{3}$ (b) $-\frac{2}{3}$ (c) $-\frac{1}{2}$ (d) $10x^3 + 4x^2 - 5x - 2$

page 156   Exercise 5H

1. $-\frac{4}{9}, \frac{1}{9}$ **2.** 2   **3.** 3   **4.** (i) 4 (ii) 8 (iii) $\frac{1}{16}$ **5.** $-1, 0$
6. 1·53   **7.** 4   **8.** (i)(a) $\log\left(\dfrac{x}{y}\right)$ (b) 0 (c) 2 (d) $\log 2$ (ii) 8·48
9. 2, 16, 1·78   **10.** (i) 3·16 (ii) 1·33 (iii) 0·16   **11.** $x = 2, y = 8; x = 8, y = 2$
12. $x = 9, y = 81; x = 27, y = 27$ **13.** (a) $-16 < a < 0$ (b) ± 4
14. (i) 2 (ii) 5 (iii) $4x^2 - 21x + 1 = 0$ **15.** (b) $[b^2 + (c - 1)^2]x^2 - b(c + 1)x + c = 0$
16. (a) $-\frac{1}{3} \leqslant k \leqslant 1$ (b) $\frac{1}{4}(2 \pm 3\sqrt{2})$
17. $x^2 - 4(k^2 - 4k + 2)x + 16 = 0, x^2 - 6x + 4 = 0$
18. $-1, -2, 3x + 2$ **19.** (a) $\frac{1}{6}$ **20.** $(3 - x)(x - 1)^2, \{x \in \mathbb{R}: x \geqslant 3 \quad \text{or} \quad x = 1\}$
21. $(2x - 1)(x + 2)(x^2 - x + 2)$

page 161   **Exercise 6A**

1. (a) $\begin{pmatrix} 5 & 0 \\ 3 & 2 \end{pmatrix}$ (b) $\begin{pmatrix} -7 & -6 \\ -1 & 0 \end{pmatrix}$ (c) $\begin{pmatrix} -2 & -6 \\ 2 & 2 \end{pmatrix}$ (d) Impossible   (e) $\begin{pmatrix} 13 & 9 \\ 3 & 1 \end{pmatrix}$

(f) $\begin{pmatrix} -12 & -6 \\ 8 & 4 \end{pmatrix}$ (g) $\begin{pmatrix} -3 & -15 \\ -1 & -5 \end{pmatrix}$ (h) $\begin{pmatrix} -2 & 0 \\ 0 & -2 \end{pmatrix}$ (i) Impossible   (j) $\begin{pmatrix} 12 & 6 \\ 12 & 6 \\ -14 & -7 \end{pmatrix}$

(k) Impossible   (l) $\begin{pmatrix} 2 & 5 \\ 5 & -2 \\ -4 & -1 \end{pmatrix}$ (m) $\frac{1}{2}\begin{pmatrix} 1 & 3 \\ -1 & -1 \end{pmatrix}$ (n) No inverse, B is singular

(o) $\begin{pmatrix} -1 & 1 \\ -3 & 1 \end{pmatrix}$ (p) $\begin{pmatrix} 1 & 2 & -3 \\ 3 & 0 & 2 \end{pmatrix}$

2. $\mathbf{AB} = \begin{pmatrix} 4 \\ 5 \end{pmatrix}$ $\mathbf{AD} = \begin{pmatrix} 4 & 4 & -1 \\ 5 & -2 & -3 \end{pmatrix}$ $\mathbf{BC} = \begin{pmatrix} -2 & 6 \\ -1 & 3 \end{pmatrix}$ $\mathbf{CD} = (1 \quad 6 \quad 1)$

$\mathbf{CA} = (8 \quad -5)$ $\mathbf{CB} = (1)$

3.

	Won	Drawn	Lost
A	4	1	3
B	3	1	4
C	2	6	0
D	2	2	4
E	3	2	3

A, C, E, B, D; C, A, E, B, D.

4. (a) $-4, 11$ (b) $-2, -3$ (c) $2, 4$ (d) $4, 3$ (e) $2, -1$ (f) $5, 19$ (g) $-1, 4$ (h) $2, -1$
(i) $-3, 2$ (j) $-3, 1\frac{1}{2}$

6. $\mathbf{X} = \begin{pmatrix} 1 & -1 \\ 3 & 4 \end{pmatrix}$ $\mathbf{Y} = \begin{pmatrix} -1 & 1 \\ 2 & -3 \end{pmatrix}$ $\mathbf{Z} = \begin{pmatrix} 2 & -1 \\ 4 & 1 \end{pmatrix}$ **7.** $4, -11$ **8.** $-\frac{1}{2}$ or 3

9. (a) $x = 3, y = -2$ (b) $x = 1, y = -\frac{2}{3}$ (c) $x = 1\frac{1}{2}, y = \frac{1}{2}$ **10.** $\begin{pmatrix} 1 & 0 & 0 \\ 0 & 1 & 0 \\ 0 & 0 & 1 \end{pmatrix}$, $-1, 2, 1.$

page 168   **Exercise 6B**

1. $(9, -3), (-3, -13), (6, -13), (3, 4)$ **2.** (a) $(4, 1)$ (b) $(5, 10)$ (c) $(3, -1)$ (d) $(-5, 2)$

3. $\begin{pmatrix} 2 & -1 \\ 3 & 4 \end{pmatrix}$, $(-5, 9), (2, -2)$

4. (a) $(5, -3), (15, -9), (13, -7), (3, -1)$ (b) 8 sq. units (c) $\frac{1}{2}\begin{pmatrix} 1 & 1 \\ 3 & 5 \end{pmatrix}$

5. $\begin{pmatrix} -1 & 0 \\ 0 & 1 \end{pmatrix}, \begin{pmatrix} 0 & 1 \\ -1 & 0 \end{pmatrix}, \begin{pmatrix} 0 & 1 \\ 1 & 0 \end{pmatrix}, \begin{pmatrix} 5 & 2 \\ 4 & -2 \end{pmatrix}$

6. (a) (i) $\begin{pmatrix} 1 & 2 \\ 0 & 1 \end{pmatrix}$ (ii) $\begin{aligned} x' &= x + 2y \\ y' &= y \end{aligned}$ (b) (i) $\begin{pmatrix} 1 & 0 \\ 3 & 1 \end{pmatrix}$ (ii) $\begin{aligned} x' &= x \\ y' &= 3x + y \end{aligned}$

7. $\begin{pmatrix} 18 & 14 \\ 6 & 5 \end{pmatrix}$, 10 sq. units, 30 sq. units

8. (a) $\begin{pmatrix} -1 & 0 \\ 0 & -1 \end{pmatrix}$ (b) $\begin{pmatrix} 3 & 0 \\ 0 & 3 \end{pmatrix}$ (c) $\begin{pmatrix} 0 & -1 \\ -1 & 0 \end{pmatrix}$ (d) $\begin{pmatrix} 1 & 0 \\ 0 & 2 \end{pmatrix}$ (e) $\begin{pmatrix} 1 & 1 \\ 0 & 1 \end{pmatrix}$

(f) $\begin{pmatrix} 0 & 1 \\ -1 & 2 \end{pmatrix}$ (g) $\begin{pmatrix} 2 & 1 \\ 1 & 2 \end{pmatrix}$ (h) $\begin{pmatrix} -1 & 2 \\ -2 & 3 \end{pmatrix}$ (i) $\begin{pmatrix} 2 & -\frac{1}{2} \\ 2 & 0 \end{pmatrix}$

9. (a) reflection in $y = x$ (b) enlargement ($\times$ 5) centre (0, 0)
(c) stretch ($\times$ 3) parallel to x-axis, y-axis fixed   (d) mapping onto (0, 0)
(e) mapping onto x-axis   (f) mapping onto $y = x$
(g) enlargement ($\times$ 3) centre (0, 0) and reflection in $y = x$
(h) shear with y-axis fixed and $(1, 0) \to (1, 2)$ (i) mapping onto $y = 4x$
(j) shear with $y = x$ fixed and $(1, 0) \to (-2, -3)$

10. $2x + 3y = 0$ **12.** (a) $\begin{pmatrix} x' \\ y' \end{pmatrix} = \begin{pmatrix} -1 & 0 \\ 0 & 1 \end{pmatrix}\begin{pmatrix} x \\ y \end{pmatrix} + \begin{pmatrix} 2 \\ 3 \end{pmatrix}$ (b) $\begin{aligned} x' &= -x + 2 \\ y' &= y + 3 \end{aligned}$

13. (a) $(-1, -1)$ (b) $(-3, 2)$ (c) $(-1, 2)$

page 176   Exercise 6C

1. (a) $(-4, -2)$ (b) $(-3, 2)$ (c) $(4\frac{1}{2}, 1)$

2. $\begin{pmatrix} 3 & 1 \\ 1 & 0 \end{pmatrix}\begin{pmatrix} x \\ y \end{pmatrix} + \begin{pmatrix} 0 \\ -1 \end{pmatrix}, \begin{pmatrix} 2 & -1 \\ -1 & 0 \end{pmatrix}\begin{pmatrix} x \\ y \end{pmatrix} + \begin{pmatrix} -3 \\ 1 \end{pmatrix}, \begin{pmatrix} 5 & -3 \\ 2 & -1 \end{pmatrix}\begin{pmatrix} x \\ y \end{pmatrix} + \begin{pmatrix} -8 \\ -4 \end{pmatrix},$

$\begin{pmatrix} 5 & 2 \\ -3 & -1 \end{pmatrix}\begin{pmatrix} x \\ y \end{pmatrix} + \begin{pmatrix} -2 \\ 1 \end{pmatrix}$

(a) $(\frac{1}{3}, -\frac{2}{3})$ (b) $(2, -1)$ (c) $(2, 0)$ (d) $(1, -1)$

3. $7y = 5x + 6$, $x = 3y + 2$ **4.** $y = 3x - 4$ **5.** $\mathbf{r} = \begin{pmatrix} -2 \\ 3 \end{pmatrix} + \lambda\begin{pmatrix} 11 \\ 1 \end{pmatrix}, \mathbf{r} = \begin{pmatrix} 1 \\ 2 \end{pmatrix} + \lambda\begin{pmatrix} -1 \\ 5 \end{pmatrix}$

6. $\begin{pmatrix} 5 & 2 \\ -2 & 1 \end{pmatrix}$, $x + 7y = 0$, $4x + y = 27$, $y = 3$ **7.** $y = x + 1$ **8.** $\mathbf{r} = \begin{pmatrix} 1 \\ -1 \end{pmatrix} + \lambda\begin{pmatrix} -2 \\ 3 \end{pmatrix}$

9. (a) $3y = 10x + 8$, $y = 5x + 21$ **13.** $\begin{pmatrix} 7 \\ -7 \end{pmatrix}$ **14.** $-2\mathbf{i} + 3\mathbf{j}$ **15.** $(-1, 1)$

16. (a) $y = x$, $y = 2x$ (b) $3y = x$, $y + x = 0$ (c) $y = \pm 2\sqrt{2}x$

17. $\begin{pmatrix} 5 & -4 \\ 2 & -1 \end{pmatrix}$, $y = x$, $2y = x$; $y = x$ **18.** 8, $y = 2x$

19. (a) $y = x + 6$, $3y = 2x + 12$ (b) $y + 4x = 1$, $y = x + 1$ (c) $y = 2$, $y = 2x - 6$
(d) $y = x - 1$ (e) $y = 3x - 9$

page 182   Exercise 6D

1. (a) $\begin{pmatrix} \frac{\sqrt{3}}{2} & -\frac{1}{2} \\ \frac{1}{2} & \frac{\sqrt{3}}{2} \end{pmatrix}$ (b) $\begin{pmatrix} \frac{1}{\sqrt{2}} & -\frac{1}{\sqrt{2}} \\ \frac{1}{\sqrt{2}} & \frac{1}{\sqrt{2}} \end{pmatrix}$ (c) $\begin{pmatrix} -\frac{1}{2} & -\frac{\sqrt{3}}{2} \\ \frac{\sqrt{3}}{2} & -\frac{1}{2} \end{pmatrix}$

2. (a) $\begin{pmatrix} -\frac{1}{2} & \frac{\sqrt{3}}{2} \\ \frac{\sqrt{3}}{2} & \frac{1}{2} \end{pmatrix}$ (b) $\begin{pmatrix} \frac{1}{2} & \frac{\sqrt{3}}{2} \\ \frac{\sqrt{3}}{2} & -\frac{1}{2} \end{pmatrix}$ (c) $\begin{pmatrix} -\frac{3}{5} & \frac{4}{5} \\ \frac{4}{5} & \frac{3}{5} \end{pmatrix}$

3. (a) rotation of 53·1° anticlockwise about origin   (b) reflection in $2y = x$
(c) rotation of 53·1° anticlockwise about origin and enlargement ($\times$ 5), centre the origin

4. (a) $\begin{pmatrix} x' \\ y' \end{pmatrix} = \begin{pmatrix} 0 & -1 \\ 1 & 0 \end{pmatrix}\begin{pmatrix} x \\ y \end{pmatrix} + \begin{pmatrix} 3 \\ 5 \end{pmatrix}$ (b) $\begin{pmatrix} x' \\ y' \end{pmatrix} = \begin{pmatrix} -1 & 0 \\ 0 & -1 \end{pmatrix}\begin{pmatrix} x \\ y \end{pmatrix} + \begin{pmatrix} 6 \\ -2 \end{pmatrix}$

(c) $\begin{pmatrix} x' \\ y' \end{pmatrix} = \begin{pmatrix} 3 & 0 \\ 0 & 3 \end{pmatrix}\begin{pmatrix} x \\ y \end{pmatrix} + \begin{pmatrix} -4 \\ 2 \end{pmatrix}$ (d) $\begin{pmatrix} x' \\ y' \end{pmatrix} = \begin{pmatrix} 0 & 1 \\ 1 & 0 \end{pmatrix}\begin{pmatrix} x \\ y \end{pmatrix} + \begin{pmatrix} -5 \\ 5 \end{pmatrix}$

(e) $\begin{pmatrix} x' \\ y' \end{pmatrix} = \begin{pmatrix} 0 & 1 \\ 1 & 0 \end{pmatrix}\begin{pmatrix} x \\ y \end{pmatrix} + \begin{pmatrix} -2 \\ 8 \end{pmatrix}$

5. (a) $x' = -y$ (b) $x' = 2x + 1$ (c) $x' = y + 5$ (d) $x' = -y + 3$
 $y' = x - 2$ $y' = 2y - 3$ $y' = -x + 3$ $y' = -x + 3$
(e) $x' = -y + 6$
 $y' = -x$

7. (a) 90° rotation anticlockwise about $(4, -1)$ (b) reflection in the line $y = x - 3$
(c) glide reflection in $y = 1$ with $(0, 1) \rightarrow (-6, 1)$ (d) enlargement ($\times 5$) centre $(1, 4)$
(e) 180° rotation about $(1, -1\frac{1}{2})$ (f) 90° rotation clockwise about $(-\frac{1}{2}, -3\frac{1}{2})$
(g) 90° rotation anticlockwise about $(\frac{1}{2}, 2\frac{1}{2})$ and an enlargement ($\times 3$) centre $(\frac{1}{2}, 2\frac{1}{2})$

8. (a) $(3, -2)$ (b) $y = x - 5$, $y + x = 1$
(c) enlargement ($\times 2$) centre $(3, -2)$ and a reflection in the line $y + x = 1$

page 183   *Exercise 6E*

1. $\frac{8}{3}$, -5

2. (a) $\begin{pmatrix} 2 & 0 & 3 & 4 \\ 14 & 0 & 21 & 28 \end{pmatrix}$ (b) $\frac{1}{2}\begin{pmatrix} 3 & -1 \\ -4 & 2 \end{pmatrix}$ (i) $\begin{pmatrix} 3 & 1 \\ 1 & 2 \end{pmatrix}$ (ii) $\begin{pmatrix} 10 & 5\frac{1}{2} \\ -10 & -5 \end{pmatrix}$

3. (a) (i) $\begin{pmatrix} -2 & 2 \\ -1 & 1 \end{pmatrix}$ (ii) $\begin{pmatrix} -1 & 2 \\ -1 & 2 \end{pmatrix}$ (iii) $\begin{pmatrix} 0 & 0 \\ 0 & 0 \end{pmatrix}$

(b) $BA = \begin{pmatrix} 1 & 0 \\ -2 & 4 \\ 2 & 2 \end{pmatrix}$ $AC = \begin{pmatrix} 2 & 0 \\ 4 & -2 \\ 0 & 6 \end{pmatrix}$ $EA = (5 \quad 1)$

4. (a) $-\frac{2}{3}$ (b) $\begin{pmatrix} 1 & 0 & 0 \\ 0 & 1 & 0 \\ 0 & 0 & 1 \end{pmatrix}$ $x = 2, y = -1, z = 3$

5. (a) $x = 3\frac{2}{3}$, $y = 6\frac{3}{4}$ (b) $\frac{1}{7}\begin{pmatrix} -2 & -3 & 8 \\ 1 & 5 & -4 \\ 3 & 1 & -5 \end{pmatrix}$

6. (c) $\begin{pmatrix} 1 & 0 \\ 0 & -1 \end{pmatrix}$ (ii) $\begin{pmatrix} 0 & 1 \\ 1 & 0 \end{pmatrix}$ (iii) $\begin{pmatrix} -1 & 0 \\ 0 & -1 \end{pmatrix}$ (iv) $(-5, 2)$ (v) $(-4, 6)$

7. $(1, 2)$, $(-2, 1)$, $\begin{pmatrix} -\frac{3}{5} & \frac{4}{5} \\ \frac{4}{5} & \frac{3}{5} \end{pmatrix}$, $\begin{pmatrix} 1 & 0 \\ 0 & 1 \end{pmatrix}$, $\begin{pmatrix} 0 & 1 \\ -1 & 0 \end{pmatrix}$, $\begin{pmatrix} \frac{4}{5} & \frac{3}{5} \\ \frac{3}{5} & -\frac{4}{5} \end{pmatrix}$

8. $\begin{pmatrix} 1 & 0 \\ 0 & -1 \end{pmatrix}$, $\begin{pmatrix} -1 & 0 \\ 0 & 1 \end{pmatrix}$, $\begin{pmatrix} -1 & 0 \\ 0 & -1 \end{pmatrix}$
(i) $(a + 6, -b)$ (ii) $(a, 4 - b)$ (iii) $(6 - a, -b)$ (iv) $(-a, b)$ (v) (a, b)
(i) no values   (ii) any points for which $b = 2$ (iii) $a = 3, b = 0$
(iv) any points for which $a = 0$ (v) all values of a and b

9. (i) $\begin{pmatrix} 1 & \dfrac{1-x}{y} \\ -1 & \dfrac{x}{y} \end{pmatrix}$ (a) $\tfrac{1}{2}, -1\tfrac{1}{2}$ (b) $\tfrac{3}{4}, -\tfrac{1}{2}$ (ii) $x = 1,\ 2y = x + 3$

10. $\begin{pmatrix} 2 & 1 \\ 2 & 3 \end{pmatrix}, y = 2x$

11. $\begin{pmatrix} 3 & 0 \\ 0 & 3 \end{pmatrix}, \begin{pmatrix} 0 & 1 \\ 1 & 0 \end{pmatrix}, \begin{pmatrix} \tfrac{4}{5} & -\tfrac{3}{5} \\ \tfrac{3}{5} & \tfrac{4}{5} \end{pmatrix}, \begin{pmatrix} -\tfrac{9}{5} & \tfrac{12}{5} \\ \tfrac{12}{5} & \tfrac{9}{5} \end{pmatrix}, (-15, 45)$

12. (i) enlargement, scale factor $\tfrac{1}{2}$, centre $(0, 0)$; reflection in the line $y = x$; rotation of 120°, anticlockwise about $(0, 0)$ (ii) 2, 3   (iii) enlargement, scale factor 2, centre $(0, 0)$; reflection in the line $y = x$; rotation of 120° clockwise about the origin.

13. $\begin{pmatrix} 1 & 0 \\ 0 & -1 \end{pmatrix}, \begin{pmatrix} \tfrac{1}{2} & \tfrac{1}{2}\sqrt{3} \\ \tfrac{1}{2}\sqrt{3} & -\tfrac{1}{2} \end{pmatrix}$, a rotation of 60° anticlockwise about $(0, 0)$, reflection in $x = y\sqrt{3}$

14. $\begin{pmatrix} 3 \\ 2 \end{pmatrix}$, 13   **15.** $\tfrac{1}{2}$ **16.** $x = \tfrac{1}{2}$, glide reflection in $x = \tfrac{1}{2}$ with $(\tfrac{1}{2}, 0) \to (\tfrac{1}{2}, 3)$

17. $(1\tfrac{2}{3}, -\tfrac{1}{3})$, 90° rotation anticlockwise about $(1\tfrac{2}{3}, -\tfrac{1}{3})$ and an enlargement, scale factor 2, centre $(1\tfrac{2}{3}, -\tfrac{1}{3})$. $\begin{matrix} x' = -y + 1 \\ y' = -x + 1 \end{matrix}$ $\begin{matrix} x' = -2x + 4 \\ y' = 2y \end{matrix}$, centre $(\tfrac{4}{3}, 0)$, scale factor 2.

18. $2y = x$

page 193 Exercise 7A

1. (a) 56 (b) 1008 (c) 600 (d) 4

2. (a) 5! (b) 3 × 5! (c) 5! (d) $\dfrac{5!}{5}$ (e) 5 × 5! (f) 9 × 5! (g) 2 × 5! (h) 36 × 5! (i) 55 × 5!

3. 18   **4.** 10!   **5.** 8!   **6.** 24   **7.** 120   **8.** 24, 256   **9.** (a) 81 (b) 256   **10.** 8!

11. 720, 120   **12.** 24   **13.** 48   **14.** 72   **15.** 12   **16.** 10080   **17.** 80640

18. (a) 14! (b) 11! (c) 11! × 4!   **19.** 1440, 240   **20.** 2520   **21.** 30240   **22.** 42

23. (a) 12 (b) 12   **24.** 120 (a) 48 (b) 72 (c) 12   **25.** 96   **26.** 1296

page 197 Exercise 7B

1. 30   **2.** 151 200   **3.** 6720   **4.** 120

5. (a) 24 (b) 64   **6.** (a) 6 (b) 16   **7.** 36   **8.** 2730

9. 2 522 520   **10.** 3360   **11.** 453 600, 10 080

12. (a) 20 160   (b) 5040   (c) 15 120   **13.** 18

14. 60, 40   **15.** 1260   **16.** 528   **17.** 42

18. 151 (a) 73 (b) 78 (c) 13 (d) 138   **19.** 19 958 400 (a) 3 628 800 (b) 302 400

20. 2 494 800 (a) 453 600 (b) 60 480   **21.** 5472   **22.** 1596, 198

page 201 Exercise 7C

1. 120   **2.** 126   **3.** 10   **4.** 330   **5.** 10   **6.** 4845

7. 22 100   **8.** 210, 210   **9.** 1960   **10.** 286   **11.** 150   **12.** 60

13. 2016   **14.** (a) 126 (b) 105   **15.** (a) 15 (b) 1   **16.** (a) 20 (b) 4   **17.** (a) 13 (b) 5

18. (a) 24 (b) 7 (c) 13   **19.** 37   **20.** 6300   **21.** (a) 5 (b) 85 (c) 365   **22.** 175

23. (a) 105 (b) 21 (c) 231   **24.** 100   **25.** (a) 1 (b) 56 (c) 1231 (d) 1230   **26.** 120

27. 3360, 1200   **28.** 30 240, 12 600   **29.** 7560, 3150 *page 205*   **Exercise 7D**

1. 630 630   **2.** 127   **3.** 63   **4.** 330   **5.** 7920   **6.** 60 060   **7.** 63
8. 64   **9.** 255   **10.** 47   **11.** 31   **12.** 255   **13.** 39
14. $(p + 1)(q + 1) 2^{n-p-q} - 1$ **15.** (a) 126 (b) 280   **16.** 5775   **17.** 126 126
18. 25 740   **19.** 23

page 206   **Exercise 7E**

1. 816   **2.** 286   **3.** 6   **4.** 360 (a) 180 (b) 240   **5.** 100   **6.** (a) 495 (b) 225 (c) 15
7. 25   **8.** 840, 96   **9.** 3528, 8624   **10.** 154   **11.** 10 080, 30   **12.** $(n-2)!(n-3)$
13. 120, 24   **14.** 480, 172 800, 462, 425
15. (a) 38 760 (i) 12 320 (ii) 5320   (b) 86 400, 172 800

page 210   **Exercise 8A**

1. (a) 10, 13   (b) 49, 64   (c) 0·00001, 0·000001,   (d) $5\frac{1}{4}$, 6
2. (a) 8, 3   (b) 23, 2   (c) 19, -3 (d) $13\frac{1}{2}$, $1\frac{1}{2}$
  (e) $-11·5, 2·5$ (f) $6\frac{1}{4}$, $\frac{1}{2}$ (g) -8, 1   (h) 6, -3
3. (a) 20 (b) 102 (c) 45 (d) 32   **4.** 56   **5.** 24   **6.** -9 **7.** 13, -15
8. (a) $2 + 3n$, 302   (b) $8 - 3n$, -292 (c) $\frac{1}{2}(18 + 7n)$, 359
9. (a) 200th (b) 411th   **10.** (a) 122nd (b) 64th
11. (a) 2, 4, 288   (b) 10, -2, -60 (c) 4·5, 1·5, 342   (d) 15, -2, 0
  (e) 7, -4, -620 (f) $-6\frac{1}{2}$, $+1\frac{1}{2}$, 21   (g) -9, 2, 96
12. (a) 5402 (b) 1000 (c) 9464 (d) $13\frac{1}{2}$
13. 2, $1\frac{1}{2}$, $87\frac{1}{2}$ **14.** $6\frac{1}{2}$, $\frac{1}{2}$, 66   **15.** 21, -2, 0   **16.** -10, 3, 18
17. -15, 3, 0   **18.** $13\frac{1}{2}$, $-3\frac{1}{2}$, $-14\frac{1}{2}$ **19.** 0, $\frac{1}{2}$, $9\frac{1}{2}$ **20.** 2160
21. 1575*d*   **22.** 20   **23.** 32   **24.** 24   **25.** 28
26. (a) $23n - 20 - 3n^2$ (b) $20 - 6n$ (c) 14, -6
27. 7, 12, 17   **28.** 2·5   **29.** $\log_e 3$, $\log_e 3^n$, $\frac{1}{2}n(n + 1) \log_e 3$
30. $\log_e(ab^n)$, $\frac{1}{2}n \log_e(a^2 b^{n+1})$

page 216   **Exercise 8B**

1. (a) 5, 2500, 12 500 (b) $\frac{1}{2}$, 3, $1\frac{1}{2}$ (c) $\frac{1}{3}$, $1\frac{2}{3}$, $\frac{5}{9}$ **2.** 59 048   **3.** $12\frac{3}{8}$, $11\frac{13}{16}$, $-\frac{9}{16}$
4. (a) $2047\frac{7}{8}$ (b) $1228\frac{7}{8}$ **5.** 3   **6.** $-\frac{1}{4}$ **7.** $\pm 1\frac{1}{2}$ **8.** 2, $\frac{3}{4}$; -2, $-\frac{3}{4}$ **9.** 31   **10.** 2186
11. (a) 14 (b) -3 (c) 4 (d) 7·5   **12.** (a) 8 (b) 12 (c) 2·5 (d) 3·5   **13.** 24
14. (a) divergent   (b) convergent, 32   (c) convergent, 56   (d) convergent, $156\frac{1}{4}$
  (e) divergent   (f) convergent, $51\frac{1}{3}$
15. (a) $|x| < 1$ (b) $|x| < 3$ (c) $|x| > 1$ (d) $-\frac{3}{2} < x < \frac{1}{2}$ **16.** $20\frac{1}{4}$
17. $2\frac{1}{2}$, $1\frac{1}{2}$ **18.** 2, 3   **19.** 6   **20.** £11 million   **21.** $\frac{2}{3}$
22. 16   **23.** $\frac{5}{8}$ **24.** $\frac{4}{5}$, 5   **25.** 8   **26.** 7
27. (a) $\frac{7}{9}$ (b) $\frac{50}{99}$ (c) $\frac{224}{495}$ (d) $\frac{40}{111}$

page 220   **Exercise 8C**

1. (a) $1 + 2 + 3 + 4 + 5$ (b) $1 + 4 + 9 + 16 + 25$ (c) $13 + 15 + 17$
  (d) $504 + 420 + 360 + 315 + 280 + 252$
  (e) $-1 + 2 - 3 + 4 - 5$ (f) $1 - 2 + 3 - 4 + 5$

2. (a) $\displaystyle\sum_{r=1}^{50} r$ (b) $\displaystyle\sum_{r=1}^{50} r^3$ (c) $\displaystyle\sum_{r=1}^{6} 5r$ (d) $\displaystyle\sum_{r=1}^{6} (2r + 1)$

(e) $\displaystyle\sum_{r=1}^{6} (-1)^{r+1} 3r$ (f) $\displaystyle\sum_{r=1}^{13} (-1)^r r^2$

4. (a) $\displaystyle\sum_{r=1}^{100} r = 5050$ (b) $\displaystyle\sum_{r=1}^{22} r^2 = 3795$ (c) $\displaystyle\sum_{r=1}^{15} r^3 = 14\,400$

(d) $\displaystyle\sum_{r=1}^{51} (2r - 1) = 2601$ (e) $\displaystyle\sum_{r=1}^{20} r(r + 2) = 3290$ (f) $\displaystyle\sum_{r=1}^{13} r(2r - 1) = 1547$

5. (a) $90\,000$ (b) $14\,400$ (c) $75\,600$ **6.** (a) $\frac{1}{3}n(n + 1)(n + 2)$ (b) 8120

7. (a) $\frac{1}{4}n(n + 1)(n + 2)(n + 3)$ (b) $53\,130$ **8.** (a) $\frac{1}{6}n(4n^2 + 15n + 17)$ (b) 6390

10. (a) $n(2n + 1)$ (b) $\frac{1}{6}(n + 1)(n + 2)(2n + 3)$ (c) $n^2(2n + 1)^2$ (d) $4n(n + 3)$ (e) $2n^2(n + 1)^2$

(f) $8n^2(2n + 1)^2$ (g) $\frac{1}{3}n(n + 1)(n + 5)$ (h) $\frac{1}{12}n(n + 1)(n + 2)(3n + 5)$

(i) $\frac{1}{3}n(n + 1)(4n - 1)$ (j) $\frac{1}{6}n(n + 1)(3n^2 + n - 1)$

page 226 **Exercise 8D**

1. (a) $27 + 27x + 9x^2 + x^3$ (b) $125 + 150x + 60x^2 + 8x^3$

(c) $16 + 32x + 24x^2 + 8x^3 + x^4$ (d) $16 - 32x + 24x^2 - 8x^3 + x^4$

(e) $32y^5 + 80y^4x + 80y^3x^2 + 40y^2x^3 + 10yx^4 + x^5$

(f) $32x^5 - 240x^4y + 720x^3y^2 - 1080x^2y^3 + 810xy^4 - 243y^5$

(g) $x^4 - 4x^2 + 6 - \dfrac{4}{x^2} + \dfrac{1}{x^4}$ (h) $x^5 - 10x^3 + 40x - \dfrac{80}{x} + \dfrac{80}{x^3} - \dfrac{32}{x^5}$

2. (a) $1 + 12x + 54x^2 + 108x^3 + 81x^4$ (b) $16x^4 + 32x^3y + 24x^2y^2 + 8xy^3 + y^4$

(c) $64 - 576x + 2160x^2 - 4320x^3 + 4860x^4 - 2916x^5 + 729x^6$

(d) $32x^5 + 240x^3 + 720x + \dfrac{1080}{x} + \dfrac{810}{x^3} + \dfrac{243}{x^5}$

3. $1 + 12x + 66x^2 + 220x^3$ **4.** $a^{12} + 12a^{11}x + 66a^{10}x^2 + 220a^9x^3$

5. $a^{10} - 30a^9x + 405a^8x^2 - 3240a^7x^3$ **6.** $1 + 10x + \dfrac{95}{2}x^2 + \dfrac{285}{2}x^3 + \dfrac{4845}{16}x^4$

7. $32 + 80x + 80x^2 + 40x^3 + 10x^4 + x^5$ (a) $40\cdot84101$ (b) $24\cdot76099$

8. (a) $1 - 8x + 25x^2 - 30x^3$ (b) $1 + 3x - 3x^2 - 25x^3$ (c) $1 - 16x + 113x^2 - 464x^3$

(d) $1 + 3x + 3x^2 + 2x^3$ (e) $1 + 16x + 111x^2 + 434x^3$ (f) $1 - 14x + 76x^2 - 195x^3$

9. (a) $1 + 6x + 21x^2 + 50x^3$ (b) $1 + 10x + 35x^2 + 40x^3$ (c) $1 - 16x + 144x^2 - 896x^3$

10. $1 - 11x + 55x^2 - 165x^3 + 330x^4, 0\cdot801$ **11.** $1 + 15x + 105x^2 + 455x^3, 1\cdot161$

12. $1 + 24x + 264x^2, 1\cdot3$ **13.** $128 + 448x + 672x^2 + 560x^3, 132\cdot548$

14. (a) $1\cdot0937$ (b) $0\cdot9860837$ (c) $0\cdot9044$ (d) $973\cdot9$

15. $\frac{1}{2}, 1, 5, 11\frac{1}{4}$ **16.** $16, \frac{1}{8}$ **17.** $1 - x - 11x^2$

page 228 **Exercise 8E**

1. $^{12}C_8 \times 4^4 \times x^8$ **2.** $^{8}C_5 \times 3^3 \times 2^5 \times x^5$ **3.** $-^9C_3 \times 2^6 \times x^3$

4. (a) $^8C_3 \times 5^5 \times 3^3$ (b) $-^7C_3 \times 7^4 \times 2^3$ **5.** $^7C_5 \times 3^2 \times 2^5 x^{10}$

6. (a) $^{10}C_5 \times 3^5$ (b) $^{12}C_8 \times 4^{10}$ (c) $^6C_4 \times 3^2 \times 2^4$ (d) $2^4(2 \times {}^{10}C_5 + {}^{10}C_6)$

page 231   Exercise 8F

1. (a) $|x| < \frac{1}{2}$ (b) all x (c) $|x| < \frac{1}{2}$ (d) $|x| < 2$

2. (a) $1 - x + x^2 - x^3, |x| < 1$ (b) $1 + \frac{1}{3}x - \frac{1}{9}x^2 + \frac{5}{81}x^3, |x| < 1$

(c) $1 - x + \frac{3}{2}x^2 - \frac{5}{2}x^3, |x| < \frac{1}{2}$ (d) $1 + 4x + 12x^2 + 32x^3, |x| < \frac{1}{2}$

(e) $1 - 6x + 27x^2 - 108x^3, |x| < \frac{1}{3}$ (f) $1 + 3x + \frac{3}{2}x^2 - \frac{1}{2}x^3, |x| < \frac{1}{2}$

(g) $2 + \frac{1}{4}x - \frac{1}{64}x^2 + \frac{1}{512}x^3, |x| < 4$ (h) $\frac{1}{10} + \frac{1}{40}x + \frac{3}{320}x^2 + \frac{1}{256}x^3, |x| < 2$

3. $1 - 7x + 30x^2 - 104x^3 + 320x^4, |x| < \frac{1}{2}$ **4.** $2, -9, 29, -82$

5. $1 - x + 7x^2 - 13x^3$ **6.** $1 + \frac{1}{2}x - \frac{1}{8}x^2 + \frac{1}{16}x^3, |x| < 1, 1\cdot0392$

7. (a) $-\frac{1}{x^3} - \frac{6}{x^4} - \frac{24}{x^5} - \frac{80}{x^6}, |x| > 2$ (b) $\frac{1}{2x} - \frac{1}{4x^2} + \frac{1}{8x^3} - \frac{1}{16x^4}, |x| > \frac{1}{2}$

(c) $\frac{1}{9x^2} + \frac{8}{27x^3} + \frac{16}{27x^4} + \frac{256}{243x^5}, |x| > \frac{4}{3}$

8. (a) $0\cdot97980$ (b) $10\cdot1980$ (c) $2\cdot0199$ (d) $1\cdot01943$ (e) $2\cdot05828$

9. $1 - \frac{3}{2}x + \frac{15}{8}x^2 - \frac{51}{16}x^3, |x| < \frac{1}{2}$

10. $1 + x + \frac{x^2}{2} + \frac{x^3}{2}, |x| < 1, 1\cdot224$

11. $3, 1, \frac{9}{2}, -\frac{27}{16}$ **12.** $\frac{2}{3}, -\frac{3}{2}$

page 232   Exercise 8G

1. $27, -6, 0$ **2.** $-144, 48$ **3.** (i) -22 (ii) -30

4. (i) $10, 4$ (ii) $r = 2, a = 4; r = -2, a = -12; 4092, 4092$

5. (i) $\frac{1}{2}n(3n + 1), 20, 26$ (ii) $\frac{1}{2}(3^n - 1), 10$ **6.** (a) 2800 (b) $1\cdot6, 9$ **7.** $a + ab + b^2N, 1250$

8. 2660 **9.** (a) $6n - 1$ **10.** $32 - 80x + 80x^2, 32\cdot808$ **11.** $256 - 3072x + 16128x^2, 253$

12. $-\dfrac{32 \times 20!}{15!5!}$ **13.** $\dfrac{15}{112}$ **14.** -20

15. (a) $1 + x - x^2 + \frac{5}{3}x^3, |x| < \frac{1}{3}, 2\cdot01980$ (b) $-3, 4$

16. $1 - x + \frac{3}{2}x^2 - \frac{5}{2}x^3, 1 + x + \frac{3}{2}x^2 + \frac{5}{2}x^3, 1\cdot0010015$

17. (a) $-\frac{1}{2}, 5, |x| < \frac{1}{3}$ (b) $1 + \frac{1}{2}x - \frac{1}{8}x^2$ for $|x| < 1, 1\cdot0392$

18. $\frac{3}{2}, \frac{1}{2}, \frac{1970}{1393}$

page 237   Exercise 9A

1. $\frac{1}{2}$ **2.** (a) $\frac{1}{2}$ (b) $\frac{5}{6}$ (c) $\frac{2}{3}$ **3.** (a) $\frac{1}{3}$ (b) $\frac{2}{3}$ (c) $\frac{1}{10}$ (d) $\frac{4}{5}$

4. (a) $\frac{1}{4}$ (b) $\frac{1}{2}$ (c) $\frac{1}{4}$ (d) $\frac{1}{5}$ (e) $\frac{17}{20}$ **5.** (a) $\frac{3}{4}$ (b) $\frac{1}{4}$ (c) $\frac{5}{8}$ **6.** (a) $\frac{1}{4}$ (b) $\frac{1}{2}$

7. (a) $\frac{1}{4}$ (b) $\frac{5}{12}$ **8.** $\frac{3}{13}$ **9.** (a) $\frac{1}{4}$ (b) $\frac{1}{13}$ (c) $\frac{1}{2}$ (d) $\frac{3}{4}$ **10.** (a) $\frac{7}{12}$ (b) $\frac{1}{4}$ (c) $\frac{5}{8}$ (d) $\frac{7}{24}$

11. (a) $\frac{8}{15}$ (b) $\frac{4}{5}$ (c) $\frac{3}{5}$ (d) $\frac{2}{15}$ **12.** (a) $\frac{1}{2}$ (b) $\frac{2}{3}$ (c) $\frac{3}{5}$ (d) $\frac{3}{10}$

page 244   Exercise 9B

1. (a) $\frac{14}{15}$ (b) $\frac{1}{5}$ **2.** (a) False (b) True **3.** (a) $\frac{1}{8}$ (b) $\frac{5}{8}$ **4.** (a) $\frac{1}{8}$ (b) $\frac{3}{8}$

5. (a) $\frac{1}{2}$ (b) $\frac{5}{6}$ **6.** (a) No (b) Yes **7.** (a) No (b) No **8.** $\frac{2}{13}$ **9.** $\frac{1}{2}$

10. $\frac{1}{4}$ **11.** $\frac{1}{16}$ **12.** (a) $\frac{1}{4}$ (b) $\frac{1}{4}$ **13.** $\frac{3}{32}$ **14.** $\frac{1}{4}$ **15.** (a) $\frac{1}{6}$ (b) $\frac{5}{36}$ (c) $\frac{11}{36}$

16. (a) $\frac{1}{12}$ (b) $\frac{5}{6}$ **17.** (a) $\frac{1}{4}$ (b) $\frac{1}{6}$ **18.** (a) $\frac{1}{16}$ (b) $\frac{1}{26}$ (c) $\frac{1}{8}$ (d) $\frac{1}{4}$ (e) $\frac{1}{169}$

19. (a) 1 (b) $\frac{5}{8}$ **20.** (a) $\frac{3}{8}$ (b) $\frac{3}{8}$ **21.** (a) $\frac{1}{8}$ (b) $\frac{1}{64}$ (c) $\frac{3}{64}$

22. (a) $\frac{13}{204}$ (b) $\frac{1}{102}$ (c) $\frac{1}{51}$ (d) $\frac{1}{17}$ **23.** (a) $\frac{1}{36}$ (b) $\frac{11}{18}$ (c) $\frac{7}{18}$

24. (a) $\frac{1}{28}$ (b) $\frac{3}{28}$ **25.** $\frac{7}{12}$ **26.** $\frac{3}{13}$ **27.** $\frac{8}{13}$ **28.** $\frac{19}{100}$

29. (a) $\frac{5}{126}$ (b) $\frac{10}{21}$ **30.** (a) $\frac{1}{816}$ (b) $\frac{7}{102}$ (c) $\frac{7}{34}$

page 246 **Exercise 9C**

1. $\frac{15}{56}$ 2. $\frac{3}{4}$ 3. (a) $\frac{11}{31}$ (b) $\frac{11}{663}$ 4. (a) $\frac{25}{32}$ (b) $\frac{1}{9}$ 5. (a) $\frac{22}{43}$ (b) $\frac{6}{11}$
6. (a) $\frac{121}{11}$ (b) $\frac{40}{8}$ 7. (a) $\frac{49}{156}$ (b) $\frac{40}{48}$ 8. (a) $\frac{17}{32}$ (b) $\frac{1}{17}$
9. (a) $\frac{20}{43}$ (b) $\frac{10}{19}$ 10. (a) $\frac{13}{90}$ (b) $\frac{9}{13}$

page 250 **Exercise 9D**

1. $\frac{2}{3}$ 2. $\frac{4}{35}$ 3. $\frac{1}{6}$ 4. $\frac{2}{5}$ 5. (a) $\frac{1}{55}$ (b) $\frac{32}{495}$ (c) $\frac{92}{99}$
6. (a) $\frac{3}{54145}$ (b) $\frac{184}{54145}$ (c) $\frac{1}{3185}$ 7. (a) $\frac{143}{3196}$ (b) $\frac{55}{11186}$ (c) $\frac{1111}{22372}$
8. (a) $\frac{1}{128}$ (b) $\frac{7}{128}$ (c) $\frac{127}{128}$ 9. (a) $\frac{5}{512}$ (b) $\frac{1}{1024}$ (c) $\frac{1}{1024}$
10. (a) $\frac{63}{256}$ (b) $\frac{9}{128}$ 11. (a) $\frac{1}{1024}$ (b) $\frac{15}{1024}$ (c) $\frac{135}{512}$ (d) $\frac{1}{64}$ (e) $\frac{27}{256}$
12. $\frac{79}{4096}$ 13. (a) $\frac{64}{729}$ (b) $\frac{1}{729}$ (c) $\frac{656}{729}$ 14. 0·000248
15. (a) $\frac{1}{64}$ (b) $\frac{3}{32}$ (c) $\frac{11}{32}$ 16. (a) 0·206 (b) 0·343 (c) 0·816
17. (a) $\frac{1024}{6561}$ (b) $\frac{256}{6561}$ (c) $\frac{1120}{6561}$ (d) $\frac{16}{2187}$ (e) $\frac{32}{6561}$

page 251 **Exercise 9E**

1. (a) $\frac{1}{17}$ (b) $\frac{25}{102}$ (c) $\frac{10}{34}$ 2. (i) $\frac{1}{221}$ (ii) $\frac{1}{663}$ (iii) $\frac{1}{17}$ (iv) $\frac{13}{102}$
3. (i) $\frac{2}{15}$ (ii) $\frac{14}{15}$ (iii) $\frac{3}{5}$ 4. (a) $\frac{1}{9}$ (b) $\frac{1}{6}$ (c) $\frac{13}{18}$

5.

score	4	6	8	9	11	14
P	$\frac{4}{25}$	$\frac{8}{25}$	$\frac{4}{25}$	$\frac{4}{25}$	$\frac{4}{25}$	$\frac{1}{25}$

£1·20 loss

6. (a)(i) $\frac{5}{18}$ (ii) $\frac{1}{3}$ (b) $\frac{12}{35}$ 7. (i) $\frac{1}{60}$ (ii) $\frac{1}{3}$ (iii) $\frac{3}{5}$ (iv) $\frac{3}{20}$
8. (a) $\frac{4}{49}$ (b) $\frac{1}{7}$ (c) $\frac{9}{49}$ (d) $\frac{13}{49}$ 9. (i) 40320 (ii) $\frac{3}{28}$
10. $\frac{5}{12}$ 11. (i) $\frac{7}{60}$ (ii) $\frac{11}{20}$ 12. (i) $\frac{1}{1000}$ (ii) $\frac{41}{100}$ (iii) $\frac{121}{1000}$
13. (a) $\frac{25}{216}$ (b) $\frac{625}{1296}$ (c) $\frac{4 \times 5^2}{6^7}$ (d) $\frac{5^3}{6^5}$ 14. (i) 0·02 (ii) 0·45
15. $\frac{1}{13}, \frac{1}{2}, \frac{1}{26}, \frac{7}{13}$, (a) $\frac{144}{2197}$ (b) $\frac{3}{8}$ (c) $\frac{15}{16}$; 7 times   16. (i) $\frac{2}{3}$
17. (i) 0·206 (ii) 0·343 (iii) 0·816 (iv) 0·451   18. (i) 0·678 (ii) 0·322; $\frac{1}{15}$
19. (i) $\frac{768}{3125}$ (ii) $\frac{2816}{3125}$ (iii) $\frac{768}{15625}$
20. (i) 0·663 (ii) 0·00542 (iii) 0·000372; 0·0000106
21. (i) $\frac{7}{24}$ (ii) $\frac{145}{288}$ (iii) $\frac{101}{288}$ (iv) $\frac{1025}{3456}$
22. $\frac{3}{10}, \frac{9}{10}, \frac{7}{10}, \frac{3}{5}$, (i) not independent. $P(A \cap B) \neq P(A) \times P(B)$
      (ii) not mutually exclusive. $P(A \cup B) \neq P(A) + P(B)$
23. (a) $\frac{27}{343}$ (b) $\frac{8}{343}$ (c) $\frac{10}{343}$ (d) $\frac{6}{19}$ (e) $\frac{56}{343}$

page 258 **Exercise 10A**

1. 7, 3, −3   2. 1, $\frac{1}{2}$, 0

3. (a) 5 (b) 9 (c) $6x$ (d) $3x^2$ (e) $2x + 3$ (f) $5 - 2x$ (g) $-\dfrac{1}{x^2}$ (h) $-\dfrac{2}{x^3}$

4. $3 - 4x$, -13, 7   5. $4x + 5$, 1, −3   6. $3x^2 - 2$, 1, −2, 1

page 261   Exercise 10B

1. $5x^4$ **2.** $3x^2$ **3.** $24x$ **4.** $20x^3$ **5.** 16

6. $6x$ **7.** $27x^2$ **8.** $12x^3$ **9.** 0   **10.** 2

11. $35x^6$ **12.** 0   **13.** $\frac{5}{3}x^{2/3}$ **14.** $\frac{3}{4}x^{-1/4}$ **15.** $\frac{2}{5}x^{-3/5}$

16. $\frac{3}{5}x^{-2/5}$ **17.** $4x^{-1/3}$ **18.** $2x^{-3/4}$ **19.** $\frac{1}{2}x^{-1/2}$ **20.** $\frac{3}{2}\sqrt{x}$

21. $2x^{-1/2}$ **22.** $-\dfrac{5}{x^2}$ **23.** $-\dfrac{6}{x^3}$ **24.** $-4x^{-3/2}$ **25.** $-\dfrac{6}{x^4}$

26. $2x + 7$ **27.** $1 - 14x$ **28.** $12x - 7$ **29.** 1   **30.** $3x^2 + 14x$

31. $18x^5 - 4x + 6$ **32.** $6x + 7 - \dfrac{1}{x^2}$ **33.** $3 + \dfrac{5}{x^2} - \dfrac{12}{x^3}$ **34.** $5 + \dfrac{3}{2}x^{-3/2}$ **35.** $2x + 4$

36. $2x + 2$ **37.** $4x + 7$ **38.** 4   **39.** 3   **40.** 34

41. 8   **42.** 13   **43.** 20   **44.** 2   **45.** 6

46. (4, 16)   **47.** (−4, 16)   **48.** (3, 2)   **49.** (1, 4)   **50.** (−1, 3)

51. (−3, 5), (1, −11)   **52.** $(\frac{1}{3}, \frac{22}{27})$, (−1, 2)   **53.** (2, 6), (−2, −6)

54. (a) 5   (b) $3x^2 + 4$ (c) 7   (d) $6x$ (e) 6

55. (a) 24   (b) $6x - \dfrac{24}{x^2}$ (c) 6   (d) $6 + \dfrac{48}{x^3}$ (e) 12   **56.** $1\frac{1}{2}$ **57.** 4

58. (a) 12   (b) $30x$ (c) $\dfrac{6}{x^3}$ **60.** 6   **61.** (−3, 0), (4, 0), −7, 7

62. 0, 8   **63.** 1, 5   **64.** (4, −5)   **65.** $(\frac{3}{4}, \frac{3}{16})$ **66.** (1, 4)

67. (a) $y = 6x - 9$, $6y + x = 57$ (b) $y = 4x + 7$, $4y + x = 11$

(c) $y = 5x - 29$, $5y + x + 15 = 0$ (d) $y + 3x = 6$, $3y = x + 8$

68. (2, 11)   **69.** 1, $-2\frac{1}{2}$

page 265   Exercise 10C

1. Min at (2, −8)   **2.** Max at (3, 7)   **3.** Max at $(-\frac{1}{3}, 7\frac{5}{27})$, Min at (1, 6)

4. Max at (−3, 22), Min at (1, −10)   **5.** Max at (−1, 3), Min at (1, −1)   **6.** Infl at (1, 0)

7. Max at $(\frac{2}{3}, \frac{14}{27})$, Min at (4, −18)   **8.** Infl at (−2, 4)

9. 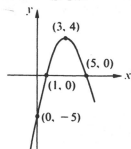   **10.** 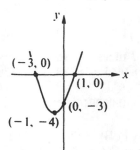   **11.**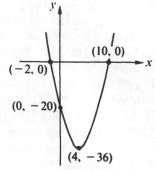

12. 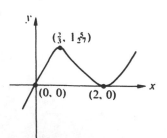   **13.** 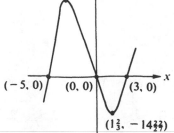   **14.**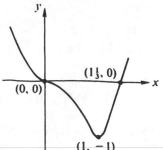

page 268   **Exercise 10D**

1. Min at $(-5, -15)$ **2.** Max at $(1\frac{1}{2}, 7)$ **3.** Max at $(-2, 28)$ Min at $(4, -80)$
4. Min at $(-3, -156)$, Max at $(5,100)$ **5.** Max at $(-2, 32)$, Min $(2, -32)$ **6.** Min at $(0, 3)$
7. Min at $(-2, -32)$, Max at $(0, 0)$, Min at $(2, -32)$
8. Min at $(-3, -162)$, Infl at $(0, 0)$, Max at $(3, 162)$

9.

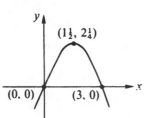
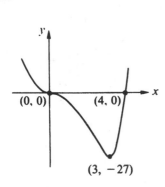

10

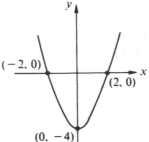
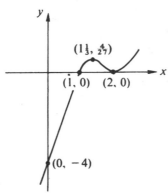

11.

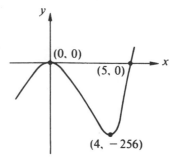

12.

13.

14.

15. (a) Min $(2, 2)$, Max $(-1, 15\frac{1}{2})$ (b) $15\frac{1}{2}$ (c) $-10\frac{1}{2}$
16. (a) Min $(-3, -81)$, Max $(2, 44)$ (b) $-145, 44$
17. $\frac{3}{4}, 1\frac{1}{4}$ **18.** $37\frac{1}{2}, 2\frac{1}{2}$ **19.** 250, 25, 10   **20.** 10, 1, 2   **21.** 90 000 m², 300 m square
22. 2 m, 4 m³   **23.** 30 cm, 30 cm   **24.** 15 cm   **25.** 20 km/h

page 271   **Exercise 10E**

1. 0·2   **2.** 0·03   **3.** 0·02   **4.** 8%   **5.** 40π cm³   **6.** 6%
7. 0·01 cm   **8.** (800 ± 16) cm²   **9.** 2%   **10.** $1\cdot8\pi$ cm²   **11.** 126·5   **12.** 4·02

page 273   **Exercise 10F**

1. $(0, 0)$ horizontal, $(1, -1)$ **2.** $(0, 0)$ **3.** $(\frac{3}{2}, \frac{11}{4})$ **4.** $(\frac{1}{3}, 8\frac{25}{27})$
5. $(1, -3)$ horizontal   **6.** $(1, 0)$ horizontal   **7.** $(4, 2)$ **8.** $(0, -5)$ horizontal
9. $(\frac{1}{3}, \frac{16}{27})$

page 273   *Exercise 10G*

1. 1   **2.** $y = 8x - 11$, $(1\frac{3}{8}, 0)$ **3.** $2y + x + 3 = 0$, $(-3, 0)$, $(0, -1\frac{1}{2})$
4. (a) $18x + 9\sqrt{x} + 1$ (b) $\pm\frac{1}{2}$ **5.** 6, 15   **6.** -3, max   **7.** $-2, 8, 10$
8.

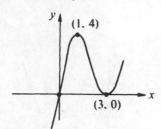

9. (i) (a) 5 cm × 5 cm × 5 cm   (b) 125 cm³

   (ii) $8x - \dfrac{27}{x^2}$, $1\frac{1}{2}$, min

10. $300r - \pi r^3$, $300 - 3\pi r^2$, $-6\pi r$, $\dfrac{10}{\sqrt{\pi}}$, $2 : 1$

12. $6x^2 - 14x$ (i) $y + 4x = 11$ (ii) $0\cdot12$ decrease
13. $\frac{31}{30}$ **14.** (i) $(2, 4)$, 4   (ii) 1%

page 281   *Exercise 11A*

1. (a), (c), (e)
2. $A(0, 6)$, $B(3, 0)$; $C(-3, 0)$, $D(1, 0)$, $E(0, -3)$, $F(-1, -4)$; $G(-4, 0)$, $H(2, 0)$, $I(0, 8)$, $J(-1, 9)$
3. $A(-2, 0)$, $a = 2$; $B(2, 0)$, $b = 1$; $C(-2, 0)$, $D(2, 0)$, $c = -3$
4. (a)   (b)   (c)   (d)

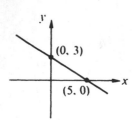
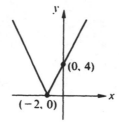
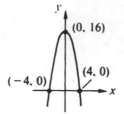
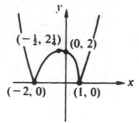

5.   **6.** (a)   (b)   (c)

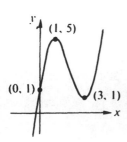
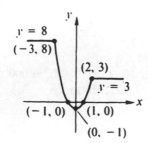
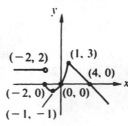
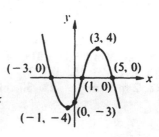

7.   **8.**   **9.**

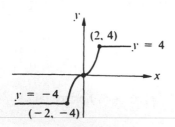
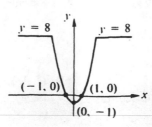
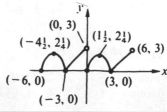

10.
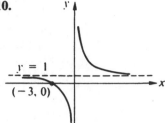

11.
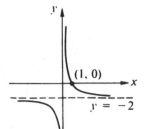

12.
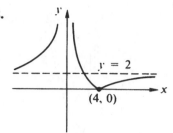

13.
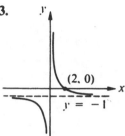

14.
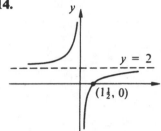

15.
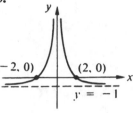

16.
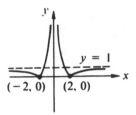

page 284   Exercise 11B

1. A$(-2, 0)$, B$(2, 0)$, $a = 2$; C$(\frac{1}{2}, 0)$, $b = 2$; D$(-1, 0)$, $c = 1$

2.
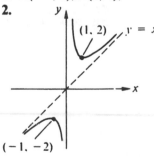

3.
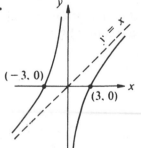

4.
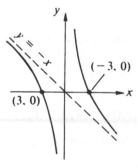

5.
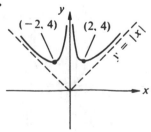

6.
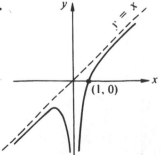

7.
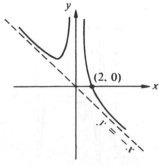

8.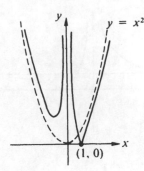

9.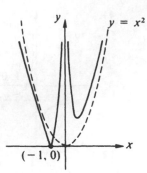

page 287   Exercise 11C

1. (c) $y = 5 + 2x - x^2$, $y = 4x - x^2$, $y = 3 - 2x - x^2$
2. (c) $y = x^2 - 2x - 3$, $y = 2x - x^2$, $y = x^2 + 6x + 8$
4. (a) $y = x^2 + x + 5$ (b) $y = x^2 + x - 2$ (c) $y = x^2 - x + 1$ (d) $y = -x^2 - x - 1$
    (e) $y = x^2 - 3x + 3$ (f) $y = x^2 + 5x + 7$ (g) $y = 2x^2 + 2x + 4$ (h) $y = 2x^2 + 2x + 6$
5. (a) translated $+2$ units in the direction of Oy,
    (b) reflection in the y-axis,
    (c) translation $+3$ units in the direction of Ox,
    (d) stretched parallel to the y-axis, scale factor 2,
    (e) stretched parallel to the x-axis, scale factor 2,
    (f) translation of -2 units in the direction of Ox followed by translation of -3 units in the direction
        of Oy,
    (g) stretched parallel to the x-axis, scale factor $\frac{1}{2}$ followed by translation of -3 units in the direction
        of Oy,
    (h) translation of $+1$ units in the direction of Ox, stretched parallel to the y-axis, scale factor 2,
        followed by translation $+5$ in direction of Oy.
6. $y = x^2 - 5x + 6$

page 290   Exercise 11D

(Only the sketches of $1/f(x)$ are shown here)

1.   **2.**   **3.**   **4.**

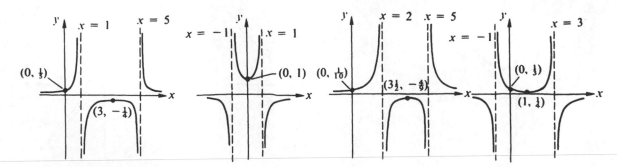

5.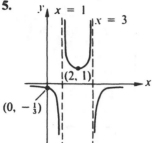

6.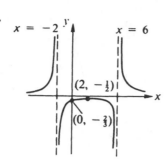

7.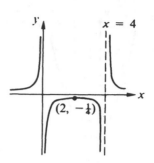

8.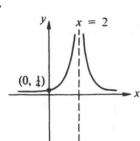

9.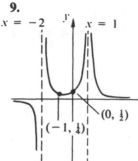

10.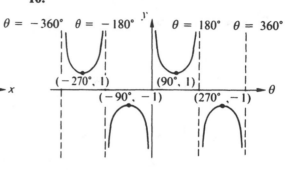

11.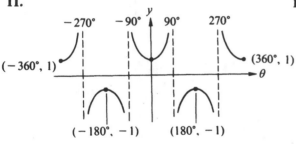

12.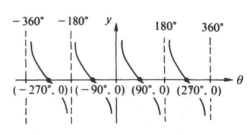

13. (a) $\{y \in \mathbb{R}: y \leqslant -\frac{4}{25}$ or $y > 0\}$ (b) $\{y \in \mathbb{R}: y \leqslant -\frac{8}{25}$ or $y > 0\}$

(c) $\{y \in \mathbb{R}: y < 0$ or $y \geqslant 1\}$ (d) $\{y \in \mathbb{R}: y \leqslant -\frac{24}{25}$ or $y > 0\}$

page 291   Exercise 11E

1. (i)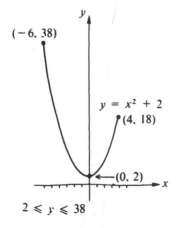

(ii)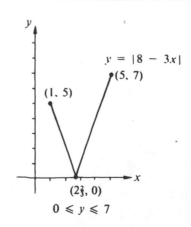

2. (a)

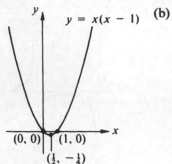
$y = x(x - 1)$
$(0, 0)$ $(1, 0)$
$(\frac{1}{2}, -\frac{1}{4})$

(b)

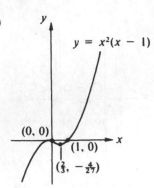
$y = x^2(x - 1)$
$(0, 0)$
$(1, 0)$
$(\frac{2}{3}, -\frac{4}{27})$

(c)

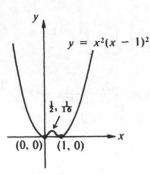
$y = x^2(x - 1)^2$
$\frac{1}{2}, \frac{1}{16}$
$(0, 0)$ $(1, 0)$

3. $m < -10, m > 2$

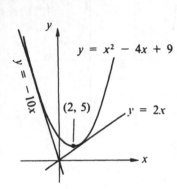
$y = x^2 - 4x + 9$
$y = -10x$
$(2, 5)$
$y = 2x$

4. (i) $-1, 5, -7, 3$ **(iii)** $2\frac{1}{3}$
(iv)

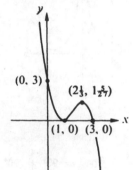
$(0, 3)$
$(2\frac{1}{3}, 1\frac{5}{27})$
$(1, 0)$ $(3, 0)$

5. (i)

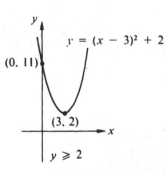
$y = (x - 3)^2 + 2$
$(0, 11)$
$(3, 2)$
$y \geqslant 2$

(ii)

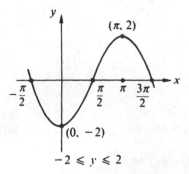
$(\pi, 2)$
$-\frac{\pi}{2}$ $\frac{\pi}{2}$ π $\frac{3\pi}{2}$
$(0, -2)$
$-2 \leqslant y \leqslant 2$

6.

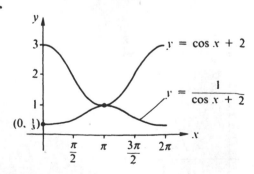
3
$y = \cos x + 2$
2
$y = \dfrac{1}{\cos x + 2}$
1
$(0, \frac{1}{3})$
$\frac{\pi}{2}$ π $\frac{3\pi}{2}$ 2π

7. $1 - \dfrac{9}{x^2}$, min 6, max -6,

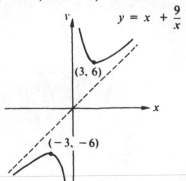
$y = x + \dfrac{9}{x}$
$(3, 6)$
$(-3, -6)$

8. (a)

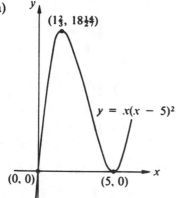

$(1\frac{2}{3}, 18\frac{14}{27})$

$y = x(x - 5)^2$

$(0, 0)$ $(5, 0)$

(b)

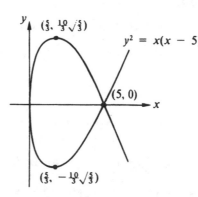

$(\frac{5}{3}, \frac{10}{3}\sqrt{\frac{5}{3}})$

$y^2 = x(x - 5)^2$

$(5, 0)$

$(\frac{5}{3}, -\frac{10}{3}\sqrt{\frac{5}{3}})$

9.

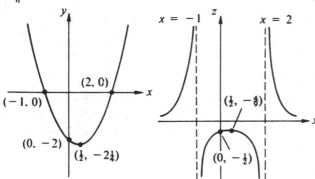

$(2, 0)$

$(-1, 0)$

$(0, -2)$ $(\frac{1}{2}, -2\frac{1}{4})$

$x = -1$ $x = 2$

$(\frac{1}{2}, -\frac{4}{3})$

$(0, -\frac{1}{2})$

$\{z \in \mathbb{R}: z > 0 \quad \text{or} \quad z \leqslant -\frac{4}{9}\}$

10.

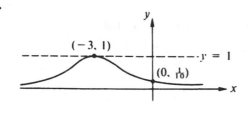

$(-3, 1)$ $y = 1$

$(0, \frac{1}{10})$

11. (i)

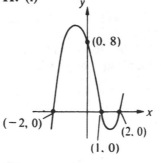

$(0, 8)$

$(-2, 0)$

$(2, 0)$

$(1, 0)$

(ii)

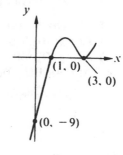

$(1, 0)$

$(3, 0)$

$(0, -9)$

$y = 2(1 - x)(x - 3)^2$, all roots are +ve, $k > 18$.

12. $b^2 - 4ac < 0$,

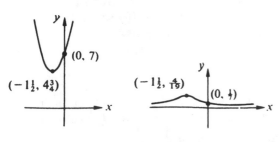

$(0, 7)$

$(-1\frac{1}{2}, 4\frac{3}{4})$

$(-1\frac{1}{2}, \frac{4}{19})$

$(0, \frac{4}{7})$

$(1, 0)$

$(-5, 0)$

$(0, -5)$

$(-2, -9)$

$x = -5$ $x = 1$

$(-2, -\frac{1}{6})$

$(0, -\frac{1}{5})$

$\{z \in \mathbb{R}: z > 0 \quad \text{or} \quad z \leqslant -\frac{1}{9}\}$

page 296   Exercise 12A

1. (a) $x^3 + c$ (b) $x^2 + c$ (c) $\frac{1}{4}x^4 + c$ (d) $\frac{2}{5}x^5 + c$

   (e) $5x + c$ (f) $\frac{1}{2}x^6 + c$ (g) $\frac{2}{3}x^{3/2} + c$ (h) $-\frac{6}{x} + c$

   (i) $\frac{2}{x^2} + c$ (j) $2\sqrt{x} + c$ (k) $\frac{1}{2}x^4 + x^3 + c$ (l) $\frac{5}{2}x^2 + x + c$

   (m) $x^2 + 3x^3 + c$ (n) $x^5 - 3x^2 + c$ (o) $2x^4 - 4x^3 + c$ (p) $2x^2 - x^3 + c$
   (q) $x^3 - 3x^2 + c$ (r) $\frac{2}{5}x^5 - 4x^2 + c$ (s) $x^3 + x^2 - x + c$ (t) $\frac{1}{3}x^3 - 4x^2 + 12x + c$
2. (a) $2x^4 + c$ (b) $6x^2 + c$ (c) $\frac{5}{3}x^3 + c$ (d) $7x + c$ (e) $7x - x^2 + c$

   (f) $-\frac{3}{x^2} + c$ (g) $\frac{3}{x^4} + c$ (h) $2x^{3/2} + c$ (i) $2x^{5/2} + c$ (j) $x^3 - \frac{6}{x} + c$

   (k) $x^4 + x^3 + x^2 + x + c$ (l) $2x^3 - 2x^4 + c$ (m) $\frac{x^5}{5} + \frac{x^3}{3} - \frac{1}{x} - \frac{1}{3x^3} + c$

3. (a) $6x^2 + c$ (b) $\frac{1}{4}x^4 + \frac{1}{2}x^2 + c$ (c) $\frac{1}{3}x^3 + \frac{1}{2}x^2 + c$

   (d) $\frac{1}{3}x^3 + x^2 - 24x + c$ (e) $-\frac{5}{3x^3} + c$ (f) $2x^5 + 2x^4 + \frac{6}{x} + c$

   (g) $\frac{x^3}{3} - \frac{1}{x} + c$ (h) $2\sqrt{x} - 2x^{3/2} + c$ (i) $\frac{x^3}{3} + \frac{1}{x} + c$

   4. $y = 3x^2 + 1$ 5. $y = x^3 - 2x + 10$ 6. $y = 8 + 2x - x^2$
   7. $s = 2t^3 + 6t^2 + t - 1$ 8. $V = 7h^2 - 4h + 1$ 9. $A = 5p - 2p^2 + 1$
   10. $y = x^3 + 8x - 1$ 11. $y = x^2 - 3x - 10, (-2, 0)$ 12. $y = x^2 - 4x + 7$
   13. $y = 7 + 2x^2 - x^3$ 14. $y = x^3 - 2x^2 + 5$ 15. $y = 5x^3 - 6x + 4$

page 303   Exercise 12B

1. (a) 24 (b) 8 (c) 15 (d) 2 (e) 4 (f) $17\frac{1}{3}$ (g) $\frac{1}{2}$ (h) 2 (i) 32 (j) $13\frac{3}{4}$
2. (a) A$(-1, 0)$, B$(0, 4)$, C$(4, 0)$, $20\frac{2}{3}$ sq. units
   (b) A$(-1, 0)$, B$(0, -5)$, C$(5, 0)$, 36 sq. units
   (c) A$(0, 4)$, B$(1, 0)$, C$(4, 0)$, $6\frac{1}{3}$ sq. units

3. (a) $\frac{1}{x}$ not defined for $x = 0$,   (b) $\frac{1}{x^2}$ not defined for $x = 0$,

   (c) $\frac{1}{x - 1}$ not defined for $x = 1$,   (d) $\frac{1}{x^2 - 1}$ not defined for $x = -1$

   (e) $\sqrt{x}$ not defined for $x < 0$
   4. 26 sq. units   5. $3\frac{3}{4}$ sq. units   6. $12\frac{2}{3}$ sq. units   7. $31\frac{1}{2}$ sq. units
   8. $10\frac{2}{3}$ sq. units   9. 36 sq. units   10. 3 sq. units   11. 13 sq. units
   12. $7\frac{2}{3}$ sq. units   13. 8 sq. units   14. $1\frac{1}{3}$ sq. units
   15. (a) $5\frac{1}{3}$ sq. units   (b) $25\frac{1}{4}$ sq. units   (c) 2 sq. units   16. $10\frac{2}{3}$ sq. units
   17. $2\frac{2}{3}$ sq. units   18. $\frac{1}{6}$ sq. units   19. A$(-2, 6)$, B$(2, 10)$, $10\frac{2}{3}$ sq. units
   20. A$(-4, 0)$, B$(-1, 0)$, C$(2, 0)$, D$(0, -8)$, E$(-5, -28)$, F$(3, 28)$, 64 sq. units
   21. $21\frac{1}{3}$ sq. units   22. $41\frac{2}{3}$ sq. units

page 307   Exercise 12C

1. $\frac{32}{5}\pi$ cu. units   2. $\frac{56}{3}\pi$ cu. units   3. 13π cu. units   4. $1\frac{28}{15}\pi$ cu. units
5. $1\frac{8}{15}\pi$ cu. units   6. $\frac{16}{9}\pi$ cu. units   7. $\frac{512}{15}\pi$ cu. units   8. $\frac{1}{105}\pi$ cu. units

9. $\frac{5}{6}\pi$ cu. units   10. $\frac{32}{5}\pi$ cu. units   11. $y = \frac{rx}{h}, V = \frac{1}{3}\pi r^2 h$

12. 8π cu. units   13. $\frac{15}{4}\pi$ cu. units   14. $\frac{56}{5}\pi$ cu. units

page 312   Exercise 12D

1. $15t^2 - 1, 30t$ **2.** (a) 0, 0, 8m/s² (b) 8 m, 4 m/s, -4 m/s²
3. $a = 6t - 8, s = t^3 - 4t^2 + 3$ **4.** (a) 1, 3 (b) 6 m/s² (c) $5\frac{1}{3}$ m
5. $V = t - \frac{1}{2}t^2 + 1, S = \frac{1}{2}t^2 - \frac{1}{6}t^3 + t + 2$ **6.** 1, 3, 10 m, 6 m   **7.** $4\frac{1}{2}$ m
8. $s = \frac{1}{4}t^3 + 9t - 18$, 182 m   **9.** $(0, 1\cdot6)$ **10.** $(1\cdot5, 0\cdot9)$ **11.** $(0, 1\frac{22}{23})$
12. $(2\frac{1}{4}, 2\frac{7}{10})$ **13.** $(2\frac{2}{5}, 7\frac{4}{5})$ **14.** $(3\frac{1}{3}, 0)$ **15.** $(1\frac{36}{55}, 0)$

page 314   Exercise 12E

1. $1\frac{1}{3}$ **2.** 11   **3.** 16   **4.** 6   **5.** $\frac{1}{5}$ **6.** $6\frac{1}{3}$ **7.** $2\frac{2}{3}$ **8.** 7
9. (a) 9 m/s (b) 12 m/s²   **10.** 4·5N   **11.** 95J

page 314   Exercise 12F

1. (i) $-\dfrac{1}{2x^2} + c$ (ii) $\frac{2}{3}\sqrt{x}(x + 3) + c$ **2.** 2 sq. units
3. (a) $\frac{1}{30}x^2(15 - 24\sqrt{x} + 10x) + c$ (b) $4\frac{17}{24}$
4. $y = x^3 + 2x^2 - 4x - 4$, max at $(-2, 4)$ min at $(\frac{2}{3}, -5\frac{13}{27})$
5. (i) 8 (ii) $42\frac{1}{2}$ m/s   **6.** $10\frac{2}{3}$ sq. units   **7.** (i) $(-1, 0)$ (ii) $\frac{32}{10}\pi$ cubic units
8.   **9.** $3y = x + 6, 3y + x = 9, (1\frac{1}{2}, 2\frac{1}{2}), (1, 0), (2, 0)$

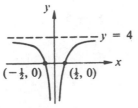

$4\frac{1}{2}$ sq. units

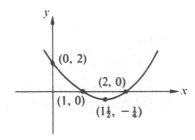

10. $\frac{16}{15}\pi$ cu. units, $\frac{2}{3}\pi$ cu. units   **11.** $y = 6x + 15$, (b) $\frac{408}{5}\pi$ cu. units
12. (i) $\frac{14}{9}$ (ii) 2 sq. units
13.

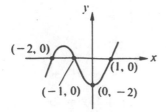

, (a) $\frac{5}{12}$ sq. units   (b) $(-\frac{2}{3}, -\frac{20}{27})$, $27y + 63x + 62 = 0$.

14. $\frac{15}{2}\pi, \frac{53}{15}\pi$

page 319   Exercise 13A

1. $12(3x + 2)^3$

2. $10(2x - 3)^4$

3. $-28(1 - 4x)^6$

4. $45(2 + 9x)^4$

5. $10x(3 + x^2)^4$

6. $-24x^2(1 - x^3)^7$

7. $48x^3(2x^4 + 1)^5$

8. $20x(x^2 + 5)^9$

9. $\dfrac{-9}{(1 + 3x)^4}$

10. $\dfrac{8x}{(1 - 4x^2)^2}$

11. $\dfrac{5}{2\sqrt{(5x - 3)}}$

12. $\dfrac{-3x}{(x^2 + 1)^{5/2}}$

13. $\dfrac{2}{(1 - 2x)^2}$

14. $-\dfrac{2x}{(x^2 + 3)^2}$

15. $\dfrac{x + 2}{\sqrt{(x^2 + 4x + 1)}}$

16. $-\dfrac{x}{(x^2 - 5)^{3/2}}$

17. $-\dfrac{1}{2\sqrt{x}(1 + \sqrt{x})^2}$

18. $40x(1 + (1 + x^2)^5)^3(1 + x^2)^4$

19. $\dfrac{1}{4\sqrt{x}\sqrt{(2 + \sqrt{x})}}$

20. $-\dfrac{9x}{(3x^2 + 1)^{5/2}}$

21. (a) $y = 12x - 23$ (b) $12y = 14 - x$

22. (a) $y = 4x - 9$ (b) $4y + x + 2 = 0$

23. (a) $8y = x + 3$ (b) $4y + 32x + 31 = 0$

24. (a) $3y = x + 5$ (b) $y + 3x = 15$

25. (a) Max at $(0, 1)$ (b) Min at $(2\frac{1}{2}, 0)$ (c) Infl at $(2\frac{1}{2}, 0)$ (d) Max at $(-2, -\frac{1}{4})$

26. (a) 5 m   (b) 0.2 m/s   (c) -0.008 m/s²

page 321   Exercise 13B

1. $(4 + 3x)^8 + c$

2. $(2 + 7x)^6 + c$

3. $(3 - 2x)^7 + c$

4. $(x^2 + 4)^5 + c$

5. $\frac{1}{18}(2 + 3x)^6 + c$

6. $\frac{1}{10}(1 + 2x)^5 + c$

7. $-\frac{1}{24}(1 - 6x)^4 + c$

8. $\frac{1}{10}(3 + 4x)^5 + c$

9. $-\frac{3}{8}(1 - 2x)^5 + c$

10. $\frac{1}{4}(1 + x^2)^4 + c$

11. $\frac{1}{15}(x^3 - 1)^5 + c$

12. $\frac{1}{16}(x^4 + 6)^4 + c$

13. $\frac{1}{5}(x^2 + x + 3)^5 + c$

14. $\frac{1}{5}(x^3 - x + 4)^5 + c$

15. $\frac{1}{6}(4 + x + x^2)^6 + c$

16. $\frac{1}{16}(x^2 - 2x + 4)^8 + c$ **17.** 1342

18. -7776

19. 210.1

20. 1302

21. $121\frac{1}{3}$

22. 20

23. $221\frac{2}{3}$

page 324   Exercise 13C

1. 77

2. $5(2t + 3)^3$

3. $4\pi r$ cm²/s

4. 40π cm²/s

5. $\dfrac{3}{2\pi}$ cm/s

6. 3 cm²/s

7. 0.1 cm/s

8. (a) $\frac{8}{3}\pi r$ cm²/s   (b) $\frac{9}{10}\pi r^2$ cm³/s   **9.** 0.25 cm/s

10. (a) 0.01 cm/s   (b) 0.1π cm²/s

11. $2s(s^2 - 2)$

page 327   Exercise 13D

1. $y = x^4 + 4$

2. $y = x^2 - 8x + 6$

3. $y = \dfrac{3}{x - 2}$

4. $x + 2y = 11$

5. $y = x^2 - 6x$

6. $9y = (x + 7)(x - 8)$

7. $y = \dfrac{x}{2x + 1}$

8. $x + 3y = 7xy$

9. $x^2 + 4y^2 = 4$

10. $x^2 = 4y^2(1 - y^2)$

11. $5x^2 + 5y^2 - 8xy = 9$

12. $\dfrac{1}{2}t$

13. $\dfrac{1}{t}$

14. $\dfrac{1}{2}t$

15. $\dfrac{1}{4t^{3/2}}$

16. $-\dfrac{1}{3t^2}$

17. $\dfrac{3}{2t + 3}$

18. $\left(\dfrac{t - 1}{t + 1}\right)^2$

19. $-2t(t + 1)^2$

20. (a) $y = 6x - 2$ (b) $6y + x = 99$

21. (a) $y + 8x = 5$ (b) $16y = 2x + 15$

22. (a) $y = x + 3$ (b) $y + x + 5 = 0$

23. (a) $y = 6x - 5$ (b) $6y + x = 44$

24. $(49, -9)$

25. (a) 2   (b) $\dfrac{3}{16t^5}$ (c) $\dfrac{3}{4(t + 1)}$

26.

t	-4	-3	-2	-1	0	1	2	3	4
x	32	18	8	2	0	2	8	18	32
y	-16	-12	-8	-4	0	4	8	12	16

27.

θ	0	$\dfrac{\pi}{6}$	$\dfrac{\pi}{3}$	$\dfrac{\pi}{2}$	$\dfrac{2\pi}{3}$	$\dfrac{5\pi}{6}$	π	$\dfrac{7\pi}{6}$	$\dfrac{4\pi}{3}$	$\dfrac{3\pi}{2}$	$\dfrac{5\pi}{3}$	$\dfrac{11\pi}{6}$	2π
x	0	2	3·46	4	3·46	2	0	-2	$-3·46$	-4	$-3·46$	-2	0
y	1	0·87	0·5	0	$-0·5$	$-0·87$	-1	$-0·87$	$-0·5$	0	0·5	0·87	1

28.

t	-2	$-1\frac{1}{2}$	-1	$-\frac{1}{2}$	0	$\frac{1}{2}$	1	$1\frac{1}{2}$	2
x	-12	$-3\frac{3}{4}$	0	$\frac{3}{4}$	0	$-\frac{3}{4}$	0	$3\frac{3}{4}$	12
y	16	9	4	1	0	1	4	9	16

page 330 **Exercise 13E**

1. $(16x + 5)(2x + 5)^6$ **2.** $2x(9x + 5)(2x + 5)^6$ **3.** $3x^3(3x + 4)(x + 3)^4$

4. $2x^3(7x^2 + 6)(x^2 + 3)^4$ **5.** $x^2(9 - 23x^4)(3 - x^4)^4$

6. $2x(x^4 + 1)(5x^4 - 28x^2 + 1)$ **7.** $3(2x^2 + 1)^2(3x - 1)^6(26x^2 - 4x + 7)$

8. $3(30x + 11)(2x - 1)^5(3x + 5)^8$ **9.** $(19 - 16x)(2x + 1)^2(4 - x)^4$

10. $\dfrac{3(x + 2)}{2\sqrt{(x + 3)}}$ **11.** $\dfrac{x(5x + 12)}{2\sqrt{(x + 3)}}$ **12.** $\dfrac{3(3x + 1)(x + 3)^3}{2\sqrt{x}}$

13. $\dfrac{2(x + 3)(3x^2 + 3x + 1)}{\sqrt{(2x^2 + 1)}}$ **14.** $\dfrac{6}{(x + 3)^2}$ **15.** $-\dfrac{15}{(x - 5)^2}$

16. $\dfrac{11}{(2x + 3)^2}$ **17.** $-\dfrac{13}{(3x + 2)^2}$ **18.** $\dfrac{2(2 + x)(2 - x)}{(x^2 + 4)^2}$ **19.** $\dfrac{(x - 3)(x + 1)}{(x^2 + 3)^2}$

20. $\dfrac{(x - 8)(x + 2)}{(x^2 + 16)^2}$ **21.** $\dfrac{x^4(2x + 5)}{(x + 1)^4}$ **22.** $\dfrac{3(3x + 2)}{(2 - x)^5}$

23. $\dfrac{-6(2x + 11)(6x + 5)}{(2x - 3)^4}$ **24.** $\dfrac{4x - 7}{(2x - 1)^{3/2}}$

25. $60(2x + 3)^3(2x + 1)$ **26.** (a) $\dfrac{1}{(x + 1)^2}$ (b) 1 (c) $-\dfrac{2}{(x + 1)^3}$ (d) -2

27. $\left(\dfrac{t + 6}{2 - t}\right)^2$ **28.** $\dfrac{3(t + 1)^2}{t(t + 2)^3}$ **29.** (a) $y = 24x - 44$ (b) $24y + x = 98$

30. (a) $y = 5x - 8$ (b) $5y + x = 12$ **31.** (a) $2y = 3x - 7$ (b) $3y + 2x + 4 = 0$

32. (a) $3y = 8x - 4$ (b) $8y + 3x = 38$ **33.** Max at $(1, 256)$, Min at $(5, 0)$.

34. Min at $(0, 0)$, Max at $(1, 16)$, Min at $(3, 0)$ **35.** Max at $(-2, \frac{1}{4})$, Min at $(4, -\frac{1}{8})$

36. Infl at $(0, 0)$, Min at $(5, 115\frac{20}{27})$ **37.** $4\left(\dfrac{t + 1}{t - 1}\right)^3$ **38.** $\dfrac{10}{9}\left(\dfrac{t + 3}{t - 2}\right)^3$

page 332 **Exercise 13F**

1. $-\dfrac{x}{y}$

2. $\dfrac{2(1 - x)}{y}$

3. $\dfrac{2(2 - 3x)}{3y^2 - 2}$

4. $\dfrac{3(x^2 + 1)}{5(y + 1)}$

5. $\dfrac{4x + 3}{7 - 4y}$

6. $-\dfrac{(2x + y)}{(x + 2y)}$

7. $\dfrac{3(x^2 + y)}{2y - 3x}$

8. $\dfrac{2x^2 + y^2}{y(y - 2x)}$

9. $\dfrac{5 - 6x - y}{6 + x + 4y}$

10. $\frac{1}{3}$

11. $-2\frac{2}{3}$

12. (a) $9y + 7x + 2 = 0$ (b) $7y = 9x - 16$ **13.** (a) $y + 3x + 9 = 0$ (b) $3y = x - 7$

14. (a) $y + 3x + 2 = 0$ (b) $3y = x + 4$

page 333 **Exercise 13G**

1. $\dfrac{7}{9}$ **2.** (a) $(0, 0)(4, 8)$ (b) 36 **3.** $\dfrac{1 - 3x}{(1 + x^2)^{3/2}}, \dfrac{1}{3}$, max

4. $\dfrac{6x}{(2x^2 + 1)^2}, x > 0, 0, -1$.   **5.** $\frac{3}{25}\pi$ cm^2/s   **6.** $\dfrac{3}{4\pi r^2}$ cm/s, $\dfrac{1}{2}$

7. (i) 1 cm   (ii) 1 sec after first observation   (iii) $\dfrac{3}{4}\pi$ cm^2/sec

8. (i) $x^2 + 4y^2 = 4$ (ii) $x = y^2 - 3y + 2$

9. (i) $y + 12x = 0, y + 3x + 3 = 0$ (ii) $(\frac{1}{3}, -4)$

10. $\frac{1}{2}(1 + t)^3(3t - 1), -3t(1 + t)^4, 2y = 135x + 27$ **11.** $y + x + 4 = 0, y + 2x = 4$

12. $(\frac{1}{4}, 2\frac{1}{4})$ **13.**

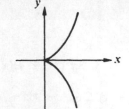

$108\frac{4}{5}$ sq. units

page 342 **Exercise 14A**

1. (a)

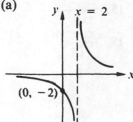

(b)

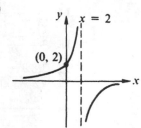

(c)

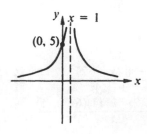

2. (a)

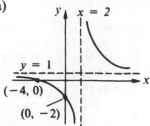

(b)

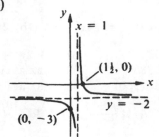

(c)

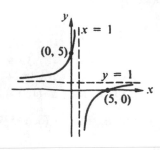

(d)

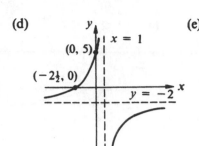

(e)

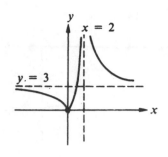

(f)

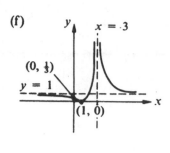

3.

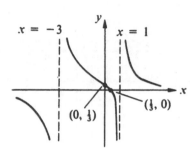

4. $-6 \leqslant y \leqslant 2$

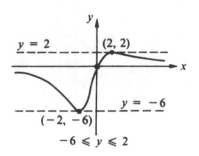

5.

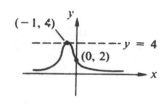

6.

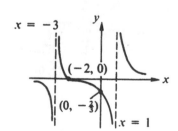

7.

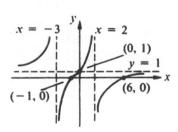

8.

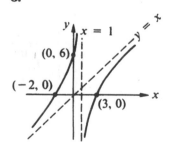

9. $y \leqslant -3, y > 0$

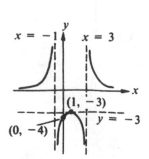

10. $y \leqslant 2$

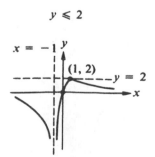

11. $y \in \mathbb{R}$

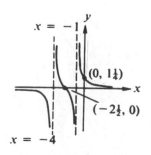

12. $y \leqslant 1, y \geqslant 4$

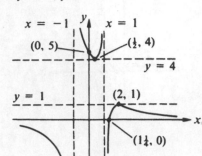

13. $y \in \mathbb{R}$

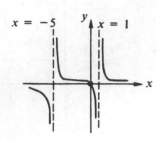

14. $-2 \leqslant y \leqslant 2$

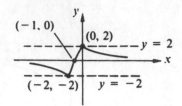

15. $y \leqslant -9, y \geqslant -1$.

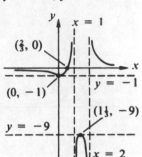

16. $y \geqslant -1$

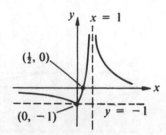

17. $y \in \mathbb{R}$

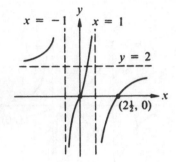

18. $y < 1$

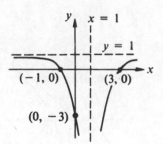

19. $y \leqslant -2, y \geqslant 6$

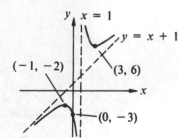

20. $y \geqslant 0$

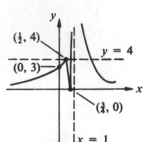

21. $k \geqslant 4$

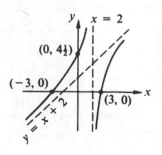

22. $k \geqslant 2$

23. $0 \leqslant k \leqslant 3$; $f(x) \leqslant 1, f(x) \geqslant 9$

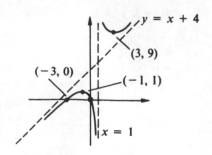

page 348   **Exercise 14B**

1. (a) $\{x \in \mathbb{R}: x < 2 \ \text{ or } \ x > 5\}$ (b) $\{x \in \mathbb{R}: -1 < x < 2 \ \text{ or } \ x > 3\}$ **2.** $x < 1, x > 2$
3. $-2 < x < 3$ **4.** $-5 < x < 0, x > 3$ **5.** $x < 0, 1\frac{1}{2} < x < 5$ **6.** $\frac{1}{2} < x < 4$
7. $x < -2, 2 < x < 4$ **8.** $x < -3, -1 < x < 1, x > 3$ **9.** $-2 < x < -\frac{1}{2}, \frac{1}{2} < x < 2$
10. $x < -2, x > 2$ **11.** $2 < x < 4$ **12.** $-6 < x < 2\frac{1}{2}$ **13.** $-2 < x < 1, x > 2$
14. $x < -1, 0 < x < 3$ **15.** $x > 1$ but $\neq 3$ **16.** $x > -1$ **17.** $x < -2, 0 < x < 3$
18. $-2 < x < 0, x > 3$ **19.** $x < -2, 1 < x < 5$ **20.** $x < -2, 0 < x < 4, x > 8$
21. $-10 < x < -1, x > 2$ **22.** $x < -5, -2 < x < 1$ **23.** $x < -4, -1 < x < 2, x > 6$
24. $x < -6, -1 < x < \frac{1}{2}, x > 2$ **25.** $-4 < x < 2, 5 < x < 8$ **26.** $0 < x < 2$
27. $x < 0, x > 3$
28. (a) $\{x \in \mathbb{R}: -7 < x < 1\}$ (b) $\{x \in \mathbb{R}: x > 3\}$ (c) $\{x \in \mathbb{R}: x < -2 \ \text{ or } \ x > 0\}$
29. $x < -1, x > 4$ **30.** $-2 < x < 2$ **31.** $x < -2, x > 4$ **32.** $-3 < x < -1$
33. $x < -4, x > -2$ **34.** $-1\frac{1}{2} < x < \frac{1}{2}$ **35.** $x > 1$ **36.** $-2 < x < 2$
37. $x < -6, x > -4\frac{1}{2}$ **38.** $0 < x < 1\frac{1}{2}$ **39.** $x < 2, x > 6$ **40.** $-2 < x < -\frac{2}{3}$
41. $x < 2, x > 6$ **42.** $0 < x < 2, x > 4$ **43.** $x < -4, x > 0$

page 349   **Exercise 14C**

1.

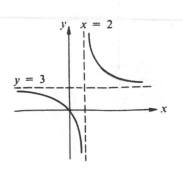

(i) $y + x = 1$,
(ii) $y + x + 3 = 0$.

2.

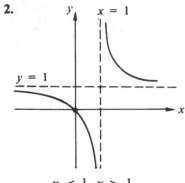

$x < \frac{1}{2}, x > 1$.

3.

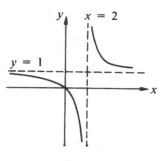

$y = \dfrac{x}{x + 2}$.

4. $x = 2, y = 3$,

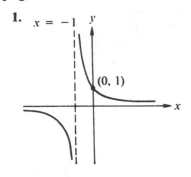

5. $-\frac{1}{2}$,

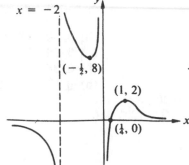

6.

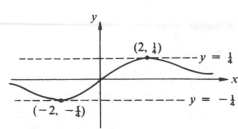

7. $x = 1, y = x + 1$,

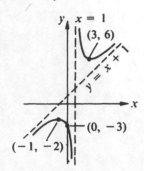

(a) $x = 2$
(b) $x < 0$
(c) $x < 3$ except $x = 2, x > 4$

8. (i) $x < \frac{1}{2}$ (ii) $x < -1, x > 2$ (iii) $x < -1, \frac{1}{2} < x < 2$

9. $-4 < x < 0$

10. $-2 < x < 0$

12. (a) $0 < x < 3$ (b) $-1 < x < 1, x > 2$

13. $\{x \in \mathbb{R}: |x| > 2\}$

14. (i) $\{x \in \mathbb{R}: x < -1 \quad \text{or} \quad x > 4\}$
     (ii) $\{x \in \mathbb{R}: -1 < x < 0 \quad \text{or} \quad x > 4\}$

15.

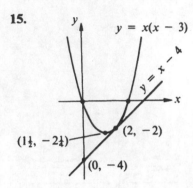

16.

$-7 \leqslant x \leqslant -3$,
$(5 - 3\sqrt{6}) \leqslant x \leqslant (5 + 3\sqrt{6})$ with $x \neq \pm 2$

page 353   Exercise 15A

1. (a) $2 \sin 4\theta \cos \theta$ (b) $2 \cos 3\theta \cos 2\theta$ (c) $2 \cos 4\theta \sin 3\theta$ (d) $-2 \sin 3\theta \sin 2\theta$
  (e) $2 \cos 5\theta \cos 2\theta$ (f) $2 \sin \frac{1}{2}\theta \cos \frac{3}{2}\theta$ (g) $2 \sin 2\theta \sin \theta$ (h) $-2 \cos 5\theta \sin \theta$
2. (a) $\sin 6\theta + \sin 2\theta$ (b) $\sin 5\theta - \sin 3\theta$ (c) $\cos 7\theta - \cos \theta$ (d) $\sin 7\theta + \sin 3\theta$
  (e) $\sin 7\theta - \sin 3\theta$ (f) $\cos 7\theta + \cos \theta$ (g) $\cos 4\theta - \cos \theta$ (h) $\cos 5\theta - \cos 7\theta$

3. (a) $\sqrt{\dfrac{3}{2}}$ (b) $\dfrac{1}{\sqrt{2}}$ (c) $-\sqrt{\dfrac{3}{2}}$ (d) $\dfrac{(1 + \sqrt{2})}{4}$

5. (a) $0, 90°, 180°$ (b) $0, 30°, 90°, 150°, 180°$
  (c) $15°, 30°, 75°, 90°, 150°$ (d) $0, 30°, 60°, 120°, 150°, 180°$
  (e) $15°, 45°, 75°, 135°$ (f) $0, 36°, 45°, 108°, 120°, 135°, 180°$
6. (a) $\pm 150°, \pm 90°, \pm 30°$ (b) $\pm 180°, \pm 120°, \pm 90°, \pm 60°, 0$
  (c) $\pm 180°, \pm 150°, \pm 120°, \pm 60°, \pm 30°, 0$
  (d) $-150°, -90°, -60°, -30°, 30°, 90°, 120°, 150°$

page 356   Exercise 15B

1. (a) $5 \cos(x + 53 \cdot 1°)$ (b) $\sqrt{29} \cos(x + 68 \cdot 2°)$ (c) $\sqrt{13} \cos(x + 33 \cdot 7°)$
2. (a) $13 \cos(x - 67 \cdot 4°)$ (b) $\sqrt{5} \cos(x - 63 \cdot 4°)$ (c) $\sqrt{13} \cos(x - 33 \cdot 7°)$
3. (a) $25 \sin(x + 73 \cdot 7°)$ (b) $\sqrt{10} \sin(x + 18 \cdot 4°)$ (c) $\sqrt{34} \sin(x + 31°)$
4. (a) $5 \sin(x - 53 \cdot 1°)$ (b) $\sqrt{58} \sin(x - 66 \cdot 8°)$ (c) $\sqrt{74} \sin(x - 54 \cdot 5°)$

5. (a) $\sqrt{2}$, 45° (b) 5, 53·1° (c) 17, 28·1° (d) 5, 323·1° (e) $\sqrt{13}$, 123·7° (f) $\sqrt{29}$, 21·8°
6. (a) 118·1°, 323·1° (b) −46·4°, −36·9°, 313·6°, 323·1°
7. (a) 11°, 225·2° (b) 244·1°, 352° (c) 36·9°, 270°
  (d) 77·9°, 325·7° (e) 15·7°, 282·4° (f) 142·5°, 172·5°, 322·5°, 352·5°
8. (a) −50·5°, 103·6° (b) −17·6°, 130·2° (c) 24·6°, 171·7°
  (d) −169·1°, 51° (e) −153·9°, 78·1° (f) ±180°, −29·6°, 0, 103·3°

page 360   Exercise 15C

1. (a) 0 (b) 60° (c) 30° (d) −30° (e) 120° (f) −60° (g) 60° (h) −60°
2. (a) $\dfrac{\pi}{2}$ (b) $-\dfrac{\pi}{2}$ (c) $\dfrac{\pi}{3}$ (d) $\dfrac{5\pi}{6}$ (e) $\dfrac{\pi}{4}$ (f) $\dfrac{\pi}{6}$ (g) $-\dfrac{\pi}{4}$ (h) $\dfrac{3\pi}{4}$

3.

x	0	0·25	0·5	0·75	1	1·25	1·5	1·75	2	2·25	2·5	2·75	3	3·14
y	1	0·97	0·88	0·73	0·54	0·32	0·07	−0·18	−0·42	−0·63	−0·80	−0·92	−0·99	−1

4. (a) $\frac{56}{65}$ (b) $\frac{220}{221}$ (c) $\frac{4}{3}$ (d) $-\frac{13}{85}$ (e) $\frac{297}{425}$ **5.** (a) $\dfrac{\pi}{2}$ (b) $\dfrac{\pi}{4}$
8. $\frac{1}{6}$ **9.** 0·425 **10.** $\frac{1}{3}$ **11.** $\frac{1}{6}$ **12.** $\sqrt{(6-4\sqrt{2})}$

page 365   Exercise 15D

1. $2n\pi \pm \dfrac{\pi}{3}$ **2.** $n\pi - \dfrac{\pi}{4}$ **3.** $n\pi + (-1)^{n+1}\dfrac{\pi}{3}$ **4.** $\dfrac{2n\pi}{3}$ **5.** $\dfrac{n\pi}{2} - \dfrac{\pi}{12}$

6. $\dfrac{n\pi}{5} + (-1)^n\dfrac{\pi}{30}$ **7.** $n\pi + \dfrac{\pi}{2}, n\pi + \dfrac{\pi}{4}$ **8.** $n\pi + (-1)^{n+1}\dfrac{\pi}{6}, 2n\pi + \dfrac{\pi}{2}$

9. $\dfrac{n\pi}{5}, \dfrac{n\pi}{2}$ **10.** $\dfrac{n\pi}{6}$ **11.** $n\pi, (2n+1)\dfrac{\pi}{8}$ **12.** $\dfrac{n\pi}{2}$

13. $\dfrac{1}{3}\left(2n\pi - \dfrac{\pi}{2}\right), \dfrac{1}{7}\left(2n\pi - \dfrac{\pi}{2}\right)$ **14.** $\dfrac{1}{7}\left(2n\pi + \dfrac{\pi}{2}\right), 2n\pi + \dfrac{\pi}{2}$ **15.** $n\pi - \dfrac{\pi}{3}, \dfrac{n\pi}{3} - \dfrac{\pi}{18}$

16. $\dfrac{n\pi}{4} + \dfrac{\pi}{16}$ **17.** $180°n + (-1)^n 5.7°$ **18.** $180°n \pm 26.6°$
19. $360°n \pm 120°, 360°n \pm 48.2°$ **20.** $180°n, 360°n \pm 75.5°$
21. $360°n + 115.3°, 360°n - 41.6°$ **22.** $360°n + 46.4°, 360°n - 90°$

page 366   Exercise 15E

1.

sin θ	cos θ	tan θ
0·1	0·995	0·1
0·0998	0·995	0·1003

sin θ	cos θ	tan θ
0·02	0·9998	0·02
0·02	0·9998	0·02

2. (a) 2 (b) 1 (c) 4 (d) 2
3. (a) $-\dfrac{1}{2}\theta$ (b) $\dfrac{1}{2\theta}$ (c) 4θ (d) $\dfrac{1}{4}\sqrt{2}(2 + 2\theta - \theta^2)$ (e) $\dfrac{\sqrt{3}}{2} - \dfrac{1}{2}\theta - \dfrac{\sqrt{3}}{4}\theta^2$ (f) $1 - 2\theta + 4\theta^2$
4. (a) 0·035 (b) 0·0035 (c) 0·999 847

page 370   **Exercise 15F**

1. $\cos x$ **2.** $-\sin x$ **3.** $\sec^2 x$ **4.** $-\operatorname{cosec} x \cot x$ **5.** $\sec x \tan x$ **6.** $-\operatorname{cosec}^2 x$
7. $2 \cos 2x$ **8.** $-5 \sin 5x$ **9.** $-7 \sin 7x$ **10.** $9 \cos 9x$ **11.** $3 \sec^2 3x$

12. $5 \sec 5x \tan 5x$ **13.** $-6 \operatorname{cosec} 6x \cot 6x$ **14.** $-2 \operatorname{cosec}^2 2x$ **15.** $-\dfrac{\pi}{180} \sin x°$

16. $\dfrac{\pi}{90} \sec^2 2x°$ **17.** $-2x \sin (x^2 + 1)$ **18.** $4x \cos (2x^2 + 1)$ **19.** $3x^2 \cos (x^3)$

20. $8x \sec (4x^2 + 1) \tan (4x^2 + 1)$ **21.** $-4x \operatorname{cosec}^2 (2x^2 + 1)$ **22.** $-3 \operatorname{cosec}^2 (3x + 4)$
23. $\sin 2x$ **24.** $-3 \cos^2 x \sin x$ **25.** $4 \tan^3 x \sec^2 x$ **26.** $7 \sin^6 x \cos x$
27. $-4 \cos^3 x \sin x$ **28.** $6 \sec^6 x \tan x$ **29.** $6 \sin^2 2x \cos 2x$ **30.** $-12 \cos^3 3x \sin 3x$
31. $6 \sin (12x + 2)$ **32.** $6 \cos^2 (3 - 2x) \sin (3 - 2x)$ **33.** $15 \tan^2 5x \sec^2 5x$
34. $35 \sec^7 5x \tan 5x$ **35.** $2x \sin x + x^2 \cos x$ **36.** $2x \cos 3x - 3x^2 \sin 3x$
37. $3x^2 (\tan 3x + x \sec^2 3x)$ **38.** $3 \cos 3x \cos x - \sin 3x \sin x$
39. $\cos 4x \cos x - 4 \sin 4x \sin x$ **40.** $4 \cos^3 x \cos 5x$ **41.** $3 \sin^2 x \cos 4x$

42. $-4 \cos^3 x \sin 5x$ **43.** $3 \sin^2 x \sin 4x$ **44.** $\dfrac{2}{x^3}(x \cos 2x - \sin 2x)$

45. $\dfrac{-2 \cos^3 x}{x^3}(2x \sin x + \cos x)$ **46.** $-\dfrac{\cos^2 5x}{6x^2}(15x \sin 5x + \cos 5x)$

47. (a) $-3\sqrt{3}$ (b) 3 (c) 1 (d) $\frac{3}{4}$

48. (a) $\left(\dfrac{\pi}{6}, \dfrac{\sqrt{3}}{2}\right)\left(\dfrac{5\pi}{6}, -\dfrac{\sqrt{3}}{2}\right)$ (b) $\left(\dfrac{\pi}{3}, \dfrac{9\sqrt{3}}{8}\right)$ **49.** $y = x + 3, \ y = 3 - x$

50. (a) max at $\left(\dfrac{2\pi}{3}, 8\right)$ min at $(2\pi, -8)$ (b) min at $(0, 1)$, max at $\left(\dfrac{\pi}{2}, \dfrac{\pi}{2}\right)$, min at $\left(\dfrac{3\pi}{2}, -\dfrac{3\pi}{2}\right)$

(c) max at $\left(\dfrac{\pi}{4}, \sqrt{2}\right)$, min at $\left(\dfrac{5\pi}{4}, -\sqrt{2}\right)$

51. (a) $-\cot \theta, \ -\operatorname{cosec}^3 \theta$ (b) $\frac{1}{4} \sec \theta, \ -\frac{1}{16} \sec^3 \theta$

page 373   **Exercise 15G**

1. $-\cos x + c$ **2.** $\sec x + c$ **3.** $-\frac{1}{3} \cos 3x + c$ **4.** $\frac{1}{2} \sin 2x + c$
5. $\frac{1}{2} \tan 2x + c$ **6.** $-\frac{1}{4} \cot 4x + c$ **7.** $-\frac{1}{2} \operatorname{cosec} 2x + c$ **8.** $\frac{1}{4} \sin 4x + c$
9. $-\frac{1}{2} \cos 4x + c$ **10.** $-2 \cos 2x + c$ **11.** $3 \cos (2x + 1) + c$ **12.** $2 \sin (3x - 4) + c$
13. $x + \frac{1}{2} \sin 2x + c$ **14.** $3x + 2 \cos 4x + c$ **15.** $-2 \cos^4 x + c$ **16.** $-\frac{1}{6} \cos^6 x + c$
17. $2 \sin^5 x + c$ **18.** $-\cos^6 2x + c$ **19.** $\frac{1}{10} \sin^5 2x + c$ **20.** $2 \sin^5 5x + c$
21. $\frac{1}{6} \sec^6 x + c$ **22.** $\frac{1}{7} \tan^7 x + c$ **23.** $\frac{1}{3} \cos^3 x - \cos x + c$
24. $\frac{2}{3} \cos^3 x - \cos x - \frac{1}{5} \cos^5 x + c$ **25.** $\sin x + \frac{1}{5} \sin^5 x - \frac{2}{3} \sin^3 x + c$
26. $\frac{1}{4} \sin 4x - \frac{1}{12} \sin^3 4x + c$ **27.** $\frac{1}{4} \sin 2x + \frac{1}{2}x + c$
28. $\frac{1}{2}x - \frac{1}{4} \sin 2x + c$ **29.** $\frac{1}{32} \sin 4x + \frac{3}{8}x + \frac{1}{4} \sin 2x + c$
30. $\frac{3}{8}x - \frac{1}{12} \sin 6x + \frac{1}{96} \sin 12x + c$ **31.** $\frac{1}{2} \cos 2x - \frac{1}{4} \cos 4x + c$
32. $-\frac{1}{6} \cos 6x - \frac{1}{4} \cos 4x + c$ **33.** $\frac{1}{16} \sin 8x + \frac{1}{4} \sin 2x + c$
34. $\frac{3}{2} \sin 2x - \frac{3}{4} \sin 4x + c$
35. (a) $\sin x + c$ (b) $x + c$ (c) $\frac{1}{2} \sin 2x + c$ (d) $\tan x + c$ (e) $x - \cos x + c$

(f) $\tan x + c$ **36.** $\dfrac{\pi}{2} - 1$ **37.** $\frac{1}{2}$ **38.** $-\frac{2}{3}$ **39.** $\frac{1}{16}$ **40.** $\frac{1}{4}$ **41.** $\dfrac{\pi}{8} - \dfrac{1}{4}$

42. $\frac{4}{3}$ **43.** $-\frac{1}{2}$ **44.** 2 sq. units   **45.** 3 sq. units   **46.** 4 sq. units   **47.** $\dfrac{\pi^2}{2}$ cu. units

48. $\dfrac{\pi}{2}(2 + \pi)$ cu. units

page 376   **Exercise 15H**

1. $\dfrac{1}{\sqrt{(1-x^2)}}$ **2.** $\dfrac{a}{a^2+x^2}$ **3.** $\dfrac{1}{\sqrt{(16-x^2)}}$ **4.** $\dfrac{-3}{\sqrt{(1-9x^2)}}$ **5.** $\dfrac{4}{1+16x^2}$

6. $\dfrac{6}{\sqrt{(1-36x^2)}}$ **7.** $\dfrac{1}{\sqrt{(x(1-x))}}$ **8.** $\dfrac{-3}{2-6x+9x^2}$ **9.** $\dfrac{2}{\sqrt{(2-x^2)}}$

10. $\sin^{-1}x + \dfrac{x}{\sqrt{(1-x^2)}}$ **11.** $\tan^{-1}x + \dfrac{x}{1+x^2}$ **12.** $2x\tan^{-1}x + 1$

13. $\sin^{-1}\left(\dfrac{x}{2}\right)+c$ **14.** $\sin^{-1}\left(\dfrac{x}{4}\right)+c$ **15.** $\tan^{-1}\left(\dfrac{x}{3}\right)+c$ **16.** $\dfrac{1}{5}\tan^{-1}\left(\dfrac{x}{5}\right)+c$

17. $\sin^{-1}\left(\dfrac{x}{7}\right)+c$ **18.** $\dfrac{1}{7}\tan^{-1}\left(\dfrac{x}{7}\right)+c$ **19.** $\sin^{-1}\left(\dfrac{x}{5}\right)+c$ **20.** $\dfrac{1}{5}\tan^{-1}\left(\dfrac{x}{10}\right)+c$

21. $\dfrac{\pi}{3}$ **22.** $\dfrac{\pi}{4}$ **23.** $\dfrac{7}{12}\pi$ **24.** $\dfrac{\pi}{3}$

page 377   **Exercise 15I**

1. $\cos x$ **2.** $\frac{1}{4}\pi, \frac{1}{2}\pi, \frac{3}{4}\pi$ **3.** (a) $7-4\sqrt{3}$ (b) $90°, 270°$

4. (a) $2\cos(x-53\cdot1°)$ (i) Max 2, Min -2 (ii) $11\cdot7°, 94\cdot5°$
(c) $10°, 50°, 90°, 130°, 170°, 250°, 270°, 290°$

5. (a) $\dfrac{1-t^2}{1+t^2}$, $20\cdot8°, 122\cdot3°$ **6.** $3\cdot2, 18\cdot4, 108\cdot4°$ (or $288\cdot4°$)   **7.** $2n\pi, n\pi + \dfrac{\pi}{4}$

8. (i) $\frac{1}{18}\pi(6n+(-1)^{n+1})$ (ii) $360°n + 36\cdot9°, (4n-1)90°$

9. (a) $180°n + (-1)^n 41\cdot8°, 360°n - 90°$ (b) $3 - \frac{5}{2}\cos(2x - 36\cdot9°)$, max $5\frac{1}{2}$, min $\frac{1}{2}$

10. $\dfrac{2-3t\tan 3t}{2\tan 3t + 3t}$ **11.** $-\cos x - \dfrac{1}{3}\cos^3 x + c$ **12.** $-\dfrac{1}{\pi}$

13. (a) (i) $2\sec^2 2x$ (ii) $9\sin^2 3x \cos 3x$ (iii) $5\cos x(1+\sin x)^4$

14.

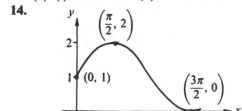

15. (a) (i) $\sin 2x + 2x\cos 2x$

(ii) $3\sec^2\left(3x + \dfrac{\pi}{4}\right)$

(iii) $-2\sin x(1+\cos x)$

(b) (i) $\frac{3}{4}$ (ii) $-\frac{1}{3}$

16. (ii) $\frac{1}{4}(\sqrt{17}-1)$

17. $x\sin^{-1}mx + \dfrac{1}{m}\sqrt{(1-m^2x^2)} + c$

18. (i) $(\frac{1}{6}\pi, \frac{3}{16}\sqrt{3}), (\frac{1}{2}\pi, 0), (\frac{5}{6}\pi, -\frac{3}{16}\sqrt{3})$ (ii)

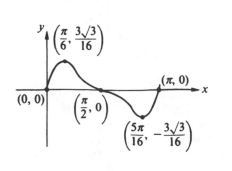

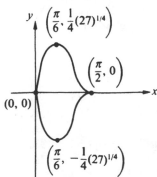

page 381   Exercise 16A

1. (a) $\frac{1}{3}$ (b) $\frac{3}{4}$ (c) $5\frac{1}{2}$ **2.** 26·6°, 45°, 108·4°   **3.** $\frac{3}{5}$, 3   **4.** 36°   **5.** 27°
6. $2y = x + 2$, $y + 2x = 16$

page 383   Exercise 16B

1. (a) $1\frac{2}{3}$ (b) $\frac{8}{13}$ (c) $5\sqrt{2}$ (d) $\sqrt{5}$ (e) $\frac{3}{2}\sqrt{5}$ **2.** A, C, $\frac{4}{13}\sqrt{13}$ **3.** -2, $-7\frac{1}{4}$
4. (a) $3y = 4x + 10$ (b) 5 units (c) $2\frac{1}{3}$ units (d) $5\frac{1}{2}$ sq. units
5. $x + 4y + 1 = 0$, $y = 4x + 1$
6. (a) $y = 2x + 1$, $x + 2y + 2 = 0$ (b) $x + 3y = 1$, $y = 3x + 2$
    (c) $7x + 4y = 18$, $7y = 4x - 1$

page 390   Exercise 16C

1. (b) (c) (f)   **2.** (a) (0, 0), $\sqrt{10}$ (b) (3, 0), 5   (c) (2, -1), $3\sqrt{2}$
    (d) (-1, 1), 2   (e) (3, -1), 4   (f) (-2, 0), $\sqrt{10}$ (g) ($-\frac{1}{2}$, 0), 5
    (h) ($\frac{1}{2}$, $-\frac{1}{2}$), 1   **3.** (2, 3)   **8.** $4\sqrt{2}$ **9.** $\sqrt{10}$
10. (a) (i) $3y = x - 20$ (ii) $y + 3x = 0$ (b) (i) $y + 5x = 17$ (ii) $5y = x + 7$
    (c) (i) $4y + 2x = 7$ (ii) $y = 2x - 2$
11. (a) $x^2 + y^2 - 4x - 4y + 3 = 0$ (b) $x^2 + y^2 + 9x + y = 22$
    (c) $2x^2 + 2y^2 + x - 10y + 2 = 0$
13. (a) external (b) external (c) internal (d) external   **14.** $3x + y = 15$ **15.** $y = 7x - 37$
16. $(y - 3)\sin\theta + (x - 2)\cos\theta = 4$, $y\sqrt{3} + x = 10 + 3\sqrt{3}$
17. $x^2 + y^2 - 4x - 22y + 115 = 0$
18. $x^2 + y^2 - 2x + 2y = 8$ **19.** $x^2 + y^2 - 8x - 4y + 15 = 0$

page 395   Exercise 16D

1. (a) (1, 0) (b) (-2, 0) (c) (5, 0) (d) ($2\frac{1}{4}$, 0)
2. (a) $x = -3$ (b) $x = 3$ (c) $x = -5$ (d) $x = \frac{1}{2}$
3. (a)   (b)   (c)

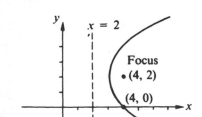

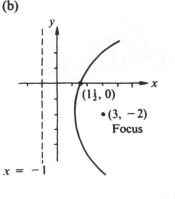

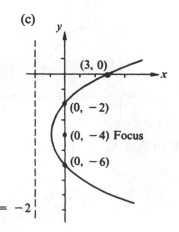

4. (1, 4)   **5.** ($-\frac{1}{4}$, -1)   **6.** (2, 8) (18, 24)   **7.** (1, -2), (9, 6), $y + x + 1 = 0$, $3y = x + 9$
8. $2y = x + 8$, $y + 2x + 1 = 0$ **9.** $2y = x + 12$, $2y + 3x + 4 = 0$
10. $yt = x + at^2$ **11.** $y + tx = 2at + at^3$; 0, ± 2
12. (a) $[a(2 + p^2), 0]$ (b) $[0, ap(2 + p^2)]$ **13.** $4y^2 + 1 = 4(x + y)$
14. $y(p + q) = 2x + 2apq$ **15.** -1, 3   **19.** $y^2 = 2ax + 4a^3$

page 400   Exercise 16E

1. (a) $(\pm 2\sqrt{2}, 0)$, $\frac{8}{9}$ (b) $(\pm 3, 0)$, $\frac{9}{25}$ (c) $(\pm 2\sqrt{3}, 0)$, $\frac{3}{4}$
2. (a) $(\pm 5, 0)$, $\frac{23}{16}$ (b) $(\pm\sqrt{10}, 0)$, 10 (c) $(\pm 2\sqrt{17}, 0)$, 17
3. (a) $2y + \sqrt{3}\,x = 4$ (b) $3y = 2x + 25$ (c) $4y + 15x + 9 = 0$ (d) $y + 2\sqrt{2}\,x = 2$
   (e) $y + x + 6 = 0$ (f) $y + 4x = 16$
4. $bx\cos\theta + ay\sin\theta = ab$, $(0, b\cosec\theta)$
5. (a) $yt_1{}^2 + x = 2ct_1$ (b) $yt_2 = xt_2{}^3 - ct_2{}^4 + c$ ·
6. (a) $yt_1t_2 + x = c(t_1 + t_2)$ (b) $yt^2 + x = 2ct$

page 404   Exercise 16F

1. $A(2, \frac{1}{3}\pi)$, $B(3, \frac{1}{2}\pi)$, $C(1\frac{1}{2}, \frac{2}{3}\pi)$, $D(1, \pi)$, $E(3, \frac{7}{6}\pi)$, $F(2\frac{1}{2}, \frac{3}{2}\pi)$, $G(1, \frac{5}{3}\pi)$, $H(2, \frac{11}{6}\pi)$
2. $P(2\sqrt{3}, 2)$, $Q(0, 5)$, $R(0, -2)$, $S(-5, 5)$, $T(-3, -3\sqrt{3})$, $U(0, -3)$
3. (a) $r = 4\sec\theta$ (b) $r = 4\sin\theta$ (c) $r^2 = \cosec 2\theta$
   (d) $r = 8\cot\theta\cosec\theta$ (e) $r = 2\cos\theta$ (f) $r = 4\cos\theta$
   (g) $r^2 = 8\sec 2\theta$ (h) $r^2 = \dfrac{4}{1 - \sin 2\theta}$
4. (a) $y = 1$ (b) $x^2 + y^2 = 9$ (c) $y = \sqrt{3}\,x$ (d) $x^2 + y^2 = y$
   (e) $x^2 + y^2 = x + y$ (f) $x^2 - y^2 = 1$ (g) $x^4 + x^2y^2 = 9y^2$ (h) $x^2 = 9 + 6y$

5.   6.   7.

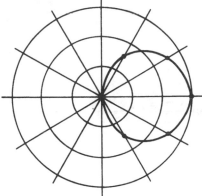 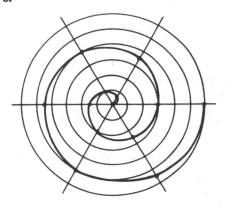 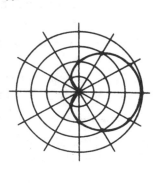

page 405   Exercise 16G

1. $(1, 1\frac{1}{3})$, $(3, 0)$, $(2, \frac{2}{3})$, $78\cdot7°$ 2. $3ty + 2x = t^2(2 + 3t^2)$, $\frac{1}{8}$
3. (i) $(1\frac{1}{2}, 0)$ (ii) $2\frac{1}{2}$ units (iii) $(0,2)$ $(0, -2)$, $(-1, 0)$, $(4, 0)$; $(3, 2)$, $(5\frac{2}{3}, 0)$
4. $(-1, 4\frac{1}{2})$, $\frac{1}{2}\sqrt{65}$ 5. 10 units, 26 units
6. $\frac{1}{4}(6 \pm \sqrt{6})$, $\frac{23}{25}$ 7. $(x - 1)^2 + y^2 = 1$, $(x - 5)^2 + (y + 4)^2 = 25$, $(1, -1)$
8. $x^2 + y^2 = 3ax$, $y^2 = 3ax$ 9. $(4a, 0)$, $2a$, $(2a, 0)$, $(6a, 0)$, $60°$, $\sqrt{3}\,y = \pm x$, $\frac{4}{3}a^2\sqrt{3}$
10. $(\frac{81}{4}a, -9a)$ 11. $[apq, a(p + q)]$
12. (b) $2ay = x^2 + a^2$ 13. (i) $4a$ (ii) $y^2 = 2ax - 8a^2$
14. $4x^2 + 4y^2 - 8x + 3 = 0$ 15. $2m + c = 3$
16. $\dfrac{ab}{\sin 2\theta}$ 17. $a^2 - b^2 = 2ax - 2by$ (ii) $(0, -b)$, $\left(\dfrac{3}{5}\sqrt{3}\,b, \dfrac{4}{5}b\right)$
18. $x + yt^2 = 6t$, $(-1\frac{1}{2}, -6)$, $(3\frac{3}{4}, 2\frac{2}{3})$, $\frac{21}{10}\sqrt{89}$
21. $r = 4\cos\theta$, $(4, 0)$, $(2\sqrt{2}, \frac{1}{4}\pi)$, $\theta = -0\cdot142$ rads, $\theta = \dfrac{\pi}{4}$ rads

page 412   Exercise 17A

1. (a) 11   (b) 7   (c) 3   (d) 12   (e) 16   (f) 81°   (g) 40°

2. (a) $\sqrt{11}$ (b) $\sqrt{6}$ (c) 76°   **3.** $\frac{3}{13}i + \frac{4}{13}j + \frac{12}{13}k$ **4.** $\begin{pmatrix} \frac{4}{9} \\ \frac{4}{9} \\ -\frac{7}{9} \end{pmatrix}$

5. $i + j - 2k$ is one example   **6.** $\frac{\sqrt{3}}{3}(i + j + k)$ is one example

7. **c** and **d**   **8.** $14 : -2 : 5, \frac{14}{13}, -\frac{2}{13}, \frac{5}{13}$ **9.** $1 : 2 : -2$

10. (a) $2i - 3j + 6k$ or any multiple thereof   (b) $\frac{2}{7}, -\frac{3}{7}, \frac{6}{7}$ (c) $\frac{1}{7}(2i - 3j + 6k)$

15. 36°, 68°, 76°   **17.** $4a + b - 2c$ **18.** $3a - b + 2c$

page 414   Exercise 17B

1. (a) $3i + 8tj$ (b) $3i + 2j - 10tk$ (c) $2j$ (d) $6i + 9t^2j - 6k$

2. (a) $\cos\theta i - \sin\theta j$ (b) $-2\sin\theta i + 2j$ (c) $\cos\theta i + \sin\theta j + 2\theta k$
   (d) $3\cos 3\theta j + 3\cos^2\theta \sin\theta k$

3. $16i - 4j + k, 24i - 2j, 24i$ **4.** $10i + 28j + 6k$ **5.** $2\pi i$ m/s, 2π m/s

6. $2i + (2t^2 + 1)j - 4k, 2ti + (\frac{2}{3}t^3 + t)j + (3 - 4t)k$

page 420   Exercise 17C

1. $r = i + j + k + \lambda(2i + 3j - k)$ **2.** $r = \begin{pmatrix} -1 \\ 2 \\ 1 \end{pmatrix} + \lambda\begin{pmatrix} 1 \\ 2 \\ 3 \end{pmatrix}$

3. $2i + 5j + 3k$ or multiples thereof   **5.** B and C   **6.** F   **7.** 6, 8

8. (a) $3i + 3j - 3k$ (b) $r = 2i - j + k + \lambda(3i + 3j - 3k)$

9. (a) $\dfrac{x - 2}{2} = \dfrac{y - 3}{3} = \dfrac{z + 1}{1}$ (b) $\dfrac{x - 3}{3} = \dfrac{y + 1}{2} = \dfrac{z - 2}{-4}$ (c) $\dfrac{x - 2}{2} = \dfrac{y - 1}{-1} = \dfrac{z - 1}{-1}$

10. $r = 2i + 5j + 4k + \lambda(3i - 2j - k)$

11. (a) $r = 2i + 2j - k + \lambda(3i + 2j + 4k)$ (b) $r = 3i - 2j + 3k + \lambda(i + 4j - k)$

12. $-4i + 2j + 3k$ **13.** $\begin{pmatrix} 9 \\ 3 \\ 0 \end{pmatrix}$

14. (a) $(5, 0, 1)$ (b) lines do not intersect   (c) $(4, 5, 9)$ (d) $(12, -3, 3)$

15. (a) parallel (b) non parallel coplanar   (c) skew   (d) non parallel coplanar

16. 79°   **17.** 69°   **18.** (a) $i + 4j + 7k$ (b) $-4i + 4j - 3k$ (c) $5\sqrt{5}$ units

20. (a) $\sqrt{5}$ units   (b) $3\sqrt{2}$ units   (c) $\sqrt{11}$ units   (d) $4\sqrt{6}$ units

21. (a) $\frac{1}{3}\sqrt{21}$ units   (b) $\frac{1}{2}\sqrt{14}$ units

22. $\frac{2}{5}\sqrt{35}$ units   **23.** (a) $\sqrt{3}$ units   (b) $\frac{3}{7}\sqrt{21}$ units

page 429   Exercise 17D

4. $r . (i - 2j - k) = 7$ **5.** $r . (2i + 3j - k) = 5$ **6.** $r = 2i + 3j - k + \lambda(2i - j + 3k)$

7. (a) and (c)   **8.** (a) and (b)

9. (a) $r . (4i + 2j + 3k) = 4$ (b) $r . (2i - 3j + 4k) = 5$ **10.** $x + 2y - 3z = 0$

11. $3y + z = 10$ **12.** $3x + 2y + z = 6$ **13.** $\frac{2}{3}$ units   **14.** 3 units

15. 2 units   **16.** $\begin{pmatrix} 12 \\ 5 \\ 7 \end{pmatrix}$ **17.** $5i - 5k$ **20.** $(1, -5, 1)$

21. (2, 14, 6) **22.** $\begin{pmatrix} 4 \\ -1 \\ -5 \end{pmatrix}$, 7 units **23.** (a) 79° (b) 11° **24.** (a) 22° (b) 72°

25. (a) parallel, not in plane (b) intersects, $7\mathbf{i} - 4\mathbf{j} + 6\mathbf{k}$ (c) parallel, in plane

26. $-\mathbf{i} + 6\mathbf{j} - 8\mathbf{k}$

27. Various forms including $\mathbf{r} = \mathbf{i} + 2\mathbf{j} + 2\mathbf{k} + \lambda(\mathbf{i} - 5\mathbf{j} - 7\mathbf{k}) + \mu(-3\mathbf{i} + 6\mathbf{j} + 3\mathbf{k})$,
$3x + 2y - z = 5$

28. 66° **29.** (a) $\mathbf{r} \cdot (3\mathbf{i} - 5\mathbf{j} + 4\mathbf{k}) = 10$ (b) $(\sqrt{2} - \tfrac{2}{3})$ **30.** (a) 3 units (b) 3 units

31. (b) $\mathbf{r} \cdot (6\mathbf{i} - 2\mathbf{j} + \mathbf{k}) = 11$ (c) $\tfrac{14}{41}\sqrt{41}$ **32.** $\tfrac{7}{26}\sqrt{26}$ units

33. $x = \dfrac{2y + 5}{7} = \dfrac{2z - 13}{-17}$ **34.** $\begin{pmatrix} 5 \\ 3 \\ -1 \end{pmatrix}$, $5x - y + 3z = 19$

35. (a) $\mathbf{r} \cdot \begin{pmatrix} 2 \\ -1 \\ 3 \end{pmatrix} = 3$ (b) $\mathbf{r} = \begin{pmatrix} 5 \\ -4 \\ 1 \end{pmatrix} + \lambda\begin{pmatrix} 2 \\ -1 \\ 3 \end{pmatrix}$ (c) $\begin{pmatrix} 3 \\ -3 \\ -2 \end{pmatrix}$

page 437 **Exercise 17E**

1. (a) (8, 3, 4) (b) (3, −5, 3) (c) (5, 4, 18)

2. (a) $\begin{pmatrix} -1 & 0 & 0 \\ 0 & 1 & 0 \\ 0 & 0 & 1 \end{pmatrix}$ (b) $\begin{pmatrix} 1 & 0 & 0 \\ 0 & -1 & 0 \\ 0 & 0 & 1 \end{pmatrix}$ (c) $\begin{pmatrix} 1 & 0 & 0 \\ 0 & -1 & 0 \\ 0 & 0 & -1 \end{pmatrix}$ (d) $\begin{pmatrix} 3 & 0 & 0 \\ 0 & 3 & 0 \\ 0 & 0 & 3 \end{pmatrix}$

(e) $\begin{pmatrix} 2 & 0 & 0 \\ 0 & 1 & 0 \\ 0 & 0 & 1 \end{pmatrix}$ (f) $\begin{pmatrix} 1 & 0 & 0 \\ 0 & 0 & -1 \\ 0 & 1 & 0 \end{pmatrix}$ (g) $\begin{pmatrix} 0 & 0 & 0 \\ 0 & 1 & 0 \\ 0 & 0 & 1 \end{pmatrix}$ (h) $\begin{pmatrix} 0 & 1 & 0 \\ 1 & 0 & 0 \\ 0 & 0 & 1 \end{pmatrix}$

(i) $\begin{pmatrix} 0 & -1 & 0 \\ -1 & 0 & 0 \\ 0 & 0 & 1 \end{pmatrix}$ (j) $\begin{pmatrix} 1 & 0 & 0 \\ 0 & 1 & 0 \\ 2 & 0 & 1 \end{pmatrix}$ (k) $\dfrac{1}{2}\begin{pmatrix} 1 & 1 & 0 \\ 1 & 1 & 0 \\ 0 & 0 & 2 \end{pmatrix}$

4. Reflection in the plane $y = z$; all points mapped onto themselves i.e. the identity transformation; reflection in the plane $y = z$.

5. 90° rotation about y-axis with $(1, 0, 0) \to (0, 0, -1)$; 180° rotation about y-axis; 90° rotation about y-axis with $(1, 0, 0) \to (0, 0, 1)$; all points mapped onto themselves.

6. (a) $\mathbf{r} = \lambda\begin{pmatrix} 1 \\ 3 \\ -2 \end{pmatrix}$, $x = \dfrac{y}{3} = -\dfrac{z}{2}$ (b) $\mathbf{r} = \lambda\begin{pmatrix} 1 \\ 2 \\ -3 \end{pmatrix}$, $x = \dfrac{y}{2} = -\dfrac{z}{3}$

(c) $\mathbf{r} = \lambda\begin{pmatrix} 1 \\ -2 \\ 1 \end{pmatrix}$, $x = -\dfrac{y}{2} = z$

7. (a) $x - y + z = 0$ (b) $x + y + z = 0$ **8.** (a) 1 (b) 2 (c) 0 (d) −5 (e) 4 (f) 24

9. (a) $\begin{pmatrix} -2 & -1 & 1 \\ -1 & -1 & 1 \\ 3 & 2 & -1 \end{pmatrix}$ (b) $\dfrac{1}{2}\begin{pmatrix} 2 & -2 & 4 \\ -4 & 6 & -8 \\ 2 & -3 & 5 \end{pmatrix}$ (c) singular matrix, no inverse

(d) $\dfrac{1}{13}\begin{pmatrix} 6 & -2 & -5 \\ -5 & 6 & 2 \\ 3 & -1 & 4 \end{pmatrix}$ (e) $\dfrac{1}{8}\begin{pmatrix} 7 & -5 & 1 \\ -5 & 7 & -3 \\ -1 & 3 & 1 \end{pmatrix}$ (f) $\dfrac{1}{3}\begin{pmatrix} 1 & -3 & 2 \\ 0 & 3 & -3 \\ 1 & -9 & 11 \end{pmatrix}$

10. (a) $\begin{pmatrix} 8 & 4 & 1 \\ 3 & 4 & -2 \\ 7 & 4 & 1 \end{pmatrix}$ (b) $\frac{1}{3}\begin{pmatrix} 3 & 0 & -3 \\ -4 & 1 & 5 \\ -5 & -1 & 7 \end{pmatrix}$ (c) 30 cubic units,   (d) 120 cubic units

11. $\begin{pmatrix} -5 & 3 & -2 \\ 6 & -4 & 3 \\ 2 & -1 & 1 \end{pmatrix}$, 3, -2, 2   **12.** $\frac{1}{10}\begin{pmatrix} 4 & 5 & -1 \\ -8 & -5 & 7 \\ -2 & -5 & 3 \end{pmatrix}$, -1, 4, 2

13. (a) $\begin{pmatrix} -3 & 1 & -4 \\ -6 & 1 & -8 \\ 5 & -1 & 7 \end{pmatrix}$, (2, 1, -3)

page 445   Exercise 17F

1. (a) $\begin{pmatrix} -3 & 1 & 1 \\ 2 & -1 & 0 \\ 7 & -2 & -2 \end{pmatrix}$ (b) $\begin{pmatrix} 3 & 1 & 1 \\ -11 & -3 & -4 \\ 13 & 4 & 5 \end{pmatrix}$ (c) $\frac{1}{7}\begin{pmatrix} 7 & -14 & 7 \\ -3 & 8 & -2 \\ 1 & -5 & 3 \end{pmatrix}$

(d) singular matrix, no inverse   (e) $\frac{1}{3}\begin{pmatrix} 11 & -2 & -5 \\ 6 & 0 & -3 \\ -7 & 1 & 4 \end{pmatrix}$ (f) $\frac{1}{6}\begin{pmatrix} 10 & -2 & -3 \\ 6 & 0 & -3 \\ -4 & 2 & 3 \end{pmatrix}$

2. (a) 5, -3, 1   (b) -1, 3, 2   (c) 1, -3, 2   **3.** (a) -3, 1, 4   (b) 4, 5, 1   (c) 1, -1, 4

4. $\begin{pmatrix} 3 & 3 & -1 \\ 3 & 2 & 1 \\ 2 & 1 & 0 \end{pmatrix}$, $\begin{pmatrix} -3 & 1 & 4 \\ -2 & -1 & 2 \\ -2 & 0 & 4 \end{pmatrix}$, $\begin{pmatrix} 1 & 0 & -1 \\ 2 & 1 & -1 \\ 1 & 1 & 1 \end{pmatrix}$ **5.** 4

6. (a) Equations inconsistent, 2 of the planes are parallel.
  (b) Planes have a common line, $x = -\lambda$, $y = -1 - 2\lambda$, $z = \lambda$.
  (c) $x = 1$, $y = 1$, $z = -1$. The three planes have a common point (1, 1, -1).
7. (a) Equations inconsistent. 2 coincident planes and a parallel plane.
  (b) 2 coincident planes and one other, meeting in a line. $x = 2 - 2\lambda$, $y = 1$, $z = \lambda$.
8. (a) Equations inconsistent. Each plane parallel to line of intersection of the other two.
  (b) The three planes have a common line, $x = \frac{3}{4}(1 - \lambda)$, $y = -\frac{1}{4}(7 + \lambda)$, $z = \lambda$.
9. (a) 4; $x = \lambda$, $y = \mu$, $z = \frac{1}{3}(2 - \lambda - 2\mu)$ (b) -8; $x = \frac{1}{3}(6 - \lambda)$, $y = \frac{1}{3}(2\lambda - 7)$, $z = \lambda$
  (c) -5; $x = -\frac{7}{3}\lambda$, $y = -\frac{1}{3}(3 + 2\lambda)$, $z = \lambda$ (d) 5; $x = \lambda$, $y = \lambda - 1$, $z = 1$
10. (a) -9 (b) 7, the line $\dfrac{x - 3}{3} = \dfrac{y + 1}{-3} = z$ is common to all three planes.

11. 2, $\mathbf{r} = \begin{pmatrix} 3 \\ -4 \\ 0 \end{pmatrix} + \lambda\begin{pmatrix} -2 \\ 3 \\ 1 \end{pmatrix}$ **12.** $\frac{1}{4}\begin{pmatrix} 1 & -8 & 6 & -1 \\ 2 & 4 & -4 & 2 \\ -1 & 4 & -2 & 1 \\ -1 & 0 & 2 & -3 \end{pmatrix}$, $x = 1$, $y = -3$, $z = -2$, $t = 4$.

page 447   Exercise 17G

1. $-\mathbf{i} - 5\mathbf{j} + 2\mathbf{k}$, 42·95°   **2.** 90°, $\frac{7}{2}\sqrt{26}$ sq. units
3. (i) $4\mathbf{i} + 18\mathbf{j} - 12\mathbf{k}$, $4\mathbf{i} + 6\mathbf{j}$ (ii) $4\mathbf{i} + 6\mathbf{j} + \lambda(-\mathbf{j} + \mathbf{k})$ (iii) $4\mathbf{i} + 2\mathbf{j} + 4\mathbf{k}$ (iv) 88°

4. 54·7°, $\begin{pmatrix} 1 \\ 3 \\ -2 \end{pmatrix}$, $\mathbf{r} . \begin{pmatrix} 4 \\ 3 \\ -3 \end{pmatrix} = 19$

5. (i) $\begin{pmatrix} 1 \\ 2 \\ 3 \end{pmatrix} + t\begin{pmatrix} 3 \\ 4 \\ -5 \end{pmatrix}$, $(7, 10, -7)$ (ii) $\frac{2}{5}\sqrt{6}$ (iii) $x - 2y - z = -6$

6. (a) $3\mathbf{j} + 4\mathbf{k}$ (b) $\mathbf{r} \cdot (3\mathbf{j} + 4\mathbf{k}) = 0$ (c) $\frac{1}{26}\sqrt{26}$

7. (i) $2x - 2y - z = -9$ (ii) $90°$ (iii) $(5, 0, 1), (-3, 8, 5)$

8. $3\mathbf{i} - \mathbf{j} - 2\mathbf{k}, 2\mathbf{j} + \mathbf{k}, \mathbf{i} - \mathbf{j}, 31\cdot5°$, not coplanar.

9. (a) $\frac{1}{2}\sqrt{14}$ (b) $\frac{13}{14}$ (c) $x + y - 5z = -1$ (d) $x = y - 1 = \frac{z - 2}{-5}$ (e) $\left(\frac{8}{27}, \frac{35}{27}, \frac{14}{27}\right)$

10. $\begin{pmatrix} 24 & -23 & -21 \\ -8 & 8 & 7 \\ -1 & 1 & 1 \end{pmatrix}$ (i) $\begin{pmatrix} 0 \\ 0 \\ 0 \end{pmatrix}$ (ii) $\begin{pmatrix} -47 \\ 16 \\ 2 \end{pmatrix}$ (iii) $\begin{pmatrix} -85 \\ 29 \\ 4 \end{pmatrix}$

11. $2, 1, 1, 3, 0, 0; \frac{1}{3}\begin{pmatrix} 0 & 0 & 3 \\ 0 & 1 & -1 \\ 3 & -1 & -5 \end{pmatrix}$

12. $(x + a + y)(x - a)(y - a); 0, \frac{1}{2}\pi, \frac{3}{2}\pi, \pi.$

13. (i) $x = 1, y = 6\frac{2}{3}, z = 7\frac{1}{3}$ (ii) $x = \frac{1}{13}(49 - 5\lambda), y = \frac{1}{13}(14\lambda - 15), z = \lambda.$

14. $x = 3 - 4\lambda, y = \lambda - 2, z = \lambda; x = 1, y = -1\frac{1}{2}, z = \frac{1}{2}$ (a) no solutions
(b) $x = \lambda, y = \mu, z = \frac{1}{3}(8 - 2\lambda + \mu).$

15. $L: x = -2\lambda, y = \frac{1}{2}\lambda, z = \lambda.$ The 3 equations all represent the same plane. $\therefore$ Any point in the plane satisfies the equations.

16. $-1, -2, 3, \frac{1}{2}\sqrt{2}$ units   **17.** $x = 1 - 3\lambda, y = 2 + 2\lambda, z = \lambda$

page 454   Exercise 18A

1. $\frac{1}{x - 2} + \frac{3}{x - 3}$

2. $\frac{4}{x - 7} - \frac{2}{4 - x}$

3. $\frac{3}{x} - \frac{5}{1 - x}$

4. $\frac{4}{x - 4} - \frac{3}{x + 3}$

5. $\frac{4}{x + 4} + \frac{2}{x}$

6. $\frac{1}{x + 2} - \frac{2}{x + 3} + \frac{3}{x - 3}$

7. $\frac{5}{x - 2} + \frac{2x + 3}{x^2 + 4}$

8. $\frac{3}{x - 2} + \frac{2 - 3x}{x^2 + 1}$

9. $\frac{3}{x - 3} + \frac{2}{x^2 + 2}$

10. $\frac{5}{x} - \frac{4}{3 + 2x^2}$

11. $\frac{1}{x - 2} - \frac{2x + 1}{3 + 2x^2}$

12. $\frac{3}{x + 3} - \frac{2}{x^2 + x + 5}$

13. $\frac{1}{x + 5} - \frac{2}{(x + 5)^2}$

14. $\frac{2}{x - 3} + \frac{7}{(x - 3)^2}$

15. $\frac{2}{x} + \frac{1}{x - 1} + \frac{5}{(x - 1)^2}$

16. $\frac{2}{x} + \frac{3}{x^2} - \frac{4}{x - 1}$

17. $\frac{1}{3(x + 1)} - \frac{1}{3(x - 2)} + \frac{1}{(x - 2)^2}$

18. $\frac{4}{x - 4} + \frac{7}{(x - 4)^2} - \frac{3}{2x - 3}$

19. $1 + \frac{5}{x + 2} - \frac{3}{x - 1}$

20. $3 - \frac{1}{x} + \frac{15}{x - 3}$

21. $3 - \frac{5}{x + 3} + \frac{2x - 1}{x^2 + 4}$

22. $x - 2 + \frac{4}{x + 1} + \frac{2}{x + 4}$

23. $2 - \frac{3}{2(x + 1)} + \frac{7}{2(x - 1)}$

24. $x - 2 + \frac{3}{x} + \frac{3}{x^2} - \frac{5}{x - 1}$

25 $\frac{2}{2x - 3} - \frac{1}{x + 2}$

26. $\frac{1}{2x - 1} - \frac{2}{x + 3} + \frac{5}{(x + 3)^2}$

27. $\frac{5}{2(x + 1)} - \frac{1}{2(x - 1)}$

28. $\dfrac{2}{x-1} - \dfrac{5}{2(2x-1)} + \dfrac{1}{2(2x-1)^2}$

29. $1 + \dfrac{1}{x-3} - \dfrac{1}{2x+5}$

30. $2 + \dfrac{x-5}{x^2+4} + \dfrac{5}{x-3}$

31. $\dfrac{1}{x-3} - \dfrac{2}{x+7} + \dfrac{2}{2x-3}$

32. $\dfrac{2}{1+2x} + \dfrac{3-x}{x^2-x+2}$

33. $2 - \dfrac{1}{2(x+3)} - \dfrac{3}{2(x-1)} + \dfrac{12}{(x-1)^2}$

34. $\dfrac{2}{1+x} + \dfrac{3}{(1+x)^2} - \dfrac{5}{(1+x)^3} + \dfrac{1}{x-2}$

35. $\dfrac{6}{x} - \dfrac{4}{x-2} + \dfrac{1}{x+2}$

36. $\dfrac{3}{x+3} + \dfrac{4}{(x+3)^2} + \dfrac{1-2x}{x^2+4}$

page 456　*Exercise 18B*

1. $\dfrac{1}{1-x} + \dfrac{1}{1+2x}$, $2 - x + 5x^2 - 7x^3 + \ldots + [(-2)^r + 1]\,x^r + \ldots, \; |x| < \dfrac{1}{2}$

2. $\dfrac{3}{1+x} - \dfrac{2}{1+3x}$, $1 + 3x - 15x^2 + 51x^3 - \ldots + (-1)^r 3[1 - 2\times 3^{r-1}]\,x^r + \ldots, \; |x| < \dfrac{1}{3}$

3. $\dfrac{1}{1-3x} - \dfrac{1}{1+5x}$, $8x - 16x^2 + 152x^3 - \ldots + [3^r - (-5)^r]\,x^r + \ldots, \; |x| < \dfrac{1}{5}$

4. $\dfrac{1}{1-x} - \dfrac{1}{3-x}$, $\dfrac{2}{3} + \dfrac{8x}{9} + \dfrac{26}{27}x^2 + \dfrac{80}{81}x^3 + \ldots + \left(1 - \dfrac{1}{3^{r+1}}\right)x^r + \ldots, \; |x| < 1$

5. $\dfrac{3}{2(3+2x)} + \dfrac{1}{2(2x-1)}$, $-\dfrac{4}{3}x - \dfrac{16}{9}x^2 - \dfrac{112}{27}x^3 - \ldots + 2^{r-1}\left(\left(\dfrac{-1}{3}\right)^r - 1\right)x^r + \ldots, \; |x| < \dfrac{1}{2}$

6. $\dfrac{3}{3+x} - \dfrac{1}{3+4x}$, $\dfrac{2}{3} + \dfrac{x}{9} - \dfrac{13}{27}x^2 + \dfrac{61}{81}x^3 - \ldots + \left(-\dfrac{1}{3}\right)^r\left(1 - \dfrac{4^r}{3}\right)x^r + \ldots, \; |x| < \dfrac{3}{4}$

7. $\dfrac{2}{(1+x)^2} - \dfrac{1}{1+2x}$, $1 - 2x + 2x^2 - 0x^3 + \ldots + (-1)^r\,2(r+1 - 2^{r-1})\,x^r + \ldots, \; |x| < \dfrac{1}{2}$

8. $\dfrac{1}{1-4x} + \dfrac{1}{1-x} + \dfrac{2}{2+3x}$, $3 + \dfrac{7}{2}x + \dfrac{77}{4}x^2 + \dfrac{493}{8}x^3 + \ldots + \left[4^r + 1 + \left(\dfrac{-3}{2}\right)^r\right]x^r + \ldots, \; |x| < \dfrac{1}{4}$

9. $\dfrac{4}{2+x} - \dfrac{4}{(2+x)^2} - \dfrac{3}{1+x}$, $-2 + 3x - \dfrac{13}{4}x^2 + \dfrac{13}{4}x^3 - \ldots + \left(-\dfrac{1}{2}\right)^r(1 - r - 3\times 2^r)x^r + \ldots, \; |x| < 1$

10. $\dfrac{x-2}{1+x^2} + \dfrac{5}{1-5x}$, $3 + 26x + 127x^2 + 624x^3 + \ldots, \; |x| < \dfrac{1}{5}$

11. $\dfrac{3x+5}{1+x+x^2} + \dfrac{8}{2-3x}$, $9 + 4x + 6x^2 + \dfrac{37}{2}x^3 + \ldots$

page 458　*Exercise 18C*

1. $n(n+2)$　　**2.** (b) $\dfrac{1}{3}n(n+1)(n+2)$　　**3.** (a) $\dfrac{1}{x} - \dfrac{1}{x+1}$ (c) $\dfrac{15}{16}$

4. (a) $\dfrac{1}{3x-1} - \dfrac{1}{3x+2}$ (c) $\dfrac{5}{32}; \dfrac{1}{6}$　　**5.** (a) $\dfrac{1}{x} - \dfrac{1}{x+2}$ (b) $\dfrac{3}{4}$ (c) $\dfrac{36}{55}$

6. (c) 1770　　**7.** (a) $1 - \dfrac{1}{4n+1}$　　(b) $\dfrac{1}{3} - \dfrac{n+1}{(2n+1)(2n+3)}$

　　(c) $\dfrac{1}{4} - \dfrac{1}{2(n+1)(n+2)}$ (d) $\dfrac{7}{6} - \dfrac{3n+7}{(n+1)(n+2)(n+3)}$

8. (b) Yes, $2\tfrac{1}{2}$　　**9.** (b) Yes, $\tfrac{5}{24}$

page 463 Exercise 18D

1. (a) 3 (b) -4 (c) $3 + 4i$ (d) 25 (e) $-7 + 24i$
2. (a) $5 + 7i$ (b) $-13 + 5i$ (c) -4 (d) 1 (e) -1 (f) 13
   (g) $\frac{1}{2}$ (h) i
3. (a) $7 + i$ (b) $32 - 7i$ (c) $1 - 7i$ (d) $11 - 13i$ (e) $11 - 2i$ (f) 5
   (g) 78 (h) $7 + i$
4. (a) $6 - 2i$ (b) $2 - 2i$ (c) $-1 + i$ (d) $\frac{1}{5} + \frac{2}{5}i$ (e) $2 + i$ (f) $\frac{4}{5} + \frac{3}{5}i$
   (g) $-\frac{1}{2} + \frac{5}{2}i$ (h) $\frac{1}{5} - \frac{7}{5}i$ (i) $\frac{5}{13} + \frac{12}{13}i$ (j) $\frac{6}{13} + \frac{17}{13}i$ 5. i

6. (a) $x = \pm 5i$ (b) $x = \pm 4i$ (c) $x = \pm\frac{3i}{2}$ (d) $x = -1 \pm 2i$

   (e) $x = 2 \pm i$ (f) $x = \dfrac{-1 \pm \sqrt{7}i}{4}$

7. (a) $\pm(3 + 2i)$ (b) $\pm(4 + i)$ (c) $\pm(4 - 3i)$
8. (a) $x^2 + 9 = 0$ (b) $x^2 - 2x + 5 = 0$ (c) $x^2 - 4x + 5 = 0$
   (d) $x^2 - 4x + 13 = 0$ (e) $x^2 - 6x + 25 = 0$ (f) $x^2 - 6x + 34 = 0$
10. (a) $x = 1, 3 \pm 2i$ (b) $x = 2, -\frac{1}{2} \pm \frac{1}{2}i$ (c) $x = -1, -1 \pm \sqrt{2}i$

    (d) $x = 1, 2, 1 \pm \frac{1}{2}i$ (e) $x = 1, -1, \dfrac{-4 \pm 3i}{5}$

11. $5, 6, 2; x = 2 \pm 2i, -\frac{2}{3} \pm \frac{1}{3}i$ 12. $a = 1, b = -8, c = 27, d = -38, e = 26$

13. $a = b = c = 1, x = 1, \dfrac{-1 \pm \sqrt{3}i}{2}$

page 467 Exercise 18E

1. $3 + 2i, -1 + 3i, -3 - 4i, 4 - i$ 2.

3. $5\left(\cos \dfrac{\pi}{6} + i \sin \dfrac{\pi}{6}\right), 4\left(\cos \dfrac{3\pi}{4} + i \sin \dfrac{3\pi}{4}\right)$

   $3\left[\cos\left(-\dfrac{\pi}{3}\right) + i \sin\left(-\dfrac{\pi}{3}\right)\right], 3\left[\cos\left(-\dfrac{2\pi}{3}\right) + i \sin\left(-\dfrac{2\pi}{3}\right)\right]$

4. (a) $\sqrt{2}, \dfrac{\pi}{4}$ (b) $2\sqrt{2}, \dfrac{3\pi}{4}$

   (c) $2, \pi$ (d) $3, -\dfrac{\pi}{2}$ (e) $2\sqrt{2}, -\dfrac{3\pi}{4}$

5. (a) $5, \dfrac{\pi}{2}$ (b) $7, 0$ (c) $2, -\dfrac{\pi}{2}$ (d) $3, \pi$

   (e) $2, \dfrac{\pi}{3}$ (f) $10, -\dfrac{\pi}{6}$ (g) $5, -0.93$ (h) $13, 1.97$

6. $z_1 = 5i, z_2 = 4 - 4i, z_3 = -4, z_4 = -6 + 6\sqrt{3}i, z_5 = 2 - 2\sqrt{3}i, z_6 = -6 + 6i$

7. (a) $1, -\dfrac{\pi}{2}$ (b) $\sqrt{2}, -\dfrac{3\pi}{4}$ (c) $\dfrac{\sqrt{10}}{5}, 1.25$ (d) $5\sqrt{2}, 1.43$

8. (b) (i) 6 (ii) $\dfrac{5\pi}{12}$ (iii) 9 (iv) 4 (v) $\dfrac{\pi}{3}$ (vi) $\dfrac{\pi}{2}$

9. (b) (i) $\dfrac{1}{3}$ (ii) $-\dfrac{7\pi}{12}$ (iii) 3 (iv) $\dfrac{7\pi}{12}$ 10. $1 + \sqrt{3}i, -4\sqrt{2} + 4\sqrt{2}i$

12. 64 13. $-64i$ 14. $1, \dfrac{1}{2}(-1 \pm i\sqrt{3})$ 15. $\dfrac{\tan \theta(3 - \tan^2 \theta)}{1 - 3 \tan^2 \theta}$

17. $\dfrac{6 \tan \theta - 20 \tan^3 \theta + 6 \tan^5 \theta}{1 - 15 \tan^2 \theta + 15 \tan^4 \theta - \tan^6 \theta}$

page 472   Exercise 18F

1. Circle centre (0, 2) radius 4; $x^2 + (y - 2)^2 = 4^2$
2. Circle centre (1, 3) radius 5; $(x - 1)^2 + (y - 3)^2 = 5^2$
3. Circle centre (−2, 3) radius 4; $(x + 2)^2 + (y - 3)^2 = 4^2$
4. Perpendicular bisector of line joining (0, 0) to (0, 2); $y = 1$
5. Perpendicular bisector of line joining (0, 1) to (1, 0); $y = x$
6. Perpendicular bisector of line joining (4, −1) to (1, 2); $y = x - 2$
7. $y = 0$, i.e. the x-axis   **8.** $x = 0$, i.e. the y-axis

9.
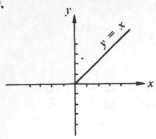

10.
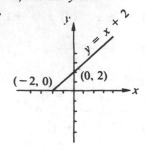

11.
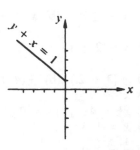

12. $2\sqrt{2}$ **13.** $-1 - i$ **14.** (a) $x^2 + y^2 = 17$
(b) $x = 1$; (1, 4), (1, −4)

15.
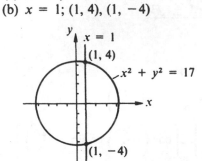

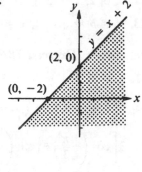

16.
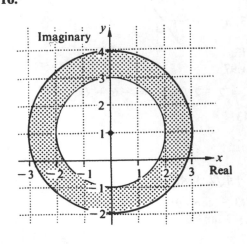

17.
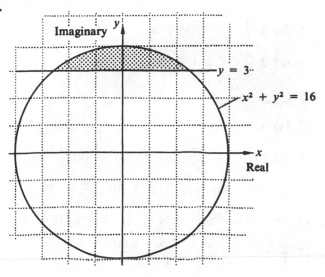

page 473 **Exercise 18G**

1. $-\dfrac{2}{5}, \dfrac{3}{5}$ **2.** $\dfrac{3}{1-3x} - \dfrac{2}{1-2x} - \dfrac{1}{(1-2x)^2}$, $x + 7x^2 + 33x^3$, $|x| < \dfrac{1}{3}$

3. $1 - x + 3x^2 - 5x^3 + 11x^4$ (a) $1\cdot0212418$ (b) $0\cdot9902951$

4. $\dfrac{1}{4(1-x)} + \dfrac{-7}{4(3+x)}$, $-\dfrac{1}{3}, \dfrac{4}{9}, \dfrac{5}{27}$ **5.** $\dfrac{1}{1-2x} + \dfrac{2}{2+x}$, $2 + \dfrac{3}{2}x + \dfrac{17}{4}x^2$

6. $-\dfrac{1}{x+1} + \dfrac{1}{(x+1)^2} + \dfrac{2}{x-1}$, $\dfrac{1}{x} + \dfrac{4}{x^2} - \dfrac{1}{x^3} + \dfrac{6}{x^4} - \dfrac{3}{x^5}$, $(2-n)$

7. $\dfrac{1}{2(r-1)} - \dfrac{1}{2(r+1)}$, $\dfrac{3}{4} - \dfrac{1}{2n} - \dfrac{1}{2(n+1)}$ **8.** $\dfrac{1}{4(2r-1)} - \dfrac{1}{4(2r+3)}$

9. $\dfrac{r+2}{(r+1)!}$, $1 - \dfrac{1}{(2n+1)!}$, $\dfrac{1}{(2n+1)!} - 1$ **10.** (i) $18 - i$, $\dfrac{6}{25} + i\dfrac{17}{25}$

11. (a) $2 - 11i$ (b) $-2 - 3i$ **12.** $-4, 13$

13. $2 - i, -4$ **14.** $x^2 - 2x + 2$, $1 \pm i$, $-1 \pm 2i$

15. $\text{Re}(z) = -3$, $\text{Im}(z) = -1$, $\sqrt{10}$, $-2\cdot82$ rads

16. $1 + 6ci - 15c^2 - 20c^3i + 15c^4 + 6c^5i - c^6$; $0, \pm\frac{1}{3}\sqrt{3}, \pm\sqrt{3}$

17. (a) $2 + 3i$, $5 - 4i$ (b) $a = 2, b = 3$; $a = -2, b = -3$; $\pm(2+3i)$

  (c) (i) circle centre $(2, 0)$ radius 3 (ii) line through $(2\frac{1}{2}, 0)$ perpendicular to real axis

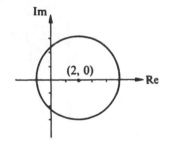

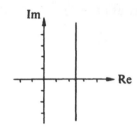

18.

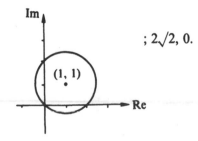

; $2\sqrt{2}, 0$. **20.** $\pm(1+i)$, $\pm(1-i)$, $(x^2 - 2x + 2)(x^2 + 2x + 2)$

21. $16\cos^5\theta - 20\cos^3\theta + 5\cos\theta$

22. $0, \dfrac{\pi}{4}, \dfrac{3\pi}{4}, \pi, \dfrac{5\pi}{4}, \dfrac{7\pi}{4}, 2\pi$

page 480 **Exercise 19A**

1. e^x **2.** $3e^{3x}$ **3.** $2e^{2x}$ **4.** $2xe^{x^2}$ **5.** $2xe^{(x^2+1)}$ **6.** $-e^{-x}$ **7.** $8e^{2x}$ **8.** $30x^2e^{2x^3}$

9. $-6e^{-2x}$ **10.** $5e^{5x+3} - 4x^{-3}$ **11.** $4xe^{x^2} + 3e^x - 4e^{-x}$ **12.** $xe^x(2 + x)$

13. $3x^2e^{2x}(3 + 2x)$ **14.** $3xe^{-x}(2 - x)$ **15.** $\dfrac{xe^x - 2e^x - 8}{x^3}$ **16.** $e^x + c$ **17.** $5e^x + c$

18. $e^{2x} + c$ **19.** $2e^{3x} + c$ **20.** $e^{x^2} + c$ **21.** $2e^{(x^2+1)} + c$ **22.** $2e^{-2x} + c$

23. $e^2x - 4e^{-x} + c$ **24.** $e(e^2 - 1)$ **25.** $1 - \dfrac{1}{e^3}$ **26.** $e^3(e^2 - 1)$ **27.** $e^3 - \dfrac{1}{e}$

28. 4 **29.** $y = e^x + x^2 - 4$

30. (a)   (b)   (c)

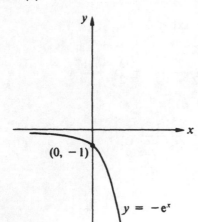

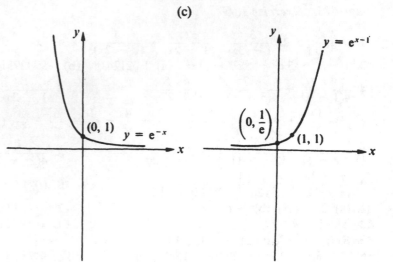

31. min at $\left(-1, -\dfrac{1}{e}\right)$ **32.** max at $(0, -1)$ **33.**

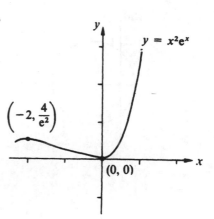

page 487   Exercise 19B

1. $\dfrac{1}{x}$ **2.** $\dfrac{1}{x}$ **3.** $\dfrac{1}{x}$ **4.** $-\dfrac{1}{x}$ **5.** $\dfrac{3}{x}$ **6.** $\dfrac{2x}{x^2 + 3}$ **7.** $\dfrac{3}{3x - 4}$ **8.** $\dfrac{2x + 2}{x^2 + 2x - 1}$

9. $\dfrac{6}{9 - x^2}$ **10.** $\dfrac{3}{x(3 - 2x)}$ **11.** $\dfrac{3x - 4}{x(3x - 2)}$ **12.** $\dfrac{2}{1 - x^2}$ **13.** $\dfrac{x + 2}{(x + 3)^2}$

14. $\dfrac{2x}{x^2 - 1} - \dfrac{1}{(x + 4)^2}$ **15.** $\dfrac{1}{x \ln 10}$ **16.** $\dfrac{1}{x \ln 3}$ **17.** $2e^x + \dfrac{1}{x}$ **18.** $\dfrac{e^x}{x}(x \ln x + 1)$

19. $\dfrac{e^{x^2}}{x}(2x^2 \ln x + 1)$ **20.** $\dfrac{1 - 2x^2 \ln x}{xe^{x^2}}$

21. (a) $2x^x(1 + \ln x)$ (b) $x2^{x^2+1} \ln 2$ (c) $\dfrac{(3x + 5)(x - 1)}{2(x + 1)^{3/2}}$ (d) $\dfrac{2}{(1 - 2x)\sqrt{(1 - 4x^2)}}$

22. $3 \ln 3 - 1$ **23.** 4   **24.** (a) $3^x \ln 3$ (b) $4^x \ln 4$

25. $a^x \ln a, \dfrac{a^x}{\ln a} + c, 39{\cdot}09$ **26.** $\ln |x| + c$ **27.** $3 \ln |x| + c$

28. $\frac{1}{2} \ln |x| + c$ **29.** $\ln (x^2 + 1) + c$ **30.** $3 \ln (x^2 + 3) + c$

31. $3 \ln |x^2 - 3| + c$ **32.** $\ln |x^2 + 3x - 1| + c$ **33.** $\frac{1}{2} \ln (x^2 - 4x + 7) + c$

34. $\frac{1}{2}e^{2x} + 3 \ln |x| + c$

35. $\ln \left[\dfrac{|3x + 4|}{(x + 1)^2} \right] + c$

36. $\ln \left[\dfrac{|x - 1|}{(x + 3)^2} \right] + c$

37. $\frac{1}{2} \ln |\sec 2x| + c$

38. $\ln |\sin x| + c$

39. $x - 3 \ln |x + 3| + c$

40. $2x + 2 \ln |x - 1| + c$

41. $x - \frac{7}{2} \ln |2x + 3| + c$

42. (a) $\dfrac{1}{x}$ undefined for $x = 0$ (b) $\dfrac{1}{x - 3}$ undefined for $x = 3$

     (c) $\dfrac{1}{2x - 1}$ undefined for $x = \frac{1}{2}$ (d) $\dfrac{2(x + 1)}{x^2 + 2x - 3}$ undefined for $x = 1$

43. (a) $1 \cdot 695$ (b) $9 \cdot 730$ (c) $1 \cdot 609$ (d) $1 \cdot 546$ (e) $-0 \cdot 693$ (f) $-0 \cdot 718$ (g) $1 \cdot 088$ (h) $1 \cdot 327$

44. (a)   (b)   (c)

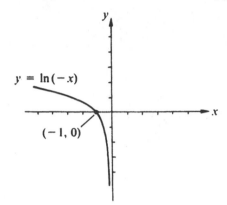
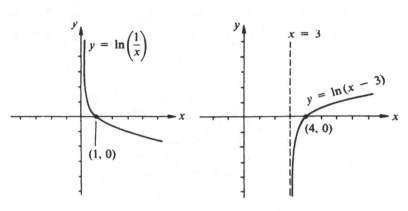

45. (a)   (b)

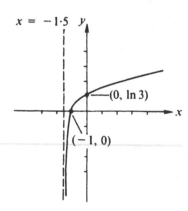
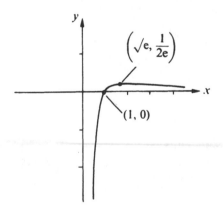

46. $\dfrac{1}{1 + x} + \dfrac{1}{1 - x}$, $\ln \left| \dfrac{1 + x}{1 - x} \right| + c$

47. $\dfrac{2x}{1 + x^2} + \dfrac{2}{1 - x}$, $\ln \left(\dfrac{x^2 + 1}{(1 - x)^2} \right) + c$

48. $\dfrac{2}{2x + 1} - \dfrac{1}{x + 2}$, $0 \cdot 182$

49. $y = \ln [3(x^2 + 1)]$

51. (a) $\frac{1}{2} \sec^2 x - \ln |\sec x| + c$

     (b) $\frac{1}{4} \sec^4 x - \sec^2 x + \ln |\sec x| + c$

page 492　Exercise 19C

1. (a) $1 - 3x + \frac{9}{2}x^2 - \frac{9}{2}x^3 + \frac{27}{8}x^4 \ldots$ (b) $1 + 2x + 2x^2 + \frac{4}{3}x^3 + \frac{2}{3}x^4 \ldots$

  (c) $1 + x^2 + \frac{1}{2}x^4 \ldots$ (d) $1 + \frac{1}{2}x + \frac{1}{8}x^2 + \frac{1}{48}x^3 + \frac{1}{384}x^4 \ldots$

  (e) $1 - \frac{1}{2}x + \frac{1}{8}x^2 - \frac{1}{48}x^3 + \frac{1}{384}x^4 \ldots$ (f) $1 + 6x + \frac{35}{2}x^2 + \frac{100}{3}x^3 + \frac{375}{8}x^4 \ldots$

  (g) $1 + 2x - \frac{16}{3}x^3 - \frac{32}{3}x^4 \ldots$ (h) $1 - x - \frac{1}{2}x^2 + \frac{1}{6}x^3 + \frac{5}{24}x^4 \ldots$

  (i) $1 + x - \frac{1}{2}x^2 - \frac{5}{6}x^3 + \frac{1}{24}x^4 \ldots$

2. (a) e^2 (b) e^{-3} (c) $e^{-1/2} - 1$ (d) $e - 2$ (e) $e^x - 1$ (f) e^{x^2}

3. (a) $2x - 2x^2 + \frac{8}{3}x^3 - 4x^4 \ldots, \; -\frac{1}{2} < x \leqslant \frac{1}{2}$

  (b) $-6x - 18x^2 - 72x^3 - 324x^4 \ldots, \; -\frac{1}{6} \leqslant x < \frac{1}{6}$

  (c) $\frac{1}{3}x - \frac{1}{18}x^2 + \frac{1}{81}x^3 - \frac{1}{324}x^4 \ldots, \; -3 < x \leqslant 3$

  (d) $\ln 2 + \frac{1}{2}x - \frac{1}{8}x^2 + \frac{1}{24}x^3 - \frac{1}{64}x^4 \ldots, \; -2 < x \leqslant 2$

  (e) $\ln 3 + \frac{2}{3}x - \frac{2}{9}x^2 + \frac{8}{81}x^3 - \frac{4}{81}x^4 \ldots, \; -\frac{3}{2} < x \leqslant \frac{3}{2}$

  (f) $-x - \frac{5}{2}x^2 - \frac{7}{3}x^3 - \frac{17}{4}x^4 \ldots, \; -\frac{1}{2} \leqslant x < \frac{1}{2}$

  (g) $3x - \frac{5}{2}x^2 + 3x^3 - \frac{17}{4}x^4 \ldots, \; -\frac{1}{2} < x \leqslant \frac{1}{2}$

  (h) $\ln 3 + \frac{4}{3}x - \frac{5}{9}x^2 + \frac{28}{81}x^3 - \frac{41}{162}x^4 \ldots, \; -1 < x \leqslant 1$

  (i) $5x - \frac{15}{2}x^2 + \frac{65}{3}x^3 - \frac{255}{4}x^4 \ldots, \; -\frac{1}{4} < x \leqslant \frac{1}{4}$

  (j) $4\ln 2 - \frac{3}{2}x + \frac{7}{16}x^2 - \frac{11}{32}x^3 + \frac{127}{512}x^4 \ldots, \; -1 < x \leqslant 1$

  (k) $x - \frac{5}{2}x^2 + \frac{7}{3}x^3 - \frac{17}{12}x^4 \ldots, \; -1 < x \leqslant 1$

  (l) $6x + 24x^2 + 66x^3 + 240x^4 \ldots, \; -\frac{1}{3} \leqslant x < \frac{1}{3}$

4. (a) $\ln\left(\frac{3}{2}\right)$ (b) $-\ln 2$ (c) $3\ln 2$ (d) $\ln 2$

5. $1{\cdot}1052, 0{\cdot}9048$ **6.** $0{\cdot}0953, -0{\cdot}1054$ **7.** (i) $0{\cdot}5108$ (ii) $1{\cdot}099$

9. (a) $\dfrac{2^r}{r!}$ (b) $\dfrac{(-1)^{r+1}}{r.2^r}$ (c) $\dfrac{1+r}{r!}$ (d) $\dfrac{2}{r(r^2-1)}$ **10.** $-3 + x^3 + \frac{1}{2}x^4$

11. $1 - x + \frac{5}{2}x^2 - \frac{29}{6}x^3, \; |x| < \frac{1}{2}$ **12.** $\frac{1}{2}x - \frac{1}{2}x^2 + \frac{5}{12}x^3 - \frac{1}{3}x^4, \; -1 < x \leqslant 1$

13. (a) $x + \frac{1}{2}x^2 + \frac{1}{3}x^3 + \frac{3}{40}x^5, \; -1 < x \leqslant 1$

  (b) $2x - 4x^2 + \frac{17}{3}x^3 - 8x^4, \; -\frac{1}{2} < x \leqslant \frac{1}{2}$

  (c) $x + \frac{23}{4}x^2 + \frac{199}{12}x^3 + \frac{1023}{32}x^4, \; -2 < x \leqslant 2$

  (d) $-x - 2x^2 - \frac{17}{6}x^3 - 4x^4, \; -\frac{1}{2} \leqslant x < \frac{1}{2}$

14. $1, 2, \frac{1}{6}$ **15.** $2, -3, -24$

page 493　Exercise 19D

1.

, $2{\cdot}08$ units

2. (a) $\dfrac{1}{e^x x^2}(x + 1)$ (b) $\dfrac{\cos x}{1 + \sin x}$ **3.** $1 + \ln x, \; x\ln x - x + c, \; 0{\cdot}386$

4. $(\frac{1}{6}\ln 4, 8)$, minimum   **5.** $y = e(2x - 1), \; ey + 1 = 0, \; 80°$

6. $\dfrac{2}{2x - 1} - \dfrac{x}{x^2 + 1}, \; \ln\left[\dfrac{|2x - 1|}{\sqrt{(x^2 + 1)}}\right] + c$

8. (i)

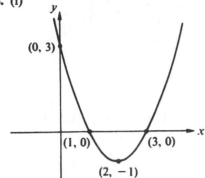

(0, 3)

(1, 0) (3, 0)

(2, −1)

(ii)

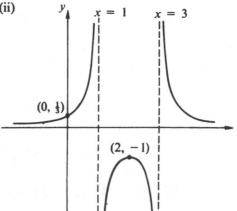

$x = 1$ $x = 3$

, 0·235 sq. units

(0, ⅓)

(2, −1)

9.

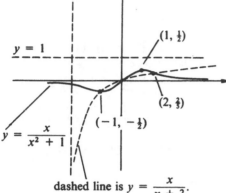

$x = -3$

$y = 1$

$(1, \frac{1}{2})$

$y = \dfrac{x}{x^2 + 1}$

$(2, \frac{2}{3})$

$(-1, -\frac{1}{2})$

dashed line is $y = \dfrac{x}{x + 3}$.

10. (i) $1 - \dfrac{x}{2} - \dfrac{x^2}{8} - \dfrac{x^3}{16}$

(ii) $-ax - \dfrac{a^2 x^2}{2} - \dfrac{a^3 x^3}{3}$, $2, \frac{3}{8}, \frac{29}{48}$

11. $-8x - 7x^2 - \frac{62}{3}x^3 \ldots \left(\dfrac{(-1)^n 2^n - 2(3)^n}{n}\right)x^n \ldots,$

$-\dfrac{1}{3} \leqslant x < \dfrac{1}{3}$

12. $y = \ln(1 + x^2) + e^{-x^2} - 1, \frac{1}{6}, -\frac{5}{24}$

13. $1 + y + \dfrac{y^2}{2} + \dfrac{y^3}{6}, 8 \ln 2, 7(\ln 2)^2, \dfrac{16}{3}(\ln 2)^3.$ **14.** $2 + 4x^2 + \frac{4}{3}x^4, \ln 2, 2, -\frac{4}{3}$

15. (a) 4·73   **(b)** $x + \dfrac{x^2}{2} - \dfrac{2x^3}{3} + \dfrac{x^4}{4} + \dfrac{x^5}{5}, -x + \dfrac{x^2}{2} + \dfrac{2x^3}{3} + \dfrac{x^4}{4} - \dfrac{x^5}{5}$

16. (i) $-6x - \dfrac{3}{2}x^2 - \dfrac{29}{2}x^3 \ldots \left[\dfrac{(-1)^n 3^n}{n!} - \dfrac{(n+1)(n+2)}{2}\right]x^n$ or leave as $(-1)^n\left[\dfrac{3^n}{n} - \binom{-3}{n}\right]x^n$

(ii) $-6x - \dfrac{3}{2}x^2 - x^3 \ldots -\dfrac{3}{n}x^n$

(iii) $1 - 6x + \frac{33}{2}x^2$

page 500   Exercise 20A

1. $-3\cos(x^2 - 4) + c$ **2.** $-\frac{1}{2}\sin(5 - x^2) + c$ **3.** $(1 + x^2)^{3/2} + c$ **4.** $\frac{1}{4}(x^2 + 6)^6 + c$

5. $\frac{1}{55}(5x - 1)(x + 2)^{10} + c$ **6.** $\frac{1}{11}(5x^2 + 3x + 1)(x - 3)^9 + c$

7. $\frac{1}{10}(6x - 1)(3x + 2)^4 + c$ **8.** $\frac{1}{8}(4x - 1)(2x + 3)^6 + c$ **9.** $(x - 3)\sqrt{(2x + 3)} + c$

10. $\sin^{-1} x + c$ **11.** $\dfrac{1}{6}\tan^{-1}\left(\dfrac{3x}{2}\right) + c$ **12.** $\dfrac{1}{2}\sin^{-1}\left(\dfrac{2x}{5}\right) + c$

13. $\dfrac{1}{3}\sin^{-1}\left(\dfrac{3x}{2}\right) + c$ **14.** $\dfrac{\sqrt{5}}{10}\tan^{-1}\left(\dfrac{x\sqrt{5}}{2}\right) + c$ **15.** $\dfrac{1}{2}\left[x\sqrt{(25 - x^2)} + 25\sin^{-1}\left(\dfrac{x}{5}\right)\right] + c$

16. $\frac{1}{4}(2x\sqrt{(1-4x^2)} + \sin^{-1} 2x) + c$ **17. (a)** $\frac{1}{4}\tan^{-1} 4x + c$ **(b)** $\frac{1}{4}\sin^{-1}\left(\frac{4x}{3}\right) + c$

(c) $\frac{\sqrt{5}}{10}\tan^{-1}\left(\frac{x\sqrt{5}}{2}\right) + c$ **(d)** $\frac{\sqrt{2}}{2}\sin^{-1}\left(x\sqrt{\frac{2}{5}}\right) + c$

18. (a) $\tan^{-1}\left(\frac{x-1}{3}\right) + c$ **(b)** $\frac{1}{2}\tan^{-1}\left(\frac{x+3}{2}\right) + c$ **(c)** $\frac{3}{4}\tan^{-1}\left(\frac{2x-3}{2}\right) + c$

19. $\frac{1}{32}(2x+5)(2x-3)^7 + c$ **20.** $\frac{1}{15}(3x-1)(2x+1)^{3/2} + c$ **21.** $\frac{1}{3}(x-1)\sqrt{(2x+1)} + c$

22. $\frac{1}{2}\sin^{-1}\left(\frac{2x}{5}\right) + c$ **23.** $\frac{1}{2}\left[x\sqrt{(9-x^2)} + 9\sin^{-1}\left(\frac{x}{3}\right)\right] + c$

24. $\tan^{-1}\left(\frac{x+1}{4}\right) + c$ **25.** $7\frac{11}{15}$ **26.** $\frac{9}{14}$ **27.** $\frac{97}{168}$ **28.** $\frac{3}{10}$

29. $\frac{2}{3}$ **30.** $\frac{1}{8}\pi$ **31.** $\frac{9}{8}\pi$ **32.** 3   **33.** $\frac{1}{3}\pi$ **34.** $2\ln 2$ **35.** $\ln 6$

page 505   Exercise 20B

1. $\frac{1}{30}(5x-1)(x+1)^5 + c$ **2.** $\frac{2}{3}(x-3)^{3/2}(x+2) + c$ **3.** $\frac{2}{3}(x-6)\sqrt{(x+3)} + c$
4. $\sin x - x\cos x + c$ **5.** $2\cos x + 2x\sin x - x^2\cos x + c$

6. $\frac{2}{9}\sin(3x-1) - \frac{2x}{3}\cos(3x-1) + c$ **7.** $\frac{x^3}{9}(3\ln x - 1) + c$

8. $\frac{x}{2}(x+2)\ln x - \frac{x^2}{4} - x + c$ **9.** $-\frac{1}{4x^2}(2\ln x + 1) + c$

10. $\frac{e^x}{2}(\sin x - \cos x) + c$ **11.** $e^x(x-1) + c$ **12.** $\frac{e^{3x}}{27}(9x^2 - 6x + 2) + c$

13. $\frac{e^x}{5}(2\sin 2x + \cos 2x) + c$ **14.** $\frac{1}{2}(2x+1)\ln(2x+1) - x + c$

15. $x\tan^{-1} x - \frac{1}{2}\ln(1+x^2) + c$ **16.** $x\sin^{-1} x + \sqrt{(1-x^2)} + c$

17. $(x^2-1)\ln(x+1) + x - \frac{x^2}{2} + c$ **18.** $\frac{(x^2-4)}{2}\ln(x-2) - x - \frac{x^2}{4} + c$

19. $-\frac{1}{42}$ **20.** $8\frac{2}{3}$ **21.** $2\frac{2}{15}$ **22.** $\frac{1}{8}(\pi-2)$ **23.** $4\ln 8 - 1$
24. $\frac{1}{4}(3 - \ln 4)$ **25.** $2(e^2 + 3)$

page 508   Exercise 20C

1. $x^3 - 6x + c$ **2.** $\ln|\sec x| + c$ **3.** $-\frac{1}{2}\cos 2x + c$ **4.** $\ln\left|\frac{x-1}{x+1}\right| + c$ **5.** $3e^x + c$

6. $\frac{1}{3}e^{3x} + c$ **7.** $-\frac{1}{x} + c$ **8.** $\ln|x| + c$ **9.** $\frac{1}{2}\sin 2x - \frac{1}{6}\sin^3 2x + c$

10. $\frac{-1}{2(2x-3)^2} + c$ **11.** $\ln|2x+3| + c$ **12.** $\sin^{-1} x + c$ **13.** $\frac{1}{56}(7x+1)(x-1)^7 + c$

14. $\frac{1}{2}\tan^{-1}\left(\frac{x}{2}\right) + c$ **15.** $-\frac{1}{(x^2-4)^4} + c$ **16.** $-\frac{1}{9}\cos 9x - \cos x + c$

17. $\frac{1}{21}(3x+7)(x+7)^6 + c$ **18.** $\sin^{-1} x - \sqrt{(1-x^2)} + c$ **19.** $x\sin x + \cos x + c$

20. $4x^3 - 10x^2 + c$ **21.** $2\tan^{-1}\left(\frac{x}{5}\right) + c$ **22.** $3\sin 3x + c$ **23.** $\frac{x^4}{4}\ln x - \frac{x^4}{16} + c$

24. $\frac{2^x}{\ln 2} + c$ **25.** $\frac{1}{5}(x-1)(2x+3)^{3/2} + c$ **26.** $\ln(x^2+4) + c$ **27.** $\ln\left(\frac{(x+1)^2}{x^2+3}\right) + c$

28. $x \ln x - x + c$ **29.** $3 \sin 5x + 5 \sin 3x + c$ **30.** $4 \tan x + c$ **31.** $2\sqrt{(x + 5)} + c$

32. $-\cos x + \cos^3 x - \frac{3}{5} \cos^5 x + \frac{1}{7} \cos^7 x + c$ **33.** $\frac{3}{8}x + \frac{1}{64} \sin 8x + \frac{1}{8} \sin 4x + c$

34. $\sec x + c$ **35.** $x - 10 \ln |x + 10| + c$ **36.** $\frac{2}{3}x^3 - \frac{1}{2}x^2 - 6x + c$

37. $-e^{-x}(x + 1) + c$ **38.** $\frac{1}{2}x^2 + x + \ln \left| \dfrac{x + 3}{2x - 1} \right| + c$ **39.** $\frac{1}{6} \tan^{-1} \left(\dfrac{2x}{3} \right) + c$

40. $-2 \cos x + c$ **41.** $\frac{1}{2} \ln |\sec 2x| + c$ **42** $\frac{1}{2} \ln (x^2 + 4x + 7) + c$

43. $\dfrac{1}{4} \cos 2x - \dfrac{1}{24} \cos 12x + c$ **44.** $x \sin^{-1} \left(\dfrac{x}{3} \right) + \sqrt{(9 - x^2)} + c$

45. $\frac{1}{4} \ln |\sec 4x + \tan 4x| + c$ **46.** $-\frac{1}{6} \operatorname{cosec}^3 (2x + 3) + c$

47. $\dfrac{1}{2} (\ln x)^2 + c$ **48.** $\dfrac{1}{2} \sin^{-1} \left(\dfrac{2x}{3}\sqrt{3} \right) + c$ **49.** $\dfrac{x}{2}\sqrt{(16 - x^2)} + 8 \sin^{-1} \left(\dfrac{x}{4} \right) + c$

50. $\frac{1}{2} \sin(x^2 + 3) + c$ **51.** $\frac{1}{2} \tan^2 x + \ln |\cos x| + c$ **52.** $\ln |x + \sin x| + c$

53. $\dfrac{1}{2}x^2 + \tan^{-1} \left(\dfrac{x}{3} \right) + \ln (x^2 + 9) + c$ **54.** $\tan (x^2 + 3) + c$

55. $\tan^{-1} \left(\dfrac{x + 1}{2} \right) + c$ **56.** $\ln \left| \dfrac{2x - 3}{x - 1} \right| + c$

page 515   Exercise 20D

1. (a) $x\dfrac{dy}{dx} = 3x^2 + y$ **(b)** $x\dfrac{dy}{dx} + y = 0$ **(c)** $\dfrac{dy}{dx} = 8x$

(d) $\dfrac{dy}{dx} = 2xy$ **(e)** $x\dfrac{dy}{dx} \ln x = y$ **(f)** $\dfrac{dy}{dx} + y \tan x = 0$

(g) $x + 4y\dfrac{dy}{dx} = 0$ **(h)** $3(x^2 + 1)\dfrac{dy}{dx} = 2xy$ **(i)** $(2y - x)\dfrac{dy}{dx} = y$

2. $y = \dfrac{1}{2}x^2 + 2x + c$ **3.** $y = \pm \sqrt{(x^3 + c)}$ **4.** $y = \dfrac{x^4}{1 + cx^4}$ **5.** $y = \dfrac{-1}{e^x + c}$

6. $y = \pm \sqrt{(2 \sin x + c)}$ **7.** $y = A - \ln |1 - x|$ **8.** $y = \cos^{-1}(x^2 + c)$ **9.** $y = Ae^{2x}$

10. $y = Ae^{x^3} - 3$ **11.** $y = \sin^{-1}(\ln |x| + c)$ **12.** $y = \tan (x^2 + c)$ **13.** $y = \ln (2x^2 + c)$

14. $y = A(x + 1)$ **15.** $y = \dfrac{Ae^x}{x + 1} - 3$ **16.** $y = \dfrac{1 + Ae^{x^3}}{1 - Ae^{x^3}}$ **17.** $y = 6x^2 - 4x - 2$

18. $y = \dfrac{1}{2 - x}$ **19.** $y = (x^2 + x + 2)^{1/3}$ **20.** $y = \sin^{-1}\left[\left(x^3 + \dfrac{7}{8} \right)^{1/3} \right]$

21. $y = 2 \ln x - 3$ **22.** $y = 2 \ln (x - 1) + 1$ **23.** $y = 4e^{2x^2}$ **24.** $y = 4e^{3x^2} - 1$

25. $y = 3(e^{2x^2 + 3x} + 1)$ **26.** $y = \ln (\cos x + e)$ **27.** $y = \dfrac{1}{2 + \ln x}$ **28.** $y = 3x - 2$

29. $y = 2x + 11$ **30.** $y = \dfrac{1 + x}{1 - x}$ **31.** $y = 2e^x \dfrac{(1 - x)}{(x + 1)}$

32. $y = 4e^{2x}(x^2 + 1) - 1$ **33. (a)** $y = x + \dfrac{A}{x}$ **(b)** $y = x + \dfrac{4}{x}$ **34.** $y = 6e^x - 4x - 4$

35. $6x^2 + 9y^2 = 12xy + 18y - 6x - 9$ **36. (a)** $x^2 + y^2 = Ax$ **(b)** $x^2 + y^2 = 10x$

37. $\ln (y^2 + 2x^2) - \sqrt{2} \tan^{-1} \left(\dfrac{y}{x\sqrt{2}} \right) = A$ **38.** $y = x \ln \left[\dfrac{|Ax|}{(x - y)^2} \right]$

39. $(y - x - 4)^3(y + 4x + 1)^2 = A$ **40.** $y = Ae^x$ **42.** 12 minutes

page 519   Exercise 20E

1. (a) $3x^4 + x^2 + c_1x + c_2$ (b) $y = \frac{1}{6}x^3 - \sin x + c_1x + c_2$

2. $y = 8e^{2x} + x^3 + 4x - 6$ **3.** $y = 5x^2 + 6x + 4$ **4.** $y = \frac{1}{6}x^3 \ln x - \frac{5}{36}x^3 + \frac{1}{3}x - \frac{11}{36}$

5. (a) $y = Ae^{5x} + Be^{-2x}$ (b) $y = (A + Bx)e^{4x}$

(c) $y = Ae^{5x} + Be^{-3x} - \frac{1}{7}e^{4x}$ (d) $y = Ae^{3x} + Be^{-x/2} + \frac{2}{11}e^{5x}$

(e) $y = e^x(A \cos 2x + B \sin 2x) + 2x + 1$ (f) $y = Ae^{-3x/2} + Be^{-4x} + \frac{1}{6}x^2 + \frac{1}{9}x - \frac{29}{27}$

8. (a) $y = A \cos 6x + B \sin 6x + \frac{1}{36}$ (b) $y = Ae^{3x} + Be^{-3x} - 2$

(c) $y = A \cos 3x + B \sin 3x + 2$ (d) $y = Ae^{3x/2} + Be^{-3x/2} - 4$

9. (a) $y = 3 \cos 2x - \sin 2x + 2$ (b) $y = 5e^x - e^{-x} - 2$

(c) $y = 2 \cos \frac{1}{2}x + 3 \sin \frac{1}{2}x + 16$ (d) $y = 5e^{x/2} + e^{-x/2} - 4$

page 520   Exercise 20F

1. (a) $\frac{2}{3}$ (b) $\frac{3}{8}$ (c) $0 \cdot 157$ **2.** (a) $\frac{61}{192}$ (b) $1 - \frac{1}{4}\pi$ (c) π (d) $\ln(\frac{4}{3})$

3. (a) $\frac{x^2}{2} \ln x - \frac{x^2}{4} + c$ (b) $\frac{2}{3}(x + 4)\sqrt{(x - 2)} + c$ **4.** $0 \cdot 59$

5. (i) $-\frac{1}{72}$ (ii) $\frac{1}{24}(3\pi + 2)$

6. (a) (i) $\frac{x^3}{3} - 4x - \frac{4}{x} + c$ (ii) $\frac{1}{2} \ln |\sin 2x| + c$ (iii) $-\sqrt{(1 - 2x)} + c$

(iv) $\frac{x^2}{2} \ln 2x - \frac{x^2}{4} + c$ (b) $\frac{2}{3}$ (c) $\frac{2}{3}$

7. $\frac{2}{4x^2 + 1}$ (i) $\frac{1}{8} \ln (4x^2 + 1) + \tan^{-1} 2x + c$ (ii) $\tan^{-1} 2x^2 + c$

8. $(\pi^2 - 2\pi)$ cubic units   **9.** $y = \frac{2x}{5 - 2x}$ **11.** $y = \frac{2 + \sin x}{2 - \sin x}$

12. (i) $y + 2x = 12$ (ii) $y = \frac{8x}{x + 1}$ **13.** $e^{3/y} = A\left(\frac{x + 1}{x - 2}\right)$, $2e^{3(1 - y)/y} = \frac{x + 1}{x - 2}$

14. (a) $1 \cdot 84$ (b) 1 (c) $y = 2(e^x - x - 1)$

15. (i) $y = \tan^{-1}(\ln x + \sqrt{3})$ (ii) $\frac{dy}{dx} = -y^2(1 + 3x^2)$

16. $\frac{dx}{dt} = k(m - x)$, $\ln\left(\frac{m}{m - x}\right) = kt$, $k = 1$ (i) $\frac{7}{8}m$ (ii) $2 \ln 2$

17. $4y^3 = 6x - 3 \sin 2x + 4$ (ii) $4y = 3 \cos 2x - 3 \sin 2x + 1$

18. (a) $y = x - \frac{1}{4x^2}$, (b) (i) $y = Ae^{-x} + Be^{-2x} + 3e^{-3x} - \frac{1}{2}$, (ii) $-\frac{1}{2}$

page 526   Exercise 21A

1. 16   **2.** $1 \cdot 724$ **3.** (a) $0 \cdot 6956$ (b) $0 \cdot 6938$ **4.** $51 \cdot 67$ **5.** $0 \cdot 6932$

6. (a) $1 \cdot 1051$ (b) $1 \cdot 1071$ **7.** (a) $1 \cdot 470$ (b) $1 \cdot 463$ **8.** (a) $0 \cdot 9968$ (b) 1

9. (a) $0 \cdot 6435$ (b) $0 \cdot 6445$ (c) $0 \cdot 6435$ (d) $0 \cdot 6431$

page 529   **Exercise 21B**

1. (a) $1 - \dfrac{x^2}{2!} + \dfrac{x^4}{4!} \ldots\ldots$ (b) $1 + x + \dfrac{x^2}{2!} + \dfrac{x^3}{3!} + \dfrac{x^4}{4!} + \dfrac{x^5}{5!} \ldots$

(c) $1 + nx + \dfrac{n(n-1)}{2!}x^2 + \dfrac{n(n-1)(n-2)}{3!}x^3 + \dfrac{n(n-1)(n-2)(n-3)}{4!}x^4$

$\quad + \dfrac{n(n-1)(n-2)(n-3)(n-4)}{5!} x^5 \ldots$

(d) $x - \dfrac{x^2}{2} + \dfrac{x^3}{3} - \dfrac{x^4}{4} + \dfrac{x^5}{5} \ldots\ldots$

(e) $1 + x \ln a + \dfrac{(x \ln a)^2}{2!} + \dfrac{(x \ln a)^3}{3!} + \dfrac{(x \ln a)^4}{4!} + \dfrac{(x \ln a)^5}{5!} + \ldots\ldots$

(f) $x + \dfrac{x^3}{6} + \dfrac{3x^5}{40} \ldots\ldots$; 0.995004, 0.2014

2. $x + \dfrac{x^3}{3} + \dfrac{2x^5}{15} \ldots\ldots$ **3.** $\dfrac{4}{5} + \dfrac{3}{5}x - \dfrac{2}{5}x^2 - \dfrac{1}{10}x^3 + \dfrac{1}{30}x^4 \ldots\ldots$, 0.8559

4. $1 + 2x + 2x^2 + \frac{8}{3}x^3 \ldots\ldots$, 1.0355

page 535   **Exercise 21C**

1. -1.84, 1.15 **2.** (a) 0.45 (b) x_1 is undefined   (c) -4.45 **3.** -1.67
4. 2.236 **5.** 3.4495 **6.** 2.414 **7.** -2.104 **8.** 1.557
9. 1.7632 **10.** 0.79206 **11.** 1.4973

page 537   **Exercise 21D**

3. $\frac{1}{5}e^x(\cos 2x + 2 \sin 2x) + c$ **4.** $i \ln (4 \pm \sqrt{15})$ **5.** $i \ln (1 + \sqrt{2})$

page 537   **Exercise 21E**

1.

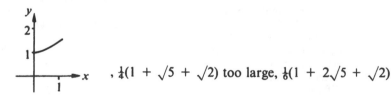

$, \frac{1}{4}(1 + \sqrt{5} + \sqrt{2})$ too large, $\frac{1}{6}(1 + 2\sqrt{5} + \sqrt{2})$

3. $\frac{2}{3}\pi(e^{-9} + 4e^{-4} + 2e^{-1} + 2)$ **4.** 2.76

5. 1.17, 1 **6.** $1 - \dfrac{x^2}{2!} + \dfrac{x^4}{4!} - \dfrac{x^6}{6!}$, $t - \dfrac{t^2}{2} + \dfrac{t^3}{3}$, $-\dfrac{x^2}{2} - \dfrac{x^4}{12} - \dfrac{x^6}{45}$, 0.0639

7. $1 + \dfrac{x^2}{2} + \dfrac{5x^4}{24}$, 1.00502 **8.** 0.37 **9.** 2.93 **10.** 0.53

11.

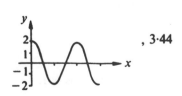

$, 3.44$

12. max at $(0, 1)$, min at $(-1, 0)$, min at $(1, 0)$, 0.64; $(0.64, 0.39)$, $(-0.64, 0.39)$

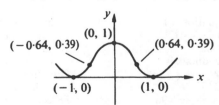

Index